PROBLEMS
AND
SOLUTIONS
IN
INTRODUCTORY
AND
ADVANCED MATRIX
CALCULUS

—— Second Edition ——

Willi-Hans Steeb
Yorick Hardy

University of Johannesburg, South Africa &
University of South Africa, South Africa

PROBLEMS
AND
SOLUTIONS
IN
INTRODUCTORY
AND
ADVANCED MATRIX
CALCULUS

— Second Edition —

W⊝ World Scientific

NEW JERSEY · LONDON · SINGAPORE · BEIJING · SHANGHAI · HONG KONG · TAIPEI · CHENNAI · TOKYO

Published by

World Scientific Publishing Co. Pte. Ltd.

5 Toh Tuck Link, Singapore 596224

USA office: 27 Warren Street, Suite 401-402, Hackensack, NJ 07601

UK office: 57 Shelton Street, Covent Garden, London WC2H 9HE

Library of Congress Cataloging-in-Publication Data

Names: Steeb, W.-H. | Hardy, Yorick, 1976–

Title: Problems and solutions in introductory and advanced matrix calculus.

Description: Second edition / by Willi-Hans Steeb (University of Johannesburg,
    South Africa & University of South Africa, South Africa),
    Yorick Hardy (University of Johannesburg, South Africa & University of South Africa,
    South Africa). | New Jersey : World Scientific, 2016. |
    Includes bibliographical references and index.

Identifiers: LCCN 2016028706| ISBN 9789813143784 (hardcover : alk. paper) |
    ISBN 9789813143791 (pbk. : alk. paper)

Subjects: LCSH: Matrices--Problems, exercises, etc. | Calculus. | Mathematical physics.

Classification: LCC QA188 .S664 2016 | DDC 512.9/434--dc23

LC record available at https://lccn.loc.gov/2016028706

British Library Cataloguing-in-Publication Data

A catalogue record for this book is available from the British Library.

Printed in Singapore

Preface

The purpose of this book is to supply a collection of problems in introductory and advanced matrix problems together with their detailed solutions which will prove to be valuable to undergraduate and graduate students as well as to research workers in these fields. Each chapter contains an introduction with the essential definitions and explanations to tackle the problems in the chapter. If necessary, other concepts are explained directly with the present problems. Thus the material in the book is self-contained. The topics range in difficulty from elementary to advanced. Students can learn important principles and strategies required for problem solving. Lecturers will also find this text useful either as a supplement or text, since important concepts and techniques are developed in the problems.

A large number of problems are related to applications. Applications include wavelets, linear integral equations, Kirchhoff's laws, global positioning systems, Floquet theory, octonians, random walks, entanglement, tensor decomposition, hyperdeterminant, matrix-valued differential forms, Kronecker product and images. A number of problems useful in quantum physics and graph theory are also provided. Advanced topics include groups and matrices, Lie groups and matrices and Lie algebras and matrices. Exercises for matrix-valued differential forms are also included.

In this second edition new problems for braid groups, mutually unbiased bases, vec operator, spectral theorem, binary matrices, nonnormal matrices, wavelets, fractals, matrices and integration are added. Each chapter also contains supplementary problems. Furthermore a number of Maxima and SymbolicC++ programs are added for solving problems. Applications in mathematical and theoretical physics are emphasized.

The book can also be used as a text for linear and multilinear algebra or matrix theory. The material was tested in the first author's lectures given around the world.

Note to the Readers

The International School for Scientific Computing (ISSC) provides certificate courses for this subject. Please contact the authors if you want to do this course or other courses of the ISSC.

e-mail addresses of the first author:

steebwilli@gmail.com
steeb_wh@yahoo.com

e-mail address of the second author:

yorickhardy@gmail.com

Home page of the first author: http://issc.uj.ac.za

Contents

Notation

$:=$	is defined as
$\in$	belongs to (a set)
$\notin$	does not belong to (a set)
$\cap$	intersection of sets
$\cup$	union of sets
$\emptyset$	empty set
$T \subset S$	subset T of set S
$S \cap T$	the intersection of the sets S and T
$S \cup T$	the union of the sets S and T
$f(S)$	image of set S under mapping f
$f \circ g$	composition of two mappings $(f \circ g)(x) = f(g(x))$
$\mathbb{N}$	set of natural numbers
$\mathbb{N}_0$	set of natural numbers including 0
$\mathbb{Z}$	set of integers
$\mathbb{Q}$	set of rational numbers
$\mathbb{R}$	set of real numbers
$\mathbb{R}^+$	set of nonnegative real numbers
$\mathbb{C}$	set of complex numbers
$\mathbb{R}^n$	n-dimensional Euclidean space
	space of column vectors with n real components
$\mathbb{C}^n$	n-dimensional complex linear space
	space of column vectors with n complex components
$\mathcal{H}$	Hilbert space
S_n	symmetric group on a set of n symbols
i	$\sqrt{-1}$
$\Re(z)$	real part of the complex number z
$\Im(z)$	imaginary part of the complex number z
$|z|$	modulus of complex number z
	$|x + iy| = (x^2 + y^2)^{1/2}, \; x, y \in \mathbb{R}$
$\mathbf{x}$	column vector in $\mathbb{C}^n$
$\mathbf{x}^T$	transpose of $\mathbf{x}$ (row vector)
$\mathbf{0}$	zero (column) vector
$\|\cdot\|$	norm
$\mathbf{x} \cdot \mathbf{y} \equiv \mathbf{x}^* \mathbf{y}$	scalar product (inner product) in $\mathbb{C}^n$
$\mathbf{x} \times \mathbf{y}$	vector product in $\mathbb{R}^3$

A, B, C	$m \times n$ matrices
P	$n \times n$ permutation matrix
Π	$n \times n$ projection matrix
U	$n \times n$ unitary matrix
$\text{vec}(A)$	vectorization of matrix A
$\det(A)$	determinant of a square matrix A
$\text{tr}(A)$	trace of a square matrix A
$\text{Pf}(A)$	Pfaffian of square matrix A
$\text{rank}(A)$	rank of matrix A
A^T	transpose of matrix A
$\overline{A}$	conjugate of matrix A
A^*	conjugate transpose of matrix A
A^{-1}	inverse of square matrix A (if it exists)
I_n	$n \times n$ unit matrix
I	unit operator
0_n	$n \times n$ zero matrix
AB	matrix product of $m \times n$ matrix A and $n \times p$ matrix B
$A \bullet B$	Hadamard product (entry-wise product) of $m \times n$ matrices A and B
$[A, B] := AB - BA$	commutator for square matrices A and B
$[A, B]_+ := AB + BA$	anticommutator for square matrices A and B
$A \otimes B$	Kronecker product of matrices A and B
$A \oplus B$	Direct sum of matrices A and B
δ_{jk}	Kronecker delta with $\delta_{jk} = 1$ for $j = k$ and $\delta_{jk} = 0$ for $j \neq k$
λ	eigenvalue
ϵ	real parameter
t	time variable
$\hat{H}$	Hamilton operator

The *elementary matrices* E_{jk} with $j = 1, \ldots, m$ and $k = 1, \ldots, n$ are defined as 1 at entry (j, k) and 0 otherwise.

The Pauli spin matrices are defined as

$$\sigma_1 := \begin{pmatrix} 0 & 1 \\ 1 & 0 \end{pmatrix}, \quad \sigma_2 := \begin{pmatrix} 0 & -i \\ i & 0 \end{pmatrix}, \quad \sigma_3 := \begin{pmatrix} 1 & 0 \\ 0 & -1 \end{pmatrix}.$$

Chapter 1

Basic Operations

Let $\mathbb{F}$ be a field, for example the set of real numbers $\mathbb{R}$ or the set of complex numbers $\mathbb{C}$. Let m, n be two integers ≥ 1. An array A of numbers in $\mathbb{F}$

$$\begin{pmatrix} a_{11} & a_{12} & a_{13} & \cdots & a_{1n} \\ a_{21} & a_{22} & a_{23} & \cdots & a_{2n} \\ \vdots & \vdots & \vdots & \ddots & \vdots \\ a_{m1} & a_{m2} & a_{m3} & \cdots & a_{mn} \end{pmatrix} = (a_{ij})$$

is called an $m \times n$ *matrix* with entry a_{ij} in the i-th row and j-th column. A *row vector* is a $1 \times n$ matrix. A *column vector* is an $n \times 1$ matrix. We have a *zero matrix*, in which $a_{ij} = 0$ for all i, j. I_n denotes the $n \times n$ identity matrix with the diagonal elements equal to 1 and 0 otherwise.

Let $A = (a_{ij})$ and $B = (b_{ij})$ be two $m \times n$ matrices. We define $A + B$ to be the $m \times n$ matrix whose entry in the i-th row and j-th column is $a_{ij} + b_{ij}$. Matrix multiplication is only defined between two matrices if the number of columns of the first matrix is the same as the number of rows of the second matrix. If A is an $m \times n$ matrix and B is an $n \times p$ matrix, then the matrix product AB is an $m \times p$ matrix defined by

$$(AB)_{ij} = \sum_{r=1}^{n} a_{ir} b_{rj}$$

for each pair i and j, where $(AB)_{ij}$ denotes the (i, j)-th entry in AB. Let $A = (a_{ij})$ and $B = (b_{ij})$ be two $m \times n$ matrices with entries in some field. Then their Hadamard product is the entrywise product of A and B, that is the $m \times n$ matrix $A \bullet B$ whose (i, j)-th entry is $a_{ij} b_{ij}$.

Let V be a vector space and let

$$B = \{\mathbf{v}_1, \mathbf{v}_2, \ldots, \mathbf{v}_n\}$$

be a finite subset of V. The set B is a linearly independent set in V if

$$c_1\mathbf{v}_1 + c_2\mathbf{v}_2 + \cdots + c_n\mathbf{v}_n = \mathbf{0}$$

only admits the trivial solution

$$c_1 = c_2 = \cdots = c_n = 0$$

for the scalars $c_1, c_2, \ldots, c_n$. We define the span of B by

$$\text{span}(B) = \{c_1\mathbf{v}_1 + c_2\mathbf{v}_2 + \cdots + c_n\mathbf{v}_n : \text{for scalar } c_1, c_2, \ldots, c_n\}.$$

If $\text{span}(B) = V$, then B is said to span V (or B is a spanning set for V). If B is both linearly independent and a spanning set for V then B is said to be a basis for V, and we write $\dim(V) = |B|$. In this case, the dimension $\dim(V)$ of V is the number of elements $|B| = n$ of B. A vector space which has no finite basis is infinite dimensional. Every basis for a finite dimensional vector space has the same number of elements.

The set of $m \times n$ matrices forms an mn dimensional vector space, for example the 2×2 matrices over $\mathbb{C}$ form vector space over $\mathbb{C}$ with standard basis

$$\left\{ \begin{pmatrix} 1 & 0 \\ 0 & 0 \end{pmatrix}, \begin{pmatrix} 0 & 1 \\ 0 & 0 \end{pmatrix}, \begin{pmatrix} 0 & 0 \\ 1 & 0 \end{pmatrix}, \begin{pmatrix} 0 & 0 \\ 0 & 1 \end{pmatrix} \right\}.$$

Another basis for the 2×2 matrices over $\mathbb{C}$ is given in terms of the Pauli spin matrices

$$\left\{ \begin{pmatrix} 1 & 0 \\ 0 & 1 \end{pmatrix}, \begin{pmatrix} 0 & 1 \\ 1 & 0 \end{pmatrix}, \begin{pmatrix} 0 & -i \\ i & 0 \end{pmatrix}, \begin{pmatrix} 1 & 0 \\ 0 & -1 \end{pmatrix} \right\}.$$

However, this is not a basis for the 2×2 matrices over $\mathbb{R}$. For the 2×2 matrices over $\mathbb{R}$ we have the same standard basis as for the 2×2 matrices over $\mathbb{C}$. Another basis for the 2×2 matrices over $\mathbb{R}$ is

$$\left\{ \begin{pmatrix} 1 & 0 \\ 0 & 1 \end{pmatrix}, \begin{pmatrix} 0 & 1 \\ 1 & 0 \end{pmatrix}, \begin{pmatrix} 0 & -1 \\ 1 & 0 \end{pmatrix}, \begin{pmatrix} 1 & 0 \\ 0 & -1 \end{pmatrix} \right\}.$$

For the vector space $\mathbb{C}^2$ an orthonormal basis frequently used is the Hadamard basis

$$\frac{1}{\sqrt{2}} \begin{pmatrix} 1 \\ 1 \end{pmatrix}, \qquad \frac{1}{\sqrt{2}} \begin{pmatrix} 1 \\ -1 \end{pmatrix}.$$

Problem 1. Let e_1, e_2, e_3 be the *standard basis* in $\mathbb{R}^3$

$$e_1 = \begin{pmatrix} 1 \\ 0 \\ 0 \end{pmatrix}, \quad e_2 = \begin{pmatrix} 0 \\ 1 \\ 0 \end{pmatrix}, \quad e_3 = \begin{pmatrix} 0 \\ 0 \\ 1 \end{pmatrix}.$$

(i) Consider the normalized vectors

$$a = \frac{1}{\sqrt{3}}(e_1 + e_2 + e_3), \quad b = \frac{1}{\sqrt{3}}(-e_1 - e_2 + e_3),$$

$$c = \frac{1}{\sqrt{3}}(-e_1 + e_2 - e_3), \quad d = \frac{1}{\sqrt{3}}(e_1 - e_2 - e_3).$$

These vectors are the unit vectors giving the direction of the four bonds of an atom in the *diamond lattice*. Show that the four vectors are linearly dependent.
(ii) Find the scalar products $a^T b$, $b^T c$, $c^T d$, $d^T a$. Discuss.

Solution 1. (i) From $c_1 a + c_2 b + c_3 c + c_4 d = 0$ we find that $c_1 = c_2 = c_3 = c_4 = 1$ is a nonzero solution. Thus the vectors are linearly dependent.
(ii) We find $a^T b = -1/3$, $b^T c = -1/3$, $c^T d = -1/3$, $d^T a = -1/3$.

Problem 2. (i) Consider the normalized vector v in $\mathbb{R}^3$ and the permutation matrix P, respectively

$$v = \frac{1}{\sqrt{3}} \begin{pmatrix} 1 \\ 1 \\ -1 \end{pmatrix}, \quad P = \begin{pmatrix} 0 & 1 & 0 \\ 0 & 0 & 1 \\ 1 & 0 & 0 \end{pmatrix}.$$

Are the three vectors v, Pv, $P^2 v$ linearly independent?
(ii) Consider the 4×4 symmetric matrix A and the vector b in $\mathbb{R}^4$

$$A = \begin{pmatrix} 0 & 1 & 0 & 0 \\ 1 & 0 & 1 & 0 \\ 0 & 1 & 0 & 1 \\ 0 & 0 & 1 & 0 \end{pmatrix}, \quad b = \begin{pmatrix} 1 \\ 1 \\ 1 \\ 1 \end{pmatrix}.$$

Are the vectors b, Ab, $A^2 b$, $A^3 b$ in $\mathbb{R}^4$ linearly independent? Show that the matrix A is invertible. Look at the column vectors of the matrix A

$$\begin{pmatrix} 0 \\ 1 \\ 0 \\ 0 \end{pmatrix}, \quad \begin{pmatrix} 1 \\ 0 \\ 1 \\ 0 \end{pmatrix}, \quad \begin{pmatrix} 0 \\ 1 \\ 0 \\ 1 \end{pmatrix}, \quad \begin{pmatrix} 0 \\ 0 \\ 1 \\ 0 \end{pmatrix}.$$

Find the inverse of A.

Solution 2. (i) Yes the three vectors

$$v = \frac{1}{\sqrt{3}} \begin{pmatrix} 1 \\ 1 \\ -1 \end{pmatrix}, \quad Pv = \frac{1}{\sqrt{3}} \begin{pmatrix} 1 \\ -1 \\ 1 \end{pmatrix}, \quad P^2 v = \frac{1}{\sqrt{3}} \begin{pmatrix} -1 \\ 1 \\ 1 \end{pmatrix}$$

are linearly independent, i.e. they form a basis in the vector space $\mathbb{R}^3$.
(ii) We obtain the four vectors

$$\mathbf{b}, \quad A\mathbf{b} = \begin{pmatrix} 1 \\ 2 \\ 2 \\ 1 \end{pmatrix}, \quad A^2\mathbf{b} = \begin{pmatrix} 2 \\ 3 \\ 3 \\ 2 \end{pmatrix}, \quad A^3\mathbf{b} = \begin{pmatrix} 3 \\ 5 \\ 5 \\ 3 \end{pmatrix}.$$

Thus the vectors $\mathbf{b}$ and $A\mathbf{b}$ are linearly independent. However $A^2\mathbf{b} = A\mathbf{b} + \mathbf{b}$, $A^3\mathbf{b} = A^2\mathbf{b} + A\mathbf{b}$. Thus $\mathbf{b}$, $A\mathbf{b}$ and $A^3\mathbf{b}$ are linearly dependent. The column vectors in A are linearly independent. Thus the inverse of A exists and is given by

$$A^{-1} = \begin{pmatrix} 0 & 1 & 0 & -1 \\ 1 & 0 & 0 & 0 \\ 0 & 0 & 0 & 1 \\ -1 & 0 & 1 & 0 \end{pmatrix}.$$

Note that $\det(A) = 1$ and thus $\det(A^{-1}) = 1$.

Problem 3. (i) Consider the Hilbert space $M_2(\mathbb{R})$ of the 2×2 matrices over $\mathbb{R}$. Show that the matrices

$$\begin{pmatrix} 1 & 0 \\ 0 & 0 \end{pmatrix}, \quad \begin{pmatrix} 1 & 1 \\ 0 & 0 \end{pmatrix}, \quad \begin{pmatrix} 1 & 1 \\ 1 & 0 \end{pmatrix}, \quad \begin{pmatrix} 1 & 1 \\ 1 & 1 \end{pmatrix}$$

are linearly independent.
(ii) Use the Gram-Schmidt orthonormalization technique to find an orthonormal basis for $M_2(\mathbb{R})$.

Solution 3. (i) From

$$a\begin{pmatrix} 1 & 0 \\ 0 & 0 \end{pmatrix} + b\begin{pmatrix} 1 & 1 \\ 0 & 0 \end{pmatrix} + c\begin{pmatrix} 1 & 1 \\ 1 & 0 \end{pmatrix} + d\begin{pmatrix} 1 & 1 \\ 1 & 1 \end{pmatrix} = \begin{pmatrix} 0 & 0 \\ 0 & 0 \end{pmatrix}$$

we find that the only solution is $a = b = c = d = 0$.
(ii) We obtain the *standard basis*

$$\begin{pmatrix} 1 & 0 \\ 0 & 0 \end{pmatrix}, \quad \begin{pmatrix} 0 & 1 \\ 0 & 0 \end{pmatrix}, \quad \begin{pmatrix} 0 & 0 \\ 1 & 0 \end{pmatrix}, \quad \begin{pmatrix} 0 & 0 \\ 0 & 1 \end{pmatrix}.$$

Problem 4. Consider the vector space of 2×2 matrices over $\mathbb{R}$ and the matrices

$$A_1 = \begin{pmatrix} 1 & 1 \\ 0 & 0 \end{pmatrix}, \quad A_2 = \begin{pmatrix} 0 & 1 \\ 0 & 1 \end{pmatrix}, \quad A_3 = \begin{pmatrix} 0 & 0 \\ 1 & 1 \end{pmatrix}, \quad A_4 = \begin{pmatrix} 1 & 0 \\ 1 & 0 \end{pmatrix}.$$

(i) Are these four 2×2 matrices linearly independent?
(ii) An $n \times n$ matrix M is called *normal* if $MM^* = M^*M$. Which of the four matrices A_j ($j = 1, 2, 3, 4$) are normal matrices?

Solution 4. (i) No. We have

$$\begin{pmatrix} 1 & 1 \\ 0 & 0 \end{pmatrix} = \begin{pmatrix} 0 & 1 \\ 0 & 1 \end{pmatrix} + \begin{pmatrix} 1 & 0 \\ 1 & 0 \end{pmatrix} - \begin{pmatrix} 0 & 0 \\ 1 & 1 \end{pmatrix}.$$

(ii) None of these matrices is normal.

Problem 5. Let σ_1, σ_2, σ_3 be the *Pauli spin matrices*

$$\sigma_1 = \begin{pmatrix} 0 & 1 \\ 1 & 0 \end{pmatrix}, \quad \sigma_2 = \begin{pmatrix} 0 & -i \\ i & 0 \end{pmatrix}, \quad \sigma_3 = \begin{pmatrix} 1 & 0 \\ 0 & -1 \end{pmatrix}$$

and $\sigma_0 = I_2$. Consider the vector space of 2×2 matrices over $\mathbb{C}$.
(i) Show that the 2×2 matrices

$$\frac{1}{\sqrt{2}} I_2, \quad \frac{1}{\sqrt{2}} \sigma_1, \quad \frac{1}{\sqrt{2}} \sigma_2, \quad \frac{1}{\sqrt{2}} \sigma_3$$

are linearly independent.
(ii) Show that any 2×2 complex matrix has a unique representation of the form

$$a_0 I_2 + i a_1 \sigma_1 + i a_2 \sigma_2 + i a_3 \sigma_3$$

for some $a_0, a_1, a_2, a_3 \in \mathbb{C}$.

Solution 5. (i) From the condition

$$c_0 \frac{1}{\sqrt{2}} I_2 + c_1 \frac{1}{\sqrt{2}} \sigma_1 + c_2 \frac{1}{\sqrt{2}} \sigma_2 + c_3 \sigma_3 = 0_2$$

we find the four equations $c_0 + c_3 = 0$, $c_1 - c_2 i = 0$, $c_1 + c_2 i = 0$, $c_0 - c_3 = 0$
with the only solution $c_0 = c_1 = c_2 = c_3 = 0$.
(ii) Since

$$a_0 I_2 + i a_1 \sigma_1 + i a_2 \sigma_2 + i a_3 \sigma_3 = \begin{pmatrix} a_0 + i a_3 & i a_1 + a_2 \\ i a_1 - a_2 & a_0 - i a_3 \end{pmatrix}$$

we obtain

$$a_0 I_2 + i a_1 \sigma_1 + i a_2 \sigma_2 + i a_3 \sigma_3 = \begin{pmatrix} \alpha & \beta \\ \gamma & \delta \end{pmatrix}$$

where $\alpha, \beta, \gamma, \delta \in \mathbb{C}$. Thus

$$a_0 = \frac{\alpha + \delta}{2}, \quad a_1 = \frac{\beta + \gamma}{2i}, \quad a_2 = \frac{\beta - \gamma}{2}, \quad a_3 = \frac{\alpha - \delta}{2i}.$$

Problem 6. Consider the normalized vector $\mathbf{v}_0 = (1 \ \ 0 \ \ 0)^T$ in $\mathbb{R}^3$. Find
three normalized vectors $\mathbf{v}_1, \mathbf{v}_2, \mathbf{v}_3$ such that

$$\sum_{j=0}^{3} \mathbf{v}_j = \mathbf{0}, \qquad \mathbf{v}_j^T \mathbf{v}_k = -\frac{1}{3} \quad (j \neq k).$$

Solution 6.   Owing to the form of $\mathbf{v}_0$ we have due to the second condition that $v_{1,1} = v_{2,1} = v_{3,1} = -1/3$. We obtain

$$\mathbf{v}_1 = \begin{pmatrix} -1/3 \\ 2\sqrt{2}/3 \\ 0 \end{pmatrix}, \qquad \mathbf{v}_2 = \begin{pmatrix} -1/3 \\ -\sqrt{2}/3 \\ \sqrt{6}/3 \end{pmatrix}, \qquad \mathbf{v}_3 = \begin{pmatrix} -1/3 \\ -\sqrt{2}/3 \\ -\sqrt{6}/3 \end{pmatrix}.$$

Problem 7.   (i) Find four normalized vectors $\mathbf{v}_1$, $\mathbf{v}_2$, $\mathbf{v}_3$, $\mathbf{v}_4$ in $\mathbb{R}^3$ such that

$$\mathbf{v}_j^T \mathbf{v}_k = \frac{4}{3}\delta_{jk} - \frac{1}{3} = \begin{cases} 1 & \text{for } j = k \\ -1/3 & \text{for } j \neq k \end{cases}.$$

(ii) Calculate the vector and the matrix, respectively

$$\sum_{j=1}^{4} \mathbf{v}_j, \qquad \frac{3}{4}\sum_{j=1}^{4} \mathbf{v}_j \mathbf{v}_j^T.$$

Solution 7.   (i) Such a quartet of vectors consists of the vectors pointing from the center of a cube to nonadjacent corners. We may picture these four vectors as the normal vectors for the faces of the *tetrahedron* that is defined by the other four corners of the cube. Owing to the conditions the four vectors are normalized. Thus

$$\mathbf{v}_1 = \frac{1}{\sqrt{3}}\begin{pmatrix} 1 \\ 1 \\ 1 \end{pmatrix}, \quad \mathbf{v}_2 = \frac{1}{\sqrt{3}}\begin{pmatrix} 1 \\ -1 \\ -1 \end{pmatrix}, \quad \mathbf{v}_3 = \frac{1}{\sqrt{3}}\begin{pmatrix} -1 \\ 1 \\ -1 \end{pmatrix}, \quad \mathbf{v}_4 = \frac{1}{\sqrt{3}}\begin{pmatrix} -1 \\ -1 \\ 1 \end{pmatrix}.$$

(ii) We have

$$\sum_{j=1}^{4} \mathbf{v}_j = \mathbf{0}, \qquad \frac{3}{4}\sum_{j=1}^{4} \mathbf{v}_j \mathbf{v}_j^T = I_3.$$

Problem 8.   Find the set of all four (column) vectors $\mathbf{u}_1$, $\mathbf{u}_2$, $\mathbf{v}_1$, $\mathbf{v}_2$ in $\mathbb{R}^2$ such that the following conditions are satisfied $\mathbf{v}_1^T \mathbf{u}_2 = 0$, $\mathbf{v}_2^T \mathbf{u}_1 = 0$, $\mathbf{v}_1^T \mathbf{u}_1 = 1$, $\mathbf{v}_2^T \mathbf{u}_2 = 1$.

Solution 8.   We obtain the following four conditions

$$v_{11}u_{21} + v_{12}u_{22} = 0, \quad v_{21}u_{11} + v_{22}u_{12} = 0, \quad v_{11}u_{11} + v_{12}u_{12} = 1, \quad v_{21}u_{21} + v_{22}u_{22} = 1.$$

We have four equations with eight unknowns. We obtain three solutions. The first one is

$$u_{11} = \frac{r_4}{r_1 r_4 - r_2 r_3}, \quad u_{12} = -\frac{r_3}{r_1 r_4 - r_2 r_3}, \quad u_{21} = -\frac{r_2}{r_1 r_4 - r_2 r_3}, \quad u_{22} = \frac{r_1}{r_1 r_4 - r_2 r_3}$$

$$v_{11} = r_1, \quad v_{12} = r_2, \quad v_{21} = r_3, \quad v_{22} = r_4$$

where r_1, r_2, r_3, r_4 are arbitrary real constants with $r_1 r_4 - r_2 r_3 \neq 0$. The second solution is

$$u_{11} = r_5, \quad u_{12} = \frac{1}{r_6}, \quad u_{21} = \frac{1}{r_7}, \quad u_{22} = 0,$$

$$v_{11} = 0, \quad v_{12} = r_6, \quad v_{21} = r_7, \quad v_{22} = -r_5 r_6 r_7$$

where r_5, r_6, r_7 are arbitrary real constants with $r_6 \neq 0$ and $r_7 \neq 0$. The third solution is

$$u_{11} = -\frac{r_{10}}{r_8 r_9}, \quad u_{12} = \frac{1}{r_8}, \quad u_{21} = \frac{1}{r_9}, \quad u_{22} = 0,$$

$$v_{11} = 0, \quad v_{12} = r_8, \quad v_{21} = r_9, \quad v_{22} = r_{10}$$

where r_8, r_9, r_{10} are arbitrary real constants with $r_8 \neq 0$ and $r_9 \neq 0$.

Problem 9. Consider the 2×2 matrices

$$A = \begin{pmatrix} a_{11} & a_{12} \\ a_{12} & a_{11} \end{pmatrix}, \quad C = \begin{pmatrix} 0 & 1 \\ 1 & 0 \end{pmatrix}$$

where $a_{11}, a_{12} \in \mathbb{R}$. Can the expression $A^3 + 3AC(A+C) + C^3$ be simplified for computation?

Solution 9. Yes. Since $[A, C] = 0_2$, i.e. $AC = CA$ we have $(A + C)^3$.

Problem 10. Let A, B be 2×2 matrices. Let $AB = 0_2$ and $BA = 0_2$. Can we conclude that at least one of the two matrices is the 2×2 zero matrix? Prove or disprove.

Solution 10. Consider the matrices

$$A = \begin{pmatrix} 1 & 0 \\ 0 & 0 \end{pmatrix}, \quad B = \begin{pmatrix} 0 & 0 \\ 0 & 1 \end{pmatrix}.$$

Then $AB = 0_2$ and $BA = 0_2$ but both are not the zero matrix. Another example is

$$C = \begin{pmatrix} 1 & 1 \\ 1 & 1 \end{pmatrix}, \quad D = \begin{pmatrix} 1 & -1 \\ -1 & 1 \end{pmatrix}.$$

Note that the rank of all these matrices is 1.

Problem 11. Let A, C be $n \times n$ matrices over $\mathbb{R}$. Let $\mathbf{x}$, $\mathbf{y}$, $\mathbf{b}$, $\mathbf{d}$ be column vectors in $\mathbb{R}^n$. Write the system of equations $(A + iC)(\mathbf{x} + i\mathbf{y}) = (\mathbf{b} + i\mathbf{d})$ as a $2n \times 2n$ set of real equations.

Solution 11. We have $(A + iC)(\mathbf{x} + i\mathbf{y}) = A\mathbf{x} - C\mathbf{y} + i(A\mathbf{y} + C\mathbf{x})$. Thus we can write

$$\begin{pmatrix} A & -C \\ C & A \end{pmatrix} \begin{pmatrix} \mathbf{x} \\ \mathbf{y} \end{pmatrix} = \begin{pmatrix} \mathbf{b} \\ \mathbf{d} \end{pmatrix}.$$

Problem 12. Let A, B be $n \times n$ symmetric matrices over $\mathbb{R}$, i.e. $A^T = A$, $B^T = B$. What is the condition on A, B such that AB is symmetric?

Solution 12.   We have $(AB)^T = B^T A^T = BA$. Thus the condition is that $AB = BA$, i.e. $[A, B] = 0_n$.

Problem 13.   (i) Compute the matrix product

$$(x_1 \quad x_2 \quad x_3) \begin{pmatrix} 4 & -1 & 2 \\ -1 & 0 & 1 \\ 2 & 1 & 0 \end{pmatrix} \begin{pmatrix} x_1 \\ x_2 \\ x_3 \end{pmatrix}.$$

(ii) Write the quadratic polynomial $3x_1^2 - 8x_1 x_2 + 2x_2^2 + 6x_1 x_3 - 3x_3^2$ in matrix form.

Solution 13.   (i) We obtain the polynomial $4x_1^2 - 2x_1 x_2 + 4x_1 x_3 + 2x_2 x_3$.
(ii) We have

$$(x_1 \quad x_2 \quad x_3) \begin{pmatrix} 3 & -2 & 3 \\ -6 & 2 & 4 \\ 3 & -4 & -3 \end{pmatrix} \begin{pmatrix} x_1 \\ x_2 \\ x_3 \end{pmatrix}.$$

Problem 14.   An $n \times n$ matrix M is called *normal* if $MM^* = M^*M$. Let A, B be normal $n \times n$ matrices. Assume that $AB^* = B^*A$, $BA^* = A^*B$. Show that their sum $A + B$ is normal. Show that their product AB is normal.

Solution 14.   We have

$$(A + B)^*(A + B) = A^*A + A^*B + B^*A + B^*B = (A + B)(A + B)^*$$

and

$$(AB)^*(AB) = B^*A^*AB = (AB^*)(BA^*) = A(BB^*)A^* = (AB)(AB)^*.$$

Problem 15.   Let A be an $n \times n$ normal matrix. Show that $\ker(A) = \ker(A^*)$, where ker denotes the *kernel*.

Solution 15.   If A is normal and $\mathbf{v} \in \mathbb{C}^n$, then

$$(A\mathbf{v})^* A\mathbf{v} = \mathbf{v}^* A^* A\mathbf{v} = \mathbf{v}^* AA^* \mathbf{v} = (A^*\mathbf{v})^* A^*\mathbf{v}.$$

Thus $A\mathbf{v} = \mathbf{0}$ if and only if $A^*\mathbf{v} = \mathbf{0}$.

Problem 16.   (i) Let A be an $n \times n$ hermitian matrix. Show that A^m is a hermitian matrix for all $m \in \mathbb{N}$.
(ii) Let B be an $n \times n$ hermitian matrix. Is iB skew-hermitian?

Solution 16.   (i) We have $(A^m)^* = (A^*)^m$.
(ii) Yes. Since $i^* = -i$ we have $(iB)^* = -iB^* = -iB$. Thus iB is skew-hermitian. Can any skew-hermitian matrix be written in the form iB with B hermitian?

Problem 17. (i) Let A be an $n \times n$ matrix with $A^2 = 0_n$. Is the matrix $I_n + A$ invertible?
(ii) Let B be an $n \times n$ matrix with $B^3 = 0_n$. Show that $I_n + B$ has an inverse.

Solution 17. (i) We have $(I_n + A)(I_n - A) = I_n + A - A - A^2 = I_n$. Thus the matrix $I_n + A$ is invertible and the inverse is given by $I_n - A$.
(ii) We have

$$(I_n + B)(I_n - B + B^2) = I_n - B + B^2 + B - B^2 + B^3 = I_n.$$

Thus $I_n - B + B^2$ is the inverse of $I_n + B$.

Problem 18. Let A be an $n \times n$ matrix over $\mathbb{R}$. Assume that $A^2 = 0_n$. Find the inverse of $I_n + iA$.

Solution 18. We have $(I_n + iA)(I_n - iA) = I_n + A^2 = I_n$. Thus $I_n - iA$ is the inverse of $I_n + iA$.

Problem 19. Let A, B be $n \times n$ matrices and c a constant. Assume that the inverses of $(A - cI_n)$ and $(A + B - cI_n)$ exist. Show that

$$(A - cI_n)^{-1} B (A + B - cI_n)^{-1} \equiv (A - cI_n)^{-1} - (A + B - cI_n)^{-1}.$$

Solution 19. Multiplying the identity from the right with $(A + B - cI_n)$ yields

$$(A - cI_n)^{-1} B \equiv (A - cI_n)^{-1}(A + B - cI_n) - I_n.$$

Multiplying this identity from the left with $(A - cI_n)$ yields $B \equiv B$. Reversing this process provides the identity.

Problem 20. (i) A 3×3 matrix M over $\mathbb{R}$ is orthogonal if and only if the columns of M form an orthonormal basis in $\mathbb{R}^3$. Show that the matrix

$$\begin{pmatrix} \sqrt{3}/3 & 0 & -\sqrt{6}/3 \\ \sqrt{3}/3 & \sqrt{2}/2 & \sqrt{6}/6 \\ \sqrt{3}/3 & -\sqrt{2}/2 & \sqrt{6}/6 \end{pmatrix}$$

is orthogonal.
(ii) Represent the nonnormal 3×3 matrix

$$A = \begin{pmatrix} 1 & 0 & 1 \\ 0 & 2 & 0 \\ -1 & 0 & -1 \end{pmatrix} \qquad \text{(relative to the natural basis)}$$

relative to the orthonormal basis in $\mathbb{R}^3$

$$\left\{ \frac{1}{\sqrt{2}} \begin{pmatrix} 1 \\ 0 \\ 1 \end{pmatrix}, \begin{pmatrix} 0 \\ 1 \\ 0 \end{pmatrix}, \frac{1}{\sqrt{2}} \begin{pmatrix} 1 \\ 0 \\ -1 \end{pmatrix} \right\}.$$

Solution 20. (i) The three normalized column vectors are

$$\mathbf{v}_1 = \begin{pmatrix} \sqrt{3}/3 \\ \sqrt{3}/3 \\ \sqrt{3}/3 \end{pmatrix}, \qquad \mathbf{v}_2 = \begin{pmatrix} 0 \\ \sqrt{2}/2 \\ -\sqrt{2}/2 \end{pmatrix}, \qquad \mathbf{v}_3 = \begin{pmatrix} -\sqrt{6}/3 \\ \sqrt{6}/6 \\ \sqrt{6}/6 \end{pmatrix}.$$

The scalar products are $\mathbf{v}_1^T \mathbf{v}_2 = 0$, $\mathbf{v}_1^T \mathbf{v}_3 = 0$, $\mathbf{v}_2^T \mathbf{v}_3 = 0$.
(ii) We form the orthogonal matrix

$$O = \begin{pmatrix} 1/\sqrt{2} & 0 & 1/\sqrt{2} \\ 0 & 1 & 0 \\ 1/\sqrt{2} & 0 & -1/\sqrt{2} \end{pmatrix}$$

from the orthonormal basis. Then $O = O^{-1} = O^T$ and

$$OAO^{-1} = \begin{pmatrix} 0 & 0 & 0 \\ 0 & 2 & 0 \\ 2 & 0 & 0 \end{pmatrix}.$$

Note that A and OAO^{-1} are nonnormal and have the eigenvalues 0 (2 times) and 2.

Problem 21. Consider the *rotation matrix*

$$R(\theta) = \begin{pmatrix} \cos(\theta) & -\sin(\theta) \\ \sin(\theta) & \cos(\theta) \end{pmatrix}.$$

Let n be a positive integer. Calculate $R^n(\theta)$.

Solution 21. Since

$$\begin{pmatrix} \cos(\alpha) & -\sin(\alpha) \\ \sin(\alpha) & \cos(\alpha) \end{pmatrix} \begin{pmatrix} \cos(\beta) & -\sin(\beta) \\ \sin(\beta) & \cos(\beta) \end{pmatrix} \equiv \begin{pmatrix} \cos(\alpha+\beta) & -\sin(\alpha+\beta) \\ \sin(\alpha+\beta) & \cos(\alpha+\beta) \end{pmatrix}$$

we obtain

$$R^n(\theta) = \begin{pmatrix} \cos(n\theta) & -\sin(n\theta) \\ \sin(n\theta) & \cos(n\theta) \end{pmatrix}.$$

Problem 22. (i) Consider the 2×2 matrix

$$A = \begin{pmatrix} 1 & \alpha \\ 1 & \beta \end{pmatrix}$$

where $\alpha, \beta \in \mathbb{R}$. Find the condition on α, β such that the inverse matrix exists. Find the inverse in this case.
(ii) Let $t \neq 0$ and $s \in \mathbb{R}$. Find the inverse of the matrix

$$M = \begin{pmatrix} t & s \\ 0 & 1 \end{pmatrix}.$$

Solution 22. (i) We have $\det(A) = \beta - \alpha$. Thus the inverse exists if $\alpha \neq \beta$. In this case the inverse is given by

$$A^{-1} = \frac{1}{\beta - \alpha} \begin{pmatrix} \beta & -\alpha \\ -1 & 1 \end{pmatrix}.$$

(ii) We find

$$M^{-1} = \begin{pmatrix} t^{-1} & -t^{-1}s \\ 0 & 1 \end{pmatrix}.$$

Problem 23. Let

$$A = \begin{pmatrix} 0 & 1 & 0 & 0 \\ -1 & 0 & 0 & 0 \\ 0 & 0 & 0 & 1 \\ 0 & 0 & -1 & 0 \end{pmatrix}, \qquad B = \begin{pmatrix} 0 & 0 & 1 & 0 \\ 0 & 0 & 0 & 1 \\ -1 & 0 & 0 & 0 \\ 0 & -1 & 0 & 0 \end{pmatrix}.$$

Can one find a 4×4 permutation matrix such that $A = PBP^T$?

Solution 23. We find the permutation matrix

$$P = P^T = P^{-1} = \begin{pmatrix} 1 & 0 & 0 & 0 \\ 0 & 0 & 1 & 0 \\ 0 & 1 & 0 & 0 \\ 0 & 0 & 0 & 1 \end{pmatrix}.$$

Problem 24. (i) Consider the 3×3 permutation matrix

$$P = \begin{pmatrix} 0 & 1 & 0 \\ 0 & 0 & 1 \\ 1 & 0 & 0 \end{pmatrix}.$$

Find all 3×3 matrices A such that $PAP^T = A$.
(ii) Consider the 4×4 permutation matrix

$$P = \begin{pmatrix} 0 & 1 & 0 & 0 \\ 0 & 0 & 1 & 0 \\ 0 & 0 & 0 & 1 \\ 1 & 0 & 0 & 0 \end{pmatrix}.$$

Find all 4×4 matrices A such that $PAP^T = A$.

Solution 24. (i) From the 9 conditions of $PAP^T = A$ we obtain

$$a_{11} = a_{22} = a_{33}, \quad a_{12} = a_{23} = a_{31}, \quad a_{13} = a_{21} = a_{32}.$$

Thus

$$A = \begin{pmatrix} a_{11} & a_{12} & a_{13} \\ a_{13} & a_{11} & a_{12} \\ a_{12} & a_{13} & a_{11} \end{pmatrix}.$$

(ii) From the 16 conditions of $PAP^T = A$ we find $a_{11} = a_{22} = a_{33} = a_{44}$ and

$$a_{12} = a_{23} = a_{34} = a_{41}, \quad a_{13} = a_{24} = a_{31} = a_{42}, \quad a_{14} = a_{21} = a_{32} = a_{43}.$$

Thus

$$A = \begin{pmatrix} a_{11} & a_{12} & a_{13} & a_{14} \\ a_{14} & a_{11} & a_{12} & a_{13} \\ a_{13} & a_{14} & a_{11} & a_{12} \\ a_{12} & a_{13} & a_{14} & a_{11} \end{pmatrix}.$$

Problem 25.   (i) Let π be a permutation on $\{1, 2, \ldots, n\}$. The matrix P_π for which $p_{i*} = e_{\pi(i)*}$ is called the *permutation matrix* associated with π, where p_{i*} is the i-th row of P_π and $e_{ij} = 1$ if $i = j$ and 0 otherwise. Let $n = 4$ and $\pi = (3\ 2\ 4\ 1)$. Find P_π.
(ii) Find the corresponding permutation matrix for the permutation

$$\begin{pmatrix} 1 & 2 & 3 & 4 \\ 3 & 4 & 1 & 2 \end{pmatrix}.$$

Solution 25.   (i) P_π is obtained from the $n \times n$ identity matrix I_n by applying π to the rows of I_n. We find

$$P_\pi = \begin{pmatrix} 0 & 0 & 1 & 0 \\ 0 & 1 & 0 & 0 \\ 0 & 0 & 0 & 1 \\ 1 & 0 & 0 & 0 \end{pmatrix}$$

and $p_{1*} = e_{3*} = e_{\pi(1)*}$, $p_{2*} = e_{2*} = e_{\pi(2)*}$, $p_{3*} = e_{4*} = e_{\pi(3)*}$, $p_{4*} = e_{1*} = e_{\pi(4)*}$.
(ii) We have two options: For $(1\ \ 2\ \ 3\ \ 4)\,P = (3\ \ 4\ \ 1\ \ 2)$ the permutation matrix P is given by

$$P = \begin{pmatrix} 0 & 0 & 1 & 0 \\ 0 & 0 & 0 & 1 \\ 1 & 0 & 0 & 0 \\ 0 & 1 & 0 & 0 \end{pmatrix}.$$

For

$$Q \begin{pmatrix} 1 \\ 2 \\ 3 \\ 4 \end{pmatrix} = \begin{pmatrix} 3 \\ 4 \\ 1 \\ 2 \end{pmatrix}$$

the permutation matrix Q is given by $Q = P$.

Problem 26.   (i) Find an invertible $2n \times 2n$ matrix T such that

$$T^{-1} \begin{pmatrix} 0_n & I_n \\ -I_n & 0_n \end{pmatrix} T = \begin{pmatrix} 0_n & -I_n \\ I_n & 0_n \end{pmatrix}.$$

(ii) Show that the $2n \times 2n$ matrix

$$R = \frac{1}{\sqrt{2}} \begin{pmatrix} I_n & I_n \\ I_n & -I_n \end{pmatrix}$$

is invertible. Find the inverse.

Solution 26.   (i) We find

$$T = T^{-1} = \begin{pmatrix} 0_n & I_n \\ I_n & -I_n \end{pmatrix}.$$

(ii) Since $R^2 = I_{2n}$ we have $R^{-1} = R$.

Problem 27.   Let M be an $2n \times 2n$ matrix with $n \geq 1$. Then M can be written in *block form*

$$M = \begin{pmatrix} A & B \\ C & D \end{pmatrix}$$

where A, B, C, D are $n \times n$ matrices. Assume that M^{-1} exists and that the $n \times n$ matrix D is also nonsingular. Find M^{-1} using this condition.

Solution 27.   We have

$$MM^{-1} = \begin{pmatrix} A & B \\ C & D \end{pmatrix} \begin{pmatrix} E & F \\ G & H \end{pmatrix} = \begin{pmatrix} I_n & 0_n \\ 0_n & I_n \end{pmatrix}.$$

It follows that

$$AE + BG = I_n, \quad AF + BH = 0_n, \quad CE + DG = 0_n, \quad CF + DH = I_n.$$

Since D is nonsingular we obtain from the last two equations that

$$G = -D^{-1}CE, \qquad H = D^{-1} - D^{-1}CF.$$

It follows that $(A - BD^{-1}C)E = I_n$. Thus E and $A - BD^{-1}C$ are invertible and we obtain $E = (A - BD^{-1}C)^{-1}$. For F we find

$$F = -EBD^{-1} = -(A - BD^{-1}C)^{-1}BD^{-1}.$$

Problem 28.   (i) Let A be an $m \times n$ matrix with $m \geq n$. Assume that A has rank n. Show that there exists an $m \times n$ matrix B such that the $n \times n$ matrix B^*A is nonsingular. The matrix B can be chosen such that $B^*A = I_n$.
(ii) An $n^2 \times n$ matrix J is called a *selection matrix* such that J^T is the $n \times n^2$ matrix

$$[E_{11} \ E_{22} \ \dots \ E_{nn}]$$

where E_{ii} is the $n \times n$ matrix of zeros except for a 1 in the (i, i)-th position. Find J for $n = 2$ and calculate $J^T J$. Calculate $J^T J$ for arbitrary n.

Solution 28. (i) If $m = n$, the A is nonsingular and we choose $B = (A^*)^{-1}$. If $n < m$, then there exists an $m \times (m - n)$ matrix C such that the $m \times m$ matrix $Q = [A\ C]$ is nonsingular. If we write $(P^*)^{-1}$ as $[B\ D]$, where B is $m \times n$ and D is $m \times (m - n)$, then $B^*A = I_n$, $B^*C = 0_{n \cdot (m-n)}$, $D^*A = 0_{m-(n \cdot n)}$ and $D^*C = I_{m-n}$.

(ii) We have

$$J^T = \begin{pmatrix} 1 & 0 & 0 & 0 \\ 0 & 0 & 0 & 1 \end{pmatrix} \Rightarrow J = \begin{pmatrix} 1 & 0 \\ 0 & 0 \\ 0 & 0 \\ 0 & 1 \end{pmatrix}.$$

Therefore $J^T J = I_2$. For the general case we find $J^T J = I_n$.

Problem 29. Let A be an arbitrary $n \times n$ matrix over $\mathbb{R}$. Can we conclude that A^2 is positive semidefinite?

Solution 29. No we cannot conclude in general that A^2 is positive semidefinite. For example

$$A = \begin{pmatrix} 0 & 1 \\ -1 & 0 \end{pmatrix} \Rightarrow A^2 = \begin{pmatrix} -1 & 0 \\ 0 & -1 \end{pmatrix}.$$

Problem 30. An $n \times n$ matrix Π is a *projection matrix* if $\Pi^* = \Pi$, $\Pi^2 = \Pi$.

(i) Let Π_1 and Π_2 be projection matrices. Is $\Pi_1 + \Pi_2$ a projection matrix?

(ii) Let Π_1 and Π_2 be projection matrices. Is $\Pi_1 \Pi_2$ a projection matrix?

(iii) Let Π be a projection matrix. Is $I_n - \Pi$ a projection matrix? Calculate $\Pi(I_n - \Pi)$.

(iv) Is

$$\Pi = \frac{1}{3} \begin{pmatrix} 1 & 1 & 1 \\ 1 & 1 & 1 \\ 1 & 1 & 1 \end{pmatrix}$$

a projection matrix?

(v) Let $\epsilon \in [0, 1]$. Show that the 2×2 matrix

$$\Pi(\epsilon) = \begin{pmatrix} \epsilon & \sqrt{\epsilon - \epsilon^2} \\ \sqrt{\epsilon - \epsilon^2} & 1 - \epsilon \end{pmatrix}$$

is a projection matrix. What are the eigenvalues of $\Pi(\epsilon)$? Be clever.

Solution 30. (i) Obviously $(\Pi_1 + \Pi_2)^* = \Pi_1^* + \Pi_2^* = \Pi_1 + \Pi_2$. We have

$$(\Pi_1 + \Pi_2)^2 = \Pi_1^2 + \Pi_1 \Pi_2 + \Pi_2 \Pi_1 + \Pi_2^2 = \Pi_1 + \Pi_1 \Pi_2 + \Pi_2 \Pi_1 + \Pi_2.$$

Thus $\Pi_1 + \Pi_2$ is a projection matrix only if $\Pi_1 \Pi_2 = 0_n$, where we used that from $\Pi_1 \Pi_2 = 0_n$ we can conclude that $\Pi_2 \Pi_1 = 0_n$. From $\Pi_1 \Pi_2 = 0_n$ it follows that $(\Pi_1 \Pi_2)^* = \Pi_2^* \Pi_1^* = \Pi_2 \Pi_1 = 0_n$.

(ii) We have $(\Pi_1 \Pi_2)^* = \Pi_2^* \Pi_1^* = \Pi_2 \Pi_1$ and $(\Pi_1 \Pi_2)^2 = \Pi_1 \Pi_2 \Pi_1 \Pi_2$. Thus we see that $\Pi_1 \Pi_2$ is a projection matrix if and only if $\Pi_1 \Pi_2 = \Pi_2 \Pi_1$.

(iii) We have $(I_n - \Pi)^* = I_n^* - \Pi^* = I_n - \Pi$ and $(I_n - \Pi)^2 = I_n - \Pi$. Thus $I_n - \Pi$ is a projection matrix. We have

$$\Pi(I_n - \Pi) = \Pi - \Pi^2 = \Pi - \Pi = 0_n.$$

(iv) We find $\Pi^* = \Pi$ and $\Pi^2 = \Pi$. Thus Π is a projection matrix.
(v) The matrix is symmetric over $\mathbb{R}$. We have $\Pi^2(\epsilon) = \Pi(\epsilon)$, $\Pi^T(\epsilon) = \Pi(\epsilon)$. Thus $\Pi(\epsilon)$ is a projection matrix. The eigenvalues are 1 and 0.

Problem 31. Let $m \geq 1$ and $N \geq 2$. Assume that $N > m$. Let X be an $N \times m$ matrix over $\mathbb{R}$ such that $X^*X = I_m$.
(i) We define $\Pi := XX^*$. Calculate Π^2, Π^* and $\text{tr}(\Pi)$.
(ii) Give an example for such a matrix X, where $m = 1$ and $N = 2$.

Solution 31. (i) We have

$$\Pi^2 = (XX^*)(XX^*) = X(X^*X)X^* = XI_mX^* = XX^* = \Pi$$

and $\Pi^* = (XX^*)^* = XX^* = \Pi$. Thus Π is an $N \times N$ projection matrix. We have $\text{tr}(\Pi) = m$.
(ii) An example is

$$X = \frac{1}{\sqrt{2}} \begin{pmatrix} 1 \\ 1 \end{pmatrix} \Rightarrow X^* = \frac{1}{\sqrt{2}} (1 \quad 1) \Rightarrow XX^* = \frac{1}{2} \begin{pmatrix} 1 & 1 \\ 1 & 1 \end{pmatrix}.$$

Any normalized vector in $\mathbb{R}^2$ can be used.

Problem 32. (i) A square matrix A is called *idempotent* if $A^2 = A$. Find all 2×2 matrices A over the real numbers which are idempotent and $a_{ij} \neq 0$ for $i, j = 1, 2$.
(ii) Let A, B be $n \times n$ idempotent matrices. Show that $A + B$ are idempotent if and only if $AB = BA = 0_n$.

Solution 32. (i) From $A^2 = A$ we obtain

$$a_{11}^2 + a_{12}a_{21} = a_{11}, \ a_{12}(a_{11}+a_{22}) = a_{12}, \ a_{21}(a_{11}+a_{22}) = a_{21}, \ a_{12}a_{21}+a_{22}^2 = a_{22}.$$

Since $a_{ij} \neq 0$ we obtain $a_{11} + a_{22} = 1$ and $a_{21} = (a_{11} - a_{11}^2)/a_{12}$. Thus the matrix is

$$A = \begin{pmatrix} a_{11} & a_{12} \\ (a_{11} - a_{11}^2)/a_{12} & 1 - a_{11} \end{pmatrix}$$

with a_{11} arbitrary and a_{12} arbitrary and nonzero.
(ii) We have

$$(A + B)^2 = A^2 + B^2 + AB + BA = A + B + AB + BA.$$

Thus $(A + B)^2 = A + B$ if $AB + BA = 0_n$. Now suppose that $AB + BA = 0_n$. We have

$$AB + ABA = A^2B + ABA = A(AB + BA) = 0_n,$$
$$ABA + BA = ABA + BA^2 = (AB + BA)A = 0_n.$$

This implies $AB - BA = AB + ABA - (ABA + BA) = 0_n$. Thus $AB = BA$ and

$$AB = \frac{1}{2}(AB + AB) = \frac{1}{2}(AB + BA) = 0_n.$$

Problem 33.   Let A, B be $n \times n$ matrices over $\mathbb{C}$. Assume that $A + B$ is invertible. Show that

$$(A + B)^{-1}A = I_n - (A + B)^{-1}B, \qquad A(A + B)^{-1} = I_n - B(A + B)^{-1}.$$

Solution 33.   Since $A + B$ is invertible we have

$$(A + B)^{-1}A + (A + B)^{-1}B = (A + B)^{-1}(A + B) = I_n.$$

Consequently $(A + B)^{-1}A = I_n - (A + B)^{-1}B$. Analogously

$$A(A + B)^{-1} + B(A + B)^{-1} = (A + B)(A + B)^{-1} = I_n.$$

Thus $A(A + B)^{-1} = I_n - B(A + B)^{-1}$.

Problem 34.   (i) Find 2×2 matrices A over $\mathbb{C}$ such that $A^2 = -I_2$, $A^* = -A$.
(ii) Find 2×2 matrices C, D such that $CD = 0_2$ and $DC \neq 0_2$.

Solution 34.   (i) Solutions are

$$A = \begin{pmatrix} i & 0 \\ 0 & i \end{pmatrix}, \qquad A = \begin{pmatrix} i & 0 \\ 0 & -i \end{pmatrix}.$$

Extend to 3×3 matrices.
(ii) An example is

$$C = \begin{pmatrix} 0 & 1 \\ 0 & 0 \end{pmatrix}, \qquad D = \begin{pmatrix} 1 & 0 \\ 0 & 0 \end{pmatrix}.$$

Problem 35.   Find the 2×2 matrices F and F' from the two equations

$$\begin{pmatrix} 0 & 0 \\ 1 & 0 \end{pmatrix} = \frac{1}{2}(F + iF'), \qquad \begin{pmatrix} 0 & 1 \\ 0 & 0 \end{pmatrix} = \frac{1}{2}(F - iF').$$

Find the anticommutator of F and F', i.e. $[F, F']_+ \equiv FF' + F'F$.

Solution 35.   Adding and subtracting the two equations yields

$$F = \begin{pmatrix} 0 & 1 \\ 1 & 0 \end{pmatrix} = \sigma_1, \qquad F' = \begin{pmatrix} 0 & i \\ -i & 0 \end{pmatrix} = -\sigma_2.$$

We obtain $[F, F']_+ = 0_2$.

Problem 36. (i) Find all 2×2 matrices X such that

$$X \begin{pmatrix} 0 & 1 \\ 1 & 0 \end{pmatrix} = \begin{pmatrix} 0 & 1 \\ 1 & 0 \end{pmatrix} X.$$

(ii) Impose the additional conditions such that $\text{tr}(X) = 0$ and $\det(X) = +1$.

Solution 36. (i) We obtain $x_{11} = x_{22}$ and $x_{12} = x_{21}$. Thus the matrix is

$$X = \begin{pmatrix} x_{11} & x_{12} \\ x_{12} & x_{11} \end{pmatrix}$$

with x_{11}, x_{12} arbitrary.
(ii) The condition $\text{tr}(X) = 0$ yields $x_{11} = 0$. The condition $\det(X) = +1$ yields $x_{12} = \pm i$.

Problem 37. Let $\mathbf{v}$ be a normalized (column) vector in $\mathbb{C}^n$. Consider the $n \times n$ matrix

$$A = \mathbf{v}\mathbf{v}^* - \frac{1}{2}I_n.$$

(i) Find A^* and AA^*.
(ii) Is the matrix A invertible?

Solution 37. (i) Since $(\mathbf{v}\mathbf{v}^*)^* = \mathbf{v}\mathbf{v}^*$ we have $A = A^*$, i.e. the matrix is hermitian. Since $\mathbf{v}\mathbf{v}^*\mathbf{v}\mathbf{v}^* = \mathbf{v}\mathbf{v}^*$ we obtain

$$AA^* = \frac{1}{4}I_n.$$

(ii) Yes. We find $A^{-1} = 4\mathbf{v}\mathbf{v}^* - 2I_n$.

Problem 38. Find all 2×2 matrices

$$A = \begin{pmatrix} a_{11} & a_{12} \\ 0 & a_{22} \end{pmatrix}$$

over $\mathbb{R}$ with $a_{12} \neq 0$ such that $A^2 = I_2$.

Solution 38. From $A^2 = I_2$ we obtain the three conditions

$$a_{11}^2 = 1, \quad a_{22}^2 = 1, \quad a_{11}a_{12} + a_{12}a_{22} = 0$$

and the two possible solutions

$$A = \begin{pmatrix} 1 & a_{12} \\ 0 & -1 \end{pmatrix}, \quad A = \begin{pmatrix} -1 & a_{12} \\ 0 & 1 \end{pmatrix}, \quad a_{12} \text{ arbitrary.}$$

Problem 39. Let $z = x + iy$ with $x, y \in \mathbb{R}$ and thus $\bar{z} = x - iy$. Consider the maps

$$x + iy \leftrightarrow \begin{pmatrix} x & -y \\ y & x \end{pmatrix} = A(x, y), \quad 1 \leftrightarrow \begin{pmatrix} 1 \\ 0 \end{pmatrix}, \quad i \leftrightarrow \begin{pmatrix} 0 \\ 1 \end{pmatrix}.$$

(i) Calculate z^2 and $A^2(x, y)$. Discuss.
(ii) Find the eigenvalues and normalized eigenvectors of $A(x, y)$.
(iii) Calculate $A(x, y) \otimes A(x, y)$ and find the eigenvalues and normalized eigen-vectors.

Solution 39.   (i) We obtain $z^2 = x^2 - y^2 + 2ixy$ and

$$A^2(x, y) = \begin{pmatrix} x^2 - y^2 & -2xy \\ 2xy & x^2 - y^2 \end{pmatrix}.$$

(ii) The eigenvalues depending on x, y are $\lambda_1 = z = x + iy$, $\lambda_2 = \bar{z} = x - iy$ with the corresponding normalized eigenvectors

$$\mathbf{v}_1 = \frac{1}{\sqrt{2}} \begin{pmatrix} 1 \\ -i \end{pmatrix}, \qquad \mathbf{v}_2 = \frac{1}{\sqrt{2}} \begin{pmatrix} 1 \\ -i \end{pmatrix}$$

which are independent of x and y.
(iii) We have

$$A(x, y) \otimes A(x, y) = \begin{pmatrix} x^2 & -xy & -xy & y^2 \\ xy & x^2 & -y^2 & -xy \\ xy & -y^2 & x^2 & -xy \\ y^2 & xy & xy & x^2 \end{pmatrix}$$

with the eigenvalues z^2, $z\bar{z}$, $\bar{z}z$, $\bar{z}\bar{z}$. Note that $z\bar{z} = \bar{z}z$.

Problem 40.   Let $A \in \mathbb{C}^{n \times m}$ with $n \geq m$ and $\text{rank}(A) = m$.
(i) Show that the matrix $m \times m$ matrix A^*A is invertible.
(ii) We set $\Pi = A(A^*A)^{-1}A$. Show that Π is a projection matrix, i.e. $\Pi^2 = \Pi$ and $\Pi = \Pi^*$.

Solution 40.   (i) To show that A^*A is invertible it suffices to show that $N(A^*A) = \{\mathbf{0}\}$. From $A^*A\mathbf{v} = \mathbf{0}$ we obtain $\mathbf{v}^*A^*A\mathbf{v} = 0$. Thus $(A\mathbf{v})^*A\mathbf{v} = 0$ and therefore $A\mathbf{v} = \mathbf{0}$. Hence $\mathbf{v} = \mathbf{0}$.
(ii) We have

$$\Pi^2 = (A(A^*A)^{-1}A)(A(A^*A)^{-1}A) = A(A^*A)^{-1} = A(A^*A)^{-1}A^* = \Pi$$

and

$$\Pi^* = A(A^*A)^{-1} = A((A^*A)^*)^{-1} = A(A^*A)^{-1}A = \Pi.$$

Problem 41.   Let A be an $n \times n$ hermitian matrix and Π be an $n \times n$ projection matrix. Then $\Pi A \Pi$ is again a hermitian matrix. Is this still true if A is a normal matrix, i.e. $AA^* = A^*A$?

Solution 41.   This is not true in general. Consider

$$A = \begin{pmatrix} 0 & 0 & 1 \\ 1 & 0 & 0 \\ 0 & 1 & 0 \end{pmatrix}, \qquad \Pi = \begin{pmatrix} 0 & 0 & 0 \\ 0 & 1 & 0 \\ 0 & 0 & 1 \end{pmatrix}.$$

Then $A^*A = AA^*$ and

$$\Pi A\Pi = \begin{pmatrix} 0 & 0 & 0 \\ 0 & 0 & 0 \\ 0 & 1 & 0 \end{pmatrix}, \qquad (\Pi A\Pi)^* = \begin{pmatrix} 0 & 0 & 0 \\ 0 & 0 & 1 \\ 0 & 0 & 0 \end{pmatrix}.$$

Thus $(\Pi A\Pi)(\Pi A\Pi)^* \neq (\Pi A\Pi)^*(\Pi A\Pi)$.

Problem 42. An $n \times n$ matrix C of the form

$$C = \begin{pmatrix} c_0 & c_{n-1} & \cdots & c_2 & c_1 \\ c_1 & c_0 & c_{n-1} & & c_2 \\ \vdots & c_1 & c_0 & \ddots & \vdots \\ c_{n-2} & & \ddots & \ddots & c_{n-1} \\ c_{n-1} & c_{n-2} & \cdots & c_1 & c_0 \end{pmatrix}$$

is called a *circulant matrix*. Show that the set of all $n \times n$ circulant matrices form an n-dimensional vector space.

Solution 42. A circulant matrix is fully specified by the n-dimensional vector

$$\mathbf{c} = \begin{pmatrix} c_0 & c_1 & c_2 & \cdots & c_{n-1} \end{pmatrix}^T$$

which appears in the first column of the matrix C. The other columns are rotations of it. The last row is in reverse order. The sum of two $n \times n$ circulant matrices is again a circulant matrix and multiplying a circulant matrix with a constant provides again a circulant matrix. Thus the $n \times n$ circulant matrices form an n-dimensional vector space.

Problem 43. Two $n \times n$ matrices A, B are called *similar* if there exists an invertible $n \times n$ matrix S such that $A = SBS^{-1}$. If A is similar to B, then B is also similar to A, since $B = S^{-1}AS$.
(i) Show that the two 2×2 matrices

$$A = \begin{pmatrix} 0 & 1 \\ 0 & 0 \end{pmatrix}, \qquad B = \begin{pmatrix} 0 & 0 \\ 1 & 0 \end{pmatrix} = A^T$$

are similar.
(ii) Consider the two 2×2 invertible matrices

$$A = \begin{pmatrix} 1 & 0 \\ 2 & 1 \end{pmatrix}, \qquad B = \begin{pmatrix} 1 & 0 \\ 0 & 1 \end{pmatrix}.$$

Are the matrices similar?
(iii) Consider the two 2×2 invertible matrices

$$C = \begin{pmatrix} 1 & 0 \\ 0 & -1 \end{pmatrix}, \qquad D = \begin{pmatrix} 0 & 1 \\ 1 & 0 \end{pmatrix}.$$

Are the matrices similar?

Solution 43.   (i) We have

$$S = \begin{pmatrix} s_{11} & s_{12} \\ s_{21} & s_{22} \end{pmatrix} \Rightarrow S^{-1} = \frac{1}{\det(S)} \begin{pmatrix} s_{22} & -s_{12} \\ -s_{21} & s_{11} \end{pmatrix}.$$

Then $PBP^{-1} = A$ provides the three conditions

$$s_{12}s_{22} = 0, \quad s_{22}^2 = 0, \quad s_{12}^2 = -\det(S).$$

Thus $s_{22} = 0$ and we arrive at $s_{12}^2 = s_{12}s_{21}$. Since $s_{12} \neq 0$ we have $s_{12} = s_{21}$. Note that s_{11} is arbitrary. Thus

$$S = \begin{pmatrix} s_{11} & s_{12} \\ s_{12} & 0 \end{pmatrix} \Rightarrow S^{-1} = \begin{pmatrix} 0 & 1/s_{12} \\ 1/s_{12} & -s_{11}/s_{12}^2 \end{pmatrix}.$$

A special case is $s_{11} = 0$, $s_{12} = 1$.

(ii) From $A = S^{-1}BS$ we obtain $SA = BS$. Then from $SA = BS$ we obtain

$$\begin{pmatrix} s_{11} + 2s_{12} & s_{12} \\ s_{21} + 2s_{22} & s_{22} \end{pmatrix} = \begin{pmatrix} s_{11} & s_{12} \\ s_{21} & s_{22} \end{pmatrix}.$$

It follows that $s_{12} = 0$ and $s_{22} = 0$. Thus S is not invertible and therefore A and B are not similar.

(iii) The matrices C and D are similar. We find $s_{11} = s_{21}$, $s_{12} = -s_{22}$. Since S must be invertible, all four matrix elements are nonzero. For example, we can select

$$S = \frac{1}{\sqrt{2}} \begin{pmatrix} 1 & 1 \\ 1 & -1 \end{pmatrix}.$$

Problem 44.   An $n \times n$ matrix is called *nilpotent* if some power of it is the zero matrix, i.e. there is a positive integer p such that $A^p = 0_n$.

(i) Show that every nonzero nilpotent matrix is nondiagonalizable.

(ii) Consider the 4×4 matrix

$$N = \begin{pmatrix} 0 & 0 & 0 & 0 \\ 1 & 0 & 0 & 0 \\ 0 & 1 & 0 & 0 \\ 0 & 0 & 1 & 0 \end{pmatrix}.$$

Show that the matrix is nilpotent.

(iii) Is the matrix product of two $n \times n$ nilpotent matrices nilpotent?

Solution 44.   (i) Assume that there is an invertible matrix S such that $SAS^{-1} = D$, where D is a diagonal matrix. Then

$$0_n = SA^pS^{-1} = SAS^{-1}SAS^{-1}\cdots SAS^{-1} = D^p.$$

Thus $D^p = 0_n$ and therefore D must be the zero matrix. Consequently A is the zero matrix.

(ii) We find that N^4 is the 4×4 zero matrix. Thus the matrix N is nilpotent.

(iii) This is not true in general. Consider

$$A = \begin{pmatrix} 0 & 1 \\ 0 & 0 \end{pmatrix}, \quad B = \begin{pmatrix} 0 & 0 \\ 1 & 0 \end{pmatrix} \Rightarrow AB = \begin{pmatrix} 1 & 0 \\ 0 & 0 \end{pmatrix}.$$

Problem 45. A matrix A for which $A^p = 0_n$, where p is a positive integer, is called *nilpotent*. If p is the least positive integer for which $A^p = 0_n$ then A is said to be nilpotent of index p. Find all 2×2 matrices A over the real numbers which are nilpotent with $p = 2$, i.e. $A^2 = 0_2$.

Solution 45. From $A^2 = 0_2$ we obtain the four equations

$$a_{11}^2 + a_{12}a_{21} = 0, \ a_{12}(a_{11} + a_{22}) = 0, \ a_{21}(a_{11} + a_{22}) = 0, \ a_{12}a_{21} + a_{22}^2 = 0.$$

Thus we have to consider the cases $a_{11} + a_{22} \neq 0$ and $a_{11} + a_{22} = 0$. If $a_{11} + a_{22} \neq 0$, then $a_{12} = a_{21} = 0$ and therefore $a_{11} = a_{22} = 0$. Thus we have the 2×2 zero matrix. If $a_{11} + a_{22} = 0$ we have $a_{11} = -a_{22}$ and $a_{11} \neq 0$, otherwise we would find the zero matrix again. Thus $a_{12}a_{21} = -a_{11}^2 = -a_{22}^2$ and for this case we find the solution

$$A = \begin{pmatrix} a_{11} & a_{12} \\ -a_{11}^2/a_{12} & -a_{11} \end{pmatrix}$$

where a_{11}, a_{12} are arbitrary with $a_{12} \neq 0$.

Problem 46. Let

$$\mathbf{a} = \begin{pmatrix} a_1 \\ a_2 \\ a_3 \end{pmatrix}, \quad \mathbf{b} = \begin{pmatrix} b_1 \\ b_2 \\ b_3 \end{pmatrix}$$

be vectors in $\mathbb{R}^3$. The *vector product* $\times$ is defined as

$$\mathbf{a} \times \mathbf{b} := \begin{pmatrix} a_2b_3 - a_3b_2 \\ a_3b_1 - a_1b_3 \\ a_1b_2 - a_2b_1 \end{pmatrix}.$$

(i) Show that we can find a 3×3 matrix $S(\mathbf{a})$ such that $\mathbf{a} \times \mathbf{b} = S(\mathbf{a})\mathbf{b}$.
(ii) Express the *Jacobi identity*

$$\mathbf{a} \times (\mathbf{b} \times \mathbf{c}) + \mathbf{c} \times (\mathbf{a} \times \mathbf{b}) + \mathbf{b} \times (\mathbf{c} \times \mathbf{a}) = \mathbf{0}$$

using the matrices $S(\mathbf{a})$, $S(\mathbf{b})$ and $S(\mathbf{c})$.

Solution 46. (i) The matrix $S(\mathbf{a})$ is the skew-symmetric matrix

$$S(\mathbf{a}) = \begin{pmatrix} 0 & -a_3 & a_2 \\ a_3 & 0 & -a_1 \\ -a_2 & a_1 & 0 \end{pmatrix}.$$

(ii) Using the result from (i) we obtain

$$\mathbf{a} \times (S(\mathbf{b})\mathbf{c}) + \mathbf{c} \times (S(\mathbf{a})\mathbf{b}) + \mathbf{b} \times (S(\mathbf{c})\mathbf{a}) = \mathbf{0}$$
$$S(\mathbf{a})(S(\mathbf{b})\mathbf{c}) + S(\mathbf{c})(S(\mathbf{a})\mathbf{b}) + S(\mathbf{b})(S(\mathbf{c})\mathbf{a}) = \mathbf{0}$$
$$(S(\mathbf{a})S(\mathbf{b}))\mathbf{c} + (S(\mathbf{c})S(\mathbf{a}))\mathbf{b} + (S(\mathbf{b})S(\mathbf{c}))\mathbf{a} = \mathbf{0}$$

where we used that matrix multiplication is associative.

Problem 47. Let

$$
\mathbf{x} = \begin{pmatrix} x_1 \\ x_2 \\ x_3 \end{pmatrix}, \qquad \mathbf{y} = \begin{pmatrix} y_1 \\ y_2 \\ y_3 \end{pmatrix}
$$

be two normalized vectors in $\mathbb{R}^3$. Assume that $\mathbf{x}^T \mathbf{y} = 0$, i.e. the vectors are orthogonal. Is the vector $\mathbf{x} \times \mathbf{y}$ a unit vector again? We have

$$
x_1^2 + x_2^2 + x_3^2 = 1, \qquad y_1^2 + y_2^2 + y_3^2 = 1 \tag{1}
$$

and

$$
\mathbf{x}^T \mathbf{y} = x_1 y_1 + x_2 y_2 + x_3 y_3 = 0. \tag{2}
$$

Solution 47. The square of the norm $\|\mathbf{x} \times \mathbf{y}\|^2$ of the vector $\mathbf{x} \times \mathbf{y}$ is given by

$$
x_1^2(y_2^2 + y_3^2) + x_2^2(y_1^2 + y_3^2) + x_3^2(y_1^2 + y_2^2) - 2(x_2 x_3 y_2 y_3 + x_1 x_3 y_1 y_3 + x_1 x_2 y_1 y_2).
$$

From (1) it follows that $x_3^2 = 1 - x_1^2 - x_2^2$, $y_3^2 = 1 - y_1^2 - y_2^2$. From (2) it follows that $x_3 y_3 = -x_1 y_1 - x_2 y_2$ and

$$
2x_1 y_1 x_2 y_2 = x_3^2 y_3^2 - x_1^2 y_1^2 - x_2^2 y_2^2 = 1 - y_1^2 - y_2^2 - x_1^2 - x_2^2 + x_1^2 y_2^2 + x_2^2 y_1^2.
$$

Inserting these equations into $\|\mathbf{x} \times \mathbf{y}\|$ gives $\|\mathbf{x} \times \mathbf{y}\|^2 = 1$. Thus the vector is normalized.

Problem 48. Consider the three linear independent normalized column vectors in $\mathbb{R}^3$

$$
\mathbf{a}_1 = \frac{1}{\sqrt{2}} \begin{pmatrix} 1 \\ 0 \\ 1 \end{pmatrix}, \qquad \mathbf{a}_2 = \begin{pmatrix} 0 \\ 1 \\ 0 \end{pmatrix}, \qquad \mathbf{a}_3 = \frac{1}{\sqrt{2}} \begin{pmatrix} 1 \\ 0 \\ -1 \end{pmatrix}.
$$

(i) Find the "volume" $V_{\mathbf{a}} := \mathbf{a}_1^T(\mathbf{a}_2 \times \mathbf{a}_3)$.

(ii) From the three vectors $\mathbf{a}_1$, $\mathbf{a}_2$, $\mathbf{a}_3$ we form the 3×3 orthogonal matrix

$$
A = \begin{pmatrix} 1/\sqrt{2} & 0 & 1/\sqrt{2} \\ 0 & 1 & 0 \\ 1/\sqrt{2} & 0 & -1/\sqrt{2} \end{pmatrix}.
$$

Find the determinant and trace. Discuss.

(iii) Find the vectors

$$
\mathbf{b}_1 = \frac{1}{V_{\mathbf{a}}} \mathbf{a}_2 \times \mathbf{a}_3, \qquad \mathbf{b}_2 = \frac{1}{V_{\mathbf{a}}} \mathbf{a}_3 \times \mathbf{a}_1, \qquad \mathbf{b}_3 = \frac{1}{V_{\mathbf{a}}} \mathbf{a}_1 \times \mathbf{a}_2.
$$

Are the three vectors linearly independent?

Solution 48. (i) We have

$$
V_{\mathbf{a}} = (1/\sqrt{2} \quad 0 \quad 1/\sqrt{2})\,((-1/\sqrt{2} \quad 0 \quad -1/\sqrt{2}))^T = -1.
$$

(ii) Since the vectors are linearly independent the determinant must be nonzero. We find $\det(A) = -1$ and $\operatorname{tr}(A) = 1$. Note that

$$\det(A) = \mathbf{a}_1^T(\mathbf{a}_2 \times \mathbf{a}_3).$$

(iii) We obtain $\mathbf{b}_1 = \mathbf{a}_1$, $\mathbf{b}_2 = \mathbf{a}_2$, $\mathbf{b}_3 = \mathbf{a}_3$. Thus the vectors are linearly independent.

Problem 49. Let $\mathbf{v}_1$, $\mathbf{v}_2$, $\mathbf{v}_3$ be three normalized linearly independent vectors in $\mathbb{R}^3$. Give an interpretation of A defined by

$$\cos\left(\frac{1}{2}A\right) = \frac{1 + \mathbf{v}_1 \cdot \mathbf{v}_2 + \mathbf{v}_2 \cdot \mathbf{v}_3 + \mathbf{v}_3 \cdot \mathbf{v}_1}{\sqrt{2(1 + \mathbf{v}_1 \cdot \mathbf{v}_2)(1 + \mathbf{v}_2 \cdot \mathbf{v}_3)(1 + \mathbf{v}_3 \cdot \mathbf{v}_1)}}$$

where $\cdot$ denotes the scalar product. Consider first the case where $\mathbf{v}_1$, $\mathbf{v}_2$, $\mathbf{v}_3$ denote the standard basis in $\mathbb{R}^3$.

Solution 49. The quantity A is the directed area of *spherical triangles* on the unit sphere S^2 with corners given by the vectors $\mathbf{v}_1$, $\mathbf{v}_2$, $\mathbf{v}_3$, whose sign is given by

$$\operatorname{sign}(A) = \operatorname{sgn}(\mathbf{v}_1 \cdot (\mathbf{v}_2 \times \mathbf{v}_3)).$$

For the standard basis we find $\cos(\frac{1}{2}A) = \frac{1}{\sqrt{2}}$.

Problem 50. Let $\mathbf{v}$ be a column vector in $\mathbb{R}^n$ and $\mathbf{v} \neq \mathbf{0}$. Let

$$A := \frac{\mathbf{v}\mathbf{v}^T}{\mathbf{v}^T\mathbf{v}}$$

where T denotes the transpose, i.e. $\mathbf{v}^T$ is a row vector. Calculate A^2.

Solution 50. Obviously $\mathbf{v}\mathbf{v}^T$ is a nonzero $n \times n$ matrix and $\mathbf{v}^T\mathbf{v}$ is a nonzero real number. We find

$$A^2 = \left(\frac{\mathbf{v}\mathbf{v}^T}{\mathbf{v}^T\mathbf{v}}\right)^2 = \frac{(\mathbf{v}\mathbf{v}^T)(\mathbf{v}\mathbf{v}^T)}{(\mathbf{v}^T\mathbf{v})^2} = \frac{\mathbf{v}(\mathbf{v}^T\mathbf{v})\mathbf{v}^T}{(\mathbf{v}^T\mathbf{v})^2} = \frac{\mathbf{v}\mathbf{v}^T}{\mathbf{v}^T\mathbf{v}} = A$$

where we used that matrix multiplication is *associative*.

Problem 51. (i) A square matrix A over $\mathbb{C}$ is called *skew-hermitian* if $A = -A^*$. Show that such a matrix is *normal*, i.e. we have $AA^* = A^*A$.
(ii) Let A be an $n \times n$ skew-hermitian matrix over $\mathbb{C}$, i.e. $A^* = -A$. Let U be an $n \times n$ *unitary matrix*, i.e. $U^* = U^{-1}$. Show that $B := U^*AU$ is a skew-hermitian matrix.

Solution 51. (i) We have $AA^* = -A^*A^* = (-A^*)(-A) = A^*A$.
(ii) We have $A^* = -A$. Thus from $B = U^*AU$ and $U^{**} = U$ it follows that

$$B^* = (U^*AU)^* = U^*A^*U = U^*(-A)U = -U^*AU = -B.$$

Problem 52. Let A, X, Y be $n \times n$ matrices. Assume that $XA = I_n$, $AY = I_n$. Show that $X = Y$.

Solution 52. We have $X = XI_n = X(AY) = (XA)Y = I_nY = Y$.

Problem 53. (i) Let A, B be $n \times n$ matrices. Assume that A is nonsingular, i.e. A^{-1} exists. Show that if $BA = 0_n$, then $B = 0_n$.
(ii) Let A, B be $n \times n$ matrices and $A + B = I_n$, $AB = 0_n$. Show that $A^2 = A$ and $B^2 = B$.

Solution 53. (i) We have $B = BI_n = B(AA^{-1}) = (BA)A^{-1} = 0_nA^{-1} = 0_n$.
(ii) Multiplying $A + B = I_n$ with A we obtain $A^2 + AB = A$ and therefore $A^2 = A$. Multiplying $A + B = I_n$ with B yields $AB + B^2 = B$ and therefore $B^2 = B$.

Problem 54. Find a 2×2 matrix A over $\mathbb{R}$ such that

$$A\begin{pmatrix} 1 \\ 0 \end{pmatrix} = \frac{1}{\sqrt{2}}\begin{pmatrix} 1 \\ 1 \end{pmatrix}, \qquad A\begin{pmatrix} 0 \\ 1 \end{pmatrix} = \frac{1}{\sqrt{2}}\begin{pmatrix} 1 \\ -1 \end{pmatrix}.$$

Solution 54. We find $a_{11} = a_{12} = a_{21} = 1/\sqrt{2}$, $a_{22} = -1/\sqrt{2}$. Thus

$$A = \frac{1}{\sqrt{2}}\begin{pmatrix} 1 & 1 \\ 1 & -1 \end{pmatrix}.$$

Problem 55. Consider the vector space $\mathbb{R}^4$. Find all pairwise orthogonal vectors (column vectors) $\mathbf{x}_1, \dots, \mathbf{x}_p$, where the entries of the column vectors can only be $+1$ or -1. Calculate the matrix

$$\sum_{j=1}^{p} \mathbf{x}_j\mathbf{x}_j^T$$

and find the eigenvalues and eigenvectors of this matrix.

Solution 55. The number of vectors p cannot exceed 4 since that would imply $\dim(\mathbb{R}^4) > 4$. A solution is

$$\mathbf{x}_1 = \begin{pmatrix} 1 \\ 1 \\ 1 \\ 1 \end{pmatrix}, \quad \mathbf{x}_2 = \begin{pmatrix} 1 \\ -1 \\ 1 \\ -1 \end{pmatrix}, \quad \mathbf{x}_3 = \begin{pmatrix} 1 \\ -1 \\ -1 \\ 1 \end{pmatrix}, \quad \mathbf{x}_4 = \begin{pmatrix} 1 \\ 1 \\ -1 \\ -1 \end{pmatrix}.$$

Thus

$$\sum_{j=1}^{4} \mathbf{x}_j\mathbf{x}_j^T = \begin{pmatrix} 4 & 0 & 0 & 0 \\ 0 & 4 & 0 & 0 \\ 0 & 0 & 4 & 0 \\ 0 & 0 & 0 & 4 \end{pmatrix}.$$

The eigenvalue is 4 with multiplicity 4. The eigenvectors are all $\mathbf{x} \in \mathbb{R}^4$ with $\mathbf{x} \neq \mathbf{0}$. Another solution is given by

$$\mathbf{x}_1 = \begin{pmatrix} 1 \\ 1 \\ 1 \\ -1 \end{pmatrix}, \quad \mathbf{x}_2 = \begin{pmatrix} 1 \\ 1 \\ -1 \\ 1 \end{pmatrix}, \quad \mathbf{x}_3 = \begin{pmatrix} 1 \\ -1 \\ 1 \\ 1 \end{pmatrix}, \quad \mathbf{x}_4 = \begin{pmatrix} -1 \\ 1 \\ 1 \\ 1 \end{pmatrix}.$$

Problem 56. The $(n+1) \times (n+1)$ *Hadamard matrix* $H(n)$ of any dimension is generated recursively as follows

$$H(n) = \begin{pmatrix} H(n-1) & H(n-1) \\ H(n-1) & -H(n-1) \end{pmatrix}$$

where $n = 1, 2, \ldots$ and $H(0)$ is the 1×1 matrix $H(0) = (1)$. The column vectors in the matrix $H(n)$ form a basis in $\mathbb{R}^n$. They are pairwise orthogonal to each other. Find $H(1)$, $H(2)$, $H(3)$ and $H(3)H^T(3)$. Find the eigenvalues of $H(1)$ and $H(2)$.

Solution 56. We find

$$H(1) = \begin{pmatrix} 1 & 1 \\ 1 & -1 \end{pmatrix}, \qquad H(2) = \begin{pmatrix} 1 & 1 & 1 & 1 \\ 1 & -1 & 1 & -1 \\ 1 & 1 & -1 & -1 \\ 1 & -1 & -1 & 1 \end{pmatrix}$$

and

$$H(3) = \begin{pmatrix} 1 & 1 & 1 & 1 & 1 & 1 & 1 & 1 \\ 1 & -1 & 1 & -1 & 1 & -1 & 1 & -1 \\ 1 & 1 & -1 & -1 & 1 & 1 & -1 & -1 \\ 1 & -1 & -1 & 1 & 1 & -1 & -1 & 1 \\ 1 & 1 & 1 & 1 & -1 & -1 & -1 & -1 \\ 1 & -1 & 1 & -1 & -1 & 1 & -1 & 1 \\ 1 & 1 & -1 & -1 & -1 & -1 & 1 & 1 \\ 1 & -1 & -1 & 1 & -1 & 1 & 1 & -1 \end{pmatrix}.$$

We find $H(3)H^T(3) = 8I_8$. The eigenvalues of $H(1)$ are $\sqrt{2}$ and $-\sqrt{2}$. The eigenvalues of $H(2)$ are 2 (2 times) and -2 (2 times).

Problem 57. A *Hadamard matrix* is an $n \times n$ matrix H with entries in $\{-1, +1\}$ such that any two distinct rows or columns of H have inner product 0. Construct a 4×4 Hadamard matrix starting from the column vector $\mathbf{x}_1 = (1\ 1\ 1\ 1)^T$.

Solution 57. The column vector $\mathbf{x}_2 = (1\ 1\ -1\ -1)^T$ is perpendicular to the vector $\mathbf{x}_1$. Next the column vector

$$\mathbf{x}_3 = (-1\ 1\ 1\ -1)^T$$

is perpendicular to $\mathbf{x}_1$ and $\mathbf{x}_2$. Finally the column vector

$$\mathbf{x}_4 = (1 \ -1 \ 1 \ -1)^T$$

is perpendicular to $\mathbf{x}_1$, $\mathbf{x}_2$ and $\mathbf{x}_3$. Thus we obtain the 4×4 Hadamard matrix

$$H = \begin{pmatrix} 1 & 1 & -1 & 1 \\ 1 & 1 & 1 & -1 \\ 1 & -1 & 1 & 1 \\ 1 & -1 & -1 & -1 \end{pmatrix}.$$

Problem 58.   Consider the 3×3 symmetric matrix over $\mathbb{R}$

$$A = \begin{pmatrix} 2 & 2 & -2 \\ 2 & 2 & -2 \\ -2 & -2 & 6 \end{pmatrix}.$$

(i) Let X be an $m \times n$ matrix. The *column rank* of X is the maximum number of linearly independent columns. The *row rank* is the maximum number of linearly independent rows. The row rank and the column rank of X are equal (called the *rank* of X). Find the rank of A and denote it by k.
(ii) Locate a $k \times k$ submatrix of A having rank k.
(iii) Find 3×3 permutation matrices P and Q such that in the matrix PAQ the submatrix from (ii) is in the upper left portion of A.

Solution 58.   (i) The vectors in the first two columns are linearly dependent. Thus the rank of A is 2.
(ii) A 2×2 submatrix having rank 2 is

$$B = \begin{pmatrix} 2 & -2 \\ -2 & 6 \end{pmatrix}.$$

(iii) We have

$$P = Q = \begin{pmatrix} 1 & 0 & 0 \\ 0 & 0 & 1 \\ 0 & 1 & 0 \end{pmatrix} \quad \Rightarrow \quad PAQ = \begin{pmatrix} 2 & -2 & 2 \\ -2 & 6 & -2 \\ 2 & -2 & 2 \end{pmatrix}.$$

Problem 59.   Find all $n \times n$ matrices A over $\mathbb{R}$ that satisfy the equation $A^T A = 0_n$.

Solution 59.   From $A^T A = 0_n$ we find $\sum_{i=1}^{n} a_{ij} a_{ij} = 0$, where $j = 1, \ldots, n$. Thus A must be the zero matrix.

Problem 60.   A square matrix A such that $A^2 = I_n$ is called *involutory*.
(i) Find all 2×2 matrices over the real numbers which are involutory. Assume that $a_{ij} \neq 0$ for $i, j = 1, 2$.

(ii) Show that an $n \times n$ matrix A is involutory if and only if $(I_n - A)(I_n + A) = 0_n$.

Solution 60. (i) From $A^2 = I_2$ we obtain

$$a_{11}^2 + a_{12}a_{21} = 1, \ a_{12}(a_{11} + a_{22}) = 0, \ a_{21}(a_{11} + a_{22}) = 0, \ a_{12}a_{21} + a_{22}^2 = 1.$$

Since $a_{ij} \neq 0$ we have $a_{11} + a_{22} = 0$ and $a_{21} = (1 - a_{11}^2)/a_{12}$. Then the matrix is given by

$$A = \begin{pmatrix} a_{11} & a_{12} \\ (1 - a_{11}^2)/a_{12} & -a_{11} \end{pmatrix}.$$

(ii) Suppose that $(I_n - A)(I_n + A) = 0_n$. Then $(I_n - A)(I_n + A) = I_n - A^2 = 0_n$. Thus $A^2 = I_n$ and A is involutory. Suppose that A is involutory. Then $A^2 = I_n$ and

$$0_n = I_n - A^2 = (I_n - A)(I_n + A).$$

Problem 61. (i) Let A be an $n \times n$ symmetric matrix over $\mathbb{R}$. Let P be an arbitrary $n \times n$ matrix over $\mathbb{R}$. Show that $P^T A P$ is symmetric.
(ii) Let B be an $n \times n$ skew-symmetric matrix over $\mathbb{R}$, i.e. $B^T = -B$. Let P be an arbitrary $n \times n$ matrix over $\mathbb{R}$. Show that $P^T B P$ is skew-symmetric.

Solution 61. (i) Using that $A^T = A$ and $(P^T)^T = P$ we have $(P^T A P)^T = P^T A^T (P^T)^T = P^T A P$. Thus $P^T A P$ is symmetric.
(ii) Using $B^T = -B$ and $(P^T)^T = P$ we have $(P^T B P)^T = P^T B^T (P^T)^T = -P^T B P$. Thus $P^T B P$ is skew-symmetric.

Problem 62. Let A be an $m \times n$ matrix. The *column rank* of A is the maximum number of linearly independent columns. The *row rank* is the maximum number of linearly independent rows. The row rank and the column rank of A are equal (called the rank of A). Find the rank of the 4×4 matrix

$$A = \begin{pmatrix} 1 & 2 & 3 & 4 \\ 5 & 6 & 7 & 8 \\ 9 & 10 & 11 & 12 \\ 13 & 14 & 15 & 16 \end{pmatrix}.$$

Solution 62. The *elementary transformations* do not change the rank of a matrix. We subtract the third column from the fourth column, the second column from the third column and the first column from the second column, i.e.

$$\begin{pmatrix} 1 & 2 & 3 & 1 \\ 5 & 6 & 7 & 1 \\ 9 & 10 & 11 & 1 \\ 13 & 14 & 15 & 1 \end{pmatrix} \sim \begin{pmatrix} 1 & 2 & 1 & 1 \\ 5 & 6 & 1 & 1 \\ 9 & 10 & 1 & 1 \\ 13 & 14 & 1 & 1 \end{pmatrix} \sim \begin{pmatrix} 1 & 1 & 1 & 1 \\ 5 & 1 & 1 & 1 \\ 9 & 1 & 1 & 1 \\ 13 & 1 & 1 & 1 \end{pmatrix}.$$

From the last matrix we see (three columns are the same) that the rank of A is 2. It follows that two eigenvalues must be 0.

Problem 63. A *Cartan matrix* A is a square matrix whose elements a_{ij} satisfy the following conditions:
1. a_{ij} is an integer, one of $\{-3, -2, -1, 0, 2\}$
2. $a_{jj} = 2$ for all diagonal elements of A
3. $a_{ij} \leq 0$ off of the diagonal
4. $a_{ij} = 0$ iff $a_{ji} = 0$
5. There exists an invertible diagonal matrix D such that DAD^{-1} gives a symmetric and positive definite quadratic form.
Give a 2×2 non-diagonal Cartan matrix and its eigenvalues.

Solution 63. For $n = 2$ the only possible Cartan matrix is

$$A = \begin{pmatrix} 2 & -1 \\ -1 & 2 \end{pmatrix}.$$

The first four conditions are obvious for the matrix A. The last condition can be seen from

$$\mathbf{x}^T A \mathbf{x} = (x_1 \quad x_2) \begin{pmatrix} 2 & -1 \\ -1 & 2 \end{pmatrix} \begin{pmatrix} x_1 \\ x_2 \end{pmatrix} = 2(x_1^2 - x_1 x_2 + x_2^2) \geq 0$$

for $\mathbf{x} \neq \mathbf{0}$. That the symmetric matrix A is positive definite can also be seen from the eigenvalues of A which are 3 and 1.

Problem 64. Let A, B, C, D be $n \times n$ matrices over $\mathbb{R}$. Assume that AB^T and CD^T are symmetric and $AD^T - BC^T = I_n$, where T denotes transpose. Show that $A^T D - C^T B = I_n$.

Solution 64. From the assumptions we have

$$AB^T = (AB^T)^T = BA^T, \quad CD^T = (CD^T)^T = DC^T, \quad AD^T - BC^T = I_n.$$

Taking the transpose of the third equation we have $DA^T - CB^T = I_n$. These four equations can be written in the form of block matrices in the identity

$$\begin{pmatrix} A & B \\ C & D \end{pmatrix} \begin{pmatrix} D^T & -B^T \\ -C^T & A^T \end{pmatrix} = \begin{pmatrix} I_n & 0_n \\ 0_n & I_n \end{pmatrix}.$$

Thus the matrices are $(2n) \times (2n)$ matrices. If X, Y are $m \times m$ matrices with $XY = I_m$, the identity matrix, then $Y = X^{-1}$ and $YX = I_m$ too. Applying this to the matrix equation with $m = 2n$ we obtain

$$\begin{pmatrix} D^T & -B^T \\ -C^T & A^T \end{pmatrix} \begin{pmatrix} A & B \\ C & D \end{pmatrix} = \begin{pmatrix} I_n & 0_n \\ 0_n & I_n \end{pmatrix}.$$

Equating the lower right blocks shows that $-C^T B + A^T D = I_n$.

Problem 65. Let n be a positive integer. Let A_n be the $(2n + 1) \times (2n + 1)$ skew-symmetric matrix for which each entry in the first n subdiagonals below the main diagonal is 1 and each of the remaining entries below the main diagonal is −1. Give A_1 and A_2. Find the rank of A_n.

Solution 65.   We have

$$A_1 = \begin{pmatrix} 0 & -1 & 1 \\ 1 & 0 & -1 \\ -1 & 1 & 0 \end{pmatrix}, \quad A_2 = \begin{pmatrix} 0 & -1 & -1 & 1 & 1 \\ 1 & 0 & -1 & -1 & 1 \\ 1 & 1 & 0 & -1 & -1 \\ -1 & 1 & 1 & 0 & -1 \\ -1 & -1 & 1 & 1 & 0 \end{pmatrix}.$$

We use induction on n to prove that $\mathrm{rank}(A_n) = 2n$. The rank of A_1 is 2 since the first vector in the matrix A_1 is a linear combination of the second and third vectors and the second and third vectors are linearly independent. Suppose $n \geq 2$ and that the $\mathrm{rank}(A_{n-1}) = 2(n-1)$ is known. Adding multiples of the first two rows of A_n to the other rows transforms A_n to a matrix of the form

$$\begin{pmatrix} 0 & -1 & \\ 1 & 0 & * \\ \mathbf{0} & & -A_{n-1} \end{pmatrix}$$

in which $\mathbf{0}$ and $*$ represent blocks of size $(2n-1) \times 2$ and $2 \times (2n-1)$, respectively. Thus $\mathrm{rank}(A_n) = 2 + \mathrm{rank}(A_{n-1}) = 2 + 2(n-1) = 2n$.

Problem 66.   A vector $\mathbf{u} = (u_1, u_2, \dots, u_n)$ is called a *probability vector* if the components are nonnegative and their sum is 1.
(i) Is the vector $\mathbf{u} = (1/2, 0, 1/4, 1/4)$ a probability vector?
(ii) Can the vector $\mathbf{v} = (2, 3, 5, 1, 0)$ be "normalized" so that we obtain a probability vector?

Solution 66.   (i) Since all the components are nonnegative and the sum of the entries is 1 we find that $\mathbf{u}$ is a probability vector.
(ii) All the entries in $\mathbf{v}$ are nonnegative but the sum is 11. Thus we can construct the probability vector

$$\widetilde{\mathbf{v}} = \frac{1}{11}(2, 3, 5, 1, 0).$$

Problem 67.   An $n \times n$ matrix $P = (p_{ij})$ is called a *stochastic matrix* if each of its rows is a probability vector, i.e. if each entry of P is nonnegative and the sum of the entries in each row is 1. Let A and B be two stochastic $n \times n$ matrices. Is the matrix product AB also a stochastic matrix?

Solution 67.   Yes. From matrix multiplication we have for the (ij)-th entry of the product

$$(AB)_{ij} = \sum_{k=1}^{n} a_{ik} b_{kj}.$$

It follows that

$$\sum_{j=1}^{n}(AB)_{ij} = \sum_{j=1}^{n}\sum_{k=1}^{n}a_{ik}b_{kj} = \sum_{k=1}^{n}a_{ik}\sum_{j=1}^{n}b_{kj} = 1$$

since

$$\sum_{k=1}^{n}a_{ik} = 1, \qquad \sum_{j=1}^{n}b_{kj} = 1.$$

Problem 68.   The *numerical range*, also known as the *field of values*, of an $n \times n$ matrix A over the complex numbers, is defined as

$$F(A) := \{\, \mathbf{z}^* A \mathbf{z} \; : \; \|\mathbf{z}\| = \mathbf{z}^* \mathbf{z} = 1, \; \mathbf{z} \in \mathbb{C}^n \,\}.$$

The *Toeplitz-Hausdorff convexity theorem* tells us that the numerical range of a square matrix is a convex compact subset of the complex plane.
(i) Find the numerical range for the 2×2 matrices

$$B = \begin{pmatrix} 1 & 0 \\ 0 & 0 \end{pmatrix}, \qquad C = \begin{pmatrix} 0 & 0 \\ 1 & 1 \end{pmatrix}.$$

(ii) Let $\alpha \in \mathbb{R}$ and the 5×5 symmetric matrix

$$A = \begin{pmatrix} \alpha & 1 & 0 & 0 & 0 \\ 1 & \alpha & 1 & 0 & 0 \\ 0 & 1 & \alpha & 1 & 0 \\ 0 & 0 & 1 & \alpha & 1 \\ 0 & 0 & 0 & 1 & \alpha \end{pmatrix}.$$

Show that the set $F(A)$ lies on the real axis. Show that $|\mathbf{z}^* A \mathbf{z}| \le \alpha + 8$.

Solution 68.   (i) Let

$$A = \begin{pmatrix} a_{11} & a_{12} \\ a_{21} & a_{22} \end{pmatrix}, \qquad \mathbf{z} = \begin{pmatrix} e^{i\phi}\cos(\theta) \\ e^{i\chi}\sin(\theta) \end{pmatrix}, \qquad \phi, \chi, \theta \in \mathbb{R}.$$

Therefore $\mathbf{z}$ is an arbitrary complex vector of length 1, i.e. $\|\mathbf{z}\| = 1$. Then

$$\mathbf{z}^* A \mathbf{z} = (\, e^{-i\phi}\cos(\theta) \quad e^{-i\chi}\sin(\theta) \,) \begin{pmatrix} a_{11} & a_{12} \\ a_{21} & a_{22} \end{pmatrix} \begin{pmatrix} e^{i\phi}\cos(\theta) \\ e^{i\chi}\sin(\theta) \end{pmatrix}$$

$$= a_{11}\cos^2(\theta) + (a_{12}e^{i(\chi-\phi)} + a_{21}e^{i(\phi-\chi)})\sin(\theta)\cos(\theta) + a_{22}\sin^2(\theta).$$

Thus for the matrix B we have $\mathbf{z}^T B \mathbf{z} = \cos^2(\theta)$, where $\cos^2(\theta) \in [0, 1]$ for all $\theta \in \mathbb{R}$. For the matrix C we have

$$\mathbf{z}^* C \mathbf{z} = e^{i(\phi-\chi)}\sin(\theta)\cos(\theta) + \sin^2(\theta).$$

Thus the numerical range $F(C)$ is the closed elliptical disc in the complex plane with foci at $(0,0)$ and $(1,0)$, minor axis 1, and major axis $\sqrt{2}$.

(ii) Since $\mathbf{z}^* = (\bar{z}_1, \bar{z}_2, \bar{z}_3, \bar{z}_4, \bar{z}_5)$ with $\mathbf{z}^*\mathbf{z} = 1$ and applying matrix multiplication we obtain

$$\mathbf{z}^* A\mathbf{z} = \alpha + \bar{z}_1 z_2 + \bar{z}_2 z_1 + \bar{z}_2 z_3 + \bar{z}_3 z_2 + \bar{z}_3 z_4 + \bar{z}_4 z_3 + \bar{z}_4 z_5 + \bar{z}_5 z_4.$$

Let $z_j = r_j e^{i\theta_j}$ with $0 \le r_j \le 1$. For $j \ne k$ we have

$$\bar{z}_j z_k + z_j \bar{z}_k = 2 r_j r_k \cos(\theta_j - \theta_k)$$

with $0 \le r_j \le 1$. It follows that $\mathbf{z}^* A\mathbf{z}$ is given by

$$\alpha + 2(r_1 r_2 \cos(\theta_{12}) + r_2 r_3 \cos(\theta_{23}) + r_3 r_4 \cos(\theta_{34}) + r_4 r_5 \cos(\theta_{45}))$$

where $\theta_{12} = \theta_1 - \theta_2$, $\theta_{23} = \theta_2 - \theta_3$, $\theta_{34} = \theta_3 - \theta_4$, $\theta_{45} = \theta_4 - \theta_5$. Thus the set $F(A)$ lies on the real axis. Since $0 \le r_j \le 1$ and $|\cos(\theta)| \le 1$ we obtain $|\mathbf{z}^* A\mathbf{z}| \le \alpha + 8$.

Problem 69.   Let A be an $n \times n$ matrix over $\mathbb{C}$ and $F(A)$ the field of values. Let U be an $n \times n$ unitary matrix.
(i) Show that $F(U^* AU) = F(A)$.
(ii) Apply the theorem to the two 2×2 matrices which are unitarily equivalent

$$A_1 = \begin{pmatrix} 0 & 1 \\ 1 & 0 \end{pmatrix}, \qquad A_2 = \begin{pmatrix} 1 & 0 \\ 0 & -1 \end{pmatrix}.$$

Solution 69.   (i) Since a unitary matrix leaves invariant the surface of the Euclidean unit ball, the complex numbers that comprise the sets $F(U^* AU)$ and $F(A)$ are the same. If $\mathbf{z} \in \mathbb{C}^n$ and $\mathbf{z}^*\mathbf{z} = 1$, we have

$$\mathbf{z}^* (U^* AU)\mathbf{z} = \mathbf{w}^* A\mathbf{w} \in F(A)$$

where $\mathbf{w} = U\mathbf{z}$, so that $\mathbf{w}^*\mathbf{w} = \mathbf{z}^* U^* U\mathbf{z} = \mathbf{z}^*\mathbf{z} = 1$. Thus $F(U^* AU) \subset F(A)$. The reverse containment is obtained similarly.
(ii) For A_1 we have

$$\mathbf{z}^* A_1 \mathbf{z} = \bar{z}_1 z_2 + \bar{z}_2 z_1 = 2 r_1 r_2 \cos(\theta_2 - \theta_1)$$

with the constraints $0 \le r_1, r_2 \le 1$ and $r_1^2 + r_2^2 = 1$. For A_2 we find

$$\mathbf{z}^* A_2 \mathbf{z} = \bar{z}_1 z_1 - \bar{z}_2 z_2 = r_1^2 - r_2^2$$

with the constraints $0 \le r_1, r_2 \le 1$ and $r_1^2 + r_2^2 = 1$. Both define the same set, namely the interval $[-1, 1]$.

Problem 70.   Let A be an $m \times n$ matrix over $\mathbb{C}$. The *Moore-Penrose pseudo inverse matrix* A^+ is the unique $n \times m$ matrix which satisfies

$$AA^+ A = A, \quad A^+ AA^+ = A^+, \quad (AA^+)^* = AA^+, \quad (A^+ A)^* = A^+ A.$$

We also have that

$$\mathbf{x} = A^+ \mathbf{b} \tag{1}$$

is the shortest length least square solution to the problem $A\mathbf{x} = \mathbf{b}$.
(i) Show that if $(A^*A)^{-1}$ exists, then $A^+ = (A^*A)^{-1}A^*$.
(ii) Let

$$A = \begin{pmatrix} 1 & 3 \\ 2 & 4 \\ 3 & 5 \end{pmatrix}.$$

Find the Moore-Penrose matrix inverse A^+ of A.

Solution 70.   (i) Suppose that $(A^*A)^{-1}$ exists we have

$$A\mathbf{x} = \mathbf{b} \Rightarrow A^*A\mathbf{x} = A^*\mathbf{b} \Rightarrow \mathbf{x} = (A^*A)^{-1}A^*\mathbf{b}.$$

Using (1) we obtain $A^+ = (A^*A)^{-1}A^*$.
(ii) We have

$$A^*A = \begin{pmatrix} 1 & 2 & 3 \\ 3 & 4 & 5 \end{pmatrix} \begin{pmatrix} 1 & 3 \\ 2 & 4 \\ 3 & 5 \end{pmatrix} = \begin{pmatrix} 14 & 26 \\ 26 & 50 \end{pmatrix}.$$

Since $\det(A^*A) \neq 0$ the inverse of A^*A exists and is given by

$$(A^*A)^{-1} = \frac{1}{12} \begin{pmatrix} 25 & -13 \\ -13 & 7 \end{pmatrix}.$$

Thus

$$A^+ = (A^*A)^{-1}A^* = \frac{1}{12} \begin{pmatrix} 25 & -13 \\ -13 & 7 \end{pmatrix} \begin{pmatrix} 1 & 2 & 3 \\ 3 & 4 & 5 \end{pmatrix} = \frac{1}{12} \begin{pmatrix} -14 & -2 & 10 \\ 8 & 2 & -4 \end{pmatrix}.$$

Problem 71.   The *Fibonacci numbers* are defined by the recurrence relation (linear difference equation of second order with constant coefficients) $s_{n+2} = s_{n+1} + s_n$, where $n = 0, 1, \dots$ and $s_0 = 0$, $s_1 = 1$. Write this recurrence relation in matrix form. Find s_6, s_5, and s_4.

Solution 71.   We have $s_2 = 1$. We can write

$$\begin{pmatrix} s_{n+1} & s_n \\ s_n & s_{n-1} \end{pmatrix} = \begin{pmatrix} 1 & 1 \\ 1 & 0 \end{pmatrix}^n, \qquad n = 1, 2, \dots \quad .$$

Thus

$$\begin{pmatrix} 1 & 1 \\ 1 & 0 \end{pmatrix}^5 = \begin{pmatrix} 8 & 5 \\ 5 & 3 \end{pmatrix}.$$

It follows that $s_6 = 8$, $s_5 = 5$ and $s_4 = 3$.

Problem 72.   Assume that $A = A_1 + iA_2$ is a nonsingular $n \times n$ matrix, where A_1 and A_2 are real $n \times n$ matrices. Assume that A_1 is also nonsingular. Find the inverse of A using the inverse of A_1.

Solution 72. We have the identity $(A_1 + iA_2)(I_n - iA_1^{-1}A_2) \equiv A_1 + A_2 A_1^{-1} A_2$. Thus we find the inverse

$$A^{-1} = (A_1 + A_2 A_1^{-1} A_2)^{-1} - iA_1^{-1}A_2(A_1 + A_2 A_1^{-1} A_2)^{-1}.$$

Problem 73. Let A and B be $n \times n$ matrices over $\mathbb{R}$. Assume that $A \neq B$, $A^3 = B^3$ and $A^2 B = B^2 A$. Is $A^2 + B^2$ invertible?

Solution 73. We have $(A^2 + B^2)(A - B) = A^3 - B^3 - A^2 B + B^2 A = 0_n$. Since $A \neq B$, we can conclude that $A^2 + B^2$ is not invertible.

Problem 74. Let A be a symmetric 2×2 matrix over $\mathbb{R}$

$$A = \begin{pmatrix} a_{00} & a_{01} \\ a_{10} & a_{11} \end{pmatrix}.$$

Thus $a_{01} = a_{10}$. Assume that $a_{00}a_{01} = a_{01}^2$, $a_{00}a_{11} = a_{01}a_{11}$. Find all matrices A that satisfy these conditions.

Solution 74. We obtain as trivial solutions

$$\begin{pmatrix} 1 & 1 \\ 1 & 1 \end{pmatrix}, \quad \begin{pmatrix} 0 & 0 \\ 0 & 0 \end{pmatrix}$$

and the non-trivial solutions

$$\begin{pmatrix} 1 & 1 \\ 1 & 0 \end{pmatrix}, \quad \begin{pmatrix} 0 & 0 \\ 0 & 1 \end{pmatrix}, \quad \begin{pmatrix} 1 & 0 \\ 0 & 0 \end{pmatrix}.$$

Problem 75. Consider the invertible 2×2 matrix

$$J = \begin{pmatrix} 0 & 1 \\ -1 & 0 \end{pmatrix} \Rightarrow J^{-1} = \begin{pmatrix} 0 & -1 \\ 1 & 0 \end{pmatrix}.$$

Find all nonzero 2×2 matrices A such that $AJ = JA$.

Solution 75. We have $A = JAJ^{-1}$ with

$$A = \begin{pmatrix} a_{11} & a_{12} \\ -a_{12} & a_{11} \end{pmatrix}.$$

Problem 76. Consider the column vector in $\mathbb{R}^8$

$$\mathbf{x} = (20.0 \ 6.0 \ 4.0 \ 2.0 \ 10.0 \ 6.0 \ 8.0 \ 4.0)^T$$

where ^{T} denotes transpose. Consider the matrices

$$H_1 = \frac{1}{2} \begin{pmatrix} 1 & 1 & 0 & 0 & 0 & 0 & 0 & 0 \\ 0 & 0 & 1 & 1 & 0 & 0 & 0 & 0 \\ 0 & 0 & 0 & 0 & 1 & 1 & 0 & 0 \\ 0 & 0 & 0 & 0 & 0 & 0 & 1 & 1 \end{pmatrix}$$

$$G_1 = \frac{1}{2} \begin{pmatrix} 1 & -1 & 0 & 0 & 0 & 0 & 0 & 0 \\ 0 & 0 & 1 & -1 & 0 & 0 & 0 & 0 \\ 0 & 0 & 0 & 0 & 1 & -1 & 0 & 0 \\ 0 & 0 & 0 & 0 & 0 & 0 & 1 & -1 \end{pmatrix}$$

$$H_2 = \frac{1}{2} \begin{pmatrix} 1 & 1 & 0 & 0 \\ 0 & 0 & 1 & 1 \end{pmatrix}, \qquad G_2 = \frac{1}{2} \begin{pmatrix} 1 & -1 & 0 & 0 \\ 0 & 0 & 1 & -1 \end{pmatrix}$$

$$H_3 = \frac{1}{2} (1 \quad 1), \qquad G_3 = \frac{1}{2} (1 \quad -1).$$

(i) Calculate the vectors

$$H_1 \mathbf{x}, \qquad G_1 \mathbf{x}$$

$$H_2 H_1 \mathbf{x}, \quad G_2 H_1 \mathbf{x}, \quad H_2 G_1 \mathbf{x}, \quad G_2 G_1 \mathbf{x}$$

$$H_3 H_2 H_1 \mathbf{x}, \quad G_3 H_2 H_1 \mathbf{x}, \quad H_3 G_2 H_1 \mathbf{x}, \quad G_3 G_2 H_1 \mathbf{x},$$

$$H_3 H_2 G_1 \mathbf{x}, \quad G_3 H_2 G_1 \mathbf{x}, \quad H_3 G_2 G_1 \mathbf{x}, \quad G_3 G_2 G_1 \mathbf{x}.$$

(ii) Calculate $H_j H_j^T$, $G_j G_j^T$, $H_j G_j^T$ for $j = 1, 2, 3$.

(iii) How can we reconstruct the original vector $\mathbf{x}$ from the vector

$$(H_3 H_2 H_1 \mathbf{x}, G_3 H_2 H_1 \mathbf{x}, H_3 G_2 H_1 \mathbf{x}, G_3 G_2 H_1 \mathbf{x}, H_3 H_2 G_1 \mathbf{x},$$
$$G_3 H_2 G_1 \mathbf{x}, H_3 G_2 G_1 \mathbf{x}, G_3 G_2 G_1 \mathbf{x}).$$

The problem plays a role in *wavelet theory*.

Solution 76.   (i) We find

$$H_1 \mathbf{x} = \begin{pmatrix} 13.0 \\ 3.0 \\ 8.0 \\ 6.0 \end{pmatrix}, \qquad G_1 \mathbf{x} = \begin{pmatrix} 7.0 \\ 1.0 \\ 2.0 \\ 2.0 \end{pmatrix}.$$

Thus we have the vector $(13.0 \ 3.0 \ 8.0 \ 6.0 \ 7.0 \ 1.0 \ 2.0 \ 2.0)^T$. Next we find

$$H_2 H_1 \mathbf{x} = \begin{pmatrix} 8.0 \\ 7.0 \end{pmatrix}, \qquad G_2 H_1 \mathbf{x} = \begin{pmatrix} 5.0 \\ 1.0 \end{pmatrix},$$

$$H_2 G_1 \mathbf{x} = \begin{pmatrix} 4.0 \\ 2.0 \end{pmatrix}, \qquad G_2 G_1 \mathbf{x} = \begin{pmatrix} 3.0 \\ 0.0 \end{pmatrix}.$$

Thus we have the vector $(8.0 \cdot 7.0 \ 5.0 \ 1.0 \ 4.0 \ 2.0 \ 3.0 \ 0.0)^T$. Finally we have

$$H_3 H_2 H_1 \mathbf{x} = 7.5, \ G_3 H_2 H_1 \mathbf{x} = 0.5, \ H_3 G_2 H_1 \mathbf{x} = 3.0, \ G_3 G_2 H_1 \mathbf{x} = 2.0$$

$$H_3 H_2 G_1 \mathbf{x} = 3.0, \quad G_3 H_2 G_1 \mathbf{x} = 1.0, \quad H_3 G_2 G_1 \mathbf{x} = 1.5, \quad G_3 G_2 G_1 \mathbf{x} = 1.5.$$

Thus we obtain the vector $(7.5 \ 0.5 \ 3.0 \ 2.0 \ 3.0 \ 1.0 \ 1.5 \ 1.5)$.

(ii) We find

$$H_1 H_1^T = \frac{1}{2} I_4, \quad G_1 G_1^T = \frac{1}{2} I_4, \quad H_1 G_1^T = 0_4$$

$$H_2 H_2^T = \frac{1}{2} I_2, \quad G_2 G_2^T = \frac{1}{2} I_2, \quad H_2 G_2^T = 0_2.$$

$$H_3 H_3^T = \frac{1}{2}, \quad G_3 G_3^T = \frac{1}{2}, \quad H_3 G_3^T = 0.$$

(iii) Let $\mathbf{w} = (7.5\ 0.5\ 3.0\ 2.0\ 3.0\ 1.0\ 1.5\ 1.5)^T$. Consider the 4×8 matrices

$$X_1 = \begin{pmatrix} 1 & 1 & 0 & 0 & 0 & 0 & 0 & 0 \\ 1 & -1 & 0 & 0 & 0 & 0 & 0 & 0 \\ 0 & 0 & 1 & 1 & 0 & 0 & 0 & 0 \\ 0 & 0 & 1 & -1 & 0 & 0 & 0 & 0 \end{pmatrix}$$

$$Y_1 = \begin{pmatrix} 0 & 0 & 0 & 0 & 1 & 1 & 0 & 0 \\ 0 & 0 & 0 & 0 & 1 & -1 & 0 & 0 \\ 0 & 0 & 0 & 0 & 0 & 0 & 1 & 1 \\ 0 & 0 & 0 & 0 & 0 & 0 & 1 & -1 \end{pmatrix}.$$

Then

$$X_1 \mathbf{w} = \begin{pmatrix} 8 \\ 7 \\ 5 \\ 1 \end{pmatrix}, \quad Y_1 \mathbf{w} = \begin{pmatrix} 4 \\ 2 \\ 3 \\ 0 \end{pmatrix}.$$

Now let

$$X_2 = Y_2 = \begin{pmatrix} 1 & 0 & 1 & 0 \\ 1 & 0 & -1 & 0 \\ 0 & 1 & 0 & 1 \\ 0 & 1 & 0 & -1 \end{pmatrix}.$$

Then

$$X_2(X_1 \mathbf{w}) = \begin{pmatrix} 13 \\ 3 \\ 8 \\ 6 \end{pmatrix}, \quad Y_2(Y_1 \mathbf{w}) = \begin{pmatrix} 7 \\ 1 \\ 2 \\ 2 \end{pmatrix}.$$

Thus the original vector $\mathbf{x}$ is reconstructed from

$$\begin{pmatrix} X_2(X_1 \mathbf{w}) \\ Y_2(Y_1 \mathbf{w}) \end{pmatrix}.$$

The odd entries come from $X_2(X_1\mathbf{w}) + Y_2(Y_1\mathbf{w})$. The even ones come from $X_2(X_1\mathbf{w}) - Y_2(Y_1\mathbf{w})$.

Problem 77. Given a signal as the column vector

$$\mathbf{x} = (3.0\ 0.5\ 2.0\ 7.0)^T.$$

The *pyramid algorithm* (for *Haar wavelets*) is as follows: The first two entries $(3.0\ 0.5)^T$ in the signal give an average of $(3.0+0.5)/2 = 1.75$ and a difference average of $(3.0-0.5)/2 = 1.25$. The second two entries $(2.0\ 7.0)$ give an average of $(2.0+7.0)/2 = 4.5$ and a difference average of $(2.0-7.0)/2 = -2.5$. Thus we end up with a vector

$$(1.75\ 1.25\ 4.5\ -2.5)^T.$$

Now we take the average of 1.75 and 4.5 providing $(1.75+4.5)/2 = 3.125$ and the difference average $(1.75-4.5)/2 = -1.375$. Thus we end up with the vector

$$\mathbf{y} = (3.125\ -1.375\ 1.25\ -2.5)^T.$$

(i) Find a 4×4 matrix A such that

$$\mathbf{x} \equiv \begin{pmatrix} 3.0 \\ 0.5 \\ 2.0 \\ 7.0 \end{pmatrix} = A\mathbf{y} \equiv A \begin{pmatrix} 3.125 \\ -1.375 \\ 1.25 \\ -2.5 \end{pmatrix}.$$

(ii) Show that the inverse of A exists. Then find the inverse.

Solution 77.   (i) Since we can write

$$\begin{pmatrix} 3.0 \\ 0.5 \\ 2.0 \\ 7.0 \end{pmatrix} = 3.125 \begin{pmatrix} 1 \\ 1 \\ 1 \\ 1 \end{pmatrix} - 1.375 \begin{pmatrix} 1 \\ 1 \\ -1 \\ -1 \end{pmatrix} + 1.25 \begin{pmatrix} 1 \\ -1 \\ 0 \\ 0 \end{pmatrix} - 2.5 \begin{pmatrix} 0 \\ 0 \\ 1 \\ -1 \end{pmatrix}$$

we obtain the matrix

$$A = \begin{pmatrix} 1 & 1 & 1 & 0 \\ 1 & 1 & -1 & 0 \\ 1 & -1 & 0 & 1 \\ 1 & -1 & 0 & -1 \end{pmatrix}.$$

(ii) All the column vectors in the matrix A are nonzero and all the pairwise scalar products are equal to 0. Thus the column vectors form a basis (not normalized) in $\mathbb{R}^n$. Thus the matrix is invertible. The inverse matrix is given by

$$A^{-1} = \begin{pmatrix} 1/4 & 1/4 & 1/4 & 1/4 \\ 1/4 & 1/4 & -1/4 & -1/4 \\ 1/2 & -1/2 & 0 & 0 \\ 0 & 0 & 1/2 & -1/2 \end{pmatrix}.$$

Problem 78.   The 8×8 matrix

$$H = \begin{pmatrix} 1 & 1 & 1 & 0 & 1 & 0 & 0 & 0 \\ 1 & 1 & 1 & 0 & -1 & 0 & 0 & 0 \\ 1 & 1 & -1 & 0 & 0 & 1 & 0 & 0 \\ 1 & 1 & -1 & 0 & 0 & -1 & 0 & 0 \\ 1 & -1 & 0 & 1 & 0 & 0 & 1 & 0 \\ 1 & -1 & 0 & 1 & 0 & 0 & -1 & 0 \\ 1 & -1 & 0 & -1 & 0 & 0 & 0 & 1 \\ 1 & -1 & 0 & -1 & 0 & 0 & 0 & -1 \end{pmatrix}$$

plays a role in the *discrete wavelet transform*. Show that the matrix is invertible without calculating the determinant. Find the inverse.

Solution 78.   (i) All the column vectors in the matrix H are nonzero. All the pairwise scalar products of the column vectors are 0. Thus the matrix has maximum rank, i.e. $\text{rank}(H) = 8$ and the column vectors form a basis (not normalized) in $\mathbb{R}^n$. The inverse matrix is given by

$$H^{-1} = \frac{1}{8} \begin{pmatrix} 1 & 1 & 1 & 1 & 1 & 1 & 1 & 1 \\ 1 & 1 & 1 & 1 & -1 & -1 & -1 & -1 \\ 2 & 2 & -2 & -2 & 0 & 0 & 0 & 0 \\ 0 & 0 & 0 & 0 & 2 & 2 & -2 & -2 \\ 4 & -4 & 0 & 0 & 0 & 0 & 0 & 0 \\ 0 & 0 & 4 & -4 & 0 & 0 & 0 & 0 \\ 0 & 0 & 0 & 0 & 4 & -4 & 0 & 0 \\ 0 & 0 & 0 & 0 & 0 & 0 & 4 & -4 \end{pmatrix}.$$

Problem 79. (i) For $n = 4$ the transform matrix for the *Daubechies wavelet* is given by

$$D_4 = \begin{pmatrix} c_0 & c_1 & c_2 & c_3 \\ c_3 & -c_2 & c_1 & -c_0 \\ c_2 & c_3 & c_0 & c_1 \\ c_1 & -c_0 & c_3 & -c_2 \end{pmatrix}, \qquad \begin{pmatrix} c_0 \\ c_1 \\ c_2 \\ c_3 \end{pmatrix} = \frac{1}{4\sqrt{2}} \begin{pmatrix} 1 + \sqrt{3} \\ 3 + \sqrt{3} \\ 3 - \sqrt{3} \\ 1 - \sqrt{3} \end{pmatrix}.$$

Is D_4 orthogonal? Prove or disprove.

(ii) For $n = 8$ the transform matrix for the Daubechies wavelet is given by

$$D_8 = \begin{pmatrix} c_0 & c_1 & c_2 & c_3 & 0 & 0 & 0 & 0 \\ c_3 & -c_2 & c_1 & -c_0 & 0 & 0 & 0 & 0 \\ 0 & 0 & c_0 & c_1 & c_2 & c_3 & 0 & 0 \\ 0 & 0 & c_3 & -c_2 & c_1 & -c_0 & 0 & 0 \\ 0 & 0 & 0 & 0 & c_0 & c_1 & c_2 & c_3 \\ 0 & 0 & 0 & 0 & c_3 & -c_2 & c_1 & -c_0 \\ c_2 & c_3 & 0 & 0 & 0 & 0 & c_0 & c_1 \\ c_1 & -c_0 & 0 & 0 & 0 & 0 & c_3 & -c_2 \end{pmatrix}.$$

Is D_8 orthogonal? Prove or disprove.

Solution 79. (i) We find $D_4 D_4^T = I_4$. Thus $D_4^{-1} = D_4^T$ and D_4 is orthogonal.
(ii) We find $D_8 D_8^T = I_8$. Thus $D_8^{-1} = D_8^T$ and D_8 is orthogonal.

Problem 80. Let A, B be $n \times n$ matrices over $\mathbb{C}$. Assume that A and $A + B$ are invertible. Then $(A + B)^{-1} \equiv A^{-1} - A^{-1} B (A + B)^{-1}$. Apply the identity to $A = \sigma_1$, $B = \sigma_3$.

Solution 80. We have

$$A + B = \begin{pmatrix} 1 & 1 \\ 1 & -1 \end{pmatrix} \Rightarrow (A + B)^{-1} = \frac{1}{2} \begin{pmatrix} 1 & 1 \\ 1 & -1 \end{pmatrix}$$

and $A^{-1} = \sigma_1$. Thus

$$\frac{1}{2} \begin{pmatrix} 1 & 1 \\ 1 & -1 \end{pmatrix} \equiv \begin{pmatrix} 0 & 1 \\ 1 & 0 \end{pmatrix} - \begin{pmatrix} 0 & 1 \\ 1 & 0 \end{pmatrix} \begin{pmatrix} 1 & 0 \\ 0 & -1 \end{pmatrix} \frac{1}{2} \begin{pmatrix} 1 & 1 \\ 1 & -1 \end{pmatrix}.$$

Problem 81.   Let A be a positive definite $n \times n$ matrix over $\mathbb{R}$. Let $\mathbf{x} \in \mathbb{R}^n$. Show that $A + \mathbf{x}\mathbf{x}^T$ is positive definite.

Solution 81.   For all vectors $\mathbf{y} \in \mathbb{R}^n$, $\mathbf{y} \neq \mathbf{0}$, we have $\mathbf{y}^T A \mathbf{y} > 0$. We have

$$\mathbf{y}^T \mathbf{x}\mathbf{x}^T \mathbf{y} = \left(\sum_{j=1}^{n} y_j x_j \right) \left(\sum_{j=1}^{n} x_j y_j \right) = \left(\sum_{j=1}^{n} x_j y_j \right)^2 \geq 0$$

and therefore we have $\mathbf{y}^T (A + \mathbf{x}\mathbf{x}^T) \mathbf{y} > 0$ for all $\mathbf{y} \in \mathbb{R}^n$, $\mathbf{y} \neq \mathbf{0}$.

Supplementary Problems

Problem 1.   (i) Are the four 2×2 nonnormal matrices

$$\frac{1}{\sqrt{2}} \begin{pmatrix} 1 & 1 \\ 0 & 0 \end{pmatrix}, \quad \frac{1}{\sqrt{2}} \begin{pmatrix} 0 & 0 \\ 1 & -1 \end{pmatrix}, \quad \frac{1}{\sqrt{2}} \begin{pmatrix} 1 & -1 \\ 0 & 0 \end{pmatrix}, \quad \frac{1}{\sqrt{2}} \begin{pmatrix} 0 & 0 \\ 1 & 1 \end{pmatrix}$$

linearly independent?

(ii) Show that the nine 3×3 matrices

$$\begin{pmatrix} 1 & 0 & 0 \\ 0 & 0 & 0 \\ 0 & 0 & 0 \end{pmatrix}, \begin{pmatrix} 1 & 1 & 0 \\ 0 & 0 & 0 \\ 0 & 0 & 0 \end{pmatrix}, \begin{pmatrix} 1 & 1 & 1 \\ 0 & 0 & 0 \\ 0 & 0 & 0 \end{pmatrix}, \begin{pmatrix} 1 & 1 & 1 \\ 0 & 0 & 1 \\ 0 & 0 & 0 \end{pmatrix}, \begin{pmatrix} 1 & 1 & 1 \\ 0 & 0 & 1 \\ 0 & 0 & 1 \end{pmatrix},$$

$$\begin{pmatrix} 1 & 1 & 1 \\ 0 & 0 & 1 \\ 0 & 1 & 1 \end{pmatrix}, \begin{pmatrix} 1 & 1 & 1 \\ 0 & 0 & 1 \\ 1 & 1 & 1 \end{pmatrix}, \begin{pmatrix} 1 & 1 & 1 \\ 1 & 0 & 1 \\ 1 & 1 & 1 \end{pmatrix}, \begin{pmatrix} 1 & 1 & 1 \\ 1 & 1 & 1 \\ 1 & 1 & 1 \end{pmatrix}$$

form a basis in the vector space of the 3×3 matrices. This basis is called the *spiral basis* (and can be extended to any dimension). Which of these matrices are nonnormal?

(iii) Do the eight 3×3 skew-hermitian matrices

$$\Gamma_1 = \frac{1}{\sqrt{2}} \begin{pmatrix} 0 & i & 0 \\ i & 0 & 0 \\ 0 & 0 & 0 \end{pmatrix}, \quad \Gamma_2 = \frac{1}{\sqrt{2}} \begin{pmatrix} 0 & 1 & 0 \\ -1 & 0 & 0 \\ 0 & 0 & 0 \end{pmatrix},$$

$$\Gamma_3 = \frac{1}{\sqrt{2}} \begin{pmatrix} i & 0 & 0 \\ 0 & -i & 0 \\ 0 & 0 & 0 \end{pmatrix}, \quad \Gamma_4 = \frac{1}{\sqrt{2}} \begin{pmatrix} 0 & 0 & i \\ 0 & 0 & 0 \\ i & 0 & 0 \end{pmatrix},$$

$$\Gamma_5 = \frac{1}{\sqrt{2}} \begin{pmatrix} 0 & 0 & 1 \\ 0 & 0 & 0 \\ -1 & 0 & 0 \end{pmatrix}, \quad \Gamma_6 = \frac{1}{\sqrt{2}} \begin{pmatrix} 0 & 0 & 0 \\ 0 & 0 & i \\ 0 & i & 0 \end{pmatrix},$$

$$\Gamma_7 = \frac{1}{\sqrt{2}} \begin{pmatrix} 0 & 0 & 0 \\ 0 & 0 & 1 \\ 0 & -1 & 0 \end{pmatrix}, \quad \Gamma_8 = \frac{1}{\sqrt{6}} \begin{pmatrix} i & 0 & 0 \\ 0 & i & 0 \\ 0 & 0 & -2i \end{pmatrix}$$

together with the 3×3 unit matrix form an orthonormal basis in the vector space of 3×3 matrices over the complex number. Find all pairwise scalar products $\langle A, B \rangle := \mathrm{tr}(AB^*)$. Discuss.

Problem 2. (i) One can describe a *tetrahedron* in the vector space $\mathbb{R}^3$ by specifying vectors $\mathbf{v}_1$, $\mathbf{v}_2$, $\mathbf{v}_3$, $\mathbf{v}_4$ normal to its faces with lengths equal to the faces' area. Give an example.

(ii) Consider a tetrahedron defined by the triple of linearly independent vectors $\mathbf{v}_j \in \mathbb{R}^3$, $j = 1, 2, 3$. Show that the normal vectors to the faces defined by two of these vectors, normalized to the area of the face, is given by

$$\mathbf{n}_1 = \frac{1}{2}\mathbf{v}_2 \times \mathbf{v}_3, \quad \mathbf{n}_2 = \frac{1}{2}\mathbf{v}_3 \times \mathbf{v}_1, \quad \mathbf{n}_3 = \frac{1}{2}\mathbf{v}_1 \times \mathbf{v}_2.$$

Show that

$$\mathbf{v}_1 = \frac{2}{3V}\mathbf{n}_2 \times \mathbf{n}_3, \quad \mathbf{v}_2 = \frac{2}{3V}\mathbf{n}_3 \times \mathbf{n}_1, \quad \mathbf{v}_3 = \frac{2}{3V}\mathbf{n}_1 \times \mathbf{n}_2$$

where V is the volume of the tetrahedron given by

$$V = \frac{1}{3!}(\mathbf{v}_1 \times \mathbf{v}_2) \cdot \mathbf{v}_3 = \sqrt{\frac{2}{9}(\mathbf{n}_1 \times \mathbf{n}_2) \cdot \mathbf{n}_3}.$$

Apply it to normalized vectors which form an orthonormal basis in $\mathbb{R}^3$

$$\mathbf{n}_1 = \frac{1}{\sqrt{2}}\begin{pmatrix} 1 \\ 0 \\ 1 \end{pmatrix}, \quad \mathbf{n}_2 = \begin{pmatrix} 0 \\ 1 \\ 0 \end{pmatrix}, \quad \mathbf{n}_3 = \frac{1}{\sqrt{2}}\begin{pmatrix} 1 \\ 0 \\ -1 \end{pmatrix}.$$

Problem 3. (i) Show that any rank one positive semidefinite $n \times n$ matrix M can be written as $M = \mathbf{v}\mathbf{v}^T$, where $\mathbf{v}$ is some (column) vector in $\mathbb{C}^n$.

(ii) Let $\mathbf{u}$, $\mathbf{v}$ be normalized (column) vectors in $\mathbb{C}^n$. Let A be an $n \times n$ positive semidefinite matrix over $\mathbb{C}$. Show that $(\mathbf{u}^*\mathbf{v})(\mathbf{u}^*A\mathbf{v}) \geq 0$.

(iii) Show that if hermitian matrices S and T are positive semidefinite and commute ($ST = TS$), then their product ST is also positive semidefinite. We have to show that

$$(ST\mathbf{v})^*\mathbf{v} \geq 0$$

for all $\mathbf{v} \in \mathbb{C}^n$. Note that if $S = 0_n$, then obviously $(ST\mathbf{u})^*\mathbf{u} = 0$. So assume that $S \neq 0_n$, i.e. $\|S\| > 0$.

Problem 4. Show that the symmetric matrices

$$\begin{pmatrix} 1 & 1 \\ 1 & 1 \end{pmatrix}, \quad \begin{pmatrix} 1 & 0 & 1 \\ 0 & 0 & 0 \\ 1 & 0 & 1 \end{pmatrix}, \quad \begin{pmatrix} 1 & 0 & 0 & 1 \\ 0 & 0 & 0 & 0 \\ 0 & 0 & 0 & 0 \\ 1 & 0 & 0 & 1 \end{pmatrix}$$

are positive semidefinite. Extend to the $n \times n$ case.

Problem 5.   Let A be a hermitian $n \times n$ matrix and $A \neq 0_n$. Show that $A^m \neq 0_n$ for all $m \in \mathbb{N}$.

Problem 6.   Let $A \in \mathbb{R}^{m \times n}$ be a nonzero matrix. Let $\mathbf{x} \in \mathbb{R}^n$, $\mathbf{y} \in \mathbb{R}^m$ be vectors such that $c := \mathbf{y}^T A \mathbf{x} \neq 0$. Show that the matrix $B := A - c^{-1} A \mathbf{x} \mathbf{y}^T A$ has *rank* exactly one less than the rank of A.

Problem 7.   (i) Consider the two-dimensional Euclidean space and let $\mathbf{e}_1$, $\mathbf{e}_2$ be the standard basis. Consider the seven vectors

$$\mathbf{v}_0 = \mathbf{0}, \quad \mathbf{v}_1 = \frac{1}{2}\mathbf{e}_1 + \frac{\sqrt{3}}{2}\mathbf{e}_2, \quad \mathbf{v}_2 = -\frac{1}{2}\mathbf{e}_1 - \frac{\sqrt{3}}{2}\mathbf{e}_2,$$

$$\mathbf{v}_3 = -\frac{1}{2}\mathbf{e}_1 + \frac{\sqrt{3}}{2}\mathbf{e}_2, \quad \mathbf{v}_4 = \frac{1}{2}\mathbf{e}_1 - \frac{\sqrt{3}}{2}\mathbf{e}_2, \quad \mathbf{v}_5 = -\mathbf{e}_1, \quad \mathbf{v}_6 = \mathbf{e}_1.$$

Find the distance between the vectors and select the vectors pairs with the shortest distance.

(ii) Consider the vectors in $\mathbb{R}^2$

$$\mathbf{v}_k = \begin{pmatrix} \cos((k-1)\pi/4) \\ \sin((k-1)\pi/4) \end{pmatrix}, \quad k = 1, 3, 5, 7$$

and

$$\mathbf{v}_k = \sqrt{2}\begin{pmatrix} \cos((k-1)\pi/4) \\ \sin(((k-1)\pi/4) \end{pmatrix}, \quad k = 2, 4, 6, 8$$

which play a role for the lattice Boltzmann model. Find the angles between the vectors.

Problem 8.   Show that the 4×4 matrices

$$\Pi = \frac{1}{2}\begin{pmatrix} 1 & 1/\sqrt{2} & 0 & -1/\sqrt{2} \\ 1/\sqrt{2} & 1 & 1/\sqrt{2} & 0 \\ 0 & 1/\sqrt{2} & 1 & 1/\sqrt{2} \\ -1/\sqrt{2} & 0 & 1/\sqrt{2} & 1 \end{pmatrix}, \quad \tilde{\Pi} = I_4 - \Pi$$

are projection matrices. Describe the subspaces of $\mathbb{R}^4$ they project into.

Problem 9.   (i) The vectors $\mathbf{u}$, $\mathbf{v}$, $\mathbf{w}$ point to the vertices of a equilateral *triangle*

$$\mathbf{u} = \begin{pmatrix} 1/\sqrt{3} \\ 0 \end{pmatrix}, \quad \mathbf{v} = \begin{pmatrix} -1/(2\sqrt{3}) \\ 1/2 \end{pmatrix}, \quad \mathbf{w} = \begin{pmatrix} -1/(2\sqrt{3}) \\ -1/2 \end{pmatrix}.$$

Find the area of this triangle.

(ii) Let $\mathbf{u}$, $\mathbf{v}$ be (column) vectors in $\mathbb{R}^n$. Show that

$$A = \sqrt{|(\mathbf{u}^T\mathbf{u})(\mathbf{v}^T\mathbf{v}) - (\mathbf{u}^T\mathbf{v})^2|}$$

calculates the area spanned by the vectors $\mathbf{u}$ and $\mathbf{v}$.

Problem 10. Assume that two planes in $\mathbb{R}^3$ given by

$$kx_1 + \ell x_2 + mx_3 + n = 0, \qquad k'x_1 + \ell'x_2 + m'x_3 + n' = 0$$

be the mirror images with respect to a third plane in $\mathbb{R}^3$ given by

$$ax_1 + bx_2 + cx_3 + d = 0.$$

Show that $(r^2 = a^2 + b^2 + c^2)$

$$\begin{pmatrix} k' \\ \ell' \\ m' \end{pmatrix} = \frac{1}{r^2} \begin{pmatrix} a^2 - b^2 - c^2 & 2ab & 2ac \\ 2ab & -a^2 + b^2 - c^2 & 2bc \\ 2ac & 2bc & -a^2 - b^2 + c^2 \end{pmatrix} \begin{pmatrix} k \\ \ell \\ m \end{pmatrix}.$$

Problem 11. (i) Let A be a hermitian $n \times n$ matrix over $\mathbb{C}$ with $A^2 = I_n$. Find the matrix $(A^{-1} + iI_n)^{-1}$.
(ii) Let B be an $n \times n$ matrix over $\mathbb{C}$. Find the condition on B such that

$$(I_n + iB)(I_n - iB) = I_n.$$

This means that $(I_n - iB)$ is the inverse of $(I_n + iB)$.

Problem 12. Consider the vector space $\mathbb{R}^d$. Suppose that $\{\mathbf{v}_j\}_{j=1}^d$ and $\{\mathbf{w}_k\}_{k=1}^d$ are two bases in $\mathbb{R}^d$. Then there is an invertible $d \times d$ matrix

$$T = (t_{jk})_{j,k=1}^d$$

so that

$$\mathbf{v}_j = \sum_{k=1}^d t_{jk} \mathbf{w}_k, \quad j = 1, \ldots, d.$$

The bases $\{\mathbf{v}_j\}_{j=1}^d$ and $\{\mathbf{w}_k\}_{k=1}^d$ are said to have the same *orientation* if $\det(T) > 0$. If $\det(T) < 0$, then they have the opposite orientation. Consider the two bases in $\mathbb{R}^2$

$$\left\{ \mathbf{v}_1 = \begin{pmatrix} 1 \\ 0 \end{pmatrix}, \ \mathbf{v}_2 = \begin{pmatrix} 0 \\ 1 \end{pmatrix} \right\}, \qquad \left\{ \mathbf{w}_1 = \frac{1}{\sqrt{2}} \begin{pmatrix} 1 \\ 1 \end{pmatrix}, \ \mathbf{w}_2 = \frac{1}{\sqrt{2}} \begin{pmatrix} 1 \\ -1 \end{pmatrix} \right\}.$$

Find the orientation.

Problem 13. (i) Let $\omega > 0$ (fixed) and $t \geq 0$. Show that the 4×4 matrix

$$S(t) = \frac{1}{4} \begin{pmatrix} 1 + 3e^{-\omega t} & 1 - e^{-\omega t} & 1 - e^{-\omega t} & 1 - e^{-\omega t} \\ 1 - e^{-\omega t} & 1 + 3e^{-\omega t} & 1 - e^{-\omega t} & 1 - e^{-\omega t} \\ 1 - e^{-\omega t} & 1 - e^{-\omega t} & 1 + 3e^{-\omega t} & 1 - e^{-\omega t} \\ 1 - e^{-\omega t} & 1 - e^{-\omega t} & 1 - e^{-\omega t} & 1 + 3e^{-\omega t} \end{pmatrix}$$

is a stochastic matrix (probabilistic matrix). Find the eigenvalues of $S(t)$. Let

$$\pi = \frac{1}{2} (1 \quad 0 \quad 0 \quad 1).$$

Find the vectors $\pi S(t)$, $(\pi S(t))S(t)$ etc.. Is $S(t_1 + t_2) = S(t_1)S(t_2)$?
(ii) Let $\epsilon \in [0, 1]$. Consider the stochastic matrix

$$S(\epsilon) = \begin{pmatrix} \epsilon & 1 - \epsilon \\ 1 - \epsilon & \epsilon \end{pmatrix}.$$

Let $n = 1, 2, \ldots$. Show that

$$S^n(\epsilon) = \frac{1}{2} \begin{pmatrix} 1 + (2\epsilon - 1)^n & 1 - (2\epsilon - 1)^n \\ 1 - (2\epsilon - 1)^n & 1 + (2\epsilon - 1)^n \end{pmatrix}.$$

Problem 14.   Can one find 4×4 matrices A and B such that

$$A \begin{pmatrix} 0 & 0 & 0 & 1 \\ 0 & 0 & 1 & 0 \\ 0 & 1 & 0 & 0 \\ 1 & 0 & 0 & 0 \end{pmatrix} \quad B = \begin{pmatrix} 1 & 0 & 0 & 1 \\ 0 & 0 & 0 & 0 \\ 0 & 0 & 0 & 0 \\ 1 & 0 & 0 & 1 \end{pmatrix}?$$

Of course at least one of the matrices A and B must be singular. The underlying field $\mathbb{F}$ could be $\mathbb{R}$ or $\mathrm{char}(\mathbb{F}) = 2$.

Problem 15.   (i) Find all 2×2 matrices such that $A^2 \neq 0_2$ but $A^3 = 0_2$.
(ii) Find all non-hermitian 2×2 matrices B such that $BB^* = I_2$.
(iii) Find all hermitian 2×2 matrices H such that $H^2 = 0_2$.
(iv) Find all 2×2 matrices C over $\mathbb{R}$ such that $C^T C + C C^T = I_2$, $C^2 = 0_2$.
(v) Find all 2×2 matrices A_1, A_2, A_3 such that $A_1 A_2 = A_2 A_3$, $A_3 A_1 = A_2 A_3$.
(vi) Find all 2×2 matrices A and B such that $A^2 = B^2 = I_2$, $AB = -I_2$.

Problem 16.   Let A be an $m \times n$ matrix and B be a $p \times q$ matrix. Then the *direct sum* of A and B, denoted by $A \oplus B$, is the $(m+p) \times (n+q)$ matrix defined by

$$A \oplus B := \begin{pmatrix} A & 0 \\ 0 & B \end{pmatrix}.$$

Let A_1, A_2 be $m \times m$ matrices and B_1, B_2 be $n \times n$ matrices. Show that

$$(A_1 \oplus B_1)(A_2 \oplus B_2) = (A_1 A_2) \oplus (B_1 B_2).$$

Problem 17.   Consider the symmetric matrix A and the normalized vector $\mathbf{v}$

$$A = \begin{pmatrix} 0 & 1 & 0 \\ 1 & 0 & 1 \\ 0 & 1 & 0 \end{pmatrix}, \qquad \mathbf{v} = \frac{1}{\sqrt{3}} \begin{pmatrix} 1 \\ 1 \\ 1 \end{pmatrix}.$$

Calculate $S_j = \mathbf{v}^* A^j \mathbf{v}$ for $j = 1, 2, 3$. Can A be reconstructed from S_1, S_2, S_3 and the information that A is real symmetric?

Problem 18.   The *standard simplex* Δ_n is defined by the set in $\mathbb{R}^n$

$$\Delta_n := \{ (x_1, \ldots, x_n)^T : x_j \geq 0, \sum_{j=1}^{n} x_j = 1 \}.$$

Consider n affinely independent points $B_1, \ldots, B_n \in \Delta_n$. They span an $(n-1)$-simplex denoted by $\Lambda = \mathrm{Con}(B_1, \ldots, B_n)$, that is

$$\Lambda = \mathrm{Con}(B_1, \ldots, B_n) = \{\, \lambda_1 B_1 + \cdots + \lambda_n B_n \, : \, \sum_{j=1}^{n} \lambda_j = 1, \ \lambda_1, \ldots, \lambda_n \geq 0 \,\}.$$

The set corresponds to an invertible $n \times n$ matrix $[\Lambda]$ whose columns are $B_1, \ldots, B_n$. Conversely, consider the matrix $C = (b_{jk})$, where $C_k = (b_{1k}, \ldots, b_{nk})^T$ $(k = 1, \ldots, n)$. If $\det(C) \neq 0$ and the sum of the entries in each column is 1, then the matrix C corresponds to an $(n-1)$-simplex $\mathrm{Con}(B_1, \ldots, B_n)$ in Δ_n. Let C_1 and C_2 be $n \times n$ matrices with nonnegative entries and all the columns of each matrix add up to 1.
(i) Show that $C_1 C_2$ and $C_2 C_1$ are also such matrices.
(ii) Are the $n^2 \times n^2$ matrices $C_1 \otimes C_2$, $C_2 \otimes C_1$ such matrices?

Problem 19. (i) Let S, T be $n \times n$ matrices over $\mathbb{C}$ with $S^2 = I_n$, $(TS)^2 = I_n$. Thus S and T are invertible. Show that $STS^{-1} = T^{-1}$, $ST^{-1}S = T$.
(ii) Let A be an invertible $n \times n$ matrix over $\mathbb{C}$ and B be an $n \times n$ matrix over $\mathbb{C}$. We define the $n \times n$ matrix $D := A^{-1}BA$. Show that $D^n = A^{-1}B^n A$ applying $AA^{-1} = I_n$.

Problem 20. Find all invertible 2×2 matrices S such that

$$\begin{pmatrix} 0 & 1 \\ 0 & 0 \end{pmatrix} S \begin{pmatrix} 0 & 0 \\ 1 & 0 \end{pmatrix} = \begin{pmatrix} 0 & 0 \\ 1 & 0 \end{pmatrix} S \begin{pmatrix} 0 & 1 \\ 0 & 0 \end{pmatrix}.$$

Problem 21. (i) Consider the 3×2 matrix

$$A = \begin{pmatrix} 1 & 1 \\ 1 & 1 \\ 1 & 1 \end{pmatrix}.$$

Can one find 2×3 matrices B such that $AB = I_2$?
(ii) Consider the 3×2 matrix

$$X = \begin{pmatrix} 1 & 2 \\ 1 & 1 \\ 1 & 1 \end{pmatrix}.$$

Can one find 2×3 matrices Y such that $XY = I_2$?

Problem 22. Let $\mathbf{x} \in \mathbb{R}^3$ and $\times$ be the vector product. Find all solutions of

$$\begin{pmatrix} 1 \\ 0 \\ 0 \end{pmatrix} \times \begin{pmatrix} x_1 \\ x_2 \\ x_3 \end{pmatrix} = \begin{pmatrix} 0 \\ 0 \\ 1 \end{pmatrix} \quad \text{and} \quad \begin{pmatrix} 0 \\ 1 \\ 0 \end{pmatrix} \times \begin{pmatrix} y_1 \\ y_2 \\ y_3 \end{pmatrix} = \begin{pmatrix} 1/\sqrt{2} \\ 0 \\ 1/\sqrt{2} \end{pmatrix}.$$

Problem 23. (i) Let A be an 3×3 matrix over $\mathbb{R}$ and $\mathbf{u}, \mathbf{v} \in \mathbb{R}^3$. Find the conditions on A, $\mathbf{u}$, $\mathbf{v}$ such that

$$A(\mathbf{u} \times \mathbf{v}) = (A\mathbf{u}) \times (A\mathbf{v}).$$

(ii) Consider the three vectors $\mathbf{v}_1$, $\mathbf{v}_2$, $\mathbf{v}_3$ in $\mathbb{R}^3$. Show that $\mathbf{v}_1 \cdot (\mathbf{v}_2 \times \mathbf{v}_3) = 0$ if $\mathbf{v}_1$, $\mathbf{v}_2$, $\mathbf{v}_3$ are linearly dependent.

(iii) Let $\mathbf{u}, \mathbf{v} \in \mathbb{R}^3$. Show that $(\mathbf{u} \times \mathbf{v}) \cdot (\mathbf{u} \times \mathbf{v}) \equiv (\mathbf{u} \cdot \mathbf{u})(\mathbf{v} \cdot \mathbf{v}) - (\mathbf{u} \cdot \mathbf{v})^2$.

(iv) Let $\mathbf{a}$, $\mathbf{b}$, $\mathbf{c}$, $\mathbf{d}$ be vectors in $\mathbb{R}^3$. Show that (*Lagrange identity*)

$$(\mathbf{a} \times \mathbf{b}) \cdot (\mathbf{c} \times \mathbf{d}) \equiv \det \begin{pmatrix} \mathbf{a} \cdot \mathbf{c} & \mathbf{b} \cdot \mathbf{c} \\ \mathbf{a} \cdot \mathbf{d} & \mathbf{b} \cdot \mathbf{d} \end{pmatrix}.$$

Problem 24. (i) Consider four nonzero vectors $\mathbf{v}_1$, $\mathbf{v}_2$, $\mathbf{v}_3$, $\mathbf{v}_4$ in $\mathbb{R}^3$. Let

$$\mathbf{w} := (\mathbf{v}_1 \times \mathbf{v}_2) \times (\mathbf{v}_3 \times \mathbf{v}_4) \neq \mathbf{0}.$$

Find $\mathbf{w} \times (\mathbf{v}_1 \times \mathbf{v}_2)$ and $\mathbf{w} \times (\mathbf{v}_3 \times \mathbf{v}_4)$. Discuss. Note that $\mathbf{v}_1$ and $\mathbf{v}_2$ span a plane in $\mathbb{R}^3$ and $\mathbf{v}_3$ and $\mathbf{v}_4$ span a plane in $\mathbb{R}^3$.

(ii) Let $\mathbf{u}$, $\mathbf{v}$, $\mathbf{w}$ be vectors in $\mathbb{R}^3$. Show that (*Jacobi identity*)

$$\mathbf{u} \times (\mathbf{v} \times \mathbf{w}) + \mathbf{w} \times (\mathbf{u} \times \mathbf{v}) + \mathbf{v} \times (\mathbf{w} \times \mathbf{u}) = \mathbf{0}.$$

(iii) Consider the normalized column vectors

$$\mathbf{v}_1 = \begin{pmatrix} 0 & 1 & 0 \end{pmatrix}^T, \qquad \mathbf{v}_2 = \begin{pmatrix} 1 & 0 & 1 \end{pmatrix}^T.$$

Do the three vectors $\mathbf{v}_1$, $\mathbf{v}_2$, $\mathbf{v}_1 \times \mathbf{v}_2$ form an orthonormal basis in $\mathbb{R}^3$?

(iv) Let $\mathbf{v}_1$, $\mathbf{v}_2$, $\mathbf{v}_3$, $\mathbf{v}_4$ be column vectors in $\mathbb{R}^3$. Show that

$$(\mathbf{v}_1 \times \mathbf{v}_2)^T (\mathbf{v}_3 \times \mathbf{v}_4) \equiv (\mathbf{v}_1^T \mathbf{v}_3)(\mathbf{v}_2^T \mathbf{v}_4) - (\mathbf{v}_2^T \mathbf{v}_3)(\mathbf{v}_1^T \mathbf{v}_4).$$

A special case with $\mathbf{v}_3 = \mathbf{v}_1$, $\mathbf{v}_4 = \mathbf{v}_2$ is

$$(\mathbf{v}_1 \times \mathbf{v}_2)^T (\mathbf{v}_1 \times \mathbf{v}_2) \equiv (\mathbf{v}_1^T \mathbf{v}_1)(\mathbf{v}_2^T \mathbf{v}_2) - (\mathbf{v}_2^T \mathbf{v}_1)(\mathbf{v}_1^T \mathbf{v}_2).$$

Problem 25. Consider the coordinates

$$\mathbf{p}_1 = (x_1, y_1, z_1)^T, \qquad \mathbf{p}_2 = (x_2, y_2, z_2)^T, \qquad \mathbf{p}_3 = (x_3, y_3, z_3)^T$$

with $\mathbf{p}_1 \neq \mathbf{p}_2$, $\mathbf{p}_2 \neq \mathbf{p}_3$, $\mathbf{p}_3 \neq \mathbf{p}_1$. We form the vectors

$$\mathbf{v}_{21} = \begin{pmatrix} x_2 - x_1 \\ y_2 - y_1 \\ z_2 - z_1 \end{pmatrix}, \qquad \mathbf{v}_{31} = \begin{pmatrix} x_3 - x_1 \\ y_3 - y_1 \\ z_3 - z_1 \end{pmatrix}.$$

What does $\frac{1}{2}|\mathbf{v}_{21} \times \mathbf{v}_{31}|$ calculate? Apply it to $\mathbf{p}_1 = (0,0,0)^T$, $\mathbf{p}_2 = (1,0,1)^T$, $\mathbf{p}_3 = (1,1,1)^T$.

Problem 26. Consider the 4×4 *Haar matrix*

$$K = \frac{1}{2} \begin{pmatrix} 1 & 1 & 1 & 1 \\ 1 & 1 & -1 & -1 \\ \sqrt{2} & -\sqrt{2} & 0 & 0 \\ 0 & 0 & \sqrt{2} & -\sqrt{2} \end{pmatrix}.$$

Find all 4×4 hermitian matrices H such that $KHK^T = H$.

Problem 27. Consider the real symmetric 3×3 matrix

$$A = \begin{pmatrix} -1/2 & -\sqrt{3}/6 & \sqrt{6}/3 \\ -\sqrt{3}/6 & -5/6 & -\sqrt{2}/3 \\ \sqrt{6}/3 & -\sqrt{2}/3 & 1/3 \end{pmatrix}.$$

Show that $A^T = A^{-1}$ by showing that the column of the matrix are normalized and pairwise orthonormal.

Problem 28. Let $X \in \mathbb{R}^{n \times n}$. Show that X can be written as

$$X = A + S + cI_n$$

where A is antisymmetric $(A^T = -A)$, S is symmetric $(S^T = S)$ with $\operatorname{tr}(S) = 0$ and $c \in \mathbb{R}$.

Problem 29. Let $c_j > 0$ for $j = 1, \ldots, n$. Show that the $n \times n$ matrices

$$\left(\frac{\sqrt{c_j c_k}}{c_j + c_k} \right), \qquad \left(\frac{1/c_j + 1/c_k}{\sqrt{c_j c_k}} \right)$$

$(k = 1, \ldots, n)$ are positive definite.

Problem 30. Let $a, b \in \mathbb{R}$ and $c^2 := a^2 + b^2$ with $c^2 > 0$. Consider the 2×2 matrix

$$M(a, b) = \frac{1}{c} \begin{pmatrix} a & b \\ b & -a \end{pmatrix}.$$

Show that $M(a, b)M(a, b) = I_2$. Thus $M(a, b)$ can be considered as a *square root* of I_2.

Problem 31. Show that a nonzero 3×3 matrix A over $\mathbb{R}$ which satisfies

$$A^2 A^T + A^T A^2 = 2A, \qquad AA^T A = 2A, \qquad A^3 = 0_3$$

is given by

$$A = \sqrt{2} \begin{pmatrix} 0 & 0 & 0 \\ 1 & 0 & 0 \\ 0 & 1 & 0 \end{pmatrix} \Rightarrow A^T = \sqrt{2} \begin{pmatrix} 0 & 1 & 0 \\ 0 & 0 & 1 \\ 0 & 0 & 0 \end{pmatrix}.$$

Problem 32. Let $\mathbf{x}$ be a normalized column vector in $\mathbb{R}^n$, i.e. $\mathbf{x}^T \mathbf{x} = 1$. A matrix T is called a *Householder matrix* if $T := I_n - 2\mathbf{x}\mathbf{x}^T$. Show that

$$T^2 = (I_n - 2\mathbf{x}\mathbf{x}^T)(I_n - 2\mathbf{x}\mathbf{x}^T) = I_n - 4\mathbf{x}\mathbf{x}^T + 4\mathbf{x}\mathbf{x}^T = I_n.$$

Problem 33.   Let A be an $n \times n$ matrix over $\mathbb{R}$. Show that there exists nonnull vectors $\mathbf{x}_1$, $\mathbf{x}_2$ in $\mathbb{R}^n$ such that

$$\frac{\mathbf{x}_1^T A \mathbf{x}_1}{\mathbf{x}_1^T \mathbf{x}_1} \leq \frac{\mathbf{x}^T A \mathbf{x}}{\mathbf{x}^T \mathbf{x}} \leq \frac{\mathbf{x}_2^T A \mathbf{x}_2}{\mathbf{x}_2^T \mathbf{x}_2}$$

for every nonnull vector $\mathbf{x}$ in $\mathbb{R}^n$.

Problem 34.   A *generalized Kronecker delta* can be defined as follows

$$\delta_{I,J} := \begin{cases} 1 & \text{if } J = (j_1, \dots, j_r) \text{ is an even permutation of } I = (i_1, \dots, i_r) \\ -1 & \text{if } J \text{ is an odd permutation of } I \\ 0 & \text{if } J \text{ is not a permutation of } I \end{cases}$$

Show that $\delta_{126,621} = -1$, $\delta_{126,651} = 0$, $\delta_{125,512} = 1$. Give a computer algebra implementation.

Problem 35.   (i) Let $j_0, j_1, k_0, k_1 \in \{0, 1\}$. Consider the tensor

$$T^{j_0, j_1}_{k_0, k_1}.$$

Give a $1-1$ map that maps $T^{j_0, j_1}_{k_0, k_1}$ to a $2^2 \times 2^2$ matrix $S = (s_{\ell_0, \ell_1})$ with $\ell_0, \ell_1 = 0, 1, 2, 3$, i.e.

$$\begin{pmatrix} t^{00}_{00} & t^{00}_{01} & t^{00}_{10} & t^{00}_{11} \\ t^{01}_{00} & t^{01}_{01} & t^{01}_{10} & t^{01}_{11} \\ t^{10}_{00} & t^{10}_{01} & t^{10}_{10} & t^{10}_{11} \\ t^{11}_{00} & t^{11}_{01} & t^{11}_{10} & t^{11}_{11} \end{pmatrix} \mapsto \begin{pmatrix} s_{00} & s_{01} & s_{02} & s_{03} \\ s_{10} & s_{11} & s_{12} & s_{13} \\ s_{20} & s_{21} & s_{22} & s_{23} \\ s_{30} & s_{31} & s_{32} & s_{33} \end{pmatrix}.$$

(ii) Let $j_0, j_1, j_2, k_0, k_1, k_2 \in \{0, 1\}$. Consider the tensor

$$T^{j_0, j_1, j_2}_{k_0, k_1, k_2}.$$

Give a $1-1$ map that maps $T^{j_0, j_1, j_2}_{k_0, k_1, k_2}$ to a $2^3 \times 2^3$ matrix $S = (s_{\ell_0, \ell_1})$ with $\ell_0, \ell_1 = 0, 1, \dots, 2^3 - 1$.

(iii) Let $n \geq 2$ and $j_0, j_1, \dots, j_n, k_0, k_1, \dots, k_n \in \{0, 1\}$. Consider the tensor

$$T^{j_0, j_1, \dots, j_{n-1}}_{k_0, k_1, \dots, k_{n-1}}.$$

Give a $1-1$ map that maps the tensor to a $2^n \times 2^n$ matrix $S = (s_{\ell_0, \ell_1})$ with $\ell_0, \ell_1 = 0, 1, \dots, 2^n - 1$. Give a SymbolicC++ implementation. The user provides n.

Chapter 2

Linear Equations

Let A be an $m \times n$ matrix over a field $\mathbb{F}$. Let $b_1, \ldots, b_m$ be elements of the field $\mathbb{F}$. The system of equations

$$a_{11}x_1 + a_{12}x_2 + \cdots + a_{1n}x_n = b_1$$
$$a_{21}x_1 + a_{22}x_2 + \cdots + a_{2n}x_n = b_2$$
$$\vdots$$
$$a_{m1}x_1 + a_{m2}x_2 + \cdots + a_{mn}x_n = b_m$$

is called a system of linear equations. We also write

$$A\mathbf{x} = \mathbf{b}$$

where $\mathbf{x}$ and $\mathbf{b}$ are considered as column vectors. The system is said to be homogeneous if all the numbers $b_1, \ldots, b_m$ are equal to 0. The number n is called the number of unknowns, and m is called the number of equations. The system of homogeneous equations also admits the trivial solution $x_1 = x_2 = \cdots = x_n = 0$.

A homogeneous system of m linear equations in n unknowns with $n > m$ admits a non-trivial solution. An underdetermined linear system is either inconsistent or has infinitely many solutions.

An important special case is $m = n$. Then for the system of linear equations $A\mathbf{x} = \mathbf{b}$ we investigate the cases A^{-1} exists and A^{-1} does not exist. If A^{-1} exists we can write the solution as $\mathbf{x} = A^{-1}\mathbf{b}$.

If $m > n$, then we have an overdetermined system and it can happen that no solution exists. One solves these problems in the least-square sense.

Problem 1.   Let
$$A = \begin{pmatrix} 1 & 1 \\ 2 & -1 \end{pmatrix}, \qquad \mathbf{b} = \begin{pmatrix} 1 \\ 5 \end{pmatrix}.$$
Find the solutions of the system of linear equations $A\mathbf{x} = \mathbf{b}$.

Solution 1.   Since A is invertible ($\det(A) = -3$) we have the unique solution $\mathbf{x} = A^{-1}\mathbf{b}$. From the equations $x_1 + x_2 = 1$, $2x_1 - x_2 = 5$ we obtain by addition of the two equations $3x_1 = 6$ and thus $x_1 = 2$. It follows that $x_2 = -1$.

Problem 2.   Let $\alpha, b_1, b_2 \in \mathbb{R}$. Solve the system of linear equations
$$\begin{pmatrix} \cos(\alpha) & -\sin(\alpha) \\ \sin(\alpha) & \cos(\alpha) \end{pmatrix} \begin{pmatrix} x_1 \\ x_2 \end{pmatrix} = \begin{pmatrix} b_1 \\ b_2 \end{pmatrix}.$$

Solution 2.   The inverse of the matrix on the left hand side exists for all α and is found by the substitution $\alpha \mapsto -\alpha$. Thus the solution is
$$\begin{pmatrix} x_1 \\ x_2 \end{pmatrix} = \begin{pmatrix} \cos(\alpha) & \sin(\alpha) \\ -\sin(\alpha) & \cos(\alpha) \end{pmatrix} \begin{pmatrix} b_1 \\ b_2 \end{pmatrix}.$$

Problem 3.   (i) Let $\epsilon \in \mathbb{R}$ and
$$A = \begin{pmatrix} 1 & 1 \\ 2 & 2 \end{pmatrix}, \qquad \mathbf{b} = \begin{pmatrix} 3 \\ \epsilon \end{pmatrix}.$$
Find the condition on ϵ so that there is a solution of $A\mathbf{x} = \mathbf{b}$.
(ii) Consider the system of three linear equations
$$\begin{pmatrix} 1 & 1 & 1 \\ 1 & 2 & 4 \\ 1 & 4 & 10 \end{pmatrix} \begin{pmatrix} x_1 \\ x_2 \\ x_3 \end{pmatrix} = \begin{pmatrix} 1 \\ \epsilon \\ \epsilon^2 \end{pmatrix}$$
with $\epsilon \in \mathbb{R}$. Find the condition on ϵ so that there is a solution.

Solution 3.   (i) From $A\mathbf{x} = \mathbf{b}$ we obtain $x_1 + x_2 = 3$, $2x_1 + 2x_2 = \epsilon$. Multiplying the first equation by 2 and then subtracting from the second equation yields $6 = \epsilon$. Thus if $\epsilon \neq 6$ there is no solution. If $\epsilon = 6$ the line $x_1 + x_2 = 3$ is the solution.
(ii) The matrix has no inverse, i.e. the determinant is equal to 0. Consider the three equations
$$x_1 + x_2 + x_3 = 1, \quad x_1 + 2x_2 + 4x_3 = \epsilon, \quad x_1 + 4x_2 + 10x_3 = \epsilon^2.$$
Subtracting the first equation from the second and third yields
$$x_2 + 3x_3 = \epsilon - 1, \quad 3x_2 + 9x_3 = \epsilon^2 - 1.$$
Inserting x_2 from the first equation into the second equation yields $\epsilon^2 - 3\epsilon + 2 = 0$ with the solutions $\epsilon = 1$ and $\epsilon = 2$.

Problem 4.   Find all solutions of the system of linear equations

$$\begin{pmatrix} \cos(\theta) & -\sin(\theta) \\ -\sin(\theta) & -\cos(\theta) \end{pmatrix} \begin{pmatrix} x_1 \\ x_2 \end{pmatrix} = \begin{pmatrix} x_1 \\ x_2 \end{pmatrix}, \qquad \theta \in \mathbb{R}$$

with $\mathbf{x} \neq \mathbf{0}$. What type of equation is this?

Solution 4.   We obtain

$$\begin{pmatrix} x_1 \\ x_2 \end{pmatrix} = \begin{pmatrix} \cos(\theta/2) \\ -\sin(\theta/2) \end{pmatrix}$$

using the identities $\sin(\theta) \equiv 2\sin(\theta/2)\cos(\theta/2)$, $\cos(\theta) \equiv 2\cos^2(\theta/2) - 1$. This is an eigenvalue equation with eigenvalue 1.

Problem 5.   Find all solutions of the system of linear equations

$$\begin{pmatrix} 5 & -2 & -4 \\ -2 & 2 & 2 \\ -4 & 2 & 5 \end{pmatrix} \begin{pmatrix} x_1 \\ x_2 \\ x_3 \end{pmatrix} = \begin{pmatrix} x_1 \\ x_2 \\ x_3 \end{pmatrix}.$$

Solution 5.   We obtain

$$\begin{pmatrix} 4 & -2 & -4 \\ -2 & 1 & 2 \\ -4 & 2 & 4 \end{pmatrix} \begin{pmatrix} x_1 \\ x_2 \\ x_3 \end{pmatrix} = \begin{pmatrix} 0 \\ 0 \\ 0 \end{pmatrix}.$$

Therefore $4x_1 - 2x_2 - 4x_3 = 0$, $-2x_1 + x_2 + 2x_3 = 0$, $-4x_1 + 2x_2 + 4x_3 = 0$. These equations are pairwise linearly dependent. We have $2x_1 - x_2 - 2x_3 = 0$. This equation defines a plane. We have an eigenvalue equation with eigenvalue 1 if at least one of the x_j is nonzero.

Problem 6.   Let $A \in \mathbb{R}^{n \times n}$ and $\mathbf{x}, \mathbf{b} \in \mathbb{R}^n$. Consider the linear equation $A\mathbf{x} = \mathbf{b}$. Show that it can be written as $\mathbf{x} = T\mathbf{x}$, i.e. find $T\mathbf{x}$.

Solution 6.   Let $C = I_n - A$. Then we can write $\mathbf{x} = C\mathbf{x} + \mathbf{b}$. Thus $\mathbf{x} = T\mathbf{x}$ with $T\mathbf{x} := C\mathbf{x} + \mathbf{b}$.

Problem 7.   If the system of linear equations $A\mathbf{x} = \mathbf{b}$ admits no solution we call the equations inconsistent. If there is a solution, the equations are called consistent. Let $A\mathbf{x} = \mathbf{b}$ be a system of m linear equations in n unknowns and suppose that the rank of A is m. Show that in this case $A\mathbf{x} = \mathbf{b}$ is consistent.

Solution 7.   Since $[A|\mathbf{b}]$ is an $m \times (n+1)$ matrix we have $m \geq \text{rank}[A|\mathbf{b}]$. We have $\text{rank}[A|\mathbf{b}] \geq \text{rank}(A)$ and by assumption $\text{rank}(A) = m$. Thus

$$m \geq \text{rank}[A|\mathbf{b}] \geq \text{rank}(A) = m.$$

Hence $\text{rank}[A|\mathbf{b}] = m$ and therefore the system of equations $A\mathbf{x} = \mathbf{b}$ are consistent.

Problem 8. Find all solutions of the linear system of three equations and four unknowns

$$x_1 + 2x_2 - 4x_3 + x_4 = 3, \quad 2x_1 - 3x_2 + x_3 + 5x_4 = -4, \quad 7x_1 - 10x_3 + 13x_4 = 0.$$

Solution 8. We obtain

$$x_1 = \frac{10}{7}t + \frac{13}{28}, \quad x_2 = \frac{9}{7}t + \frac{39}{28}, \quad x_3 = t, \quad x_4 = -\frac{1}{4}, \quad t \in \mathbb{R}.$$

Problem 9. (i) Solve the linear equation

$$(\, x_1 \quad x_2 \quad x_3 \,) \begin{pmatrix} 1 & 2 & 3 \\ 1 & 2 & 3 \\ 1 & 2 & 3 \end{pmatrix} = (\, x_1 \quad x_2 \quad x_3 \,).$$

(ii) Find $x, y \in \mathbb{R}$ such that

$$\begin{pmatrix} 2 & -3 \\ -1 & 2 \end{pmatrix} = \begin{pmatrix} 3 & 0 \\ 0 & 1 \end{pmatrix} \begin{pmatrix} 2/3 & x \\ y & 2 \end{pmatrix}.$$

Solution 9. (i) We obtain

$$(\, x_1 \quad x_2 \quad x_3 \,) \begin{pmatrix} 0 & 2 & 3 \\ 1 & 1 & 3 \\ 1 & 2 & 2 \end{pmatrix} = (\, 0 \quad 0 \quad 0 \,).$$

The rank of the matrix on the left-hand side is 3 and thus the inverse exists. Hence the solution is $x_1 = x_2 = x_3 = 0$.
(ii) We obtain

$$\begin{pmatrix} 2 & -3 \\ -1 & 2 \end{pmatrix} = \begin{pmatrix} 2 & 3x \\ y & 2 \end{pmatrix}.$$

Thus $x = y = -1$.

Problem 10. Show that the *curve fitting problem*

j	0	1	2	3	4
t_j	-1.0	-0.5	0.0	0.5	1.0
y_j	1.0	0.5	0.0	0.5	2.0

by a quadratic polynomial of the form $p(t) = a_2 t^2 + a_1 t + a_0$ leads to an overdetermined linear system.

Solution 10. From the interpolation conditions $p(t_j) = y_j$ with $j = 0, 1, \ldots, 4$ we obtain the overdetermined linear system

$$\begin{pmatrix} 1 & t_0 & t_0^2 \\ 1 & t_1 & t_1^2 \\ 1 & t_2 & t_2^2 \\ 1 & t_3 & t_3^2 \\ 1 & t_4 & t_4^2 \end{pmatrix} \begin{pmatrix} a_0 \\ a_1 \\ a_2 \end{pmatrix} = \begin{pmatrix} y_0 \\ y_1 \\ y_2 \\ y_3 \\ y_4 \end{pmatrix}.$$

Problem 11. Consider the overdetermined linear system $A\mathbf{x} = \mathbf{b}$. Find an $\hat{\mathbf{x}}$ such that

$$\|A\hat{\mathbf{x}} - \mathbf{b}\|_2 = \min_{\mathbf{x}} \|A\mathbf{x} - \mathbf{b}\|_2 \equiv \min_{\mathbf{x}} \|\mathbf{r}(\mathbf{x})\|_2$$

with the *residual vector* $\mathbf{r}(\mathbf{x}) := \mathbf{b} - A\mathbf{x}$ and $\|.\|_2$ denotes the Euclidean norm.

Solution 11. From

$$\|\mathbf{r}(\mathbf{x})\|_2^2 = \mathbf{r}^T\mathbf{r} = (\mathbf{b} - A\mathbf{x})^T(\mathbf{b} - A\mathbf{x}) = \mathbf{b}^T\mathbf{b} - 2\mathbf{x}^T A^T\mathbf{b} + \mathbf{x}^T A^T A\mathbf{x}$$

where we used that $\mathbf{x}^T A^T\mathbf{b} = \mathbf{b}^T A\mathbf{x}$, and the necessary condition

$$\nabla \|\mathbf{r}(\mathbf{x})\|_2^2|_{\mathbf{x}=\hat{\mathbf{x}}} = 0$$

we obtain $A^T A\hat{\mathbf{x}} - A^T\mathbf{b} = 0$. This system is called *normal equations*. We can also write this system as

$$A^T(\mathbf{b} - A\hat{\mathbf{x}}) \equiv A^T\mathbf{r}(\hat{\mathbf{x}}) = 0.$$

This justifies the name normal equations.

Problem 12. Consider the overdetermined linear system $A\mathbf{x} = \mathbf{b}$ with

$$A = \begin{pmatrix} 1 & 1 \\ 1 & 2 \\ 1 & 3 \\ 1 & 4 \\ 1 & 5 \\ 1 & 6 \\ 1 & 7 \\ 1 & 8 \\ 1 & 9 \\ 1 & 10 \end{pmatrix}, \quad \mathbf{x} = \begin{pmatrix} x_1 \\ x_2 \end{pmatrix}, \quad \mathbf{b} = \begin{pmatrix} 444 \\ 458 \\ 478 \\ 493 \\ 506 \\ 516 \\ 523 \\ 531 \\ 543 \\ 571 \end{pmatrix}.$$

Solve this linear system in the least squares sense (see the previous problem) by the normal equations method.

Solution 12. From the normal equations $A^T A\hat{\mathbf{x}} = A^T\mathbf{b}$ we obtain

$$\begin{pmatrix} 10 & 55 \\ 55 & 385 \end{pmatrix} \begin{pmatrix} \hat{x}_1 \\ \hat{x}_2 \end{pmatrix} = \begin{pmatrix} 5063 \\ 28898 \end{pmatrix}$$

with $\det(A^T A) \neq 0$. The solution is approximately

$$\begin{pmatrix} \hat{x}_1 \\ \hat{x}_2 \end{pmatrix} = \begin{pmatrix} 436.2 \\ 12.7455 \end{pmatrix}.$$

Problem 13. An underdetermined linear system is either inconsistent or has infinitely many solutions. Consider the underdetermined linear system $H\mathbf{x} = \mathbf{y}$, where H is an $n \times m$ matrix with $m > n$ and

$$\mathbf{x} = \begin{pmatrix} x_1 \\ x_2 \\ \vdots \\ x_m \end{pmatrix}, \qquad \mathbf{y} = \begin{pmatrix} y_1 \\ y_2 \\ \vdots \\ y_n \end{pmatrix}.$$

Assume that $H\mathbf{x} = \mathbf{y}$ has infinitely many solutions. Let Q be the $n \times m$ matrix

$$Q = \begin{pmatrix} 1 & 0 & \cdots & 0 & 0 & \cdots & 0 \\ 0 & 1 & \cdots & 0 & 0 & \cdots & 0 \\ \vdots & \vdots & \ddots & \vdots & \vdots & & \vdots \\ 0 & 0 & \cdots & 1 & 0 & \cdots & 0 \end{pmatrix}.$$

We define $\hat{\mathbf{x}} := Q\mathbf{x}$. Find

$$\min_{\mathbf{x}} \|Q\mathbf{x} - \mathbf{y}\|_2^2$$

subject to the constraint $\|H\mathbf{x} - \mathbf{y}\|_2^2 = 0$. Assume that $(\lambda H^T H + Q^T Q)^{-1}$ exists for all $\lambda > 0$. Apply the *Lagrange multiplier method*.

Solution 13. We have

$$V(\mathbf{x}) = \|Q\mathbf{x} - \mathbf{y}\|_2^2 + \lambda \|H\mathbf{x} - \mathbf{y}\|_2^2$$

where λ is the Lagrange multiplier. Thus $V(\mathbf{x}) \to \min$ if λ is sufficiently large. The derivative of $V(\mathbf{x})$ with respect to the unknown $\mathbf{x}$ is

$$\frac{\partial}{\partial \mathbf{x}} V(\mathbf{x}) = 2\lambda H^T(H\mathbf{x} - \mathbf{y}) + 2Q^T(Q\mathbf{x} - \mathbf{y}).$$

Thus

$$(\lambda H^T H + Q^T Q)\mathbf{x} = (\lambda H^T + Q^T)\mathbf{y}.$$

It follows that $\hat{\mathbf{x}} = \left(\lambda H^T H + Q^T Q\right)^{-1} (\lambda H + Q)^T \mathbf{y}$.

Problem 14. Show that solving the system of nonlinear equations with the unknowns x_1, x_2, x_3, x_4

$$(x_1 - 1)^2 + (x_2 - 2)^2 + x_3^2 = a^2(x_4 - b_1)^2$$
$$(x_1 - 2)^2 + x_2^2 + (x_3 - 2)^2 = a^2(x_4 - b_2)^2$$
$$(x_1 - 1)^2 + (x_2 - 1)^2 + (x_3 - 1)^2 = a^2(x_4 - b_3)^2$$
$$(x_1 - 2)^2 + (x_2 - 1)^2 + x_3^2 = a^2(x_4 - b_4)^2$$

leads to a linear underdetermined system. Solve this system with respect to x_1, x_2 and x_3.

Solution 14. Expanding all the squares and rearranging that the linear terms are on the left-hand side, yields

$$2x_1 + 4x_2 - 2a^2 b_1 x_4 = 5 - a^2 b_1^2 + x_1^2 + x_2^2 + x_3^2 - a^2 x_4^2$$
$$4x_1 + 4x_3 - 2a^2 b_2 x_4 = 8 - a^2 b_2^2 + x_1^2 + x_2^2 + x_3^2 - a^2 x_4^2$$
$$2x_1 + 2x_2 + 2x_3 - 2a^2 b_3 x_4 = 3 - a^2 b_3^2 + x_1^2 + x_2^2 + x_3^2 - a^2 x_4^2$$
$$4x_1 + 2x_2 - 2a^2 b_4 x_4 = 5 - a^2 b_4^2 + x_1^2 + x_2^2 + x_3^2 - a^2 x_4^2.$$

The quadratic terms in all the equations are the same. Thus by subtracting the first equation from each of the other three, we obtain an underdetermined system of three linear equations

$$2x_1 - 4x_2 + 4x_3 - 2a^2(b_2 - b_1)x_4 = 3 + a^2(b_1^2 - b_2^2)$$
$$-2x_2 + 2x_3 - 2a^2(b_3 - b_1)x_4 = -2 + a^2(b_1^2 - b_3^2)$$
$$2x_1 - 2x_2 - 2a^2(b_4 - b_1)x_4 = a^2(b_1^2 - b_4^2).$$

Solving these equations we obtain

$$x_1 = a^2(b_1 + b_2 - 2b_3)x_4 + \frac{a^2}{2}(-b_1^2 - b_2^2 + 2b_3^2) + \frac{7}{2}$$

$$x_2 = a^2(2b_1 + b_2 - 2b_3 - b_4)x_4 + \frac{a^2}{2}(-2b_1^2 - b_2^2 + b_3^2 + b_4^2) + \frac{7}{2}$$

$$x_3 = a^2(b_1 + b_2 - b_3 - b_4)x_4 + \frac{a^2}{2}(-b_1^2 - b_2^2 + b_3^2 + b_4^2) + \frac{5}{2}.$$

Inserting these solutions into one of the nonlinear equations provides a quadratic equation for x_4. Such a system of equations plays a role in the *Global Positioning System* (GPS), where x_4 plays the role of time.

Problem 15. Let A be an $m \times n$ matrix over $\mathbb{R}$. We define

$$N_A := \{ \mathbf{x} \in \mathbb{R}^n \ : \ A\mathbf{x} = \mathbf{0} \}.$$

N_A is called the *kernel* of A and $\nu(A) := \dim(N_A)$ is called the *nullity* of A. If N_A only contains the zero vector, then $\nu(A) = 0$.
(i) Let

$$A = \begin{pmatrix} 1 & 2 & -1 \\ 2 & -1 & 3 \end{pmatrix}.$$

Find N_A and $\nu(A)$.
(ii) Let

$$A = \begin{pmatrix} 2 & -1 & 3 \\ 4 & -2 & 6 \\ -6 & 3 & -9 \end{pmatrix}.$$

Find N_A and $\nu(A)$.

Solution 15. (i) From

$$\begin{pmatrix} 1 & 2 & -1 \\ 2 & -1 & 3 \end{pmatrix} \begin{pmatrix} x_1 \\ x_2 \\ x_3 \end{pmatrix} = \begin{pmatrix} 0 \\ 0 \end{pmatrix}$$

we find the system of linear equations $x_1 + 2x_2 - x_3 = 0$, $2x_1 - x_2 + 3x_3 = 0$. Eliminating x_3 yields $x_1 = -x_2$. It follows that $x_3 = -x_1$. Thus N_A is spanned by the vector $(-1 \ \ 1 \ \ 1)^T$. Therefore $\nu(A) = 1$.
(ii) From

$$\begin{pmatrix} 2 & -1 & 3 \\ 4 & -2 & 6 \\ -6 & 3 & -9 \end{pmatrix} \begin{pmatrix} x_1 \\ x_2 \\ x_3 \end{pmatrix} = \begin{pmatrix} 0 \\ 0 \\ 0 \end{pmatrix}$$

we find the system of linear equations

$$2x_1 - x_2 + 3x_3 = 0, \quad 4x_1 - 2x_2 + 6x_3 = 0, \quad -6x_1 + 3x_2 - 9x_3 = 0.$$

The three equations are the same. Thus from $2x_1 - x_2' + 3x_3 = 0$ we find that

$$\left\{ \begin{pmatrix} 1 \\ 2 \\ 0 \end{pmatrix}, \begin{pmatrix} 0 \\ 3 \\ 1 \end{pmatrix} \right\}$$

is a basis for N_A and $\nu(A) = 2$.

Problem 16. (i) Let $x_1, x_2, x_3 \in \mathbb{Z}$. Find all solutions of the system of linear equations

$$7x_1 + 5x_2 - 5x_3 = 8, \qquad 17x_1 + 10x_2 - 15x_3 = -42.$$

(ii) Find all positive solutions.

Solution 16. (i) Eliminating x_2 yields $3x_1 - 5x_3 = -58$ or $3x_1 \equiv -58 \pmod 5$. The solution is $x_1 \equiv 4 \pmod 5$ or $x_1 = 4 + 5s$, $s \in \mathbb{Z}$. Thus using $3x_1 - 5x_3 = -58$ we obtain $x_3 = 14 + 3s$ and using $7x_1 + 5x_2 - 5x_3 = 8$ we find $x_2 = 10 - 4s$.
(ii) For x_2 positive we have $s \leq 2$, for x_1 positive we have $0 \leq s$ and x_3 remains positive. Thus the solution set for positive x_1, x_2, x_3 is (4, 10, 14), (9, 6, 17), (14, 2, 20). We can write

$$\begin{pmatrix} x_1 \\ x_2 \\ x_3 \end{pmatrix} = \begin{pmatrix} 4 \\ 10 \\ 14 \end{pmatrix} + s \begin{pmatrix} 5 \\ -4 \\ 3 \end{pmatrix}$$

where $s = 0, 1, 2$.

Problem 17. Consider the inhomogeneous linear *integral equation*

$$\int_0^1 (\alpha_1(x)\beta_1(y) + \alpha_2(x)\beta_2(y))\varphi(y)dy + f(x) = \varphi(x) \qquad (1)$$

for the unknown function φ, $f(x) = x$ and

$$\alpha_1(x) = x, \quad \alpha_2(x) = \sqrt{x}, \quad \beta_1(y) = y, \quad \beta_2(y) = \sqrt{y}.$$

Thus α_1 and α_2 are continuous in $[0, 1]$ and likewise for β_1 and β_2. We define

$$B_1 := \int_0^1 \beta_1(y)\varphi(y)dy, \qquad B_2 := \int_0^1 \beta_2(y)\varphi(y)dy$$

and

$$a_{\mu\nu} := \int_0^1 \beta_\mu(y)\alpha_\nu(y)dy, \qquad b_\mu := \int_0^1 \beta_\mu(y)f(y)dy$$

where $\mu, \nu = 1, 2$. Show that the integral equation can be cast into a system of linear equations for B_1 and B_2. Solve this system of linear equations and thus find a solution of the integral equation.

Solution 17. Using B_1 and B_2 equation (1) can be written as

$$\varphi(x) = \alpha_1(x)B_1 + \alpha_2(x)B_2 + f(x) \tag{2}$$

or

$$\varphi(y) = \alpha_1(y)B_1 + \alpha_2(y)B_2 + f(y). \tag{3}$$

We insert (2) into the left-hand side and (3) into the right-hand side of

$$\varphi(x) = \alpha_1(x) \int_0^1 \beta_1(y)\varphi(y)dy + \alpha_2(x) \int_0^1 \beta_2(y)\varphi(y)dy + f(x).$$

Using $a_{\mu\nu}$ and b_μ we find

$$\alpha_1(x)B_1 + \alpha_2(x)B_2 = \alpha_1(x)B_1 a_{11} + \alpha_1(x)B_2 a_{12} + \alpha_1(x)b_1$$
$$+ \alpha_2(x)B_1 a_{21} + \alpha_2(x)B_2 a_{22} + \alpha_2(x)b_2.$$

Now α_1 and α_2 are linearly independent. Comparing the coefficients of α_1 and α_2 we obtain the system of linear equations for B_1 and B_2

$$B_1 = a_{11}B_1 + a_{12}B_2 + b_1, \quad B_2 = a_{21}B_1 + a_{22}B_2 + b_2$$

or in matrix form

$$\begin{pmatrix} 1 - a_{11} & -a_{12} \\ -a_{21} & 1 - a_{22} \end{pmatrix} \begin{pmatrix} B_1 \\ B_2 \end{pmatrix} = \begin{pmatrix} b_1 \\ b_2 \end{pmatrix}.$$

Since

$$a_{11} = \int_0^1 y^2 dy = \frac{1}{3}, \quad a_{12} = \int_0^1 y^{3/2} dy = \frac{2}{5}, \quad a_{21} = \int_0^1 y^{3/2} dy = \frac{2}{5},$$

$$a_{22} = \int_0^1 y dy = \frac{1}{2}, \quad b_1 = \int_0^1 y^2 dy = \frac{1}{3}, \quad b_2 = \int_0^1 y^{2/3} dy = \frac{2}{5}$$

we obtain

$$\begin{pmatrix} 2/3 & -2/5 \\ -2/5 & 1/2 \end{pmatrix} \begin{pmatrix} B_1 \\ B_2 \end{pmatrix} = \begin{pmatrix} 1/3 \\ 2/5 \end{pmatrix}$$

with the unique solution $B_1 = 49/26$, $B_2 = 30/13$. Since $\varphi(x) = \alpha_1(x)B_1 + \alpha_2(x)B_2 + x$ we obtain the solution of the integral equation as

$$\varphi(x) = \frac{75}{26}x + \frac{30}{13}\sqrt{x}.$$

Problem 18. Consider the area-preserving map of the two-dimensional torus (modulo 1)

$$\begin{pmatrix} x_1' \\ x_2' \end{pmatrix} = A \begin{pmatrix} x_1 \\ x_2 \end{pmatrix}, \quad A = \begin{pmatrix} 4 & 15 \\ 1 & 4 \end{pmatrix}$$

where $\det(A) = 1$ (area-preserving). Consider a rational point on the torus

$$\begin{pmatrix} x_1 \\ x_2 \end{pmatrix} = \begin{pmatrix} n_1/p \\ n_2/p \end{pmatrix}$$

where p is a prime number (except 2, 3, 5) and n_1, n_2 are integers between 0 and $p - 1$. One finds that the orbit has the following property. It is periodic and its period T depends on p alone. Consider $p = 7$, $n_1 = 2$, $n_2 = 3$. Find the orbit and the period T.

Solution 18.   We have

$$\begin{pmatrix} x_1 \\ x_2 \end{pmatrix} = \begin{pmatrix} 4 & 15 \\ 1 & 4 \end{pmatrix} \begin{pmatrix} 2/7 \\ 3/7 \end{pmatrix} = \begin{pmatrix} 53/7 \\ 14/7 \end{pmatrix} \mapsto \begin{pmatrix} 4/7 \\ 0 \end{pmatrix}.$$

Repeating this process we find the orbit

$$\begin{pmatrix} 2/7 \\ 3/7 \end{pmatrix}, \begin{pmatrix} 4/7 \\ 0 \end{pmatrix}, \begin{pmatrix} 2/7 \\ 4/7 \end{pmatrix}, \begin{pmatrix} 5/7 \\ 4/7 \end{pmatrix}, \begin{pmatrix} 3/7 \\ 0 \end{pmatrix}, \begin{pmatrix} 5/7 \\ 3/7 \end{pmatrix}, \begin{pmatrix} 2/7 \\ 3/7 \end{pmatrix}.$$

Thus the period is $T = 6$.

Problem 19.   *Gordan's theorem* tells us the following. Let A be an $m \times n$ matrix over $\mathbb{R}$ and $\mathbf{c}$ be an n-vector in $\mathbb{R}^n$. Then exactly one of the following systems has a solution:
System 1: $A\mathbf{x} < \mathbf{0}$ for some $\mathbf{x} \in \mathbb{R}^n$.
System 2: $A^T \mathbf{p} = \mathbf{0}$ and $\mathbf{p} \geq \mathbf{0}$ for some $\mathbf{p} \in \mathbb{R}^m$.
(i) Let

$$A = \begin{pmatrix} 1 & 0 & 1 \\ 0 & 1 & 0 \\ 1 & 0 & 1 \end{pmatrix}.$$

Find out whether system (1) or system (2) has a solution.
(ii) Let

$$A = \begin{pmatrix} 2 & 1 & 1 \\ 1 & 2 & 1 \\ 1 & 1 & 2 \end{pmatrix}.$$

Find out whether system (1) or system (2) has a solution.

Solution 19.   (i) We test system (1). We have

$$\begin{pmatrix} 1 & 0 & 1 \\ 0 & 1 & 0 \\ 1 & 0 & 1 \end{pmatrix} \begin{pmatrix} x_1 \\ x_2 \\ x_3 \end{pmatrix} < \begin{pmatrix} 0 \\ 0 \\ 0 \end{pmatrix}.$$

Thus we have the system of inequalities $x_2 < 0$, $x_1 + x_3 < 0$. Obviously this system has solutions. For example $x_1 = x_3 = -1$ and $x_2 = -1/2$. Thus system (2) has no solution.
(ii) From

$$\begin{pmatrix} 2 & 1 & 1 \\ 1 & 2 & 1 \\ 1 & 1 & 2 \end{pmatrix} \begin{pmatrix} p_1 \\ p_2 \\ p_3 \end{pmatrix} = \begin{pmatrix} 0 \\ 0 \\ 0 \end{pmatrix}$$

we find that $p_1 = p_2 = p_3 = 0$. Thus system (1) has a solution.

Problem 20. *Farkas' theorem* tells us the following. Let A be an $m \times n$ matrix over $\mathbb{R}$ and c be an n-vector in $\mathbb{R}^n$. Then exactly one of the following systems has a solution:

System 1: $Ax \leq 0$ and $c^T x > 0$ for some $x \in \mathbb{R}^n$.

System 2: $A^T y = c$ and $y \geq 0$ for some $y \in \mathbb{R}^m$.

Let

$$A = \begin{pmatrix} 1 & 0 & 1 \\ 0 & 1 & 0 \\ 1 & 0 & 1 \end{pmatrix}, \qquad c = \begin{pmatrix} 1 \\ 1 \\ 1 \end{pmatrix}.$$

Find out whether system (1) or system (2) has a solution.

Solution 20. We test system (1). We have

$$\begin{pmatrix} 1 & 0 & 1 \\ 0 & 1 & 0 \\ 1 & 0 & 1 \end{pmatrix} \begin{pmatrix} x_1 \\ x_2 \\ x_3 \end{pmatrix} \leq \begin{pmatrix} 0 \\ 0 \\ 0 \end{pmatrix}, \qquad (1 \quad 1 \quad 1) \begin{pmatrix} x_1 \\ x_2 \\ x_3 \end{pmatrix} > 0.$$

Thus we have the system of inequalities

$$x_1 + x_3 \leq 0, \quad x_2 \leq 0, \quad x_1 + x_3 \leq 0, \qquad x_1 + x_2 + x_3 > 0.$$

Since $x_1 + x_3 \leq 0$ and $x_2 \leq 0$ this system has no solution. Thus after Farkas' theorem system (2) has a solution.

Problem 21. Let A be an $n \times n$ matrix. Consider the linear equation $Ax = 0$. If the matrix A has rank r, then there are $n - r$ linearly independent solutions of $Ax = 0$. Let $n = 3$ and

$$A = \begin{pmatrix} 0 & 1 & 1 \\ 0 & 0 & 1 \\ 0 & 0 & 0 \end{pmatrix}.$$

Find the rank of A and the linearly independent solutions.

Solution 21. The rank of A is 2. Thus we have one $3 - 2 = 1$ linearly independent solution. We obtain $x = (t \quad 0 \quad 0)^T$, $t \in \mathbb{R}$.

Problem 22. Consider the curve described by the equation

$$2x^2 + 4xy - y^2 + 4x - 2y + 5 = 0 \tag{1}$$

relative to the natural basis (standard basis $e_1 = (1 \quad 0)^T$, $e_2 = (0 \quad 1)^T$).
(i) Write the equation in matrix form.
(ii) Find an orthogonal change of basis so that the equation relative to the new basis has no crossterms, i.e. no $x'y'$ term. This change of coordinate system does not change the origin.

Solution 22. (i) Equation (1) in matrix form is

$$(x \quad y) \begin{pmatrix} 2 & 2 \\ 2 & -1 \end{pmatrix} \begin{pmatrix} x \\ y \end{pmatrix} + (4 \quad -2) \begin{pmatrix} x \\ y \end{pmatrix} + 5 = 0.$$

(ii) The symmetric 2×2 matrix given in (i) admits the eigenvalues 3 and -2 with the corresponding normalized eigenvectors

$$\frac{1}{\sqrt{5}} \begin{pmatrix} 2 \\ 1 \end{pmatrix}, \qquad \frac{1}{\sqrt{5}} \begin{pmatrix} -1 \\ 2 \end{pmatrix}.$$

Thus the coordinates (x, y) of a point relative to the natural basis and (x', y') relative to the basis consisting of these normalized eigenvectors are related by

$$\begin{pmatrix} 2/\sqrt{5} & -1/\sqrt{5} \\ 1/\sqrt{5} & 2/\sqrt{5} \end{pmatrix} \begin{pmatrix} x' \\ y' \end{pmatrix} = \begin{pmatrix} x \\ y \end{pmatrix}$$

or

$$(x' \quad y') \begin{pmatrix} 2/\sqrt{5} & 1/\sqrt{5} \\ -1/\sqrt{5} & 2/\sqrt{5} \end{pmatrix} = (x \quad y).$$

Relative to the new basis the quadratic function becomes $3x'^2 - 2y'^2$. To determine the equation of the original curve relative to the new basis, we substitute

$$x = 2x'/\sqrt{5} - y'/\sqrt{5}, \qquad y = x'/\sqrt{5} + 2y'/\sqrt{5}$$

in the original equation (1) and obtain $3x'^2 - 2y'^2 + 6x'/\sqrt{5} - 8y'/\sqrt{5} + 5 = 0$.

Problem 23.   Consider the 2×2 matrix

$$\begin{pmatrix} b & -a \\ a & b \end{pmatrix}$$

with $a, b \in \mathbb{R}$ and positive determinant, i.e. $a^2 + b^2 > 0$. Solve the equation

$$\begin{pmatrix} b & -a \\ a & b \end{pmatrix} \begin{pmatrix} x_1 \\ y_1 \end{pmatrix} = \begin{pmatrix} b & a \\ -a & b \end{pmatrix} \begin{pmatrix} x_0 \\ y_0 \end{pmatrix}$$

for the vector $(x_1 \ y_1)^T$ with a given vector $(x_0 \ y_0)^T$.

Solution 23.   We have

$$\begin{pmatrix} b & -a \\ a & b \end{pmatrix}^{-1} = \frac{1}{a^2 + b^2} \begin{pmatrix} b & a \\ -a & b \end{pmatrix}. \tag{1}$$

Thus

$$\begin{pmatrix} x_1 \\ y_1 \end{pmatrix} = \begin{pmatrix} b & -a \\ a & b \end{pmatrix}^{-1} \begin{pmatrix} b & a \\ -a & b \end{pmatrix} \begin{pmatrix} x_0 \\ y_0 \end{pmatrix}.$$

Using equation (1) we obtain

$$\begin{pmatrix} x_1 \\ y_1 \end{pmatrix} = \frac{1}{a^2 + b^2} \begin{pmatrix} (b^2 - a^2)x_0 + 2aby_0 \\ -2abx_0 + (b^2 - a^2)y_0 \end{pmatrix}.$$

Problem 24.   Let V be a vector space over a field $\mathbb{F}$. Let W be a subspace of V. We define an *equivalence relation* $\sim$ on V by stating that $v_1 \sim v_2$ if $v_1 - v_2 \in W$.

The *quotient space* V/W is the set of equivalence classes $[v]$ where $v_1 - v_2 \in W$. Thus we can say that v_1 is equivalent to v_2 modulo W if $v_1 = v_2 + w$ for some $w \in W$. Let

$$V = \mathbb{R}^2 = \left\{ \begin{pmatrix} x_1 \\ x_2 \end{pmatrix} : x_1, x_2 \in \mathbb{R} \right\}$$

and the subspace

$$W = \left\{ \begin{pmatrix} x_1 \\ 0 \end{pmatrix} : x_1 \in \mathbb{R} \right\}.$$

(i) Is

$$\begin{pmatrix} 3 \\ 0 \end{pmatrix} \sim \begin{pmatrix} 1 \\ 0 \end{pmatrix}, \qquad \begin{pmatrix} 4 \\ 1 \end{pmatrix} \sim \begin{pmatrix} -3 \\ 1 \end{pmatrix}, \qquad \begin{pmatrix} 3 \\ 0 \end{pmatrix} \sim \begin{pmatrix} 4 \\ 1 \end{pmatrix} ?$$

(ii) Give the quotient space V/W.

Solution 24. (i) We have

$$\begin{pmatrix} 3 \\ 0 \end{pmatrix} \sim \begin{pmatrix} 1 \\ 0 \end{pmatrix} \quad \text{since} \quad \begin{pmatrix} 3 \\ 0 \end{pmatrix} - \begin{pmatrix} 1 \\ 0 \end{pmatrix} = \begin{pmatrix} 2 \\ 0 \end{pmatrix} \in W$$

$$\begin{pmatrix} 4 \\ 1 \end{pmatrix} \sim \begin{pmatrix} -3 \\ 1 \end{pmatrix} \quad \text{since} \quad \begin{pmatrix} 4 \\ 1 \end{pmatrix} - \begin{pmatrix} -3 \\ 1 \end{pmatrix} = \begin{pmatrix} 7 \\ 0 \end{pmatrix} \in W$$

$$\begin{pmatrix} 3 \\ 0 \end{pmatrix} \nsim \begin{pmatrix} 4 \\ 1 \end{pmatrix} \quad \text{since} \quad \begin{pmatrix} 3 \\ 0 \end{pmatrix} - \begin{pmatrix} 4 \\ 1 \end{pmatrix} = \begin{pmatrix} -1 \\ -1 \end{pmatrix} \notin W.$$

(ii) Thus the elements of the quotient space consists of straight lines parallel to the x_1 axis.

Problem 25. Suppose that V is a vector space over a field $\mathbb{F}$ and $U \subset V$ is a subspace. We define an equivalence relation $\sim$ on V by $x \sim y$ iff $x - y \in U$. Let $V/U = V/\sim$. Define addition and scalar multiplication on V/U by $[x] + [y] = [x] + [y]$, $c[x] = [cx]$, where $c \in \mathbb{F}$ and

$$[x] = \{ y \in V : y \sim x \}.$$

(i) Show that these operations do not depend on which representative x we choose.
(ii) Let $V = \mathbb{C}^2$ and the subspace $U = \{ (x_1, x_2) : x_1 = 2x_2 \}$. Find V/U.

Solution 25. (i) Suppose $[x] = [x']$ and $[y] = [y']$. We want to show that $[x + y] = [x' + y']$, i.e. that $x + y \sim x' + y'$. We have

$$(x + y) - (x' + y') = (x - x') + (y - y')$$

and $x - x' \in U$, $y - y' \in U$. Thus $(x+y) - (x' + y') \in U$ which means $x + y \sim x' + y'$ which means $[x + y] = [x' + y']$. Similarly $(cx) - (cx') = c(x - x') \in U$ so that $[cx] = [cx']$. All the vector space axioms are satisfied. Therefore V/U is a vector space over the field $\mathbb{F}$ (quotient vector space).
(ii) We have $(x_1, x_2) \sim (x_1', x_2')$ iff $x_1 - x_1' = 2(x_2 - x_2')$ or $x_1 - 2x_2 = x_1' - 2x_2'$.

Problem 26. Let $b > a$. Consider the system of linear equations

$$
\begin{pmatrix}
1 & 1 & 1 & \cdots & 1 \\
x_0 & x_1 & x_2 & \cdots & x_n \\
x_0^2 & x_1^2 & x_2^2 & \cdots & x_n^2 \\
\vdots & \vdots & \vdots & \ddots & \vdots \\
x_0^n & x_1^n & x_2^n & \cdots & x_n^n
\end{pmatrix}
\begin{pmatrix}
w_0 \\ w_1 \\ w_2 \\ \vdots \\ w_n
\end{pmatrix}
=
\begin{pmatrix}
b - a \\
(b^2 - a^2)/2 \\
(b^3 - a^3)/3 \\
\vdots \\
(b^{n+1} - a^{n+1})/(n+1)
\end{pmatrix}.
$$

Let $n = 2$, $a = 0$, $b = 1$, $x_0 = 0$, $x_1 = 1/2$, $x_2 = 1$. Find w_0, w_1, w_2.

Solution 26. From

$$
\begin{pmatrix}
1 & 1 & 1 \\
0 & 1/2 & 1 \\
0 & 1/4 & 1
\end{pmatrix}
\begin{pmatrix}
w_0 \\ w_1 \\ w_2
\end{pmatrix}
=
\begin{pmatrix}
1 \\ 1/2 \\ 1/3
\end{pmatrix}
$$

we find $w_0 = \frac{1}{6}$, $w_1 = \frac{2}{3}$, $w_2 = \frac{1}{6}$.

Problem 27. Let Y, X, A, B, C, E $n \times n$ matrices over $\mathbb{R}$. Consider the system of matrix equations $Y + CE + DX = 0_n$, $AE + BX = 0_n$. Assume that A has an inverse. Eliminate the matrix E and solve the system for Y.

Solution 27. From the first equation we obtain $Y = -CE - DX$. The second equation provides $E = -A^{-1}BX$. Inserting this equation into $Y = -CE - DX$ we obtain

$$
Y = (CA^{-1}B - D)X.
$$

Problem 28. For the *three-body problem* the following linear transformation plays a role

$$
X(x_1, x_2, x_3) = \frac{1}{3}(x_1 + x_2 + x_3)
$$

$$
x(x_1, x_2, x_3) = \frac{1}{\sqrt{2}}(x_1 - x_2)
$$

$$
y(x_1, x_2, x_3) = \frac{1}{\sqrt{6}}(x_1 + x_2 - 2x_3).
$$

(i) Find the inverse transformation.
(ii) Introduce *polar coordinates* $x(r, \phi) = r \sin(\phi)$, $y(r, \phi) = r \cos(\phi)$ and

$$
r^2 = \frac{1}{3}((x_1 - x_2)^2 + (x_2 - x_3)^2 + (x_3 - x_1)^2).
$$

Express $(x_1 - x_2)$, $(x_2 - x_3)$, $(x_3 - x_1)$ using these three coordinates.

Solution 28. (i) In matrix form we have

$$
\begin{pmatrix} X \\ x \\ y \end{pmatrix}
=
\begin{pmatrix}
1/3 & 1/3 & 1/3 \\
1/\sqrt{2} & -1/\sqrt{2} & 0 \\
1/\sqrt{6} & 1/\sqrt{6} & -2/\sqrt{6}
\end{pmatrix}
\begin{pmatrix} x_1 \\ x_2 \\ x_3 \end{pmatrix}.
$$

Calculating the inverse of the matrix on the right-hand side yields

$$\begin{pmatrix} 1 & 1/\sqrt{2} & 1/\sqrt{6} \\ 1 & -1/\sqrt{2} & 1/\sqrt{6} \\ 1 & 0 & -2/\sqrt{6} \end{pmatrix}.$$

(ii) We find

$$x_1 - x_2 = \sqrt{2}r\sin(\phi),$$
$$x_2 - x_3 = \sqrt{2}r\sin(\phi + 2\pi/3),$$
$$x_3 - x_1 = \sqrt{2}r\sin(\phi + 4\pi/3).$$

Problem 29. Let $\alpha \in [0, 2\pi)$. Find all solutions of the linear equation

$$\begin{pmatrix} \cos(\alpha) & \sin(\alpha) \\ \sin(\alpha) & \cos(\alpha) \end{pmatrix} \begin{pmatrix} x_1 \\ x_2 \end{pmatrix} = \begin{pmatrix} 1 \\ 1 \end{pmatrix}.$$

Thus x_1 and x_2 depend on α.

Solution 29. The determinant of the matrix is given by $\cos(2\alpha)$. Hence the determinant is equal to 0 if $\alpha = \pi/4$ or $\alpha = 3\pi/4$. Note that $\cos(\pi/4) = 1/\sqrt{2}$, $\sin(\pi/4) = 1/\sqrt{2}$, $\cos(3\pi/4) = -1/\sqrt{2}$, $\sin(3\pi/4) = 1/\sqrt{2}$. Thus if $\alpha \neq \pi/4$ and $\alpha \neq 3\pi/4$ the inverse of the matrix exists and we find the solution

$$\begin{pmatrix} x_1 \\ x_2 \end{pmatrix} = \frac{1}{\cos^2(\alpha) - \sin^2(\alpha)} \begin{pmatrix} \cos(\alpha) & -\sin(\alpha) \\ -\sin(\alpha) & \cos(\alpha) \end{pmatrix}.$$

If $\alpha = \pi/4$ we have the matrix

$$\frac{1}{\sqrt{2}} \begin{pmatrix} 1 & 1 \\ 1 & 1 \end{pmatrix}$$

and we find the solution $(x_1 + x_2)/\sqrt{2} = 1$. If $\alpha = 3\pi/4$ we have the matrix

$$\frac{1}{\sqrt{2}} \begin{pmatrix} -1 & 1 \\ 1 & -1 \end{pmatrix}$$

and the solution is $(x_1 - x_2)/\sqrt{2} = 1$.

Problem 30. Consider the partial differential equation (*Laplace equation*)

$$\frac{\partial^2 u}{\partial x^2} + \frac{\partial^2 u}{\partial y^2} = 0 \quad \text{on} \quad [0,1] \times [0,1]$$

with the boundary conditions

$$u(x,0) = 1, \quad u(x,1) = 2, \quad u(0,y) = 1, \quad u(1,y) = 2.$$

Apply the *central difference scheme*

$$\left(\frac{\partial^2 u}{\partial x^2}\right)_{j,k} \approx \frac{u_{j-1,k} - 2u_{j,k} + u_{j+1,k}}{(\Delta x)^2}, \qquad \left(\frac{\partial^2 u}{\partial y^2}\right)_{j,k} \approx \frac{u_{j,k-1} - 2u_{j,k} + u_{j,k+1}}{(\Delta y)^2}$$

and then solve the linear equation. Consider the cases $\Delta x = \Delta y = 1/3$ and $\Delta x = \Delta y = 1/4$.

Solution 30.   Case $\Delta x = \Delta y = 1/3$: We have the 16 grid points (j,k) with $j,k = 0,1,2,3$ and four interior grid points $(1,1)$, $(2,1)$, $(1,2)$, $(2,2)$.
For $j = k = 1$ we have $u_{0,1} + u_{2,1} + u_{1,0} + u_{1,2} - 4u_{1,1} = 0$.
For $j = 1$, $k = 2$ we have $u_{0,2} + u_{2,2} + u_{1,1} + u_{1,3} - 4u_{1,2} = 0$.
For $j = 2$, $k = 1$ we have $u_{1,1} + u_{3,1} + u_{2,0} + u_{2,3} - 4u_{2,1} = 0$.
For $j = 2$, $k = 2$ we have $u_{1,2} + u_{3,2} + u_{2,1} + u_{2,3} - 4u_{2,2} = 0$.
From the boundary conditions we find that

$$u_{0,1} = 1, \quad u_{1,0} = 1, \quad u_{0,2} = 1, \quad u_{1,3} = 2,$$

$$u_{3,1} = 2, \quad u_{2,0} = 1, \quad u_{3,2} = 2, \quad u_{2,3} = 2.$$

Thus we find the linear equation in matrix form $A\mathbf{u} = \mathbf{b}$

$$\begin{pmatrix} -4 & 1 & 1 & 0 \\ 1 & -4 & 0 & 1 \\ 1 & 0 & -4 & 1 \\ 0 & 1 & 1 & -4 \end{pmatrix} \begin{pmatrix} u_{1,1} \\ u_{1,2} \\ u_{2,1} \\ u_{2,2} \end{pmatrix} = \begin{pmatrix} -2 \\ -3 \\ -3 \\ -4 \end{pmatrix}.$$

The matrix A is invertible with

$$A^{-1} = -\frac{1}{12}\begin{pmatrix} 7/2 & 1 & 1 & 1/2 \\ 1 & 7/2 & 1/2 & 1 \\ 1 & 1/2 & 7/2 & 1 \\ 1/2 & 1 & 1 & 7/2 \end{pmatrix}$$

and thus we obtain the solution

$$u_{1,1} = \frac{5}{4}, \quad u_{1,2} = \frac{3}{2}, \quad u_{2,1} = \frac{3}{2}, \quad u_{2,2} = \frac{7}{4}.$$

Case $\Delta x = \Delta y = 1/4$. We have 25 grid points and 9 interior grid points. The equations for these nine grid points are as follows

$$u_{0,1} + u_{2,1} + u_{1,0} + u_{1,2} - 4u_{1,1} = 0$$
$$u_{0,2} + u_{2,2} + u_{1,1} + u_{1,3} - 4u_{1,2} = 0$$
$$u_{0,3} + u_{2,3} + u_{1,2} + u_{1,4} - 4u_{1,3} = 0$$
$$u_{1,1} + u_{3,1} + u_{2,0} + u_{2,2} - 4u_{2,1} = 0$$
$$u_{1,2} + u_{3,2} + u_{2,1} + u_{2,3} - 4u_{2,2} = 0$$
$$u_{1,3} + u_{3,3} + u_{2,2} + u_{2,4} - 4u_{2,3} = 0$$
$$u_{2,1} + u_{4,1} + u_{3,0} + u_{3,2} - 4u_{3,1} = 0$$
$$u_{2,2} + u_{4,2} + u_{3,1} + u_{3,3} - 4u_{3,2} = 0$$
$$u_{2,3} + u_{4,3} + u_{3,2} + u_{3,4} - 4u_{3,3} = 0.$$

From the boundary conditions we find the 12 values

$$u_{0,1} = 1, \quad u_{1,0} = 1, \quad u_{0,2} = 1, \quad u_{0,3} = 1, \quad u_{1,4} = 2, \quad u_{2,0} = 1,$$

$$u_{2,4} = 2, \quad u_{4,1} = 2, \quad u_{3,0} = 1, \quad u_{4,2} = 2, \quad u_{4,3} = 2, \quad u_{3,4} = 2.$$

Inserting the boundary values the linear equation in matrix form $A\mathbf{u} = \mathbf{b}$ is given by

$$\begin{pmatrix} -4 & 1 & 0 & 1 & 0 & 0 & 0 & 0 & 0 \\ 1 & -4 & 1 & 0 & 1 & 0 & 0 & 0 & 0 \\ 0 & 1 & -4 & 0 & 0 & 1 & 0 & 0 & 0 \\ 1 & 0 & 0 & -4 & 1 & 0 & 1 & 0 & 0 \\ 0 & 1 & 0 & 1 & -4 & 1 & 0 & 1 & 0 \\ 0 & 0 & 1 & 0 & 1 & -4 & 0 & 0 & 1 \\ 0 & 0 & 0 & 1 & 0 & 0 & -4 & 1 & 0 \\ 0 & 0 & 0 & 0 & 1 & 0 & 1 & -4 & 1 \\ 0 & 0 & 0 & 0 & 0 & 1 & 0 & 1 & -4 \end{pmatrix} \begin{pmatrix} u_{1,1} \\ u_{1,2} \\ u_{1,3} \\ u_{2,1} \\ u_{2,2} \\ u_{2,3} \\ u_{3,1} \\ u_{3,2} \\ u_{3,3} \end{pmatrix} = \begin{pmatrix} -2 \\ -1 \\ -3 \\ -1 \\ 0 \\ -2 \\ -3 \\ -2 \\ -4 \end{pmatrix}.$$

The matrix is invertible and thus the solution is $\mathbf{u} = A^{-1}\mathbf{b}$.

Problem 31. (i) The equation of a line in the Euclidean space $\mathbb{R}^2$ passing through the points (x_1, y_1) and (x_2, y_2) is given by

$$(y - y_1)(x_2 - x_1) = (y_2 - y_1)(x - x_1).$$

Apply this equation to the points in $\mathbb{R}^2$ given by $(x_1, y_1) = (1, 1/2)$, $(x_2, y_2) = (1/2, 1)$. Consider the unit square with the corner points $(0, 0)$, $(0, 1)$, $(1, 0)$, $(1, 1)$ and the map

$$(0, 0) \to 0, \quad (0, 1) \to 0, \quad (1, 0) \to 0, \quad (1, 1) \to 1.$$

We can consider this as a 2 input AND-gate. Show that the line constructed above classifies this map.

(ii) The equation of a plane in $\mathbb{R}^3$ passing through the points (x_1, y_1, z_1), (x_2, y_2, z_2), (x_3, y_3, z_3) in $\mathbb{R}^3$ is given by

$$\det \begin{pmatrix} x - x_1 & y - y_1 & z - z_1 \\ x_2 - x_1 & y_2 - y_1 & z_2 - z_1 \\ x_3 - x_1 & y_3 - y_1 & z_3 - z_1 \end{pmatrix} = 0.$$

Apply this equations to the points $(1, 1, 1/2)$, $(1, 1/2, 1)$, $(1/2, 1, 1)$. Consider the unit cube in $\mathbb{R}^3$ with the corner points (vertices)

$$(0, 0, 0), \ (0, 0, 1), \ (0, 1, 0), \ (0, 1, 1), \ (1, 0, 0), \ (1, 0, 1), \ (1, 1, 0), \ (1, 1, 1)$$

and the map where all corner points are mapped to 0 except for $(1, 1, 1)$ which is mapped to 1. We can consider this as a 3 input AND-gate. Show that the plane constructed in (i) separates these solutions.

Solution 31. (i) The line is given by $x + y = 3/2$.

(ii) The plane is given by $x + y + z = 5/2$.

This can be extend to any dimension. For a hypercube in $\mathbb{R}^4$ and the points

$$(1,1,1,1/2), \ (1,1,1/2,1), \ (1,1/2,1,1), \ (1/2,1,1,1)$$

we obtain the hyperplane $x_1 + x_2 + x_3 + x_4 = 7/2$ which separates the corner points $(1,1,1,1)$ from all the other points.

Problem 32. Find the system of linear equations for a and b given by

$$\frac{x+15}{(x+3)(x-1)} = \frac{a}{x+3} + \frac{b}{x-1}.$$

Solve the system of linear equations.

Solution 32. We have

$$\frac{a}{x+3} + \frac{b}{x-1} = \frac{a(x-1) + b(x+3)}{(x+3)(x-1)} = \frac{(a+b)x + (-a+3b)}{(x+3)(x-1)}.$$

Comparing coefficients leads to the system of linear equations given in matrix form by

$$\begin{pmatrix} 1 & 1 \\ -1 & 3 \end{pmatrix} \begin{pmatrix} a \\ b \end{pmatrix} = \begin{pmatrix} 1 \\ 15 \end{pmatrix}.$$

The matrix is invertible and the solution is $a = -3$, $b = 4$.

Problem 33. Let $k = 1, 2, 3, 4$ and $x_0 = 1$, $x_5 = 0$. Solve $x_{k-1} - 2x_k + x_{k+1} = 0$.

Solution 33. Inserting $k = 1, 2, 3, 4$ we obtain the system of linear equations

$$-2x_1 + x_2 = -1, \quad x_1 - 2x_2 + x_3 = 0, \quad x_2 - 2x_3 + x_4 = 0, \quad x_3 - 2x_4 = 0$$

or in matrix form

$$\begin{pmatrix} -2 & 1 & 0 & 0 \\ 1 & -2 & 1 & 0 \\ 0 & 1 & -2 & 1 \\ 0 & 0 & 1 & -2 \end{pmatrix} \begin{pmatrix} x_1 \\ x_2 \\ x_3 \\ x_4 \end{pmatrix} = \begin{pmatrix} -1 \\ 0 \\ 0 \\ 0 \end{pmatrix}.$$

The matrix is invertible.

Problem 34. Consider the polynomial $p(x) = a + bx + cx^2$. Find a, b, c from the conditions $p(0) = 0$, $p(1) = 1$, $p(2) = 0$.

Solution 34. We obtain the three linear equations $0 = a$, $1 = a + b + c$, $0 = a + 2b + 4c$ with the solution $a = 0$, $b = 2$, $c = -1$.

Problem 35. Let $A \in \mathbb{C}^{n \times m}$ with $n \geq m$ and $\mathrm{rank}(A) = m$.
(i) Show that the $m \times m$ matrix $A^* A$ is invertible.

(ii) We set $\Pi := A(A^*A)^{-1}A$. Show that Π is a projection matrix, i.e. $\Pi^2 = \Pi$ and $\Pi = \Pi^*$.

Solution 35. (i) To show that A^*A is invertible it suffices to show that $N(A^*A) = \{0\}$. From $A^*Av = 0$ we obtain

$$A^*Av = 0 \Rightarrow v^*A^*Av = 0 \Rightarrow (Av)^*(Av) = 0 \Rightarrow Av = 0.$$

Thus $v = 0$ since $\text{rank}(A) = m$. Hence $N(A^*A) = \{0\}$ and A^*A is invertible.
(ii) We have

$$\Pi^2 = A(A^*A)^{-1}A^*A(A^*A)^{-1}A^* = A(A^*A)^{-1}A^* = \Pi$$

and $\Pi^* = (A(A^*A)^{-1}A^*)^* = A((A^*A)^*)^{-1}A^* = A(A^*A)^{-1}A^* = \Pi$.

Problem 36. Let $a, b, c \in \mathbb{R}$ and $abc \neq 0$. Find the solution of the system of linear equations

$$\begin{pmatrix} 0 & c & b \\ c & 0 & a \\ b & a & 0 \end{pmatrix} \begin{pmatrix} \cos(\alpha) \\ \cos(\beta) \\ \cos(\gamma) \end{pmatrix} = \begin{pmatrix} a \\ b \\ c \end{pmatrix}.$$

Solution 36. Since

$$\det \begin{pmatrix} 0 & c & b \\ c & 0 & a \\ b & a & 0 \end{pmatrix} = 2abc \neq 0$$

the inverse of the matrix exists and is given by

$$\frac{1}{2abc} \begin{pmatrix} -a^2 & ab & ac \\ ab & -b^2 & bc \\ ac & bc & -c^2 \end{pmatrix}.$$

Thus the solution is

$$\begin{pmatrix} \cos(\alpha) \\ \cos(\beta) \\ \cos(\gamma) \end{pmatrix} = \frac{1}{2abc} \begin{pmatrix} -a^2 & ab & ac \\ ab & -b^2 & bc \\ ac & bc & -c^2 \end{pmatrix} \begin{pmatrix} a \\ b \\ c \end{pmatrix}.$$

Problem 37. Let A be an invertible $n \times n$ matrix over $\mathbb{R}$. Consider the system of linear equation $Ax = b$ or

$$\sum_{j=1}^{n} a_{ij}x_j = b_i, \qquad i = 1, \ldots, n.$$

Let $A = C - R$. This is called a *splitting* of the matrix A and R is the defect matrix of the splitting. Consider the iteration

$$Cx^{(k+1)} = Rx^{(k)} + b, \qquad k = 0, 1, 2, \ldots.$$

Let

$$A = \begin{pmatrix} 4 & -1 & 0 \\ -1 & 4 & -1 \\ 0 & -2 & 4 \end{pmatrix}, \quad C = \begin{pmatrix} 4 & 0 & 0 \\ 0 & 4 & 0 \\ 0 & 0 & 4 \end{pmatrix}, \quad b = \begin{pmatrix} 3 \\ 2 \\ 2 \end{pmatrix}, \quad x^{(0)} = \begin{pmatrix} 0 \\ 0 \\ 0 \end{pmatrix}.$$

The iteration converges if $\rho(C^{-1}R) < 1$, where $\rho(C^{-1}R)$ denotes the *spectral radius* of $C^{-1}R$. Show that $\rho(C^{-1}R) < 1$. Perform the iteration.

Solution 37.   We have

$$C^{-1}R = \frac{1}{4} \begin{pmatrix} 0 & 1 & 0 \\ 1 & 0 & 1 \\ 0 & 2 & 0 \end{pmatrix}$$

with the eigenvalues $\sqrt{3}/4$, $-\sqrt{3}/4$, 0. Thus $\rho(C^{-1}R) < 1$. We have

$$C\mathbf{x}^{(1)} = R\mathbf{x}^{(0)} + \mathbf{b} \Rightarrow C\mathbf{x}^{(1)} = \mathbf{b} = \begin{pmatrix} 3 \\ 2 \\ 2 \end{pmatrix} \Rightarrow \mathbf{x}^{(1)} = \begin{pmatrix} 3/4 \\ 1/2 \\ 1/2 \end{pmatrix}.$$

The second iteration yields

$$C\mathbf{x}^{(2)} = R\mathbf{x}^{(1)} + \mathbf{b} = \begin{pmatrix} 7/2 \\ 13/4 \\ 3 \end{pmatrix} \Rightarrow \mathbf{x}^{(2)} = \begin{pmatrix} 7/8 \\ 13/16 \\ 3/4 \end{pmatrix}.$$

The series converges to $(1\ 1\ 1)^T$ the solution of the linear system $A\mathbf{x} = \mathbf{b}$.

Problem 38.   Let A be an $n \times n$ matrix over $\mathbb{R}$ and let $\mathbf{b} \in \mathbb{R}^n$. Consider the linear equation $A\mathbf{x} = \mathbf{b}$. Assume that $a_{jj} \neq 0$ for $j = 1, \ldots, n$. We define the diagonal matrix $D = \text{diag}(a_{jj})$. Then the linear equation $A\mathbf{x} = \mathbf{b}$ can be written as

$$\mathbf{x} = B\mathbf{x} + \mathbf{c}$$

with $B := -D^{-1}(A - D)$, $\mathbf{c} := D^{-1}\mathbf{b}$. The *Jacobi method* for the solution of the linear equation $A\mathbf{x} = \mathbf{b}$ is given by

$$\mathbf{x}^{(k+1)} = B\mathbf{x}^{(k)} + \mathbf{c}, \quad k = 0, 1, \ldots$$

where $\mathbf{x}^{(0)}$ is any initial vector in $\mathbb{R}^n$. The sequence converges if

$$\rho(B) := \max_{j=1,\ldots,n} |\lambda_j(B)| < 1$$

where $\rho(B)$ is the *spectral radius* of B. Let

$$A = \begin{pmatrix} 2 & 1 & 0 \\ 1 & 2 & 1 \\ 0 & 1 & 2 \end{pmatrix}.$$

(i) Show that the Jacobi method can applied for this matrix.

(ii) Find the solution of the linear equation with $\mathbf{b} = (1\ 1\ 1)^T$.

Solution 38. (i) We have

$$B = -D^{-1}(A - D) = -\frac{1}{2}\begin{pmatrix} 1 & 0 & 0 \\ 0 & 1 & 0 \\ 0 & 0 & 1 \end{pmatrix}\begin{pmatrix} 0 & 1 & 0 \\ 1 & 0 & 1 \\ 0 & 1 & 0 \end{pmatrix} = -\frac{1}{2}\begin{pmatrix} 0 & 1 & 0 \\ 1 & 0 & 1 \\ 0 & 1 & 0 \end{pmatrix}.$$

The eigenvalues of B are 0, $1/\sqrt{2}$, $-1/\sqrt{2}$. Thus $\rho(B) = 1/\sqrt{2} < 1$ and the Jacobi method can be applied.

(ii) The solution is $\mathbf{x} = (1/2\ 0\ 1/2)^T$.

Problem 39. *Kirchhoff's current law* states that the algebraic sum of all the currents flowing into a junction is 0. *Kirchhoff's voltage law* states that the algebraic sum of all the voltages around a closed circuit is 0. Use Kirchhoff's laws and *Ohm's law* $(V = RI)$ to setting up the system of linear equations for the circuit depicted in the figure.

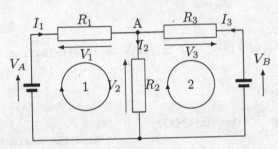

Given the voltages V_A, V_B and the resistors R_1, R_2, R_3. Find I_1, I_2, I_3.

Solution 39. From Kirchhoff's voltage law we find for loop 1 and loop 2 that

$$V_1 + V_2 = V_A, \qquad V_2 + V_3 = V_B.$$

From Kirchhoff's current law for node A we obtain

$$I_1 - I_2 + I_3 = 0.$$

Thus the linear system can be written in matrix form

$$\begin{pmatrix} R_1 & R_2 & 0 \\ 0 & R_2 & R_3 \\ 1 & -1 & 1 \end{pmatrix}\begin{pmatrix} I_1 \\ I_2 \\ I_3 \end{pmatrix} = \begin{pmatrix} V_A \\ V_B \\ 0 \end{pmatrix}.$$

The determinant of the matrix on the left-hand side is given by

$$R_1R_2 + R_2R_3 + R_1R_3.$$

Since $R_1, R_2, R_3 > 0$ the inverse matrix exists. The solution is given by

$$I_1 = \frac{(R_2 + R_3)V_A - R_2V_B}{R_1R_2 + R_1R_3 + R_2R_3}, \qquad I_2 = \frac{R_3V_A + R_1V_B}{R_1R_2 + R_2R_3 + R_1R_3},$$

$$I_3 = \frac{-R_2 V_A + (R_1 + R_2)V_B}{R_1 R_2 + R_2 R_3 + R_1 R_3}.$$

Problem 40.   Write a C++ program that implements Gauss elimination to solve linear equations. Apply it to the system

$$\begin{pmatrix} 1 & 1 & 1 \\ 8 & 4 & 2 \\ 27 & 9 & 3 \end{pmatrix} \begin{pmatrix} x_1 \\ x_2 \\ x_3 \end{pmatrix} = \begin{pmatrix} 1 \\ 5 \\ 14 \end{pmatrix}.$$

Solution 40.   The program is

```
// Gauss elimination + back substitution

#include <iostream>
#include <cmath>
using namespace std;

const int n = 3;

void gauss(double C[n][n+1]) // in place Gauss elimination
{
 int i, j, k, m;
 for(i=0;i<n-1;i++)
 {
 double temp, max = fabs(C[m=i][i]);
 for(j=i+1;j<n;j++)                     // find the maximum
 if(fabs(C[j][i]) > max) max = fabs(C[m=j][i]);
 for(j=i;j<=n;j++)                      // swap the rows
 { temp = C[i][j]; C[i][j] = C[m][j]; C[m][j] = temp; }
 if(C[i][i] != 0)              // no elimination necessary
 for(j=i+1;j<n;j++)                     // elimination
 {
 double r = C[j][i]/C[i][i];
 C[j][i] = 0.0;                // perform the elimination
 for(k=i+1;k<=n;k++) C[j][k] = C[j][k]-r*C[i][k]; }
 }
}

int solve(double A[n][n],double b[n],double x[n]) // solve Ax = b
{
 int i, j;
 double C[n][n+1];
 for(i=0;i<n;i++)               // 1. Construct augmented matrix
 { for(j=0;j<n;j++) C[i][j] = A[i][j]; C[i][n] = b[i]; }
 gauss(C);                      // 2. Gauss elimination
 for(i=n-1;i>=0;i--)            // 3. Back substitution
 {
 double sum = C[i][n];         // Initial value from the last column
 if(C[i][i]==0.0) return 0; // no solution/no unique solution
 for(j=i+1;j<n;j++) sum -= C[i][j]*x[j]; // Substitution
```

```
x[i] = sum/C[i][i];          // divide by the leading coefficient
}
return 1;
}

int main(void)
{
  double A[n][n] = {{ 1.0,1.0,1.0 },{ 8.0,4.0,2.0 },{ 27.0,9.0,3.0 }};
  double b[n] = { 1.0,5.0,14.0 };
  double x[n];
  if(solve(A,b,x))
  { cout << "("; for(int i=0;i<n;i++) cout << x[i] << " ";
    cout << ")" << endl; }
  else cout << "No solution or no unique solution" << endl;
  return 0;
}
```

The output is (0.333333 0.5 0.166667) compared to (1/3 1/2 1/6).

Supplementary Problems

Problem 1. Let $\mathbf{n}$ and $\mathbf{p}$ be vectors in $\mathbb{R}^n$ with $\mathbf{n} \neq \mathbf{0}$. The set of all vectors $\mathbf{x}$ in $\mathbb{R}^n$ which satisfy the equation $\mathbf{n} \cdot (\mathbf{x} - \mathbf{p}) = 0$ is called a *hyperplane* through the point $\mathbf{p} \in \mathbb{R}$. We call $\mathbf{n}$ a normal vector for the hyperplane and call $\mathbf{n} \cdot (\mathbf{x} - \mathbf{p}) = 0$ a normal equation for the hyperplane. Find $\mathbf{n}$ and $\mathbf{p}$ in $\mathbb{R}^4$ such that we obtain the hyperplane given by

$$x_1 + x_2 + x_3 + x_4 = \frac{7}{2}.$$

Any hyperplane of the Euclidean space $\mathbb{R}^n$ has exactly two unit normal vectors.

Problem 2. The hyperplanes $x_1 + x_2 + x_3 + x_4 = 1$, $x_1 + x_2 + x_3 + x_4 = 2$ in $\mathbb{R}^4$ do not intersect. Find the shortest distance.

Problem 3. (i) Find the solutions of the equation

$$\begin{pmatrix} 1 & 2 \\ 3 & 4 \end{pmatrix} \begin{pmatrix} x_{11} & x_{12} \\ x_{21} & x_{22} \end{pmatrix} = \begin{pmatrix} 0 & 0 \\ 0 & 0 \end{pmatrix}.$$

(ii) Find the solutions of the equation

$$\begin{pmatrix} 1 & 2 \\ 3 & 4 \end{pmatrix} \begin{pmatrix} x_{11} & x_{12} \\ x_{21} & x_{22} \end{pmatrix} = \begin{pmatrix} 1 & 0 \\ 0 & 1 \end{pmatrix}.$$

Problem 4. Let A be a given 3×3 matrix over $\mathbb{R}$ with $\det(A) \neq 0$. Is the transformation

$$x_1'(x_1, x_2) = \frac{a_{11}x_1 + a_{12}x_2 + a_{13}}{a_{31}x_1 + a_{32}x_2 + a_{33}}, \quad x_2'(x_1, x_2) = \frac{a_{21}x_1 + a_{22}x_2 + a_{23}}{a_{31}x_1 + a_{32}x_2 + a_{33}}$$

invertible? If so find the inverse.

Problem 5.   (i) Consider the linear equation written in matrix form

$$\begin{pmatrix} 1 & 0 & 1 \\ 0 & 1 & 1 \\ 1 & 0 & 2 \end{pmatrix} \begin{pmatrix} x_1 \\ x_2 \\ x_3 \end{pmatrix} = \begin{pmatrix} 1 \\ 2 \\ 1 \end{pmatrix}.$$

The determinant of the 3×3 matrix is nonzero $(+1)$ and the inverse matrix is

$$\begin{pmatrix} 2 & 0 & -1 \\ 1 & 1 & -1 \\ -1 & 0 & 1 \end{pmatrix}.$$

Apply two different methods (Gauss elimination and the Leverrier's method) to find the solution. Compare the two methods and discuss.
(ii) Apply the Gauss-Seidel method to solve the linear system

$$\begin{pmatrix} 4 & -1 & -1 & 0 \\ -1 & 4 & 0 & -1 \\ -1 & 0 & 4 & -1 \\ 0 & -1 & -1 & 4 \end{pmatrix} \begin{pmatrix} x_1 \\ x_2 \\ x_3 \\ x_4 \end{pmatrix} = \begin{pmatrix} 1 \\ 0 \\ 0 \\ 0 \end{pmatrix}$$

and show that the solution is given by

$$x_1 = \frac{7}{24}, \quad x_2 = \frac{1}{12}, \quad x_3 = \frac{1}{12}, \quad x_4 = \frac{1}{24}.$$

(iii) Show that the solution of the system of linear equations

$$x_1 + x_2 + x_3 = 0, \quad x_1 + 2x_2 + x_3 = 1, \quad 2x_1 + x_2 + x_3 = 2$$

is given by $x_1 = 2$, $x_2 = 1$, $x_3 = -3$.

Problem 6.   (i) Given the 2×2 matrix A. Find all 2×2 matrices X such that $AX = XA$. From $AX = XA$ we obtain

$$\begin{pmatrix} a_{11}x_{11} + a_{12}x_{21} & a_{11}x_{12} + a_{12}x_{22} \\ a_{21}x_{11} + a_{22}x_{21} & a_{21}x_{12} + a_{22}x_{22} \end{pmatrix} = \begin{pmatrix} a_{11}x_{11} + a_{21}x_{12} & a_{12}x_{11} + a_{22}x_{12} \\ a_{11}x_{21} + a_{21}x_{22} & a_{12}x_{21} + a_{22}x_{22} \end{pmatrix}.$$

Thus we obtain the four equations

$$a_{12}x_{21} = a_{21}x_{12}, \quad a_{11}x_{12} + a_{12}x_{12} = a_{12}x_{11} + a_{22}x_{12}$$
$$a_{21}x_{12} = a_{12}x_{21}, \quad a_{21}x_{11} + a_{22}x_{21} = a_{11}x_{21} + a_{21}x_{22}.$$

(ii) Show that the solution of the equation

$$\begin{pmatrix} 0 & 1 \\ 1 & 0 \end{pmatrix} \begin{pmatrix} x_{11} & x_{12} \\ x_{21} & x_{22} \end{pmatrix} = \begin{pmatrix} x_{11} & x_{12} \\ x_{21} & x_{22} \end{pmatrix} \begin{pmatrix} 0 & 1 \\ 1 & 0 \end{pmatrix}$$

is given by $x_{11} = x_{22}$, $x_{12} = x_{21}$.

Problem 7.   Let $\alpha, b_1, b_2 \in \mathbb{R}$. Solve the system of linear equations

$$\begin{pmatrix} \cosh(\alpha) & \sinh(\alpha) \\ \sinh(\alpha) & \cosh(\alpha) \end{pmatrix} \begin{pmatrix} x_1 \\ x_2 \end{pmatrix} = \begin{pmatrix} b_1 \\ b_2 \end{pmatrix}.$$

Chapter 3

Kronecker Product

Let A be an $m \times n$ matrix and B be an $r \times s$ matrix. The *Kronecker product* of A and B is defined as the $(m \cdot r) \times (n \cdot s)$ matrix

$$A \otimes B := \begin{pmatrix} a_{11}B & a_{12}B & \cdots & a_{1n}B \\ a_{21}B & a_{22}B & \cdots & a_{2n}B \\ \vdots & \vdots & \ddots & \vdots \\ a_{m1}B & a_{m2}B & \cdots & a_{mn}B \end{pmatrix}.$$

The Kronecker product is associative

$$(A \otimes B) \otimes C = A \otimes (B \otimes C).$$

Let A and B be $m \times n$ matrices and C a $p \times q$ matrix. Then

$$(A + B) \otimes C = A \otimes C + B \otimes C.$$

Let $c \in \mathbb{C}$. Then $(cA) \otimes B = c(A \otimes B) = A \otimes (cB)$. We have

$$(A \otimes B)^T = A^T \otimes B^T, \qquad (A \otimes B)^* = A^* \otimes B^*.$$

Let A be an $m \times n$ matrix, B be a $p \times q$ matrix, C be an $n \times r$ matrix and D be a $q \times s$ matrix. Then

$$(A \otimes B)(C \otimes D) = (AC) \otimes (BD).$$

Let A be an $n \times n$ matrix and B an $m \times m$ matrix. Then we have

$$\mathrm{tr}(A \otimes B) = \mathrm{tr}(A)\mathrm{tr}(B), \quad \det(A \otimes B) = (\det(A))^m (\det(B))^n.$$

The *Kronecker sum* is defined as $A \oplus_K B := A \otimes I_m + I_n \otimes B$.

Problem 1. Consider the Pauli spin matrices

$$\sigma_1 := \begin{pmatrix} 0 & 1 \\ 1 & 0 \end{pmatrix}, \qquad \sigma_3 := \begin{pmatrix} 1 & 0 \\ 0 & -1 \end{pmatrix}.$$

Find $\sigma_1 \otimes \sigma_3$ and $\sigma_3 \otimes \sigma_1$. Discuss. Find $\operatorname{tr}(\sigma_1 \otimes \sigma_3)$ and $\operatorname{tr}(\sigma_3 \otimes \sigma_1)$.

Solution 1. We obtain

$$\sigma_1 \otimes \sigma_3 = \begin{pmatrix} 0 & 0 & 1 & 0 \\ 0 & 0 & 0 & -1 \\ 1 & 0 & 0 & 0 \\ 0 & -1 & 0 & 0 \end{pmatrix}, \qquad \sigma_3 \otimes \sigma_1 = \begin{pmatrix} 0 & 1 & 0 & 0 \\ 1 & 0 & 0 & 0 \\ 0 & 0 & 0 & -1 \\ 0 & 0 & -1 & 0 \end{pmatrix}.$$

We see that $\sigma_1 \otimes \sigma_3 \neq \sigma_3 \otimes \sigma_1$. We find $\operatorname{tr}(\sigma_1 \otimes \sigma_3) = \operatorname{tr}(\sigma_3 \otimes \sigma_1) = 0$.

Problem 2. Let

$$\mathbf{x}_1 = \frac{1}{\sqrt{2}} \begin{pmatrix} 1 \\ 1 \end{pmatrix}, \qquad \mathbf{x}_2 = \frac{1}{\sqrt{2}} \begin{pmatrix} 1 \\ -1 \end{pmatrix}.$$

Thus $\{\mathbf{x}_1, \mathbf{x}_2\}$ forms an orthonormal basis in $\mathbb{R}^2$. Calculate

$$\mathbf{x}_1 \otimes \mathbf{x}_1, \quad \mathbf{x}_1 \otimes \mathbf{x}_2, \quad \mathbf{x}_2 \otimes \mathbf{x}_1, \quad \mathbf{x}_2 \otimes \mathbf{x}_2$$

and interpret the result.

Solution 2. We obtain

$$\mathbf{x}_1 \otimes \mathbf{x}_1 = \frac{1}{2} \begin{pmatrix} 1 \\ 1 \\ 1 \\ 1 \end{pmatrix}, \qquad \mathbf{x}_1 \otimes \mathbf{x}_2 = \frac{1}{2} \begin{pmatrix} 1 \\ -1 \\ 1 \\ -1 \end{pmatrix},$$

$$\mathbf{x}_2 \otimes \mathbf{x}_1 = \frac{1}{2} \begin{pmatrix} 1 \\ 1 \\ -1 \\ -1 \end{pmatrix}, \qquad \mathbf{x}_2 \otimes \mathbf{x}_2 = \frac{1}{2} \begin{pmatrix} 1 \\ -1 \\ -1 \\ 1 \end{pmatrix}.$$

Thus we find an orthonormal basis in $\mathbb{R}^4$ from an orthonormal basis in $\mathbb{R}^2$.

Problem 3. Given the orthonormal basis

$$\mathbf{x}_1 = \begin{pmatrix} e^{i\phi} \cos(\theta) \\ \sin(\theta) \end{pmatrix}, \qquad \mathbf{x}_2 = \begin{pmatrix} -\sin(\theta) \\ e^{-i\phi} \cos(\theta) \end{pmatrix}$$

in the vector space $\mathbb{C}^2$. Use this orthonormal basis to find an orthonormal basis in $\mathbb{C}^4$.

Solution 3. A basis in $\mathbb{C}^4$ is given by $\{\mathbf{x}_1 \otimes \mathbf{x}_1, \mathbf{x}_1 \otimes \mathbf{x}_2, \mathbf{x}_2 \otimes \mathbf{x}_1, \mathbf{x}_2 \otimes \mathbf{x}_2\}$ since

$$(\mathbf{x}_j^* \otimes \mathbf{x}_k^*)(\mathbf{x}_m \otimes \mathbf{x}_n) = \delta_{jm}\delta_{kn}$$

where $j, k, m, n = 1, 2$.

Problem 4. (i) An orthonormal basis in $\mathbb{R}^3$ is given by

$$\mathbf{v}_1 = \frac{1}{\sqrt{2}} \begin{pmatrix} 1 \\ 0 \\ 1 \end{pmatrix}, \quad \mathbf{v}_2 = \begin{pmatrix} 0 \\ 1 \\ 0 \end{pmatrix}, \quad \mathbf{v}_3 = \frac{1}{\sqrt{2}} \begin{pmatrix} 1 \\ 0 \\ -1 \end{pmatrix}.$$

Use this orthonormal basis and the Kronecker product to find an orthonormal basis in $\mathbb{R}^9$.

(ii) A basis in the vector space $M_2(\mathbb{C})$ of 2×2 matrices over $\mathbb{C}$ is given by the Pauli spin matrices (including the 2×2 identity matrix) $\sigma_0 = I_2, \sigma_1, \sigma_2, \sigma_3$. Use this basis and the Kronecker product to find a basis in the vector space $M_4(\mathbb{C}^4)$ of all 4×4 matrices over $\mathbb{C}$.

Solution 4. (i) An orthonormal basis in $\mathbb{C}^9$ is given by the nine vectors

$$\mathbf{v}_j \otimes \mathbf{v}_k, \quad j, k = 1, 2, 3.$$

(ii) A basis in the vector space of 4×4 matrices is given by the 16 matrices

$$\sigma_j \otimes \sigma_k, \quad j, k = 0, 1, 2, 3.$$

Problem 5. Consider the two normalized vectors

$$\frac{1}{\sqrt{2}} \begin{pmatrix} 1 \\ 0 \\ 0 \\ 1 \end{pmatrix}, \quad \frac{1}{\sqrt{2}} \begin{pmatrix} 1 \\ 0 \\ 0 \\ -1 \end{pmatrix}$$

in the Hilbert space $\mathbb{R}^4$. Show that the vectors obtained by applying the matrix $(I_2 \otimes \sigma_1)$ together with the original ones form an orthonormal basis in $\mathbb{R}^4$.

Solution 5. We have

$$(I_2 \otimes \sigma_1) \frac{1}{\sqrt{2}} \begin{pmatrix} 1 \\ 0 \\ 0 \\ 1 \end{pmatrix} = \frac{1}{\sqrt{2}} \begin{pmatrix} 0 \\ 1 \\ 1 \\ 0 \end{pmatrix}, \quad (I_2 \otimes \sigma_1) \frac{1}{\sqrt{2}} \begin{pmatrix} 1 \\ 0 \\ 0 \\ -1 \end{pmatrix} = \frac{1}{\sqrt{2}} \begin{pmatrix} 0 \\ 1 \\ -1 \\ 0 \end{pmatrix}.$$

The four vectors (*Bell basis*)

$$\frac{1}{\sqrt{2}} \begin{pmatrix} 1 \\ 0 \\ 0 \\ 1 \end{pmatrix}, \quad \frac{1}{\sqrt{2}} \begin{pmatrix} 0 \\ 1 \\ 1 \\ 0 \end{pmatrix}, \quad \frac{1}{\sqrt{2}} \begin{pmatrix} 1 \\ 0 \\ 0 \\ -1 \end{pmatrix}, \quad \frac{1}{\sqrt{2}} \begin{pmatrix} 0 \\ 1 \\ -1 \\ 0 \end{pmatrix}.$$

are linearly independent, normalized and pairwise orthogonal.

Problem 6. Let A be an $m \times m$ matrix and B be an $n \times n$ matrix. The underlying field is $\mathbb{C}$. Let I_m, I_n be the $m \times m$ and $n \times n$ unit matrix, respectively.

(i) Show that $\mathrm{tr}(A \otimes B) = \mathrm{tr}(A)\mathrm{tr}(B)$.

(ii) Show that $\mathrm{tr}(A \otimes I_n + I_m \otimes B) = n\mathrm{tr}(A) + m\mathrm{tr}(B)$.

Solution 6. (i) We have

$$\mathrm{tr}(A \otimes B) = \sum_{j=1}^{m}\sum_{k=1}^{n} a_{jj}b_{kk} = \left(\sum_{j=1}^{m} a_{jj}\right)\left(\sum_{k=1}^{n} b_{kk}\right) = \mathrm{tr}(A)\mathrm{tr}(B).$$

(ii) Since the trace operation is linear and $\mathrm{tr}(I_n) = n$ we find

$$\mathrm{tr}(A \otimes I_n + I_m \otimes B) = \mathrm{tr}(A \otimes I_n) + \mathrm{tr}(I_m \otimes B) = n\,\mathrm{tr}(A) + m\,\mathrm{tr}(B).$$

Problem 7. (i) Let A be an arbitrary $n \times n$ matrix over $\mathbb{C}$. Show that

$$\exp(A \otimes I_n) \equiv \exp(A) \otimes I_n.$$

(ii) Let A be an $n \times n$ matrix and I_m be the $m \times m$ identity matrix. Show that

$$\sin(A \otimes I_m) \equiv \sin(A) \otimes I_m.$$

(iii) Let A be an $n \times n$ matrix and B be an $m \times m$ matrix. Is

$$\sin(A \otimes I_m + I_n \otimes B) \equiv (\sin(A)) \otimes (\cos(B)) + (\cos(A)) \otimes (\sin(B))?$$

Prove or disprove.

Solution 7. (i) Using the expansion of the exponential function

$$\exp(A \otimes I_n) = \sum_{k=0}^{\infty} \frac{(A \otimes I_n)^k}{k!} = I_n \otimes I_n + \frac{1}{1!}(A \otimes I_n) + \frac{1}{2!}(A \otimes I_n)^2 + \cdots$$

and $(A \otimes I_n)^k = A^k \otimes I_n$ $(k \in \mathbb{N})$ we find the identity.

(ii) Let $k \in \mathbb{N}$. Since $(A \otimes I_m)^k = A^k \otimes I_m$ we find the identity.

(iii) Since $[A \otimes I_m, I_n \otimes B] = 0_{n \times m}$ and using the result

$$\sin(A \otimes I_m) = \sin(A) \otimes I_m, \qquad \cos(I_n \otimes B) = I_n \otimes \cos(B)$$

we find the identity.

Problem 8. Let A, B be arbitrary $n \times n$ matrices over $\mathbb{C}$. Show that

$$\exp(A \otimes I_n + I_n \otimes B) \equiv \exp(A) \otimes \exp(B).$$

Solution 8. The proof of this identity relies on $[A \otimes I_n, I_n \otimes B] = 0_{n^2}$ and

$$(A \otimes I_n)^r (I_n \otimes B)^s \equiv (A^r \otimes I_n)(I_n \otimes B^s) \equiv A^r \otimes B^s, \qquad r, s \in \mathbb{N}.$$

Thus

$$\exp(A \otimes I_n + I_n \otimes B) = \sum_{j=0}^{\infty} \sum_{k=0}^{j} \frac{1}{j!} \binom{j}{k} (A \otimes I_n)^k (I_n \otimes B)^{j-k}$$

$$= \sum_{j=0}^{\infty} \sum_{k=0}^{j} \frac{1}{j!} \binom{j}{k} (A^k \otimes B^{j-k})$$

$$= \left(\sum_{j=0}^{\infty} \frac{A^j}{j!} \right) \otimes \left(\sum_{k=0}^{\infty} \frac{B^k}{k!} \right)$$

$$= \exp(A) \otimes \exp(B).$$

Problem 9. Let A and B be arbitrary $n \times n$ matrices over $\mathbb{C}$. Is $e^{A \otimes B} = e^A \otimes e^B$?

Solution 9. Obviously this is not true in general. Let $A = B = I_n$. Then

$$e^{A \otimes B} = e^{I_{n^2}} \quad \text{and} \quad e^A \otimes e^B = e^{I_n} \otimes e^{I_n} \neq e^{I_{n^2}}.$$

Problem 10. The underlying field is $\mathbb{C}$. Let A be an $m \times m$ matrix with eigenvalue λ and corresponding normalized eigenvector $\mathbf{v}$. Let B be an $n \times n$ matrix with eigenvalue μ and corresponding normalized eigenvector $\mathbf{u}$. Let ϵ_1, ϵ_2 and ϵ_3 be real parameters. Find an eigenvalue and the corresponding normalized eigenvectors of the matrix

$$\epsilon_1 A \otimes B + \epsilon_2 A \otimes I_n + \epsilon_3 I_m \otimes B.$$

Solution 10. We have

$$(A \otimes B)(\mathbf{v} \otimes \mathbf{u}) = (A\mathbf{v}) \otimes (B\mathbf{u}),$$

$$(A \otimes I_n)(\mathbf{v} \otimes \mathbf{u}) = (A\mathbf{v}) \otimes \mathbf{u}, \qquad (I_m \otimes B)(\mathbf{v} \otimes \mathbf{u}) = \mathbf{v} \otimes (B\mathbf{u}).$$

Thus a normalized eigenvector of the matrix $\epsilon_1 A \otimes B + \epsilon_2 A \otimes I_n + \epsilon_3 I_m \otimes B$ is $\mathbf{u} \otimes \mathbf{v}$ and the corresponding eigenvalue is given by $\epsilon_1 \lambda \mu + \epsilon_2 \lambda + \epsilon_3 \mu$.

Problem 11. Let A_j $(j = 1, \dots, k)$ be matrices of size $m_j \times n_j$. We introduce the notation

$$\otimes_{j=1}^{k} A_j = (\otimes_{j=1}^{k-1} A_j) \otimes A_k = A_1 \otimes A_2 \otimes \cdots \otimes A_k.$$

Consider the *elementary matrices*

$$E_{00} = \begin{pmatrix} 1 & 0 \\ 0 & 0 \end{pmatrix}, \quad E_{10} = \begin{pmatrix} 0 & 0 \\ 1 & 0 \end{pmatrix}, \quad E_{01} = \begin{pmatrix} 0 & 1 \\ 0 & 0 \end{pmatrix}, \quad E_{11} = \begin{pmatrix} 0 & 0 \\ 0 & 1 \end{pmatrix}.$$

(i) Calculate $\otimes_{j=1}^{n}(E_{00} + E_{01} + E_{11})$ for $k = 1$, $k = 2$, $k = 3$ and $k = 8$. Give an interpretation of the result when each entry in the matrix represents a pixel (1 for black and 0 for white). This means we use the Kronecker product for representing images.

(ii) Calculate

$$\left(\otimes_{j=1}^{k}(E_{00} + E_{01} + E_{10} + E_{11})\right) \otimes \begin{pmatrix} 0 & 1 \\ 1 & 0 \end{pmatrix}$$

for $k = 2$ and give an interpretation as an image, i.e. each entry 0 is identified with a black pixel and an entry 1 with a white pixel. Discuss the case for arbitrary k.

Solution 11.   (i) We obtain

$$E_{00} + E_{01} + E_{11} = \begin{pmatrix} 1 & 1 \\ 0 & 1 \end{pmatrix}$$

$$(E_{00} + E_{01} + E_{11}) \otimes (E_{00} + E_{01} + E_{11}) = \begin{pmatrix} 1 & 1 & 1 & 1 \\ 0 & 1 & 0 & 1 \\ 0 & 0 & 1 & 1 \\ 0 & 0 & 0 & 1 \end{pmatrix}.$$

For $k = 3$ we obtain the 8×8 matrix

$$\begin{pmatrix} 1 & 1 & 1 & 1 & 1 & 1 & 1 & 1 \\ 0 & 1 & 0 & 1 & 0 & 1 & 0 & 1 \\ 0 & 0 & 1 & 1 & 0 & 0 & 1 & 1 \\ 0 & 0 & 0 & 1 & 0 & 0 & 0 & 1 \\ 0 & 0 & 0 & 0 & 1 & 1 & 1 & 1 \\ 0 & 0 & 0 & 0 & 0 & 1 & 0 & 1 \\ 0 & 0 & 0 & 0 & 0 & 0 & 1 & 1 \\ 0 & 0 & 0 & 0 & 0 & 0 & 0 & 1 \end{pmatrix}.$$

For n larger and larger the image approaches the *Sierpinski triangle*. For $k = 8$ we have an 256×256 matrix which provides a good approximation for the Sierpinski triangle.

(ii) Note that

$$E_{00} + E_{01} + E_{10} + E_{11} = \begin{pmatrix} 1 & 1 \\ 1 & 1 \end{pmatrix}.$$

Then for $n = 2$ we obtain the 8×8 matrix

$$\begin{pmatrix} 0 & 1 & 0 & 1 & 0 & 1 & 0 & 1 \\ 1 & 0 & 1 & 0 & 1 & 0 & 1 & 0 \\ 0 & 1 & 0 & 1 & 0 & 1 & 0 & 1 \\ 1 & 0 & 1 & 0 & 1 & 0 & 1 & 0 \\ 0 & 1 & 0 & 1 & 0 & 1 & 0 & 1 \\ 1 & 0 & 1 & 0 & 1 & 0 & 1 & 0 \\ 0 & 1 & 0 & 1 & 0 & 1 & 0 & 1 \\ 1 & 0 & 1 & 0 & 1 & 0 & 1 & 0 \end{pmatrix}$$

which is a checkerboard. For all n we find a checkerboard pattern.

Problem 12. (i) Let A and X be $n \times n$ matrices over $\mathbb{C}$. Assume that $[X, A] = 0_n$. Calculate the commutator $[X \otimes I_n + I_n \otimes X, A \otimes A]$.

(ii) Let A, B, C be $n \times n$ matrices. Assume that $[A, B] = 0_n$, $[A, C] = 0_n$. Let

$$X := I_n \otimes A + A \otimes I_n, \qquad Y := I_n \otimes B + B \otimes I_n + A \otimes C.$$

Calculate the commutator $[X, Y]$.

Solution 12. (i) We have

$$[X \otimes I_n + I_n \otimes X, A \otimes A] = (XA) \otimes A + A \otimes (XA) - (AX) \otimes A - A \otimes (AX).$$

Since $AX = XA$ we obtain $[X \otimes I_n + I_n \otimes X, A \otimes A] = 0_{n^2}$.

(ii) We have

$$\begin{aligned}
[X, Y] &= I_n \otimes (AB - BA) + (AB - BA) \otimes I_n + A \otimes (AC - CA) \\
&= I_n \otimes [A, B] + [A, B] \otimes I_n + A \otimes [A, C] \\
&= 0_{n^2}.
\end{aligned}$$

Problem 13. A square matrix is called a *stochastic matrix* if each entry is nonnegative and the sum of the entries in each row is 1. Let A, B be $n \times n$ stochastic matrices. Is $A \otimes B$ a stochastic matrix?

Solution 13. The answer is yes. For example the sum of the first row of the matrix $A \otimes B$ is

$$a_{11} \sum_{j=1}^{n} b_{1j} + a_{12} \sum_{j=1}^{n} b_{1j} + \cdots + a_{1n} \sum_{j=1}^{n} b_{1j} = \sum_{j=1}^{n} a_{1j} = 1.$$

Analogously for the other rows.

Problem 14. Let X be an $m \times m$ and Y be an $n \times n$ matrix. The *direct sum* is the $(m + n) \times (m + n)$ matrix

$$X \oplus Y := \begin{pmatrix} X & 0 \\ 0 & Y \end{pmatrix}.$$

Let A be an $n \times n$ matrix, B be an $m \times m$ matrix and C be an $p \times p$ matrix. Then we have the identity

$$(A \oplus B) \otimes C \equiv (A \otimes C) \oplus (B \otimes C).$$

Is $A \otimes (B \oplus C) = (A \otimes B) \oplus (A \otimes C)$ true?

Solution 14. This is not true in general. For example, let

$$A = \begin{pmatrix} 2 & 1 \\ 0 & 3 \end{pmatrix}, \quad B = \begin{pmatrix} 1 & 1 \\ 1 & 1 \end{pmatrix}, \quad C = \begin{pmatrix} 0 & 1 \\ 1 & 0 \end{pmatrix}.$$

Then $A \otimes (B \oplus C) \neq (A \otimes B) \oplus (A \otimes C)$.

Problem 15. Let A, B be $n \times n$ matrices. Then the direct sum $A \oplus B$ is the $2n \times 2n$ matrix

$$A \oplus B = \begin{pmatrix} A & 0_n \\ 0_n & B \end{pmatrix}.$$

(i) Show that $A \oplus B$ can also be written using the Kronecker product and addition of matrices.

(ii) Let C be a 2×2 matrix. Can $C \oplus C$ be written as a Kronecker product of two 2×2 matrices?

Solution 15. (i) Yes. We have

$$A \oplus B \equiv \begin{pmatrix} 1 & 0 \\ 0 & 0 \end{pmatrix} \otimes A + \begin{pmatrix} 0 & 0 \\ 0 & 1 \end{pmatrix} \otimes B.$$

(ii) Yes. We have $C \oplus C \equiv I_2 \otimes C$.

Problem 16. Let A, B be 2×2 matrices, C a 3×3 matrix and D a 1×1 matrix. Find the condition on these matrices such that $A \otimes B = C \oplus D$, where $\oplus$ denotes the *direct sum*. We assume that D is nonzero.

Solution 16. Since

$$A \otimes B = \begin{pmatrix} a_{11}b_{11} & a_{11}b_{12} & a_{12}b_{11} & a_{12}b_{12} \\ a_{11}b_{21} & a_{11}b_{22} & a_{12}b_{21} & a_{12}b_{22} \\ a_{21}b_{11} & a_{21}b_{12} & a_{22}b_{11} & a_{22}b_{12} \\ a_{21}b_{21} & a_{21}b_{22} & a_{22}b_{21} & a_{22}b_{22} \end{pmatrix}, \quad C \oplus D = \begin{pmatrix} c_{11} & c_{12} & c_{13} & 0 \\ c_{21} & c_{22} & c_{23} & 0 \\ c_{31} & c_{32} & c_{33} & 0 \\ 0 & 0 & 0 & d \end{pmatrix}$$

we obtain from the last row and column of the two matrices $A \otimes B$

$$a_{12}b_{12} = a_{12}b_{22} = a_{22}b_{12} = 0, \quad a_{21}b_{21} = a_{21}b_{22} = a_{22}b_{21} = 0$$

and $a_{22}b_{22} = d$. Since $d \neq 0$ we have $a_{22} \neq 0$ and $b_{22} \neq 0$. Thus it follows that

$$a_{12} = a_{21} = b_{12} = b_{21} = 0.$$

Therefore $c_{12} = c_{21} = c_{23} = c_{32} = c_{13} = c_{31} = 0$ and

$$c_{11} = a_{11}b_{11}, \quad c_{22} = a_{11}b_{22}, \quad c_{33} = a_{22}b_{11}, \quad d = a_{22}b_{22}.$$

Problem 17. Let A be an 2×2 matrix over $\mathbb{C}$. Solve the equation $A \oplus A = A \otimes A$.

Solution 17. The equation yields the sixteen conditions

$$a_{11} = a_{11}^2, \quad a_{22} = a_{22}^2,$$

$$a_{12}(a_{11} - 1) = 0, \quad a_{21}(a_{11} - 1) = 0, \quad a_{22}(a_{11} - 1) = 0,$$

$$a_{11}(a_{22} - 1) = 0, \quad a_{12}(a_{22} - 1) = 0, \quad a_{21}(a_{22} - 1) = 0,$$

$$a_{12}a_{11} = a_{12}^2 = a_{12}a_{21} = a_{12}a_{22} = 0, \quad a_{11}a_{21} = a_{12}a_{21} = a_{21}^2 = a_{21}a_{22} = 0.$$

Thus $a_{12} = a_{21} = 0$. For a_{11} and a_{22} we find the two options $a_{11} = a_{22} = 0$ and $a_{11} = a_{22} = 1$. Thus either we have the 2×2 zero matrix or the 2×2 identity matrix as solution.

Problem 18. With each $m \times n$ matrix Y we associate the column vector vec(Y) of length $m \times n$ defined by

$$\text{vec}(Y) := (y_{11}, \ldots, y_{m1}, y_{12}, \ldots, y_{m2}, \ldots, y_{1n}, \ldots, y_{mn})^T.$$

(i) Let A be an $m \times n$ matrix, B an $p \times q$ matrix, and C an $m \times q$ matrix. Let X be an unknown $n \times p$ matrix. Show that the *matrix equation* $AXB = C$ is equivalent to the system of qm equations in np unknowns given by

$$(B^T \otimes A)\text{vec}(X) = \text{vec}(C)$$

that is, vec(AXB) = $(B^T \otimes A)$vec(X).
(ii) Let A, B, D be $n \times n$ matrices and I_n the $n \times n$ identity matrix. Use the result from (i) to prove that $AX + XB = D$ can be written as

$$((I_n \otimes A) + (B^T \otimes I_n))\text{vec}(X) = \text{vec}(D).$$

Solution 18. (i) For a given matrix Y, let Y_k denote the k-th column of Y. Let $B = (b_{ij})$. Then

$$(AXB)_k = AXB_k = A\left(\sum_{i=1}^{p} b_{ik}X_i\right) = (b_{1k}A \ b_{2k}A \ \ldots \ b_{pk}A)\text{vec}(X)$$

$$= (B_k^T \otimes A)\text{vec}(X).$$

Thus

$$\text{vec}(AXB) = \begin{pmatrix} B_1^T \otimes A \\ \vdots \\ B_q^T \otimes A \end{pmatrix} \text{vec}(X) = (B^T \otimes A)\text{vec}(X)$$

since the transpose of a column of B is a row of B^T. Therefore

$$\text{vec}(C) = \text{vec}(AXB) = (B^T \otimes A)\text{vec}(X).$$

(ii) Utilizing the result from (i) and that the vec operation is linear we have

$$\text{vec}(AX + XB) = \text{vec}(AX) + \text{vec}(XB) = \text{vec}(AXI_n) + \text{vec}(I_nXB)$$
$$= (I_n \otimes A)\text{vec}(X) + (B^T \otimes I_n)\text{vec}(X)$$
$$= \text{vec}(D).$$

Problem 19.   Let A, B, C, D be symmetric $n \times n$ matrices over $\mathbb{R}$. Assume that these matrices commute with each other. Consider the $4n \times 4n$ matrix

$$H = \begin{pmatrix} A & B & C & D \\ -B & A & D & -C \\ -C & -D & A & B \\ -D & C & -B & A \end{pmatrix}.$$

(i) Calculate HH^T and express the result using the Kronecker product.
(ii) Assume that $A^2 + B^2 + C^2 + D^2 = 4nI_n$.

Solution 19.   (i) We find $H^T H = (A^2 + B^2 + C^2 + D^2) \otimes I_4$.
(ii) For the special case with $A^2 + B^2 + C^2 + D^2 = 4nI_n$ we find that H is a Hadamard matrix, since $HH^T = 4nI_n \otimes I_4 = 4nI_{4n}$. This is the *Williamson construction* of Hadamard matrices.

Problem 20.   (i) Can the 4×4 matrix

$$C = \begin{pmatrix} 1 & 0 & 0 & 1 \\ 0 & 1 & 1 & 0 \\ 0 & 1 & -1 & 0 \\ 1 & 0 & 0 & -1 \end{pmatrix}$$

be written as the Kronecker product of two 2×2 matrices A and B, i.e. $C = A \otimes B$?
(ii) Can the 4×4 matrix

$$A = \frac{1}{2} \begin{pmatrix} 0 & 0 & -1 & 1 \\ 0 & 0 & 1 & -1 \\ -1 & 1 & 0 & 0 \\ 1 & -1 & 0 & 0 \end{pmatrix}$$

be written as a Kronecker product of 2×2 matrices?

Solution 20.   (i) From $C = A \otimes B$ we find the 16 conditions

$$\begin{array}{llll} a_{11}b_{11} = c_{11} & a_{11}b_{12} = c_{12} & a_{11}b_{21} = c_{21} & a_{11}b_{22} = c_{22} \\ a_{12}b_{11} = c_{13} & a_{12}b_{12} = c_{14} & a_{12}b_{21} = c_{23} & a_{12}b_{22} = c_{24} \\ a_{21}b_{11} = c_{31} & a_{21}b_{12} = c_{32} & a_{21}b_{21} = c_{41} & a_{21}b_{22} = c_{42} \\ a_{22}b_{11} = c_{33} & a_{22}b_{12} = c_{34} & a_{22}b_{21} = c_{43} & a_{22}b_{22} = c_{44}. \end{array}$$

Since $c_{12} = c_{13} = c_{21} = c_{24} = c_{31} = c_{34} = c_{42} = c_{43} = 0$ and

$$c_{11} = c_{14} = c_{22} = c_{23} = c_{32} = c_{41} = 1, \qquad c_{33} = c_{44} = -1,$$

we obtain

$$\begin{array}{llll} a_{11}b_{11} = 1 & a_{11}b_{12} = 0 & a_{11}b_{21} = 0 & a_{11}b_{22} = 1 \\ a_{12}b_{11} = 0 & a_{12}b_{12} = 1 & a_{12}b_{21} = 1 & a_{12}b_{22} = 0 \\ a_{21}b_{11} = 0 & a_{21}b_{12} = 1 & a_{21}b_{21} = 1 & a_{21}b_{22} = 0 \\ a_{22}b_{11} = -1 & a_{22}b_{12} = 0 & a_{22}b_{21} = 0 & a_{22}b_{22} = -1. \end{array}$$

Thus from $a_{11}b_{11} = 1$ we obtain that $a_{11} \neq 0$. Then from $a_{11}b_{12} = 0$ we obtain $b_{12} = 0$. However $a_{21}b_{12} = 1$. This is a contradiction. Thus C cannot be written as a Kronecker product of 2×2 matrices.

(ii) Yes. We have

$$A = \frac{1}{2} \begin{pmatrix} 0 & 1 \\ 1 & 0 \end{pmatrix} \otimes \begin{pmatrix} -1 & 1 \\ 1 & -1 \end{pmatrix}.$$

Problem 21. Let $\mathbf{x}, \mathbf{y}, \mathbf{z} \in \mathbb{R}^n$. We define a *wedge product*

$$\mathbf{x} \wedge \mathbf{y} := \mathbf{x} \otimes \mathbf{y} - \mathbf{y} \otimes \mathbf{x}.$$

Show that $(\mathbf{x} \wedge \mathbf{y}) \wedge \mathbf{z} + (\mathbf{z} \wedge \mathbf{x}) \wedge \mathbf{y} + (\mathbf{y} \wedge \mathbf{z}) \wedge \mathbf{x} = 0$.

Solution 21. We have

$$(\mathbf{x} \wedge \mathbf{y}) \wedge \mathbf{z} = \mathbf{x} \otimes \mathbf{y} \otimes \mathbf{z} - \mathbf{y} \otimes \mathbf{x} \otimes \mathbf{z} - \mathbf{z} \otimes \mathbf{x} \otimes \mathbf{y} + \mathbf{z} \otimes \mathbf{y} \otimes \mathbf{x}$$

$$(\mathbf{z} \wedge \mathbf{x}) \wedge \mathbf{y} = \mathbf{z} \otimes \mathbf{x} \otimes \mathbf{y} - \mathbf{x} \otimes \mathbf{z} \otimes \mathbf{y} - \mathbf{y} \otimes \mathbf{z} \otimes \mathbf{x} + \mathbf{y} \otimes \mathbf{x} \otimes \mathbf{z}$$

$$(\mathbf{y} \wedge \mathbf{z}) \wedge \mathbf{x} = \mathbf{y} \otimes \mathbf{z} \otimes \mathbf{x} - \mathbf{z} \otimes \mathbf{y} \otimes \mathbf{x} - \mathbf{x} \otimes \mathbf{y} \otimes \mathbf{z} + \mathbf{x} \otimes \mathbf{z} \otimes \mathbf{y}.$$

Adding these equations the result follows.

Problem 22. Consider the vectors $\mathbf{v}$ and $\mathbf{w}$ in $\mathbb{C}^2$

$$\mathbf{v} = \begin{pmatrix} v_1 \\ v_2 \end{pmatrix}, \quad \mathbf{w} = \begin{pmatrix} w_1 \\ w_2 \end{pmatrix}.$$

(i) Calculate $\mathbf{v} \otimes \mathbf{w}$, $\mathbf{v}^T \otimes \mathbf{w}$, $\mathbf{v} \otimes \mathbf{w}^T$, $\mathbf{v}^T \otimes \mathbf{w}^T$.
(ii) Is $\text{tr}(\mathbf{v}^T \otimes \mathbf{w}) = \text{tr}(\mathbf{v} \otimes \mathbf{w}^T)$?
(iii) Is $\det(\mathbf{v}^T \otimes \mathbf{w}) = \det(\mathbf{v} \otimes \mathbf{w}^T)$?

Solution 22. (i) We have

$$\mathbf{v} \otimes \mathbf{w} = \begin{pmatrix} v_1 w_1 \\ v_1 w_2 \\ v_2 w_1 \\ v_2 w_2 \end{pmatrix}, \quad \mathbf{v}^T \otimes \mathbf{w} = \begin{pmatrix} v_1 w_1 & v_2 w_1 \\ v_1 w_2 & v_2 w_2 \end{pmatrix},$$

$$\mathbf{v} \otimes \mathbf{w}^T = \begin{pmatrix} v_1 w_1 & v_1 w_2 \\ v_2 w_1 & v_2 w_2 \end{pmatrix}, \quad \mathbf{v}^T \otimes \mathbf{w}^T = \begin{pmatrix} v_1 w_1 & v_1 w_2 & v_2 w_1 & v_2 w_2 \end{pmatrix}.$$

(ii) Yes we have $\text{tr}(\mathbf{v}^T \otimes \mathbf{w}) = \text{tr}(\mathbf{v} \otimes \mathbf{w}^T) = v_1 w_1 + v_2 w_2$.
(iii) Yes we have $\det(\mathbf{v}^T \otimes \mathbf{w}) = \det(\mathbf{v} \otimes \mathbf{w}^T) = v_1 w_1 v_2 w_2 - v_1 w_2 v_2 w_1$.

Problem 23. Let A, B be *skew-hermitian matrices*, i.e. $A^* = -A$ and $B^* = -B$. Is $A \otimes B$ skew-hermitian?

Solution 23. We have $(A \otimes B)^* = A^* \otimes B^* = (-A) \otimes (-B) = A \otimes B$. Thus $A \otimes B$ is not skew-hermitian in general. However, $A \otimes B$ is hermitian.

Problem 24. Let A, B be $n \times n$ normal matrices over $\mathbb{C}$, i.e. $AA^* = A^*A$ and $BB^* = B^*B$. Show that $A \otimes B$ is a normal matrix.

Solution 24. We have

$$
\begin{aligned}
(A \otimes B)(A \otimes B)^* &= (A \otimes B)(A^* \otimes B^*) = (AA^* \otimes BB^*) \\
&= (A^*A \otimes B^*B) = (A^* \otimes B^*)(A \otimes B) \\
&= (A \otimes B)^*(A \otimes B).
\end{aligned}
$$

Problem 25. Let $\mathbf{z} = (\, z_0 \quad z_1 \quad \cdots \quad z_N \,)^T$ be a (column) vector in $\mathbb{C}^{N+1}$ and assume that $\mathbf{z}$ is nonzero. Then

$$
\mathbf{z}^* = (\, \bar{z}_0 \quad \bar{z}_1 \quad \cdots \quad \bar{z}_N \,)
$$

i.e. $\mathbf{z}^*$ is the transpose and conjugate complex of $\mathbf{z}$. Consider the $(N+1) \times (N+1)$ matrix

$$
\Pi := I_{N+1} - \frac{1}{\mathbf{z}^*\mathbf{z}}(\mathbf{z} \otimes \mathbf{z}^*).
$$

Show that $\Pi^* = \Pi$, $\Pi^2 = \Pi$. This means that Π is a *projection matrix*.

Solution 25. We have

$$
\mathbf{z} \otimes \mathbf{z}^* = \begin{pmatrix} z_0 \\ z_1 \\ \vdots \\ z_N \end{pmatrix} \otimes (\bar{z}_0, \bar{z}_1, \ldots, \bar{z}_N) = \begin{pmatrix} z_0\bar{z}_0 & z_0\bar{z}_1 & \cdots & z_0\bar{z}_N \\ z_1\bar{z}_0 & z_1\bar{z}_1 & \cdots & z_1\bar{z}_N \\ \vdots & & \ddots & \vdots \\ z_N\bar{z}_0 & z_N\bar{z}_1 & \cdots & z_N\bar{z}_N \end{pmatrix}
$$

and

$$
\mathbf{z}^* \otimes \mathbf{z} = (\bar{z}_0, \bar{z}_1, \ldots, \bar{z}_N) \otimes \begin{pmatrix} z_0 \\ z_1 \\ \vdots \\ z_N \end{pmatrix} = \begin{pmatrix} \bar{z}_0 z_0 & \bar{z}_1 z_0 & \cdots & \bar{z}_N z_0 \\ \bar{z}_0 z_1 & \bar{z}_1 z_1 & \cdots & \bar{z}_N z_1 \\ \vdots & & \ddots & \vdots \\ \bar{z}_0 z_N & \bar{z}_1 z_N & \cdots & \bar{z}_N z_N \end{pmatrix}.
$$

Thus $\mathbf{z} \otimes \mathbf{z}^* = \mathbf{z}^* \otimes \mathbf{z}$. Since $(\mathbf{z} \otimes \mathbf{z}^*)^* = \mathbf{z}^* \otimes \mathbf{z}^{**} = \mathbf{z}^* \otimes \mathbf{z}$ and $I_{N+1} = I_{N+1}^*$ we obtain $\Pi = \Pi^*$. Using $\mathbf{z} \otimes \mathbf{z}^* = \mathbf{z}\mathbf{z}^*$ and $\mathbf{z} \otimes \mathbf{z}^* = \mathbf{z}^* \otimes \mathbf{z}$ we find

$$
\begin{aligned}
\Pi^2 &= \left(I_{N+1} - \frac{1}{\mathbf{z}^*\mathbf{z}}(\mathbf{z} \otimes \mathbf{z}^*) \right)\left(I_{N+1} - \frac{1}{\mathbf{z}^*\mathbf{z}}(\mathbf{z} \otimes \mathbf{z}^*) \right) \\
&= I_{N+1} - \frac{2}{\mathbf{z}^*\mathbf{z}}(\mathbf{z} \otimes \mathbf{z}^*) + \frac{1}{(\mathbf{z}^*\mathbf{z})^2}(\mathbf{z}^*\mathbf{z})(\mathbf{z}\mathbf{z}^*) = I_{N+1} - \frac{1}{\mathbf{z}^*\mathbf{z}}(\mathbf{z} \otimes \mathbf{z}^*) \\
&= \Pi.
\end{aligned}
$$

Problem 26. Let A, B be $n \times n$ *tridiagonal matrices* with $n \geq 3$. Is $A \otimes B$ a tridiagonal matrix?

Solution 26. This is obviously not true in general. For example, consider the 3×3 matrix

$$A = \begin{pmatrix} a_{11} & a_{12} & 0 \\ a_{21} & a_{22} & a_{23} \\ 0 & a_{32} & a_{33} \end{pmatrix}$$

and the 9×9 matrix $A \otimes A$.

Problem 27. Consider the 2×2 matrix A over $\mathbb{C}$ with $\det(A) = 1$.
(i) Find the inverse of A.
(ii) Let $\mathbf{v}$ be a normalized vector in $\mathbb{C}^2$. Is the vector $A\mathbf{v}$ normalized?
(iii) Let

$$E = \begin{pmatrix} 0 & 1 \\ -1 & 0 \end{pmatrix}.$$

Calculate AEA^T.
(iv) Let B be a 2×2 matrix over $\mathbb{C}$ with $\det(B) = 1$. Find the 4×4 matrix $(A \otimes B)(E \otimes E)(A \otimes B)^T$.

Solution 27. (i) The inverse matrix of A is given by

$$A^{-1} = \begin{pmatrix} a_{22} & -a_{12} \\ -a_{21} & a_{11} \end{pmatrix}.$$

(ii) In general the vector $A\mathbf{v}$ is not normalized, for example

$$\begin{pmatrix} 2 & 1 \\ 1 & 1 \end{pmatrix} \begin{pmatrix} 1 \\ 0 \end{pmatrix} = \begin{pmatrix} 2 \\ 1 \end{pmatrix}.$$

(iii) Utilizing $\det(A) = 1$ we find

$$AEA^T = \begin{pmatrix} -a_{12} & a_{11} \\ -a_{22} & a_{21} \end{pmatrix} \begin{pmatrix} a_{11} & a_{21} \\ a_{12} & a_{22} \end{pmatrix} = \begin{pmatrix} 0 & \det(A) \\ -\det(A) & 0 \end{pmatrix} = E.$$

(iv) We obtain

$$(A \otimes B)(E \otimes E)(A \otimes B)^T = (AEA^T) \otimes (BEB^T) = E \otimes E.$$

Problem 28. Let A be an $n \times n$ hermitian matrix with $A^2 = I_n$. Let

$$\Pi_1 = \frac{1}{2}(I_n + A) \otimes I_n, \qquad \Pi_2 = \frac{1}{2}(I_n - A) \otimes I_n.$$

Show that Π_1 and Π_2 are projection matrices. Calculate the product $\Pi_1\Pi_2$.

Solution 28. Since $A = A^*$ we have $\Pi_1 = \Pi_1^*$ and $\Pi_2 = \Pi_2^*$. Since $A^2 = I_n$ we have

$$\Pi_1^2 = \Pi_1, \qquad \Pi_2^2 = \Pi_2.$$

Since $(I_n + A)(I_n - A) = 0_n$ we obtain $\Pi_1\Pi_2 = 0_{n^2}$.

Problem 29.   Is the Kronecker product of two *circulant matrices* again a circulant matrix? Let A, B be two $n \times n$ circulant matrices. Is $A \otimes B = B \otimes A$?

Solution 29.   The Kronecker product of two circulant matrices is not a circulant matrix in general as we see by calculating the Kronecker product of two circulant matrices

$$
\begin{pmatrix} c_0 & c_1 \\ c_1 & c_0 \end{pmatrix} \otimes \begin{pmatrix} d_0 & d_1 \\ d_1 & d_0 \end{pmatrix} = \begin{pmatrix} c_0 d_0 & c_0 d_1 & c_1 d_0 & c_1 d_1 \\ c_0 d_1 & c_0 d_0 & c_1 d_1 & c_1 d_0 \\ c_1 d_0 & c_1 d_1 & c_0 d_0 & c_0 d_1 \\ c_1 d_1 & c_1 d_0 & c_0 d_1 & c_0 d_0 \end{pmatrix}.
$$

In general we do not have $A \otimes B = B \otimes A$ for two $n \times n$ circulant matrices.

Problem 30.   A *Hankel matrix* is a matrix where for each antidiagonal the element is the same. For example, Hankel matrices are

$$
\begin{pmatrix} 17 & 16 & 15 & 24 \\ 16 & 15 & 24 & 33 \\ 15 & 24 & 33 & 2 \\ 24 & 33 & 2 & 41 \end{pmatrix}, \qquad \begin{pmatrix} 72 & 60 & 55 & 43 & 30 & 21 \\ 60 & 55 & 43 & 30 & 21 & 10 \\ 55 & 43 & 30 & 21 & 10 & 8 \end{pmatrix}.
$$

Given two square Hankel matrices. Is the Kronecker product again a Hankel matrix?

Solution 30.   Let

$$
A = \begin{pmatrix} 1 & 2 & 3 \\ 2 & 3 & 4 \\ 3 & 4 & 1 \end{pmatrix}, \qquad B = \begin{pmatrix} 2 & 1 & 4 \\ 1 & 4 & 5 \\ 4 & 5 & 8 \end{pmatrix}.
$$

We see that $A \otimes B$ is not a Hankel matrix.

Problem 31.   Let A be an invertible $n \times n$ matrix. Then the matrix $(A^{-1} \otimes I_n)(I_n \otimes A)$ also has an inverse. Find the inverse.

Solution 31.   We have $(A^{-1} \otimes I_n)(I_n \otimes A) = A^{-1} \otimes A$. Thus we find that the inverse is given by $A \otimes A^{-1}$ since

$$
(A^{-1} \otimes A)(A \otimes A^{-1}) = I_n \otimes I_n.
$$

Note that (identity) $A \otimes A^{-1} \equiv (A \otimes I_n)(I_n \otimes A^{-1})$.

Problem 32.   (i) Let $\mathbf{x}$, $\mathbf{y}$ be vectors in $\mathbb{R}^2$. Can one find a 4×4 permutation matrix P such that $P(\mathbf{x} \otimes \mathbf{y}) = \mathbf{y} \otimes \mathbf{x}$?
(ii) Let A, B be 2×2 matrices. Can one find a 4×4 permutation matrix such that

$$
P(A \otimes B)P^{-1} = B \otimes A?
$$

(iii) Consider the vectors $\mathbf{u}$ and $\mathbf{v}$ in $\mathbb{C}^3$

$$\mathbf{u} = \begin{pmatrix} u_1 \\ u_2 \\ u_3 \end{pmatrix}, \qquad \mathbf{v} = \begin{pmatrix} v_1 \\ v_2 \\ v_3 \end{pmatrix}.$$

Find the 9×9 permutation matrix P such that $P(\mathbf{u} \otimes \mathbf{v}) = \mathbf{v} \otimes \mathbf{u}$.

Solution 32. (i) We have

$$\mathbf{x} \otimes \mathbf{y} = \begin{pmatrix} x_1 y_1 \\ x_1 y_2 \\ x_2 y_1 \\ x_2 y_2 \end{pmatrix}, \qquad \mathbf{x} \otimes \mathbf{y} = \begin{pmatrix} y_1 x_1 \\ y_1 x_2 \\ y_2 x_1 \\ y_2 x_2 \end{pmatrix} \Rightarrow P = \begin{pmatrix} 1 & 0 & 0 & 0 \\ 0 & 0 & 1 & 0 \\ 0 & 1 & 0 & 0 \\ 0 & 0 & 0 & 1 \end{pmatrix}.$$

(ii) We find the same permutation matrix as in (i). Note that $P = P^{-1} = P^T$.
(iii) We find the 9×9 matrix (written as a direct sum)

$$P = (1) \oplus \begin{pmatrix} 0 & 0 & 1 & 0 & 0 & 0 & 0 \\ 0 & 0 & 0 & 0 & 0 & 1 & 0 \\ 1 & 0 & 0 & 0 & 0 & 0 & 0 \\ 0 & 0 & 0 & 1 & 0 & 0 & 0 \\ 0 & 0 & 0 & 0 & 0 & 0 & 1 \\ 0 & 1 & 0 & 0 & 0 & 0 & 0 \\ 0 & 0 & 0 & 0 & 1 & 0 & 0 \end{pmatrix} \oplus (1).$$

Problem 33. Let A, B be symmetric matrices over $\mathbb{R}$.
(i) What is the condition on A, B such that AB is symmetric?
(ii) What is condition on A, B such that $A \otimes B$ is symmetric?

Solution 33. (i) We have $(AB)^T = B^T A^T = BA$. Thus the condition is that $AB = BA$.
(ii) We have $(A \otimes B)^T = A^T \otimes B^T = A \otimes B$.

Problem 34. Let A, B be 2×2 matrices over $\mathbb{R}$. Find the conditions on A and B such that

$$(A \otimes I_2) \frac{1}{\sqrt{2}} \begin{pmatrix} 1 \\ 0 \\ 0 \\ 1 \end{pmatrix} = (I_2 \otimes B) \frac{1}{\sqrt{2}} \begin{pmatrix} 1 \\ 0 \\ 0 \\ 1 \end{pmatrix}.$$

Solution 34. We find $a_{11} = b_{11}$, $a_{12} = b_{21}$, $a_{21} = b_{21}$, $a_{22} = b_{22}$.

Problem 35. Let σ_1, σ_2, σ_3 be the Pauli spin matrices.
(i) Consider the Hilbert space $\mathbb{C}^4$. Show that the matrices

$$\Pi_1 = \frac{1}{2}(I_2 \otimes I_2 + \sigma_1 \otimes \sigma_1), \qquad \Pi_2 = \frac{1}{2}(I_2 \otimes I_2 - \sigma_1 \otimes \sigma_1)$$

are projection matrices in $\mathbb{C}^4$.

(ii) Find $\Pi_1\Pi_2$. Discuss.

(iii) Let e_1, e_2, e_3, e_4 be the standard basis in $\mathbb{C}^4$. Calculate $\Pi_1 e_j$, $\Pi_2 e_j$, $j = 1, 2, 3, 4$ and show that we obtain 2 two-dimensional Hilbert spaces under these projections.

(iv) Do the matrices $I_2 \otimes I_2$, $\sigma_1 \otimes \sigma_1$ form a group under matrix multiplication?

(v) Show that the matrices

$$\Pi_j = \frac{1}{2}(I_2 \otimes I_2 + \sigma_j \otimes \sigma_j), \qquad j = 1, 2, 3$$

are projection matrices.

(vi) Is

$$\frac{1}{2}(I_2 \otimes I_2 - \frac{1}{2}(i\sigma_1 \otimes i\sigma_1 + i\sigma_2 \otimes i\sigma_2))$$

a projection matrix?

Solution 35.   (i) We have $\Pi_1^* = \Pi_1$, $\Pi_2^* = \Pi_2$ and $\Pi_1^2 = \Pi_1$, $\Pi_2^2 = \Pi_2$. Thus Π_1 and Π_2 are projection matrices.

(ii) We obtain $\Pi_1\Pi_2 = 0_4$. Let $\mathbf{u}$ be an arbitrary vector in $\mathbb{C}^4$. Then the vectors $\Pi_1\mathbf{u}$ and $\Pi_2\mathbf{u}$ are perpendicular.

(iii) We obtain

$$\Pi_1 e_1 = \Pi_1 e_4 = \frac{1}{2}\begin{pmatrix} 1 \\ 0 \\ 0 \\ 1 \end{pmatrix}, \qquad \Pi_1 e_2 = \Pi_1 e_3 = \frac{1}{2}\begin{pmatrix} 0 \\ 1 \\ 1 \\ 0 \end{pmatrix},$$

$$\Pi_2 e_1 = -\Pi_2 e_4 = \frac{1}{2}\begin{pmatrix} 1 \\ 0 \\ 0 \\ -1 \end{pmatrix}, \qquad \Pi_2 e_2 = -\Pi_2 e_3 = \frac{1}{2}\begin{pmatrix} 0 \\ 1 \\ -1 \\ 0 \end{pmatrix}.$$

Thus Π_1 projects into a two-dimensional Hilbert space spanned by the normalized vectors

$$\left\{ \frac{1}{\sqrt{2}}\begin{pmatrix} 1 \\ 0 \\ 0 \\ 1 \end{pmatrix}, \quad \frac{1}{\sqrt{2}}\begin{pmatrix} 0 \\ 1 \\ 1 \\ 0 \end{pmatrix} \right\}.$$

The projection matrix Π_2 projects into a two-dimensional Hilbert space spanned by the normalized vectors

$$\left\{ \frac{1}{\sqrt{2}}\begin{pmatrix} 1 \\ 0 \\ 0 \\ -1 \end{pmatrix}, \quad \frac{1}{\sqrt{2}}\begin{pmatrix} 0 \\ 1 \\ -1 \\ 0 \end{pmatrix} \right\}.$$

The four vectors we find are the *Bell basis*.

(iv) Yes. Matrix multiplication yields

$$(I_2 \otimes I_2)(I_2 \otimes I_2) = I_2 \otimes I_2, \quad (I_2 \otimes I_2)(\sigma_1 \otimes \sigma_1) = \sigma_1 \otimes \sigma_1,$$

$$(\sigma_1 \otimes \sigma_1)(I_2 \otimes I_2) = \sigma_1 \otimes \sigma_1, \quad (\sigma_1 \otimes \sigma_1)(\sigma_1 \otimes \sigma_1) = I_2 \otimes I_2.$$

The neutral element is $I_2 \otimes I_2$. The inverse element of $\sigma_1 \otimes \sigma_1$ is $\sigma_1 \otimes \sigma_1$.
(v) Since $\sigma_j^2 = I_2$ we obtain $\Pi_j^2 = \Pi_j$ and $\Pi_j^* = \Pi_j$ for $j = 1, 2, 3$. Thus the Π_j's are projection matrices.
(vi) We find the projection matrix

$$\begin{pmatrix} 1/2 & 0 & 0 & 0 \\ 0 & 1/2 & 1/2 & 0 \\ 0 & 1/2 & 1/2 & 0 \\ 0 & 0 & 0 & 1/2 \end{pmatrix}.$$

Problem 36. Let A be an $n \times n$ matrix. Show that the $2n \times 2n$ matrix

$$B = \begin{pmatrix} I_n & I_n \\ A & I_n + A \end{pmatrix}$$

can be expressed using the Kronecker product, the $n \times n$ identity matrix I_n, the matrix A and the 2×2 matrices

$$C = \begin{pmatrix} 1 & 1 \\ 0 & 1 \end{pmatrix}, \qquad D = \begin{pmatrix} 0 & 0 \\ 1 & 1 \end{pmatrix}.$$

Solution 36. We have $B = C \otimes I_n + D \otimes A$.

Problem 37. Let E_{jk} $(j, k = 1, \ldots, n)$ be the $n \times n$ matrices with 1 at entry (j, k) and 0 otherwise. Let $n = 2$. Find the 4×4 matrices

$$X = E_{11} \otimes E_{11} + E_{12} \otimes E_{12} + E_{21} \otimes E_{21} + E_{22} \otimes E_{22}$$

and

$$Y = E_{11} \otimes E_{11} + E_{21} \otimes E_{12} + E_{12} \otimes E_{21} + E_{22} \otimes E_{22}.$$

Are the matrices invertible?

Solution 37. We obtain

$$X = \begin{pmatrix} 1 & 0 & 0 & 1 \\ 0 & 0 & 0 & 0 \\ 0 & 0 & 0 & 0 \\ 1 & 0 & 0 & 1 \end{pmatrix}, \qquad Y = \begin{pmatrix} 1 & 0 & 0 & 0 \\ 0 & 0 & 1 & 0 \\ 0 & 1 & 0 & 0 \\ 0 & 0 & 0 & 1 \end{pmatrix}.$$

The matrix Y is invertible, but not the matrix X.

Problem 38. Let A, B, C be 2×2 matrices. Can we find 8×8 permutation matrices P and Q such that $P(A \otimes B \otimes C)Q = C \otimes B \otimes A$?

Solution 38.   We find $P = Q$ with

$$P = (1) \oplus \begin{pmatrix} 0 & 0 & 0 & 1 & 0 & 0 \\ 0 & 1 & 0 & 0 & 0 & 0 \\ 0 & 0 & 0 & 0 & 0 & 1 \\ 1 & 0 & 0 & 0 & 0 & 0 \\ 0 & 0 & 0 & 0 & 1 & 0 \\ 0 & 0 & 1 & 0 & 0 & 0 \end{pmatrix} \oplus (1).$$

Problem 39.   Let P_1 be an $m \times m$ permutation matrix and P_2 an $n \times n$ permutation matrix. Then $P_1 \otimes P_2$ is an $mn \times mn$ permutation matrix. Let K_1 and K_2 be $m \times m$ and $n \times n$ matrices, respectively such that $P_1 = e^{K_1}$ and $P_2 = e^{K_2}$. Find K such that $P_1 \otimes P_2 = e^K$, where K is an $mn \times mn$ matrix.

Solution 39.   We have $P_1 \otimes P_2 = e^{K_1} \otimes e^{K_2}$. Since

$$e^{K_1} \otimes e^{K_2} \equiv e^{K_1 \otimes I_n + I_m \otimes K_2}$$

we obtain $K = K_1 \otimes I_n + I_m \otimes K_2$.

Problem 40.   Let A and B be two complex valued matrices, not necessarily of the same size. If $B \neq 0$ then the operation $\oslash$ given by

$$(A \otimes B) \oslash B = A$$

is well defined, where $\otimes$ is the Kronecker product. The operation $\oslash$ is called a right *Kronecker quotient*. Show that the nearest Kronecker product

$$M \oslash B := A \qquad (\|M - A \otimes B\| \text{ is a minimum})$$

defines a Kronecker quotient, where $\| \cdot \|$ denotes the Frobenius norm. Show that

$$M \oslash B = \frac{\mathrm{tr}_2((I \otimes B^*)M)}{\|B\|^2}$$

where tr_2 denotes the partial trace over matrices with the same size as B^*B, and I is an appropriate identity matrix. Are there any other Kronecker quotients?

Solution 40.   Clearly, $\|A \otimes B - C \otimes B\|$ attains a minimum when $C = A$. We need to show that no other matrix C can attain the minimum. Suppose $\|A \otimes B - C \otimes B\| = 0$. Then

$$\begin{aligned} \|A \otimes B - C \otimes B\|^2 &= \mathrm{tr}((A \otimes B - C \otimes B)^*(A \otimes B - C \otimes B)) \\ &= \|A - C\|^2 \|B\|^2 \\ &= 0 \end{aligned}$$

if and only if $\|A - C\| = 0$ (since $B \neq 0$ by assumption). In other words, the minimum 0 is only attained when $A = C$. Let $\{B_1, \ldots, B_m\}$ be an orthogonal basis (with respect to the Frobenius inner product) for the vector space of matrices of the same size as B, with $B_1 = B$. In other words $\operatorname{tr}(B_j B_k^*) = \operatorname{tr}(B_k^* B_j) = \delta_{j,k} \|B_j\|^2$. We may write

$$M = \sum_{j=1}^{m} M_j \otimes B_j$$

so that

$$\|M - A \otimes B\|^2 = \operatorname{tr}((M - A \otimes B)^*(M - A \otimes B))$$

$$= \|M_1 - A\|^2 \|B\|^2 + \sum_{j=2}^{m} \|M_j\|^2 \|B_j\|^2$$

which achieves a minimum when $\|A - M_1\|$ is minimized, i.e. $A = M_1$ where

$$M_1 = \frac{\operatorname{tr}_2((I \otimes B^*)M)}{\|B\|^2}.$$

There are many other Kronecker quotients, for example

$$M \oslash B = \frac{1}{|\{(i,j) : (B)_{i,j} \neq 0\}|} \sum_{i,j:(B)_{i,j} \neq 0} \frac{\operatorname{tr}_2((I \otimes E_{ij})^* M)}{(B)_{i,j}}$$

defines a Kronecker quotient, where E_{ij} is the matrix of the same size as B with a 1 in row i and column j and 0 in all other entries.

Problem 41. Let $d \geq 2$. Consider the vector space $V = \mathbb{C}^d$. Let $V \times V$ be the Cartesian product of V with itself. Let $\mathbf{u}$, $\mathbf{v}$, $\mathbf{w}$ be elements of V. The Kronecker product (tensor product) is defined by the relations

$$(\mathbf{u} + \mathbf{v}) \otimes \mathbf{w} = \mathbf{u} \otimes \mathbf{w} + \mathbf{v} \otimes \mathbf{w}, \quad \mathbf{u} \otimes (\mathbf{v} + \mathbf{w}) = \mathbf{u} \otimes \mathbf{v} + \mathbf{u} \otimes \mathbf{w},$$

$$(c\mathbf{u}) \otimes \mathbf{v} = \mathbf{u} \otimes (c\mathbf{v}) = c(\mathbf{u} \otimes \mathbf{v}).$$

Let $V \otimes V \equiv V^{\otimes^2} = \operatorname{span}\{\mathbf{u} \otimes \mathbf{v} : \mathbf{u}, \mathbf{v} \in V\}$, where the span is taken in $\mathbb{C}^{d^2}$. Let $\{\mathbf{e}_j\}_{j=1}^{d}$ be a basis in the vector space V. Then $\{\mathbf{e}_j \otimes \mathbf{e}_k\}$ $(j, k = 1, \ldots, d)$ is a basis in the vector space $V \otimes V$ with dimension d^2. We define the k-th tensor power of the vector space V denoted by $V^{\otimes^k}$ given by the elements

$$\mathbf{u}_1 \otimes \mathbf{u}_2 \otimes \cdots \otimes \mathbf{u}_k.$$

Thus the dimension of the vector space $V^{\otimes^k}$ is d^k. Let $k \leq d$. Then the *wedge product* (also called *Grassmann product, exterior product*) is defined by

$$\mathbf{v}_1 \wedge \cdots \wedge \mathbf{v}_k := \sum_{\pi \in S_k} \sigma(\pi) \mathbf{v}_{\pi(1)} \otimes \cdots \otimes \mathbf{v}_{\pi(k)}$$

where S_k is the permutation group and σ is its alternating representation, i.e. $\sigma(\pi) = +1$ for even permutations and $\sigma(\pi) = -1$ for odd permutations. The complex vector space spanned by elements of the form

$$\mathbf{v}_1, \ldots, \mathbf{v}_n$$

is denoted by $\wedge^k V$ and has dimension $\binom{d}{k}$. Give a computer algebra implementation with SymbolicC++ of the wedge product. Apply it to the case $d = 3$ with the basis

$$\mathbf{v}_1 = \frac{1}{\sqrt{2}} \begin{pmatrix} 1 \\ 0 \\ 1 \end{pmatrix}, \quad \mathbf{v}_2 = \begin{pmatrix} 0 \\ 1 \\ 0 \end{pmatrix}, \quad \mathbf{v}_3 = \frac{1}{\sqrt{2}} \begin{pmatrix} 1 \\ 0 \\ -1 \end{pmatrix}.$$

Calculate $\mathbf{v}_1 \wedge \mathbf{v}_2$, $\mathbf{v}_2 \wedge \mathbf{v}_3$, $\mathbf{v}_3 \wedge \mathbf{v}_1$, $\mathbf{v}_1 \wedge \mathbf{v}_2 \wedge \mathbf{v}_3$.

Solution 41.   The `vector` class of the Standard Template Library of C++ is utilized and kron (Kronecker product) from SymbolicC++.

```
/* wedge.cpp */

#include <iostream>
#include <vector>
#include "symbolicc++.h"
using namespace std;

Symbolic wedge(const vector<Symbolic> &v)
{
 Symbolic term, sum = 0;
 int sign, i, k, l, n = v.size(); i = 0;
 vector<int> j(n,-1); j[0] = -1;
 while(i >= 0)
 {
 if(++j[i]==n) { j[i--] = -1; continue; }
 if(i < 0) break;
 for(k=0;k<i;++k) if(j[k]==j[i]) break;
 if(k!=i) continue; ++i;
 if(i==n)
 {
 for(sign=1,k=0;k<n;k++)
 {
 if(k==0) term = v[j[0]]; else term = kron(term,v[j[k]]);
 for(l=k+1;l<n;l++) if(j[l]<j[k]) sign = -sign;
 }
 sum += sign*term; --i;
 }
 }
 return sum;
}

int main(void)
{
 vector<Symbolic> w1(2), w2(3);
 Symbolic v1 = (Symbolic(1),0,1)/sqrt(Symbolic(2));
 Symbolic v2 = (Symbolic(0),1,0);
 Symbolic v3 = (Symbolic(1),0,-1)/sqrt(Symbolic(2));
 cout << "v1 = " << v1 << endl;
 cout << "v2 = " << v2 << endl;
 cout << "v3 = " << v3 << endl;
```

```
w1[0] = v1; w1[1] = v2; cout << "v1 ^ v2 = " << wedge(w1) << endl;
w1[0] = v2; w1[1] = v3; cout << "v2 ^ v3 = " << wedge(w1) << endl;
w1[0] = v3; w1[1] = v1; cout << "v3 ^ v1 = " << wedge(w1) << endl;
w2[0] = v1; w2[1] = v2; w2[2] = v3;
cout << "v1 ^ v2 ^ v3 = " << wedge(w2) << endl;
return 0;
}
```

Supplementary Problems

Problem 1. (i) Given the standard basis $\{e_1, e_2\}$ in $\mathbb{C}^2$. Is

$$\frac{1}{\sqrt{2}}(e_1 \otimes e_1 + e_2 \otimes e_2), \quad \frac{1}{\sqrt{2}}(e_1 \otimes e_1 - e_2 \otimes e_2),$$

$$\frac{1}{\sqrt{2}}(e_1 \otimes e_2 + e_2 \otimes e_1), \quad \frac{1}{\sqrt{2}}(e_1 \otimes e_2 - e_2 \otimes e_1)$$

a basis in $\mathbb{C}^4$?

(ii) Let e_1, e_2 be the standard basis in $\mathbb{C}^2$. The eight vectors

$$e_1 \otimes e_1 \otimes e_1, \quad e_1 \otimes e_1 \otimes e_2, \quad e_1 \otimes e_2 \otimes e_1, \quad e_1 \otimes e_2 \otimes e_2,$$

$$e_2 \otimes e_1 \otimes e_1, \quad e_2 \otimes e_1 \otimes e_2, \quad e_2 \otimes e_2 \otimes e_1, \quad e_2 \otimes e_2 \otimes e_2$$

provides the standard basis in $\mathbb{R}^8$. Do the 8 vectors

$$\frac{1}{\sqrt{2}}(e_1 \otimes e_1 \otimes e_1 \pm e_2 \otimes e_2 \otimes e_2), \quad \frac{1}{\sqrt{2}}(e_1 \otimes e_1 \otimes e_2 \pm e_2 \otimes e_2 \otimes e_1),$$

$$\frac{1}{\sqrt{2}}(e_1 \otimes e_2 \otimes e_1 \pm e_2 \otimes e_1 \otimes e_2), \quad \frac{1}{\sqrt{2}}(e_1 \otimes e_2 \otimes e_2 \pm e_2 \otimes e_1 \otimes e_1)$$

form an orthonormal basis in $\mathbb{R}^8$?

(iii) Let the vectors v_1, v_2 form an orthonormal basis in $\mathbb{C}^2$. Then the four vectors

$$v_1 \otimes v_1, \quad v_1 \otimes v_2, \quad v_2 \otimes v_1, \quad v_2 \otimes v_2$$

form an orthonormal basis in $\mathbb{C}^4$. Do the four vectors

$$v_1 \otimes v_1 + v_2 \otimes v_2, \ v_1 \otimes v_1 - v_2 \otimes v_2, \ v_1 \otimes v_2 + v_2 \otimes v_1, \ v_1 \otimes v_2 - v_2 \otimes v_1$$

form a basis in $\mathbb{C}^4$? Can the four vectors be written as a Kronecker product of two vectors in $\mathbb{C}^2$?

(iv) The three vectors

$$e_0 = \begin{pmatrix} 0 \\ 1 \\ 0 \end{pmatrix}, \quad e_+ = \frac{1}{\sqrt{2}}\begin{pmatrix} 1 \\ 0 \\ 1 \end{pmatrix}, \quad e_- = \frac{1}{\sqrt{2}}\begin{pmatrix} 1 \\ 0 \\ -1 \end{pmatrix}$$

form an orthonormal basis in $\mathbb{R}^3$. Do the 9 vectors

$$\mathbf{v}_{0,0} = \frac{1}{\sqrt{3}}(\mathbf{e}_+ \otimes \mathbf{e}_- + \mathbf{e}_- \otimes \mathbf{e}_+ - \mathbf{e}_0 \otimes \mathbf{e}_0)$$

$$\mathbf{v}_{1,0} = \frac{1}{\sqrt{2}}(\mathbf{e}_+ \otimes \mathbf{e}_- - \mathbf{e}_- \otimes \mathbf{e}_+)$$

$$\mathbf{v}_{1,\pm 1} = \pm \frac{1}{\sqrt{2}}(\mathbf{e}_\pm \otimes \mathbf{e}_0 - \mathbf{e}_0 \otimes \mathbf{e}_\pm)$$

$$\mathbf{v}_{2,0} = \frac{1}{\sqrt{6}}(\mathbf{e}_+ \otimes \mathbf{e}_- + \mathbf{e}_- \otimes \mathbf{e}_+ + 2\mathbf{e}_0 \otimes \mathbf{e}_0)$$

$$\mathbf{v}_{2,\pm 1} = \frac{1}{\sqrt{2}}(\mathbf{e}_\pm \otimes \mathbf{e}_0 + \mathbf{e}_0 \otimes \mathbf{e}_\pm)$$

$$\mathbf{v}_{2,\pm 2} = \mathbf{e}_\pm \otimes \mathbf{e}_\pm$$

form an orthonormal basis in $\mathbb{R}^9$? These vectors play a role for the π-mesons.

Problem 2.   Show that the conditions on $x_1, x_2, y_1, y_2 \in \mathbb{R}$ such that

$$\begin{pmatrix} x_1 \\ x_2 \end{pmatrix} \otimes \begin{pmatrix} y_1 \\ y_2 \end{pmatrix} = \begin{pmatrix} y_1 \\ y_2 \end{pmatrix} \otimes \begin{pmatrix} x_1 \\ x_2 \end{pmatrix}$$

is given by $x_1 y_2 = x_2 y_1$.

Problem 3.   The 6×6 *primary cyclic matrix* S is given by

$$S = \begin{pmatrix} 0 & 1 & 0 & 0 & 0 & 0 \\ 0 & 0 & 1 & 0 & 0 & 0 \\ 0 & 0 & 0 & 1 & 0 & 0 \\ 0 & 0 & 0 & 0 & 1 & 0 \\ 0 & 0 & 0 & 0 & 0 & 1 \\ 1 & 0 & 0 & 0 & 0 & 0 \end{pmatrix} \Rightarrow S^T = \begin{pmatrix} 0 & 0 & 0 & 0 & 0 & 1 \\ 1 & 0 & 0 & 0 & 0 & 0 \\ 0 & 1 & 0 & 0 & 0 & 0 \\ 0 & 0 & 1 & 0 & 0 & 0 \\ 0 & 0 & 0 & 1 & 0 & 0 \\ 0 & 0 & 0 & 0 & 1 & 0 \end{pmatrix}.$$

Let $\mathbf{e}_j$ $(j = 1, \ldots, 6)$ be the standard basis in $\mathbb{R}^6$. Then

$$S^T \mathbf{e}_1 = \mathbf{e}_2, \quad S^T \mathbf{e}_2 = \mathbf{e}_3, \quad S^T \mathbf{e}_3 = \mathbf{e}_4, \quad S^T \mathbf{e}_4 = \mathbf{e}_5, \quad S^T \mathbf{e}_5 = \mathbf{e}_6, \quad S^T \mathbf{e}_6 = \mathbf{e}_1$$

and $S\mathbf{e}_1 = \mathbf{e}_6$, $S\mathbf{e}_2 = \mathbf{e}_1$, $S\mathbf{e}_3 = \mathbf{e}_2$, $S\mathbf{e}_4 = \mathbf{e}_3$, $S\mathbf{e}_5 = \mathbf{e}_4$, $S\mathbf{e}_6 = \mathbf{e}_5$. Consider the Hadamard matrix

$$U = \frac{1}{\sqrt{2}} \begin{pmatrix} 1 & 1 \\ 1 & -1 \end{pmatrix}.$$

Apply $S^T(I_3 \otimes U)$ and $S(I_3 \otimes U)$ to the standard basis. Discuss.

Problem 4.   (i) We know that if A and B normal matrices, then $A \otimes B$ is normal. Can we find two matrices C and D which are nonnormal but $C \otimes D$ is normal?

(ii) Consider the invertible nonnormal 2×2 matrix M

$$M = \begin{pmatrix} 1 & 1 \\ 0 & 1 \end{pmatrix} \Rightarrow M^{-1} = \begin{pmatrix} 1 & -1 \\ 0 & 1 \end{pmatrix}.$$

Is the matrix $M \otimes M^{-1}$ nonnormal?

Problem 5. Let A, B be invertible matrices over $\mathbb{R}$. Then

$$(AB)^{-1} = B^{-1}A^{-1} \qquad (A \otimes B)^{-1} = A^{-1} \otimes B^{-1}.$$

Find the conditions on 2×2 matrices over $\mathbb{R}$ such that $A^{-1} \otimes B^{-1} = B^{-1} \otimes A^{-1}$.

Problem 6. Consider the set of 2×2 matrices

$$A = \begin{pmatrix} 1 & 1 \\ 0 & 0 \end{pmatrix}, \quad B = \begin{pmatrix} 0 & 1 \\ 0 & 1 \end{pmatrix}, \quad C = \begin{pmatrix} 0 & 0 \\ 1 & 1 \end{pmatrix}, \quad D = \begin{pmatrix} 1 & 0 \\ 1 & 0 \end{pmatrix}.$$

(i) Are the matrices linearly independent?
(ii) Are the matrices $A \otimes A$, $B \otimes B$, $C \otimes C$, $D \otimes D$ linearly independent?
(iii) Show that the 2×2 matrices B and D are similar. Show that the matrices $B \otimes D$ and $D \otimes B$ are similar.

Problem 7. Let $\oplus$ be the direct sum.
(i) Let A be an 2×2 matrix over $\mathbb{C}$. Can $A \otimes A$ be written as

$$A \otimes A = B_1 \oplus B_2 + C_1 \oplus C_2 + D_1 \oplus D_2$$

where B_1, B_2 are 2×2 matrices, C_1, D_2 are 1×1 matrices and C_2, D_1 are 3×3 matrices?
(ii) Let A, B, C, D be $n \times n$ matrices. Is

$$(A \oplus B) \otimes (C \oplus D) = (A \otimes (C \oplus D)) \oplus (B \otimes (C \oplus D))?$$

(iii) Let A, B be 2×2 matrices over $\mathbb{R}$. Can $A \otimes B$ be written as

$$A \otimes B = C \oplus D + 1 \oplus E + F \oplus 1?$$

Here C, D are 2×2 matrices over $\mathbb{R}$ and E, F are 3×3 matrices over $\mathbb{R}$.

Problem 8. Let $J = i\sigma_2$, where σ_2 is the second Pauli spin matrix.
(i) Show that any 2×2 matrix $A \in SL(2, \mathbb{C})$ satisfies $A^T J A = J$, where A^T is the transpose of A.
(ii) Let A be a 2×2 matrix with $A \in SL(2, \mathbb{C})$. Show that

$$(A \otimes A)^T (J \otimes J)(A \otimes A) = J \otimes J \text{ and } (A \oplus A)^T (J \oplus J)(A \oplus A) = J \oplus J.$$

Problem 9. (i) Find the conditions on the 2×2 matrices A and B such that

$$(A \otimes B) \otimes (A \otimes B) = (A \otimes A) \otimes (B \otimes B).$$

(ii) Let A, B be 2×2 matrices. Find the conditions on A and B such that

$$[A \otimes B, B \otimes A] = 0_4 \iff (AB) \otimes (BA) = (BA) \otimes (AB).$$

Problem 10.   Consider the 4×4 matrix

$$M = \begin{pmatrix} 16 & 3 & 2 & 13 \\ 5 & 10 & 11 & 8 \\ 9 & 6 & 7 & 12 \\ 4 & 15 & 14 & 1 \end{pmatrix}.$$

The matrix M is a so-called *magic square*, since its row sums, column sums, principal diagonal sum, principal counter diagonal sums are equal. Is $M \otimes M$ a magic square? Prove or disprove. Find the eigenvalues of M.

Problem 11.   Let $\{ \mathbf{e}_j \}_{1 \leq j \leq N}$ be the standard basis in $\mathbb{C}$ and $S = \{1, \dots, N\}$ be the index set. For each ordered index subset $I = \{j_1, j_2, \dots, j_p\} \subset S$, one defines $\mathbf{e}_I$ as

$$\mathbf{e}_I = \sum_{\sigma \in S_p} (\mathrm{sgn}(\sigma)) \mathbf{e}_{j_{\sigma(1)}} \otimes \mathbf{e}_{j_{\sigma(2)}} \otimes \cdots \otimes \mathbf{e}_{j_{\sigma(p)}}.$$

Let $n = 3$ and $p = 2$. Find all $\mathbf{e}_I$'s.

Problem 12.   Let A, B be 2×2 matrices over $\mathbb{C}$.
(i) Assume that $A \otimes B = B \otimes A$. Can we conclude that $AB = BA$? Prove or disprove.
(ii) Assume that $AB = BA$. Can be conclude that $A \otimes B = B \otimes A$?

Problem 13.   Let

$$T = \sum_{j,k,\ell=0}^{1} t_{jk\ell} \mathbf{v}_j \otimes \mathbf{v}_k \otimes \mathbf{v}_\ell.$$

(i) Assume that

$$\mathbf{v}_0 = \begin{pmatrix} 1 \\ 0 \end{pmatrix}, \qquad \mathbf{v}_1 = \begin{pmatrix} 0 \\ 1 \end{pmatrix}.$$

Find the coefficients $t_{jk\ell}$ such that

$$T = \frac{1}{\sqrt{2}} (1 \ 0 \ 0 \ 0 \ 0 \ 0 \ 0 \ 1)^T.$$

(ii) Assume that

$$\mathbf{v}_0 = \frac{1}{\sqrt{2}} \begin{pmatrix} 1 \\ 1 \end{pmatrix}, \qquad \mathbf{v}_1 = \frac{1}{\sqrt{2}} \begin{pmatrix} 1 \\ -1 \end{pmatrix}.$$

Find the coefficients $t_{jk\ell}$ such that

$$T = \frac{1}{\sqrt{2}} (1 \ 0 \ 0 \ 0 \ 0 \ 0 \ 0 \ 1)^T.$$

Problem 14.   (i) Consider the vectors

$$\mathbf{u} = \begin{pmatrix} u_1 \\ u_2 \\ u_3 \end{pmatrix}, \qquad \mathbf{v} = \begin{pmatrix} v_1 \\ v_2 \end{pmatrix}$$

in $\mathbb{C}^3$ and $\mathbb{C}^2$, respectively. Find the 6×6 permutation matrix P such that $P(\mathbf{u} \otimes \mathbf{v}) = \mathbf{v} \otimes \mathbf{u}$.

(ii) Find all non-trivial solutions of the equation

$$\begin{pmatrix} x_1 \\ x_2 \end{pmatrix} \otimes \begin{pmatrix} y_1 \\ y_2 \\ y_3 \end{pmatrix} = \begin{pmatrix} y_1 \\ y_2 \\ y_3 \end{pmatrix} \otimes \begin{pmatrix} x_1 \\ x_2 \end{pmatrix}.$$

From the equation we obtain the four conditions

$$x_1 y_2 = y_1 x_2, \quad x_1 y_3 = y_2 x_1, \quad x_2 y_1 = y_2 x_2, \quad x_2 y_2 = y_3 x_1.$$

A case study has to be done starting with $x_1 = 0$ and $x_1 \neq 0$.

(iii) Consider the standard basis $\mathbf{e}_1$, $\mathbf{e}_2$, $\mathbf{e}_3$ in $\mathbb{R}^3$. Find a 8×8 permutation matrix T such that

$$T(\mathbf{e}_1 \otimes \mathbf{e}_2 \otimes \mathbf{e}_3) = \mathbf{e}_3 \otimes \mathbf{e}_2 \otimes \mathbf{e}_1.$$

Problem 15. Consider the invertible $2n \times 2n$ matrix

$$C = i\sigma_2 \otimes I_n \equiv \begin{pmatrix} 0_n & I_n \\ -I_n & 0_n \end{pmatrix}.$$

The defining equation for an element $S \in Sp(2n)$ is given by

$$(S^{-1})^* = S = C(S^{-1})^T C^{-1}.$$

Find the inverse of the matrix C.

Problem 16. Show that

$$\begin{pmatrix} 0 & 1 \\ 1 & 0 \end{pmatrix} \otimes \begin{pmatrix} 1 & 0 \\ 0 & 1 \end{pmatrix} = \begin{pmatrix} 0 \\ 0 \\ 1 \end{pmatrix} \oplus (1 \quad 0 \quad 0) + (0 \quad 0 \quad 1) \oplus \begin{pmatrix} 1 \\ 0 \\ 0 \end{pmatrix}$$

where $\oplus$ denotes the direct sum.

Problem 17. Let $\sigma_0 = I_2$, σ_1, σ_2, σ_3 be the Pauli spin matrices. These matrices form an orthogonal basis in the Hilbert space of 2×2 matrices.

(i) Show that any 2×2 matrix can be written as linear combination

$$\sum_{j=0}^{3} c_j \sigma_j, \quad c_j \in \mathbb{C}.$$

Show that any 4×4 matrix can be written as linear combination

$$\sum_{j_1=0}^{3} \sum_{j_2=0}^{3} c_{j_1 j_2} \sigma_{j_1} \otimes \sigma_{j_2}, \quad c_{j_1 j_2} \in \mathbb{C}.$$

Show that any $2^n \times 2^n$ matrix can be written as linear combination

$$\sum_{j_1=0}^{3} \sum_{j_2=0}^{3} \cdots \sum_{j_n=0}^{3} c_{j_1 j_2 \ldots j_n} \sigma_{j_1} \otimes \sigma_{j_2} \otimes \cdots \otimes \sigma_{j_n}, \quad c_{j_1 j_2 \ldots j_n} \in \mathbb{C}.$$

(ii) Consider the 256 sixteen times sixteen matrices

$$T_{j_0 j_1 j_2 j_3} := \sigma_{j_0} \otimes \sigma_{j_1} \otimes \sigma_{j_2} \otimes \sigma_{j_3}$$

where $j_0, j_1, j_2, j_3 \in \{0, 1, 2, 3\}$. Let

$$|\psi\rangle = \frac{1}{4} (1 \quad 1 \quad 1 \quad 1 \quad 1 \quad 1 \quad 1 \quad 1 \quad 1 \quad 1 \quad 1 \quad 1 \quad 1 \quad 1 \quad 1 \quad 1)^T$$

Find $\langle \psi | T_{j_0 j_1 j_2 j_3} | \psi \rangle = t_{j_0 j_1 j_2 j_3}$.

Problem 18. Let A be an invertible $n \times n$ matrix with $A^2 = I_n$. Consider

$$X_1 = I_n \otimes A \otimes A^{-1}, \quad X_2 = A \otimes I_n \otimes A^{-1}, \quad X_3 = A \otimes A^{-1} \otimes I_n.$$

(i) Find X_1^2, X_2^2, X_3^2, $X_1 X_2$, $X_1 X_3$, $X_2 X_3$.
(ii) Find the commutators $[X_1, X_2]$, $[X_2, X_3]$, $[X_3, X_2]$.

Problem 19. Let A, B be given nonzero 2×2 matrices. Let X be a 2×2 matrix. Find the solution of the equation $X \otimes A = B \otimes X$.

Problem 20. Let A be an $m \times m$ matrix and B be an $n \times n$ matrix. The *Kronecker sum* is defined by

$$A \oplus_K B := A \otimes I_n + I_m \otimes B.$$

Is $(A \oplus_K B) \otimes C = (A \otimes C) \oplus_K (B \otimes C)$? Prove or disprove.

Problem 21. Let A be $m \times m$ over $\mathbb{C}$, and B be $n \times n$ over $\mathbb{C}$. Solve the equation

$$A \otimes B = A \otimes I_n + I_m \otimes B.$$

Obviously $A = 0_m$, $B = 0_n$ is a solution. Are there nonzero solutions? Could be the vec-operator of any use?

Problem 22. Let A be an 2×2 matrix over $\mathbb{C}$. Show that the condition on A such that $A \otimes I_2 = I_2 \otimes A$ is given by $a_{12} = a_{21} = 0$, $a_{11} = a_{22}$.

Problem 23. (i) Let $\mathbf{u}, \mathbf{v}$ be normalized column vectors in $\mathbb{C}^n$ with $\mathbf{u}^* \mathbf{v} = 0$. Show that $\mathbf{u} \otimes \mathbf{v} - \mathbf{v} \otimes \mathbf{u}$ cannot be written as

$$\mathbf{u} \otimes \mathbf{v} - \mathbf{v} \otimes \mathbf{u} = \mathbf{a} \otimes \mathbf{b}$$

where $\mathbf{a}, \mathbf{b} \in \mathbb{C}^n$. This problem plays a role for *entanglement*.
(ii) Let $\mathbf{u}, \mathbf{v} \in \mathbb{C}^n$ and linearly independent. Can one find $\mathbf{x}, \mathbf{y} \in \mathbb{C}^n$ such that

$$\mathbf{x} \otimes \mathbf{y} = \mathbf{u} \otimes \mathbf{v} - \mathbf{v} \otimes \mathbf{u}?$$

(iii) Let A, B be nonzero $n \times n$ matrices over $\mathbb{C}$. Assume that A and B are linearly independent. Show that $A \otimes B - B \otimes A$ cannot be written as $A \otimes B - B \otimes A = X \otimes Y$, where X, Y are $n \times n$ matrices over $\mathbb{C}$.

Problem 24. Let $n \in \mathbb{N}$. The *Fibonacci numbers* are provided by the entries of the matrix

$$\begin{pmatrix} 1 & 1 \\ 1 & 0 \end{pmatrix}^n, \qquad n = 1, 2, \dots.$$

Find the sequence provided by the 4×4 matrix

$$\left(\begin{pmatrix} 1 & 1 \\ 1 & 0 \end{pmatrix} \otimes \begin{pmatrix} 1 & 1 \\ 1 & 0 \end{pmatrix} \right)^n, \qquad n = 1, 2, \dots.$$

Compare with the sequence of the Fibonacci numbers. Discuss.

Problem 25. Let A be an $n \times n$ matrix over $\mathbb{R}$. The matrix A is called symmetric if $A = A^T$. If the matrix A is symmetric about the 'northeast-to-south-west' diagonal, i.e. $a_{j,k} = a_{n-k+1,n-j+1}$ it is called persymmetric. Let X, Y be $n \times n$ matrices over $\mathbb{R}$ which are symmetric and persymmetric. Show that $X \otimes Y$ is again symmetric and persymmetric.

Problem 26. Let A be an $n \times n$ matrix over $\mathbb{C}$ and B be an $m \times m$ matrix over $\mathbb{C}$. Assume that $A \otimes B$ is positive semidefinite. Can we conclude that A and B are positive semidefinite?

Problem 27. Let A, B be $n \times n$ positive semidefinite matrices. Show that

$$\mathrm{tr}(A \otimes B) \leq \frac{1}{4}(\mathrm{tr}(A) + \mathrm{tr}(B))^2, \quad \mathrm{tr}(A \otimes B) \leq \frac{1}{2}\mathrm{tr}(A \otimes A + B \otimes B).$$

Problem 28. Let U be a 2×2 unitary matrix and I_2 be the 2×2 identity matrix. Show that the 4×4 matrix

$$V = \begin{pmatrix} 0 & 0 \\ 0 & 1 \end{pmatrix} \otimes U + \begin{pmatrix} e^{i\alpha} & 0 \\ 0 & 0 \end{pmatrix} \otimes I_2, \qquad \alpha \in \mathbb{R}$$

is unitary. Utilize $UU^* = I_2$ and

$$\begin{pmatrix} 0 & 0 \\ 0 & 1 \end{pmatrix} \begin{pmatrix} e^{i\alpha} & 0 \\ 0 & 0 \end{pmatrix} = \begin{pmatrix} e^{i\alpha} & 0 \\ 0 & 0 \end{pmatrix} \begin{pmatrix} 0 & 0 \\ 0 & 1 \end{pmatrix} = \begin{pmatrix} 0 & 0 \\ 0 & 0 \end{pmatrix}.$$

Problem 29. Let A, B be $n \times n$ matrices over $\mathbb{R}$. Let $\mathbf{u}$, $\mathbf{v}$ be column vectors in $\mathbb{R}^n$. Show that

$$A(\mathbf{u} \otimes \mathbf{v}^T)B \equiv (A\mathbf{u}) \otimes (\mathbf{v}^T B)?$$

It can also be written as $A(\mathbf{u}\mathbf{v}^T)B \equiv (A\mathbf{u})(\mathbf{v}^T B)$ since matrix multiplication is associative.

Problem 30. Consider the invertible 4×4 matrix

$$T = \begin{pmatrix} 1 & 0 & 0 & 1 \\ 1 & 0 & 0 & -1 \\ 0 & 1 & 1 & 0 \\ 0 & -i & i & 0 \end{pmatrix} \Rightarrow T^{-1} = \frac{1}{2}\begin{pmatrix} 1 & 1 & 0 & 0 \\ 0 & 0 & 1 & i \\ 0 & 0 & 1 & -i \\ 1 & -1 & 0 & 0 \end{pmatrix}.$$

Let

$$R(\alpha) = \begin{pmatrix} \cos(\alpha) & -\sin(\alpha) \\ \sin(\alpha) & \cos(\alpha) \end{pmatrix}.$$

Show that $T(R(\alpha) \otimes R(\beta)^T)T^{-1}$ is given by

$$\begin{pmatrix} \cos(\alpha - \beta) & 0 & 0 & i\sin(\alpha - \beta) \\ 0 & \cos(\alpha + \beta) & -\sin(\alpha + \beta) & 0 \\ 0 & \sin(\alpha + \beta) & \cos(\alpha + \beta) & 0 \\ i\sin(\alpha - \beta) & 0 & 0 & \cos(\alpha - \beta) \end{pmatrix}.$$

Problem 31. Let $\mathbf{v}$ be a normalized (column) vector in $\mathbb{R}^2$. Is the matrix

$$I_2 - \mathbf{v} \otimes \mathbf{v}^T$$

invertible?

Chapter 4

Traces, Determinants and Hyperdeterminants

A function on $n \times n$ matrices $\det : \mathbb{C}^{n \times n} \to \mathbb{C}$ is called a determinant function if and only if it satisfies the following conditions:
1) det is linear in each row if the other rows of the matrix are held fixed.
2) If the $n \times n$ matrix A has two identical rows then $\det(A) = 0$.
3) If I_n is the $n \times n$ identity matrix, then $\det(I_n) = 1$.
Let A, B be $n \times n$ matrices and $c \in \mathbb{C}$. Then we have

$$\det(AB) = \det(A)\det(B), \qquad \det(cA) = c^n \det(A).$$

The determinant of A is the product of the eigenvalues of A, $\det(A) = \lambda_1 \cdot \lambda_2 \cdot \ldots \cdot \lambda_n$. Let A be an $n \times n$ matrix. Then the *trace* is defined as

$$\operatorname{tr}(A) := \sum_{j=1}^{n} a_{jj}.$$

The trace is independent of the underlying basis. The trace is the sum of the eigenvalues (counting multiplicities) of A, i.e.

$$\operatorname{tr}(A) = \sum_{j=1}^{n} \lambda_j.$$

Let A, B, C be $n \times n$ matrices. Then (*cyclic invariance*)

$$\operatorname{tr}(ABC) = \operatorname{tr}(CAB) = \operatorname{tr}(BCA).$$

The trace and determinant of a square matrix A are related by the identity

$$\det(\exp(A)) \equiv \exp(\operatorname{tr}(A)).$$

Problem 1.   Consider the *Hadamard matrix*

$$U = \frac{1}{\sqrt{2}} \begin{pmatrix} 1 & 1 \\ 1 & -1 \end{pmatrix}.$$

Find $\mathrm{tr}(U)$, $\mathrm{tr}(U^2)$, $\det(U)$ and $\det(U^2)$.

Solution 1.   We obtain $\mathrm{tr}(U) = 0$, $\mathrm{tr}(U^2) = 2$, $\det(U) = -1$, $\det(U^2) = 1$. Can the matrix U be reconstructed from this information?

Problem 2.   Consider the 2×2 nonnormal matrix

$$A = \begin{pmatrix} 0 & 1 \\ 0 & 0 \end{pmatrix}.$$

Can we find an invertible 2×2 matrix Q such that $Q^{-1}AQ$ is a diagonal matrix?

Solution 2.   The answer is no. Let

$$\tilde{A} = Q^{-1}AQ, \qquad \tilde{A} = \begin{pmatrix} a & 0 \\ 0 & b \end{pmatrix}$$

where $a, b \in \mathbb{C}$. Taking the trace of $\tilde{A}$ we find $a + b = \mathrm{tr}(Q^{-1}AQ) = \mathrm{tr}(A) = 0$. Taking the determinant of $\tilde{A}$ we obtain

$$ab = \det(Q^{-1}AQ) = \det(Q^{-1})\det(A)\det(Q) = \det(A) = 0.$$

Thus from $a + b = 0$ and $ab = 0$ we find $a = b = 0$. It follows that $Q\tilde{A}Q^{-1} = A$. Therefore A is the zero matrix which contradicts the assumption for the matrix A. Thus no invertible Q can be found. This means that A is not diagonalizable.

Problem 3.   Let A be a 2×2 matrix over $\mathbb{R}$. Assume that $\mathrm{tr}(A) = 0$ and $\mathrm{tr}(A^2) = 0$. Can we conclude that A is the 2×2 zero matrix?

Solution 3.   No we cannot conclude that A is the 2×2 zero matrix. The nonnormal matrix

$$A = \begin{pmatrix} 0 & 0 \\ 1 & 0 \end{pmatrix}$$

satisfies $\mathrm{tr}(A) = 0$ and $\mathrm{tr}(A^2) = 0$. What happens if we also assume that A is normal?

Problem 4.   For an integer $n \geq 3$, let $\theta := 2\pi/n$. Find the determinant of the $n \times n$ matrix $A + I_n$, where the matrix $A = (a_{jk})$ has the entries $a_{jk} = \cos(j\theta + k\theta)$ for all $j, k = 1, \ldots, n$.

Solution 4.   The determinant of a square matrix is the product of its eigenvalues. We compute the determinant of $I_n + A$ by calculating the eigenvalues of $I_n + A$. The eigenvalues of $I_n + A$ are obtained by adding 1 to each of the eigenvalues of A. Thus we calculate the eigenvalues of A and then add 1. We show that the eigenvalues of A are $n/2, -n/2, 0, \dots, 0$, where 0 occurs with multiplicity $n-2$. We define column vectors $\mathbf{v}^{(m)}$, $0 \le m \le n-1$, componentwise by $v_k^{(m)} = e^{ikm\theta}$. We form a matrix from the column vectors $\mathbf{v}^{(m)}$. Its determinant is a Vandermonde product and hence is nonzero. Thus the vectors $\mathbf{v}^{(m)}$ form a basis in $\mathbb{C}^n$. Since $\cos(z) \equiv (e^{iz} + e^{-iz})/2$ for any $z \in \mathbb{C}$ we obtain

$$(A\mathbf{v}^{(m)})_j = \sum_{k=1}^{n} \cos(j\theta + k\theta)e^{ikm\theta} = \frac{e^{ij\theta}}{2}\sum_{k=1}^{n} e^{ik(m+1)\theta} + \frac{e^{-ij\theta}}{2}\sum_{k=1}^{n} e^{ik(m-1)\theta}$$

where $j = 1, \dots, n$. Since

$$\sum_{k=1}^{n} e^{ik\ell\theta} = 0$$

for integer ℓ unless $n|\ell$, we conclude that $A\mathbf{v}^{(m)} = \mathbf{0}$ for $m = 0$ and for $2 \le m \le n-1$. We also find

$$(A\mathbf{v}^{(1)})_j = \frac{n}{2}e^{-ij\theta} = \frac{n}{2}(\mathbf{v}^{(n-1)})_j, \qquad (A\mathbf{v}^{(n-1)})_j = \frac{n}{2}e^{ij\theta} = \frac{n}{2}(\mathbf{v}^{(1)})_j.$$

Thus

$$A(\mathbf{v}^{(1)} \pm \mathbf{v}^{(n-1)}) = \pm\frac{n}{2}(\mathbf{v}^{(1)} \pm \mathbf{v}^{(n-1)}).$$

Consequently

$$\{\mathbf{v}^{(0)}, \mathbf{v}^{(2)}, \mathbf{v}^{(3)}, \dots, \mathbf{v}^{(n-2)}, \mathbf{v}^{(1)} + \mathbf{v}^{(n-1)}, \mathbf{v}^{(1)} - \mathbf{v}^{(n-1)}\}$$

is a basis for $\mathbb{C}^n$ of eigenvectors of A with the corresponding eigenvalues. Since the determinant of $I_n + A$ is the product of $(1 + \lambda)$ over all eigenvalues λ of A, we obtain

$$\det(I_n + A) = (1 + n/2)(1 - n/2) = 1 - n^2/4.$$

Problem 5. Let α, β, γ, δ be real numbers. Is the 2×2 matrix

$$U = e^{i\alpha} \begin{pmatrix} e^{-i\beta/2} & 0 \\ 0 & e^{i\beta/2} \end{pmatrix} \begin{pmatrix} \cos(\gamma/2) & -\sin(\gamma/2) \\ \sin(\gamma/2) & \cos(\gamma/2) \end{pmatrix} \begin{pmatrix} e^{-i\delta/2} & 0 \\ 0 & e^{i\delta/2} \end{pmatrix}$$

unitary? What the determinant of U?

Solution 5. Each of the three matrices on the right-hand side is unitary and $e^{i\alpha}$ is unitary. The product of two unitary matrices is again a unitary matrix. Thus U is unitary. The determinant of each of the three matrices on the right-hand side is 1. Thus $\det(U) = e^{2i\alpha}$.

Problem 6. Let A and B be two $n \times n$ matrices over $\mathbb{C}$. If there exists a nonsingular $n \times n$ matrix X such that $A = XBX^{-1}$, then A and B are said to be *similar matrices*. Show that the spectra (eigenvalues) of two similar matrices are equal.

Solution 6.   We have

$$\det(A - \lambda I_n) = \det(XBX^{-1} - X\lambda I_n X^{-1}) = \det(X(B - \lambda I_n)X^{-1})$$
$$= \det(X)\det(B - \lambda I_n)\det(X^{-1})$$
$$= \det(B - \lambda I_n).$$

Problem 7.   Let A and B be $n \times n$ matrices over $\mathbb{C}$. Show that the matrices AB and BA have the same set of eigenvalues.

Solution 7.   Consider first the case that A is invertible. Then we have $AB = A(BA)A^{-1}$. Thus AB and BA are similar and therefore have the same set of eigenvalues. If A is singular we apply the *continuity argument*: If A is singular, consider $A + \epsilon I_n$. We choose $\delta > 0$ such that $A + \epsilon I_n$ is invertible for all ϵ, $0 < \epsilon < \delta$. Thus $(A + \epsilon I_n)B$ and $B(A + \epsilon I_n)$ have the same set of eigenvalues for every $\epsilon \in (0, \delta)$. We equate their characteristic polynomials to obtain

$$\det(\lambda I_n - (A + \epsilon I_n)B) = \det(\lambda I_n - B(A + \epsilon I_n)), \quad 0 < \epsilon < \delta.$$

Since both sides are analytic functions of ϵ we find with $\epsilon \to 0^+$ that

$$\det(\lambda I_n - AB) = \det(\lambda I_n - BA).$$

Problem 8.   Let A, B be $n \times n$ matrices. Assume that $[A, B] = A$. What can be said about the trace of A?

Solution 8.   Since $\mathrm{tr}([A, B]) = 0$ we have $\mathrm{tr}(A) = 0$.

Problem 9.   An $n \times n$ matrix A is called *reducible* if there is a permutation matrix P such that

$$P^T A P = \begin{pmatrix} B & C \\ 0 & D \end{pmatrix}$$

where B and D are square matrices of order at least 1. An $n \times n$ matrix A is called *irreducible* if it is not reducible. Show that the $n \times n$ *primary permutation matrix*

$$A := \begin{pmatrix} 0 & 1 & 0 & \cdots & 0 \\ 0 & 0 & 1 & \cdots & 0 \\ \vdots & \vdots & \vdots & \ddots & \vdots \\ 0 & 0 & 0 & \cdots & 1 \\ 1 & 0 & 0 & \cdots & 0 \end{pmatrix}$$

is irreducible.

Solution 9.   Suppose the matrix A is reducible. Let

$$P^T A P = J_1 \oplus J_2 \oplus \cdots \oplus J_k, \quad k \geq 2$$

where P is some permutation matrix and the J_j are irreducible matrices of order $< n$. Here $\oplus$ denotes the direct sum. The rank of $A - I_n$ is $n - 1$ since

$\det(A - I_n) = 0$ and the submatrix of size $n - 1$ by deleting the last row and the last column from $A - I_n$ is nonsingular. It follows that

$$\text{rank}(P^T AP - I_n) = \text{rank}(P^T(A - I_n)P) = n - 1.$$

By using the above decomposition, we obtain

$$\text{rank}(P^T AP - I_n) = \sum_{j=1}^{k} \text{rank}(J_j - I_n) \leq (n - k) < (n - 1).$$

This is a contradiction. Thus A is irreducible.

Problem 10. We define a linear *bijection*, h, between $\mathbb{R}^4$ and $\mathbf{H}(2)$, the set of complex 2×2 hermitian matrices, by

$$(t, x, y, z) \to \begin{pmatrix} t + x & y - iz \\ y + iz & t - x \end{pmatrix}.$$

We denote the matrix on the right-hand side by H.
(i) Show that the matrix can be written as a linear combination of the Pauli spin matrices σ_1, σ_2, σ_3 and the identity matrix I_2.
(ii) Find the inverse map.
(iii) Calculate the determinant of 2×2 hermitian matrix H. Discuss.

Solution 10. (i) We have $H = tI_2 + x\sigma_3 + y\sigma_1 + z\sigma_2$.
(ii) Consider

$$\begin{pmatrix} a & c \\ c^* & b \end{pmatrix} = \begin{pmatrix} t + x & y - iz \\ y + iz & t - x \end{pmatrix}.$$

Compare the entries of the 2×2 matrix we obtain

$$t = \frac{a + b}{2}, \quad x = \frac{a - b}{2}, \quad y = \frac{c + c^*}{2}, \quad z = \frac{c^* - c}{2i}.$$

(iii) We obtain $\det(H) = t^2 - x^2 - y^2 - z^2$. This is the *Lorentz metric*. Let U be a unitary 2×2 matrix. Then $\det(UHU^*) = \det(H)$.

Problem 11. Let A be an $n \times n$ invertible matrix over $\mathbb{C}$. Assume that A can be written as $A = B + iB$, where B has only real coefficients. Show that B^{-1} exists and

$$A^{-1} = \frac{1}{2}(B^{-1} - iB^{-1}).$$

Solution 11. Since $A = (1 + i)B$ and $\det(A) \neq 0$ we have $(1 + i)^n \det(B) \neq 0$. Thus $\det(B) \neq 0$ and B^{-1} exists. We have

$$(1 + i)B\frac{1}{2}(1 - i)B^{-1} = I_n.$$

Problem 12. (i) Let A be an invertible matrix. Assume that $A = A^{-1}$. What are the possible values for $\det(A)$?

(ii) Let B be a skew-symmetric matrix over $\mathbb{R}$, i.e. $A^T = -B$ and of order $2n-1$. Show that $\det(B) = 0$.

(iii) Show that if a square matrix C is hermitian, i.e. $C^* = C$ then $\det(C)$ is a real number.

Solution 12.   (i) Since $1 = \det(I_n) = \det(AA^{-1}) = \det(A)\det(A^{-1})$ and by assumption $\det(A) = \det(A^{-1})$ we have $1 = (\det(A))^2$. Thus $\det(A)$ is either $+1$ or -1.

(ii) From $B^T = -B$ we obtain

$$\det(B^T) = \det(-B) = (-1)^{2n-1}\det(B) = -\det(B).$$

Since $\det(B) = \det(B^T)$ we obtain $\det(B) = -\det(B)$ and therefore $\det(B) = 0$.

(iii) Since C is hermitian we have $C^* = C$ or $\bar{C} = C^T$. Furthermore $\det(C) = \det(C^T)$. Thus

$$\det(\bar{C}) = \det(C^T) = \det(C).$$

Now if $\det(C) = x + iy$ then $\det(\bar{C}) = x - iy$ with $x, y \in \mathbb{R}$. Thus $x + iy = x - iy$ and therefore $y = 0$. Thus $\det(C)$ is a real number.

Problem 13.   (i) Let A, B are 2×2 matrices over $\mathbb{R}$. Let $H := A + iB$. Express $\det(H)$ as a sum of determinants.

(ii) Let A, B are 2×2 matrices over $\mathbb{R}$. Let $H := A + iB$. Assume that H is hermitian. Show that $\det(H) = \det(A) - \det(B)$.

Solution 13.   (i) Since

$$H = \begin{pmatrix} a_{11} + ib_{11} & a_{12} + ib_{12} \\ a_{21} + ib_{21} & b_{22} + ib_{22} \end{pmatrix}$$

we find

$$\det(H) = \det(A) - \det(B) + i\det\begin{pmatrix} a_{11} & a_{12} \\ b_{21} & b_{22} \end{pmatrix} + i\det\begin{pmatrix} b_{11} & b_{12} \\ a_{21} & a_{22} \end{pmatrix}.$$

(ii) Since H is hermitian, i.e. $H^* = H$ with $H^* = A^T - iB^T$ we have $A = A^T$ and $B = -B^T$. Thus $a_{12} = a_{21}$, $b_{11} = b_{22} = 0$ and $b_{12} = -b_{21}$. It follows that

$$\det\begin{pmatrix} a_{11} + ib_{11} & a_{12} + ib_{12} \\ a_{21} + ib_{21} & a_{22} + ib_{22} \end{pmatrix} = \det(A) - \det(B).$$

Problem 14.   (i) Let A, B, and C be $n \times n$ matrices. Calculate

$$\det\begin{pmatrix} A & 0_n \\ C & B \end{pmatrix}.$$

(ii) Let A, B, C, D be $n \times n$ matrices. Assume that $DC = CD$, i.e. C and D commute and $\det(D) \neq 0$. Consider the $(2n) \times (2n)$ matrix

$$M = \begin{pmatrix} A & B \\ C & D \end{pmatrix}.$$

Show that

$$\det(M) = \det(AD - BC). \tag{1}$$

We know that

$$\det \begin{pmatrix} U & 0_n \\ X & Y \end{pmatrix} = \det(U)\det(Y), \quad \det \begin{pmatrix} U & V \\ 0_n & Y \end{pmatrix} = \det(U)\det(Y) \tag{2}$$

where U, V, X, Y are $n \times n$ matrices.

Solution 14. (i) Obviously we find

$$\det \begin{pmatrix} A & 0_n \\ C & B \end{pmatrix} = \det(A)\det(B).$$

(ii) We have the identity

$$\begin{pmatrix} A & B \\ C & D \end{pmatrix} \begin{pmatrix} D & 0_n \\ -C & I_n \end{pmatrix} = \begin{pmatrix} AD - BC & B \\ CD - DC & D \end{pmatrix} = \begin{pmatrix} AD - BC & B \\ 0_n & D \end{pmatrix} \tag{3}$$

where we used that $CD = DC$. Applying the determinant to the right and left-hand side of (3) and using (2) and $\det(D) \neq 0$ we obtain identity (1).

Problem 15. Let A, B be $n \times n$ matrices. We have the identity

$$\det \begin{pmatrix} A & B \\ B & A \end{pmatrix} \equiv \det(A + B)\det(A - B).$$

Use this identity to calculate the determinant of the left-hand side using the right-hand side, where

$$A = \begin{pmatrix} 2 & 3 \\ 1 & 7 \end{pmatrix}, \quad B = \begin{pmatrix} 0 & 2 \\ 4 & 6 \end{pmatrix}.$$

Solution 15. We have

$$A + B = \begin{pmatrix} 2 & 5 \\ 5 & 13 \end{pmatrix}, \quad A - B = \begin{pmatrix} 2 & 1 \\ -3 & 1 \end{pmatrix}.$$

Therefore $\det(A+B) = 1$ and $\det(A-B) = 5$. Finally $\det(A+B)\det(A-B) = 5$.

Problem 16. (i) Let A, B be $n \times n$ matrices. Show that

$$\operatorname{tr}((A + B)(A - B)) \equiv \operatorname{tr}(A^2) - \operatorname{tr}(B^2).$$

(ii) Let A, B, C be $n \times n$ matrices. Show that $\operatorname{tr}([A, B]C) = \operatorname{tr}(A[B, C])$.

Solution 16. (i) We have

$$(A + B)(A - B) = A^2 - AB + BA - B^2 = A^2 + [B, A] - B^2.$$

Since $\text{tr}([B, A]) = 0$ and the trace is linear we obtain the identity.

(ii) Using *cyclic invariance* for the trace we have

$$\text{tr}([A, B]C) = \text{tr}((AB - BA)C) = \text{tr}(ABC) - \text{tr}(BAC)$$
$$= \text{tr}(ABC) - \text{tr}(ACB) = \text{tr}(ABC - ACB)$$
$$= \text{tr}(A[B, C]).$$

Problem 17.   Let A, B be $n \times n$ positive definite matrices. Show that $\text{tr}(AB) > 0$.

Solution 17.   Let U be a unitary matrix such that

$$U^* AU = D = \text{diag}(d_1, d_2, \ldots, d_n).$$

Obviously, d_1, d_2, $\ldots$, d_n are the eigenvalues of A. Then $\text{tr}(A) = \text{tr}(D)$ and

$$\text{tr}(AB) = \text{tr}(U^* AUU^* BU) = \text{tr}(DC)$$

where $C = U^* BU$. Now C is positive definite and therefore its diagonal entries c_{ii} are real and positive and $\text{tr}(C) = \text{tr}(B)$. The diagonal entries of DC are $d_i c_{ii}$ and therefore

$$\text{tr}(DC) = \sum_{i=1}^{n} d_i c_{ii} > 0.$$

Problem 18.   Let

$$A = \begin{pmatrix} \dfrac{1}{x_1 + y_1} & \dfrac{1}{x_1 + y_2} \\ \dfrac{1}{x_2 + y_1} & \dfrac{1}{x_2 + y_2} \end{pmatrix}$$

where we assume that $x_i + y_j \neq 0$ for $i, j = 1, 2$. Show that

$$\det(A) = \frac{(x_1 - x_2)(y_1 - y_2)}{(x_1 + y_1)(x_1 + y_2)(x_2 + y_1)(x_2 + y_2)}.$$

Solution 18.   We have

$$\det(A) = \frac{(x_2 + y_1)(x_1 + y_2) - (x_1 + y_1)(x_2 + y_2)}{(x_1 + y_1)(x_1 + y_2)(x_2 + y_1)(x_2 + y_2)}$$
$$= \frac{x_1 y_1 - x_1 y_2 + x_2 y_2 - x_2 y_1}{(x_1 + y_1)(x_1 + y_2)(x_2 + y_1)(x_2 + y_2)}$$
$$= \frac{(x_1 - x_2)(y_1 - y_2)}{(x_1 + y_1)(x_1 + y_2)(x_2 + y_1)(x_2 + y_2)}.$$

For the general case with the matrix

$$A = \begin{pmatrix} \dfrac{1}{x_1 + y_1} & \dfrac{1}{x_1 + y_2} & \dfrac{1}{x_1 + y_3} & \cdots & \dfrac{1}{x_1 + y_n} \\ \dfrac{1}{x_2 + y_1} & \dfrac{1}{x_2 + y_2} & \dfrac{1}{x_2 + y_3} & \cdots & \dfrac{1}{x_2 + y_n} \\ \vdots & \vdots & \vdots & \ddots & \vdots \\ \dfrac{1}{x_n + y_1} & \dfrac{1}{x_n + y_2} & \dfrac{1}{x_n + y_3} & \cdots & \dfrac{1}{x_n + y_n} \end{pmatrix}$$

we find the determinant (*Cauchy determinant*)

$$\det(A) = \frac{\prod_{i>j}^{n}(x_i - x_j)(y_i - y_j)}{\prod_{i,j=1}^{n}(x_i + y_j)}.$$

Problem 19. For a 3×3 matrix we can use the *rule of Sarrus* to calculate the determinant. Let

$$A = \begin{pmatrix} a_{11} & a_{12} & a_{13} \\ a_{21} & a_{22} & a_{23} \\ a_{31} & a_{32} & a_{33} \end{pmatrix}.$$

Write the first two columns again to the right of the matrix to obtain

$$\begin{pmatrix} a_{11} & a_{12} & a_{13} & | & a_{11} & a_{12} \\ a_{21} & a_{22} & a_{23} & | & a_{21} & a_{22} \\ a_{31} & a_{32} & a_{33} & | & a_{31} & a_{32} \end{pmatrix}.$$

Now look at the diagonals. The product of the diagonals sloping down to the right have a plus sign, the ones up to the left have a negative sign. This leads to the determinant

$$\det(A) = a_{11}a_{22}a_{33} + a_{12}a_{23}a_{31} + a_{13}a_{21}a_{32} - a_{31}a_{22}a_{13} - a_{32}a_{23}a_{11} - a_{33}a_{21}a_{12}.$$

(i) Use this rule to calculate the determinant of the *rotational matrix*

$$R(\theta) = \begin{pmatrix} \cos(\theta) & 0 & -\sin(\theta) \\ 0 & 1 & 0 \\ \sin(\theta) & 0 & \cos(\theta) \end{pmatrix}.$$

(ii) Use the rule to find the determinant of the matrix

$$A(\alpha) = \begin{pmatrix} 0 & \cos(\alpha) & \sin(\alpha) \\ \cos(\alpha) & \sin(\alpha) & 0 \\ 0 & \cos(\alpha) & \sin(\alpha) \end{pmatrix}.$$

Solution 19. (i) We have

$$\begin{pmatrix} \cos(\theta) & 0 & -\sin(\theta) & | & \cos(\theta) & 0 \\ 0 & 1 & 0 & | & 0 & 1 \\ \sin(\theta) & 0 & \cos(\theta) & | & \sin(\theta) & 0 \end{pmatrix}.$$

Thus $\det(R(\theta))$ is given by

$$\cos(\theta)\cos(\theta) + 0 + 0 - \sin(\theta)(-\sin(\theta)) - 0 - 0 = \cos^2(\theta) + \sin^2(\theta) = 1.$$

(ii) We find $\det(A(\alpha)) = 0$.

Problem 20. Let A, S be $n \times n$ matrices. Assume that S is invertible and assume that $S^{-1}AS = \rho S$, where $\rho \neq 0$. Show that A is invertible.

Solution 20. From $S^{-1}AS = \rho S$ we obtain $\det(S^{-1}AS) = \det(\rho S)$. Thus

$$\det(S^{-1})\det(A)\det(S) = \det(\rho S).$$

It follows that $\det(A) = \det(\rho S) = \rho^n \det(S)$. Since $\rho^n \neq 0$ and $\det(S) \neq 0$ it follows that $\det(A) \neq 0$ and therefore A^{-1} exists.

Problem 21. Let A be an invertible $n \times n$ matrix. Let $c = 2$. Can we find an invertible matrix S such that $SAS^{-1} = cA$?

Solution 21. From $SAS^{-1} = cA$ it follows that

$$\det(SAS^{-1}) = \det(cA) \implies \det(A) = c^n \det(A).$$

Since $\det(A) \neq 0$ we have $c^n = 1$. Thus such an S does not exist.

Problem 22. The determinant of an $n \times n$ *circulant matrix* is given by

$$\det \begin{pmatrix} a_1 & a_2 & a_3 & \cdots & a_n \\ a_n & a_1 & a_2 & \cdots & a_{n-1} \\ \vdots & \vdots & \vdots & \ddots & \vdots \\ a_3 & a_4 & a_5 & \cdots & a_2 \\ a_2 & a_3 & a_4 & \cdots & a_1 \end{pmatrix} = (-1)^{n-1} \prod_{j=0}^{n-1} \left(\sum_{k=1}^{n} \zeta^{jk} a_k \right) \tag{1}$$

where $\zeta := \exp(2\pi i/n)$. Find the determinant of the circulant $n \times n$ matrix

$$C = \begin{pmatrix} 1 & 4 & 9 & \cdots & n^2 \\ n^2 & 1 & 4 & \cdots & (n-1)^2 \\ \vdots & \vdots & \vdots & \ddots & \vdots \\ 9 & 16 & 25 & \cdots & 4 \\ 4 & 9 & 16 & \cdots & 1 \end{pmatrix}$$

using equation (1).

Solution 22. Applying (1) we have

$$\det(C) = (-1)^{n-1} \prod_{j=0}^{n-1} \left(\sum_{k=1}^{n} \zeta^{jk} k^2 \right).$$

Now

$$\sum_{k=1}^{n} k^2 x^k = \frac{n^2 x^{n+3} - (2n^2 + 2n - 1)x^{n+2} + (n^2 + 2n + 1)x^{n+1} - x^2 - x}{(x-1)^3}$$

for $x \neq 1$ and

$$\sum_{k=1}^{n} k^2 x^k = \frac{n(n+1)(2n+1)}{6}$$

for $x = 1$. For the product we have

$$\prod_{j=1}^{n-1}(\zeta^j - a) = (-1)^{n-1}\sum_{k=0}^{n-1} a_k = \begin{cases} (-1)^{n-1}\frac{a^n-1}{a-1} & \text{if } a \neq 1 \\ (-1)^{n-1}n & \text{if } a = 1. \end{cases}$$

It follows that

$$\prod_{j=1}^{n-1}\zeta^j = (-1)^{n-1}, \qquad \prod_{j=1}^{n-1}(\zeta^j - 1) = (-1)^{n-1}n$$

and

$$\prod_{j=1}^{n-1}\left(\zeta^j - \frac{n+2}{n}\right) = (-1)^{n-1}\frac{(n+2)^n - n^n}{2n^{n-1}}.$$

Finally

$$\det(C) = (-1)^{n-1}\frac{n^{n-2}(n+1)(2n+1)((n+2)^n - n^n)}{12}.$$

Problem 23. Let A be a nonzero 2×2 matrix over $\mathbb{R}$. Let B_1, B_2, B_3, B_4 be 2×2 matrices over $\mathbb{R}$ and assume that

$$\det(A + B_j) = \det(A) + \det(B_j) \qquad \text{for} \quad j = 1, 2, 3, 4.$$

Show that there exist real numbers c_1, c_2, c_3, c_4, not all zero, such that

$$c_1 B_1 + c_2 B_2 + c_3 B_3 + c_4 B_4 = 0_2. \tag{1}$$

Solution 23. Let

$$A = \begin{pmatrix} a_{11} & a_{12} \\ a_{21} & a_{22} \end{pmatrix}, \qquad B_j = \begin{pmatrix} b_{11}^{(j)} & b_{12}^{(j)} \\ b_{21}^{(j)} & b_{22}^{(j)} \end{pmatrix}$$

with $j = 1, 2, 3, 4$. If $\det(A + B_j) = \det(A) + \det(B_j)$, it follows that

$$a_{22}b_{11}^{(j)} - a_{21}b_{12}^{(j)} - a_{12}b_{21}^{(j)} + a_{11}b_{22}^{(j)} = 0. \tag{2}$$

Since A is a nonzero matrix the solution space to (1) for fixed j is a three-dimensional vector space. Any four vectors in a three-dimensional space are linearly dependent. Thus there must exist c_1, c_2, c_3, c_4, not all 0, for which equation (1) is true.

Problem 24. An $n \times n$ matrix Q is *orthogonal* if Q is real and $Q^T Q = Q^T Q = I_n$ i.e. $Q^{-1} = Q^T$.
(i) Find the determinant of an orthogonal matrix.
(ii) Let $\mathbf{u}$, $\mathbf{v}$ be two vectors in $\mathbb{R}^3$ and $\mathbf{u} \times \mathbf{v}$ denotes the *vector product* of $\mathbf{u}$ and $\mathbf{v}$

$$\mathbf{u} \times \mathbf{v} := \begin{pmatrix} u_2 v_3 - u_3 v_2 \\ u_3 v_1 - u_1 v_3 \\ u_1 v_2 - u_2 v_1 \end{pmatrix}.$$

Let Q be a 3×3 orthogonal matrix. Calculate $(Q\mathbf{u}) \times (Q\mathbf{v})$.

Solution 24. (i) We have $\det(Q) = \det(Q^T) = \det(Q^{-1}) = 1/\det(Q)$. Thus $(\det(Q))^2 = 1$ and therefore $\det(Q) = \pm 1$.
(ii) We have $(Q\mathbf{u}) \times (Q\mathbf{v}) = Q(\mathbf{u} \times \mathbf{v}) \det(Q)$.

Problem 25. (i) Calculate the determinant of the $n \times n$ matrix

$$A = \begin{pmatrix} 1 & 1 & 1 & \cdots & 1 & 1 \\ 1 & 0 & 1 & \cdots & 1 & 1 \\ 1 & 1 & 0 & \cdots & 1 & 1 \\ \vdots & \vdots & \vdots & \ddots & \vdots & \vdots \\ 1 & 1 & 1 & \cdots & 0 & 1 \\ 1 & 1 & 1 & \cdots & 1 & 0 \end{pmatrix}.$$

(ii) Find the determinant of the matrix

$$C = \begin{pmatrix} 0 & 1 & 1 & \cdots & 1 & 1 \\ 1 & 0 & 1 & \cdots & 1 & 1 \\ 1 & 1 & 0 & \cdots & 1 & 1 \\ \vdots & \vdots & \vdots & \ddots & \vdots & \vdots \\ 1 & 1 & 1 & \cdots & 0 & 1 \\ 1 & 1 & 1 & \cdots & 1 & 0 \end{pmatrix}.$$

Solution 25. (i) Subtracting the second row from the first row in A $(n \geq 2)$ we find the matrix

$$B = \begin{pmatrix} 0 & 1 & 0 & \cdots & 0 & 0 \\ 1 & 0 & 1 & \cdots & 1 & 1 \\ 1 & 1 & 0 & \cdots & 1 & 1 \\ \vdots & \vdots & \vdots & \ddots & \vdots & \vdots \\ 1 & 1 & 1 & \cdots & 0 & 1 \\ 1 & 1 & 1 & \cdots & 1 & 0 \end{pmatrix}.$$

Thus $\det(A) = \det(B)$. Let $s_n := \det(B)$. Now an expansion yields $s_n = -s_{n-1}$ with the initial value $s_1 = 1$ from the matrix A. It follows that

$$s_n = \det(A) = (-1)^{n-1}, \qquad n = 1, 2, \ldots$$

(ii) Using the result from (i) we find $\det(C) = (-1)^{n-1}(n-1)$.

Problem 26. Let A be a 2×2 matrix over $\mathbb{R}$ with $\det(A) \neq 0$. Is $(A^T)^{-1} = (A^{-1})^T$?

Solution 26. We have

$$A^{-1} = \frac{1}{\det(A)} \begin{pmatrix} a_{22} & -a_{12} \\ -a_{21} & a_{11} \end{pmatrix}, \qquad (A^{-1})^T = \frac{1}{\det(A)} \begin{pmatrix} a_{22} & -a_{21} \\ -a_{12} & a_{11} \end{pmatrix}$$

and

$$A^T = \begin{pmatrix} a_{11} & a_{21} \\ a_{12} & a_{22} \end{pmatrix}, \qquad (A^T)^{-1} = \frac{1}{\det(A)} \begin{pmatrix} a_{22} & -a_{21} \\ -a_{12} & a_{11} \end{pmatrix}.$$

Thus $(A^T)^{-1} = (A^{-1})^T$. This is also true for $n \times n$ matrices.

Problem 27. Find all 2×2 matrices A over $\mathbb{C}$ that satisfy the three conditions

$$\text{tr}(A) = 0, \quad A = A^*, \quad A^2 = I_2.$$

Solution 27. From the first condition we find $a_{11} = -a_{22}$. From the second condition we find $a_{12} = \bar{a}_{21}$ and that a_{11} and a_{22} are real. Inserting these two conditions into the third one we have

$$A^2 = \begin{pmatrix} a_{11} & a_{12} \\ \bar{a}_{12} & -a_{11} \end{pmatrix}^2 = \begin{pmatrix} a_{11}^2 + a_{12}\bar{a}_{12} & 0 \\ 0 & a_{11}^2 + a_{12}\bar{a}_{12} \end{pmatrix} = \begin{pmatrix} 1 & 0 \\ 0 & 1 \end{pmatrix}.$$

Thus $a_{11}^2 + a_{12}\bar{a}_{12} = 1$. Setting $a_{12} = r_{12}e^{i\phi}$ we have $a_{12}\bar{a}_{12} = r_{12}^2$ with $r_{12} \geq 0$. Thus $-1 \leq a_{11} \leq +1$ and $0 \leq r_{12} \leq 1$ with the constraint $a_{11}^2 + r_{12}^2 = 1$.

Problem 28. Let A be an $n \times n$ matrix with $A^2 = I_n$. Let B be a matrix with $AB = -BA$, i.e. $[A, B]_+ = 0_n$. Show that $\text{tr}(B) = 0$. Find $\text{tr}(A \otimes B)$.

Solution 28. We have

$$\text{tr}(B) = \text{tr}(I_n B) = \text{tr}(A^2 B) = -\text{tr}(ABA) = -\text{tr}(A^2 B) = -\text{tr}(B).$$

Thus $\text{tr}(B) = 0$. Using this result we have $\text{tr}(A \otimes B) = \text{tr}(A)\text{tr}(B) = 0$.

Problem 29. (i) Consider the two 2×2 matrices

$$A = \begin{pmatrix} a_{11} & 1 \\ a_{21} & 0 \end{pmatrix}, \qquad B = \begin{pmatrix} a_{11} & 0 \\ a_{21} & 1 \end{pmatrix}.$$

The first column of the matrices A and B agree, but the second column of the two matrices differ. Is $\det(A + B) = 2(\det(A) + \det(B))$?

Solution 29. We have $\det(A) = -a_{21}$ and $\det(B) = a_{11}$ and

$$A + B = \begin{pmatrix} 2a_{11} & 1 \\ 2a_{21} & 1 \end{pmatrix}$$

with $\det(A + B) = 2a_{11} - 2a_{21}$. Thus the equation holds.

Problem 30.   Let A, B be 2×2 matrices. Assume that $\det(A) = 0$ and $\det(B) = 0$. Can we conclude that $\det(A + B) = 0$?

Solution 30.   We cannot conclude that $\det(A+B) = 0$ in general. For example

$$A = \begin{pmatrix} 0 & 1 \\ 0 & -1 \end{pmatrix}, \quad B = \begin{pmatrix} 1 & 0 \\ 1 & 0 \end{pmatrix} \Rightarrow A + B = \begin{pmatrix} 1 & 1 \\ 1 & -1 \end{pmatrix}.$$

Thus $\det(A) = \det(B) = 0$, but $\det(A + B) = -2$.

Problem 31.   Consider the symmetric 3×3 matrix

$$A = \begin{pmatrix} n_1 & 1 & 0 \\ 1 & n_2 & 1 \\ 0 & 1 & n_3 \end{pmatrix}.$$

Find all positive integers n_1, n_2, n_3 such that $\det(A) = 1$ and A is positive definite.

Solution 31.   The five solutions are $\{ 131, \quad 312, \quad 221, \quad 122, \quad 213 \}$. Thus the number of solutions is 5, which is the *Catalan number* C_3. Consider the 4×4 matrix

$$B = \begin{pmatrix} n_1 & 1 & 0 & 0 \\ 1 & n_2 & 1 & 0 \\ 0 & 1 & n_3 & 1 \\ 0 & 0 & 1 & n_4 \end{pmatrix}$$

with $n_1, n_2, n_3, n_4 \in \mathbb{N}$ and show that there are 14 solutions with $C_4 = 14$.

Problem 32.   The oriented volume of an n-simplex in n-dimensional Euclidean space with vertices $\mathbf{v}_0, \mathbf{v}_1, \ldots, \mathbf{v}_n$ is given by

$$\frac{1}{n!} \det(S)$$

where S is the $n \times n$ matrix $S := (\mathbf{v}_1 - \mathbf{v}_0 \; \mathbf{v}_2 - \mathbf{v}_0 \; \cdots \; \mathbf{v}_{n-1} - \mathbf{v}_0 \; \mathbf{v}_n - \mathbf{v}_0)$. Thus each column of the $n \times n$ matrix is the difference between the vectors representing two vertices.
(i) Let

$$\mathbf{v}_0 = \begin{pmatrix} 0 \\ 0 \end{pmatrix}, \qquad \mathbf{v}_1 = \begin{pmatrix} 1 \\ 1 \end{pmatrix}, \qquad \mathbf{v}_2 = \begin{pmatrix} 1/2 \\ 1 \end{pmatrix}.$$

Find the oriented volume.
(ii) Let

$$\mathbf{v}_0 = \begin{pmatrix} 0 \\ 0 \\ 0 \end{pmatrix}, \qquad \mathbf{v}_1 = \begin{pmatrix} 1 \\ 0 \\ 0 \end{pmatrix}, \qquad \mathbf{v}_2 = \begin{pmatrix} 0 \\ 1 \\ 0 \end{pmatrix}, \qquad \mathbf{v}_3 = \begin{pmatrix} 0 \\ 0 \\ 1 \end{pmatrix}.$$

Find the oriented volume.

Solution 32. (i) We have

$$\frac{1}{2} \det \begin{pmatrix} 1 & 1/2 \\ 1 & 1 \end{pmatrix} = \frac{1}{4}.$$

(ii) We find

$$\frac{1}{6} \det \begin{pmatrix} 1 & 0 & 0 \\ 0 & 1 & 0 \\ 0 & 0 & 1 \end{pmatrix} = \frac{1}{6}.$$

Problem 33. The area A of a *triangle* given by the coordinates of its vertices

$$(x_0, y_0), \qquad (x_1, y_1), \qquad (x_2, y_2)$$

is

$$A = \frac{1}{2} \det \begin{pmatrix} x_0 & y_0 & 1 \\ x_1 & y_1 & 1 \\ x_2 & y_2 & 1 \end{pmatrix}.$$

(i) Let $(x_0, y_0) = (0,0)$, $(x_1, y_1) = (1,0)$, $(x_2, y_2) = (0,1)$. Find A.

(ii) A *tetrahedron* is a polyhedron composed of four triangular faces, three of which meet at each vertex. A tetrahedron can be defined by the coordinates of the vertices

$$(x_0, y_0, z_0), \qquad (x_1, y_1, z_1), \qquad (x_2, y_2, z_2), \qquad (x_3, y_3, z_3).$$

The volume V of the tetrahedron is given by (actually $|V|$)

$$V = \frac{1}{6} \det \begin{pmatrix} x_0 & y_0 & z_0 & 1 \\ x_1 & y_1 & z_1 & 1 \\ x_2 & y_2 & z_2 & 1 \\ x_3 & y_3 & z_3 & 1 \end{pmatrix}.$$

Let

$$(x_0, y_0, z_0) = (0,0,0), \quad (x_1, y_1, z_1) = (0,0,1),$$
$$(x_2, y_2, z_2) = (0,1,0), \quad (x_3, y_3, z_3) = (1,0,0).$$

Find the volume V.

(iii) Let

$$(x_0, y_0, z_0) = (+1,+1,+1), \quad (x_1, y_1, z_1) = (-1,-1,+1),$$
$$(x_2, y_2, z_2) = (-1,+1,-1), \quad (x_3, y_3, z_3) = (+1,-1,-1).$$

Find the volume V.

Solution 33. (i) We obtain

$$A = \frac{1}{2} \det \begin{pmatrix} 0 & 0 & 1 \\ 1 & 0 & 1 \\ 0 & 1 & 1 \end{pmatrix} = \frac{1}{2}.$$

(ii) We find

$$V = \frac{1}{6} \det \begin{pmatrix} 0 & 0 & 0 & 1 \\ 0 & 0 & 1 & 1 \\ 0 & 1 & 0 & 1 \\ 1 & 0 & 0 & 1 \end{pmatrix} = \frac{1}{6}.$$

(iii) We find

$$V = \frac{1}{6} \det \begin{pmatrix} 1 & 1 & 1 & 1 \\ -1 & -1 & 1 & 1 \\ -1 & 1 & -1 & 1 \\ 1 & -1 & -1 & 1 \end{pmatrix} = -\frac{8}{3}.$$

Problem 34. Let A, B be $n \times n$ matrices over $\mathbb{C}$. Assume that $\text{tr}(AB) = 0$.
(i) Can we conclude that $\text{tr}(AB^*) = 0$?
(ii) Consider the case that B is skew-hermitian.

Solution 34. (i) We cannot conclude that $\text{tr}(AB^*) = 0$ follows from $\text{tr}(AB) = 0$. For example, for $n = 2$ let

$$A = \begin{pmatrix} 0 & 1 \\ 0 & 0 \end{pmatrix}, \qquad B = \begin{pmatrix} 0 & 1 \\ 0 & 0 \end{pmatrix}.$$

Then we have $\text{tr}(AB) = 0$, but $\text{tr}(AB^*) = 1$.
(ii) If B is skew-hermitian we have $B^* = -B$. Thus $\text{tr}(AB^*) = -\text{tr}(AB) = 0$.

Problem 35. Let A, B be $n \times n$ matrices over $\mathbb{C}$. Is $\text{tr}(AB^*) = \text{tr}(A^*B)$?

Solution 35. This is not true in general. For example let

$$A = \begin{pmatrix} 2i & 0 \\ 0 & -i \end{pmatrix}, \qquad B = \begin{pmatrix} 3 & 0 \\ 0 & 4 \end{pmatrix}.$$

Then $\text{tr}(AB^*) = 2i$ and $\text{tr}(A^*B) = -2i$. In general we have $(\text{tr}(AB^*))^* = \text{tr}(A^*B)$. If $A = B$, then $\text{tr}(AA^*) = \text{tr}(A^*A)$.

Problem 36. Let

$$A = (\mathbf{a}_1, \mathbf{a}_2, \ldots, \mathbf{a}_{n-1}, \mathbf{u}), \qquad B = (\mathbf{a}_1, \mathbf{a}_2, \ldots, \mathbf{a}_{n-1}, \mathbf{v})$$

be $n \times n$ matrices, where the first $n - 1$ columns $\mathbf{a}_1, \ldots, \mathbf{a}_{n-1}$ are the same and for the last column $\mathbf{u} \neq \mathbf{v}$. Show that

$$\det(A + B) = 2^{n-1}(\det(A) + \det(B)).$$

Solution 36. This identity is true since determinants are multilinear (alternating) maps (Grassmann product) for their columns, i.e.

$$\begin{aligned} \det(A + B) &= \det(2\mathbf{a}_1, 2\mathbf{a}_2, \ldots, 2\mathbf{a}_{n-1}, \mathbf{u} + \mathbf{v}) \\ &= \det(2\mathbf{a}_1, 2\mathbf{a}_2, \ldots, 2\mathbf{a}_{n-1}, \mathbf{u}) + \det(2\mathbf{a}_1, \mathbf{a}_2, \ldots, 2\mathbf{a}_{n-1}, \mathbf{v}) \\ &= 2^{n-1} \det(\mathbf{a}_1, \mathbf{a}_2, \ldots, \mathbf{a}_{n-1}, \mathbf{u}) + 2^{n-1} \det(\mathbf{a}_1, \mathbf{a}_2, \ldots, \mathbf{a}_{n-1}, \mathbf{v}) \\ &= 2^{n-1} \det(A) + 2^{n-1} \det(B). \end{aligned}$$

As a consequence we have that if $\det(A) = \det(B) = 0$, then $\det(A + B) = 0$.

Problem 37. Find all 2×2 matrices g over $\mathbb{C}$ such that $\det(g) = 1$, $\eta g^* \eta = g^{-1}$, where η is the diagonal matrix $\eta = \mathrm{diag}(1, -1)$.

Solution 37. We have

$$g = \begin{pmatrix} \bar{u}_2 & u_1 \\ \bar{u}_1 & u_2 \end{pmatrix}, \quad \det(g) = u_2\bar{u}_2 - u_1\bar{u}_1 = 1$$

since

$$\eta g^* \eta = \begin{pmatrix} 1 & 0 \\ 0 & -1 \end{pmatrix} \begin{pmatrix} u_2 & u_1 \\ \bar{u}_1 & \bar{u}_2 \end{pmatrix} \begin{pmatrix} 1 & 0 \\ 0 & -1 \end{pmatrix} = \begin{pmatrix} u_2 & -u_1 \\ -\bar{u}_1 & \bar{u}_2 \end{pmatrix} = g^{-1}.$$

From $u_1 = r_1 e^{i\phi_1}$, $u_2 = r_2 e^{i\phi_2}$ we obtain $r_2^2 - r_1^2 = 1$.

Problem 38. An $n \times n$ *tridiagonal matrix* ($n \geq 3$) has nonzero elements only in the main diagonal, the first diagonal below this, and the first diagonal above the main diagonal. The determinant of an $n \times n$ tridiagonal matrix can be calculated by the recursive formula

$$\det(A) = a_{n,n} \det[A]_{\{1,\dots,n-1\}} - a_{n,n-1}a_{n-1,n} \det[A]_{\{1,\dots,n-2\}}$$

where $\det[A]_{\{1,\dots,k\}}$ denotes the k-th principal minor, that is, $[A]_{\{1,\dots,k\}}$ is the submatrix by the first k rows and columns of A. The cost of computing the determinant of a tridiagonal matrix using this recursion is linear in n, while the cost is cubic for a general matrix. Apply this recursion relation to calculate the determinant of the 4×4 matrix

$$A = \begin{pmatrix} 0 & 1 & 0 & 0 \\ 1 & 1 & 2 & 0 \\ 0 & 2 & 2 & 3 \\ 0 & 0 & 3 & 3 \end{pmatrix}.$$

Solution 38. We have

$$\det(A) = 3 \det \begin{pmatrix} 0 & 1 & 0 \\ 1 & 1 & 2 \\ 0 & 2 & 2 \end{pmatrix} - 9 \det \begin{pmatrix} 0 & 1 \\ 1 & 1 \end{pmatrix} = 3 \left(2 \det \begin{pmatrix} 0 & 1 \\ 1 & 1 \end{pmatrix} \right) + 9 = 3.$$

Problem 39. Consider the symmetric 3×3 matrix

$$A(\alpha) = \begin{pmatrix} \alpha & 1 & 1 \\ 1 & \alpha & 1 \\ 1 & 1 & \alpha \end{pmatrix}, \quad \alpha \in \mathbb{R}.$$

(i) Find the maxima and minima of the function $f(\alpha) = \det(A(\alpha))$.

(ii) For which values of α is the matrix noninvertible?

Solution 39.   (i) We obtain $f(\alpha) = \det(A(\alpha)) = \alpha^3 - 3\alpha + 2$. Therefore

$$\frac{df(\alpha)}{d\alpha} = 3\alpha^2 - 3.$$

From $3\alpha^2 - 3 = 0$ we find $\alpha^2 = 1$ or $\alpha_1 = 1$, $\alpha_2 = -1$. Since $d^2 f(\alpha)/d\alpha^2 = 6\alpha$ we have $d^2 f(\alpha = \alpha_1)/d\alpha^2 = 6$ and $d^2 f(\alpha = \alpha_2)/d\alpha^2 = -6$. Thus at $\alpha_1 = 1$ we have a (local) minimum and at $\alpha_2 = -1$ we have a (local) maximum. Furthermore $f(\alpha = 1) = 0$ and $f(\alpha = -1) = 4$.
(ii) Solving $\alpha^3 - 3\alpha + 2 = 0$ we find that the matrix is noninvertible for $\alpha = 1$ and $\alpha = -2$.

Problem 40.   The *Pascal matrix* of order n is defined as

$$P_n := \left(\frac{(i+j-2)!}{(i-1)!(j-1)!} \right), \qquad i,j = 1,\ldots,n.$$

Thus

$$P_2 = \begin{pmatrix} 1 & 1 \\ 1 & 2 \end{pmatrix}, \quad P_3 = \begin{pmatrix} 1 & 1 & 1 \\ 1 & 2 & 3 \\ 1 & 3 & 6 \end{pmatrix}, \quad P_4 = \begin{pmatrix} 1 & 1 & 1 & 1 \\ 1 & 2 & 3 & 4 \\ 1 & 3 & 6 & 10 \\ 1 & 4 & 10 & 20 \end{pmatrix}.$$

Find the determinant of P_2, P_3, P_4. Find the inverse of P_2, P_3, P_4.

Solution 40.   The determinant of P_2 is given by 1 and the inverse takes the form

$$P_2^{-1} = \begin{pmatrix} 2 & -1 \\ -1 & 1 \end{pmatrix}.$$

The determinant of P_3 is also given by 1 and the inverse takes the form

$$P_3^{-1} = \begin{pmatrix} 3 & -3 & 1 \\ -3 & 5 & -2 \\ 1 & -2 & 1 \end{pmatrix}.$$

The determinant of P_4 is also given by 1 and the inverse takes the form

$$P_4^{-1} = \begin{pmatrix} 4 & -6 & 4 & -1 \\ -6 & 14 & -11 & 3 \\ 4 & -11 & 10 & -3 \\ -1 & 3 & -3 & 1 \end{pmatrix}.$$

Problem 41.   Let A, B be $n \times n$ matrices over $\mathbb{R}$. Assume that A is invertible. Let t be a nonzero real number. Show that

$$\det(A + tB) = t^n \det(A) \det(A^{-1}B + t^{-1}I_n).$$

Solution 41. We have

$$\det(A + tB) = t^n \det(t^{-1}A + B) = t^n \det(A(A^{-1}B + t^{-1}I_n))$$
$$= t^n \det(A) \det(A^{-1}B + t^{-1}I_n).$$

Problem 42. Let A be an $n \times n$ invertible matrix over $\mathbb{R}$. Show that A^T is also invertible. Is $(A^T)^{-1} = (A^{-1})^T$?

Solution 42. Since $\det(A^T) = \det(A)$ and $\det(A) \neq 0$ we conclude that $\det(A^T) \neq 0$. Thus A^T is invertible. We find $(A^T)^{-1}A^T = I_n$ and

$$(AA^{-1})^T = (A^{-1})^T A^T = I_n.$$

Thus $(A^{-1})^T A^T = (A^T)^{-1} A^T$ and therefore $(A^T)^{-1} = (A^{-1})^T$.

Problem 43. Calculate the determinant of the 4×4 matrix

$$A = \begin{pmatrix} 1 & 0 & 0 & 1 \\ 0 & 1 & 1 & 0 \\ 0 & 1 & -1 & 0 \\ 1 & 0 & 0 & -1 \end{pmatrix}$$

using the *exterior product* $\wedge$ (*wedge product*). This means calculate

$$\begin{pmatrix} 1 \\ 0 \\ 0 \\ 1 \end{pmatrix} \wedge \begin{pmatrix} 0 \\ 1 \\ 1 \\ 0 \end{pmatrix} \wedge \begin{pmatrix} 0 \\ 1 \\ -1 \\ 0 \end{pmatrix} \wedge \begin{pmatrix} 1 \\ 0 \\ 0 \\ -1 \end{pmatrix}.$$

Solution 43. Let e_j ($j = 1, 2, 3, 4$) be the standard basis in $\mathbb{R}^4$. Then

$$(e_1 + e_4) \wedge (e_2 + e_3) \wedge (e_2 - e_3) \wedge (e_1 - e_4) = 4e_1 \wedge e_2 \wedge e_3 \wedge e_4$$

utilizing that $e_j \wedge e_k = -e_k \wedge e_j$. Thus the determinant is 4.

Problem 44. The 3×3 diagonal matrices over $\mathbb{R}$ with trace equal to 0 form a vector space. Provide a basis for this vector space. Using the scalar product $\text{tr}(AB^T)$ for $n \times n$ matrices A, B over $\mathbb{R}$ the elements of the basis should be orthogonal to each other.

Solution 44. Owing to the constraint that the trace of the matrices is equal to 0 the dimension of the vector space is 2. We select

$$V_1 = \begin{pmatrix} 1 & 0 & 0 \\ 0 & -1 & 0 \\ 0 & 0 & 0 \end{pmatrix}, \qquad V_2 = \begin{pmatrix} 1 & 0 & 0 \\ 0 & 1 & 0 \\ 0 & 0 & -2 \end{pmatrix}.$$

Then V_1, V_2 are linearly independent and $\text{tr}(V_1 V_2^T) = 0$.

Problem 45. The *Hilbert-Schmidt norm* of an $n \times n$ matrix over $\mathbb{C}$ is defined by

$$\|A\|_2 := \sqrt{\text{tr}(A^*A)}.$$

Another norm is the *trace norm* defined by

$$\|A\|_1 := \text{tr}\sqrt{(A^*A)}.$$

Calculate the two norms for the 2×2 matrix

$$A = \begin{pmatrix} 0 & -2i \\ i & 0 \end{pmatrix}.$$

Solution 45. We have

$$A^* = \begin{pmatrix} 0 & -i \\ 2i & 0 \end{pmatrix} \Rightarrow A^*A = \begin{pmatrix} 1 & 0 \\ 0 & 4 \end{pmatrix}.$$

Thus $\|A\|_2 = \sqrt{5}$, $\|A\|_1 = 3$.

Problem 46. Let A be a 3×3 matrix over $\mathbb{R}$. Consider the 3×3 permutation matrix

$$P = \begin{pmatrix} 0 & 0 & 1 \\ 0 & 1 & 0 \\ 1 & 0 & 0 \end{pmatrix}.$$

Assume that $AP = A$. Show that A cannot be invertible.

Solution 46. We have $\det(AP) = \det(A) \Rightarrow \det(A)\det(P) = \det(A)$. Since $\det(P) = -1$ we obtain $-\det(A) = \det(A)$ and therefore A is not invertible. We also have

$$[P, A] = \begin{pmatrix} a_{31} - a_{13} & a_{32} - a_{12} & a_{33} - a_{11} \\ a_{21} - a_{23} & 0 & a_{23} - a_{21} \\ a_{11} - a_{33} & a_{12} - a_{32} & a_{13} - a_{31} \end{pmatrix}.$$

The determinant of this matrix is 0.

Problem 47. Let $A = (a_{ij})$ be a $2n \times 2n$ skew-symmetric matrix. The *Pfaffian* is defined as

$$\text{Pf}(A) := \frac{1}{2^n n!} \sum_{\sigma \in S_{2n}} \text{sgn}(\sigma) \prod_{j=1}^{n} a_{\sigma(2j-1), \sigma(2j)}$$

where S_{2n} is the symmetric group and $\text{sgn}(\sigma)$ is the signature of permutation σ. Consider the case with $n = 2$, i.e.

$$A = \begin{pmatrix} 0 & a_{12} & a_{13} & a_{14} \\ -a_{12} & 0 & a_{23} & a_{24} \\ -a_{13} & -a_{23} & 0 & a_{34} \\ -a_{14} & -a_{24} & -a_{34} & 0 \end{pmatrix}.$$

Calculate $\text{Pf}(A)$.

Solution 47. We have $n = 2$ and $4! = 24$ permutations. Thus

$$\text{Pf}(A) = \frac{1}{8} \sum_{\sigma \in S_4} \text{sgn}(\sigma) \prod_{j=1}^{2} a_{\sigma(2j-1),\sigma(2j)} = \frac{1}{8} \sum_{\sigma \in S_4} \text{sgn}(\sigma) a_{\sigma(1)\sigma(2)} a_{\sigma(3)\sigma(4)}$$

$$= a_{12}a_{34} - a_{13}a_{24} + a_{23}a_{14}$$

where we have taken into account that for $j > k$ we have $a_{jk} = -a_{kj}$. The summation over all permutations can be avoided. Let Π be the set of all partitions of the set $\{1, 2, \ldots, 2n\}$ into pairs without regard to order. There are $2n - 1$ such partitions. An element $\alpha \in \Pi$ can be written as

$$\alpha = \{(i_1, j_1), (i_2, j_2), \ldots, (i_n, j_n)\}$$

with $i_k < j_k$ and $i_1 < i_2 < \cdots < i_n$. Let

$$\pi = \begin{pmatrix} 1 & 2 & 3 & 4 & \cdots & 2n \\ i_1 & j_1 & i_2 & j_2 & \cdots & j_n \end{pmatrix}$$

be a corresponding permutation. Given a partition α we define the number

$$A_\alpha = \text{sgn}(\pi) a_{i_1 j_1} a_{i_2 j_2} \cdots a_{i_n j_n}.$$

The Pfaffian of A is then given by

$$\text{Pf}(A) = \sum_{\alpha \in \Pi} A_\alpha.$$

For $n = 2$ we have $(1, 2)(3, 4)$, $(1, 3)(2, 4)$, $(1, 4)(2, 3)$, i.e. we have 3 partitions.

Problem 48. Let A be a skew-symmetric $2n \times 2n$ matrix. For the Pfaffian we have the properties

$$(\text{Pf}(A))^2 = \det(A), \qquad \text{Pf}(BAB^T) = \det(B)\text{Pf}(A)$$

$$\text{Pf}(\lambda A) = \lambda^n \text{Pf}(A), \qquad \text{Pf}(A^T) = (-1)^n \text{Pf}(A)$$

where B is an arbitrary $2n \times 2n$ matrix. Let J be a $2n \times 2n$ skew-symmetric matrix with $\text{Pf}(J) \neq 0$. Let B be a $2n \times 2n$ matrix such that $B^T J B = J$. Show that $\det(B) = 1$.

Solution 48. Using the property such that $\text{Pf}(BAB^T) = \det(B)\text{Pf}(A)$, where A is a $2n \times 2n$ skew-symmetric matrix we obtain

$$\text{Pf}(J) = \text{Pf}(B^T J B) = \det(B)\text{Pf}(J).$$

Since $\text{Pf}(J) \neq 0$, we find that $\det(B) = 1$.

Problem 49. Find all 2×2 invertible matrices A such that $A + A^{-1} = I_2$.

Solution 49. Since $\det(A) = ad - bc \neq 0$ we have

$$A = \begin{pmatrix} a & b \\ c & d \end{pmatrix} \Rightarrow A^{-1} = \frac{1}{\det(A)} \begin{pmatrix} d & -b \\ -c & a \end{pmatrix}.$$

Thus from the condition we obtain the four equations

$$b\left(1 - \frac{1}{\det(A)}\right) = 0, \quad c\left(1 - \frac{1}{\det(A)}\right) = 0, \quad a + \frac{1}{\det(A)} = 1, \quad d + \frac{1}{\det(A)} = 1.$$

Thus we have to do a case study. Case $b \neq 0$. Then $\det(A) = 1$ and we obtain $a + d = 1$. If $b = 0$, then $\det(A) = ad \neq 0$. Thus a and d must be nonzero.

Problem 50.   Let V_1 be a hermitian $n \times n$ matrix. Let V_2 be a positive semidefinite $n \times n$ matrix. Let k be a positive integer. Show that $\mathrm{tr}((V_2 V_1)^k)$ can be written as $\mathrm{tr}(V^k)$, where $V := V_2^{1/2} V_1 V_2^{1/2}$.

Solution 50.   Since V_2 is positive semidefinite the square root of V_2 exists. Thus applying cyclic invariance of the trace we have

$$\mathrm{tr}((V_2 V_1)^k) = \mathrm{tr}(V_2^{1/2} V_1 V_2^{1/2} V_2^{1/2} V_1 \cdots V_2^{1/2} V_1 V_2^{1/2}) = \mathrm{tr}(V^k).$$

Problem 51.   Consider the 2×2 matrix

$$M = \begin{pmatrix} \cosh(r) - \sinh(r)\cos(2\theta) & -\sinh(r)\sin(2\theta) \\ -\sinh(r)\sin(2\theta) & \cosh(r) + \sinh(r)\cos(2\theta) \end{pmatrix}.$$

Find the determinant of M. Thus show that the inverse of M exists. Find the inverse of M.

Solution 51.   We have

$$\det(M) = \cosh^2(r) - \sinh^2(r)\cos^2(2\theta) - \sinh^2(r)\sin^2(2\theta) = 1.$$

The inverse of M are obtained by replacing $r \to -r$ and $\theta \to -\theta$ We obtain

$$M^{-1} = \begin{pmatrix} \cosh(r) + \sinh(r)\cos(2\theta) & \sinh(r)\sin(2\theta) \\ \sinh(r)\sin(2\theta) & \cosh(r) - \sinh(r)\cos(2\theta) \end{pmatrix}.$$

Problem 52.   Consider the 2×2 matrix C over $\mathbb{C}$. Calculate $\det(CC^*)$ and show that $\det(CC^*) \geq 0$.

Solution 52.   We have

$$\begin{aligned}
\det(CC^*) &= \det(C)\det(C^*) = (c_{11}c_{22} - c_{12}c_{21})(\bar{c}_{11}\bar{c}_{22} - \bar{c}_{12}\bar{c}_{21}) \\
&= c_{11}\bar{c}_{11}c_{22}\bar{c}_{22} + c_{12}\bar{c}_{12}c_{21}\bar{c}_{21} - c_{11}c_{22}\bar{c}_{12}\bar{c}_{21} - c_{12}c_{21}\bar{c}_{11}\bar{c}_{22}.
\end{aligned}$$

Setting $(r_{jk} \geq 0)$ $c_{11} = r_{11}e^{i\phi_{11}}$, $c_{22} = r_{22}e^{i\phi_{22}}$, $c_{12} = r_{12}e^{i\phi_{12}}$, $c_{21} = r_{21}e^{i\phi_{21}}$ yields

$$\det(CC^*) = r_{11}^2 r_{22}^2 + r_{12}^2 r_{21}^2 - 2r_{11}r_{22}r_{12}r_{21}\cos(\alpha)$$

which cannot be negative since $|\cos(\alpha)| \leq 1$.

Problem 53. (i) Let $z \in \mathbb{C}$. Find the determinant of

$$A = \begin{pmatrix} 1 & z \\ \bar{z} & z\bar{z} \end{pmatrix}.$$

Is the matrix

$$\Pi_2 = I_2 - \frac{1}{1+z\bar{z}}A$$

a projection matrix?

(ii) Let $z_1, z_2 \in \mathbb{C}$. Find the determinant of

$$B = \begin{pmatrix} 1 & z_1 & z_2 \\ \bar{z}_1 & z_1\bar{z}_1 & z_2\bar{z}_1 \\ \bar{z}_2 & z_1\bar{z}_2 & z_2\bar{z}_2 \end{pmatrix}.$$

Is the matrix

$$\Pi_3 = I_3 - \frac{1}{1+z_1\bar{z}_1 + z_2\bar{z}_2}B$$

a projection matrix?

Solution 53. (i) We find $\det(A) = z\bar{z} - z\bar{z} = 0$. We have $A^2 = (1 + z\bar{z})A$. Thus

$$\Pi_2^2 = (I_2 - \frac{1}{1+z\bar{z}}A)^2 = I_2 - \frac{2}{1+z\bar{z}}A + \frac{1}{1+z\bar{z}}A = \Pi_2.$$

We also have $\Pi_2 = \Pi_2^*$. Thus Π_2 is a projection matrix.

(ii) We have $\det(B) = 0$. We find $\Pi_3^2 = \Pi_3$ and $\Pi_3^* = \Pi_3$. Thus Π_3 is a projection matrix.

Problem 54. Consider the *golden mean number* $\tau = (\sqrt{5}-1)/2$ and the 2×2 matrix

$$F = \begin{pmatrix} \tau & \sqrt{\tau} \\ \sqrt{\tau} & -\tau \end{pmatrix}.$$

Find $\mathrm{tr}(F)$ and $\det(F)$. Since $\det(F) \neq 0$ we have an inverse. Find F^{-1}.

Solution 54. We have $\mathrm{tr}(F) = 0$ and $\det(F) = -\tau^2 - \tau = -1$. The inverse matrix is $F^{-1} = F$.

Problem 55. Let A be an $n \times n$ matrix and B be an invertible $n \times n$ matrix. Show that

$$\det(I_n + A) = \det(I_n + BAB^{-1}).$$

Solution 55. We have

$$\det(I_n + A) = \det(B)\det(B^{-1})\det(I_n + A) = \det(B)\det(I_n + A)\det(B^{-1})$$
$$= \det(B(I_n + A)B^{-1}) = \det(BB^{-1} + BAB^{-1})$$
$$= \det(I_n + BAB^{-1}).$$

Problem 56.   Let A be an 2×2 matrix. Show that

$$\det(I_2 + A) = 1 + \operatorname{tr}(A) + \det(A).$$

Solution 56.   We have

$$\det(I_2 + A) = \det \begin{pmatrix} 1 + a_{11} & a_{12} \\ a_{21} & 1 + a_{22} \end{pmatrix} = 1 + a_{11} + a_{22} + a_{11}a_{22} - a_{12}a_{21}$$
$$= 1 + \operatorname{tr}(A) + \det(A).$$

Can the result extended to $\det(I_3 + B)$, where B is a 3×3 matrix?

Problem 57.   Let $\{\, e_j \,:\, j = 1, 2, 3 \,\}$ be the three orthonormal vectors in $\mathbb{Z}^3$

$$e_1 = \begin{pmatrix} 1 \\ 0 \\ 0 \end{pmatrix}, \quad e_2 = \begin{pmatrix} 0 \\ 1 \\ 0 \end{pmatrix}, \quad e_3 = \begin{pmatrix} 0 \\ 0 \\ 1 \end{pmatrix}.$$

We consider the face-centered cubic lattice as a sublattice of $\mathbb{Z}^3$ generated by the three primitive vectors $e_1 + e_2$, $e_1 + e_3$, $e_2 + e_3$. Form the 3×3 matrix $(e_1 + e_2 \; e_1 + e_3 \; e_2 + e_3)$ and show that this matrix has an inverse. Find the inverse.

Solution 57.   The determinant of the matrix

$$\begin{pmatrix} 1 & 1 & 0 \\ 1 & 0 & 1 \\ 0 & 1 & 1 \end{pmatrix}$$

is -2 and thus the inverse exists. The inverse is given by

$$\frac{1}{2} \begin{pmatrix} 1 & 1 & -1 \\ 1 & -1 & 1 \\ -1 & 1 & 1 \end{pmatrix}.$$

Problem 58.   Let $n \geq 2$. Consider the $n \times n$ symmetric tridiagonal matrix over $\mathbb{R}$

$$A_n = \begin{pmatrix} c & 1 & 0 & 0 & \cdots & & & & 0 \\ 1 & c & 1 & 0 & \cdots & & & & 0 \\ 0 & 1 & c & 1 & \cdots & & & & 0 \\ \vdots & & & & \ddots & & & & \vdots \\ 0 & 0 & 0 & & \cdots & 1 & c & 1 & 0 \\ 0 & 0 & 0 & & \cdots & 0 & 1 & c & 1 \\ 0 & 0 & 0 & & \cdots & 0 & 0 & 1 & c \end{pmatrix}$$

where $c \in \mathbb{R}$. Find the determinant of A_n.

Solution 58.   We obtain the linear difference equation with constant coefficients

$$A_{j+1} = cA_j - A_{j-1}, \qquad j = 1, \ldots, n - 1$$

with the initial conditions $A_0 = 1$, $A_1 = c$. The solution of this linear difference equation is

$$A_n = \sum_{j=0}^{[n/2]} (-1)^j \binom{k-j}{j} c^{k-2j}$$

where $[n/2]$ denotes the integer part of $n/2$. If $c \leq -2$, one can set $c = -2\cosh(\alpha)$ with $\alpha \in \mathbb{R}$. Then we have

$$A_n = (-1)^n \frac{\sinh((n+1)\alpha)}{\sinh(\alpha)}.$$

If $c \geq 2$ and $c = 2\cosh(\alpha)$, then

$$A_n = \frac{\sinh((n+1)\alpha)}{\sinh(\alpha)}.$$

For $-2 < c < 2$ and $c = 2\cos(\alpha)$ we have

$$A_n = (-1)^n \frac{\sin((n+1)\alpha)}{\sin(\alpha)}.$$

Problem 59. Find a nonzero 2×2 matrix V over $\mathbb{C}$ such that $V^2 = \text{tr}(V)V$. Can such a matrix be invertible?

Solution 59. Let $z \in \mathbb{C}$. Then

$$V = \begin{pmatrix} e^z & 1 \\ 1 & e^{-z} \end{pmatrix} \quad \Rightarrow \quad V^2 = (e^z + e^{-z})V.$$

Such a matrix cannot be invertible. We can assume that $\text{tr}(V) \neq 0$. Suppose that V is invertible. Then from $V^2 = \text{tr}(V)V$ it follows that $V = \text{tr}(V)I_2$. Taking the trace yields $\text{tr}(V) = 2\text{tr}(V)$. Thus we have a contradiction.

Problem 60. Consider the $(n+1) \times (n+1)$ matrix over $\mathbb{C}$

$$A = \begin{pmatrix} 1 & 0 & 0 & \cdots & 0 & z_1 \\ 0 & 1 & 0 & \cdots & 0 & z_2 \\ 0 & 0 & 1 & \cdots & 0 & z_3 \\ \vdots & \vdots & \vdots & \ddots & \vdots & \vdots \\ 0 & 0 & 0 & \cdots & 1 & z_n \\ z_1 & z_2 & z_3 & \cdots & z_n & 1 \end{pmatrix}.$$

Find the determinant. What is the condition on the z_j's such that A is invertible?

Solution 60. We find

$$\det(A) = 1 - \sum_{j=1}^{n} z_j^2.$$

For A^{-1} to exist we need $\sum_{j=1}^{n} z_j^2 \neq 1$.

Problem 61. Let $z_k = x_k + iy_k$, where $x_k, y_k \in \mathbb{R}$ and $k = 1, \ldots, n$. Find the $2n \times 2n$ matrix A such that

$$\begin{pmatrix} z_1 \\ \vdots \\ z_n \\ \bar{z}_1 \\ \vdots \\ \bar{z}_n \end{pmatrix} = A \begin{pmatrix} x_1 \\ \vdots \\ x_n \\ y_1 \\ \vdots \\ y_n \end{pmatrix}.$$

Find the determinant of the matrix A.

Solution 61. The $2n \times 2n$ matrix A can be written in block form

$$A = \begin{pmatrix} I_n & iI_n \\ I_n & -iI_n \end{pmatrix}$$

with $\det(A) = (-2i)^n$.

Problem 62. Let A, B be $n \times n$ matrices over $\mathbb{C}$. We define the product

$$A \diamond B := \frac{1}{2}(AB + BA) - \frac{1}{n}\operatorname{tr}(AB)I_n.$$

(i) Find the trace of $A \diamond B$.
(ii) Is the product commutative?

Solution 62. (i) Since the trace is a linear operation and $\operatorname{tr}(I_n) = n$ we find $\operatorname{tr}(A \diamond B) = 0$. Thus $A \diamond B$ is an element of the Lie algebra $s\ell(n, \mathbb{C})$.
(ii) Yes. We have $A \diamond B = B \diamond A$. Is the product associative?

Problem 63. Let A be an $n \times n$ invertible matrix over $\mathbb{C}$. Let $\mathbf{x}, \mathbf{y} \in \mathbb{C}^n$. Then we have the identity

$$\det(A + \mathbf{x}\mathbf{y}^*) \equiv \det(A)(1 + \mathbf{y}^* A^{-1} \mathbf{x}).$$

Can we conclude that $A + \mathbf{x}\mathbf{y}^*$ is also invertible?

Solution 63. This is not true in general. Consider for example

$$A = \frac{1}{\sqrt{2}} \begin{pmatrix} 1 & 1 \\ 1 & -1 \end{pmatrix}, \quad \mathbf{x} = \begin{pmatrix} 1 \\ 1 \end{pmatrix}, \quad \mathbf{y} = \frac{1}{\sqrt{2}} \begin{pmatrix} -1 \\ -1 \end{pmatrix}.$$

Then

$$A + \mathbf{x}\mathbf{y}^* = \begin{pmatrix} 0 & 0 \\ 0 & -\sqrt{2} \end{pmatrix} \quad \Rightarrow \quad \det(A + \mathbf{x}\mathbf{y}^*) = 0.$$

Problem 64. Let $n \geq 1$. Consider the $2^n \times 2^n$ matrices A and B. Assume that $A^2 = I_{2^n}$, $B^2 = I_{2^n}$ and $[A, B]_+ = 0_{2^n}$. Show that

$$\det(A + B) = 2^{2^n/2}.$$

Solution 64. We have

$$\det(A+B)\det(A+B) = \det((A+B)(A+B)) = \det(A^2 + B^2 + AB + BA)$$
$$= \det(I_{2^n} + I_{2^n}) = \det(2I_{2^n}) = 2^{2^n}.$$

Hence the result follows.

Problem 65. Let A, B be $n \times n$ matrices over $\mathbb{C}$. Assume that B is invertible. Show that there exists $c \in \mathbb{C}$ such that $A + cB$ is not invertible. Hint. Start of with the identity $A + cB \equiv (AB^{-1} + cI_n)B$ and apply the determinant.

Solution 65. From $A + cB \equiv (AB^{-1} + cI_n)B$ we obtain

$$\det(A + cB) = \det(AB^{-1} + cI_n)\det(B)$$

where $\det(B) \neq 0$. If we choose $-c$ equal to an eigenvalue of AB^{-1}, then $\det(AB^{-1} + cI_n) = 0$ and $A + cB$ is not invertible.

Problem 66. The *permanent* of an $n \times n$ matrix A is defined as

$$\text{Per}(A) := \sum_{\sigma \in S_n} \prod_{j=1}^{n} a_{j,\sigma(j)}$$

where S_n is the set of all permutation of n elements. Find the permanent of a 2×2 matrix A.

Solution 66. We obtain $\text{Per}(A) = a_{11}a_{22} + a_{12}a_{21}$.

Problem 67. Let A be a 4×4 matrix we write as

$$A = \begin{pmatrix} A_1 & A_2 \\ A_3 & A_4 \end{pmatrix}$$

where A_1, A_2, A_3, A_4 are the (block) 2×2 matrices in the matrix A. Consider the map

$$\begin{pmatrix} A_1 & A_2 \\ A_3 & A_4 \end{pmatrix} \mapsto \begin{pmatrix} A_1 & A_2^T \\ A_3^T & A_4 \end{pmatrix}$$

where T denotes the transpose. Is the trace preserved under this map? Is the determinant preserved under this map?

Solution 67. The trace is obviously preserved. The determinant is not preserved in general. For example consider the case

$$A = \begin{pmatrix} 1 & 0 & 0 & 0 \\ 0 & 0 & 1 & 0 \\ 0 & 1 & 0 & 0 \\ 0 & 0 & 0 & 1 \end{pmatrix} \mapsto \begin{pmatrix} 1 & 0 & 0 & 1 \\ 0 & 0 & 0 & 0 \\ 0 & 0 & 0 & 0 \\ 1 & 0 & 0 & 1 \end{pmatrix}.$$

The matrix A is invertible but not the matrix after the mapping.

Problem 68.   Let H be a 4×4 matrix and U, V be 2×2 unitary matrices. Calculate

$$\det((U \otimes V)H(U^* \otimes V^*)), \qquad \mathrm{tr}((U \otimes V)H(U^* \otimes V^*)).$$

Solution 68.   Since $UU^* = I_2$ and $VV^* = I_2$ we have

$$
\begin{aligned}
\det((U \otimes V)H(U^* \otimes V^*)) &= \det(U \otimes V)\det(H)\det(U^* \otimes V^*) \\
&= \det((U \otimes V)(U^* \otimes V^*))\det(H) \\
&= \det(I_2 \otimes I_2)\det(H) \\
&= \det(H).
\end{aligned}
$$

Using *cyclic invariance* of the trace we obtain

$$
\begin{aligned}
\mathrm{tr}((U \otimes V)H(U^* \otimes V^*)) &= \mathrm{tr}((U^* \otimes V^*)(U \otimes V)\mathrm{tr}(H)) \\
&= \mathrm{tr}((U^*U) \otimes (V^*V)H) = \mathrm{tr}((I_2 \otimes I_2)H) \\
&= \mathrm{tr}(H).
\end{aligned}
$$

Problem 69.   Let A, B be $n \times n$ matrices over $\mathbb{C}$. Consider the expression

$$(A \otimes B - B \otimes A) \oplus (-[A, B])$$

where $\oplus$ denotes the direct sum. Find the trace of this expression. This expression plays a role for universal enveloping algebras.

Solution 69.   With $\mathrm{tr}(A \otimes B) = \mathrm{tr}(A)\mathrm{tr}(B)$ we have

$$
\begin{aligned}
\mathrm{tr}((A \otimes B - B \otimes A) \oplus (-[A, B])) &= \mathrm{tr}(A \otimes B - B \otimes A) + \mathrm{tr}(-[A, B]) \\
&= \mathrm{tr}(A \otimes B) - \mathrm{tr}(B \otimes A) - \mathrm{tr}(AB - BA) \\
&= 0.
\end{aligned}
$$

Problem 70.   Let A be an 2×2 matrix over $\mathbb{R}$. Let $\mathrm{tr}(A) = c_1$, $\mathrm{tr}(A^2) = c_2$. Can $\det(A)$ be calculated from c_1, c_2?

Solution 70.   Yes we can calculate $\det(A)$ from c_1, c_2. We have

$$\mathrm{tr}(A) = a_{11} + a_{22} = c_1, \qquad \mathrm{tr}(A^2) = a_{11}^2 + a_{22}^2 + 2a_{12}a_{21} = c_2.$$

Thus we have

$$(a_{11} + a_{22})^2 = a_{11}^2 + a_{22}^2 + 2a_{11}a_{22} = c_1^2.$$

Inserting this expression into $\det(A^2) = c_2$ yields $c_1^2 - 2a_{11}a_{22} + 2a_{12}a_{21} = c_2$ or

$$\det(A) = \frac{1}{2}(c_1^2 - c_2).$$

Can this be extended to 3×3 matrices, i.e. given $\mathrm{tr}(A) = c_1$, $\mathrm{tr}(A^2) = c_2$, $\mathrm{tr}(A^3) = c_3$? Find $\det(A)$ using only c_1, c_2, c_3.

Problem 71. Let J be the $2n \times 2n$ matrix

$$J := \begin{pmatrix} 0_n & I_n \\ -I_n & 0_n \end{pmatrix}.$$

We define symplectic **G**-reflectors to be those $2n \times 2n$ symplectic matrices that have a $(2n-1)$-dimensional fixed-point subspace. Any symplectic **G**-reflector can be expressed as

$$G = I_{2n} + \beta \mathbf{u}\mathbf{u}^T J \tag{1}$$

for some $0 \neq \beta \in \mathbb{F}$, $\mathbf{0} \neq \mathbf{u} \in \mathbb{F}^{2n}$ and $\mathbf{u}$ is considered as a column vector. The underlying field is $\mathbb{F}$. Conversely, any G given by (1) is always a symplectic **G**-reflector. Show that $\det(G) = +1$.

Solution 71. Let $G = I_{2n} + \beta \mathbf{u}\mathbf{u}^T J$ be an arbitrary symplectic **G**-reflector. Consider the continuous path of matrices given by

$$G(t) = I_{2n} + (1-t)\beta \mathbf{u}\mathbf{u}^T J$$

with $0 \le t \le 1$. Then $G(0) = G$. Now $G(t)$ is a symplectic **G**-reflector for $0 \le t < 1$, so $\det(G(t)) = \pm 1$ for all $t < 1$. However

$$\lim_{t \to 1} G(t) = I_{2n}$$

so by continuity $\det(G) = +1$ for all t, in particular for $t = 0$. We can also prove it as follows. Any symplectic **G**-reflector is the square of another symplectic **G**-reflector. Since $\mathbf{u}^T J \mathbf{u} = 0$ for all $\mathbf{u} \in \mathbb{F}^{2n}$, we have

$$G = I_{2n} + \beta \mathbf{u}\mathbf{u}^T J = \left(I_{2n} + \frac{1}{2}\beta \mathbf{u}\mathbf{u}^T J \right)^2 = S^2.$$

Thus $\det(G) = \det(S^2) = (\det(S))^2 = (\pm 1)^2 = +1$.

Problem 72. Let A, B be positive semidefinite $n \times n$ matrices. Then

$$\det(A + B) \ge \det(A) + \det(B), \qquad \mathrm{tr}(AB) \le \mathrm{tr}(A)\mathrm{tr}(B).$$

Let

$$A = \frac{1}{3}\begin{pmatrix} 1 & 1 & 1 \\ 1 & 1 & 1 \\ 1 & 1 & 1 \end{pmatrix}, \qquad B = \frac{1}{3}\begin{pmatrix} 1 & 0 & 0 \\ 0 & 1 & 0 \\ 0 & 0 & 1 \end{pmatrix}.$$

Both matrices are *density matrices*, with A a pure state and B a mixed state. Calculate the left- and right-hand side of the two inequality. Discuss.

Solution 72. We have $\det(A) = 0$, $\det(B) = 1/27$ and $\det(A + B) = 4/27$. We have $\mathrm{tr}(A) = 1$, $\mathrm{tr}(B) = 1$ and $\mathrm{tr}(AB) = 1/3$.

Problem 73.   Let A be an $n \times n$ positive semidefinite matrix. Let B be an $n \times n$ positive definite matrix. Then we have *Klein's inequality*

$$\mathrm{tr}(A(\ln(A) - \ln(B))) \geq \mathrm{tr}(A - B).$$

(i) Let

$$A = \begin{pmatrix} 1/2 & -1/2 \\ -1/2 & 1/2 \end{pmatrix}, \qquad B = \begin{pmatrix} 1/2 & 0 \\ 0 & 1/2 \end{pmatrix}.$$

Calculate the left-hand side and the right-hand side of the inequality.
(ii) When is the inequality an equality?

Solution 73.   (i) Obviously the right-hand side of the inequality is equal to 0. The unitary matrix

$$U = \frac{1}{\sqrt{2}} \begin{pmatrix} 1 & 1 \\ 1 & -1 \end{pmatrix}$$

puts A into diagonal form, i.e.

$$\tilde{A} = U^{-1}AU = \begin{pmatrix} 0 & 0 \\ 0 & 1 \end{pmatrix}$$

and $U^{-1}BU = B$. Thus

$$\begin{aligned}
\mathrm{tr}(A(\ln(A) - \ln(B))) &= \mathrm{tr}(UA(\ln(A) - \ln(B))U^{-1}) \\
&= \mathrm{tr}(UAU^{-1}\ln(UAU^{-1}) - UAU^{-1}\ln(UBU^{-1})) \\
&= \mathrm{tr}(\tilde{A}\ln(\tilde{A}) - \tilde{A}\ln(B)).
\end{aligned}$$

Since $0\ln(0) = 0$ and $0\ln(1) = 0$ we have $\mathrm{tr}(\tilde{A}\ln(\tilde{A})) = 0$, $\mathrm{tr}(\tilde{A}\ln(B)) = -\ln(2)$. Thus Klein's inequality provides $\ln 2 > 0$.
(ii) We have equality if and only if $A = B$.

Problem 74.   Consider the two polynomials

$$p_1(x) = a_0 + a_1 x + \cdots + a_n x^n, \qquad p_2(x) = b_0 + b_1 x + \cdots + b_m x^m$$

where $n = \deg(p_1)$ and $m = \deg(p_2)$. Assume that $n > m$. Let $r(x) := p_2(x)/p_1(x)$. We expand $r(x)$ in powers of $1/x$, i.e.

$$r(x) = \frac{c_1}{x} + \frac{c_2}{x^2} + \cdots.$$

From the coefficients $c_1, c_2, \ldots, c_{2n-1}$ we can form an $n \times n$ *Hankel matrix*

$$H_n = \begin{pmatrix} c_1 & c_2 & \cdots & c_n \\ c_2 & c_3 & \cdots & c_{n+1} \\ \vdots & \vdots & \ddots & \vdots \\ c_n & c_{n+1} & \cdots & c_{2n-1} \end{pmatrix}.$$

The determinant of this matrix is proportional to the *resultant* of the two polynomials. If the resultant vanishes, then the two polynomials have a non-trivial greatest common divisor. Apply this theorem to the polynomials

$$p_1(x) = x^3 + 6x^2 + 11x + 6, \qquad p_2(x) = x^2 + 4x + 3.$$

Solution 74. Since $n = 3$ we need the coefficients $c_1, c_2, \ldots, c_5$. Straightforward division yields

$$(x^2 + 4x + 3) : (x^3 + 6x^2 + 11x + 6) = \frac{1}{x} - \frac{2}{x^2} + \frac{4}{x^3} - \frac{8}{x^4} + \frac{16}{x^5} + \cdots .$$

Thus $c_1 = 1$, $c_2 = -2$, $c_3 = 4$, $c_4 = -8$, $c_5 = 16$. Thus the Hankel matrix is

$$H_3 = \begin{pmatrix} 1 & -2 & 4 \\ -2 & 4 & -8 \\ 4 & -8 & 16 \end{pmatrix}.$$

We find $\det(H_3) = 0$. Thus p_1, p_2 have a greatest common divisor. Obviously

$$(x^2 + 4x + 3)(x + 2) = x^3 + 6x^2 + 11x + 6.$$

Problem 75. Consider the 5×5 symmetric matrix over $\mathbb{R}$

$$A = \begin{pmatrix} 1 & 0 & 0 & 0 & 1 \\ 0 & 1 & 1 & 1 & 0 \\ 0 & 1 & 1 & 1 & 0 \\ 0 & 1 & 1 & 1 & 0 \\ 1 & 0 & 0 & 0 & 1 \end{pmatrix}.$$

The *rank* is the number of linearly independent row (or column) vectors of A.
(i) Find the rank of the matrix. Explain.
(ii) Find the determinant and trace of the matrix.
(iii) Find all eigenvalues of the matrix.
(iv) Find one eigenvector.
(v) Is the matrix positive semidefinite?

Solution 75. (i) We have $\text{rank}(A) = 2$. There are only two unique row vectors in A and they are linearly independent.
(ii) Since $\text{rank}(A) = 2 < 5$ the determinant must be 0. The trace is given by $\text{tr}(A) = 5$.
(iii) Since $\text{rank}(A) = 2$ three of the five eigenvalues must be 0. The other two eigenvalues are 2 and 3.
(iv) Obviously $(1 \ \ 0 \ \ 0 \ \ 0 \ \ -1)^T$ is an eigenvector.
(v) From (iii) we find that the matrix is positive semidefinite.

Problem 76. Let A be an $n \times n$ matrix over $\mathbb{R}$. Assume that A^{-1} exists. Let $\mathbf{u}, \mathbf{v} \in \mathbb{R}^n$, where $\mathbf{u}$, $\mathbf{v}$ are considered as column vectors.
(i) Show that if $\mathbf{v}^T A^{-1} \mathbf{u} = -1$, then $A + \mathbf{u}\mathbf{v}^T$ is not invertible.
(ii) Assume that $\mathbf{v}^T A^{-1} \mathbf{u} \neq -1$. Show that

$$(A + \mathbf{u}\mathbf{v}^T)^{-1} = A^{-1} - \frac{A^{-1}\mathbf{u}\mathbf{v}^T A^{-1}}{1 + \mathbf{v}^T A^{-1}\mathbf{u}}. \tag{1}$$

Solution 76.   (i) From $\mathbf{v}^T A^{-1} \mathbf{u} = -1$ it follows that $\mathbf{u} \neq \mathbf{0}$, $\mathbf{v} \neq \mathbf{0}$ and $A^{-1} \mathbf{u} \neq \mathbf{0}$. Now we have

$$(A + \mathbf{u}\mathbf{v}^T)A^{-1}\mathbf{u} = AA^{-1}\mathbf{u} + \mathbf{u}\mathbf{v}^T A^{-1}\mathbf{u} = \mathbf{u} + \mathbf{u}(\mathbf{v}^T A^{-1}\mathbf{u}) = \mathbf{u} - \mathbf{u}$$
$$= \mathbf{0}$$

where we used $\mathbf{v}^T A^{-1} \mathbf{u} = -1$. Now $(A + \mathbf{u}\mathbf{v}^T)(A^{-1}\mathbf{u}) = \mathbf{0}$ is an eigenvalue equation with eigenvalue 0 and eigenvector $A^{-1}\mathbf{u}$. Since

$$\det(A + \mathbf{u}\mathbf{v}^T) = \lambda_1 \cdot \lambda_2 \cdot \ldots \cdot \lambda_n$$

where $\lambda_1, \ldots, \lambda_n$ are the eigenvalues of $A + \mathbf{u}\mathbf{v}^T$, we have $\det(A + \mathbf{u}\mathbf{v}^T) = 0$. Thus the matrix $A + \mathbf{u}\mathbf{v}^T$ is not invertible.

(ii) Since $\mathbf{v}^T A^{-1}\mathbf{u} \in \mathbb{R}$ we have the identity

$$I_n = I_n + \mathbf{u}\mathbf{v}^T A^{-1}(\mathbf{v}^T A^{-1}\mathbf{u}) - \mathbf{u}(\mathbf{v}^T A^{-1}\mathbf{u})\mathbf{v}^T A^{-1}.$$

Thus

$$I_n = I_n + \mathbf{u}\mathbf{v}^T A^{-1} - \mathbf{u}\mathbf{v}^T A^{-1} + \mathbf{u}\mathbf{v}^T A^{-1}\mathbf{v}^T A^{-1}\mathbf{u} - \mathbf{u}\mathbf{v}^T A^{-1}\mathbf{u}\mathbf{v}^T A^{-1}$$

and therefore

$$(A + \mathbf{u}\mathbf{v}^T)(A + \mathbf{u}\mathbf{v}^T)^{-1} = (A + \mathbf{u}\mathbf{v}^T)A^{-1} - \frac{(A + \mathbf{u}\mathbf{v}^T)A^{-1}\mathbf{u}\mathbf{v}^T A^{-1}}{1 + \mathbf{v}^T A^{-1}\mathbf{u}}.$$

Multiplying with $(A + \mathbf{u}\mathbf{v}^T)^{-1}$ provides equation (1).

Problem 77.   The *collocation polynomial* $p(x)$ for unequally-spaced arguments $x_0, x_1, \ldots, x_n$ can be found by the determinant method

$$\det \begin{pmatrix} p(x) & 1 & x & x^2 & \cdots & x^n \\ y_0 & 1 & x_0 & x_0^2 & \cdots & x_0^n \\ y_1 & 1 & x_1 & x_1^2 & \cdots & x_1^n \\ \vdots & \vdots & \vdots & \vdots & \ddots & \vdots \\ y_n & 1 & x_n & x_n^2 & \cdots & x_n^n \end{pmatrix} = 0$$

where $p(x_k) = y_k$ for $k = 0, 1, \ldots, n$. Apply it to $(n = 2)$

$$p(x_0 = 0) = 1 = y_0, \qquad p(x_1 = 1/2) = 9/4 = y_1, \qquad p(x_2 = 1) = 4 = y_2.$$

Solution 77.   We have

$$\det \begin{pmatrix} p(x) & 1 & x & x^2 \\ 1 & 1 & 0 & 0 \\ 9/4 & 1 & 1/2 & 1/4 \\ 4 & 1 & 1 & 1 \end{pmatrix} = 0.$$

It follows that $p(x) = 1 + 2x + x^2$.

Problem 78.   Let M be the *Minkowski space* endowed with the standard coordinates x_0, x_1, x_2, x_3 and the metric tensor field

$$g = -dx_0 \otimes dx_0 + dx_1 \otimes dx_1 + dx_2 \otimes dx_2 + dx_3 \otimes dx_3$$

and the quadratic form

$$q(x) = -(x_0)^2 + (x_1)^2 + (x_2)^2 + (x_3)^2.$$

Let $H(2)$ be the vector space of 2×2 hermitian matrices. Consider the map $\varphi : M \to H(2)$

$$\varphi(x) = X = \begin{pmatrix} x_0 + x_3 & x_1 - ix_2 \\ x_1 + ix_2 & x_0 - x_3 \end{pmatrix}$$

which is an isomorphism.
(i) Find the determinant of X.
(ii) Find the *Cayley transform* of X, i.e.

$$U = (X + iI_2)^{-1}(X - iI_2)$$

with the inverse

$$X = i(I_2 + U)(I_2 - U)^{-1}.$$

Solution 78.   (i) The determinant of X is given by

$$\det(X) = (x_0)^2 - ((x_1)^2 + (x_2)^2 + (x_3)^2) = -q(x).$$

(ii) For the Cayley transform of X we find

$$U = \frac{1}{-q(x) - 1 + 2ix_0} \begin{pmatrix} 1 - q(x) + 2ix_3 & 2(ix_1 + x_2) \\ 2(ix_1 - x_2) & 1 - q(x) - 2ix_3 \end{pmatrix}.$$

Problem 79.   Let A be an $n \times n$ skew-symmetric matrix over $\mathbb{R}$. Show that

$$\det(I_n + A) \geq 1$$

with equality holding if and only if $A = 0_n$.

Solution 79.   First note that

$$\det(I_n + A) = \det((I_n + A)^T) = \det(I_n - A).$$

The matrix $I_n + AA^T$ is positive definite and all its eigenvalues are greater or equal to 1. Thus $\det(I_n + AA^T) \geq 1$. Now

$$\det(I_n + AA^T) = \det(I_n - A^2) = \det((I_n - A)(I_n + A))$$
$$= \det((I_n + A)^2) \geq 1.$$

Thus $\det(I_n + A) \geq 1$ or $\det(I_n + A) \leq -1$. Since the eigenvalues of $I_n + A$ are either 1 or form conjugate pairs it follows that $\det(I_n + A) \geq 0$ and consequently $\det(I_n + A) \geq 1$.

Problem 80.   Consider the $(n-1) \times (n-1)$ symmetric matrix over $\mathbb{R}$

$$A_n = \begin{pmatrix} 3 & 1 & 1 & \cdots & 1 \\ 1 & 4 & 1 & \cdots & 1 \\ 1 & 1 & 5 & \cdots & 1 \\ \vdots & \vdots & \vdots & \ddots & \vdots \\ 1 & 1 & 1 & \cdots & n+1 \end{pmatrix}$$

with $n \geq 2$. Let $D_n = \det(A_n)$. Is the sequence $\{D_n/n!\}$ bounded?

Solution 80.   If we expand the last row we obtain the recursion

$$D_n = nD_{n-1} + (n-1)!, \qquad n = 3, 4, \ldots$$

with the initial condition $D_2 = 3$. We divide by $n!$ to obtain

$$\frac{D_n}{n!} = \frac{D_{n-1}}{(n-1)!} + \frac{1}{n}.$$

Hence

$$D_n = n! \left(1 + \frac{1}{2} + \cdots + \frac{1}{n} \right).$$

Therefore the sequence $\{D_n/n!\}$ is the nth partial sum of the *harmonic series*, which is unbounded as $n \to \infty$.

Problem 81.   The *hyperdeterminant* $\mathrm{Det}(A)$ of the three-dimensional array $A = (a_{ijk}) \in \mathbb{R}^{2 \times 2 \times 2}$ can be calculated as follows

$$\mathrm{Det}(A) = \frac{1}{4} \left(\det \left(\begin{pmatrix} a_{000} & a_{010} \\ a_{001} & a_{011} \end{pmatrix} + \begin{pmatrix} a_{100} & a_{110} \\ a_{101} & a_{111} \end{pmatrix} \right) \right.$$

$$- \det \left(\begin{pmatrix} a_{000} & a_{010} \\ a_{001} & a_{011} \end{pmatrix} - \begin{pmatrix} a_{100} & a_{110} \\ a_{101} & a_{111} \end{pmatrix} \right) \Bigg)^2$$

$$\left. - 4 \det \begin{pmatrix} a_{000} & a_{010} \\ a_{001} & a_{011} \end{pmatrix} \det \begin{pmatrix} a_{100} & a_{110} \\ a_{101} & a_{111} \end{pmatrix} \right.$$

Assume that only one of the coefficients a_{ijk} is nonzero. Calculate the hyperdeterminant.

Solution 81.   Since no terms of the form a_{ijk}^2 appear in calculation the determinants we find that the hyperdeterminant of A is 0.

Problem 82.   Let $\epsilon_{00} = \epsilon_{11} = 0$, $\epsilon_{01} = 1$, $\epsilon_{10} = -1$, i.e. we consider the 2×2 matrix

$$\epsilon = \begin{pmatrix} 0 & 1 \\ -1 & 0 \end{pmatrix}.$$

Then the determinant of a 2×2 matrix $A_2 = (a_{ij})$ with $i, j = 0, 1$ can be defined as

$$\det(A_2) := \frac{1}{2} \sum_{i=0}^{1} \sum_{j=0}^{1} \sum_{\ell=0}^{1} \sum_{m=0}^{1} \epsilon_{ij} \epsilon_{\ell m} a_{i\ell} a_{jm}.$$

Thus $\det(A_2) = a_{00}a_{11} - a_{01}a_{10}$. In analogy the *hyperdeterminant* $\mathrm{Det}(A_3)$ of the $2 \times 2 \times 2$ array $A_3 = (a_{ijk})$ with $i, j, k = 0, 1$ is defined as

$$-\frac{1}{2} \sum_{ii'=0}^{1} \sum_{jj'=0}^{1} \sum_{kk'=0}^{1} \sum_{mm'=0}^{1} \sum_{nn'=0}^{1} \sum_{pp'=0}^{1} \epsilon_{ii'} \epsilon_{jj'} \epsilon_{kk'} \epsilon_{mm'} \epsilon_{nn'} \epsilon_{pp'} a_{ijk} a_{i'j'm} a_{npk} a_{n'p'm'}.$$

Calculate $\mathrm{Det}(A_3)$.

Solution 82. There are $2^8 = 256$ terms for $\mathrm{Det}(A)$, but only 24 are nonzero. We find

$$a_{000}^2 a_{111}^2 + a_{001}^2 a_{110}^2 + a_{001}^2 a_{101}^2 + a_{100} a_{011}^2$$
$$- 2(a_{000}a_{001}a_{110}a_{111} + a_{000}a_{010}a_{101}a_{111} + a_{000}a_{100}a_{011}a_{111} + a_{001}a_{010}a_{101}a_{110}$$
$$+ a_{001}a_{100}a_{011}a_{110} + a_{010}a_{100}a_{011}a_{101})$$
$$+ 4(a_{000}a_{011}a_{101}a_{110} + a_{001}a_{010}a_{100}a_{111}).$$

Supplementary Problems

Problem 1. Let A, B be 2×2 matrices over $\mathbb{C}$. What are the conditions on A, B such that $\mathrm{tr}(AB) = \mathrm{tr}(A)\mathrm{tr}(B)$?

Problem 2. Given a 2×2 matrix M over $\mathbb{C}$. What can be said about the determinant of $M \otimes I_2 - I_2 \otimes M$?

Problem 3. (i) What is condition on $\alpha \in \mathbb{R}$ such that

$$F(\alpha) = \begin{pmatrix} 1 & \alpha & \alpha \\ \alpha & 1 & \alpha \\ \alpha & \alpha & 1 \end{pmatrix}$$

is invertible?
(ii) What is the condition on $\alpha, \beta \in \mathbb{R}$ such that

$$G(\alpha, \beta) = \begin{pmatrix} 1 & \beta & \beta & \beta \\ \beta & 1 & \alpha & \alpha \\ \beta & \alpha & 1 & \alpha \\ \beta & \alpha & \alpha & 1 \end{pmatrix}$$

is invertible?

Problem 4. Let A, B be $n \times n$ matrices over $\mathbb{C}$. Then $\mathrm{tr}([A, B]) = 0$. Consider $n = 3$ and

$$C = \begin{pmatrix} 1 & 0 & 0 \\ 0 & 0 & 0 \\ 0 & 0 & -1 \end{pmatrix}$$

with $\mathrm{tr}(C) = 0$. Construct 3×3 matrices A and B such that $[A, B] = C$.

Problem 5.   Consider the k vectors

$$\mathbf{v}_j = \begin{pmatrix} v_{1j} \\ v_{2j} \\ \vdots \\ v_{nj} \end{pmatrix} \quad j = 1, \ldots, k$$

in the Euclidean space $\mathbb{R}^n$. The *parallelepiped* determined by these vectors has the k-dimensional area

$$\sqrt{\det(V^T V)}$$

where V is the $n \times k$ matrix with $\mathbf{v}_1, \mathbf{v}_2, \ldots, \mathbf{v}_k$ as its columns. Apply it to the two vectors in $\mathbb{R}^3$

$$\mathbf{v}_1 = \frac{1}{\sqrt{2}} \begin{pmatrix} 1 \\ 0 \\ 1 \end{pmatrix}, \quad \mathbf{v}_2 = \frac{1}{\sqrt{2}} \begin{pmatrix} 1 \\ 0 \\ -1 \end{pmatrix}.$$

Problem 6.   The equation of a *hyperplane* passing through the n points $\mathbf{x}_1$, $\mathbf{x}_2, \ldots, \mathbf{x}_n$ in the Euclidean space $\mathbb{R}^n$ is given by

$$\det \begin{pmatrix} 1 & 1 & 1 & \cdots & 1 \\ \mathbf{x} & \mathbf{x}_1 & \mathbf{x}_2 & \cdots & \mathbf{x}_n \end{pmatrix} = 0.$$

Let $n = 3$. Apply it to the three vectors in $\mathbb{R}^3$

$$\mathbf{x}_1 = \frac{1}{\sqrt{2}} \begin{pmatrix} 1 \\ 0 \\ 1 \end{pmatrix}, \quad \mathbf{x}_2 = \begin{pmatrix} 0 \\ 1 \\ 0 \end{pmatrix}, \quad \mathbf{x}_3 = \frac{1}{\sqrt{2}} \begin{pmatrix} 1 \\ 0 \\ 1 \end{pmatrix}.$$

Problem 7.   Let A, B be 2×2 matrices. Show that

$$[A, B]_+ \equiv AB + BA \equiv (\mathrm{tr}(AB) - \mathrm{tr}(A)\mathrm{tr}(B))I_2 + \mathrm{tr}(A)B + \mathrm{tr}(B)A.$$

Can this identity be extended to 3×3 matrices?

Problem 8.   (i) Find all nonzero 2×2 matrices A and B such that $BA^* = \mathrm{tr}(BA^*)A$.
(ii) Find all 2×2 matrices A such that $A^2 = \mathrm{tr}(A)A$. Calculate $\det(A)$ and $\det(A^2)$ of such a matrix.
(iii) Let B be a 2×2 matrix over $\mathbb{C}$. Assume that $B^2 = I_2$ and thus $\mathrm{tr}(B^2) = 2$. What can be said about the trace of B?
(iv) Find all 2×2 matrices C with $\mathrm{tr}(C^2) - (\mathrm{tr}(C))^2 = 0$.
(v) Let D be a 2×2 matrix over $\mathbb{C}$. Find the condition on D such that $\det(D) = \mathrm{tr}(D^2)$. Find the condition on D such that $\det(D \otimes D) = \mathrm{tr}(D \otimes D)$.

Problem 9.   Let A, B be $n \times n$ matrices. Let U, V be unitary $n \times n$ matrices. Show that

$$\mathrm{tr}((U \otimes V)(A \otimes B)(U^* \otimes V^*)) = \mathrm{tr}(A)\mathrm{tr}(B)$$

applying cyclic invariance and $UU^* = I_n$, $VV^* = I_n$.

Problem 10. Let A, B be 2×2 matrices over $\mathbb{C}$ with $\text{tr}(A) = 0$ and $\text{tr}(B) = 0$. Show that the condition on A and B such that $\text{tr}(AB) = 0$ is given by

$$2a_{11}b_{11} + a_{12}b_{21} + a_{21}b_{12} = 0.$$

Problem 11. Let A, B be nonzero 2×2 matrices over $\mathbb{R}$ with $\det(A \otimes B) = 0$. Can we conclude that both A and B are noninvertible?

Problem 12. Consider the symmetric $n \times n$ band matrix $(n \geq 3)$

$$M_n = \begin{pmatrix} 1 & 1 & 0 & \cdots & 0 & 0 \\ 1 & 1 & 1 & \cdots & 0 & 0 \\ 0 & 1 & 1 & \cdots & 0 & 0 \\ \vdots & \vdots & \vdots & \ddots & \vdots & \vdots \\ 0 & 0 & 0 & \cdots & 1 & 1 \\ 0 & 0 & 0 & \cdots & 1 & 1 \end{pmatrix}$$

with the elements in the field $\mathbb{F}_2$. Show that

$$\det(M_n) = \det(M_{n-1}) - \det(M_{n-2})$$

with the initial conditions $\det(M_3) = \det(M_4) = 1$. Show that the solution is

$$\det(M_n) = \frac{2\sqrt{3}}{3} \cos\left(\frac{n\pi}{3} - \frac{\pi}{6}\right) \pmod 2.$$

Problem 13. Consider the nonnormal matrices

$$A_2 = \begin{pmatrix} 0 & 1 \\ 0 & 0 \end{pmatrix}, \qquad A_3 = \begin{pmatrix} 0 & 1 & 1 \\ 0 & 0 & 1 \\ 0 & 0 & 0 \end{pmatrix}.$$

Find $(\det(A_2^* A_2))^{1/2}$ and $(\det(A_3^* A_3))^{1/2}$. Extend to n dimensions.

Problem 14. Let A be a $n \times n$ matrix with $\det(A) = -1$. Show that $\det(A^{-1}) = -1$.

Problem 15. Let n be even. Consider a skew-symmetric $n \times n$ matrix A over $\mathbb{R}$. Let B be a symmetric $n \times n$ matrix over $\mathbb{R}$ with the entries of $B = (b_{jk})$ given by $b_{jk} = b_j b_k$ $(b_j, b_k \in \mathbb{R})$. Show that

$$\det(A + B) = \det(A).$$

Notice that $\det(B) = 0$. Let S_n be the *symmetric group*. Then we have

$$\det(A + B) = \sum_{\sigma \in S_n} (-1)^\sigma (a_{1\sigma(1)} + b_1 b_{\sigma(1)})(a_{2\sigma(2)} + b_2 b_{\sigma(2)}) \cdots (a_{n\sigma(n)} + b_n b_{\sigma(n)}).$$

Problem 16. Let $h > 0$ and $b > a$. Consider the matrix

$$M(a, b, h) = \begin{pmatrix} 1 & ah \\ 1 & bh \end{pmatrix}.$$

Show that $\det(M) = h(b - a) > 0$ and the inverse of M is given by

$$M^{-1}(a, b, h) = \begin{pmatrix} b/(b-a) & -a/(b-a) \\ -1/((b-a)h) & 1/((b-a)h) \end{pmatrix}.$$

Problem 17. The $n \times n$ permutation matrices form a group under matrix multiplications. Show that $\det(I_n - P) = 0$ for any $n \times n$ permutation matrix.

Problem 18. Let $\mathbf{u}, \mathbf{v}, \mathbf{w}$ be column vectors in $\mathbb{R}^3$. We form the 3×3 matrix

$$M = (\mathbf{u} \quad \mathbf{v} \quad \mathbf{w}).$$

Show that $\det(M) = (\mathbf{u} \times \mathbf{v}) \cdot \mathbf{w}$.

Problem 19. (i) Let $n \geq 2$. Consider the $n \times n$ matrix

$$A = \begin{pmatrix} 1 & 2 & 3 & \cdots & n-1 & n \\ 2 & 3 & 4 & \cdots & n & 1 \\ 3 & 4 & 5 & \cdots & 1 & 2 \\ \vdots & \vdots & \vdots & \ddots & \vdots & \vdots \\ n & 1 & 2 & \cdots & n-2 & n-1 \end{pmatrix}.$$

Show that $\det(A) = (-1)^{n(n-1)/2} \frac{1}{2}(n+1)n^{n-1}$.
(ii) Let $n \geq 2$. Consider the $n \times n$ matrix

$$M(x) = \begin{pmatrix} c_1 & x & x & \cdots & x & x \\ x & c_2 & x & \cdots & x & x \\ x & x & c_3 & \cdots & x & x \\ \vdots & \vdots & \vdots & \ddots & \vdots & \vdots \\ x & x & x & \cdots & x & c_n \end{pmatrix}.$$

Show that $\det(M(x)) = (-1)^n(P(x) - x dP(x)/dx)$, where

$$P(x) = (x - c_1)(x - c_2) \cdots (x - c_n).$$

Problem 20. (i) Consider the 2×2 matrix

$$C = \begin{pmatrix} 0 & 1 \\ 1 & 1 \end{pmatrix} \quad \Rightarrow \quad \operatorname{tr}(C) = 1, \ \operatorname{tr}(C^2) = 3.$$

Show that $\operatorname{tr}(C^k) = \operatorname{tr}(C^{k-1}) + \operatorname{tr}(C^{k-2})$ with $k = 3, 4, \ldots$.

Problem 21. Let A be an $n \times n$ matrix. Assume that

$$\text{tr}(A^j) = 0 \quad \text{for} \quad j = 1, \ldots, n.$$

Can we conclude that $\det(A) = 0$?

Problem 22. Let A_1, A_2, A_3 be 2×2 matrices over $\mathbb{R}$. We define

$$c_j = \text{tr}(A_j), \quad j = 1, 2, 3; \qquad c_{jk} = \text{tr}(A_j A_k), \quad j, k = 1, 2, 3.$$

(i) Given the coefficients c_j, c_{jk} reconstruct the matrices A_1, A_2, A_3.
(ii) Apply the result to the case $c_1 = c_2 = c_3 = 0$

$$c_{11} = 0, \, c_{12} = 1, \, c_{13} = 0, \, c_{21} = 1, \, c_{22} = 0, \, c_{23} = 0, \, c_{31} = 0, \, c_{32} = 0, \, c_{33} = 2$$

and show that

$$A_1 = \begin{pmatrix} 0 & 1 \\ 0 & 0 \end{pmatrix}, \quad A_2 = \begin{pmatrix} 0 & 0 \\ 1 & 0 \end{pmatrix}, \quad A_3 = \begin{pmatrix} 1 & 0 \\ 0 & -1 \end{pmatrix}.$$

Problem 23. (i) Let B be a 2×2 matrix over $\mathbb{C}$. Given

$$d_1 = \det(B), \quad d_2 = \det(B^2), \quad d_3 = \det(B^3), \quad d_4 = \det(B^4).$$

Can we reconstruct B from d_1, d_2, d_3, d_4. Does it depend on whether the matrix B is normal?
(ii) Find all 2×2 matrices A_1, A_2, A_3 over $\mathbb{C}$ such that

$$\text{tr}(A_1) = \text{tr}(A_2) = \text{tr}(A_3) = 0, \quad \text{tr}(A_1 A_2) = \text{tr}(A_2 A_3) = \text{tr}(A_3 A_1) = 0.$$

(iii) Find all 3×3 matrices B_1, B_2, B_3 over $\mathbb{C}$ such that

$$\text{tr}(B_1) = \text{tr}(B_2) = \text{tr}(B_3) = 0, \quad \text{tr}(B_1 B_2) = \text{tr}(B_2 B_3) = \text{tr}(B_3 B_1) = 0.$$

(iv) Find all 2×2 matrices C such that $\det(C) = 1$, $\text{tr}(C) = 0$. Do these matrices form a group under matrix multiplication?

Problem 24. Let A, B be $n \times n$ matrices over $\mathbb{C}$. Assume that B is invertible. Show that $\det(I_n + BAB^{-1}) = \det(I_n + A)$.

Problem 25. An $n \times n$ matrix A is called *idempotent* if $A^2 = A$. Show that $\text{rank}(A) = \text{tr}(A)$.

Problem 26. Let A be a real symmetric and positive definite $n \times n$ matrix. Then

$$\ln(\det(A)) \leq \text{tr}(A) - n.$$

For the identity matrix I_n we obtain equality. Show that for the matrix

$$A = \frac{1}{2} \begin{pmatrix} 3 & -1 \\ -1 & 3 \end{pmatrix}$$

we have $n = 2$, $\det(A) = 2$, $\mathrm{tr}(A) = 3$ and $\ln(2) < 1$.

Problem 27.   Let A, B, C be $n \times n$ matrices over $\mathbb{C}$. Let $\mathbf{a}_j$, $\mathbf{b}_j$, $\mathbf{c}_j$ ($j = 1, \ldots, n$) be the j-th column of A, B, C, respectively. Show that if for some $k \in \{1, \ldots, n\}$

$$\mathbf{c}_k = \mathbf{a}_k + \mathbf{b}_k$$

and

$$\mathbf{c}_j = \mathbf{a}_j = \mathbf{b}_j, \quad j = 1, \ldots, k-1, k+1, \ldots, n$$

then $\det(C) = \det(A) + \det(B)$.

Problem 28.   Consider the *Legendre polynomials* p_j ($j = 0, 1, \ldots$) with

$$p_0(x) = 1, \quad p_1(x) = x, \quad p_2(x) = \frac{1}{2}(3x^2 - 1),$$

$$p_3(x) = \frac{1}{2}(5x^3 - 3x), \quad p_4(x) = \frac{1}{8}(35x^4 - 30x^2 + 3).$$

Show that

$$\det \begin{pmatrix} p_0(x) & p_1(x) & p_2(x) \\ p_1(x) & p_2(x) & p_3(x) \\ p_2(x) & p_3(x) & p_4(x) \end{pmatrix} = (1 - x^2)^3 \begin{pmatrix} p_0(0) & 0 & p_2(0) \\ 0 & p_2(0) & 0 \\ p_2(0) & 0 & p_4(0) \end{pmatrix}.$$

Note that (x omitted)

$$\det \begin{pmatrix} p_0 & p_1 & p_2 \\ p_1 & p_2 & p_3 \\ p_2 & p_3 & p_4 \end{pmatrix}$$

$$= p_0(p_2 p_4 - (p_3)^2) - p_1(p_1 p_4 - p_2 p_3 + p_2(p_1 p_3 - (p_2)^2)$$

and $p_0(0) = 1$, $p_1(0) = 0$, $p_2(0) = -1/2$, $p_3(0) = 0$, $p_4(0) = 3/8$.

Problem 29.   Let A be an $n \times n$ diagonal matrix over $\mathbb{C}$. Let B be an $n \times n$ matrix over $\mathbb{C}$ with $b_{jj} = 0$ for all $j = 1, \ldots, n$. Can we conclude that all diagonal elements of the commutator $[A, B]$ are 0?

Problem 30.   Let A, B be $n \times n$ matrices over $\mathbb{R}$ with $\det(A) = 1$ and $\det(B) = 1$. This means A, B are elements of the Lie group $SL(n, \mathbb{R})$. Can we conclude that

$$\mathrm{tr}(AB) + \mathrm{tr}(AB^{-1}) = \mathrm{tr}(A)\mathrm{tr}(B)?$$

Problem 31.   Let H be a hermitian $n \times n$ matrix. Show that $\det(H + iI_n) \neq 0$.

Problem 32.   Let A be an 2×2 matrix over $\mathbb{R}$. Calculate $r = \mathrm{tr}(A^2) - (\mathrm{tr}(A))^2$. Show that the conditions on a_{jk} such that $r = 0$ are given by $a_{12}a_{21} = a_{11}a_{22}$, i.e. $\det(A) = 0$.

Problem 33. Consider a triangle embedded in $\mathbb{R}^3$. Let $\mathbf{v}_j = (x_j, y_j, z_j)$ $(j = 1, 2, 3)$ be the coordinates of the vertices. Then the area A of the triangle is given by

$$A = \frac{1}{2}\|(\mathbf{v}_2 - \mathbf{v}_1) \times (\mathbf{v}_1 - \mathbf{v}_3)\| = \frac{1}{2}\|(\mathbf{v}_3 - \mathbf{v}_1) \times (\mathbf{v}_3 - \mathbf{v}_2)\|$$

where $\times$ denotes the vector product and $\|.\|$ denotes the Euclidean norm. The area A of the triangle can also be found via

$$\frac{1}{2}\sqrt{\left(D\begin{pmatrix} x_1 & y_1 & 1 \\ x_2 & y_2 & 1 \\ x_3 & y_3 & 1 \end{pmatrix}\right)^2 + \left(D\begin{pmatrix} y_1 & z_1 & 1 \\ y_2 & z_2 & 1 \\ y_3 & z_3 & 1 \end{pmatrix}\right)^2 + \left(D\begin{pmatrix} z_1 & x_1 & 1 \\ z_2 & x_2 & 1 \\ z_3 & x_3 & 1 \end{pmatrix}\right)^2}$$

where D denotes the determinant. Consider

$$\mathbf{v}_1 = (1, 0, 0), \quad \mathbf{v}_2 = (0, 1, 0), \quad \mathbf{v}_3 = (0, 0, 1).$$

Find the area of the triangle using both expressions. Discuss. The triangle could be one of the faces of a tetrahedron.

Problem 34. Consider the 3×3 matrix M with entries

$$(M)_{jk} = x_k^{j-1}, \quad j, k = 1, 2, 3.$$

Find the determinant of this matrix.

Problem 35. The *permanent* and the *determinant* of an $n \times n$ matrix M over $\mathbb{C}$ are respectively defined as

$$\text{perm}(M) := \sum_{\pi \in S_n} \left(\prod_{j=1}^{n} M_{j,\pi(j)}\right), \quad \det(M) := \sum_{\pi \in S_n} (-1)^{sgn(\pi)} \left(\prod_{j=1}^{n} M_{j,\pi(j)}\right)$$

where S_n denotes the *symmetric group* on a set of n symbols. For an $n \times n$ matrix A and an $m \times m$ matrix B we know that

$$\det(A \otimes B) \equiv (\det(A))^m (\det(B))^n.$$

Study $\text{perm}(A \otimes B)$.

Problem 36. (i) Let A, B be positive definite $n \times n$ matrices. Show that $\text{tr}(A + B) \geq \text{tr}(A)$.
(ii) Let C, D be $n \times n$ matrices over $\mathbb{C}$. Is $\text{tr}(CC^*DD^*) \leq \text{tr}(CC^*)\text{tr}(DD^*)$?

Problem 37. Let A, B, C, D be $n \times n$ matrices. Assume that D is invertible. Consider the $(2n) \times (2n)$ matrix

$$M = \begin{pmatrix} A & B \\ C & D \end{pmatrix}.$$

Show that $\det(M) = \det(AD - BD^{-1}CD)$ using the identity

$$\begin{pmatrix} A & B \\ C & D \end{pmatrix} \begin{pmatrix} I_n & 0_n \\ -D^{-1}C & I_n \end{pmatrix} \equiv \begin{pmatrix} A - BD^{-1}C & B \\ 0_n & D \end{pmatrix}.$$

Problem 38.   Let $\mu \in \mathbb{R}$ and A be an 2×2 matrix over $\mathbb{R}$. Show that the determinant of the 4×4 matrix

$$\begin{pmatrix} -\mu I_2 & A \\ A^T & -\mu I_2 \end{pmatrix}$$

only contains even powers of μ.

Problem 39.   Consider the invertible 3×3 matrix A

$$A = \begin{pmatrix} 1/\sqrt{2} & 0 & 1/\sqrt{2} \\ 1/\sqrt{2} & 0 & -1/\sqrt{2} \\ 0 & 1 & 0 \end{pmatrix} \Rightarrow A^{-1} = \begin{pmatrix} 1/\sqrt{2} & 1/\sqrt{2} & 0 \\ 0 & 0 & 1 \\ 1/\sqrt{2} & -1/\sqrt{2} & 0 \end{pmatrix}.$$

Find a permutation matrix P such that $PA^{-1}P^{-1} = A$.

Problem 40.   Let $x, \epsilon \in \mathbb{R}$. Show that the determinant of the symmetric $n \times n$ matrix

$$A(x, \epsilon) = \begin{pmatrix} x+\epsilon & x & \cdots & x \\ x & x+\epsilon & & x \\ \vdots & & \ddots & \\ x & x & & x+\epsilon \end{pmatrix}$$

is given by $\det(A(x, \epsilon)) = \epsilon^{n-1}(nx + \epsilon)$.

Problem 41.   Show that the determinant of the $2n \times 2n$ matrix

$$\Omega = \begin{pmatrix} 0_n & I_n \\ -I_n & 0_n \end{pmatrix} \equiv \begin{pmatrix} 0 & 1 \\ -1 & 0 \end{pmatrix} \otimes I_n$$

is given by $\det(\Omega) = \left(\det \begin{pmatrix} 0 & 1 \\ -1 & 0 \end{pmatrix} \right)^n (\det(I_n))^2 = 1$.

Problem 42.   Given a $2 \times 2 \times 2$ *hypermatrix* $A = (a_{jk\ell})$, $j, k, \ell = 0, 1$ and the 2×2 matrix

$$S = \begin{pmatrix} s_{00} & s_{01} \\ s_{10} & s_{11} \end{pmatrix}.$$

The multiplication AS which is again a 2×2 hypermatrix is defined by

$$(AS)_{jk\ell} := \sum_{r=0}^{1} a_{jkr} s_{r\ell}.$$

Assume that $\det(S) = 1$, i.e. $S \in SL(2, \mathbb{C})$. Show that $\mathrm{Det}(AS) = \mathrm{Det}(A)$. This is a typical problem to apply computer algebra.

Problem 43.   The *Levi-Civita symbol* (also called completely antisymmetric constant tensor) is defined by

$$\epsilon_{j_1,j_2,\dots,j_n} := \begin{cases} +1 & \text{if } j_1, j_2, \dots, j_n \text{ is an even permutation of } 12\dots n \\ -1 & \text{if } j_1, j_2, \dots, j_n \text{ is an odd permutation of } 12\dots n \\ 0 & \text{otherwise} \end{cases}$$

Let δ_{jk} be the Kronecker delta. Show that

$$\epsilon_{j_1,j_2,\dots,j_n} \epsilon_{k_1,k_2,\dots,k_n} = \det \begin{pmatrix} \delta_{j_1 k_1} & \delta_{j_2 k_1} & \cdots & \delta_{j_n k_1} \\ \delta_{j_1 k_2} & \delta_{j_2 k_2} & \cdots & \delta_{j_n k_2} \\ \vdots & \vdots & \ddots & \vdots \\ \delta_{j_1 k_n} & \delta_{j_2 k_n} & \cdots & \delta_{j_n k_n} \end{pmatrix}.$$

Chapter 5

Eigenvalues and Eigenvectors

Let A be an $n \times n$ matrix over $\mathbb{C}$. The complex number λ is called an *eigenvalue* of A if and only if the matrix $(A - \lambda I_n)$ is singular. Let $\mathbf{x}$ be a nonzero column vector in $\mathbb{C}^n$. Then $\mathbf{x}$ is called an *eigenvector* belonging to (or associated with) the eigenvalue λ if and only if $(A - \lambda I_n)\mathbf{x} = 0$. The equation

$$A\mathbf{x} = \lambda\mathbf{x}$$

is called the *eigenvalue equation*. For an eigenvector $\mathbf{x}$ of A, $A\mathbf{x}$ is a scalar multiple of $\mathbf{x}$. The eigenvalues are found from the characteristic equation

$$\det(A - \lambda I_n) = 0.$$

From the eigenvalue equation we obtain $\mathbf{x}^* A^* = \lambda^* \mathbf{x}^*$ and thus

$$\mathbf{x}^* A^* A \mathbf{x} = \lambda\lambda^* \mathbf{x}^* \mathbf{x}.$$

If the matrix is hermitian then the eigenvalues are real. If the matrix is unitary then the absolute value of the eigenvalues is 1. If the matrix is skew-hermitian then the eigenvalues are purely imaginary. If the matrix is a projection matrix then the eigenvalues can only take the values 1 and 0. The eigenvalues of an upper or lower triangular matrix are the elements on the diagonal. Let A, B be $n \times n$ matrices. Then AB and BA have the same spectrum.

The trace of a square matrix A is the sum of its eigenvalues (counting degeneracy). The determinant of a square matrix A is the product of its eigenvalues (counting degeneracy). Let f be an analytic function $f : \mathbb{R} \to \mathbb{R}$ or $f : \mathbb{C} \to \mathbb{C}$. Then from $A\mathbf{x} = \lambda\mathbf{x}$ we obtain $f(A)\mathbf{x} = f(\lambda)\mathbf{x}$.

Problem 1. (i) Find the eigenvalues and normalized eigenvectors of the 2×2 matrix

$$A = \begin{pmatrix} \sin(\theta) & \cos(\theta) \\ -\cos(\theta) & \sin(\theta) \end{pmatrix}.$$

Are the eigenvectors orthogonal to each other?
(ii) Find the eigenvalues and normalized eigenvectors of the 2×2 matrix

$$\begin{pmatrix} \cos(\theta) & -e^{i\phi}\sin(\theta) \\ e^{-i\phi}\sin(\theta) & \cos(\theta) \end{pmatrix}.$$

(iii) Consider the normalized vector in $\mathbb{C}^3$

$$\mathbf{n} = \begin{pmatrix} \sin(\theta)\cos(\phi) \\ \sin(\theta)\sin(\phi) \\ \cos(\theta) \end{pmatrix}.$$

Calculate the 2×2 matrix $U(\theta, \phi) = \mathbf{n} \cdot \boldsymbol{\sigma} \equiv n_1\sigma_1 + n_2\sigma_2 + n_3\sigma_3$, where σ_1, σ_2, σ_3 are the Pauli spin matrices. Is the matrix $U(\theta, \phi)$ unitary? Find the trace and the determinant. Is the matrix $U(\theta, \phi)$ hermitian? Find the eigenvalues and normalized eigenvectors of $U(\theta, \phi)$.

Solution 1. (i) From the characteristic equation $\lambda^2 - 2\lambda\sin(\theta) + 1 = 0$ we obtain the eigenvalues

$$\lambda_1 = \sin(\theta) + i\cos(\theta), \qquad \lambda_2 = \sin(\theta) - i\cos(\theta).$$

The corresponding normalized eigenvectors are

$$\mathbf{x}_1 = \frac{1}{\sqrt{2}} \begin{pmatrix} -i \\ 1 \end{pmatrix}, \qquad \mathbf{x}_2 = \frac{1}{\sqrt{2}} \begin{pmatrix} i \\ 1 \end{pmatrix}.$$

We see that $\mathbf{x}_1^*\mathbf{x}_2 = 0$. Thus the eigenvectors are orthogonal to each other.
(ii) The matrix is unitary. Thus the absolute value of the eigenvalues must be 1. The eigenvalues are $e^{-i\theta}$, $e^{i\theta}$ with the corresponding normalized eigenvectors

$$\frac{1}{\sqrt{2}} \begin{pmatrix} 1 \\ ie^{-i\phi} \end{pmatrix}, \qquad \frac{1}{\sqrt{2}} \begin{pmatrix} 1 \\ -ie^{-i\phi} \end{pmatrix}.$$

(iii) We find

$$U(\theta, \phi) = \begin{pmatrix} \cos(\theta) & e^{-i\phi}\sin(\theta) \\ e^{i\phi}\sin(\theta) & -\cos(\theta) \end{pmatrix}.$$

The matrix is unitary since $U(\theta, \phi)U^*(\theta, \phi) = I_2$. The trace is zero and the determinant is -1. The matrix is also hermitian. Since the matrix is unitary, hermitian and the trace is equal to 0, we obtain the eigenvalues $\lambda_1 = +1$ and $\lambda_2 = -1$. Using the identities

$$\sin(\theta) \equiv 2\sin(\theta/2)\cos(\theta/2), \qquad \cos(\theta) \equiv \cos^2(\theta/2) - \sin^2(\theta/2)$$

we obtain the normalized eigenvectors

$$\begin{pmatrix} e^{-i\phi/2}\cos(\theta/2) \\ e^{i\phi/2}\sin(\theta/2) \end{pmatrix}, \qquad \begin{pmatrix} -e^{-i\phi/2}\sin(\theta/2) \\ e^{i\phi/2}\cos(\theta/2) \end{pmatrix}.$$

Problem 2. Find all solutions of the linear equation

$$\begin{pmatrix} \cos(\theta) & -\sin(\theta) \\ -\sin(\theta) & -\cos(\theta) \end{pmatrix} \mathbf{x} = \mathbf{x}, \qquad \theta \in \mathbb{R} \tag{1}$$

with the condition that $\mathbf{x} \in \mathbb{R}^2$ and $\mathbf{x}^T\mathbf{x} = 1$, i.e. the vector $\mathbf{x}$ must be normalized. What type of equation is (1)?

Solution 2. We find

$$\mathbf{x} = \begin{pmatrix} \cos(\theta/2) \\ -\sin(\theta/2) \end{pmatrix}.$$

Obviously (1) is an eigenvalue equation and the eigenvalue is $+1$.

Problem 3. (i) Let $\epsilon \in \mathbb{R}$. Find the eigenvalues and eigenvectors of the 2×2 matrix

$$A(\epsilon) = \frac{1}{\sqrt{1+\epsilon^2}} \begin{pmatrix} \epsilon & 1 \\ 1 & -\epsilon \end{pmatrix}.$$

For $\epsilon = 0$ we obtain the Pauli spin matrix σ_1, for $\epsilon = 1$ we have the Hadamard matrix and for $\epsilon \to \infty$ we obtain the Pauli spin matrix σ_3.
(ii) Let $\epsilon \in \mathbb{R}$. Find the eigenvalues and eigenvectors of the 2×2 matrix

$$B(\epsilon) = \begin{pmatrix} \tanh(\epsilon) & 1/\cosh(\epsilon) \\ 1/\cosh(\epsilon) & -\tanh(\epsilon) \end{pmatrix}.$$

The matrices $A(\epsilon)$ and $B(\epsilon)$ are connected via the invertible transformation $\epsilon \to \sinh(\epsilon)$.

Solution 3. (i) The eigenvalues of $A(\epsilon)$ are $+1$ and -1 and thus are independent of ϵ. The corresponding eigenvectors are

$$\begin{pmatrix} \sqrt{1+\epsilon^2} \\ 1+\epsilon^2 - \epsilon\sqrt{1+\epsilon^2} \end{pmatrix} \quad \text{and} \quad \begin{pmatrix} \sqrt{1+\epsilon^2} \\ -1-\epsilon^2 - \epsilon\sqrt{1+\epsilon^2} \end{pmatrix}.$$

(ii) The eigenvalues of $B(\epsilon)$ are $+1$ and -1 and thus are independent of ϵ. The corresponding eigenvectors are

$$\begin{pmatrix} 1/\cosh(\epsilon) \\ 1 - \tanh(\epsilon) \end{pmatrix} \quad \text{and} \quad \begin{pmatrix} 1/\cosh(\epsilon) \\ -1 - \tanh(\epsilon) \end{pmatrix}.$$

Problem 4. Let A be a 2×2 matrix. Assume that $\det(A) = 0$ and $\operatorname{tr}(A) = 0$. What can be said about the eigenvalues of A?

Solution 4. We have $\lambda_1\lambda_2 = 0$ and $\lambda_1 + \lambda_2 = 0$. From these conditions it follows that both eigenvalues must be 0. Is such a matrix normal?

Problem 5. Find all 2×2 matrices A over $\mathbb{C}$ which admit the normalized eigenvectors

$$\mathbf{v}_1 = \frac{1}{\sqrt{2}} \begin{pmatrix} 1 \\ 1 \end{pmatrix}, \qquad \mathbf{v}_2 = \frac{1}{\sqrt{2}} \begin{pmatrix} 1 \\ -1 \end{pmatrix}$$

with the corresponding eigenvalues λ_1 and λ_2.

Solution 5. From the eigenvalue equations $A\mathbf{v}_1 = \lambda_1 \mathbf{v}_1$, $A\mathbf{v}_2 = \lambda_2 \mathbf{v}_2$ we find after addition and subtracting to eliminate λ_1 and λ_2 that $a_{11} = a_{22}$, $a_{12} = a_{21}$. Inserting these equations into the eigenvalue equations we obtain the linear equations

$$\begin{pmatrix} 1 & 1 \\ 1 & -1 \end{pmatrix} \begin{pmatrix} a_{11} \\ a_{12} \end{pmatrix} = \begin{pmatrix} \lambda_1 \\ \lambda_2 \end{pmatrix} \Leftrightarrow \begin{pmatrix} a_{11} \\ a_{12} \end{pmatrix} = \frac{1}{2} \begin{pmatrix} 1 & 1 \\ 1 & -1 \end{pmatrix} \begin{pmatrix} \lambda_1 \\ \lambda_2 \end{pmatrix}.$$

Thus

$$a_{11} = a_{22} = \frac{1}{2}(\lambda_1 + \lambda_2), \qquad a_{12} = a_{21} = \frac{1}{2}(\lambda_1 - \lambda_2).$$

Since the normalized eigenvectors are pairwise orthonormal we can find A also with the *spectral theorem* $A = \lambda_1 \mathbf{v}_1 \mathbf{v}_1^* + \lambda_2 \mathbf{v}_2 \mathbf{v}_2^*$.

Problem 6. Find the eigenvalues and normalized eigenvectors of the 2×2 matrix

$$M(\theta) = \begin{pmatrix} \sin(\theta) & \cos(\theta) \\ \cos(\theta) & -\sin(\theta) \end{pmatrix}.$$

Solution 6. We have $\det(M(\theta)) = -1$ and $\operatorname{tr}(M(\theta)) = 0$. The eigenvalues are $+1$ and -1 and the corresponding normalized eigenvectors are

$$\frac{1}{\sqrt{2}} \begin{pmatrix} \sqrt{1 - \sin(\theta)} \\ \sqrt{1 + \sin(\theta)} \end{pmatrix}, \qquad \frac{1}{\sqrt{2}} \begin{pmatrix} \sqrt{1 + \sin(\theta)} \\ -\sqrt{1 - \sin(\theta)} \end{pmatrix}.$$

Extend the problem to the matrix

$$\begin{pmatrix} \sin(\theta) & e^{i\phi} \cos(\theta) \\ e^{-i\phi} \cos(\theta) & -\sin(\theta) \end{pmatrix}.$$

This matrix contains the Pauli spin matrices σ_1, σ_2, σ_3 for $\theta = 0$, $\phi = 0$; $\theta = 0$, $\phi = 3\pi/2$; $\theta = \pi/2$, $\phi = 0$.

Problem 7. Let $\mu_1, \mu_2 \in \mathbb{R}$. Consider the 2×2 matrix

$$A(\mu_1^2, \mu_2^2) = \begin{pmatrix} \mu_1^2 \cos^2(\theta) + \mu_2^2 \sin^2(\theta) & (\mu_2^2 - \mu_1^2) \cos(\theta) \sin(\theta) \\ (\mu_2^2 - \mu_1^2) \sin(\theta) \cos(\theta) & \mu_1^2 \sin^2(\theta) + \mu_2^2 \cos^2(\theta) \end{pmatrix}.$$

(i) Find the trace and determinant of $A(\mu_1^2, \mu_2^2)$.
(ii) Find the eigenvalues of $A(\mu_1^2, \mu_2^2)$.
Remark. Any positive semidefinite 2×2 matrix over $\mathbb{R}$ can be written in this form. Try the 2×2 matrix with all entries $+1$.

Solution 7. (i) Let λ_1, λ_2 be the eigenvalues $A(\mu_1^2, \mu_2^2)$. We have

$$\operatorname{tr}(A(\mu_1^2, \mu_2^2)) = \mu_1^2 + \mu_2^2 = \lambda_1 + \lambda_2.$$

(ii) We obtain $\det(A(\mu_1^2, \mu_2^2)) = \mu_1^2 \mu_2^2 = \lambda_1 \lambda_2$.

(iii) From (i) and (ii) we find that the eigenvalues are given by $\lambda_1 = \mu_1^2$, $\lambda_2 = \mu_2^2$. Thus the eigenvalues cannot be negative.

Extra. Find the eigenvalues of $A(\mu_1^2, \mu_2^2) \otimes A(\nu_1^2, \nu_2^2)$.

Problem 8.   (i) Let $\alpha \in \mathbb{R}$. Consider the matrices

$$A(\alpha) = \begin{pmatrix} \cos(\alpha) & -\sin(\alpha) \\ \sin(\alpha) & \cos(\alpha) \end{pmatrix}, \qquad B(\alpha) = \begin{pmatrix} \cos(\alpha) & \sin(\alpha) \\ \sin(\alpha) & -\cos(\alpha) \end{pmatrix}.$$

Find the trace and determinant of these matrices. Show that for the matrix $A(\alpha)$ the eigenvalues depend on α but the eigenvectors do not. Show that for the matrix $B(\alpha)$ the eigenvalues do not depend on α but the eigenvectors do.
(ii) Let $\alpha \in \mathbb{R}$. Consider the matrices

$$C(\alpha) = \begin{pmatrix} \cosh(\alpha) & \sinh(\alpha) \\ \sinh(\alpha) & \cosh(\alpha) \end{pmatrix}, \qquad D(\alpha) = \begin{pmatrix} \cosh(\alpha) & -\sinh(\alpha) \\ \sinh(\alpha) & -\cosh(\alpha) \end{pmatrix}.$$

Find the trace and determinant of these matrices. Show that for the matrix $C(\alpha)$ the eigenvalues depend on α but the eigenvectors do not. Show that for the matrix $D(\alpha)$ the eigenvalues do not depend on α but the eigenvectors do.

Solution 8.   (i) For the matrix $A(\alpha)$ we have $\det(A(\alpha)) = 1$ and $\mathrm{tr}(A(\alpha)) = 2\cos(\alpha)$. The eigenvalues are given by $\lambda_+ = e^{i\alpha}$, $\lambda_- = e^{-i\alpha}$ with the corresponding normalized eigenvectors

$$\frac{1}{\sqrt{2}} \begin{pmatrix} 1 \\ -i \end{pmatrix}, \qquad \frac{1}{\sqrt{2}} \begin{pmatrix} 1 \\ i \end{pmatrix}.$$

Thus the eigenvalues depend on α, but the eigenvectors do not depend on α. For the matrix $B(\alpha)$ we find that $\det(B(\alpha)) = -1$ and $\mathrm{tr}(B(\alpha)) = 0$. The eigenvalues are given by

$$\lambda_+ = +1, \qquad \lambda_- = -1$$

with the corresponding normalized eigenvectors

$$\begin{pmatrix} \cos(\alpha/2) \\ \sin(\alpha/2) \end{pmatrix}, \qquad \begin{pmatrix} \sin(\alpha/2) \\ -\cos(\alpha/2) \end{pmatrix}$$

where we utilized the identities

$$\sin(\alpha) \equiv 2\sin(\alpha/2)\cos(\alpha/2), \quad \cos(\alpha) \equiv \cos^2(\alpha/2) - \sin^2(\alpha/2).$$

Thus the eigenvalues do not depend on α but the eigenvectors do.
(ii) For the matrix $C(\alpha)$ we have $\det(C(\alpha)) = 1$ and $\mathrm{tr}(C(\alpha)) = 2\cosh(\alpha)$. The eigenvalues are given by $\lambda_+ = e^{\alpha}$, $\lambda_- = e^{-\alpha}$ with the corresponding normalized eigenvectors

$$\frac{1}{\sqrt{2}} \begin{pmatrix} 1 \\ 1 \end{pmatrix}, \qquad \frac{1}{\sqrt{2}} \begin{pmatrix} 1 \\ -1 \end{pmatrix}.$$

Thus the eigenvalues depend on α, but the eigenvectors do not depend on α. For the matrix $D(\alpha)$ we find that $\det(D(\alpha)) = -1$ and $\operatorname{tr}(D(\alpha)) = 0$. The eigenvalues are given by $\lambda_+ = +1$, $\lambda_- = -1$ with the corresponding normalized eigenvectors

$$\frac{1}{\sqrt{\cosh(\alpha)}} \begin{pmatrix} \cosh(\alpha/2) \\ \sinh(\alpha/2) \end{pmatrix}, \qquad \frac{1}{\sqrt{\cosh(\alpha)}} \begin{pmatrix} \sinh(\alpha/2) \\ \cosh(\alpha/2) \end{pmatrix}$$

where we utilized the identities

$$\sinh(\alpha) \equiv 2\sinh(\alpha/2)\cosh(\alpha/2), \quad \cosh(\alpha) \equiv \cosh^2(\alpha/2) + \sinh^2(\alpha/2).$$

Thus the eigenvalues do not depend on α but the eigenvectors do.

Problem 9. Let $a_{11}, a_{22} \in \mathbb{R}$ and $a_{12} \in \mathbb{C}$. Consider the 2×2 hermitian matrix

$$H = \begin{pmatrix} a_{11} & a_{12} \\ \bar{a}_{12} & a_{22} \end{pmatrix}$$

with the real eigenvalues λ_+ and λ_-. What conditions are imposed on the matrix elements of H if $\lambda_+ = \lambda_-$?

Solution 9. The eigenvalues are

$$\lambda_\pm = \frac{1}{2}(a_{11} + a_{22}) \pm \sqrt{a_{12}\bar{a}_{12} + (a_{11} - a_{22})^2/4}.$$

From $\lambda_+ = \lambda_-$ we obtain

$$\sqrt{a_{12}\bar{a}_{12} + (a_{11} - a_{22})^2/4} = 0.$$

Thus $a_{12}\bar{a}_{12} + (a_{11} - a_{22})^2/4 = 0$ and therefore $a_{12} = 0$, $a_{11} = a_{22}$.

Problem 10. (i) Let $A^T = (1/2, 1/2)^T$. Find the eigenvalues of the 2×2 matrix AA^T and the 1×1 matrix $A^T A$.
(ii) Let $\mathbf{x}$ be a nonzero column vector in $\mathbb{R}^n$. Then $\mathbf{x}\mathbf{x}^T$ is an $n \times n$ matrix and $\mathbf{x}^T\mathbf{x}$ is a real number. Show that $\mathbf{x}^T\mathbf{x}$ is an eigenvalue of $\mathbf{x}\mathbf{x}^T$ and $\mathbf{x}$ is the corresponding eigenvector.
(iii) Let $\mathbf{x}$ be a nonzero column vector in $\mathbb{R}^n$ and $n \geq 2$. Consider the $n \times n$ matrix $\mathbf{x}\mathbf{x}^T$. Find one nonzero eigenvalue and the corresponding eigenvector of this matrix.

Solution 10. (i) We have

$$AA^T = \begin{pmatrix} 1/4 & 1/4 \\ 1/4 & 1/4 \end{pmatrix}, \qquad A^T A = (1/2).$$

Thus the eigenvalues of the 2×2 matrix AA^T are $1/2$ and 0. The eigenvalue of the 1×1 matrix $A^T A$ is $1/2$. Thus the nonzero eigenvalues are the same.
(ii) Since matrix multiplication is associative we have $(\mathbf{x}\mathbf{x}^T)\mathbf{x} = \mathbf{x}(\mathbf{x}^T\mathbf{x}) = (\mathbf{x}^T\mathbf{x})\mathbf{x}$.

(iii) Since matrix multiplication is associative we have $(\mathbf{x}\mathbf{x}^T)\mathbf{x} = \mathbf{x}(\mathbf{x}^T\mathbf{x}) = (\mathbf{x}^T\mathbf{x})\mathbf{x}$. Thus $\mathbf{x}$ is an eigenvector and $\mathbf{x}^T\mathbf{x}$ is the nonzero eigenvalue.

Problem 11. Consider the Pauli spin matrix σ_1 with the eigenvalues $+1$ and -1 and the corresponding normalized eigenvectors

$$\mathbf{v}_1 = \frac{1}{\sqrt{2}}\begin{pmatrix} 1 \\ 1 \end{pmatrix}, \qquad \mathbf{v}_2 = \frac{1}{\sqrt{2}}\begin{pmatrix} 1 \\ -1 \end{pmatrix}.$$

With the two normalized eigenvectors we form the unitary matrix

$$U = \frac{1}{\sqrt{2}}\begin{pmatrix} 1 & 1 \\ 1 & -1 \end{pmatrix} \equiv \frac{1}{\sqrt{2}}(\sigma_3 + \sigma_1).$$

Find the eigenvalues and normalized eigenvectors of U.

Solution 11. The eigenvalues of U are given by $+1$ and -1 with the corresponding normalized eigenvectors

$$\mathbf{w}_1 = \frac{1}{\sqrt{4 - 2\sqrt{2}}}\begin{pmatrix} 1 \\ \sqrt{2} - 1 \end{pmatrix}, \qquad \mathbf{w}_2 = \frac{1}{\sqrt{4 + 2\sqrt{2}}}\begin{pmatrix} 1 \\ -\sqrt{2} - 1 \end{pmatrix}.$$

Problem 12. Let σ_1, σ_2, σ_3 be the Pauli spin matrices. Find the eigenvalues and normalized eigenvectors of 2×2 nonnormal matrix

$$\sigma_3 + i\sigma_1 = \begin{pmatrix} 1 & i \\ i & -1 \end{pmatrix}.$$

Solution 12. The eigenvalues are 0 (twice). There is only one (normalized) eigenvector

$$\frac{1}{\sqrt{2}}\begin{pmatrix} 1 \\ i \end{pmatrix}.$$

Problem 13. (i) Find all 2×2 matrices over $\mathbb{R}$ that admit only one eigenvector. (ii) Consider the 2×2 matrix

$$A(\epsilon) = \begin{pmatrix} 1 & 0 \\ \epsilon & 1 \end{pmatrix}, \qquad \epsilon \in \mathbb{R}.$$

Can one find a condition on the parameter ϵ so that $A(\epsilon)$ has only one eigenvector?

Solution 13. (i) A necessary condition is that the matrices admit only one eigenvalue (twice degenerate). The matrices are

$$\begin{pmatrix} a & b \\ 0 & a \end{pmatrix}, \qquad \begin{pmatrix} a & 0 \\ b & a \end{pmatrix}$$

where $b \neq 0$. The eigenvectors are

$$\begin{pmatrix} 1 \\ 0 \end{pmatrix}, \quad \begin{pmatrix} 0 \\ 1 \end{pmatrix}.$$

(ii) Obviously the eigenvalues are 1 (twice). Thus the eigenvalue equation is

$$\begin{pmatrix} 1 & 0 \\ \epsilon & 1 \end{pmatrix} \begin{pmatrix} x_1 \\ x_2 \end{pmatrix} = \begin{pmatrix} x_1 \\ x_2 \end{pmatrix}.$$

If $\epsilon \neq 0$ we have $x_2 = 0$ and therefore we find only one eigenvector. If $\epsilon = 0$ we find two linearly independent eigenvectors.

Problem 14. Find all 2×2 matrices over $\mathbb{R}$ which commute with

$$A = \begin{pmatrix} 0 & 1 \\ 0 & 0 \end{pmatrix}.$$

What is the relation between the eigenvectors of these matrices?

Solution 14. We have

$$\left[\begin{pmatrix} a & b \\ c & d \end{pmatrix}, \begin{pmatrix} 0 & 1 \\ 0 & 0 \end{pmatrix} \right] = \begin{pmatrix} 0 & 0 \\ 0 & 0 \end{pmatrix} \Rightarrow \begin{pmatrix} -c & a-d \\ 0 & c \end{pmatrix} = \begin{pmatrix} 0 & 0 \\ 0 & 0 \end{pmatrix}.$$

Hence $c = 0$ and $a = d$. Thus the matrices are of the form

$$\begin{pmatrix} a & b \\ 0 & a \end{pmatrix}; \quad a, b \in \mathbb{R}.$$

The eigenvectors of A are $\begin{pmatrix} x \\ 0 \end{pmatrix}$, $x \in \mathbb{R}/\{0\}$. The eigenvectors of

$$\begin{pmatrix} a & b \\ 0 & a \end{pmatrix}$$

are $\begin{pmatrix} x \\ 0 \end{pmatrix}$, $x \in \mathbb{R}/\{0\}$ and, when $b = 0$, $\begin{pmatrix} 0 \\ y \end{pmatrix}$, $y \in \mathbb{R}/\{0\}$. Thus the matrices share the eigenvector

$$\begin{pmatrix} x \\ 0 \end{pmatrix}$$

$x \in \mathbb{R}/\{0\}$. More generally, let A and B be $n \times n$ matrices over $\mathbb{C}$ which commute, i.e. $[A, B] = 0_n$. Let v be an eigenvector of A corresponding to the eigenvalue λ. Then

$$A(Bv) = B(Av) = \lambda(Bv).$$

Thus Bv is either 0 or Bv is also an eigenvector of A corresponding to λ. Now suppose the eigenspace of A corresponding to λ is one dimensional. Then $Bv = \mu v$ for some μ. It follows that A and B share one-dimensional eigenspaces.

Problem 15. (i) Let A, B be 2×2 matrices over $\mathbb{R}$ and vectors x, y in $\mathbb{R}^2$ such that $Ax = y$, $By = x$, $x^T y = 0$ and $x^T x = 1$, $y^T y = 1$. Show that AB and BA have an eigenvalue $+1$.

(ii) Find all 2×2 matrices A, B which satisfy the conditions given in (i). Use

$$\mathbf{x} = \begin{pmatrix} \cos(\alpha) \\ \sin(\alpha) \end{pmatrix}, \qquad \mathbf{y} = \begin{pmatrix} -\sin(\alpha) \\ \cos(\alpha) \end{pmatrix}.$$

Solution 15. (i) We have

$$B(A\mathbf{x}) = B\mathbf{y} = \mathbf{x} = (BA)\mathbf{x}, \quad A(B\mathbf{y}) = A\mathbf{x} = \mathbf{y} = (AB)\mathbf{y}.$$

Thus the matrices AB and BA have the eigenvalue $+1$.
(ii) From

$$A\mathbf{x} = \begin{pmatrix} a_{11} & a_{12} \\ a_{21} & a_{22} \end{pmatrix} \begin{pmatrix} \cos(\alpha) \\ \sin(\alpha) \end{pmatrix} = \begin{pmatrix} -\sin(\alpha) \\ \cos(\alpha) \end{pmatrix}$$

we obtain two equations for the unknown a_{11}, a_{12}, a_{21}, a_{22}

$$a_{11} \cos(\alpha) + a_{12}(1 + \sin(\alpha)) = 0, \quad (a_{21} - 1)\cos(\alpha) + a_{22} \sin(\alpha) = 0.$$

We eliminate a_{12} and a_{21}. The cases $\cos(\alpha) = 0$, $\sin(\alpha) \neq 0$ and $\sin(\alpha) = 0$, $\cos(\alpha) \neq 0$ are obvious. Assume now $\sin(\alpha) \neq 0$ and $\cos(\alpha) \neq 0$. Then we obtain

$$A = \begin{pmatrix} a_{11} & -1 - a_{11}\cos(\alpha)/\sin(\alpha) \\ 1 - a_{22}\sin(\alpha)/\cos(\alpha) & a_{22} \end{pmatrix}.$$

Similarly for the matrix B.

Problem 16. Let $A = (a_{jk})$ be a normal nonsymmetric 3×3 matrix over the real numbers. Show that

$$\mathbf{a} = \begin{pmatrix} a_1 \\ a_2 \\ a_3 \end{pmatrix} = \begin{pmatrix} a_{23} - a_{32} \\ a_{31} - a_{13} \\ a_{12} - a_{21} \end{pmatrix}$$

is an eigenvector of A.

Solution 16. Since A is normal with the underlying field $\mathbb{R}$ we have $AA^T = A^T A$. Since A is nonsymmetric the matrix

$$S := A - A^T = \begin{pmatrix} 0 & a_3 & -a_2 \\ -a_3 & 0 & a_1 \\ a_2 & -a_1 & 0 \end{pmatrix}$$

is skew-symmetric with rank 2. The *kernel* of S is therefore one-dimensional. We have $A\mathbf{a} = \mathbf{0}$. Therefore $\mathbf{a}$ is an element of the kernel. Since A is normal, we find

$$AS = A(A - A^T) = A^2 - AA^T = A^2 - A^T A = (A - A^T)A = SA.$$

If we multiply the equation $\mathbf{0} = S\mathbf{a}$ with A we obtain

$$\mathbf{0} = A\mathbf{0} = A(S\mathbf{a}) = (AS)\mathbf{a} = (SA)\mathbf{a} = S(A\mathbf{a}).$$

Thus besides **a**, $A\mathbf{a}$ also belongs to the one-dimensional kernel of S. Therefore $A\mathbf{a} = \lambda\mathbf{a}$.

Problem 17. Let $c \in \mathbb{R}$ and consider the symmetric 3×3 matrix

$$A(c) = \begin{pmatrix} c & 1 & 0 \\ 1 & c & 1 \\ 0 & 1 & c \end{pmatrix}.$$

Show that c is an eigenvalue of A and find the corresponding eigenvector. Find the two other eigenvalues and normalized eigenvectors.

Solution 17. From

$$A(c) \begin{pmatrix} x_1 \\ x_2 \\ x_3 \end{pmatrix} = c \begin{pmatrix} x_1 \\ x_2 \\ x_3 \end{pmatrix}$$

we obtain $x_2 = 0$ and $x_3 = -x_1$. Thus we have the normalized eigenvector

$$\frac{1}{\sqrt{2}} \begin{pmatrix} 1 \\ 0 \\ -1 \end{pmatrix}.$$

The other eigenvalues are $\sqrt{2}+c$ and $-\sqrt{2}+c$ with the corresponding normalized eigenvectors

$$\frac{1}{2} \begin{pmatrix} 1 \\ \sqrt{2} \\ 1 \end{pmatrix}, \quad \frac{1}{2} \begin{pmatrix} 1 \\ -\sqrt{2} \\ 1 \end{pmatrix}.$$

Problem 18. (i) Let $\alpha \in \mathbb{R}$. Find the eigenvalues of

$$A(\alpha) = \begin{pmatrix} 0 & \alpha & 0 \\ \alpha & 0 & \alpha \\ 0 & \alpha & 0 \end{pmatrix}.$$

Do the eigenvalues cross as a function of α?
(ii) Let $\alpha \in \mathbb{R}$. Find the eigenvalues and eigenvectors of the 3×3 matrix

$$B(\alpha) = \begin{pmatrix} -1 & \alpha & 0 \\ \alpha & 0 & \alpha \\ 0 & \alpha & 1 \end{pmatrix}.$$

Discuss the dependence of the eigenvalues and eigenvectors of α.
(iii) Let $\alpha \in \mathbb{R}$. Find the eigenvalues of the 4×4 symmetric matrix

$$C(\alpha) = \begin{pmatrix} -1 & \alpha & 0 & 0 \\ \alpha & -1/2 & \alpha & 0 \\ 0 & \alpha & 1/2 & \alpha \\ 0 & 0 & \alpha & 1 \end{pmatrix}.$$

Discuss the eigenvalues $\lambda_j(\alpha)$ as functions of α.

Solution 18.   (i) The eigenvalues are $\lambda_1(\alpha) = -\sqrt{2}\alpha$, $\lambda_2(\alpha) = 0$, $\lambda_3(\alpha) = +\sqrt{2}\alpha$. Thus the eigenvalues $\lambda_1(\alpha)$ and $\lambda_3(\alpha)$ cross at 0.

(ii) From $\det(\lambda I_3 - B(\alpha)) = 0$ we obtain $(-1-\lambda)(-\lambda)(-1-\lambda) + 2\lambda\alpha^2 = 0$. Thus $\lambda = 0$ is an eigenvalue independent of α. This leads to the quadratic equation

$$\lambda^2 = 1 + 2\alpha^2$$

with $\lambda_\pm = \pm\sqrt{1 + 2\alpha^2}$. We write the eigenvalues in the ordering

$$\lambda_0 = -\sqrt{1 + 2\alpha^2}, \quad \lambda_1 = 0, \quad \lambda_2 = \sqrt{1 + 2\alpha^2}.$$

For the eigenvalue $\lambda_1 = 0$ we find the normalized eigenvector

$$\frac{1}{\lambda_2} \begin{pmatrix} \alpha \\ 1 \\ -\alpha \end{pmatrix}.$$

For the eigenvalue $\lambda_2 = \sqrt{1 + 2\alpha^2}$ we obtain the normalized eigenvector

$$\frac{1}{2\lambda_2} \begin{pmatrix} \lambda_2 - 1 \\ 2\alpha \\ \lambda_2 + 1 \end{pmatrix}.$$

For the eigenvalue $\lambda_0 = -\sqrt{1 + 2\alpha^2}$ we obtain the normalized eigenvector

$$\frac{1}{2\lambda_2} \begin{pmatrix} \lambda_2 + 1 \\ -2\alpha \\ \lambda_2 - 1 \end{pmatrix}.$$

(iii) We obtain the characteristic equation

$$\lambda^4 + \left(-\frac{5}{4} - 3\alpha^2\right)\lambda^2 + \alpha^4 + \frac{1}{4} = 0.$$

Setting $\lambda^2 = z$ with $\lambda = \pm\sqrt{z}$ we obtain

$$z^2 + pz + q = 0, \quad p = -\frac{5}{4} - 3\alpha^2, \quad q = \frac{1}{4} + \alpha^4.$$

It follows that

$$z_1 = \frac{1}{2}(-p + \sqrt{p^2 - 4q}), \quad z_2 = \frac{1}{2}(-p - \sqrt{p^2 - 4q})$$

with $p^2 = 25/16 + 9\alpha^4 + 15\alpha^2/2$. Thus

$$z_1 = \frac{1}{2}\left(\frac{5}{4} + 3\alpha^2 + \sqrt{9/16 + 15\alpha^2/2 + 5\alpha^4}\right),$$

$$z_2 = \frac{1}{2}\left(\frac{5}{4} + 3\alpha^2 - \sqrt{9/16 + 15\alpha^2/2 + 5\alpha^4}\right)$$

and

$$\lambda_1(\alpha) = \sqrt{z_1(\alpha)}, \quad \lambda_2(\alpha) = -\sqrt{z_1(\alpha)},$$

$$\lambda_3(\alpha) = \sqrt{z_2(\alpha)}, \quad \lambda_4(\alpha) = -\sqrt{z_2(\alpha)}.$$

For $\alpha = 0$ we have $\lambda_1(0) = 1$, $\lambda_2(0) = -1$, $\lambda_3(0) = 1/2$, $\lambda_4(0) = -1/2$.

Problem 19. Consider the following 3×3 matrix A and vector $\mathbf{v}$ in $\mathbb{R}^3$

$$A = \begin{pmatrix} 0 & 1 & 0 \\ 1 & 0 & 1 \\ 0 & 1 & 0 \end{pmatrix}, \qquad \mathbf{v} = \begin{pmatrix} \sin(\alpha) \\ \sin(2\alpha) \\ \sin(3\alpha) \end{pmatrix}$$

where $\alpha \in \mathbb{R}$ and $\alpha \neq n\pi$ with $n \in \mathbb{Z}$. Show that using this vector we can find the eigenvalues and eigenvectors of A. Start of with $A\mathbf{v} = \lambda\mathbf{v}$.

Solution 19. We obtain

$$A\mathbf{v} = \begin{pmatrix} \sin(2\alpha) \\ \sin(\alpha) + \sin(3\alpha) \\ \sin(2\alpha) \end{pmatrix}.$$

Under the assumption that $\mathbf{v}$ is an eigenvector of A for specific α's we find the condition

$$\begin{pmatrix} \sin(2\alpha) \\ \sin(\alpha) + \sin(3\alpha) \\ \sin(2\alpha) \end{pmatrix} = \lambda \begin{pmatrix} \sin(\alpha) \\ \sin(2\alpha) \\ \sin(3\alpha) \end{pmatrix}.$$

Thus we have to solve the three equations

$$\sin(2\alpha) = \lambda\sin(\alpha), \quad \sin(\alpha) + \sin(3\alpha) = \lambda\sin(2\alpha), \quad \sin(2\alpha) = \lambda\sin(3\alpha).$$

We have to study the case $\lambda = 0$ and $\lambda \neq 0$. For $\lambda = 0$ we have solve the two equations

$$\sin(2\alpha) = 0, \qquad \sin(2\alpha)\cos(\alpha) = 0$$

with the solution $\alpha = \pi/2$. Thus the eigenvalue $\lambda = 0$ has the normalized eigenvector

$$\frac{1}{\sqrt{2}} \begin{pmatrix} 1 \\ 0 \\ -1 \end{pmatrix}.$$

For $\lambda \neq 0$ we obtain the equation $2\sin(\alpha) = \lambda^2 \sin(\alpha)$ after elimination $\sin(2\alpha)$ and $\sin(3\alpha)$. Since $\alpha \neq n\pi$ we have that $\sin(\alpha) \neq 0$. Thus $\lambda^2 = 2$ with the eigenvalues $\lambda = -\sqrt{2}$ and $\lambda = \sqrt{2}$ and the corresponding normalized eigenvectors

$$\frac{1}{2} \begin{pmatrix} 1 \\ -\sqrt{2} \\ 1 \end{pmatrix}, \qquad \frac{1}{2} \begin{pmatrix} 1 \\ \sqrt{2} \\ 1 \end{pmatrix}.$$

Can this technique be extended to the 4×4 matrix and vector $\mathbf{v} \in \mathbb{R}^4$

$$A = \begin{pmatrix} 0 & 1 & 0 & 0 \\ 1 & 0 & 1 & 0 \\ 0 & 1 & 0 & 1 \\ 0 & 0 & 1 & 0 \end{pmatrix}, \qquad \mathbf{v} = \begin{pmatrix} \sin(\alpha) \\ \sin(2\alpha) \\ \sin(3\alpha) \\ \sin(4\alpha) \end{pmatrix}$$

and even higher dimensions?

Problem 20. Consider the symmetric matrix over $\mathbb{R}$

$$A = \begin{pmatrix} 2 & 1 & -1 \\ 1 & 1 & 0 \\ -1 & 0 & 1 \end{pmatrix}.$$

Find an invertible matrix B such that $B^{-1}AB$ is a diagonal matrix. Construct the matrix B from the normalized eigenvectors of A.

Solution 20. The eigenvalues of A are given by $\lambda_1 = 1$, $\lambda_2 = 0$, $\lambda_3 = 3$. The corresponding normalized eigenvectors are given by

$$\mathbf{v}_1 = \frac{1}{\sqrt{2}} \begin{pmatrix} 0 \\ 1 \\ 1 \end{pmatrix}, \quad \mathbf{v}_{2,} = \frac{1}{\sqrt{3}} \begin{pmatrix} 1 \\ -1 \\ 1 \end{pmatrix}, \quad \mathbf{v}_3 = \frac{1}{\sqrt{6}} \begin{pmatrix} -2 \\ -1 \\ 1 \end{pmatrix}.$$

Hance

$$B = \begin{pmatrix} 0 & 1/\sqrt{3} & -2/\sqrt{6} \\ 1/\sqrt{2} & -1/\sqrt{3} & -1/\sqrt{6} \\ 1/\sqrt{2} & 1/\sqrt{3} & 1/\sqrt{6} \end{pmatrix}.$$

We have $B^{-1} = B^T$.

Problem 21. Find the eigenvalues of the 3×3 symmetric matrix over $\mathbb{R}$

$$A = \begin{pmatrix} 2 & 1 & 1 \\ 1 & 2 & 1 \\ 1 & 1 & 2 \end{pmatrix}$$

using the trace and the determinant of the matrix and the information that two eigenvalues are the same.

Solution 21. The trace is the sum of the eigenvalues and the determinant is the product of the eigenvalues. Thus for the given matrix A we have $\lambda_1 + \lambda_2 + \lambda_3 = 6$, $\lambda_1 \lambda_2 \lambda_3 = 4$. Now two eigenvalues are the same, say $\lambda_3 = \lambda_2$. Thus we obtain the two equations

$$\lambda_1 + 2\lambda_2 = 6, \qquad \lambda_1 \lambda_2^2 = 4$$

with the solution $\lambda_1 = 4$, $\lambda_2 = \lambda_3 = 1$.

Problem 22. Consider the symmetric matrix

$$A = \begin{pmatrix} a_{11} & a_{12} & a_{13} \\ a_{12} & a_{22} & a_{23} \\ a_{13} & a_{23} & a_{33} \end{pmatrix}$$

over $\mathbb{R}$. Write down the characteristic polynomial $\det(\lambda I_3 - A)$ and express it using the trace and determinant of A.

Solution 22. We obtain

$$\det(\lambda I_3 - A) = \lambda^3 + \lambda^2(-a_{11} - a_{22} - a_{33})$$
$$+ \lambda(a_{11}a_{22} + a_{11}a_{33} + a_{22}a_{33} - a_{12}^2 - a_{13}^2 - a_{23}^2)$$
$$+ a_{11}a_{23}^2 + a_{22}a_{13}^2 + a_{33}a_{12}^2 - 2a_{12}a_{13}a_{23} - a_{11}a_{22}a_{33}.$$

Thus we can write

$$\det(\lambda I_3 - A) = \lambda^3 - \mathrm{tr}(A)\lambda^2 + \lambda(a_{11}a_{22} + a_{11}a_{33} + a_{22}a_{33} - a_{12}^2 - a_{13}^2 - a_{23}^2) - \det(A).$$

If the real numbers λ_1, λ_2, λ_3 are the eigenvalues of A we have

$$\lambda_1 + \lambda_2 + \lambda_3 = \mathrm{tr}(A)$$
$$\lambda_1\lambda_2\lambda_3 = \det(A)$$
$$\lambda_1\lambda_2 + \lambda_1\lambda_3 + \lambda_2\lambda_3 = a_{11}a_{22} + a_{11}a_{33} + a_{22}a_{33} - a_{12}^2 - a_{13}^2 - a_{23}^2.$$

Problem 23. Consider the *quadratic form*

$$7x_1^2 + 6x_2^2 + 5x_3^2 - 4x_1x_2 - 4x_2x_3 + 14x_1 - 8x_2 + 10x_3 + 6 = 0.$$

Write this equation in matrix form and find the eigenvalues and normalized eigenvectors of the 3×3 matrix.

Solution 23. We have

$$(x_1 \quad x_2 \quad x_3) \begin{pmatrix} 7 & -2 & 0 \\ -2 & 6 & -2 \\ 0 & -2 & 5 \end{pmatrix} \begin{pmatrix} x_1 \\ x_2 \\ x_3 \end{pmatrix} + 2 (7 \quad -4 \quad 5) \begin{pmatrix} x_1 \\ x_2 \\ x_3 \end{pmatrix} + 6 = 0.$$

The eigenvalues and normalized eigenvectors of the symmetric 3×3 matrix are $\lambda_1 = 3$, $\lambda_2 = 6$, $\lambda_3 = 9$ with the corresponding normalized eigenvectors

$$\mathbf{v}_1 = \frac{1}{3}\begin{pmatrix} 1 \\ 2 \\ 2 \end{pmatrix}, \quad \mathbf{v}_2 = \frac{1}{3}\begin{pmatrix} 2 \\ 1 \\ -2 \end{pmatrix}, \quad \mathbf{v}_3 = \frac{1}{3}\begin{pmatrix} 2 \\ -2 \\ 1 \end{pmatrix}.$$

Problem 24. (i) Let H be an $n \times n$ matrix over $\mathbb{C}$. Assume that H is hermitian and unitary. What can be said about the eigenvalues of H?
(ii) An $n \times n$ matrix A such that $A^2 = A$ is called *idempotent*. What can be said about the eigenvalues of such a matrix?
(iii) An $n \times n$ matrix A for which $A^p = 0_n$, where p is a positive integer, is called *nilpotent*. What can be said about the eigenvalues of such a matrix?
(iv) An $n \times n$ matrix A such that $A^2 = I_n$ is called *involutory*. What can be said about the eigenvalues of such a matrix?

Solution 24. (i) Since H is hermitian the eigenvalues are real. Since H is unitary the eigenvalues lie on the unit circle in the complex plane. Now H is hermitian and unitary we have that $\lambda \in \{+1, -1\}$.

(ii) From the eigenvalue equation we obtain $A^2\mathbf{x} = A(A\mathbf{x}) = A\lambda\mathbf{x} = \lambda A\mathbf{x} = \lambda^2\mathbf{x}$. Thus since $A^2 = A$ we have $\lambda^2 = \lambda$.

(iii) From the eigenvalue equation $A\mathbf{x} = \lambda\mathbf{x}$ we obtain $A^p\mathbf{x} = \lambda^p\mathbf{x}$. Since $A^p = 0_n$ we obtain $\lambda^p = 0$. Thus all eigenvalues are 0.

(iv) From the eigenvalue equation $A\mathbf{x} = \lambda\mathbf{x}$ we obtain $A^2\mathbf{x} = \lambda^2\mathbf{x}$. Since $A^2 = I_n$ we have $\lambda^2 = 1$. Thus $\lambda = \pm 1$.

Problem 25.   (i) Let A be an $n \times n$ matrix over $\mathbb{C}$. Show that the eigenvectors corresponding to distinct eigenvalues are linearly independent.

(ii) Show that eigenvectors of a normal matrix M corresponding to distinct eigenvalues are orthogonal.

Solution 25.   (i) The proof is by contradiction. Consider the eigenvalue equations

$$Ax = \lambda_1 x, \qquad Ay = \lambda_2 y, \qquad x, y \neq 0$$

with $\lambda_1 \neq \lambda_2$. Assume that $\mathbf{x} = c\mathbf{y}$ with $c \neq 0$, i.e. we assume that the eigenvectors $\mathbf{x}$ and $\mathbf{y}$ are linearly dependent. Then from $A\mathbf{x} = \lambda_1\mathbf{x}$ we obtain $A(c\mathbf{y}) = \lambda_1(c\mathbf{y})$. Thus $c(A\mathbf{y}) = c(\lambda_1\mathbf{y})$. Using $A\mathbf{y} = \lambda_2\mathbf{y}$ we arrive at $c(\lambda_1 - \lambda_2)\mathbf{y} = \mathbf{0}$. Since $c \neq 0$ and $\mathbf{y} \neq \mathbf{0}$ we obtain $\lambda_1 = \lambda_2$. Thus we have a contradiction and therefore $\mathbf{x}$ and $\mathbf{y}$ must be linearly independent.

(ii) Let λ_1, λ_2 be the two different eigenvalues with the corresponding eigenvectors $\mathbf{v}_1$, $\mathbf{v}_2$, respectively. Using that from $M\mathbf{v} = \lambda\mathbf{v}$ follows $M^*\mathbf{v} = \bar{\lambda}\mathbf{v}$ we have

$$(\lambda_1 - \lambda_2)\mathbf{v}_1^*\mathbf{v}_2 = \lambda_1\mathbf{v}_1^*\mathbf{v}_2 - \lambda_2\mathbf{v}_1^*\mathbf{v}_2 = (M\mathbf{v}_1)^*\mathbf{v}_2 - \mathbf{v}_1^*(M^*\mathbf{v}_2) = 0.$$

Problem 26.   Let A be an $n \times n$ hermitian matrix, i.e. $A = A^*$. Assume that all n eigenvalues are different. Then the normalized eigenvectors $\{\mathbf{v}_j : j = 1, \ldots, n\}$ form an orthonormal basis in $\mathbb{C}^n$. Consider

$$\beta := (A\mathbf{x} - \mu\mathbf{x}, A\mathbf{x} - \nu\mathbf{x}) \equiv (A\mathbf{x} - \mu\mathbf{x})^*(A\mathbf{x} - \nu\mathbf{x})$$

where $(\,,\,)$ denotes the scalar product in $\mathbb{C}^n$ and μ, ν are real constants with $\mu < \nu$. Show that if no eigenvalue lies between μ and ν, then $\beta \geq 0$.

Solution 26.   First note that $\lambda_j \in \mathbb{R}$, since the matrix A is hermitian. We expand $\mathbf{x}$ with respect to the basis $\{\mathbf{v}_j : j = 1, \ldots, n\}$, i.e.

$$\mathbf{x} = \sum_{j=1}^{n} \alpha_j \mathbf{v}_j.$$

Since $A\mathbf{v}_j = \lambda_j\mathbf{v}_j$ and therefore $\mathbf{v}_j^* A = \lambda_j\mathbf{v}_j^*$ we have

$$A\mathbf{x} = \sum_{j=1}^{n} \alpha_j\lambda_j\mathbf{v}_j, \qquad \mathbf{x}^* A = \sum_{k=1}^{n} \alpha_k^*\lambda_k\mathbf{v}_k^*.$$

Since $(\mathbf{v}_j, \mathbf{v}_k) = \mathbf{v}_j^* \mathbf{v}_k = \delta_{jk}$ we find

$$\beta = \mathbf{x}^* A A \mathbf{x} - \nu \mathbf{x}^* A \mathbf{x} - \mu \mathbf{x}^* A \mathbf{x} + \mu \nu \mathbf{x}^* \mathbf{x}$$

$$= \sum_{j=1}^{n} \lambda_j^2 |\alpha_j|^2 - \nu \sum_{j=1}^{n} \lambda_j |\alpha_j|^2 - \mu \sum_{j=1}^{n} \lambda_j |\alpha_j|^2 + \mu \nu \sum_{j=1}^{n} |\alpha_j|^2$$

$$= \sum_{j=1}^{n} (\lambda_j - \mu)(\lambda_j - \nu) |\alpha_j|^2.$$

If no eigenvalue lies between μ and ν, then $(\lambda_j - \mu)$ and $(\lambda_j - \nu)$ have the same sign for all eigenvalues. Therefore $(\lambda_j - \mu)(\lambda_j - \nu)$ is nonnegative. Thus $\beta \geq 0$.

Problem 27. Let A be an arbitrary $n \times n$ matrix over $\mathbb{C}$. Let

$$H := \frac{A + A^*}{2}, \qquad S := \frac{A - A^*}{2i}.$$

Let λ be an eigenvalue of A and $\mathbf{x}$ be the corresponding normalized eigenvector (column vector).
(i) Show that $\lambda = \mathbf{x}^* H \mathbf{x} + i \mathbf{x}^* S \mathbf{x}$.
(ii) Show that the real part λ_r of the eigenvalue λ is given by $\lambda_r = \mathbf{x}^* H \mathbf{x}$ and the imaginary part λ_i is given by $\lambda_i = \mathbf{x}^* S \mathbf{x}$.

Solution 27. (i) The eigenvalue equation is given by $A\mathbf{x} = \lambda \mathbf{x}$. Then using $\mathbf{x}^* \mathbf{x} = 1$ we find

$$\lambda = \mathbf{x}^* A \mathbf{x} = \mathbf{x}^* \frac{A + A^*}{2} \mathbf{x} + \mathbf{x}^* \frac{A - A^*}{2} \mathbf{x} = \mathbf{x}^* H \mathbf{x} + i \mathbf{x}^* S \mathbf{x}.$$

(ii) Note that H and S are hermitian, i.e. $H^* = H$ and $S^* = S$. Hence

$$\lambda^* = \mathbf{x}^* H \mathbf{x} - i \mathbf{x}^* S \mathbf{x}.$$

Calculating $\lambda + \lambda^*$ and $\lambda - \lambda^*$ provides the results.

Problem 28. Let λ_1, λ_2 and λ_3 be the eigenvalues of the matrix

$$A = \begin{pmatrix} 0 & 1 & 2 \\ 0 & 0 & 1 \\ 2 & 2 & 1 \end{pmatrix}.$$

Find $\lambda_1^2 + \lambda_2^2 + \lambda_3^2$ without calculating the eigenvalues of A or A^2.

Solution 28. From the eigenvalue equation $A\mathbf{v} = \lambda \mathbf{v}$ we obtain $A^2 \mathbf{v} = \lambda^2 \mathbf{v}$. Now

$$\mathrm{tr}(A) = \lambda_1 + \lambda_2 + \lambda_3.$$

Thus $\mathrm{tr}(A^2) = \lambda_1^2 + \lambda_2^2 + \lambda_3^2$. Since the diagonal part of A^2 is given by (4 2 7) we find $\lambda_1^2 + \lambda_2^2 + \lambda_3^2 = 13$. Thus we only have to calculate the diagonal part of A^2.

Problem 29.   Consider the column vectors **u** and **v** in $\mathbb{R}^n$

$$\mathbf{u} = \begin{pmatrix} \cos(\theta) \\ \cos(2\theta) \\ \vdots \\ \cos(n\theta) \end{pmatrix}, \qquad \mathbf{v} = \begin{pmatrix} \sin(\theta) \\ \sin(2\theta) \\ \vdots \\ \sin(n\theta) \end{pmatrix}$$

where $n \geq 3$ and $\theta = 2\pi/n$.
(i) Calculate $\mathbf{u}^T\mathbf{u} + \mathbf{v}^T\mathbf{v}$.
(ii) Calculate $\mathbf{u}^T\mathbf{u} - \mathbf{v}^T\mathbf{v} + 2i\mathbf{u}^T\mathbf{v}$.
(iii) Calculate the matrix $A = \mathbf{u}\mathbf{u}^T - \mathbf{v}\mathbf{v}^T$, $A\mathbf{u}$ and $A\mathbf{v}$. Discuss.

Solution 29.   (i) We obtain

$$\mathbf{u}^T\mathbf{u} + \mathbf{v}^T\mathbf{v} = \sum_{j=1}^{n}(\cos^2(j\theta) + \sin^2(j\theta)) = \sum_{j=1}^{n} 1 = n.$$

(ii) We obtain

$$\mathbf{u}^T\mathbf{u} - \mathbf{v}^T\mathbf{v} + 2i\mathbf{u}^T\mathbf{v} = \sum_{j=1}^{n}(\cos^2(j\theta) - \sin^2(j\theta) + 2i\cos(j\theta)\sin(j\theta)) = 0.$$

Thus $\mathbf{u}^T\mathbf{u} = \mathbf{v}^T\mathbf{v}$ and $\mathbf{u}^T\mathbf{v} = 0$. Using the result of (i) we find $\mathbf{u}^T\mathbf{u} = \mathbf{v}^T\mathbf{v} = n/2$.
(iii) Using the result from (ii) we find

$$A\mathbf{u} = (\mathbf{u}\mathbf{u}^T - \mathbf{v}\mathbf{v}^T)\mathbf{u} = \mathbf{u}(\mathbf{u}^T\mathbf{u}) - \mathbf{v}(\mathbf{v}^T\mathbf{u}) = \frac{n}{2}\mathbf{u}$$

$$A\mathbf{v} = (\mathbf{u}\mathbf{u}^T - \mathbf{v}\mathbf{v}^T)\mathbf{v} = \mathbf{u}(\mathbf{u}^T\mathbf{v}) - \mathbf{v}(\mathbf{v}^T\mathbf{v}) = -\frac{n}{2}\mathbf{v}.$$

Thus we conclude that A has the eigenvalues $n/2$ and $-n/2$.

Problem 30.   (i) Let **u** be a nonzero column vector in $\mathbb{R}^n$. Consider the $n \times n$ matrix

$$A = \mathbf{u}\mathbf{u}^T - \mathbf{u}^T\mathbf{u}I_n.$$

Is **u** an eigenvector of this matrix? If so what is the eigenvalue?
(ii) Let **u**, **v** be nonzero column vectors in $\mathbb{R}^n$ and $\mathbf{u} \neq \mathbf{v}$. Consider the $n \times n$ matrix A over $\mathbb{R}$

$$A = \mathbf{u}\mathbf{u}^T + \mathbf{u}\mathbf{v}^T - \mathbf{v}\mathbf{u}^T - \mathbf{v}\mathbf{v}^T.$$

Find the nonzero eigenvalues of A and the corresponding eigenvector. Note that $\mathbf{u}^T\mathbf{v} = \mathbf{v}^T\mathbf{u}$.

Solution 30.   (i) We have

$$(\mathbf{u}\mathbf{u}^T - \mathbf{u}^T\mathbf{u}I_n)\mathbf{u} = \mathbf{u}(\mathbf{u}^T\mathbf{u}) - (\mathbf{u}^T\mathbf{u})\mathbf{u} = (\mathbf{u}^T\mathbf{u})(\mathbf{u} - \mathbf{u}) = 0.$$

Thus **u** is an eigenvector with eigenvalue 0.

(ii) Since

$$Au = (u^Tu + v^Tv)u + (-u^Tu - v^Tu)v, \quad Av = (u^Tv + v^Tv)u + (-u^Tv - v^Tv)v$$

we have the eigenvalue equation

$$A(u - v) = (u^Tu - v^Tv)(u - v)$$

since $u^Tv = v^Tu$. Thus $n - 1$ of the eigenvalues of A are 0 and the nonzero eigenvalue is $u^Tu - v^Tv$ with the eigenvector $u - v$.

Problem 31. Let A be an $n \times n$ matrix over $\mathbb{C}$. We define

$$r_j := \sum_{\substack{k=1 \\ k \neq j}}^{n} |a_{jk}|, \quad j = 1, \ldots, n.$$

(i) Show that each eigenvalue λ of A satisfies at least one of the following inequalities

$$|\lambda - a_{jj}| \leq r_j, \quad j = 1, \ldots, n.$$

In other words show that all eigenvalues of A can be found in the union of disks

$$\{ z : |z - a_{jj}| \leq r_j, \ j = 1, \ldots, n \}.$$

This is *Geršgorin disk theorem.*
(ii) Apply this theorem to the 2×2 unitary matrix

$$A = \begin{pmatrix} 0 & i \\ -i & 0 \end{pmatrix}.$$

(iii) Apply this theorem to the 3×3 nonnormal matrix

$$B = \begin{pmatrix} 1 & 2 & 3 \\ 3 & 4 & 9 \\ 1 & 1 & 1 \end{pmatrix}.$$

Solution 31. (i) From the eigenvalue equation $Ax = \lambda x$ we obtain $(\lambda I_n - A)x = 0$. Splitting the matrix A into its diagonal part and nondiagonal part we obtain

$$(\lambda - a_{jj})x_j = \sum_{\substack{k=1 \\ k \neq j}}^{n} a_{jk}x_k, \quad j = 1, \ldots, n$$

where x_j is the j-th component of the vector x. Let x_k be the largest component (in absolute value) of the vector x. Then, since $|x_j|/|x_k| \leq 1$ for $j \neq k$, we obtain

$$|\lambda - a_{kk}| \leq \sum_{\substack{j=1 \\ j \neq k}}^{n} |a_{kj}| \frac{|x_j|}{|x_k|} \leq \sum_{\substack{j=1 \\ j \neq k}}^{n} |a_{kj}|.$$

Thus the eigenvalue λ is contained in the disk $\{\lambda : |\lambda - a_{kk}| \le r_k\}$.

(ii) Since $|i| = |-i| = 1$ we obtain $r_1 = 1$, $r_2 = 1$. Thus the Geršgorin disks are

$$R_1 : \{z : |z| \le 1\}, \qquad R_2 : \{z : |z| \le 1\}.$$

Since the matrix is hermitian the eigenvalues are real and therefore we have $|x| \le 1$ in both cases for x real. The eigenvalues of the matrix A are given by 1, -1.

(iii) We obtain $r_1 = 5$, $r_2 = 12$, $r_3 = 2$. The Geršgorin disks are

$$R_1 : \{z : |z - 1| \le 5\}, \quad R_2 : \{z : |z - 4| \le 12\}, \quad R_3 : \{z : |z - 1| \le 2\}.$$

The eigenvalues of the matrix B are $\lambda_1 = 7.3067$, $\lambda_{2,3} = -0.6533 \pm 0.3473i$.

Problem 32. Let A be a normal matrix over $\mathbb{C}$, i.e. $A^*A = AA^*$. Show that if $\mathbf{x}$ is an eigenvector of A with eigenvalue λ, then $\mathbf{x}$ is an eigenvector of A^* with eigenvalue $\bar{\lambda}$.

Solution 32. Since $AA^* = A^*A$ and $A^{**} = A$ we have

$$(A\mathbf{x})^*A\mathbf{x} = \mathbf{x}^*A^*A\mathbf{x} = \mathbf{x}^*AA^*\mathbf{x} = (A^*\mathbf{x})^*A^*\mathbf{x}.$$

It follows that

$$\begin{aligned}
0 = \mathbf{0}^*\mathbf{0} &= (A\mathbf{x} - \lambda\mathbf{x})^*(A\mathbf{x} - \lambda\mathbf{x}) = (\mathbf{x}^*A^* - \bar{\lambda}\mathbf{x}^*)(A\mathbf{x} - \lambda\mathbf{x}) \\
&= \mathbf{x}^*AA^*\mathbf{x} - \bar{\lambda}\mathbf{x}^*A\mathbf{x} - \lambda\mathbf{x}^*A^*\mathbf{x} + \lambda\bar{\lambda}\mathbf{x}^*\mathbf{x} \\
&= (A^*\mathbf{x} - \bar{\lambda}\mathbf{x})^*(A^*\mathbf{x} - \bar{\lambda}\mathbf{x}).
\end{aligned}$$

We have the eigenvalue equation $A^*\mathbf{x} = \bar{\lambda}\mathbf{x}$, which implies that $\mathbf{x}$ is an eigenvector of A^* corresponding to the eigenvalue $\bar{\lambda}$.

Problem 33. (i) Show that an $n \times n$ matrix A is singular if and only if at least one eigenvalue is 0.

(ii) Let B be an invertible $n \times n$ matrix. Show that if $\mathbf{x}$ is an eigenvector of B with eigenvalue λ, then $\mathbf{x}$ is an eigenvector of B^{-1} with eigenvalue λ^{-1}.

Solution 33. (i) From the eigenvalue equation $A\mathbf{x} = \lambda\mathbf{x}$ we obtain $\det(A - \lambda I_n) = 0$. Thus if $\lambda = 0$ we obtain $\det(A) = 0$. It follows that A is singular, i.e. A^{-1} does not exist.

(ii) From $B\mathbf{x} = \lambda\mathbf{x}$ we obtain $B^{-1}B\mathbf{x} = B^{-1}(\lambda\mathbf{x})$. Thus $\mathbf{x} = \lambda B^{-1}\mathbf{x}$. Since $\det(B) \ne 0$ all eigenvalues must be nonzero. Therefore

$$B^{-1}\mathbf{x} = \frac{1}{\lambda}\mathbf{x}.$$

Problem 34. Let A be an $n \times n$ matrix over $\mathbb{R}$. Show that A and A^T have the same eigenvalues.

Solution 34. Let λ be an eigenvalue of A. Then we have

$$0 = \det(A - \lambda I_n) = \det((A^T)^T - \lambda I_n^T) = \det(A^T - \lambda I_n)^T = \det(A^T - \lambda I_n)$$

since for any $n \times n$ matrix B we have $\det(B) = \det(B^T)$.

Problem 35. Let A be an $n \times n$ real symmetric matrix and

$$Q(\mathbf{x}) := \mathbf{x}^T A \mathbf{x}.$$

The following statements hold (*maximum principle*)
1) $\lambda_1 = \max_{\|\mathbf{x}\|=1}(Q(\mathbf{x})) = Q(\mathbf{x}_1)$ is the largest eigenvalue of the matrix A and $\mathbf{x}_1$ is the eigenvector corresponding to eigenvalue λ_1.
2) (inductive statement). Let $\lambda_k = \max Q(\mathbf{x})$ subject to the constraints
a) $\mathbf{x}^T \mathbf{x}_j = 0$, $j = 1, \ldots, k-1$.
b) $\|\mathbf{x}\| = 1$.
Then $\lambda_k = Q(\mathbf{x}_k)$ is the kth eigenvalue of A, $\lambda_1 \geq \lambda_2 \geq \cdots \geq \lambda_k$ and $\mathbf{x}_k$ is the corresponding eigenvectors of A.
Apply the maximum principle to the symmetric 2×2 matrix

$$A = \begin{pmatrix} 3 & 1 \\ 1 & 3 \end{pmatrix}.$$

Solution 35. We find $Q(\mathbf{x}) = 3x_1^2 + 2x_1 x_2 + 3x_2^2$. We change to *polar coordinates* $x_1 = \cos(\theta)$, $x_2 = \sin(\theta)$. Note that $2\cos(\theta)\sin(\theta) \equiv \sin(2\theta)$. Thus

$$Q(\theta) = 3 + \sin(2\theta).$$

Clearly $\max(Q(\theta)) = 4$ which occurs at $\theta = \pi/4$ and $\min(Q(\theta)) = 2$ at $\theta = -\pi/4$. It follows that $\lambda_1 = 4$ with $\mathbf{x}_1 = (1, 1)^T$ and $\lambda_2 = 2$ with $\mathbf{x}_2 = (1, -1)^T$.

Problem 36. Let A be an $n \times n$ matrix. An $n \times n$ matrix can have at most n linearly independent eigenvectors. Now assume that A has $n+1$ eigenvectors (at least one must be linearly dependent) such that any n of them are linearly independent. Show that A is a scalar multiple of the identity matrix I_n.

Solution 36. Let $\mathbf{x}_1, \ldots, \mathbf{x}_{n+1}$ be the eigenvectors of A with eigenvalues $\lambda_1, \ldots, \lambda_{n+1}$. Since $\mathbf{x}_1, \ldots, \mathbf{x}_n$ are linearly independent, they span the vector space. Consequently

$$\mathbf{x}_{n+1} = \sum_{i=1}^{n} \alpha_i \mathbf{x}_i. \tag{1}$$

Multiplying equation (1) with λ_{n+1} we have

$$\lambda_{n+1} \mathbf{x}_{n+1} = \sum_{i=1}^{n} \alpha_i \lambda_{n+1} \mathbf{x}_i.$$

Applying A to equation (1) yields

$$\lambda_{n+1}\mathbf{x}_{n+1} = \sum_{i=1}^{n} \alpha_i \lambda_i \mathbf{x}_i \;\Rightarrow\; \sum_{i=1}^{n} \alpha_i \lambda_{n+1} \mathbf{x}_i = \sum_{i=1}^{n} \alpha_i \lambda_i \mathbf{x}_i.$$

It follows that $\alpha_i \lambda_{n+1} = \alpha_i \lambda_i$ for all $i = 1, \ldots, n$. If $\alpha_i = 0$ for some i, then $\mathbf{x}_{n+1}$ can be expressed as a linear combination of $\mathbf{x}_1, \ldots, \mathbf{x}_{i-1}, \mathbf{x}_{i+1}, \ldots, \mathbf{x}_n$, contradicting the linear independence. Therefore $\alpha_i \neq 0$ for all i, so that $\lambda_{n+1} = \lambda_i$ for all i. This implies $A = \lambda_{n+1} I_n$.

Problem 37.   An $n \times n$ *stochastic matrix* P satisfies the following conditions

$$p_{ij} \geq 0 \quad \text{for all } i, j = 1, 2, \ldots, n$$

and

$$\sum_{i=1}^{n} p_{ij} = 1 \quad \text{for all } j = 1, 2, \ldots, n.$$

Show that a stochastic matrix always has at least one eigenvalue equal to one.

Solution 37.   Assume that 1 is an eigenvalue. Then the eigenvalue equation is given by

$$\begin{pmatrix} p_{11} & p_{12} & p_{13} & \cdots & p_{1n} \\ p_{21} & p_{22} & p_{23} & \cdots & p_{2n} \\ \vdots & \vdots & \vdots & \ddots & \vdots \\ p_{n1} & p_{n2} & p_{n3} & \cdots & p_{nn} \end{pmatrix} \begin{pmatrix} x_1 \\ x_2 \\ \vdots \\ x_n \end{pmatrix} = \begin{pmatrix} x_1 \\ x_2 \\ \vdots \\ x_n \end{pmatrix}.$$

From the second condition we find the eigenvector $\mathbf{x} = (1 \;\; 1 \;\; \cdots \;\; 1)^T$. Thus 1 is an eigenvalue.

Problem 38.   The matrix difference equation

$$\mathbf{p}(t+1) = M\mathbf{p}(t), \qquad t = 0, 1, 2, \ldots$$

with the column vector (vector of probabilities)

$$\mathbf{p}(t) = (p_1(t), p_2(t), \ldots, p_n(t))^T$$

and the $n \times n$ matrix

$$M = \begin{pmatrix} (1-w) & 0.5w & 0 & \cdots & 0.5w \\ 0.5w & (1-w) & 0.5w & \cdots & 0 \\ 0 & 0.5w & (1-w) & \cdots & 0 \\ \vdots & \vdots & \vdots & \ddots & \vdots \\ 0.5w & 0 & 0 & \cdots & (1-w) \end{pmatrix}$$

plays a role in random walk in one dimension. M is called the *transition probability matrix* and w denotes the probability $w \in [0, 1]$ that at a given time step the particle jumps to either of its nearest neighbor sites and the probability that the particle does not jump either to the right of left is $(1-w)$. The matrix M is of the type known as *circulant matrix*. Such an $n \times n$ matrix is of the form

$$
C = \begin{pmatrix}
c_0 & c_1 & c_2 & \cdots & c_{n-1} \\
c_{n-1} & c_0 & c_1 & \cdots & c_{n-2} \\
c_{n-2} & c_{n-1} & c_0 & \cdots & c_{n-3} \\
\vdots & \vdots & \vdots & \ddots & \vdots \\
c_1 & c_2 & c_3 & \cdots & c_0
\end{pmatrix}
$$

with the normalized eigenvectors

$$
\mathbf{e}_j = \frac{1}{\sqrt{n}} \begin{pmatrix}
1 \\
e^{2\pi ij/n} \\
\vdots \\
e^{2(n-1)\pi ij/n}
\end{pmatrix}, \quad j = 1, \ldots, n.
$$

(i) Use this result to find the eigenvalues of the matrix C.

(ii) Use (i) to find the eigenvalues of the matrix M.

(iii) Use (ii) to find $\mathbf{p}(t)$ ($t = 0, 1, 2, \ldots$), where we expand the initial distribution vector $\mathbf{p}(0)$ in terms of the eigenvectors

$$
\mathbf{p}(0) = \sum_{k=1}^{n} a_k \mathbf{e}_k \quad \text{with} \quad \sum_{j=1}^{n} p_j(0) = 1.
$$

(iv) Assume that $\mathbf{p}(0) = \frac{1}{n}(1 \quad 1 \quad \cdots \quad 1)^T$. Give the time evolution of $\mathbf{p}(0)$.

Solution 38. (i) Calculating $C\mathbf{e}_j$ provides the eigenvalues

$$
\lambda_j = c_0 + c_1 \exp(2\pi ij/n) + c_2 \exp(4\pi ij/n) + \cdots + c_{n-1}\exp(2(n-1)\pi ij/n).
$$

(ii) We have $c_0 = 1 - w$, $c_1 = 0.5w$, $c_{n-1} = 0.5w$ and $c_2 = c_3 = \cdots = c_{n-2} = 0$. Using $e^{2\pi ij} = 1$ and the identity $\cos(\alpha) \equiv (e^{i\alpha} + e^{-i\alpha})/2$ we obtain as eigenvalues of M

$$
\lambda_j = (1-w) + w\cos(2j\pi/n), \quad j = 1, \ldots, n.
$$

(iii) Since $\mathbf{p}(0)$ is a probability vector we have $p_j(0) \geq 0$ and $\sum_{j=1}^{n} p_j(0) = 1$. The expansion coefficients a_k are given by

$$
a_k = \frac{1}{\sqrt{n}} \sum_{j=1}^{n} p_j(0) \exp(-2\pi i(j-1)k/n).
$$

Thus we obtain

$$
\mathbf{p}(t) = \sum_{k=1}^{n} a_k \lambda_k^t \mathbf{e}_k \quad t = 0, 1, 2, \ldots.
$$

(iv) Since $M\mathbf{p}(0) = \mathbf{p}(0)$ we find that the probability vector $\mathbf{p}(0)$ is a *fixed point*.

Problem 39.   Let n be a positive integer. Consider the 3×3 matrix with rows of elements summing to unity

$$M = \frac{1}{n} \begin{pmatrix} n-a-b & a & b \\ a & n-2a-c & a+c \\ c & a & n-a-c \end{pmatrix}$$

where the values of a, b, c are such that, $0 \le a$, $0 \le b$, $a+b \le n$, $2a+c \le n$. Thus the matrix is a *stochastic matrix*. Find the eigenvalues of M.

Solution 39.   A stochastic matrix always has an eigenvalue $\lambda_1 = 1$ with the corresponding eigenvector $\mathbf{u}_1 = (1 \ \ 1 \ \ 1)^T$. The other two eigenvalues are

$$\lambda_2 = 1 - \frac{1}{n}(a+b+c), \qquad \lambda_3 = 1 - \frac{1}{n}(3a+c).$$

If $b = 2a$, the eigenvalues λ_2 and λ_3 are the same.

Problem 40.   Let U be a unitary matrix and $\mathbf{x}$ an eigenvector of U with the corresponding eigenvalue λ, i.e. $U\mathbf{x} = \lambda\mathbf{x}$.
(i) Show that $U^*\mathbf{x} = \bar{\lambda}\mathbf{x}$.
(ii) Let λ, μ be distinct eigenvalues of a unitary matrix U with the corresponding eigenvectors $\mathbf{x}$ and $\mathbf{y}$, i.e. $U\mathbf{x} = \lambda\mathbf{x}$, $U\mathbf{y} = \mu\mathbf{y}$. Show that $\mathbf{x}^*\mathbf{y} = 0$.

Solution 40.   (i) From $U\mathbf{x} = \lambda\mathbf{x}$ we obtain $U^*U\mathbf{x} = \lambda U^*\mathbf{x}$. Therefore $\mathbf{x} = \lambda U^*\mathbf{x}$. Finally $U^*\mathbf{x} = \bar{\lambda}\mathbf{x}$ since $\bar{\lambda}\lambda = 1$. Now λ can be written as $e^{i\alpha}$ with $\alpha \in \mathbb{R}$ we have $\bar{\lambda} = e^{-i\alpha}$.
(ii) Using the result from (i) we have

$$\mu\mathbf{x}^*\mathbf{y} = \mathbf{x}^*(\mu\mathbf{y}) = \mathbf{x}^*U\mathbf{y} = (U^*\mathbf{x})^*\mathbf{y} = (\bar{\lambda}\mathbf{x})^*\mathbf{y} = \lambda\mathbf{x}^*\mathbf{y}.$$

Thus $(\mu - \lambda)\mathbf{x}^*\mathbf{y} = 0$. Since μ and λ are distinct it follows that $\mathbf{x}^*\mathbf{y} = 0$.

Problem 41.   Let H, H_0, V be $n \times n$ matrices over $\mathbb{C}$ and $H = H_0 + V$. Let $z \in \mathbb{C}$ and assume that z is chosen so that $(H_0 - zI_n)^{-1}$ and $(H - zI_n)^{-1}$ exist. Show that

$$(H - zI_n)^{-1} = (H_0 - zI_n)^{-1} - (H_0 - zI_n)^{-1}V(H - zI_n)^{-1}.$$

This is called the second *resolvent identity*.

Solution 41.   We have

$$H_0 + V = H$$
$$\Leftrightarrow H_0 = H - V$$
$$\Leftrightarrow (H_0 - zI_n) = (H - zI_n) - V$$
$$\Leftrightarrow (H_0 - zI_n)^{-1}(H_0 - zI_n) = (H_0 - zI_n)^{-1}(H - zI_n) - (H_0 - zI_n)^{-1}V$$
$$\Leftrightarrow I_n = (H_0 - zI_n)^{-1}(H - zI_n) - (H_0 - zI_n)^{-1}V$$
$$(H - zI_n)^{-1} = (H_0 - zI_n)^{-1} - (H_0 - zI_n)^{-1}V(H - zI_n)^{-1}.$$

Problem 42. An $n \times n$ matrix A is called a *Hadamard matrix* if each entry of A is 1 or -1 and if the rows or columns of A are orthogonal, i.e.

$$AA^T = nI_n \quad \text{or} \quad A^T A = nI_n.$$

Note that $AA^T = nI_n$ and $A^T A = nI_n$ are equivalent. Hadamard matrices H_n of order 2^n can be generated recursively by defining

$$H_1 = \begin{pmatrix} 1 & 1 \\ 1 & -1 \end{pmatrix}, \qquad H_n = \begin{pmatrix} H_{n-1} & H_{n-1} \\ H_{n-1} & -H_{n-1} \end{pmatrix}$$

for $n \geq 2$. Show that the eigenvalues of H_n are given by $+2^{n/2}$ and $-2^{n/2}$ each of multiplicity 2^{n-1}.

Solution 42. We use induction on n. The case $n = 1$ is obvious. Now for $n \geq 2$ we have

$$\det(\lambda I - H_n) = \det((\lambda I - H_{n-1})(\lambda I + H_{n-1}) - H_{n-1}^2).$$

Thus

$$\det(\lambda I - H_n) = \det(\lambda^2 I - 2H_{n-1}^2) = \det(\lambda I - \sqrt{2}H_{n-1}) \det(\lambda I + \sqrt{2}H_{n-1}).$$

This shows that each eigenvalue μ of H_{n-1} generates two eigenvalues $\pm\sqrt{2}\mu$ of H_n. The assertion then follows by the induction hypothesis, for H_{n-1} has éigenvalues $+2^{(n-1)/2}$ and $-2^{(n-1)/2}$ each of multiplicity 2^{n-2}.

Problem 43. An $n \times n$ matrix A over the complex numbers is called *positive semidefinite* (written as $A \geq 0$), if

$$\mathbf{x}^* A\mathbf{x} \geq 0 \quad \text{for all } \mathbf{x} \in \mathbb{C}^n.$$

Show that for every $A \geq 0$, there exists a unique $B \geq 0$ so that $B^2 = A$. Thus B is a *square root* of A.

Solution 43. Let $A = U^*\text{diag}(\lambda_1, \ldots, \lambda_n)U$, where U is unitary. We take

$$B = U^*\text{diag}(\lambda_1^{1/2}, \ldots, \lambda_n^{1/2})U.$$

Then the matrix B is positive semidefinite and $B^2 = A$ since $U^*U = I_n$. To show the uniqueness, suppose that C is an $n \times n$ positive semidefinite matrix satisfying $C^2 = A$. Since the eigenvalues of C are the nonnegative square roots of the eigenvalues of A, we can write

$$C = V\text{diag}(\lambda_1^{1/2}, \ldots, \lambda_n^{1/2})V^*$$

for some unitary matrix V. Then the identity $C^2 = A = B^2$ yields

$$T\text{diag}(\lambda_1, \ldots, \lambda_n) = \text{diag}(\lambda_1, \ldots, \lambda_n)T$$

where $T = UV$. This yields $t_{jk}\lambda_k = \lambda_j t_{jk}$. Thus $t_{jk}\lambda_k^{1/2} = \lambda_j^{1/2} t_{jk}$. Hence

$$T\mathrm{diag}(\lambda_1^{1/2}, \ldots, \lambda_n^{1/2}) = \mathrm{diag}(\lambda_1^{1/2}, \ldots, \lambda_n^{1/2})T.$$

Since $T = UV$ it follows that $B = C$.

Problem 44. (i) Consider the polynomial $p(x) = x^2 - sx + d$, $s, d \in \mathbb{C}$. Find a 2×2 matrix A such that its characteristic polynomial is p.
(ii) Consider the polynomial $q(x) = -x^3 + sx^2 - qx + d$, $s, q, d \in \mathbb{C}$. Find a 3×3 matrix B such that its characteristic polynomial is q.

Solution 44. (i) We obtain

$$A = \begin{pmatrix} s & d \\ -1 & 0 \end{pmatrix}.$$

(ii) We obtain

$$B = \begin{pmatrix} s & q & d \\ -1 & 0 & 0 \\ 0 & -1 & 0 \end{pmatrix}.$$

Problem 45. Calculate the eigenvalues of the 4×4 matrix

$$A = \begin{pmatrix} 1 & 0 & 0 & 1 \\ 0 & 1 & 1 & 0 \\ 0 & 1 & -1 & 0 \\ 1 & 0 & 0 & -1 \end{pmatrix}$$

by calculating the eigenvalues of A^2.

Solution 45. The matrix A is symmetric over $\mathbb{R}$. Thus the eigenvalues of A are real. Now A^2 is the diagonal matrix

$$A^2 = \mathrm{diag}(2\ 2\ 2\ 2)$$

with eigenvalue 2 (four times). Since $\mathrm{tr}(A) = 0$ we obtain $\sqrt{2}, \sqrt{2}, -\sqrt{2}, -\sqrt{2}$ as the eigenvalues of A.

Problem 46. If $\{A_j\}_{j=1}^m$ is a commuting family of matrices that is to say $A_j A_k = A_k A_j$ for every pair from the set, then there exists a unitary matrix V such that for all A_j in the set the matrix $\tilde{A}_j = V^* A_j V$ is upper triangular. Apply this to the matrices

$$A_1 = \begin{pmatrix} 1 & 1 \\ 1 & 1 \end{pmatrix}, \qquad A_2 = \begin{pmatrix} 1 & -1 \\ -1 & 1 \end{pmatrix}.$$

Solution 46. The eigenvalues of A_1 are 0 and 2 with the normalized eigenvectors

$$\frac{1}{\sqrt{2}}\begin{pmatrix} 1 \\ -1 \end{pmatrix}, \qquad \frac{1}{\sqrt{2}}\begin{pmatrix} 1 \\ 1 \end{pmatrix}.$$

The eigenvalues of A_2 are 0 and 2 with the normalized eigenvectors

$$\frac{1}{\sqrt{2}} \begin{pmatrix} 1 \\ 1 \end{pmatrix}, \qquad \frac{1}{\sqrt{2}} \begin{pmatrix} 1 \\ -1 \end{pmatrix}.$$

Thus the set of the eigenvectors are the same. The unitary matrix V is given by

$$V = \frac{1}{\sqrt{2}} \begin{pmatrix} 1 & 1 \\ -1 & 1 \end{pmatrix}$$

with

$$\tilde{A}_1 = V^* A_1 V = \begin{pmatrix} 0 & 0 \\ 0 & 2 \end{pmatrix}, \qquad \tilde{A}_2 = V^* A_2 V = \begin{pmatrix} 2 & 0 \\ 0 & 0 \end{pmatrix}.$$

Problem 47. Let A be an $n \times n$ matrix over $\mathbb{C}$. The *spectral radius* of the matrix A is the non-negative number defined by

$$\rho(A) := \max\{ |\lambda_j(A)| : 1 \le j \le n \}$$

where $\lambda_j(A)$ are the eigenvalues of A. We define the *norm* of A as

$$\|A\| := \sup_{\|\mathbf{x}\|=1} \|A\mathbf{x}\|$$

where $\|A\mathbf{x}\|$ denotes the Euclidean norm of the vector $A\mathbf{x}$.
(i) Show that $\rho(A) \le \|A\|$.
(ii) Consider the 2×2 matrix

$$A = \begin{pmatrix} 1/4 & 1/2 \\ 1/2 & 1/4 \end{pmatrix}.$$

Now $\rho(A) < 1$. If $\rho(A) < 1$, then $(I_2 - A)^{-1} = I_2 + A + A^2 + \cdots$. Calculate $(I_2 - A)^{-1}$. Calculate $(I_2 - A)(I_2 + A + A^2 + \cdots + A^k)$.

Solution 47. (i) From the eigenvalue equation $A\mathbf{x} = \lambda\mathbf{x}$ $(\mathbf{x} \ne 0)$ we obtain

$$|\lambda|\|\mathbf{x}\| = \|\lambda\mathbf{x}\| = \|A\mathbf{x}\| \le \|A\|\|\mathbf{x}\|.$$

Therefore since $\|\mathbf{x}\| \ne 0$ we obtain $|\lambda| \le \|A\|$ and $\rho(A) \le \|A\|$.
(ii) Since A is symmetric over $\mathbb{R}$ the eigenvalues are real. We find $\lambda_1 = -1/4$ and $\lambda_2 = 3/4$. Thus $\rho(A) < 1$. We obtain

$$(I_2 - A)^{-1} = \begin{pmatrix} 12/5 & 8/5 \\ 8/5 & 12/5 \end{pmatrix}.$$

We have $(I_2 - A)(I_2 + A + A^2 + \cdots + A^k) = I_2 - A^{k+1}$.

Problem 48. Let A be an $n \times n$ matrix with entries $a_{jk} \ge 0$ and with positive *spectral radius* ρ. Then there is a (column) vector $\mathbf{x}$ with $x_j \ge 0$ and a (column) vector $\mathbf{y}$ such that the following conditions hold

$$A\mathbf{x} = \rho\mathbf{x}, \quad \mathbf{y}^T A = \rho\mathbf{y}, \quad \mathbf{y}^T\mathbf{x} = 1.$$

Consider the 2×2 symmetric matrix

$$B = \begin{pmatrix} 2 & 1 \\ 1 & 2 \end{pmatrix}.$$

Show that B has a positive spectral radius. Find the vectors **x** and **y**.

Solution 48.   The eigenvalues of A are 3 and 1. Thus the spectral radius is $\rho = 3$ and the vectors are

$$\mathbf{x} = \begin{pmatrix} 1 \\ 1 \end{pmatrix}, \qquad \mathbf{y} = \begin{pmatrix} 1/2 \\ 1/2 \end{pmatrix}.$$

Problem 49.   Consider a symmetric 2×2 matrix A over $\mathbb{R}$ with $a_{11} > 0$, $a_{22} > 0$, $a_{12} < 0$ and $a_{jj} > |a_{12}|$ for $j = 1, 2$. Is the matrix A positive definite?

Solution 49.   We show that the matrix A is positive definite by calculating the eigenvalues. The eigenvalues of A are given by

$$\lambda_+ = \frac{1}{2}(\sqrt{(a_{11} - a_{22})^2 + 4a_{12}^2} + a_{11} + a_{22}),$$
$$\lambda_- = -\frac{1}{2}(\sqrt{(a_{11} - a_{22})^2 + 4a_{12}^2} - a_{11} - a_{22}).$$

Thus both eigenvalues are positive and therefore the matrix is positive semidefinite.

Problem 50.   Let A be a *positive definite* $n \times n$ matrix. Show that A^{-1} exists and is also positive definite.

Solution 50.   Since A is positive definite we have for the eigenvalues (which are all real and positive) using the ordering $0 < \lambda_1 \leq \lambda_2 \leq \cdots \leq \lambda_n$. The inverse A^{-1} has the real and positive eigenvalues

$$0 < \lambda_n^{-1} \leq \lambda_{n-1}^{-1} \leq \cdots \leq \lambda_1^{-1}.$$

Thus A^{-1} is also positive definite.

Problem 51.   Find the eigenvalues of the symmetric matrices

$$A_3 = \begin{pmatrix} 0 & 1 & 1 \\ 1 & 0 & 1 \\ 1 & 1 & 0 \end{pmatrix}, \quad A_4 = \begin{pmatrix} 0 & 1 & 0 & 1 \\ 1 & 0 & 1 & 0 \\ 0 & 1 & 0 & 1 \\ 1 & 0 & 1 & 0 \end{pmatrix}, \quad A_5 = \begin{pmatrix} 0 & 1 & 0 & 0 & 1 \\ 1 & 0 & 1 & 0 & 0 \\ 0 & 1 & 0 & 1 & 0 \\ 0 & 0 & 1 & 0 & 1 \\ 1 & 0 & 0 & 1 & 0 \end{pmatrix}.$$

Solution 51.   Since the matrices are symmetric over $\mathbb{R}$ the eigenvalues must be real. Since the trace for all three matrices is 0 the sum of the eigenvalues must be 0. For A_3 we obtain the eigenvalues -1 (twice) and 2. For A_4 we obtain the eigenvalues -2, 0 (twice) and $+2$. For A_5 we obtain

$$-\frac{1+\sqrt{5}}{2} \text{ (twice)} \quad \frac{\sqrt{5}-1}{2} \text{ (twice)}, \quad +2.$$

Extend the result to n-dimensions. The largest eigenvalue of A_n is $+2$.

Problem 52. Consider the $n \times n$ *cyclic matrix*

$$A = \begin{pmatrix} a_{11} & a_{12} & a_{13} & a_{14} & \cdots & a_{1n-1} & a_{1n} \\ a_{1n} & a_{11} & a_{12} & a_{13} & \cdots & a_{1n-2} & a_{1n-1} \\ a_{1n-1} & a_{1n} & a_{11} & a_{12} & \cdots & a_{n-3} & a_{n-2} \\ \vdots & \vdots & \vdots & \vdots & \vdots & \ddots & \vdots \\ a_{12} & a_{13} & a_{14} & a_{15} & \cdots & a_{1n} & a_{11} \end{pmatrix}$$

where $a_{jk} \in \mathbb{R}$. Show that

$$\mathbf{v} = \frac{1}{\sqrt{n}} \begin{pmatrix} \epsilon^{2k} \\ \epsilon^{4k} \\ \vdots \\ \epsilon^{2(n-1)k} \\ 1 \end{pmatrix}, \quad \epsilon \equiv e^{i\pi/n}, \quad 1 \le k \le n$$

is a normalized eigenvector of A. Find the eigenvalues.

Solution 52. We have

$$\frac{1}{n} \sum_{j=1}^{n} 1 = 1.$$

Thus the vector is normalized. Applying A to the vector yields

$$A\mathbf{v} = (a_{11} + \epsilon^{2k} a_{12} + \epsilon^{4k} a_{13} + \cdots + \epsilon^{2(n-1)k} a_{1n})\mathbf{v}.$$

Thus we have an eigenvalue equation with the n eigenvalues

$$a_{11} + \epsilon^{2k} a_{12} + \epsilon^{4k} a_{13} + \cdots + \epsilon^{2(n-1)k} a_{1n}, \quad k = 1, \ldots, n.$$

Problem 53. (i) Let $a, b \in \mathbb{R}$. Find on inspection two eigenvectors and the corresponding eigenvalues of the symmetric 4×4 matrix

$$\begin{pmatrix} a & 0 & 0 & b \\ 0 & a & 0 & b \\ 0 & 0 & a & b \\ b & b & b & 0 \end{pmatrix}.$$

(ii) Let $a, b \in \mathbb{R}$. Find on inspection two eigenvectors and the corresponding eigenvalues of the 4×4 matrix

$$\begin{pmatrix} a & 0 & 0 & b \\ 0 & a & 0 & b \\ 0 & 0 & -a & b \\ b & b & b & 0 \end{pmatrix}.$$

Solution 53. (i) Obviously $(1 \quad -1 \quad 0 \quad 0)^T$, $(1 \quad 0 \quad -1 \quad 0)^T$ are eigenvectors with the corresponding eigenvalues a and a.
(ii) Obviously $(1 \quad -1 \quad 0 \quad 0)^T$ is an eigenvector with the corresponding eigenvalue a.

Problem 54. Consider the two 4×4 permutation matrices

$$
S = \begin{pmatrix} 0 & 0 & 1 & 0 \\ 0 & 1 & 0 & 0 \\ 1 & 0 & 0 & 0 \\ 0 & 0 & 0 & 1 \end{pmatrix}, \qquad T = \begin{pmatrix} 1 & 0 & 0 & 0 \\ 0 & 0 & 0 & 1 \\ 0 & 0 & 1 & 0 \\ 0 & 1 & 0 & 0 \end{pmatrix}.
$$

Show that the two matrices have the same (normalized) eigenvectors. Find the commutator $[S, T]$.

Solution 54. We find the normalized eigenvectors

$$
\mathbf{v}_1 = \frac{1}{2}\begin{pmatrix} 1 \\ 1 \\ 1 \\ 1 \end{pmatrix}, \quad \mathbf{v}_2 = \frac{1}{\sqrt{2}}\begin{pmatrix} 1 \\ 0 \\ -1 \\ 0 \end{pmatrix}, \quad \mathbf{v}_3 = \frac{1}{\sqrt{2}}\begin{pmatrix} 0 \\ 1 \\ 0 \\ -1 \end{pmatrix}, \quad \mathbf{v}_4 = \frac{1}{2}\begin{pmatrix} 1 \\ -1 \\ 1 \\ -1 \end{pmatrix}.
$$

Obviously we find $[S, T] = 0_4$ applying the spectral theorem.

Problem 55. Let $\alpha \in \mathbb{R}$. Consider the symmetric 4×4 matrix

$$
A(\alpha) = \begin{pmatrix} 1 & \alpha & 0 & 0 \\ \alpha & 2 & 2\alpha & 0 \\ 0 & 2\alpha & 3 & \alpha \\ 0 & 0 & \alpha & 4 \end{pmatrix}.
$$

(i) Find the characteristic equation.
(ii) Show that

$$
\lambda_1 + \lambda_2 + \lambda_3 + \lambda_4 = 10
$$
$$
\lambda_1\lambda_2 + \lambda_1\lambda_3 + \lambda_1\lambda_4 + \lambda_2\lambda_3 + \lambda_2\lambda_4 + \lambda_3\lambda_4 = 35 - 6\alpha^2
$$
$$
\lambda_1\lambda_2\lambda_3 + \lambda_1\lambda_2\lambda_4 + \lambda_1\lambda_3\lambda_4 + \lambda_2\lambda_3\lambda_4 = 50 - 30\alpha^2
$$
$$
\lambda_1\lambda_2\lambda_3\lambda_4 = 24 - 30\alpha^2 + \alpha^4
$$

where λ_1, λ_2, λ_3, λ_4 denote the eigenvalues.

Solution 55. (i) From $\det(A(\alpha) - \lambda I_4) = 0$ we find

$$
\lambda^4 - 10\lambda^3 + (35 - 6\alpha^2)\lambda^2 + (-50 + 30\alpha^2)\lambda + 24 - 30\alpha^2 + \alpha^4 = 0.
$$

(ii) From

$$
\det \begin{pmatrix} \lambda_1 - \lambda & 0 & 0 & 0 \\ 0 & \lambda_2 - \lambda & 0 & 0 \\ 0 & 0 & \lambda_3 - \lambda & 0 \\ 0 & 0 & 0 & \lambda_4 - \lambda \end{pmatrix} = 0
$$

we obtain

$$\lambda^4 - \lambda^3 \Big(\sum_{j=1}^{4} \lambda_j\Big) + \lambda^2(\lambda_1\lambda_2 + \lambda_1\lambda_3 + \lambda_1\lambda_4 + \lambda_2\lambda_3 + \lambda_2\lambda_4 + \lambda_3\lambda_4)$$

$$-\lambda(\lambda_1\lambda_2\lambda_3 + \lambda_1\lambda_2\lambda_4 + \lambda_1\lambda_3\lambda_4 + \lambda_2\lambda_3\lambda_4) + \lambda_1\lambda_2\lambda_3\lambda_4 = 0.$$

Thus the result given above follows.

Problem 56. Let B be a 2×2 matrix with eigenvalues λ_1 and λ_2. Find the eigenvalues of the 4×4 matrix

$$X = \begin{pmatrix} 0 & 0 & 1 & 0 \\ 0 & 0 & 0 & 1 \\ b_{11} & b_{12} & 0 & 0 \\ b_{21} & b_{22} & 0 & 0 \end{pmatrix}.$$

Let $\mathbf{v}$ be an eigenvector of B with eigenvalue λ. What can be said about an eigenvector of the 4×4 matrix X given by eigenvector $\mathbf{v}$ and eigenvalue of B?

Solution 56. The characteristic equation $\det(X - rI_4) = 0$ is given by

$$r^4 - r^2(b_{11} + b_{22}) + (b_{11}b_{22} - b_{12}b_{21}) = 0$$

or $r^4 - r^2\mathrm{tr}(B) + \det(B) = 0$. Thus since $\mathrm{tr}(B) = \lambda_1 + \lambda_2$ and $\det(B) = \lambda_1\lambda_2$ we have

$$r^4 - r^2(\lambda_1 + \lambda_2) + \lambda_1\lambda_2 = 0.$$

This quartic equation can be reduced to a quadratic equation. We find the eigenvalues

$$r_1 = \sqrt{\lambda_1}, \quad r_2 = -\sqrt{\lambda_1}, \quad r_3 = \sqrt{\lambda_2}, \quad r_4 = -\sqrt{\lambda_2}.$$

Let $\mathbf{v}$ be an eigenvector of B with eigenvalue λ, i.e. $B\mathbf{v} = \lambda\mathbf{v}$. Then we have

$$\begin{pmatrix} 0 & 0 & 1 & 0 \\ 0 & 0 & 0 & 1 \\ b_{11} & b_{12} & 0 & 0 \\ b_{21} & b_{22} & 0 & 0 \end{pmatrix} \begin{pmatrix} v_1 \\ v_2 \\ \sqrt{\lambda}v_1 \\ \sqrt{\lambda}v_2 \end{pmatrix} = \sqrt{\lambda} \begin{pmatrix} v_1 \\ v_2 \\ \sqrt{\lambda}v_1 \\ \sqrt{\lambda}v_2 \end{pmatrix}.$$

Thus

$$\begin{pmatrix} v_1 \\ v_2 \\ \sqrt{\lambda}v_1 \\ \sqrt{\lambda}v_2 \end{pmatrix}$$

is an eigenvector of X.

Problem 57. Let A, B be two $n \times n$ matrices over $\mathbb{C}$. The set of all matrices of the form $A - \lambda B$ with $\lambda \in \mathbb{C}$ is said to be a *pencil*. The eigenvalues of the pencil are elements of the set $\lambda(A, B)$ defined by

$$\lambda(A, B) := \{\, z \in \mathbb{C} : \det(A - zB) = 0 \,\}.$$

If $\lambda \in \lambda(A, B)$ and $A\mathbf{v} = \lambda B\mathbf{v}$, $\mathbf{v} \neq \mathbf{0}$ then $\mathbf{v}$ is referred to as an eigenvector of $A - \lambda B$. Note that λ may be finite, empty or infinite. Let

$$A = \begin{pmatrix} 0 & 0 & 0 & 1 \\ 0 & 0 & 1 & 0 \\ 0 & 1 & 0 & 0 \\ 1 & 0 & 0 & 0 \end{pmatrix}, \qquad B = \frac{1}{\sqrt{2}} \begin{pmatrix} 1 & 0 & 0 & 1 \\ 0 & 1 & 1 & 0 \\ 0 & 1 & -1 & 0 \\ 1 & 0 & 0 & -1 \end{pmatrix}.$$

Find the eigenvalue of the pencil.

Solution 57.   Since B is invertible we have $\lambda(A, B) = \lambda(B^{-1}A, I_4) = \lambda(B^{-1}A)$.

Problem 58.   Let A be an $n \times n$ matrix with eigenvalues $\lambda_1, \ldots, \lambda_n$. Let $c \in \mathbb{C} \setminus \{0\}$. What are the eigenvalues of cA?

Solution 58.   We set $B = cA$. Then

$$\det(B - \mu I_n) = \det(cA - \mu I_n) = \det\left(c\left(A - \frac{\mu}{c}\right)\right) = c^n \det\left(A - \frac{\mu}{c}I_n\right).$$

Thus $\det(A - \mu I_n/c) = 0$ and therefore the eigenvalues of cA are $c\lambda_1, \ldots, c\lambda_n$.

Problem 59.   Let A, B be $n \times n$ matrices over $\mathbb{C}$. Let $\alpha, \beta \in \mathbb{C}$. Assume that $A^2 = I_n$ and $B^2 = I_n$ and $AB + BA = 0_n$. What can be said about the eigenvalues of $\alpha A + \beta B$?

Solution 59.   From the eigenvalue equation $(\alpha A + \beta B)\mathbf{u} = \lambda \mathbf{u}$ we obtain

$$(\alpha A + \beta B)^2 \mathbf{u} = \lambda(\alpha A + \beta B)\mathbf{u} = \lambda^2 \mathbf{u}.$$

Thus

$$(\alpha^2 A^2 + \beta B^2 + \alpha\beta(AB + BA))\mathbf{u} = \lambda^2 \mathbf{u}$$

or $(\alpha^2 + \beta^2)I_n \mathbf{u} = \lambda^2 \mathbf{u}$. Hence the eigenvalues can only be $\lambda = \pm\sqrt{\alpha^2 + \beta^2}$.

Problem 60.   (i) Let A, B be $n \times n$ matrices. Show that AB and BA have the same eigenvalues.
(ii) Can we conclude that every eigenvector of AB is also an eigenvector of BA?

Solution 60.   (i) If B is nonsingular, then B^{-1} exists and BA is similar to

$$B^{-1}(BA)B = AB$$

and the result follows since two similar matrices have the same eigenvalues. The same arguments holds if A is nonsingular. If both A and B are singular, then 0 is an eigenvalue of B. Let δ be the modulus of the nonzero eigenvalue of B of smallest modulus. If $0 < \epsilon < \delta$, then $B + \epsilon I$ is nonsingular and the eigenvalues of $A(B + \epsilon I)$ are the same as those of $(B + \epsilon I)A$. As $\epsilon \to 0$ both sets of eigenvalues converge to the eigenvalues of AB and BA, respectively.

(ii) No. Consider for example

$$A = \begin{pmatrix} 1 & 1 \\ 0 & 0 \end{pmatrix}, \quad B = \begin{pmatrix} 0 & 0 \\ 1 & 1 \end{pmatrix} \Rightarrow AB = \begin{pmatrix} 1 & 1 \\ 0 & 0 \end{pmatrix}, \quad BA = \begin{pmatrix} 0 & 0 \\ 1 & 1 \end{pmatrix}$$

and $(1 \ 0)^T$ is an eigenvector of AB, but not of BA.

Problem 61. (i) Show that if A is an $n \times m$ matrix and if B is an $m \times n$ matrix, then $\lambda \neq 0$ is an eigenvalue of the $n \times n$ matrix AB if and only if λ is an eigenvalue of the $m \times m$ matrix BA. Show that if $m = n$ then the conclusion is true even for $\lambda = 0$.
(ii) Let M be an $m \times n$ matrix $(m < n)$ over $\mathbb{R}$. Show that at least one eigenvalue of the $n \times n$ matrix $M^T M$ is equal to 0. Show that the eigenvalues of the $m \times m$ matrix MM^T are also eigenvalues of $M^T M$.

Solution 61. (i) If $\lambda \neq 0$ is an eigenvalue of AB and if $\mathbf{v}$ is an eigenvector of AB, then $AB\mathbf{v} = \lambda \mathbf{v} \neq \mathbf{0}$ since $\lambda \neq 0$ and $\mathbf{v} \neq \mathbf{0}$. Thus $B\mathbf{v} \neq \mathbf{0}$. Therefore $BAB\mathbf{v} = \lambda B\mathbf{v}$ and λ is an eigenvalue of BA. If A and B are square matrices and $\lambda = 0$ is an eigenvalue of AB with eigenvector $\mathbf{v}$, then we have $\det(AB) = \det(BA) = 0$. Thus $\lambda = 0$ is an eigenvalue of BA.
(ii) Since $\text{rank}(M^T M) \leq \text{rank}(M) \leq m < n$ less than n eigenvalues of $M^T M$ are nonzero, i.e. at least one eigenvalue is 0. From the eigenvalue equation $MM^T \mathbf{x} = \lambda \mathbf{x}$ we obtain $M^T M(M^T \mathbf{x}) = \lambda(M^T \mathbf{x})$.

Problem 62. We know that a hermitian matrix has only real eigenvalues. Can we conclude that a matrix with only real eigenvalues is hermitian?

Solution 62. We cannot conclude that a matrix with real eigenvalue is hermitian. For example the nonnormal matrix

$$A = \begin{pmatrix} 0 & 1 & 0 \\ 1 & 0 & 1 \\ 0 & 2 & 0 \end{pmatrix}$$

admits the eigenvalues $\sqrt{3}, -\sqrt{3}, 0$.

Problem 63. Let A be an $n \times n$ matrix over $\mathbb{C}$. Show that the eigenvalues of $A^* A$ are nonnegative.

Solution 63. Let $\mathbf{v}$ be a normalized eigenvector of $A^* A$, i.e. $A^* A\mathbf{v} = \lambda \mathbf{v}$. Then

$$0 \leq |A\mathbf{v}|^2 = \mathbf{v}^* A^* A\mathbf{v} = \lambda \mathbf{v}^* \mathbf{v} = \lambda.$$

Thus $\lambda \geq 0$.

Problem 64. Let n be a positive integer. Consider the 2×2 matrix

$$T_n = \begin{pmatrix} 2n & 4n^2 - 1 \\ 1 & 2n \end{pmatrix} \Rightarrow \det(T_n) = 1.$$

Show that the eigenvalues of T_n are real and not of absolute value 1.

Solution 64.   The eigenvalues are $\lambda_\pm = 2n \pm \sqrt{4n^2 - 1}$. Thus the eigenvalues are real and not of absolute value 1 since n is a positive integer.

Problem 65.   Let L_n be the $n \times n$ matrix

$$
L_n = \begin{pmatrix}
n-1 & -1 & -1 & \cdots & -1 \\
-1 & n-1 & -1 & \cdots & -1 \\
\vdots & \vdots & \ddots & \ddots & \vdots \\
-1 & -1 & -1 & \ddots & -1 \\
-1 & -1 & -1 & \cdots & n-1
\end{pmatrix}.
$$

Find the eigenvalues.

Solution 65.   The matrix L_n has a single eigenvalue at 0. All the other eigenvalues are equal to n. For example, for $n = 3$ we have the matrix

$$
L_3 = \begin{pmatrix}
2 & -1 & -1 \\
-1 & 2 & -1 \\
-1 & -1 & 2
\end{pmatrix}
$$

with the eigenvalues 0 and 3 (twice). We can write $L_n = nI_n - J_n$, where J_n is the $n \times n$ matrix with all entries $+1$. Obviously, J_n has rank 1 and therefore only one nonzero eigenvalue, namely n the trace of J_n. Thus the eigenvalues of L_n are $n - n = 0$ and $n - 0 = n$ ($n-1$ times degenerate).

Problem 66.   Let A be an $n \times n$ matrix over $\mathbb{C}$. Assume that $A^2 = -I_n$. What can be said about the eigenvalues of A?

Solution 66.   From the eigenvalue equation $A\mathbf{v} = \lambda\mathbf{v}$ we obtain

$$
A^2\mathbf{v} = \lambda A\mathbf{v} \Rightarrow -\mathbf{v} = \lambda^2\mathbf{v} \Rightarrow (\lambda^2 + 1)\mathbf{v} = \mathbf{0}.
$$

Thus the eigenvalues can only be $\pm i$.

Problem 67.   Consider the symmetric 6×6 matrix over $\mathbb{R}$

$$
A = \begin{pmatrix}
0 & 1 & 1 & 1 & 1 & 1 \\
1 & 0 & 1 & -1 & -1 & 1 \\
1 & 1 & 0 & 1 & -1 & -1 \\
1 & -1 & 1 & 0 & 1 & -1 \\
1 & -1 & -1 & 1 & 0 & 1 \\
1 & 1 & -1 & -1 & 1 & 0
\end{pmatrix}.
$$

This matrix plays a role in the construction of the *icosahedron* which is a regular *polyhedron* with 20 identical equilateral triangular faces, 30 edges and 12 vertices.
(i) Find the eigenvalues of this matrix.

(ii) Consider the matrix $A + \sqrt{5}I_6$. Find the eigenvalues.
(iii) The matrix $A + \sqrt{5}I_6$ induces an Euclidean structure on the quotient space $\mathbb{R}^6/\ker(A + \sqrt{5}I_6)$. Find the dimension of $\ker(A + \sqrt{5}I_6)$.

Solution 67. (i) The matrix is symmetric over $\mathbb{R}$. Thus the eigenvalues must all be real. The trace is 0 which is the sum of the eigenvalues. The rank of the matrix is 6. Thus the matrix is invertible. Thus all eigenvalues are nonzero. Now $A^2 = 5I_6$. Thus the eigenvalues are $\sqrt{5}$ (three-fold) and $-\sqrt{5}$ (three-fold).
(ii) From (i) we obtain that the eigenvalues of $A + \sqrt{5}I_6$ are 0 (three-fold) and $2\sqrt{5}$ (three-fold).
(iii) The kernel of $A + \sqrt{5}I_6$ has dimension 3. Thus the quotient space is isomorphic to $\mathbb{R}^3$.

Problem 68. Let A be an $n \times n$ matrix over $\mathbb{C}^n$. Let λ be an eigenvalue of A. A generalized eigenvector $\mathbf{x} \in \mathbb{C}^n$ of A corresponding to the eigenvalue λ is a nontrivial solution of

$$(A - \lambda I_n)^j \mathbf{x} = \mathbf{0}_n$$

for some $j \in \{1, 2, \ldots\}$, where $\mathbf{0}_n$ is the n-dimensional zero vector. For $j = 1$ we find the eigenvectors. It follows that $\mathbf{x}$ is a generalized eigenvector of A corresponding to λ if and only if

$$(A - \lambda I_n)^n \mathbf{x} = \mathbf{0}_n.$$

Find the eigenvectors and generalized eigenvectors of the nonnormal matrix

$$\begin{pmatrix} 0 & 1 & 0 \\ 0 & 0 & 0 \\ 0 & -1 & 0 \end{pmatrix}.$$

Solution 68. The eigenvalues are all 0. The eigenvectors follow from solving

$$\left(\begin{pmatrix} 0 & 1 & 0 \\ 0 & 0 & 0 \\ 0 & -1 & 0 \end{pmatrix} - 0 \begin{pmatrix} 1 & 0 & 0 \\ 0 & 1 & 0 \\ 0 & 0 & 1 \end{pmatrix} \right) \begin{pmatrix} x_1 \\ x_2 \\ x_3 \end{pmatrix} = \begin{pmatrix} 0 \\ 0 \\ 0 \end{pmatrix}$$

for x_1, x_2 and x_3. Thus $x_2 = 0$. The set of eigenvectors is given by

$$\left\{ \begin{pmatrix} u \\ 0 \\ v \end{pmatrix} : (u, v) \in \mathbb{C}^2/(0, 0) \right\}.$$

The generalized eigenvectors follow from solving

$$\left(\begin{pmatrix} 0 & 1 & 0 \\ 0 & 0 & 0 \\ 0 & -1 & 0 \end{pmatrix} - 0 \begin{pmatrix} 1 & 0 & 0 \\ 0 & 1 & 0 \\ 0 & 0 & 1 \end{pmatrix} \right)^3 \begin{pmatrix} x_1 \\ x_2 \\ x_3 \end{pmatrix} = \begin{pmatrix} 0 \\ 0 \\ 0 \end{pmatrix}$$

for x_1, x_2 and x_3. This equation is satisfied for all x_1, x_2 and x_3. The set of generalized eigenvectors is given by $\mathbb{C}^3/(0, 0, 0)$.

Problem 69. Consider the $2n \times 2n$ matrix

$$J := \begin{pmatrix} 0_n & I_n \\ -I_n & 0_n \end{pmatrix} \in \mathbb{R}^{2n \times 2n}.$$

A matrix $S \in \mathbb{R}^{2n \times 2n}$ is called a *symplectic matrix* if $S^T J S = J$.
(i) Show that symplectic matrices are nonsingular.
(ii) Show that the product of two symplectic matrices S_1 and S_2 is also symplectic.
(iii) Show that if S is symplectic S^{-1} and S^T are also symplectic.
(iv) Let S be a symplectic matrix. Show that if $\lambda \in \sigma(S)$, then $\lambda^{-1} \in \sigma(S)$, where $\sigma(S)$ denotes the spectrum of S.

Solution 69.   (i) Since $\det(J) \neq 0$, $\det(S) = \det(S^T)$ and

$$\det(S^T J S) = \det(S^T) \det(J) \det(S) = \det(J)(\det(S))^2$$

we find that $\det(S) \neq 0$. We have $S^{-1} = J S^T J^T$.
(ii) Let S_1, S_2 be symplectic matrices, i.e. $S_1^T J S_1 = J$, $S_2^T J S_2 = J$. Now

$$(S_1 S_2)^T J (S_1 S_2) = S_2^T (S_1^T J S_1) S_2 = S_2^T J S_2 = J.$$

(iii) From $S^T J S = J$ we have $(S^T J S)^{-1} = J^{-1}$. Since $J^{-1} = -J$ it follows that

$$S^{-1} J (S^{-1})^T = J.$$

(iv) We note that $J^T = J^{-1} = -J$.

Problem 70.   Let A, B be $n \times n$ matrices over $\mathbb{C}$ and $\mathbf{u}$ a nonzero vector in $\mathbb{C}^n$. Assume that $[A, B] = A$ and $A\mathbf{v} = \lambda \mathbf{v}$. Find $(AB)\mathbf{v}$.

Solution 70.   We have the identity $AB \equiv BA + [A, B]$. Thus

$$(AB)\mathbf{v} = (BA + [A, B])\mathbf{v} = BA\mathbf{v} + [A, B]\mathbf{v} = \lambda B\mathbf{v} + A\mathbf{v} = (\lambda B + A)\mathbf{v}.$$

Problem 71.   Let A, B be $n \times n$ matrices over $c \in \mathbb{C}$ with $[A, B] = 0_n$. Then $[A + cI_n, B + cI_n] = 0_n$, where $c \in \mathbb{C}$. Let $\mathbf{v}$ be an eigenvector of the $n \times n$ matrix A with eigenvalue λ. Show that $\mathbf{v}$ is also an eigenvector of $A + cI_n$, where $c \in \mathbb{C}$.

Solution 71.   We have $(A + cI_n)\mathbf{v} = A\mathbf{v} + cI_n\mathbf{v} = \lambda\mathbf{v} + c\mathbf{v} = (\lambda + c)\mathbf{v}$. Thus $\mathbf{v}$ is an eigenvector of $A + cI_n$ with eigenvalue $\lambda + c$.

Problem 72.   Let A be an $n \times n$ normal matrix over $\mathbb{C}$. How would one apply genetic algorithms to find the eigenvalues of A. This means we have to construct a *fitness function* f with the minima as the eigenvalues. The eigenvalue equation is given by $A\mathbf{x} = z\mathbf{x}$ ($z \in \mathbb{C}$ and $\mathbf{x} \in \mathbb{C}^n$ with $\mathbf{x} \neq \mathbf{0}$). The characteristic equation is $p(z) \equiv \det(A - zI_n) = 0$. What would be a fitness function? Apply it to the matrices

$$B = \begin{pmatrix} 0 & -i \\ i & 0 \end{pmatrix}, \quad C = \begin{pmatrix} 0 & i \\ i & 0 \end{pmatrix}, \quad D = \begin{pmatrix} 0 & 0 & 1 \\ 0 & 1 & 0 \\ 1 & 0 & 0 \end{pmatrix}.$$

Solution 72. Let $z = x + iy$ $(x, y \in \mathbb{R})$. A fitness function would be

$$f(x, y) = |\det(A - zI_n)|^2$$

which we have to minimize, where $f(x, y) \geq 0$ for all $x, y \in \mathbb{R}$.
For the matrix B we have $p(z) = z^2 - 1$ and the fitness function

$$f(x, y) = |x^2 - y^2 - 1 + 2ixy|^2 = (x^2 - y^2 - 1)^2 + (2xy)^2$$
$$= x^4 + y^4 + 2x^2y^2 - 2x^2 + 2y^2 + 1.$$

The minimum of f is reached for $(x = 1, y = 0)$ with $f(x = 1, y = 0) = 0$ and $(x = -1, y = 0)$ with $f(x = -1, y = 0) = 0$. Obviously since the matrix B is hermitian we can simplify the problem right at the beginning by setting $y = 0$ since the eigenvalues of a hermitian matrix are real. Thus the function we have to minimize would be $g(x) = x^4 - 2x^2 + 1$.

The matrix C is skew-hermitian. Thus the eigenvalues are purely imaginary, i.e. we can set $x = 0$ and the fitness function is $f(y) = (-y^2 + 1)^2$.

For the matrix D we have $p(z) = -z^3 + z^2 + z - 1$. The matrix is hermitian and so we can set $y = 0$ and the fitness function is $f(x) = (-x^3 + x^2 + x - 1)^2$.

Problem 73. Let $\mathbf{v}$ be a nonzero column vector in $\mathbb{R}^n$. Matrix multiplication is associative. Then we have $(\mathbf{v}\mathbf{v}^T)\mathbf{v} = \mathbf{v}(\mathbf{v}^T\mathbf{v})$. Discuss.

Solution 73. This is an eigenvalue equation with the (positive semidefinite) $n \times n$ matrix given by

$$\mathbf{v}\mathbf{v}^T = \begin{pmatrix} v_1^2 & v_1v_2 & \cdots & v_1v_n \\ v_2v_1 & v_2^2 & \cdots & v_2v_n \\ \vdots & \vdots & \ddots & \vdots \\ v_nv_1 & v_nv_2 & \cdots & v_n^2 \end{pmatrix}.$$

The eigenvector is $\mathbf{v}$ and the positive eigenvalue is $\mathbf{v}^T\mathbf{v} = \sum_{j=1}^{n} v_j^2$.

Problem 74. Let $n \geq 2$ and even. Consider an $n \times n$ hermitian matrix A. Thus the eigenvalues are real. Assume we have the information that if λ is an eigenvalue then $-\lambda$ is also an eigenvalue of A. How can the calculation of the eigenvalues be simplified with this information?

Solution 74.   The characteristic equation is

$$\lambda^n + c_{n-1}\lambda^{n-1} + c_{n-2}\lambda^{n-2} + \cdots + c_2\lambda^2 + c_1\lambda + c_0 = 0.$$

The substitution $\lambda \to -\lambda$ yields

$$\lambda^n - c_{n-1}\lambda^{n-1} + c_{n-2}\lambda^{n-2} + \cdots + c_2\lambda^2 - c_1\lambda + c_0 = 0.$$

Addition of the two equations provides

$$\lambda^n + c_{n-2}\lambda^{n-2} + \cdots + c_2\lambda^2 + c_0 = 0.$$

Study the problem for n odd. So in this case one of the eigenvalues must be 0.

Problem 75.   Let A, B be two nonzero $n \times n$ matrices over $\mathbb{C}$. Let $A\mathbf{v} = \lambda\mathbf{v}$ be the eigenvalue equation for A. Assume that $[A, B] = 0_n$. Then from $[A, B]\mathbf{v} = \mathbf{0}$ it follows that

$$[A, B]\mathbf{v} = (AB - BA)\mathbf{v} = A(B\mathbf{v}) - B(A\mathbf{v}) = \mathbf{0}.$$

Therefore $A(B\mathbf{v}) = \lambda(B\mathbf{v})$. If $B\mathbf{v} \neq \mathbf{0}$ we find that $B\mathbf{v}$ is an eigenvector of A with eigenvalue λ. Apply it to $A = \sigma_1 \otimes \sigma_2$ and $B = \sigma_3 \otimes \sigma_3$.

Solution 75.   The matrix σ_1 has the eigenvalue $+1$ with normalized eigenvector $\frac{1}{\sqrt{2}}(1 \quad 1)^T$ and σ_2 has the eigenvalue $+1$ with eigenvector $\frac{1}{\sqrt{2}}(1 \quad i)^T$. Thus

$$\frac{1}{\sqrt{2}}\begin{pmatrix} 1 \\ 1 \end{pmatrix} \otimes \frac{1}{\sqrt{2}}\begin{pmatrix} 1 \\ i \end{pmatrix} \equiv \frac{1}{2}\begin{pmatrix} 1 \\ i \\ 1 \\ i \end{pmatrix}$$

is a normalized eigenvector of $\sigma_1 \otimes \sigma_2$ with eigenvalue $+1$. Then

$$(\sigma_3 \otimes \sigma_3)\frac{1}{2}\begin{pmatrix} 1 \\ i \\ 1 \\ i \end{pmatrix} = \begin{pmatrix} 1 \\ -i \\ -1 \\ i \end{pmatrix}$$

is an eigenvector of $\sigma_1 \otimes \sigma_2$.

Problem 76.   Let A, B be two nonzero $n \times n$ matrices over $\mathbb{C}$. Let $A\mathbf{v} = \lambda\mathbf{v}$ be the eigenvalue equation for A. Assume that $[A, B]_+ = 0_n$. Then from $[A, B]_+\mathbf{v} = \mathbf{0}$ it follows that

$$[A, B]_+\mathbf{v} = (AB + BA)\mathbf{v} = A(B\mathbf{v}) + B(A\mathbf{v}) = \mathbf{0}.$$

Therefore $A(B\mathbf{v}) = -\lambda(B\mathbf{v})$. If $B\mathbf{v} \neq \mathbf{0}$ we have an eigenvalue equation with eigenvalue $-\lambda$. Apply it to $A = \sigma_1$ and $B = \sigma_2$, where $[\sigma_1, \sigma_2]_+ = 0_2$.

Solution 76.   The matrix σ_1 has the eigenvalue $+1$ with normalized eigenvector $\frac{1}{\sqrt{2}}(1 \quad 1)^T$ and σ_2 admits the eigenvalue $+1$ with normalized eigenvector

$\frac{1}{\sqrt{2}}(1 \quad i)^T$. It follows that

$$\sigma_1(\sigma_2 \frac{1}{\sqrt{2}} \begin{pmatrix} 1 \\ 1 \end{pmatrix}) = \sigma_1 \frac{1}{\sqrt{2}} \begin{pmatrix} -i \\ i \end{pmatrix} = \frac{1}{\sqrt{2}} \begin{pmatrix} i \\ -i \end{pmatrix} = -1\sigma_2 \frac{1}{\sqrt{2}} \begin{pmatrix} 1 \\ 1 \end{pmatrix}.$$

Problem 77. Consider the real symmetric 3×3 matrices

$$A = \begin{pmatrix} 0 & 0 & 1 \\ 0 & 1 & 0 \\ 1 & 0 & 0 \end{pmatrix}, \quad B = \begin{pmatrix} 0 & 1 & 1 \\ 1 & 0 & 1 \\ 1 & 1 & 0 \end{pmatrix}.$$

(i) Find the eigenvalues and normalized eigenvectors for the matrices A and B.
(ii) Find the commutator $[A, B]$.
(iii) Discuss the results from (ii) and (i) with respect to the eigenvectors.

Solution 77. (i) The eigenvalues of A are $\lambda = -1$ and $\lambda = +1$ (twice) with the corresponding normalized eigenvectors

$$\frac{1}{\sqrt{2}} \begin{pmatrix} 1 \\ 0 \\ -1 \end{pmatrix}, \quad \frac{1}{\sqrt{2}} \begin{pmatrix} 1 \\ 0 \\ 1 \end{pmatrix}, \quad \begin{pmatrix} 0 \\ 1 \\ 0 \end{pmatrix}.$$

For the matrix B the eigenvalues are $\lambda = -1$ (twice) and $\lambda = 2$ with the corresponding normalized eigenvectors

$$\frac{1}{\sqrt{2}} \begin{pmatrix} 1 \\ 0 \\ -1 \end{pmatrix}, \quad \frac{1}{\sqrt{2}} \begin{pmatrix} 0 \\ 1 \\ -1 \end{pmatrix}, \quad \frac{1}{\sqrt{3}} \begin{pmatrix} 1 \\ 1 \\ 1 \end{pmatrix}.$$

(ii) The commutator of A and B vanishes, i.e. $[A, B] = 0_3$.
(iii) The matrices A and B have an eigenvalue $\lambda = +1$ and the normalized eigenvector

$$\frac{1}{\sqrt{2}} \begin{pmatrix} 1 \\ 0 \\ -1 \end{pmatrix}$$

in common which relates that the commutator of A and B is the zero matrix.

Problem 78. Consider the Pauli spin matrix σ_2 with the eigenvalues $+1, -1$ and the corresponding normalized eigenvectors

$$\frac{1}{\sqrt{2}} \begin{pmatrix} 1 \\ i \end{pmatrix}, \quad \frac{1}{\sqrt{2}} \begin{pmatrix} 1 \\ -i \end{pmatrix}.$$

Then $\sigma_2 \otimes \sigma_2$ admits the eigenvalues $+1$ (twice) and -1 (twice). Show that there are product states as eigenvectors and two sets of entangled eigenvectors for $\sigma_2 \otimes \sigma_2$.

Solution 78. From the eigenvectors of σ_2 we find the four product states

$$\frac{1}{2} \begin{pmatrix} 1 \\ i \end{pmatrix} \otimes \begin{pmatrix} 1 \\ i \end{pmatrix}, \quad \frac{1}{2} \begin{pmatrix} 1 \\ i \end{pmatrix} \otimes \begin{pmatrix} 1 \\ -i \end{pmatrix}, \quad \frac{1}{2} \begin{pmatrix} 1 \\ -i \end{pmatrix} \otimes \begin{pmatrix} 1 \\ i \end{pmatrix}, \quad \frac{1}{2} \begin{pmatrix} 1 \\ -i \end{pmatrix} \otimes \begin{pmatrix} 1 \\ -i \end{pmatrix}.$$

The first set of entangled states (the so-called *Bell states*) are

$$\frac{1}{\sqrt{2}}\begin{pmatrix}1\\0\\0\\1\end{pmatrix}, \quad \frac{1}{\sqrt{2}}\begin{pmatrix}0\\1\\1\\0\end{pmatrix}, \quad \frac{1}{\sqrt{2}}\begin{pmatrix}1\\0\\0\\-1\end{pmatrix}, \quad \frac{1}{\sqrt{2}}\begin{pmatrix}0\\1\\-1\\0\end{pmatrix}.$$

The other set of fully entangled states are

$$\frac{1}{2}\begin{pmatrix}-1\\-1\\-1\\1\end{pmatrix}, \quad \frac{1}{2}\begin{pmatrix}-1\\1\\1\\1\end{pmatrix}, \quad \frac{1}{2}\begin{pmatrix}1\\-1\\1\\1\end{pmatrix}, \quad \frac{1}{2}\begin{pmatrix}1\\1\\-1\\1\end{pmatrix}.$$

Problem 79.   (i) Let A be an invertible $n\times n$ matrix. Given the eigenvalues and eigenvectors of A. What can be said about the eigenvalues of $A\otimes A^{-1}+A^{-1}\otimes A$? (ii) Let $c \in \mathbb{R}$ and A be a normal $n \times n$ matrix. Assume that A is invertible. The eigenvalue equation is given by $A\mathbf{v}_j = \lambda_j\mathbf{v}_j$ for $j = 1,\dots,n$ and $\lambda_j \neq 0$ for all $j = 1,\dots,n$. Find all the eigenvalues and eigenvectors of the matrix

$$A \otimes A^{-1} + c(A^{-1} \otimes A).$$

Find $\det(A \otimes A^{-1})$ and $\operatorname{tr}(A \otimes A^{-1})$.

Solution 79.   (i) Let $\mathbf{u}$ and $\mathbf{v}$ be two eigenvectors of A, i.e. $A\mathbf{u} = \lambda\mathbf{u}$ and $A\mathbf{v} = \mu\mathbf{v}$. Then

$$(A \otimes A^{-1} + A^{-1} \otimes A)(\mathbf{u} \otimes \mathbf{v}) = \lambda\mathbf{u} \otimes \frac{1}{\mu}\mathbf{v} + \frac{1}{\lambda}\mathbf{u} \otimes \mu\mathbf{v} = \left(\frac{\lambda}{\mu} + \frac{\mu}{\lambda}\right)(\mathbf{u} \otimes \mathbf{v}).$$

Thus the eigenvalue is $\lambda/\mu + \mu/\lambda$. If $\mathbf{u} = \mathbf{v}$ we have the eigenvalue 2.
(ii) We have the eigenvalue equation

$$((A \otimes A^{-1}) + c(A^{-1} \otimes A))(\mathbf{v}_j \otimes \mathbf{v}_k) = \left(\frac{\lambda_j}{\lambda_k} + c\frac{\lambda_k}{\lambda_j}\right)(\mathbf{v}_j \otimes \mathbf{v}_k).$$

Problem 80.   Let M be an $n \times n$ matrix over $\mathbb{R}$ given by

$$M = D + \mathbf{u} \otimes \mathbf{v}^T$$

where D is a diagonal matrix over $\mathbb{R}$ and $\mathbf{u}$, $\mathbf{v}$ are column vectors in $\mathbb{R}^n$. The eigenvalue equation is given by

$$M\mathbf{x} \equiv (D + \mathbf{u} \otimes \mathbf{v}^T)\mathbf{x} = \lambda\mathbf{x}.$$

Assume that $\sum_{j=1}^n u_j v_j \neq 0$. Show that

$$\sum_{k=1}^n \frac{u_k v_k}{(d_k - \lambda)} + 1 = 0.$$

where d_k are the diagonal elements of D.

Solution 80. From the eigenvalue equation we obtain

$$(D - \lambda I_n)\mathbf{x} + (\mathbf{u} \otimes \mathbf{v}^T)\mathbf{x} = \mathbf{0}.$$

Now

$$\mathbf{u} \otimes \mathbf{v}^T = \begin{pmatrix} u_1 v_1 & u_1 v_2 & \cdots & u_1 v_n \\ u_2 v_1 & \cdots & \cdots & u_2 v_n \\ \vdots & \vdots & \ddots & \vdots \\ u_n v_1 & \cdots & \cdots & u_n v_n \end{pmatrix}.$$

Thus

$$(d_j - \lambda)x_j + u_j \sum_{k=1}^n v_k x_k = 0, \qquad j = 1, 2, \ldots, n$$

or

$$x_j = -\frac{u_j \sum_{k=1}^n v_k x_k}{d_j - \lambda}.$$

We set $s := \sum_{k=1}^n v_k x_k$. It follows that

$$x_j = -\frac{s u_j}{d_j - \lambda}.$$

Multiplying this equation with v_j on the left and right-hand side, sum over j and using s with the condition $s \neq 0$ yields

$$\sum_{k=1}^n \frac{u_k v_k}{(d_k - \lambda)} + 1 = 0.$$

Problem 81. Let A be an $n \times n$ matrix over $\mathbb{R}$ with eigenvalues $\lambda_1, \ldots, \lambda_n$. Find the eigenvalues of the $2n \times 2n$ matrix

$$\begin{pmatrix} 0_n & A \\ -A & 0_n \end{pmatrix} \equiv \begin{pmatrix} 0 & 1 \\ -1 & 0 \end{pmatrix} \otimes A.$$

Is this matrix skew-symmetric over $\mathbb{R}$?

Solution 81. The eigenvalues of the 2×2 matrix

$$\begin{pmatrix} 0 & 1 \\ -1 & 0 \end{pmatrix}$$

are $i, -i$. Thus the eigenvalues of the $2n \times 2n$ matrix are $i\lambda_k, -i\lambda_k, k = 1, \ldots, n$. The $2n \times 2n$ matrix is not skew-symmetric over $\mathbb{R}$ as can be seen from the example

$$\begin{pmatrix} 0 & 0 & a_{11} & a_{12} \\ 0 & 0 & a_{21} & a_{22} \\ -a_{11} & -a_{12} & 0 & 0 \\ -a_{21} & -a_{22} & 0 & 0 \end{pmatrix}.$$

Problem 82.   Consider the nonnormal 2×2 matrix

$$A = \begin{pmatrix} 3 & 2 \\ 4 & 3 \end{pmatrix}.$$

For $j \geq 1$, let d_j be the greatest common divisor of the entries of $A^j - I_2$. Show that

$$\lim_{j \to \infty} d_j = \infty.$$

Use the eigenvalues of A and the characteristic polynomial.

Solution 82.   We have $\det(A) = 1$ and thus $1 = \lambda_1 \lambda_2$, where λ_1 and λ_2 denote the eigenvalues of A. Now $\det(A - \lambda I_2) = \lambda^2 - 6\lambda + 1 = 0$ and the eigenvalues are given by

$$\lambda_1 = 3 + 2\sqrt{2}, \qquad \lambda_2 = \frac{1}{\lambda_1} = 3 - 2\sqrt{2}.$$

Therefore there exists an invertible matrix C such that $A = CDC^{-1}$ with

$$D = \begin{pmatrix} \lambda_1 & 0 \\ 0 & 1/\lambda_1 \end{pmatrix}$$

and the entries of the matrix C are in $\mathbb{Q}(\sqrt{2})$. We choose an integer $k \geq 1$ such that the entries of kC and kC^{-1} are in $\mathbb{Z}(\sqrt{2})$. Then

$$k^2(A^j - I_2) = (kC)(D^j - I_2)(kC^{-1})$$

and

$$D^j - I_2 = (\lambda_1^j - 1) \begin{pmatrix} 1 & 0 \\ 0 & \lambda_1^{-j} \end{pmatrix}.$$

Thus $\lambda_1^j - 1$ divides $k^2 d_j$ in $\mathbb{Z}(\sqrt{2})$. Taking norms, we find that the integer $(\lambda_1^j - 1)(\lambda_1^{-j} - 1)$ divides $k^4 d_j^2$. However $|\lambda_1| > 1$, so

$$|(\lambda_1^j - 1)(\lambda_1^{-j} - 1)| \to \infty$$

as $j \to \infty$. Hence $\lim_{j \to \infty} d_j = \infty$.

Problem 83.   The 2×3 matrix

$$A = \begin{pmatrix} 0 & 1 & 0 \\ 1 & 0 & 1 \end{pmatrix}$$

is the starting matrix in the construction of the *Hironaka curve* utilizing the Kronecker product.
(i) Find A^T and the rank of A and A^T.
(ii) Find AA^T and the eigenvalues of AA^T. Find $A^T A$ and the eigenvalues of $A^T A$. Compare the eigenvalues of AA^T and $A^T A$. Discuss.
(iii) Calculate $A \otimes A$ and $A^T \otimes A^T$. Calculate the eigenvalues of

$$(A \otimes A)(A^T \otimes A^T), \qquad (A^T \otimes A^T)(A \otimes A).$$

Solution 83. (i) We have

$$A^T = \begin{pmatrix} 0 & 1 \\ 1 & 0 \\ 0 & 1 \end{pmatrix}.$$

The rank of A and A^T is two.

(ii) We have

$$AA^T = \begin{pmatrix} 1 & 0 \\ 0 & 2 \end{pmatrix}, \qquad A^T A = \begin{pmatrix} 1 & 0 & 1 \\ 0 & 1 & 0 \\ 1 & 0 & 1 \end{pmatrix}.$$

The eigenvalues of AA^T are 1 and 2 and the eigenvalues of $A^T A$ are 0, 1, 2. The eigenvalues of AA^T are also eigenvalues of $A^T A$ with the additional eigenvalue 0 for $A^T A$.

(iii) The following Maxima program

```
A: matrix([0,1,0],[1,0,1]);
AT: transpose(A);
KA: kronecker_product(A,A);
KAT: kronecker_product(AT,AT);
PKAKAT: KA . KAT;
eigenvalues(PKAKAT);
PKATKA: KAT . KA;
eigenvalues(PKATKA);
```

will do the job. Note that . is matrix multiplication in Maxima. The eigenvalues of $(A \otimes A)(A^T \otimes A^T)$ are 1,2 (twice),4. The eigenvalues of $(A^T \otimes A^T)(A \otimes A)$ are 1, 2 (twice), 4 and 0 (5 times).

Problem 84. Consider the unitary 4×4 matrix

$$U = \begin{pmatrix} 0 & 0 & -1 & 0 \\ 0 & 0 & 0 & -1 \\ 1 & 0 & 0 & 0 \\ 0 & 1 & 0 & 0 \end{pmatrix} = \begin{pmatrix} 0 & 1 \\ -1 & 0 \end{pmatrix} \otimes \begin{pmatrix} -1 & 0 \\ 0 & -1 \end{pmatrix} = -i\sigma_2 \otimes I_2.$$

(i) Find the eigenvalues and normalized eigenvectors of U.

(ii) Can the normalized eigenvectors of U be written as Kronecker product of two normalized vectors in $\mathbb{C}^2$. If so they are not entangled.

Solution 84. (i) For the eigenvalues we find i (twice) and $-i$ (twice) with the corresponding normalized eigenvectors

$$\mathbf{u}_1 = \frac{1}{\sqrt{2}} \begin{pmatrix} 1 \\ 0 \\ -i \\ 0 \end{pmatrix}, \quad \mathbf{u}_2 = \frac{1}{\sqrt{2}} \begin{pmatrix} 0 \\ 1 \\ 0 \\ -i \end{pmatrix}, \quad \mathbf{u}_3 = \frac{1}{\sqrt{2}} \begin{pmatrix} 1 \\ 0 \\ i \\ 0 \end{pmatrix}, \quad \mathbf{u}_4 = \frac{1}{\sqrt{2}} \begin{pmatrix} 0 \\ 1 \\ 0 \\ i \end{pmatrix}.$$

(ii) For the eigenvalue i we have the eigenvectors

$$\begin{pmatrix} 1 \\ 0 \\ -i \\ 0 \end{pmatrix} = \begin{pmatrix} 1 \\ -i \end{pmatrix} \otimes \begin{pmatrix} 1 \\ 0 \end{pmatrix}, \qquad \begin{pmatrix} 0 \\ 1 \\ 0 \\ -i \end{pmatrix} = \begin{pmatrix} 1 \\ -i \end{pmatrix} \otimes \begin{pmatrix} 0 \\ 1 \end{pmatrix}.$$

For the eigenvalue $-i$ we have

$$\begin{pmatrix} 1 \\ 0 \\ i \\ 0 \end{pmatrix} = \begin{pmatrix} 1 \\ i \end{pmatrix} \otimes \begin{pmatrix} 1 \\ 0 \end{pmatrix}, \qquad \begin{pmatrix} 0 \\ 1 \\ 0 \\ i \end{pmatrix} = \begin{pmatrix} 1 \\ i \end{pmatrix} \otimes \begin{pmatrix} 0 \\ 1 \end{pmatrix}.$$

Problem 85.   Consider the Hilbert space $\mathbb{C}^n$. Let A, B, C be $n \times n$ matrices acting in $\mathbb{C}^n$. We consider the *nonlinear eigenvalue problem*

$$A\mathbf{v} = \lambda B\mathbf{v} + \lambda^2 C\mathbf{v}$$

where $\mathbf{v} \in \mathbb{C}^n$ and $\mathbf{v} \neq \mathbf{0}$. Let σ_1, σ_2, σ_3 be the Pauli spin matrices. Find the solutions of the nonlinear eigenvalue problem

$$\sigma_1\mathbf{v} = \lambda\sigma_2\mathbf{v} + \lambda^2\sigma_3\mathbf{v}$$

where $\mathbf{v} \in \mathbb{C}^2$ and $\mathbf{v} \neq \mathbf{0}$.

Solution 85.   We obtain the eigenvalues

$$\lambda_1 = \sqrt{\frac{1 + \sqrt{5}}{2}}, \quad \lambda_2 = -\lambda_1, \quad \lambda_3 = \sqrt{\frac{1 - \sqrt{5}}{2}}, \quad \lambda_4 = -\lambda_3.$$

The corresponding eigenvectors are

$$\mathbf{v}_1 = \begin{pmatrix} 1 + i\lambda_1 \\ (1 + \sqrt{5})/2 \end{pmatrix}, \qquad \mathbf{v}_2 = \begin{pmatrix} 1 - i\lambda_1 \\ (1 + \sqrt{5})/2 \end{pmatrix},$$

$$\mathbf{v}_3 = \begin{pmatrix} 1 + i\lambda_3 \\ (1 - \sqrt{5})/2 \end{pmatrix}, \qquad \mathbf{v}_4 = \begin{pmatrix} 1 - i\lambda_3 \\ (1 - \sqrt{5})/2 \end{pmatrix}.$$

Problem 86.   Let A be an $n \times n$ symmetric matrix over $\mathbb{R}$. Since A is symmetric over $\mathbb{R}$ there exists a set of orthonormal eigenvectors $\mathbf{v}_1, \mathbf{v}_2, \ldots, \mathbf{v}_n$ which form an orthonormal basis in $\mathbb{R}^n$. Let $\mathbf{x} \in \mathbb{R}^n$ be a reasonably good approximation to an eigenvector, say $\mathbf{v}_1$. Calculate

$$R := \frac{\mathbf{x}^T A \mathbf{x}}{\mathbf{x}^T \mathbf{x}}.$$

The quotient is called *Rayleigh quotient*. Discuss.

Solution 86.   We can write

$$\mathbf{x} = c_1\mathbf{v}_1 + c_2\mathbf{v}_2 + \cdots + c_n\mathbf{v}_n, \qquad c_j \in \mathbb{R}.$$

Then, since $A\mathbf{v}_j = \lambda_j\mathbf{v}_j$, $j = 1, \ldots, n$ and using $\mathbf{v}_j^T\mathbf{v}_k = 0$ if $j \neq k$ we find

$$R = \frac{\mathbf{x}^T A \mathbf{x}}{\mathbf{x}^T \mathbf{x}} = \frac{(c_1\mathbf{v}_1 + \cdots + c_n\mathbf{v}_n)^T A(c_1\mathbf{v}_1 + \cdots + c_n\mathbf{v}_n)}{(c_1\mathbf{v}_1 + \cdots + c_n\mathbf{v}_n)^T(c_1\mathbf{v}_1 + \cdots + c_n\mathbf{v}_n)}$$

$$= \frac{(c_1\mathbf{v}_1 + \cdots + c_n\mathbf{v}_n)^T(c_1\lambda_1\mathbf{v}_1 + \cdots + c_n\lambda_n\mathbf{v}_n)}{c_1^2 + c_2^2 + \cdots + c_n^2}$$

$$= \frac{\lambda_1 c_1^2 + \lambda_2 c_2^2 + \cdots + \lambda_n c_n^2}{c_1^2 + c_2^2 + \cdots + c_n^2}$$

$$= \lambda_1\left(\frac{1 + (\lambda_2/\lambda_1)(c_2/c_1)^2 + \cdots + (\lambda_n/\lambda_1)(c_n/c_1)^2}{1 + (c_2/c_1)^2 + \cdots + (c_n/c_1)^2}\right).$$

Since $\mathbf{x}$ is a good approximation to $\mathbf{v}_1$, the coefficient c_1 is larger than the other c_j, $j = 2, 3, \ldots, n$. Thus the expression in the parenthesis is close to 1, which means that R_q is close to λ_1.

Supplementary Problems

Problem 1. (i) Let $\theta \in [0, \pi/2)$. Find the trace, determinant, eigenvalues and eigenvectors of the 2×2 matrices

$$M_1 = \frac{1}{2}\begin{pmatrix} \tan^2(\theta) & -\tan(\theta) \\ -\tan(\theta) & 1 \end{pmatrix}, \quad M_2 = \frac{1}{2}\begin{pmatrix} \tan^2(\theta) & \tan(\theta) \\ \tan(\theta) & 1 \end{pmatrix}.$$

(ii) Find the eigenvalues and normalized eigenvectors of the matrix ($\phi \in [0, 2\pi)$)

$$A(\phi) = \frac{1}{\sqrt{2}}\begin{pmatrix} 1 & e^{i\phi} \\ 1 & e^{-i\phi} \end{pmatrix}.$$

Is the matrix invertible? Make the decision by looking at the eigenvalues. If so find the inverse matrix.

(iii) Let $\epsilon \in \mathbb{R}$. Find the eigenvalues and eigenvectors of

$$A(\epsilon) = \frac{1}{\sqrt{1 + \epsilon^2}}\begin{pmatrix} \epsilon & 1 \\ 1 & \epsilon \end{pmatrix}.$$

For which ϵ is $A(\epsilon)$ not invertible? Let $\epsilon \in [0, 1]$. Consider the 2×2 matrix

$$B(\epsilon) = \frac{1}{\sqrt{1 + \epsilon^2}}\begin{pmatrix} 1 & \epsilon \\ \epsilon & -1 \end{pmatrix}.$$

For $\epsilon = 0$ we have the Pauli spin matrix σ_3 and for $\epsilon = 1$ we have the Hadamard matrix. Find the eigenvalues and eigenvectors of $A(\epsilon)$.

(iv) Let $\epsilon \in \mathbb{R}$. Find the eigenvalues $\lambda_+(\epsilon)$, $\lambda_-(\epsilon)$ and normalized eigenvectors of the matrix

$$A(\epsilon) = \begin{pmatrix} \sinh(\epsilon) & 1 \\ 1 & -\sinh(\epsilon) \end{pmatrix}.$$

Find the shortest distance between $\lambda_+(\epsilon)$ and $\lambda_-(\epsilon)$.

(v) Let $\theta \in \mathbb{R}$. Show that the eigenvalues of the hermitian matrix

$$A(\theta) = \begin{pmatrix} 1 & \cos(\theta) \\ \cos(\theta) & 1 \end{pmatrix}$$

are given by $\lambda_\pm = 1 \pm \cos(\theta)$ and thus cannot be negative.

(vi) Let $\tau := (1 + \sqrt{5})/2$ be the *golden ratio*. Consider the modular 2×2 matrix

$$M = \begin{pmatrix} 1 & 1 \\ 1 & 0 \end{pmatrix}.$$

Show that the eigenvalues are given by τ and $-\tau^{-1}$, where $\tau^{-1} = (\sqrt{5} - 1)/2$ and the normalized eigenvectors are

$$\mathbf{v}_1 = \frac{1}{1 + \tau^2} \begin{pmatrix} \tau \\ 1 \end{pmatrix}, \qquad \mathbf{v}_2 = \frac{1}{1 + \tau^2} \begin{pmatrix} -\tau^{-1} \\ 1 \end{pmatrix}.$$

(vii) Consider the two 2×2 matrices

$$A_\pm = \begin{pmatrix} -1/2 & \mp 3/2 \\ \pm 1/2 & -1/2 \end{pmatrix}.$$

Show that $\det(A_\pm) = 1$, $\mathrm{tr}(A_\pm) = -1$, $A_\pm^{-1} = A_\mp$, $A_\pm^3 = I_2$ and that the eigenvalues of the two matrices are the same, namely

$$\frac{1}{2}(\sqrt{3}i - 1), \qquad -\frac{1}{2}(\sqrt{3}i + 1).$$

Problem 2. (i) The symmetric 3×3 matrix over $\mathbb{R}$

$$A = \begin{pmatrix} 0 & 1 & 1 \\ 1 & 0 & 1 \\ 1 & 1 & 0 \end{pmatrix}$$

plays a role for the chemical compounds ZnS and $NaCl$. Find the eigenvalues and eigenvectors of A. Then find the inverse of A. Find all $\mathbf{v}$ such that $A\mathbf{v} = \mathbf{v}$.
(ii) Let $a, b, c \in \mathbb{R}$. Find the eigenvalues and eigenvectors of the symmetric matrix

$$\begin{pmatrix} 0 & a & b \\ a & 0 & c \\ b & c & 0 \end{pmatrix}.$$

The characteristic equation is $\lambda^3 - (a^2 + b^2 + c^2)\lambda - 2abc = 0$.

Problem 3. (i) Find the eigenvalues and eigenvectors of the orthogonal matrices

$$R(\alpha) = \begin{pmatrix} 1 & 0 & 0 \\ 0 & \cos(\alpha) & \sin(\alpha) \\ 0 & -\sin(\alpha) & \cos(\alpha) \end{pmatrix}, \qquad S(\beta) = \begin{pmatrix} \cos(\beta) & \sin(\beta) & 0 \\ -\sin(\beta) & \cos(\beta) & 0 \\ 0 & 0 & 1 \end{pmatrix}.$$

(ii) Find the eigenvalues and eigenvectors of $R(\alpha)S(\beta)$.

Problem 4. (i) Find the eigenvalues and eigenvectors of the 3×3 matrices

$$\begin{pmatrix} 0 & 0 & a_{13} \\ 1 & 0 & a_{23} \\ 0 & 1 & a_{33} \end{pmatrix}, \qquad \begin{pmatrix} 0 & a_{12} & 0 \\ a_{21} & 0 & a_{23} \\ 0 & a_{32} & a_{33} \end{pmatrix}.$$

(ii) Let A be a normal 2×2 matrix over $\mathbb{C}$ with eigenvalues λ_1, λ_2 and corresponding eigenvectors $\mathbf{v}_1$, $\mathbf{v}_2$, respectively. Let $c \in \mathbb{C}$. What can be said about the eigenvalues and eigenvectors of the 3×3 matrices

$$B_1 = \begin{pmatrix} c & 0 & 0 \\ 0 & a_{11} & a_{12} \\ 0 & a_{21} & a_{22} \end{pmatrix}, \quad B_2 = \begin{pmatrix} a_{11} & 0 & a_{12} \\ 0 & c & 0 \\ a_{21} & 0 & a_{22} \end{pmatrix}, \quad B_3 = \begin{pmatrix} a_{11} & a_{12} & 0 \\ a_{21} & a_{22} & 0 \\ 0 & 0 & c \end{pmatrix} ?$$

(iii) Let $\phi \in \mathbb{R}$. Find the eigenvalues and eigenvectors of the 3×3 matrices

$$X(\phi) = \begin{pmatrix} 0 & 0 & e^{i\phi} \\ 1 & 0 & 0 \\ 0 & 1 & 0 \end{pmatrix}, \quad Y(\phi) = \begin{pmatrix} 0 & 1 & 0 \\ 0 & 0 & 1 \\ e^{i\phi} & 0 & 0 \end{pmatrix}.$$

(iv) Consider the 3×3 matrix

$$A = \begin{pmatrix} 0 & 0 & 1 \\ 1 & 0 & 0 \\ 0 & 1 & 1 \end{pmatrix}.$$

Show that the characteristic polynomial is given by $p(\lambda) = \lambda^3 - \lambda^2 - 1$. Find the eigenvalues. Calculate $A \otimes A$, $A \otimes A \otimes A$, $A \otimes A \otimes A \otimes A$. Then identify 1 with a black square and 0 with a white square. Draw the pictures.

(v) Let $m > 0$ and $\theta \in \mathbb{R}$. Consider the three 3×3 matrices

$$M_1(\theta) = m \begin{pmatrix} 0 & \sin(\theta) & 0 \\ \sin(\theta) & 0 & \cos(\theta) \\ 0 & \cos(\theta) & 0 \end{pmatrix}, \quad M_2(\theta) = m \begin{pmatrix} 0 & 0 & \sin(\theta) \\ 0 & 0 & \cos(\theta) \\ \sin(\theta) & \cos(\theta) & 0 \end{pmatrix},$$

$$M_3(\theta) = m \begin{pmatrix} 0 & \sin(\theta) & \cos(\theta) \\ \sin(\theta) & 0 & 0 \\ \cos(\theta) & 0 & 0 \end{pmatrix}.$$

Find the eigenvalues and eigenvectors of the matrices. These matrices play a role for the Majorana neutrino.

Problem 5. Let $\epsilon \in \mathbb{R}$.
(i) Find the eigenvalues and eigenvectors of the matrices

$$\begin{pmatrix} 1 & \epsilon \\ 1 & 1 \end{pmatrix}, \quad \begin{pmatrix} 1 & 0 & \epsilon \\ 1 & 1 & 0 \\ 1 & 1 & 1 \end{pmatrix}, \quad \begin{pmatrix} 1 & 0 & 0 & \epsilon \\ 1 & 1 & 0 & 0 \\ 1 & 1 & 1 & 0 \\ 1 & 1 & 1 & 1 \end{pmatrix}.$$

Extend to $n \times n$ matrices.
(ii) Find the eigenvalues and eigenvectors of the matrices

$$\begin{pmatrix} 0 & 1 \\ \epsilon & 0 \end{pmatrix}, \quad \begin{pmatrix} 0 & 1 & 0 \\ 0 & 0 & 1 \\ \epsilon & 0 & 0 \end{pmatrix}, \quad \begin{pmatrix} 0 & 1 & 0 & 0 \\ 0 & 0 & 1 & 0 \\ 0 & 0 & 0 & 1 \\ \epsilon & 0 & 0 & 0 \end{pmatrix}.$$

Extend to the $n \times n$ case.

Problem 6.   (i) Consider the 5×4 matrix

$$A = \begin{pmatrix} 0 & 1 & 0 & 1 \\ 1 & 0 & 1 & 0 \\ 1 & 0 & 1 & 0 \\ 0 & 1 & 0 & 1 \\ 0 & 1 & 0 & 1 \end{pmatrix}.$$

Find the eigenvalues of the matrices $A^T A$ and $A A^T$ by only calculating the eigenvalues of $A^T A$.

(ii) Let say we want to calculate the eigenvalues of the 4×4 matrix

$$B = \begin{pmatrix} 1 & 0 & 1 & 0 \\ 0 & 1 & 0 & 1 \\ 1 & 0 & 1 & 0 \\ 0 & 1 & 0 & 1 \end{pmatrix}.$$

How could we utilize the facts that

$$\begin{pmatrix} 0 & 1 \\ 1 & 0 \\ 0 & 1 \\ 1 & 0 \end{pmatrix} \begin{pmatrix} 0 & 1 & 0 & 1 \\ 1 & 0 & 1 & 0 \end{pmatrix} = B, \qquad \begin{pmatrix} 0 & 1 & 0 & 1 \\ 1 & 0 & 1 & 0 \end{pmatrix} \begin{pmatrix} 0 & 1 \\ 1 & 0 \\ 0 & 1 \\ 1 & 0 \end{pmatrix} = \begin{pmatrix} 2 & 0 \\ 0 & 2 \end{pmatrix}$$

and $\mathrm{tr}(B) = 4$ and thus avoid solving $\det(B - \lambda I_4) = 0$?

Problem 7.   Let A be an $n \times n$ diagonalizable matrix with distinct eigenvalues λ_j $(j = 1, \ldots, n)$. Show that the *Vandermonde matrix*

$$V = \begin{pmatrix} 1 & 1 & \cdots & 1 \\ \lambda_1 & \lambda_2 & \cdots & \lambda_n \\ \lambda_1^2 & \lambda_2^2 & \cdots & \lambda_n^2 \\ \vdots & \vdots & \ddots & \vdots \\ \lambda_1^{n-1} & \lambda_2^{n-1} & \cdots & \lambda_n^{n-1} \end{pmatrix}$$

is nonsingular.

Problem 8.   (i) Let σ_1, σ_2, σ_3 be the Pauli spin matrices. Solve the *nonlinear eigenvalue problem*

$$\sigma_1 \mathbf{u} = \lambda \sigma_2 \mathbf{u} + \lambda^2 \sigma_3 \mathbf{u}$$

where $\mathbf{u} \in \mathbb{C}^2$ and $\mathbf{u} \neq \mathbf{0}$.

(ii) Solve the nonlinear eigenvalue problem

$$(\sigma_1 \otimes \sigma_1)\mathbf{v} = \lambda(\sigma_2 \otimes \sigma_2)\mathbf{v} + \lambda^2(\sigma_3 \otimes \sigma_3)\mathbf{v}$$

where $\mathbf{v} \in \mathbb{C}^4$ and $\mathbf{v} \neq \mathbf{0}$. Compare the result to (i). Discuss.

Problem 9. Let A, B, C, D, E, F, G, H be 2×2 matrices over $\mathbb{C}$. We define the star product

$$\begin{pmatrix} A & B \\ C & D \end{pmatrix} \star \begin{pmatrix} E & F \\ G & H \end{pmatrix} := \begin{pmatrix} E & 0_2 & 0_2 & F \\ 0_2 & A & B & 0_2 \\ 0_2 & C & D & 0_2 \\ G & 0_2 & 0_2 & H \end{pmatrix}.$$

Thus the right-hand side is an 8×8 matrix. Assume we know the eigenvalues and eigenvectors of the two 4×4 matrices on the left-hand side. What can be said about the eigenvalues and eigenvectors of the 8×8 matrix of the right-hand side?

Problem 10. Consider a 2×2 matrix $A = (a_{jk})$ over $\mathbb{R}$ with

$$\sum_{j,k=1}^{2} a_{jk} = 1, \qquad \mathrm{tr}(A) = 0.$$

What can be said about the eigenvalues of such a matrix?

Problem 11. Consider the *Hadamard matrix*

$$U = \frac{1}{\sqrt{2}} \begin{pmatrix} 1 & 1 \\ 1 & -1 \end{pmatrix} \equiv \frac{1}{\sqrt{2}} (\sigma_3 + \sigma_1).$$

The eigenvalues of the Hadamard matrix are given by $+1$ and -1 with the corresponding normalized eigenvectors

$$\frac{1}{\sqrt{8}} \begin{pmatrix} \sqrt{4 + 2\sqrt{2}} \\ \sqrt{4 - 2\sqrt{2}} \end{pmatrix}, \qquad \frac{1}{\sqrt{8}} \begin{pmatrix} \sqrt{4 - 2\sqrt{2}} \\ -\sqrt{4 + 2\sqrt{2}} \end{pmatrix}.$$

How can this information be used to find the eigenvalues and eigenvectors of the *Bell matrix*

$$B = U \star U \equiv \frac{1}{\sqrt{2}} \begin{pmatrix} 1 & 0 & 0 & 1 \\ 0 & 1 & 1 & 0 \\ 0 & 1 & -1 & 0 \\ 1 & 0 & 0 & -1 \end{pmatrix} ?$$

Problem 12. (i) Let U be an $n \times n$ unitary matrix. Assume that $U^* = -U$, i.e. the matrix is also skew-hermitian. Find the eigenvalues of such a matrix.
(ii) Let V be an $n \times n$ unitary matrix. What can be concluded about the eigenvalues of V if $V^* = V^T$?

Problem 13. Find the eigenvalues of the staircase matrices

$$\begin{pmatrix} 0 & 1 \\ 1 & 1 \end{pmatrix}, \quad \begin{pmatrix} 0 & 0 & 1 \\ 0 & 1 & 1 \\ 1 & 1 & 1 \end{pmatrix}, \quad \begin{pmatrix} 0 & 0 & 0 & 1 \\ 0 & 0 & 1 & 1 \\ 0 & 1 & 1 & 1 \\ 1 & 1 & 1 & 1 \end{pmatrix}.$$

Problem 14.   Let A, B be hermitian matrices over $\mathbb{C}$ and eigenvalues $\lambda_1, \ldots,$ λ_n and $\mu_1, \ldots, \mu_n$, respectively. Assume that $\text{tr}(AB) = 0$ (scalar product). What can be said about the eigenvalues of $A + B$?

Problem 15.   Consider the reverse-diagonal $n \times n$ unitary matrix

$$A(\phi_1, \ldots, \phi_n) = \begin{pmatrix} 0 & 0 & \cdots & 0 & e^{i\phi_1} \\ 0 & 0 & \cdots & e^{i\phi_2} & 0 \\ \vdots & \vdots & & & \\ 0 & e^{i\phi_{n-1}} & \cdots & 0 & 0 \\ e^{i\phi_n} & 0 & \cdots & 0 & 0 \end{pmatrix}$$

where $\phi_j \in \mathbb{R}$ $(j = 1, \ldots, n)$. Find the eigenvalues and eigenvectors.

Problem 16.   Let A, B be $n \times n$ matrices over $\mathbb{C}$. The two $n \times n$ matrices A and B have a common eigenvector if and only if the $n \times n$ matrix

$$\sum_{j,k=1}^{n-1} [A^j, B^k]^*[A^j, B^k]$$

is singular, i.e. the determinant is equal to 0.
(i) Apply the theorem to the 2×2 matrices

$$A = \begin{pmatrix} 0 & 1 \\ 1 & 0 \end{pmatrix}, \qquad B = \frac{1}{\sqrt{2}} \begin{pmatrix} 1 & 1 \\ 1 & -1 \end{pmatrix}.$$

(ii) Apply the theorem to the 3×3 matrices

$$A(\alpha) = \begin{pmatrix} 1 & 0 & 0 \\ 0 & \cos(\alpha) & -\sin(\alpha) \\ 0 & \sin(\alpha) & \cos(\alpha) \end{pmatrix}, \qquad B(\alpha) = \begin{pmatrix} \cos(\alpha) & -\sin(\alpha) & 0 \\ \sin(\alpha) & \cos(\alpha) & 0 \\ 0 & 0 & 1 \end{pmatrix}.$$

Problem 17.   Let $0 \leq x < 1$. Consider the $N \times N$ matrix C (correlation matrix) with the entries $C_{jk} := x^{|j-k|}$, $j, k = 1, \ldots, N$. Find the eigenvalues of C. Show that if $N \to \infty$ the distribution of its eigenvalues becomes a continuous function of $\phi \in [0, 2\pi]$

$$\lambda(\phi) = \frac{1 - x^2}{1 - 2x\cos(\phi) + x^2}.$$

Problem 18.   (i) Let A be an $n \times n$ matrix over $\mathbb{C}$. Show that A is normal if and only if there exists an $n \times n$ unitary matrix U and an $n \times n$ diagonal matrix D such that $D = U^{-1}AU$. Note that $U^{-1} = U^*$.
(ii) Let A be a normal $n \times n$ matrix over $\mathbb{C}$. Show that A has a set of n orthonormal eigenvectors. Show that if A has a set of n orthonormal eigenvectors, then A is normal.

Problem 19. (i) The $2n \times 2n$ *symplectic matrix* is defined by

$$S = \begin{pmatrix} 0_n & I_n \\ -I_n & 0_n \end{pmatrix}.$$

The matrix S is unitary and skew-hermitian. Find the eigenvalues of S from this information.

(ii) Find the eigenvalues of the $2n \times 2n$ matrix

$$R = \frac{1}{\sqrt{2}} \begin{pmatrix} I_n & I_n \\ -I_n & I_n \end{pmatrix}.$$

(iii) Let A be an $n \times n$ matrix over $\mathbb{R}$. Let $\lambda_1, \ldots, \lambda_n$ be the eigenvalues. What can be said about the eigenvalues of the $2n \times 2n$ matrix

$$\begin{pmatrix} 0_n & A \\ A^T & 0_n \end{pmatrix}?$$

Problem 20. Given the matrix

$$A = \begin{pmatrix} 1 & 2 & 3 & 4 \\ 5 & 6 & 7 & 8 \\ 9 & 10 & 11 & 12 \\ 13 & 14 & 15 & 16 \end{pmatrix}.$$

Prove or disprove that exactly two eigenvalues are 0.

Problem 21. Let A be an $n \times n$ matrix over $\mathbb{C}$. What is the condition on A such that all eigenvalues are 0 and A admits only one eigenvector?

Problem 22. Consider the $n \times n$ matrix

$$J_n = \begin{pmatrix} 0 & 1 & 0 & & \cdots & 0 \\ 0 & 0 & 1 & 0 & \cdots & 0 \\ & & \ddots & \ddots & & \\ & & & & & 1 \\ & & & & & 0 \end{pmatrix}.$$

Hence an arbitrary *Jordan block* is given by $zI_n + J_n$, where $z \in \mathbb{C}$. Find the eigenvalues of

$$\begin{pmatrix} a & b \\ -b & a \end{pmatrix} \otimes I_n + I_2 \otimes J_n.$$

Problem 23. Let I_2 be the 2×2 identity matrix, 0_2 be the 2×2 zero matrix and σ_3 be the Pauli spin matrix.

(i) Find the eigenvalues and normalized eigenvectors of the 4×4 matrix

$$A(t) = \begin{pmatrix} \cosh(2t)I_2 & \sinh(2t)\sigma_3 \\ \sinh(2t)\sigma_3 & \cosh(2t)I_2 \end{pmatrix}.$$

(ii) Find the eigenvalues and normalized eigenvectors of the 6×6 matrix

$$B(t) = \begin{pmatrix} \cosh(2t)I_2 & 0_2 & \sinh(2t)\sigma_3 \\ 0_2 & I_2 & 0_2 \\ \sinh(2t)\sigma_3 & 0_2 & \cosh(2t)I_2 \end{pmatrix}.$$

Can the results from (i) be utilized here?

(iii) Find the eigenvalues of the 6×6 matrices

$$\begin{pmatrix} \frac{1}{\sqrt{2}}I_2 & 0_2 & \frac{1}{\sqrt{2}}I_2 \\ 0_2 & I_2 & 0_2 \\ \frac{1}{\sqrt{2}}I_2 & 0_2 & -\frac{1}{\sqrt{2}}I_2 \end{pmatrix}, \quad \begin{pmatrix} 0_3 & I_3 \\ -I_3 & 0_3 \end{pmatrix}.$$

Problem 24. Let $n \geq 3$. Consider the tridiagonal $n \times n$ matrix

$$A = \begin{pmatrix} a_1 & b_2 & 0 & \cdots & 0 & 0 \\ c_2 & a_2 & b_3 & \cdots & 0 & 0 \\ 0 & c_3 & a_3 & \cdots & 0 & 0 \\ \vdots & \vdots & \vdots & \ddots & \vdots & \vdots \\ 0 & 0 & 0 & \cdots & a_{n-1} & b_n \\ 0 & 0 & 0 & \cdots & c_n & a_n \end{pmatrix}$$

with $a_1, a_j, b_j, c_j \in \mathbb{C}$ $(j = 1, \ldots, n)$. It has in general n complex eigenvalues the n roots of the characteristic polynomial $p(\lambda)$. Show that this polynomial can be evaluated by the recursive formula

$$p_k(\lambda) = (\lambda - a_k)p_{k-1} - b_k c_k p_{k-2}(\lambda), \quad k = 2, 3, \ldots, n$$
$$p_1(\lambda) = \lambda - a_1$$
$$p_0(\lambda) = 1.$$

Problem 25. Consider an $n \times n$ symmetric tridiagonal matrix over $\mathbb{R}$. Let $f_n(\lambda) := \det(A - \lambda I_n)$ and

$$f_k(\lambda) = \det \begin{pmatrix} \alpha_1 - \lambda & \beta_1 & 0 & \cdots & 0 \\ \beta_1 & \alpha_2 - \lambda & \beta_2 & \cdots & 0 \\ 0 & \beta_2 & \ddots & \cdots & 0 \\ \vdots & \vdots & \vdots & \vdots & \vdots \\ 0 & \cdots & \cdots & \alpha_{k-1} - \lambda & \beta_{k-1} \\ 0 & \cdots & 0 & \beta_{k-1} & \alpha_k - \lambda \end{pmatrix}$$

for $k = 1, 2, \ldots, n$ and $f_0(\lambda) = 1$, $f_{-1}(\lambda) = 0$. Then

$$f_k(\lambda) = (\alpha - \lambda)f_{k-1}(\lambda) - \beta_{k-1}^2 f_{k-2}(\lambda)$$

for $k = 2, 3, \ldots, n$. Find $f_4(\lambda)$ for the 4×4 matrix

$$\begin{pmatrix} 0 & \sqrt{1} & 0 & 0 \\ \sqrt{1} & 0 & \sqrt{2} & 0 \\ 0 & \sqrt{2} & 0 & \sqrt{3} \\ 0 & 0 & \sqrt{3} & 0 \end{pmatrix}.$$

Problem 26. Let $\epsilon_j \in \mathbb{R}$ with $j = 1, 2, 3, 4$. Consider the tridiagonal hermitian matrix

$$H(\epsilon_j) = \begin{pmatrix} \epsilon_1 & 1 & 0 & 0 \\ 1 & \epsilon_2 & 1 & 0 \\ 0 & 1 & \epsilon_3 & 1 \\ 0 & 0 & 1 & \epsilon_4 \end{pmatrix}.$$

Find the eigenvalues and eigenvectors. Let $\mathbf{v} = (v_1 \ \ v_2 \ \ v_3 \ \ v_4)^T$ be an eigenvectors. Find v_1/v_2, v_2/v_3, v_3/v_4.

Problem 27. Let M be an $n \times n$ invertible matrix with eigenvalues $\lambda_1, \ldots, \lambda_n$ (counting degeneracy). Find the eigenvalues of

$$M_+ = \frac{1}{2}(M + M^{-1}) \quad \text{and} \quad M_- = \frac{1}{2}(M - M^{-1}).$$

Problem 28. Let $A = (a_{jk})$ be an $n \times n$ semi-positive definite matrix with $\sum_{k=1}^n a_{jk} < 1$ for $j = 1, \ldots, n$. Show that $0 \le \lambda < 1$ for all eigenvalues λ of A.

Problem 29. Given an $n \times n$ permutation matrix P.
(i) Assume that n is even and $\operatorname{tr}(P) = 0$. Can we conclude that half of the eigenvalues of such a matrix are $+1$ and the other half are -1?
(ii) Assume that n is odd and $\operatorname{tr}(P) = 0$. Can we conclude that the eigenvalues are given by the n solutions of $\lambda^n = 1$?

Problem 30. Let A be an $n \times n$ matrix over $\mathbb{C}$. The characteristic polynomial of A is defined by

$$\Delta(\lambda) := \det(\lambda I_n - A) = \lambda^n + \sum_{k=1}^n (-1)^k \sigma(k) \lambda^{n-k}.$$

The eigenvalues $\lambda_1, \ldots, \lambda_n$ of A are the solutions of the characteristic equation $\Delta(\lambda) = 0$. The coefficients $\sigma(j)$ $(j = 1, \ldots, n)$ are given by

$$\sigma(1) = \sum_{j=1}^n \lambda_j = \operatorname{tr}(A)$$

$$\sigma(2) = \sum_{j<k} \lambda_j \lambda_k$$

$$\vdots$$

$$\sigma(n) = \prod_{j=1}^n \lambda_j = \det(A).$$

Another set of symmetric polynomials is given by the traces of powers of the matrix A, namely

$$s(j) = \sum_{k=1}^n (\lambda_k)^j = \operatorname{tr}(A^j), \quad j = 1, \ldots, n.$$

One has (so-called *Newton relation*)

$$j\sigma(j) - s(1)\sigma(j-1) + \cdots + (-1)^{j-1}s(j-1)\sigma(1) + (-1)^j s(j) = 0$$

where $j = 1, \ldots, n$. Consider the 3×3 matrix

$$A = \begin{pmatrix} 1/\sqrt{2} & 0 & 1/\sqrt{2} \\ 0 & 1 & 0 \\ 1/\sqrt{2} & 0 & -1/\sqrt{2} \end{pmatrix}.$$

Calculate $s(1)$, $s(2)$, $s(3)$ from the traces of the powers of A. Then apply the Newton relation to find $\sigma(1)$, $\sigma(2)$, $\sigma(3)$.

Problem 31. (i) Is the 3×3 *stochastic matrix*

$$\begin{pmatrix} 1/3 & 1/3 & 1/3 \\ 2/3 & 0 & 1/3 \\ 1/2 & 1/2 & 0 \end{pmatrix}$$

diagonalizable? First find the eigenvalues and normalized eigenvectors.
(ii) Find the eigenvalues and eigenvectors of the stochastic matrix

$$\begin{pmatrix} 1/2 & 1/2 & 0 \\ 0 & 1/2 & 1/2 \\ 1 & 0 & 0 \end{pmatrix}.$$

(iii) Find the eigenvalues and eigenvectors of the stochastic matrix

$$\begin{pmatrix} p_{AA} & p_{AB} & p_{AC} & 0 \\ 0 & 0 & 0 & 1 \\ 0 & 0 & 0 & 1 \\ p_{DA} & p_{DB} & p_{DC} & 0 \end{pmatrix}$$

where $p_{AA} = p_{DA} = 2 - \sqrt{3}$, $p_{AB} = p_{DB} = \sqrt{3} - \sqrt{2}$, $p_{AC} = p_{DC} = \sqrt{2} - 1$.
(iv) Find the eigenvalues of the double stochastic matrices

$$\begin{pmatrix} \sin^2(\theta) & \cos^2(\theta) \\ \cos^2(\theta) & \sin^2(\theta) \end{pmatrix}, \quad \begin{pmatrix} \sin^2(\theta) & 0 & \cos^2(\theta) \\ \cos^2(\theta) & \sin^2(\theta) & 0 \\ 0 & \cos^2(\theta) & \sin^2(\theta) \end{pmatrix},$$

$$\begin{pmatrix} \sin^2(\theta) & 0 & 0 & \cos^2(\theta) \\ \cos^2(\theta) & \sin^2(\theta) & 0 & 0 \\ 0 & \cos^2(\theta) & \sin^2(\theta) & 0 \\ 0 & 0 & \cos^2(\theta) & \sin^2(\theta) \end{pmatrix}.$$

Problem 32. (i) Let $c_j \in \mathbb{R}$. Find the eigenvalues of the matrices

$$\begin{pmatrix} 0 & 1 \\ c_1 & c_2 \end{pmatrix}, \quad \begin{pmatrix} 0 & 1 & 0 \\ 0 & 0 & 1 \\ c_1 & c_2 & c_3 \end{pmatrix}, \quad \begin{pmatrix} 0 & 1 & 0 & 0 \\ 0 & 0 & 1 & 0 \\ 0 & 0 & 0 & 1 \\ c_1 & c_2 & c_3 & c_4 \end{pmatrix}.$$

Generalize to the $n \times n$ case. Note that for the 2×2 matrix we find

$$\lambda_{1,2} = \frac{c_2}{2} \pm \frac{1}{2}\sqrt{4c_1 + c_2^2}.$$

(ii) Show that the eigenvalues the 4×4 matrix

$$A = \begin{pmatrix} 0 & a_{12} & a_{13} & a_{14} \\ a_{12} & 0 & 0 & 0 \\ a_{13} & 0 & 0 & 0 \\ a_{14} & 0 & 0 & 0 \end{pmatrix}$$

are given by 0 (2 times) and $\sqrt{a_{12}^2 + a_{13}^2 + a_{14}^2}$, $-\sqrt{a_{12}^2 + a_{13}^2 + a_{14}^2}$. Find the eigenvectors.

Problem 33. Consider the 3×3 matrices

$$A = \begin{pmatrix} 0 & 1 & 1 \\ 1 & 0 & 0 \\ 1 & 0 & 0 \end{pmatrix}, \quad B = \begin{pmatrix} 0 & 0 & 1 \\ 0 & 0 & 1 \\ 1 & 1 & 0 \end{pmatrix}.$$

Find the eigenvalues of A and B and the eigenvalues of the commutator $[A, B]$. Are the matrices A and B similar?

Problem 34. What can be said about the eigenvalues and eigenvectors of a nonzero hermitian $n \times n$ matrix A with $\det(A) = 0$ and $\mathrm{tr}(A) = 0$. Give one eigenvalue of A. Give one eigenvalue of $A \otimes A$. Consider the case $n = 3$ for the matrix A. Find all the eigenvalues and eigenvectors.

Problem 35. Let A be an $n \times n$ matrix over $\mathbb{C}$. Show that the matrix A admits the inverse matrix A^{-1} iff $A\mathbf{v} = \mathbf{0}$ implies $\mathbf{v} = \mathbf{0}$. Note that from $A\mathbf{v} = \mathbf{0}$ it follows that $A^{-1}(A\mathbf{v}) = \mathbf{0}$ and then $\mathbf{v} = \mathbf{0}$.

Problem 36. Let S and Q be $n \times n$ matrices and

$$(S - \lambda I_n)\mathbf{u} = \mathbf{0}, \qquad (Q - \mu I_n)\mathbf{v} = \mathbf{0}.$$

These are eigenvalue equations. Show that

$$(S \otimes Q - \lambda\mu(I_n \otimes I_n))(\mathbf{u} \otimes \mathbf{v}) = \mathbf{0}_{n^2}$$

and

$$(S \oplus Q - \lambda\mu(I_n \oplus I_n))\begin{pmatrix} \mathbf{u} \\ \mathbf{v} \end{pmatrix} = \mathbf{0}_{2n}.$$

Problem 37. Let $\gamma \in \mathbb{R}$ and $\gamma^2 < 1$. Consider the 2×2 matrix

$$K = \sigma_1 - i\gamma\sigma_3.$$

Show that the eigenvalues of K and K^* are given by $\pm\sqrt{1 - \gamma^2}$. Find the normalized eigenvectors.

Problem 38.   Let $n \geq 2$. Consider the $(n-1) \times n$ matrix A over $\mathbb{R}$. Then AA^T is an $(n-1) \times (n-1)$ matrix over $\mathbb{R}$ and $A^T A$ is an $n \times n$ matrix over $\mathbb{R}$. Show that the $(n-1)$ eigenvalues of AA^T are also eigenvalues of $A^T A$ and $A^T A$ additionally admits the eigenvalue 0.

Problem 39.   Let A, B be 3×3 matrices. We define the composition

$$A \Diamond B := \begin{pmatrix} a_{11} & 0 & a_{12} & 0 & a_{13} \\ 0 & b_{11} & b_{12} & b_{13} & 0 \\ a_{21} & b_{21} & a_{22}b_{22} & b_{23} & a_{23} \\ 0 & b_{31} & b_{32} & b_{33} & 0 \\ a_{31} & 0 & a_{32} & 0 & a_{33} \end{pmatrix}.$$

Let

$$M = \begin{pmatrix} 1/\sqrt{2} & 0 & 1/\sqrt{2} \\ 0 & 1 & 0 \\ 1/\sqrt{2} & 0 & -1/\sqrt{2} \end{pmatrix}.$$

Find the eigenvalues of M and $M \Diamond M$.

Problem 40.   Let $j, k \in \{0, 1\}$. Find the eigenvalues of the 2×2 matrix $M = (M_{jk})$ with $M_{jk} = e^{i\pi(j \cdot k)}$.

Problem 41.   Compare the lowest eigenvalues of the two 6×6 matrices

$$\begin{pmatrix} 0 & 1 & 0 & 0 & 0 & 0 \\ 1 & 0 & 1 & 0 & 0 & 0 \\ 0 & 1 & 0 & 1 & 0 & 0 \\ 0 & 0 & 1 & 0 & 1 & 0 \\ 0 & 0 & 0 & 1 & 0 & 1 \\ 0 & 0 & 0 & 0 & 1 & 0 \end{pmatrix}, \quad \begin{pmatrix} 0 & 1 & 0 & 0 & 0 & 1 \\ 1 & 0 & 1 & 0 & 0 & 0 \\ 0 & 1 & 0 & 1 & 0 & 0 \\ 0 & 0 & 1 & 0 & 1 & 0 \\ 0 & 0 & 0 & 1 & 0 & 1 \\ 1 & 0 & 0 & 0 & 1 & 0 \end{pmatrix}.$$

Problem 42.   Let A_n be the $n \times n$ matrices of the form

$$A_2 = \begin{pmatrix} 0 & t \\ t & r \end{pmatrix}, \quad A_3 = \begin{pmatrix} 0 & 0 & t \\ 0 & 1 & 0 \\ t & 0 & r \end{pmatrix}, \quad A_4 = \begin{pmatrix} 0 & 0 & 0 & t \\ 0 & 0 & t & 0 \\ 0 & t & r & 0 \\ t & 0 & 0 & r \end{pmatrix},$$

$$A_5 = \begin{pmatrix} 0 & 0 & 0 & 0 & t \\ 0 & 0 & 0 & t & 0 \\ 0 & 0 & 1 & 0 & 0 \\ 0 & t & 0 & r & 0 \\ t & 0 & 0 & 0 & r \end{pmatrix}.$$

Thus the even dimensional matrix A_{2n} has t along the skew-diagonal and r along the lower main diagonal. Otherwise the entries are 0. The odd dimensional matrix A_{2n+1} has t along the skew-diagonal except 1 at the centre and r along the lower main diagonal. Otherwise the entries are 0. Find the eigenvalues of these matrices.

(i) Show that $n = 2$ the eigenvalues are

$$-\frac{1}{2}\left(\sqrt{4t^2 + r^2} - r\right), \quad \frac{1}{2}\left(\sqrt{4t^2 + r^2} + r\right).$$

(ii) Show that $n = 3$ the eigenvalues are

$$-\frac{1}{2}\left(\sqrt{4t^2 + r^2} - r\right), \quad \frac{1}{2}\left(\sqrt{4t^2 + r^2} + r\right), \quad 1.$$

(iii) Show that for $n = 4$ the eigenvalues are

$$-\frac{1}{2}\left(\sqrt{4t^2 + r^2} - r\right), \quad \frac{1}{2}\left(\sqrt{4t^2 + r^2} + r\right)$$

twice each.

(iv) Show that for $n = 5$ we have the same eigenvalues as for $n = 4$ and additional the eigenvalue 1. Extend to general n.

Problem 43.   Let $n \geq 2$ and $j, k = 0, 1, \ldots, n - 1$. Consider

$$M(j, k) = \begin{cases} 1 & \text{for } j < k \\ 0 & \text{for } j = k \\ -1 & \text{for } j > k \end{cases}.$$

Write down the matrix M for $n = 2$ and find the eigenvalues and normalized eigenvectors. Write down the matrix M for $n = 3$ and find the eigenvalues and normalized eigenvectors.

Problem 44.   Find the spectrum of the 3×3 matrix $A = (a_{jk})$ $(j, k = 1, 2, 3)$

$$a_{jk} = \sqrt{j(k - 1)}.$$

Problem 45.   Let U be a unitary 4×4 matrix. Assume that each column vector (which obviously is normalized) of U can be written as the Kronecker product of two normalized vectors in $\mathbb{C}^2$.
(i) Can we conclude that the eigenvectors of such a unitary matrix can also be written a Kronecker product of two vectors in $\mathbb{C}^2$?
(ii) Can the matrix U be written as a Kronecker product of two 2×2 matrices?

Problem 46.   Let A_1 be an $n_1 \times n_1$ matrix, A_2 be an $n_2 \times n_2$ matrix, A_3 be an $n_3 \times n_3$ matrix and I_{n_j} be the $n_j \times n_j$ matrices $(j = 1, 2, 3)$. Let $\mathbf{u}$ be a normalized eigenvector of A_1 with eigenvalue λ, $\mathbf{v}$ be a normalized eigenvector of A_2 with eigenvalue μ and $\mathbf{w}$ be a normalized eigenvector with eigenvalue ν. Consider

$$\hat{H} = A_1 \otimes I_{n_2} \otimes I_{n_3} + I_{n_1} \otimes A_2 \otimes I_{n_3} + I_{n_1} \otimes I_{n_2} \otimes A_3$$
$$+ A_1 \otimes A_2 \otimes I + A_1 \otimes I \otimes A_3 + I \otimes A_2 \otimes A_3 + A_1 \otimes A_2 \otimes A_3.$$

Show that $\mathbf{u} \otimes \mathbf{v} \otimes \mathbf{w}$ is a normalized eigenvector of $\hat{H}$. Find the corresponding eigenvalue.

Problem 47. Let λ be an eigenvalue of the $n \times n$ matrix A with normalized eigenvector $\mathbf{v}$ and μ be an eigenvalue of the $n \times n$ matrix B with normalized eigenvector $\mathbf{w}$. Show that $\lambda\mu - \lambda - \mu$ is an eigenvalue of $A \otimes B - A \otimes I_n - I_n \otimes B$ with normalized eigenvector $\mathbf{v} \otimes \mathbf{w}$.

Problem 48. Let A, B be 2×2 matrices over $\mathbb{C}$. Let λ be an eigenvalue of A with corresponding normalized eigenvector $\mathbf{u}$ and μ be an eigenvalue of B with corresponding normalized eigenvector $\mathbf{v}$. Let $A \otimes B$ be the Kronecker product of A and B. Then $\lambda\mu$ is an eigenvalue of $A \otimes B$ with corresponding normalized eigenvector $\mathbf{u} \otimes \mathbf{v}$. Let $A \oplus B$ be the direct sum of A and B. Then we have

$$
\begin{pmatrix} A & 0_2 \\ 0_2 & B \end{pmatrix} \begin{pmatrix} u_1 \\ u_2 \\ 0 \\ 0 \end{pmatrix} = \lambda \begin{pmatrix} u_1 \\ u_2 \\ 0 \\ 0 \end{pmatrix}, \quad \begin{pmatrix} A & 0_2 \\ 0_2 & B \end{pmatrix} \begin{pmatrix} 0 \\ 0 \\ v_1 \\ v_2 \end{pmatrix} = \mu \begin{pmatrix} 0 \\ 0 \\ v_1 \\ v_2 \end{pmatrix}.
$$

Consequently

$$
\begin{pmatrix} A & 0_2 \\ 0_2 & B \end{pmatrix} \begin{pmatrix} u_1 \\ u_2 \\ v_1 \\ v_2 \end{pmatrix} = \begin{pmatrix} \lambda u_1 \\ \lambda u_2 \\ \mu v_1 \\ \mu v_2 \end{pmatrix}.
$$

Thus the vector $(u_1 \; u_2 \; v_1 \; v_2)^T$ would be an eigenvector of $A \oplus B$ if $\lambda = \mu$. Let $A \star B$ be the star product

$$
A \star B = \begin{pmatrix} b_{11} & 0 & 0 & b_{12} \\ 0 & a_{11} & a_{12} & 0 \\ 0 & a_{21} & a_{22} & 0 \\ b_{21} & 0 & 0 & b_{22} \end{pmatrix} \Rightarrow (A \star B) \begin{pmatrix} v_1 \\ u_1 \\ u_2 \\ v_2 \end{pmatrix} = \begin{pmatrix} \mu v_1 \\ \lambda u_1 \\ \lambda u_2 \\ \mu v_2 \end{pmatrix}.
$$

Thus the vector $(v_1 \; u_1 \; u_2 \; v_2)^T$ would be an eigenvector of $A \star B$ if $\mu = \lambda$.
(i) Study the eigenvalue problem for $A \otimes B + A \oplus B + A \star B$.
(ii) Apply the result to $A = \sigma_1$ and $B = \sigma_2$, where σ_1, σ_2 denote the Pauli spin matrices. Both admit the eigenvalues $+1$, -1.

Problem 49. (i) Given the eigenvalues of the $n \times n$ matrix M. Find the eigenvalues of

$$
e^{-M} \otimes I_n + I_n \otimes e^M.
$$

(ii) Let A and B be normal $n \times n$ matrices with eigenvalues $\lambda_1, \ldots, \lambda_n$ and $\mu_1, \ldots, \mu_n$, respectively. Find the eigenvalues of $e^{i\phi} A \otimes I_n + e^{-i\phi} I_n \otimes B$.

Problem 50. Let $\alpha \in \mathbb{R}$. Consider the symmetric 4×4 matrix

$$
A(\alpha) = \frac{1}{\sqrt{2}} \begin{pmatrix} 1+\alpha & 0 & 1-\alpha & 0 \\ 0 & 1+\alpha & 0 & 1-\alpha \\ 1-\alpha & 0 & 1+\alpha & 0 \\ 0 & 1-\alpha & 0 & 1+\alpha \end{pmatrix}.
$$

Show that an invertible matrix B and diagonal matrix D such that $A(\alpha) = B^{-1}DB$ are given by

$$B = B^{-1} = \frac{1}{2}\begin{pmatrix} 1 & 1 & 1 & 1 \\ 1 & -1 & 1 & -1 \\ 1 & 1 & -1 & -1 \\ 1 & -1 & -1 & 1 \end{pmatrix}, \quad D = \sqrt{2}\begin{pmatrix} 1 & 0 & 0 & 0 \\ 0 & 1 & 0 & 0 \\ 0 & 0 & \alpha & 0 \\ 0 & 0 & 0 & \alpha \end{pmatrix}.$$

Problem 51. Let A be an $n \times n$ matrix with eigenvalue λ and eigenvector $\mathbf{u}$, i.e. $A\mathbf{u} = \lambda\mathbf{u}$, $\mathbf{u} \neq \mathbf{0}$. Consider the $n^n \times n^n$ matrix

$$K = A \otimes I_n \otimes \cdots \otimes I_n + I_n \otimes A \otimes I_n \otimes \cdots \otimes I_n + \cdots + I_n \otimes \cdots \otimes I_n \otimes A$$

where each term has n Kronecker products. Show that

$$K(\mathbf{u} \otimes \cdots \otimes \mathbf{u}) = \lambda(\mathbf{u} \otimes \cdots \otimes \mathbf{u}) + \cdots + \lambda(\mathbf{u} \otimes \cdots \otimes \mathbf{u}) = n\lambda(\mathbf{u} \otimes \cdots \otimes \mathbf{u})$$

i.e. $n\lambda$ is an eigenvalue.

Problem 52. Consider the $(n+1) \times (n+1)$ matrix

$$A = \begin{pmatrix} 0 & \mathbf{s}^* \\ \mathbf{r} & 0_{n\times n} \end{pmatrix}$$

where $\mathbf{r}$ and $\mathbf{s}$ are $n \times 1$ vectors with complex entries, $\mathbf{s}^*$ denoting the conjugate transpose of $\mathbf{s}$. Show that the characteristic polynomial is given by

$$\det(B - \lambda I_{n+1}) = (-\lambda)^{n-1}(\lambda^2 - \mathbf{s}^*\mathbf{r}).$$

Problem 53. Let $n \geq 2$ and n even. Find the eigenvalues of the $n \times n$ matrices

$$A = \begin{pmatrix} 2 & -1 & 0 & \cdots & 0 & -1 \\ -1 & 2 & -1 & \ddots & & 0 \\ 0 & \ddots & \ddots & \ddots & \ddots & \vdots \\ \vdots & \ddots & \ddots & \ddots & \ddots & 0 \\ 0 & & \ddots & \ddots & \ddots & -1 \\ -1 & 0 & \ddots & 0 & -1 & 2 \end{pmatrix}.$$

Problem 54. Let $j, k = 0, 1, 2, 3$. Write down the 4×4 matrix A with the entries

$$a_{jk} = \cos\left(\frac{\pi j - \pi k}{2}\right)$$

and find the eigenvalues.

Problem 55. Let H be a hermitian $n \times n$ matrix and U an $n \times n$ unitary matrix such that $UHU^* = H$. We call H invariant under U. From $UHU^* = H$

it follows that $[H, U] = 0_n$. If $\mathbf{v}$ is an eigenvector of H, i.e. $H\mathbf{v} = \lambda\mathbf{v}$, then $U\mathbf{v}$ is also an eigenvector of H since

$$H(U\mathbf{v}) = (HU)\mathbf{v} = U(H\mathbf{v}) = \lambda(U\mathbf{v}).$$

The set of all unitary matrices U_j $(j = 1, \ldots, m)$ that leave a given hermitian matrix invariant, i.e. $U_j H U_j^* = H$ $(j = 1, \ldots, m)$ form a group under matrix multiplication.

(i) Find all 2×2 hermitian matrices H such that $[H, U] = 0_2$, where $U = \sigma_1$.
(ii) Find all 2×2 hermitian matrices H such that $[H, U] = 0_2$, where $U = \sigma_2$.
(iii) Find all 4×4 hermitian matrices H such that $[H, U] = 0_4$, where $U = \sigma_1 \otimes \sigma_1$.
(iv) Find all 4×4 hermitian matrices H such that $[H, U] = 0_4$, where $U = \sigma_2 \otimes \sigma_2$.

Problem 56. The *Cartan matrix* A of a rank r root system of a semi-simple Lie algebra is an $r \times r$ matrix whose entries are derived from the simple roots. The entries of the Cartan matrix are given by

$$a_{jk} = 2\frac{(\alpha_j, \alpha_k)}{(\alpha_j, \alpha_j)}$$

where $(\ ,\ )$ is the Euclidean inner product and α_j are the simple roots. The entries are independent of the choice of simple roots (up to ordering).
(i) The Cartan matrix of the Lie algebra e_8 is given by the 8×8 matrix

$$\begin{pmatrix} 2 & -1 & 0 & 0 & 0 & 0 & 0 & 0 \\ -1 & 2 & -1 & 0 & 0 & 0 & 0 & 0 \\ 0 & -1 & 2 & -1 & 0 & 0 & 0 & -1 \\ 0 & 0 & -1 & 2 & -1 & 0 & 0 & 0 \\ 0 & 0 & 0 & -1 & 2 & -1 & 0 & 0 \\ 0 & 0 & 0 & 0 & -1 & 2 & -1 & 0 \\ 0 & 0 & 0 & 0 & 0 & -1 & 2 & 0 \\ 0 & 0 & -1 & 0 & 0 & 0 & 0 & 2 \end{pmatrix}.$$

Show that the trace is equal to 16. Show that the determinant is equal to 1. Find the eigenvalues and eigenvectors of the matrix.
(ii) The Lie algebra $sl(3, \mathbb{R})$ is the rank 2 Lie algebra with *Cartan matrix*

$$C = \begin{pmatrix} 2 & -1 \\ 1 & 2 \end{pmatrix}.$$

Find the eigenvalues and normalized eigenvectors of C.
(iii) Let $\tau = 2\cos(\pi/5)$. Find the eigenvalues and eigenvectors of the three Cartan matrices

$$A_2 = \begin{pmatrix} 2 & -\tau \\ -\tau & 2 \end{pmatrix}, \quad A_3 = \begin{pmatrix} 2 & -1 & 0 \\ -1 & 2 & -\tau \\ 0 & -\tau & 2 \end{pmatrix}, \quad A_4 = \begin{pmatrix} 2 & -1 & 0 & 0 \\ -1 & 2 & -1 & 0 \\ 0 & -1 & 2 & -\tau \\ 0 & 0 & -\tau & 2 \end{pmatrix}.$$

Problem 57. Let A be an $n \times n$ matrix. Let J be the $n \times n$ matrix with 1's in the counter diagonal and 0's otherwise. Let $\text{tr}(A) = 0$, $\text{tr}(JA) = 0$. What can be said about the eigenvalues of A? Consider first the cases $n = 2$ and $n = 3$.

Problem 58. Let A be an $n \times n$ matrix over $\mathbb{C}$. Let $n \geq 1$. We define

$$\Delta(A^n) := \sum_{j=0}^{n} \binom{n}{j} A^k \otimes A^{n-k}.$$

Given the eigenvalues of A. What can be said about the eigenvalues of $\Delta(A^n)$?

Problem 59. Let A, B be $n \times n$ matrices over $\mathbb{C}$. Assume we know the eigenvalues and eigenvectors of A and B. What can be said about the eigenvalues and eigenvectors of $A \otimes B - B \otimes A$ and $A \otimes B + B \otimes A$?

Problem 60. Let U be a unitary 4×4 matrix. Assume that each column (vector) can be written as the Kronecker product of two vectors in $\mathbb{C}^2$. Can we conclude that the eigenvectors of such a unitary matrix can also be written a Kronecker product of two vectors in $\mathbb{C}^2$?

Problem 61. Let A, B be $n \times n$ normal matrices. Consider the normal matrices

$$H = A \otimes I_n + I_n \otimes B + A \otimes B, \quad K = A \otimes I_n + I_n \otimes B + B \otimes A.$$

Given the eigenvalues and eigenvectors of A, i.e. λ_j, $\mathbf{u}_j$ ($j = 1, \ldots, n$) and the eigenvalues and eigenvectors of B, i.e. μ_j, $\mathbf{v}_j$ ($j = 1, \ldots, n$). Then the eigenvalues and eigenvectors of H are given by $\lambda_j + \mu_k + \lambda_j \mu_k$, $\mathbf{u}_j \otimes \mathbf{v}_k$. Assume that

$$\text{tr}(AB^*) = 0$$

i.e. considering the Hilbert space of the $n \times n$ matrices with the scalar product $\langle X, Y \rangle := \text{tr}(XY^*)$ the matrices A and B are orthogonal. What can be said about the eigenvalues and eigenvectors of K?

Problem 62. (i) Let $a, b, c \in \mathbb{R}$. Show that the four eigenvalues of the 4×4 matrix

$$A = \begin{pmatrix} 0 & -a & -b & -c \\ a & 0 & -ic & ib \\ b & ic & 0 & -ia \\ c & -ib & ia & 0 \end{pmatrix}$$

are given by $\pm\sqrt{a^2 + b^2 + c^2}$, $\pm i\sqrt{a^2 + b^2 + c^2}$.
(ii) Let $x \in \mathbb{R}$. Show that the lowest eigenvalue of the symmetric 4×4 matrix ($x \in \mathbb{R}$)

$$\begin{pmatrix} 0 & -x\sqrt{5} & 0 & 0 \\ -x\sqrt{5} & 4 & -2x & -2x \\ 0 & -2x & 4 - 2x & -x \\ 0 & -2x & -x & 8 - 2x \end{pmatrix}$$

is given by

$$\lambda_0 = 4 - x - 2\sqrt{1 + x^2 + 2x \cos(2\pi/5)} - 2\sqrt{1 + x^2 + 2x \cos(4\pi/5)}.$$

Problem 63. Let A be an $n \times n$ normal matrix, i.e. $AA^* = A^*A$. Let $\mathbf{u}$ be an eigenvector of A, i.e. $A\mathbf{u} = \lambda\mathbf{u}$. Show that $\mathbf{u}$ is also an eigenvector of A^* with eigenvalue $\bar{\lambda}$, i.e.

$$A^*\mathbf{u} = \bar{\lambda}\mathbf{u}.$$

Apply $(A\mathbf{u})^*A\mathbf{u} = (A^*\mathbf{u})^*A^*\mathbf{u}$.

Problem 64. Let A be an $n \times n$ normal matrix. Assume that λ_j $(j = 1, \ldots, n)$ are the eigenvalues of A. Calculate

$$\prod_{k=1}^{n}(1 + \lambda_k)$$

without using the eigenvalues. Apply the identity

$$\prod_{k=1}^{n}(1 + \lambda_k) \equiv \det(I_n + A).$$

Problem 65. Let A be an $n \times n$ matrix over $\mathbb{C}$, which satisfies $A^2 \equiv AA = cA$, where $c \in \mathbb{C}$ is a constant. Obviously the equation is satisfied by the zero matrix. Assume that $A \neq 0_n$. Then we have a "type of eigenvalue equation".
(i) Is c an eigenvalue of A?
(ii) Take the determinant of both sides of the equation. Discuss. Study the cases that A is invertible and non-invertible. If the matrix A is invertible it follows that $A = cI_n$ and $c \neq 0$. Taking the determinant yields

$$(\det(A))^2 = c^n \det(A).$$

(iii) Study the case

$$A(z) = \begin{pmatrix} e^{-z} & 1 \\ 1 & e^z \end{pmatrix}, \quad z \in \mathbb{C}$$

and show that

$$A^2(z) = (e^z + e^{-z})A(z)$$

where $e^z + e^{-z}$ together with 0 are the eigenvalues of $A(z)$.
(iv) Study $(A \otimes A)^2 = c(A \otimes A)$.
(v) Let A be a 2×2 matrix and $A \star A$ be the star product. Study the case $(A \star A)^2 = c(A \star A)$.
(vi) Study the case that $A^3 = cA$.

Problem 66. The additive inverse eigenvalue problem is as follows: Let A be an $n \times n$ symmetric matrix over $\mathbb{R}$ with $a_{jj} = 0$ for $j = 1, \ldots, n$. Find a real diagonal $n \times n$ matrix D such that the matrix $A + D$ has the prescribed eigenvalues $\lambda_1, \ldots, \lambda_n$. The number of solutions for the real matrix D varies from 0 to $n!$. Consider the 2×2 matrix $A = \sigma_1$ and the prescribed eigenvalues $\lambda_1 = 2$, $\lambda_2 = 3$. Can one find a D? We have

$$A + D = \begin{pmatrix} d_{11} & 1 \\ 1 & d_{22} \end{pmatrix}.$$

Problem 67. Let A be an $n \times n$ normal matrix with pairwise different eigenvalues. Are the matrices

$$\Pi_j = \prod_{k=1,j\neq k}^{n} \frac{A - \lambda_k I_n}{\lambda_j - \lambda_k}$$

projection matrices?

Problem 68. Consider the hermitian 3×3 matrix

$$A = \begin{pmatrix} 0 & 1 & 1 \\ 1 & 0 & 1 \\ 1 & 1 & 0 \end{pmatrix}.$$

Find A^2 and A^3. We know that

$$\text{tr}(A) = \lambda_1 + \lambda_2 + \lambda_3, \quad \text{tr}(A^2) = \lambda_1^2 + \lambda_2^2 + \lambda_3^2, \quad \text{tr}(A^3) = \lambda_1^3 + \lambda_2^3 + \lambda_3^3.$$

Use Newton's method to solve this system of three equations to find the eigenvalues of A.

Problem 69. Consider an $n \times n$ permutation matrix P. Obviously $+1$ is always an eigenvalue since the column vector with all n entries equal to $+1$ is an eigenvector. Apply a *brute force method* and give a C++ implementation to figure out whether -1 is an eigenvalue. We run over all column vectors $\mathbf{v}$ of length n, where the entries can only be $+1$ or -1, where of course the cases with all entries $+1$ or all entries -1 can be omitted. Thus the number of column vectors we have to run through are $2^n - 2$. The condition then to be checked is $P\mathbf{v} = -\mathbf{v}$. If true we have an eigenvalues -1 with the corresponding eigenvector $\mathbf{v}$.

Problem 70. The *power method* is the simplest algorithm for computing eigenvectors and eigenvalues. Consider the vector space $\mathbb{R}^n$ with the Euclidean norm $\|\mathbf{x}\|$ of a vector $\mathbf{x} \in \mathbb{R}^n$. The iteration is as follows: Given a nonsingular $n \times n$ matrix M and a vector $\mathbf{x}_0$ with $\|\mathbf{x}_0\| = 1$. One defines

$$\mathbf{x}_{t+1} = \frac{M\mathbf{x}_t}{\|M\mathbf{x}_t\|}, \quad t = 0, 1, \ldots$$

This defines a dynamical system on the sphere S^{n-1}. Since M is invertible we have

$$\mathbf{x}_t = \frac{M^{-1}\mathbf{x}_{t+1}}{\|M^{-1}\mathbf{x}_{t+1}\|}, \quad t = 0, 1, \ldots$$

(i) Apply the power method to the nonnormal matrix

$$A = \begin{pmatrix} 1 & 1 \\ 0 & 1 \end{pmatrix} \quad \text{and} \quad \mathbf{x}_0 = \begin{pmatrix} 1 \\ 0 \end{pmatrix}.$$

(ii) Apply the power method to the Bell matrix

$$B = \frac{1}{\sqrt{2}} \begin{pmatrix} 1 & 0 & 0 & 1 \\ 0 & 1 & 1 & 0 \\ 0 & 1 & -1 & 0 \\ 1 & 0 & 0 & -1 \end{pmatrix} \quad \text{and} \quad \mathbf{x}_0 = \begin{pmatrix} 1 \\ 0 \\ 0 \\ 0 \end{pmatrix}.$$

(iii) Consider the 3×3 symmetric matrix over $\mathbb{R}$

$$A = \begin{pmatrix} 2 & -1 & 0 \\ -1 & 2 & -1 \\ 0 & -1 & 2 \end{pmatrix}.$$

Find the largest eigenvalue and the corresponding eigenvector using the power method. Start from the vector

$$\mathbf{v} = \begin{pmatrix} 1 \\ 1 \\ 1 \end{pmatrix} \Rightarrow A\mathbf{v} = \begin{pmatrix} 1 \\ 0 \\ 1 \end{pmatrix}, \quad A^2\mathbf{v} = \begin{pmatrix} 2 \\ -2 \\ 2 \end{pmatrix}.$$

The largest eigenvalue is $\lambda = 2 + \sqrt{2}$ with the eigenvector $(1 \ \ -\sqrt{2} \ \ 1)^T$.

Problem 71.   Let A be an $n \times n$ matrix over $\mathbb{C}$. Then any eigenvalue of A satisfies the inequality

$$|\lambda| \leq \max_{1 \leq j \leq n} \sum_{k=1}^{n} |a_{jk}|.$$

Write a C++ program that calculates the right-hand side of the inequality for a given matrix. Apply the complex class of STL. Apply it to the matrix

$$A = \begin{pmatrix} i & 0 & 0 & i \\ 0 & 2i & 2i & 0 \\ 0 & 3i & 3i & 0 \\ 4i & 0 & 0 & 4i \end{pmatrix}.$$

Chapter 6

Spectral Theorem

An $n \times n$ matrix over $\mathbb{C}$ is called *normal* if $A^*A = AA^*$. Let A be an $n \times n$ normal matrix over $\mathbb{C}$ with eigenvalues $\lambda_1, \ldots, \lambda_n$ and corresponding pairwise orthonormal eigenvectors $\mathbf{v}_j$ $(j = 1, \ldots, n)$. Then the matrix A can be written as (spectral decomposition)

$$A = \sum_{j=1}^{n} \lambda_j \mathbf{v}_j \mathbf{v}_j^*.$$

Note that
$$\mathbf{v}_j \mathbf{v}_j^*, \quad j = 1, \ldots, n$$
are projection matrices, i.e. $(\mathbf{v}_j \mathbf{v}_j^*)(\mathbf{v}_j \mathbf{v}_j^*) = \mathbf{v}_j \mathbf{v}_j^*$, $(\mathbf{v}_j \mathbf{v}_j^*)^* = \mathbf{v}_j \mathbf{v}_j^*$ and $\mathbf{v}_j^* \mathbf{v}_j = 1$ for $j = k$ and $\mathbf{v}_j^* \mathbf{v}_k = 0$ for $j \neq k$.

Let $M_n(\mathbb{C})$ be the vector space of $n \times n$ matrices over $\mathbb{C}$. Let $A \in M_n(\mathbb{C})$ with eigenvalues $\lambda_1, \ldots, \lambda_n$ counted according to multiplicity. The following statements are equivalent.
(1) The matrix A is normal.
(2) The matrix A is unitarily diagonalizable, i.e. there is an $n \times n$ unitary matrix U such that UAU^* is a diagonal matrix.
(3) There exists an orthonormal basis of n eigenvectors of A.
(4) With $A = (a_{jk})$ $(j, k = 1, \ldots, n)$

$$\sum_{j=1}^{n} \sum_{k=1}^{n} |a_{jk}|^2 = \sum_{j=1}^{n} |\lambda_j|^2.$$

The spectral theorem, for example, can be utilized to calculate $\exp(A)$ of a normal matrix or the square roots of a normal matrix.

Problem 1. Consider the vectors in $\mathbb{C}^2$ (sometimes called *spinors*)

$$\mathbf{v}_1 = \begin{pmatrix} \cos(\theta/2)e^{-i\phi/2} \\ \sin(\theta/2)e^{i\phi/2} \end{pmatrix}, \qquad \mathbf{v}_2 = \begin{pmatrix} \sin(\theta/2)e^{-i\phi/2} \\ -\cos(\theta/2)e^{i\phi/2} \end{pmatrix}.$$

(i) First show that they are normalized and orthonormal.
(ii) Assume that for the vector $\mathbf{v}_1$ is an eigenvector with the corresponding eigenvalue $+1$ and that the vector $\mathbf{v}_2$ is an eigenvector with the corresponding eigenvalue -1. Apply the spectral theorem to find the corresponding 2×2 matrix.
(iii) Since the vectors $\mathbf{v}_1$ and $\mathbf{v}_2$ form an orthonormal basis in $\mathbb{C}^2$ we can form an orthonormal basis in $\mathbb{C}^4$ via the Kronecker product

$$\mathbf{w}_{11} = \mathbf{v}_1 \otimes \mathbf{v}_1, \quad \mathbf{w}_{12} = \mathbf{v}_1 \otimes \mathbf{v}_2, \quad \mathbf{w}_{21} = \mathbf{v}_2 \otimes \mathbf{v}_1, \quad \mathbf{w}_{22} = \mathbf{v}_2 \otimes \mathbf{v}_2.$$

Assume that the eigenvalue for $\mathbf{w}_{11}$ is $+1$, for $\mathbf{w}_{12}$ -1, for $\mathbf{w}_{21}$ -1 and for $\mathbf{w}_{22}$ $+1$. Apply the spectral theorem to find the corresponding 4×4 matrix.

Solution 1. (i) Straightforward calculation yields

$$\mathbf{v}_1^*\mathbf{v}_1 = 1, \quad \mathbf{v}_2^*\mathbf{v}_2 = 1, \quad \mathbf{v}_2^*\mathbf{v}_1 = 0.$$

(ii) Since $U(\theta, \phi) = \lambda_1\mathbf{v}_1\mathbf{v}_1^* + \lambda_2\mathbf{v}_2\mathbf{v}_2^*$ we find utilizing the identities

$$\cos^2(\theta/2) - \sin^2(\theta/2) \equiv \cos(\theta), \quad 2\cos(\theta/2)\sin(\theta/2) \equiv \sin(\theta)$$

that

$$U(\theta, \phi) = \lambda_1\mathbf{v}_1\mathbf{v}_1^* + \lambda_2\mathbf{v}_2\mathbf{v}_2^* = \begin{pmatrix} \cos(\theta) & \sin(\theta)e^{-i\phi} \\ \sin(\theta)e^{i\phi} & -\cos(\theta) \end{pmatrix}.$$

(iii) For the 4×4 matrix $V(\theta, \phi)$

$$V(\theta, \phi) = \mathbf{w}_{11}\mathbf{w}_{11}^* - \mathbf{w}_{12}\mathbf{w}_{12}^* - \mathbf{w}_{21}\mathbf{w}_{21}^* + \mathbf{w}_{22}\mathbf{w}_{22}^*.$$

Problem 2. Given a 4×4 symmetric matrix A over $\mathbb{R}$ with the eigenvalues $\lambda_1 = 0$, $\lambda_2 = 1$, $\lambda_3 = 2$, $\lambda_4 = 3$ and the corresponding normalized eigenvectors

$$\mathbf{v}_1 = \frac{1}{\sqrt{2}}\begin{pmatrix} 1 \\ 0 \\ 0 \\ 1 \end{pmatrix}, \quad \mathbf{v}_2 = \frac{1}{\sqrt{2}}\begin{pmatrix} 1 \\ 0 \\ 0 \\ -1 \end{pmatrix}, \quad \mathbf{v}_3 = \frac{1}{\sqrt{2}}\begin{pmatrix} 0 \\ 1 \\ 1 \\ 0 \end{pmatrix}, \quad \mathbf{v}_4 = \frac{1}{\sqrt{2}}\begin{pmatrix} 0 \\ 1 \\ -1 \\ 0 \end{pmatrix}.$$

Reconstruct the matrix A using the spectral theorem. The eigenvectors given above are called *Bell basis*.

Solution 2. Since $\lambda_1 = 0$ we have

$$A = \mathbf{v}_2\mathbf{v}_2^T + 2\mathbf{v}_3\mathbf{v}_3^T + 3\mathbf{v}_4\mathbf{v}_4^T = \begin{pmatrix} 1/2 & 0 & 0 & -1/2 \\ 0 & 5/2 & -1/2 & 0 \\ 0 & -1/2 & 5/2 & 0 \\ -1/2 & 0 & 0 & 1/2 \end{pmatrix}.$$

Problem 3.   (i) Consider the 2×2 permutation matrix

$$\sigma_1 = \begin{pmatrix} 0 & 1 \\ 1 & 0 \end{pmatrix}.$$

Find the spectral decomposition of σ_1. Find a matrix K with $\exp(K) = \sigma_1$.
(ii) Consider the permutation matrix

$$P = \begin{pmatrix} 0 & 0 & 1 \\ 0 & 1 & 0 \\ 1 & 0 & 0 \end{pmatrix}.$$

Find the spectral decomposition of P. Find a matrix X with $\exp(X) = P$.
(iii) Consider the permutation matrix

$$P = \begin{pmatrix} 0 & 0 & 0 & 1 \\ 0 & 0 & 1 & 0 \\ 1 & 0 & 0 & 0 \\ 0 & 1 & 0 & 0 \end{pmatrix}.$$

Show that the eigenvalues form a group under multiplication. Find the spectral representation to find K such that $P = \exp(K)$.

Solution 3.   (i) The eigenvalues are given by $\lambda_1 = +1$ and $\lambda_2 = -1$ with the corresponding normalized eigenvectors

$$\mathbf{v}_1 = \frac{1}{\sqrt{2}} \begin{pmatrix} 1 \\ 1 \end{pmatrix}, \qquad \mathbf{v}_2 = \frac{1}{\sqrt{2}} \begin{pmatrix} 1 \\ -1 \end{pmatrix}.$$

Thus we find $\sigma_1 = \lambda_1 \mathbf{v}_1 \mathbf{v}_1^* + \lambda_2 \mathbf{v}_2 \mathbf{v}_2^*$. Hence

$$K = \ln(\lambda_1) \mathbf{v}_1 \mathbf{v}_1^* + \ln(\lambda_2) \mathbf{v}_2 \mathbf{v}_2^*.$$

Since $\ln(1) = 0$ and $\ln(-1) = \ln(e^{i\pi}) = i\pi$ we obtain the noninvertible matrix

$$K = \ln(\lambda_2) \mathbf{v}_2 \mathbf{v}_2^* = \frac{i\pi}{2} \begin{pmatrix} 1 & -1 \\ -1 & 1 \end{pmatrix}.$$

Note that $\mathrm{tr}(K) = i\pi$, $\det(P) = -1$ and $e^{i\pi} = -1$.
(ii) The eigenvalues are given by $\lambda_1 = +1$, $\lambda_2 = +1$ and $\lambda_3 = -1$ with the corresponding normalized eigenvectors

$$\mathbf{v}_1 = \begin{pmatrix} 0 \\ 1 \\ 0 \end{pmatrix}, \qquad \mathbf{v}_2 = \frac{1}{\sqrt{2}} \begin{pmatrix} 1 \\ 0 \\ 1 \end{pmatrix}, \qquad \mathbf{v}_3 = \frac{1}{\sqrt{2}} \begin{pmatrix} 1 \\ 0 \\ -1 \end{pmatrix}.$$

The eigenvectors are pairwise orthogonal. Thus we find the spectral decomposition of P as

$$P = \lambda_1 \mathbf{v}_1 \mathbf{v}_1^* + \lambda_2 \mathbf{v}_2 \mathbf{v}_2^* + \lambda_3 \mathbf{v}_3 \mathbf{v}_3^*.$$

Applying this result we find

$$X = \ln(\lambda_1) \mathbf{v}_1 \mathbf{v}_1^* + \ln(\lambda_2) \mathbf{v}_2 \mathbf{v}_2^* + \ln(\lambda_3) \mathbf{v}_3 \mathbf{v}_3^*.$$

Since $\ln(1) = 0$ and $\ln(-1) = \ln(e^{i\pi}) = i\pi$ we obtain the noninvertible matrix

$$X = \ln(\lambda_3)\mathbf{v}_3\mathbf{v}_3^* = i\pi\frac{1}{\sqrt{2}}\begin{pmatrix} 1 \\ 0 \\ -1 \end{pmatrix}\frac{1}{\sqrt{2}}(1 \quad 0 \quad -1) = \frac{i\pi}{2}\begin{pmatrix} 1 & 0 & -1 \\ 0 & 0 & 0 \\ -1 & 0 & 1 \end{pmatrix}.$$

Note that $\operatorname{tr}(X) = i\pi$, $\det(P) = -1$ and $e^{i\pi} = -1$.

(iii) The eigenvalues are $\lambda_1 = +1$, $\lambda_2 = -1$, $\lambda_3 = +i$, $\lambda_4 = -i$ with the corresponding eigenvectors

$$\mathbf{v}_1 = \frac{1}{2}\begin{pmatrix} 1 \\ 1 \\ 1 \\ 1 \end{pmatrix}, \quad \mathbf{v}_2 = \frac{1}{2}\begin{pmatrix} 1 \\ 1 \\ -1 \\ -1 \end{pmatrix}, \quad \mathbf{v}_3 = \frac{1}{2}\begin{pmatrix} 1 \\ -1 \\ -i \\ i \end{pmatrix}, \quad \mathbf{v}_4 = \frac{1}{2}\begin{pmatrix} 1 \\ -1 \\ i \\ -i \end{pmatrix}.$$

Obviously the numbers $+1, -1, +i, -i$ form a commutative group under multiplication. From the spectral representation of P

$$P = \lambda_1\mathbf{v}_1\mathbf{v}_1^* + \lambda_2\mathbf{v}_2\mathbf{v}_2^* + \lambda_3\mathbf{v}_3\mathbf{v}_3^* + \lambda_4\mathbf{v}_4\mathbf{v}_4^*$$

and $\ln(1) = 0$, $\ln(-1) = i\pi$, $\ln(i) = i\pi/2$, $\ln(-i) = -i\pi/2$ provides the skew-hermitian K as

$$K = \frac{\pi}{4}\begin{pmatrix} i & i & -1-i & 1-i \\ i & i & 1-i & -1-i \\ 1-i & -1-i & i & i \\ -1-i & 1-i & i & i \end{pmatrix}.$$

Problem 4. Let $z \in \mathbb{C}$ and A be an $n \times n$ normal matrix with eigenvalues $\lambda_1, \ldots, \lambda_n$ and pairwise orthonormal eigenvectors $\mathbf{v}_1, \ldots, \mathbf{v}_n$. Use the spectral decomposition to calculate $\exp(zA)$.

Solution 4. Since

$$e^{\lambda_j\mathbf{v}_j\mathbf{v}_j^*} = I_n + \mathbf{v}_j\mathbf{v}_j^*(e^{z\lambda_j} - 1)$$

and the matrices $\mathbf{v}_j\mathbf{v}_j^*$ and $\mathbf{v}_k\mathbf{v}_k^*$ $(j \neq k)$ commute we have

$$e^{zA} = (I_n + \mathbf{v}_1\mathbf{v}_1^*(e^{z\lambda_1} - 1))(I_n + \mathbf{v}_2\mathbf{v}_2^*(e^{z\lambda_2} - 1))\cdots(I_n + \mathbf{v}_n\mathbf{v}_n^*(e^{z\lambda_n} - 1)).$$

Now $\mathbf{v}_j^*\mathbf{v}_k = \delta_{jk}$ and $\sum_{j=1}^n \mathbf{v}_j\mathbf{v}_j^* = I_n$. It follows that

$$e^{zA} = \sum_{j=1}^n e^{z\lambda_j}\mathbf{v}_j\mathbf{v}_j^*.$$

Problem 5. Let A be a hermitian $n \times n$ matrix. Assume that all the eigenvalues $\lambda_1, \ldots, \lambda_n$ are pairwise different. Then the normalized eigenvectors $\mathbf{u}_j$ $(j = 1, \ldots, n)$ satisfy $\mathbf{u}_j^*\mathbf{u}_k = 0$ for $j \neq k$ and $\mathbf{u}_j^*\mathbf{u}_j = 1$. We have

$$A = \sum_{j=1}^n \lambda_j\mathbf{u}_j\mathbf{u}_j^*.$$

Let e_k $(k = 1, \ldots, n)$ be the standard basis in $\mathbb{C}^n$. Calculate U^*AU, where $U = \sum_{k=1}^{n} u_k e_k^*$.

Solution 5. From U we find

$$U^* = \sum_{\ell=1}^{n} e_\ell u_\ell^*.$$

Thus

$$U^*AU = \sum_{\ell=1}^{n} e_\ell u_\ell^* \sum_{j=1}^{n} \lambda_j u_j u_j^* \sum_{k=1}^{n} u_k e_k^* = \sum_{j=1}^{n} \lambda_j \sum_{\ell=1}^{n} \sum_{k=1}^{n} e_\ell (u_\ell^* u_j)(u_j^* u_k) e_k^*$$

$$= \sum_{j=1}^{n} \lambda_j \sum_{\ell=1}^{n} \sum_{k=1}^{n} e_\ell \delta_{\ell j} \delta_{jk} e_k$$

$$= \sum_{j=1}^{n} \lambda_j e_j e_j^*.$$

Thus U^*AU is a diagonal matrix with the eigenvalues on the diagonal.

Problem 6. Let A be a positive definite $n \times n$ matrix. Thus all the eigenvalues are real and positive. Assume that all the eigenvalues $\lambda_1, \ldots, \lambda_n$ are pairwise different. Then the normalized eigenvectors u_j $(j = 1, \ldots, n)$ satisfy $u_j^* u_k = 0$ for $j \neq k$ and $u_j^* u_j = 1$. We have (spectral theorem)

$$A = \sum_{j=1}^{n} \lambda_j u_j u_j^*.$$

Let e_k $(k = 1, \ldots, n)$ be the standard basis in $\mathbb{C}^n$. Calculate $\ln(A)$. Note that the unitary matrix

$$U = \sum_{k=1}^{n} u_k e_k^*$$

transforms A into a diagonal matrix, i.e. $\tilde{A} = U^*AU$ is a diagonal matrix.

Solution 6. From $\tilde{A} = U^*AU$ it follows that $A = U\tilde{A}U^*$. Thus

$$\ln(A) = \ln(U\tilde{A}U^*) = U \ln(\tilde{A})U^* = \sum_{j=1}^{n} \ln(\lambda_j) u_j u_j^*.$$

Problem 7. Consider the normal noninvertible 3×3 matrix

$$A = \begin{pmatrix} 0 & 1 & 0 \\ 1 & 0 & 1 \\ 0 & 1 & 0 \end{pmatrix}.$$

(i) Find the spectral decomposition.

(ii) Use this result to calculate $\exp(zA)$, where $z \in \mathbb{C}$.

Solution 7. (i) The eigenvalues are $\lambda_1 = 0$, $\lambda_2 = \sqrt{2}$, and $\lambda_3 = -\sqrt{2}$ with the corresponding normalized eigenvectors

$$
\mathbf{v}_1 = \frac{1}{\sqrt{2}} \begin{pmatrix} 1 \\ 0 \\ -1 \end{pmatrix}, \quad
\mathbf{v}_2 = \frac{1}{2} \begin{pmatrix} 1 \\ \sqrt{2} \\ 1 \end{pmatrix}, \quad
\mathbf{v}_3 = \frac{1}{2} \begin{pmatrix} 1 \\ -\sqrt{2} \\ 1 \end{pmatrix}.
$$

Thus we have the spectral decomposition of A

$$
A = \lambda_1 \mathbf{v}_1 \mathbf{v}_1^* + \lambda_2 \mathbf{v}_2 \mathbf{v}_2^* + \lambda_3 \mathbf{v}_3 \mathbf{v}_3^* = \lambda_2 \mathbf{v}_2 \mathbf{v}_2^* + \lambda_3 \mathbf{v}_3 \mathbf{v}_3^*.
$$

Note that $\mathbf{v}_1 \mathbf{v}_1^*$, $\mathbf{v}_2 \mathbf{v}_2^*$, $\mathbf{v}_3 \mathbf{v}_3^*$ are projection matrices.
(ii) The spectral decomposition of $\exp(zA)$ is

$$
e^{z\lambda_2 \mathbf{v}_2 \mathbf{v}_2^*} e^{z\lambda_3 \mathbf{v}_3 \mathbf{v}_3^*} = (I_3 + (e^{z\lambda_2} - 1)\mathbf{v}_2 \mathbf{v}_2^*)(I_3 + (e^{z\lambda_3} - 1)\mathbf{v}_3 \mathbf{v}_3^*)
$$

where

$$
\mathbf{v}_1 \mathbf{v}_1^* = \frac{1}{4} \begin{pmatrix} 1 & \sqrt{2} & 1 \\ \sqrt{2} & 2 & \sqrt{2} \\ 1 & \sqrt{2} & 1 \end{pmatrix}, \quad
\mathbf{v}_2 \mathbf{v}_2^* = \frac{1}{4} \begin{pmatrix} 1 & -\sqrt{2} & 1 \\ -\sqrt{2} & 2 & -\sqrt{2} \\ 1 & -\sqrt{2} & 1 \end{pmatrix}.
$$

Problem 8. Consider the two 3×3 permutation matrices with $P_1 = P_2^T$

$$
P_1 = \begin{pmatrix} 0 & 1 & 0 \\ 0 & 0 & 1 \\ 1 & 0 & 0 \end{pmatrix}, \quad
P_2 = \begin{pmatrix} 0 & 0 & 1 \\ 1 & 0 & 0 \\ 0 & 1 & 0 \end{pmatrix}.
$$

(i) Find the spectral decomposition of the permutation matrices.
(ii) Use the spectral decomposition to find the matrices K_1 and A_2 such that $P_1 = \exp(K_1)$, $P_2 = \exp(A_2)$.
(iii) We want to find efficiently K_1 and K_2 such that $P_1 = e^{K_1}$ and $P_2 = e^{K_2}$. We would apply the spectral decomposition theorem to find K_1, i.e.

$$
K_1 = \sum_{j=1}^{3} \ln(\lambda_j) \mathbf{v}_j \mathbf{v}_j^*
$$

where λ_j are the eigenvalues of U_1 and $\mathbf{v}_j$ are the corresponding normalized eigenvectors. But then to find K_2 we would apply the property that $U_1^2 = U_2$. Or could we actually apply that $U_2 = U_1^T$? Note that U_1, U_2, I_3 form a commutative subgroup of the group of 3×3 permutation under matrix multiplication.

Solution 8. (i) For the eigenvalues of P_1 we find

$$
\lambda_1 = 1, \quad \lambda_2 = e^{i2\pi/3}, \quad \lambda_3 = e^{i4\pi/3}.
$$

with $\lambda_1 + \lambda_2 + \lambda_3 = 0$. The corresponding normalized eigenvectors are

$$\mathbf{v}_1 = \frac{1}{\sqrt{3}} \begin{pmatrix} 1 \\ 1 \\ 1 \end{pmatrix}, \quad \mathbf{v}_2 = \frac{1}{\sqrt{3}} \begin{pmatrix} 1 \\ e^{i2\pi/3} \\ e^{i4\pi/3} \end{pmatrix}, \quad \mathbf{v}_3 = \frac{1}{\sqrt{3}} \begin{pmatrix} 1 \\ e^{i4\pi/3} \\ e^{i2\pi/3} \end{pmatrix}.$$

For the eigenvalues of P_2 we find the same as for P_1, namely

$$\lambda_1 = 1, \quad \lambda_2 = e^{i2\pi/3}, \quad \lambda_3 = e^{i4\pi/3}$$

with $\lambda_1 + \lambda_2 + \lambda_3 = 0$. The corresponding normalized eigenvectors are

$$\mathbf{w}_1 = \frac{1}{\sqrt{3}} \begin{pmatrix} 1 \\ 1 \\ 1 \end{pmatrix}, \quad \mathbf{w}_2 = \frac{1}{\sqrt{3}} \begin{pmatrix} 1 \\ e^{i4\pi/3} \\ e^{i2\pi/3} \end{pmatrix}, \quad \mathbf{w}_3 = \frac{1}{\sqrt{3}} \begin{pmatrix} 1 \\ e^{i2\pi/3} \\ e^{i4\pi/3} \end{pmatrix}.$$

Hence the spectral decomposition of P_1 and P_2 are

$$P_1 = \sum_{j=1}^{3} \lambda_j \mathbf{v}_j \mathbf{v}_j^*, \quad P_2 = \sum_{j=1}^{3} \lambda_j \mathbf{w}_j \mathbf{w}_j^*.$$

(ii) Since $\ln(1) = 0$, $\ln(e^{i2\pi/3}) = 2\pi i/3$, $\ln(e^{i4\pi/3}) = 4\pi i/3$ we obtain

$$K_1 = \frac{4\pi i}{9} \begin{pmatrix} 3/2 & e^{-i2\pi/3}/2 + e^{-i4\pi/3} & e^{-i4\pi/3}/2 + e^{-i2\pi/3} \\ e^{i2\pi/3}/2 + e^{i4\pi/3} & 3/2 & e^{-i2\pi/3}/2 + e^{i2\pi/3} \\ e^{i4\pi/3}/2 + e^{i2\pi/3} & e^{i2\pi/3}/2 + e^{-i2\pi/3} & 3/2 \end{pmatrix}.$$

Analogously we construct the matrix K_2. Note that $K_2 = K_1^T$. Also note that the trace of these matrices are $2\pi i$.

(iii) For the second and third eigenvalues we obtain the corresponding normalized eigenvectors

$$\mathbf{v}_2 = \frac{1}{\sqrt{3}} \begin{pmatrix} 1 \\ e^{i2\pi/3} \\ e^{i4\pi/3} \end{pmatrix}, \quad \mathbf{v}_3 = \frac{1}{\sqrt{3}} \begin{pmatrix} 1 \\ e^{i4\pi/3} \\ e^{i2\pi/3} \end{pmatrix}.$$

Now since $\ln(e^{i2\pi/3}) = i2\pi/3$, $\ln(e^{i4\pi/3}) = i4\pi/3$ we find for K_1

$$K_1 = i\frac{2\pi}{3} \mathbf{v}_1 \mathbf{v}_1^* + i\frac{4\pi}{3} \mathbf{v}_2 \mathbf{v}_2^*.$$

Since $U_1^2 = U_2 = e^{K_1} e^{K_1} = e^{2K_1} = e^{K_2}$, the matrix K_2 is given by $K_2 = 2K_1$.

Problem 9. Consider the unitary matrix

$$U = \begin{pmatrix} 0 & 0 & 0 & 1 \\ 0 & 0 & 1 & 0 \\ 0 & 1 & 0 & 0 \\ 1 & 0 & 0 & 0 \end{pmatrix} = \begin{pmatrix} 0 & 1 \\ 1 & 0 \end{pmatrix} \otimes \begin{pmatrix} 0 & 1 \\ 1 & 0 \end{pmatrix}.$$

Find the skew-hermitian matrix K such that $U = \exp(K)$. Utilize the spectral representation.

Solution 9. The eigenvalues of U are $\lambda_1 = 1$, $\lambda_2 = 1$, $\lambda_3 = -1$, $\lambda_4 = -1$ with the corresponding normalized eigenvectors (*Bell basis*)

$$
\mathbf{v}_1 = \frac{1}{\sqrt{2}} \begin{pmatrix} 1 \\ 0 \\ 0 \\ 1 \end{pmatrix}, \quad
\mathbf{v}_2 = \frac{1}{\sqrt{2}} \begin{pmatrix} 0 \\ 1 \\ 1 \\ 0 \end{pmatrix}, \quad
\mathbf{v}_3 = \frac{1}{\sqrt{2}} \begin{pmatrix} 1 \\ 0 \\ 0 \\ -1 \end{pmatrix}, \quad
\mathbf{v}_4 = \frac{1}{\sqrt{2}} \begin{pmatrix} 0 \\ 1 \\ -1 \\ 0 \end{pmatrix}.
$$

The eigenvectors are pairwise orthonormal. Thus we have the spectral representation of U

$$
U = \lambda_1 \mathbf{v}_1 \mathbf{v}_1^* + \lambda_2 \mathbf{v}_2 \mathbf{v}_2^* + \lambda_3 \mathbf{v}_3 \mathbf{v}_3^* + \lambda_4 \mathbf{v}_4 \mathbf{v}_4^*.
$$

Thus K is given by (since $\ln(1) = 0$)

$$
K = \ln(\lambda_3) \mathbf{v}_3 \mathbf{v}_3^* + \ln(\lambda_4) \mathbf{v}_4 \mathbf{v}_4^* = i\pi \begin{pmatrix} 1 & 0 & 0 & -1 \\ 0 & 1 & -1 & 0 \\ 0 & -1 & 1 & 0 \\ -1 & 0 & 0 & 1 \end{pmatrix}
$$

where we used $\ln(1) = 0$ and $-1 = \exp(i\pi)$.

Problem 10. Consider the Hadamard matrix

$$
U = \frac{1}{\sqrt{2}} \begin{pmatrix} 1 & 1 \\ 1 & -1 \end{pmatrix} \equiv \frac{1}{\sqrt{2}} (\sigma_3 + \sigma_1)
$$

with eigenvalues $+1$ and -1. Find the square roots of this matrix applying the spectral theorem. Since $\sqrt{1} = \pm 1$ and $\sqrt{-1} = \pm i$ four cases have to be considered

$$
(1, i), \quad (1, -i), \quad (-1, i), \quad (-1, -i).
$$

Solution 10. For the eigenvalues we find $\lambda_1 = +1$ and $\lambda_2 = -1$. The corresponding normalized eigenvectors are

$$
\mathbf{v}_1 = \frac{1}{\sqrt{8}} \begin{pmatrix} \sqrt{4 + 2\sqrt{2}} \\ \sqrt{4 - 2\sqrt{2}} \end{pmatrix}, \quad
\mathbf{v}_2 = \frac{1}{\sqrt{8}} \begin{pmatrix} \sqrt{4 - 2\sqrt{2}} \\ \sqrt{4 + 2\sqrt{2}} \end{pmatrix}.
$$

Thus $U = \lambda_1 \mathbf{v}_1 \mathbf{v}_1^* + \lambda_2 \mathbf{v}_2 \mathbf{v}_2^*$. Thus for the main branch ($\sqrt{1} = 1$, $\sqrt{-1} = i$) we have

$$
\sqrt{U} = 1 \cdot \mathbf{v}_1 \mathbf{v}_1^* + i \cdot \mathbf{v}_2 \mathbf{v}_2^*.
$$

Thus

$$
\sqrt{U} = \frac{1}{2\sqrt{2}} \begin{pmatrix} 1 + \sqrt{2} - i(1 - \sqrt{2}) & 1 - i \\ 1 - i & -1 + \sqrt{2} + i(1 + \sqrt{2}) \end{pmatrix}.
$$

Analogously we find the other three branches.

Problem 11. Let A be an $n \times n$ matrix over $\mathbb{C}$. An $n \times n$ matrix B is called a square root of A if $B^2 = A$. Find the square roots of the 2×2 identity matrix applying the spectral theorem. The eigenvalues of I_2 are $\lambda_1 = 1$ and $\lambda_2 = 1$. As normalized eigenvectors choose

$$\begin{pmatrix} e^{i\phi_1}\cos(\theta) \\ e^{i\phi_2}\sin(\theta) \end{pmatrix}, \quad \begin{pmatrix} e^{i\phi_1}\sin(\theta) \\ -e^{i\phi_2}\cos(\theta) \end{pmatrix}$$

which form an orthonormal basis in $\mathbb{C}^2$. Four cases

$$(\sqrt{\lambda_1}, \sqrt{\lambda_2}) = (1, 1), \quad (\sqrt{\lambda_1}, \sqrt{\lambda_2}) = (1, -1),$$

$$(\sqrt{\lambda_1}, \sqrt{\lambda_2}) = (-1, 1), \quad (\sqrt{\lambda_1}, \sqrt{\lambda_2}) = (-1, -1)$$

have to be studied. The first and last cases are trivial. So study the second case $(\sqrt{\lambda_1}, \sqrt{\lambda_2}) = (1, -1)$. The second case and the third case are "equivalent".

Solution 11. From the normalized eigenvectors we obtain the spectral representation of I_2 as

$$I_2 = \lambda_1 \begin{pmatrix} \cos^2(\theta) & e^{i(\phi_1-\phi_2)}\sin(\theta)\cos(\theta) \\ e^{i(\phi_2-\phi_1)}\sin(\theta)\cos(\theta) & \sin^2(\theta) \end{pmatrix}$$
$$+ \lambda_2 \begin{pmatrix} \sin^2(\theta) & -e^{i(\phi_1-\phi_2)}\sin(\theta)\cos(\theta) \\ -e^{i(\phi_2-\phi_1)}\sin(\theta)\cos(\theta) & \cos^2(\theta) \end{pmatrix}.$$

With $(\sqrt{\lambda_1}, \sqrt{\lambda_2}) = (1, -1)$ we obtain the square of I_2 for this case as

$$\begin{pmatrix} \cos(2\theta) & e^{i(\phi_1-\phi_2)}\sin(2\theta) \\ e^{i(\phi_2-\phi_1)}\sin(2\theta) & -\cos(2\theta) \end{pmatrix}$$

utilizing $\cos^2(\theta) - \sin^2(\theta) \equiv \cos(2\theta)$, $2\cos(\theta)\sin(\theta) \equiv \sin(2\theta)$. This case also includes the Pauli spin matrices σ_1, σ_2 and σ_3 as square root of I_2. The Hadamard matrix is also included. Note that the square root of a unitary matrix is always a unitary matrix.

Problem 12. Let $\lambda_1, \lambda_2, \mu_1, \mu_2 \in \mathbb{C}$. Let $\mathbf{v}_1$, $\mathbf{v}_2$ (column vectors) be an orthonormal basis in $\mathbb{C}^2$. We define the 2×2 matrices (spectral theorem)

$$A = \lambda_1 \mathbf{v}_1 \mathbf{v}_1^* + \lambda_2 \mathbf{v}_2 \mathbf{v}_2^*, \quad B = \mu_1 \mathbf{v}_1 \mathbf{v}_1^* + \mu_2 \mathbf{v}_2 \mathbf{v}_2^*.$$

Find the commutator $[A, B]$. Find the conditions on $\lambda_1, \lambda_2, \mu_1, \mu_2$ such that $[A, B] = 0_2$.

Solution 12. Since $\mathbf{v}_j^* \mathbf{v}_j = 1$ for $j = 1, 2$ and $\mathbf{v}_j^* \mathbf{v}_k = \delta_{kj}$ we obtain $[A, B] = 0_2$. Thus there is no condition on $\mu_1, \mu_2, \lambda_1, \lambda_2$.

Problem 13. Let $\mathbf{v}_0$, $\mathbf{v}_1$, $\ldots$, $\mathbf{v}_{2^n-1}$ be an orthonormal basis in $\mathbb{C}^{2^n}$. We define

$$U := \frac{1}{\sqrt{2^n}} \sum_{j=0}^{2^n-1} \sum_{k=0}^{2^n-1} e^{-i2\pi kj/2^n} \mathbf{v}_k \mathbf{v}_j^*. \tag{1}$$

Show that U is unitary. In other words show that $UU^* = I_{2^n}$, using the *completeness relation*

$$I_{2^n} = \sum_{j=0}^{2^n-1} \mathbf{v}_j \mathbf{v}_j^*.$$

Solution 13.   From the definition (1) we find

$$U^* = \frac{1}{\sqrt{2^n}} \sum_{j=0}^{2^n-1} \sum_{k=0}^{2^n-1} e^{i2\pi kj/2^n} \mathbf{v}_j \mathbf{v}_k^*$$

where $*$ denotes the adjoint. Therefore

$$UU^* = \frac{1}{2^n} \sum_{j=0}^{2^n-1} \sum_{k=0}^{2^n-1} \sum_{l=0}^{2^n-1} \sum_{m=0}^{2^n-1} e^{i2\pi(kj-lm)/2^n} \mathbf{v}_j \mathbf{v}_k^* \mathbf{v}_l \mathbf{v}_m^*$$

$$= \frac{1}{2^n} \sum_{j=0}^{2^n-1} \sum_{k=0}^{2^n-1} \sum_{m=0}^{2^n-1} e^{i2\pi(kj-km)/2^n} \mathbf{v}_j \mathbf{v}_m^*.$$

We have for $j = m$, $e^{i2\pi(kj-km)/2^n} = 1$. Thus for $j, m = 0, 1, \ldots, 2^n - 1$

$$\sum_{k=0}^{2^n-1} (e^{i2\pi(j-m)/2^n})^k = 2^n, \quad j = m$$

$$\sum_{k=0}^{2^n-1} (e^{i2\pi(j-m)/2^n})^k = \frac{1 - e^{i2\pi(j-m)}}{1 - e^{i2\pi(j-m)/2^n}} = 0, \quad j \neq m.$$

Thus

$$UU^* = \sum_{j=0}^{2^n-1} \mathbf{v}_j \mathbf{v}_j = I_{2^n}.$$

Supplementary Problems

Problem 1.   The star product of the *Hadamard matrix* U_H with itself provides the *Bell matrix*

$$B = U_H \star U_H = \frac{1}{\sqrt{2}} \begin{pmatrix} 1 & 0 & 0 & 1 \\ 0 & 1 & 1 & 0 \\ 0 & 1 & -1 & 0 \\ 1 & 0 & 0 & -1 \end{pmatrix}$$

which is a unitary matrix. Note that the eigenvalues of the Bell matrix are $+1$ (twice) and -1 (twice).

(i) Find the square roots of the Bell matrix applying the spectral theorem.

(ii) Apply the spectral theorem to find the skew-hermitian matrix K such that $B = e^K$.

Problem 2.   Let A be a normal matrix with eigenvalues $\lambda_1, \ldots, \lambda_n$ and pairwise orthonormal eigenvectors $\mathbf{a}_j$ (column vectors). Let B be a normal matrix with eigenvalues $\mu_1, \ldots, \mu_n$ and pairwise orthonormal eigenvectors $\mathbf{b}_j$ (column vectors). Then A and B can be written as

$$A = \sum_{j=1}^{n} \lambda_j \mathbf{a}_j \mathbf{a}_j^*, \quad B = \sum_{k=1}^{n} \mu_k \mathbf{b}_k \mathbf{b}_k^*.$$

Let $z \in \mathbb{C}$. Use the spectral decomposition to calculate $e^{zA} B e^{-zA}$.

Problem 3.   Consider the normalized vectors

$$\mathbf{v}_1 = \frac{1}{\sqrt{2}} \begin{pmatrix} 1 \\ 0 \\ 1 \end{pmatrix}, \quad \mathbf{v}_2 = \begin{pmatrix} 0 \\ 1 \\ 0 \end{pmatrix}, \quad \mathbf{v}_3 = \frac{1}{\sqrt{2}} \begin{pmatrix} 1 \\ 0 \\ -1 \end{pmatrix}$$

in $\mathbb{R}^3$. Find the eigenvalues and eigenvectors of the 3×3 matrix $M = 1 \cdot \mathbf{v}_1 \mathbf{v}_1^* + 0 \cdot \mathbf{v}_2 \mathbf{v}_2^* - 1 \cdot \mathbf{v}_3 \mathbf{v}_3^*$ without calculating M.

Problem 4.   Let A be an $n \times n$ normal matrix with eigenvalues $\lambda_1, \ldots, \lambda_n$ and pairwise normalized orthogonal eigenvectors $\mathbf{v}_j$ $(j = 1, \ldots, n)$. Then

$$A = \sum_{j=1}^{n} \lambda_j \mathbf{v}_j \mathbf{v}_j^*.$$

Show that $\cos(A) = \sum_{j=1}^{n} \cos(\lambda_j) \mathbf{v}_j \mathbf{v}_j^*$, $\sin(A) = \sum_{j=1}^{n} \sin(\lambda_j) \mathbf{v}_j \mathbf{v}_j^*$.

Problem 5.   The *spectral theorem* for $n \times n$ normal matrices over $\mathbb{C}$ is as follows: A matrix A is normal if and only if there exists an $n \times n$ unitary matrix U and a diagonal matrix D such that $D = U^* A U$. Use this theorem to prove that the matrix

$$A = \begin{pmatrix} 0 & 1 & 1 \\ 0 & 0 & 1 \\ 0 & 0 & 0 \end{pmatrix}$$

is nonnormal.

Problem 6.   Find all 2×2 matrices over $\mathbb{C}$ which admit the normalized eigenvectors

$$\mathbf{v}_1 = \frac{1}{\sqrt{2}} \begin{pmatrix} 1 \\ 1 \end{pmatrix}, \quad \mathbf{v}_2 = \frac{1}{\sqrt{2}} \begin{pmatrix} 1 \\ -1 \end{pmatrix}$$

with the corresponding eigenvalues λ_1 and λ_2.

Problem 7.   Consider the nonnormal 2×2 matrix

$$A = \begin{pmatrix} 1 & 1 \\ 0 & -1 \end{pmatrix}$$

with eigenvalues $\lambda_1 = +1$ and $\lambda_2 = -1$ and normalized eigenvectors

$$\mathbf{v}_1 = \begin{pmatrix} 1 \\ 0 \end{pmatrix}, \quad \mathbf{v}_2 = \frac{2}{\sqrt{5}} \begin{pmatrix} 1/2 \\ -1 \end{pmatrix}.$$

The eigenvectors are linearly independent, but not orthonormal. Calculate $\mathbf{v}_1^* \mathbf{v}_2$ and

$$B = \lambda_1 \mathbf{v}_1 \mathbf{v}_1^* + \lambda_2 \mathbf{v}_2 \mathbf{v}_2^*.$$

Find $\|A - B\|$. Discuss. Is the matrix B normal?

Chapter 7

Commutators and Anticommutators

Let A and B be $n \times n$ matrices. Then we define the *commutator* of A and B as

$$[A, B] := AB - BA.$$

For all $n \times n$ matrices A, B, C we have the *Jacobi identity*

$$[A, [B, C]] + [C, [A, B]] + [B, [C, A]] = 0_n$$

where 0_n is the $n \times n$ zero matrix. Since $\text{tr}(AB) = \text{tr}(BA)$ we have

$$\text{tr}([A, B]) = 0.$$

If

$$[A, B] = 0_n$$

we say that the matrices A and B *commute*. For example, if A and B are diagonal matrices then the commutator is the zero matrix 0_n. We define the *anticommutator* of A and B as

$$[A, B]_+ := AB + BA.$$

We have

$$\text{tr}([A, B]_+) = 2\text{tr}(AB), \qquad AB \equiv \frac{1}{2}[A, B] + \frac{1}{2}[A, B]_+.$$

The anticommutator plays a role for Fermi operators. Let A, B, C be $n \times n$ matrices. Then

$$[A, [B, C]_+] = [[A, B], C]_+ + [B, [A, C]]_+.$$

Problem 1. (i) Let A, B be $n \times n$ matrices. Assume that $[A, B] = 0_n$ and $[A, B]_+ = 0_n$. What can be said about AB and BA?
(ii) Let A, B be $n \times n$ matrices. Assume that A is invertible. Assume that $[A, B] = 0_n$. Can we conclude that $[A^{-1}, B] = 0_n$?
(iii) Let A, B be $n \times n$ matrices. Suppose that

$$[A, B] = 0_n, \qquad [A, B]_+ = 0_n$$

and that A is invertible. Show that B must be the zero matrix.

Solution 1. (i) Since $AB = BA$ and $AB = -BA$ we find $AB = 0_n$, $BA = 0_n$.
(ii) From $[A, B] = 0_n$ we have $AB = BA$. Thus $B = A^{-1}BA$ and therefore $BA^{-1} = A^{-1}B$. It follows that $A^{-1}B - BA^{-1} = [A^{-1}, B] = 0_n$.
(iii) From the conditions we obtain $AB = 0_n$, $BA = 0_n$. Thus $A^{-1}AB = 0_n$ and therefore $B = 0_n$. Analogously, $BAA^{-1} = 0_n$ also implies $B = 0_n$.

Problem 2. (i) Let A and B be symmetric $n \times n$ matrices over $\mathbb{R}$. Show that AB is symmetric if and only if A and B commute.
(ii) Let A, B be $n \times n$ hermitian matrices. Is $i[A, B]$ hermitian?
(iii) Let A, B be symmetric $n \times n$ matrices over $\mathbb{R}$. Show that $[A, B]$ is skew-symmetric over $\mathbb{R}$.

Solution 2. (i) Suppose that A and B commute, i.e. $AB = BA$. Then

$$(AB)^T = B^T A^T = BA = AB$$

and thus AB is symmetric. Suppose that AB is symmetric, i.e. $(AB)^T = AB$. Then $(AB)^T = B^T A^T = BA$. Hence $AB = BA$ and the matrices A and B commute.
(ii) Since $A^* = A$ and $B = B^*$ we have

$$(i[A, B])^* = (i(AB - BA))^* = (-i(BA - AB)) = i(AB - BA) = i[A, B].$$

Thus $i[A, B]$ is hermitian.
(iii) Using that $A^T = A$ and $B^T = B$ we have

$$([A, B])^T = (AB - BA)^T = (AB)^T - (BA)^T = B^T A^T - A^T B^T = -[A, B].$$

Problem 3. (i) Let A and B be $n \times n$ matrices over $\mathbb{C}$. Show that A and B commute if and only if $A - cI_n$ and $B - cI_n$ commute over every $c \in \mathbb{C}$.
(ii) Let A, B be $n \times n$ matrices over $\mathbb{C}$ with $[A, B] = 0_n$. Calculate

$$[A + cI_n, B + cI_n].$$

(iii) Let $\mathbf{v}$ be an eigenvector of the $n \times n$ matrix A with eigenvalue λ. Show that $\mathbf{v}$ is also an eigenvector of $A + cI_n$, where $c \in \mathbb{C}$.

Solution 3. (i) Suppose that A and B commute, i.e. $AB = BA$. Then

$$(A - cI_n)(B - cI_n) = BA - c(A + B) + c^2 I_n = (B - cI_n)(A - cI_n).$$

Thus $A - cI_n$ and $B - cI_n$ commute. Now suppose that $A - cI_n$ and $B - cI_n$ commute, i.e.

$$(A - cI_n)(B - cI_n) = (B - cI_n)(A - cI_n).$$

Thus $AB - c(A + B) + c^2 I_n = BA - c(A + B) + c^2 I_n$ and $AB = BA$.
(ii) We obtain $[A + cI_n, B + cI_n] = [A, B] + c[A, I_n] + c[I_n, B] + c^2[I_n, I_n] = 0_n$.
(iii) We have $(A + cI_n)\mathbf{v} = A\mathbf{v} + cI_n\mathbf{v} = \lambda\mathbf{v} + c\mathbf{v} = (\lambda + c)\mathbf{v}$. Thus $\mathbf{v}$ is an eigenvector of $A + cI_n$ with eigenvalue $\lambda + c$.

Problem 4. Can one find 2×2 matrices A and B such that $[A^2, B^2] = 0_2$ while $[A, B] \neq 0_2$?

Solution 4. Yes. An example is

$$A = \begin{pmatrix} 0 & 1 \\ 0 & 0 \end{pmatrix}, \quad B = \begin{pmatrix} 0 & 0 \\ 1 & 0 \end{pmatrix} \quad \Rightarrow \quad [A, B] = \begin{pmatrix} 1 & 0 \\ 0 & -1 \end{pmatrix}$$

with $A^2 = 0_2$ and $B^2 = 0_2$.

Problem 5. Let A, B, H be $n \times n$ matrices over $\mathbb{C}$ such that $[A, H] = 0_n$, $[B, H] = 0_n$. Find $[[A, B], H]$.

Solution 5. Using the *Jacobi identity* for arbitrary $n \times n$ matrices X, Y, Z

$$[[X, Y], Z] + [[Z, X], Y] + [[Y, Z], X] = 0_n$$

we obtain $[[A, B], H] = 0_n$.

Problem 6. Let A and B be $n \times n$ hermitian matrices. Suppose that

$$A^2 = I_n, \qquad B^2 = I_n \tag{1}$$

and

$$[A, B]_+ \equiv AB + BA = 0_n. \tag{2}$$

Let $\mathbf{x} \in \mathbb{C}^n$ be normalized, i.e. $\|\mathbf{x}\| = 1$. Here $\mathbf{x}$ is considered as a column vector.
(i) Show that

$$(\mathbf{x}^* A\mathbf{x})^2 + (\mathbf{x}^* B\mathbf{x})^2 \leq 1. \tag{3}$$

(ii) Give an example for the matrices A and B.

Solution 6. (i) Let $a, b \in \mathbb{R}$ and let $r^2 := a^2 + b^2$. The matrix $C = aA + bB$ is again hermitian. Then

$$C^2 = a^2 A^2 + abAB + baBA + b^2 B^2.$$

Using the properties (1) and (2) we find

$$C^2 = a^2 I_n + b^2 I_n = r^2 I_n.$$

Therefore $(\mathbf{x}^*C^2\mathbf{x}) = r^2$ and $-r \leq a(\mathbf{x}^*A\mathbf{x}) + b(\mathbf{x}^*B\mathbf{x}) \leq r$. Let

$$a = \mathbf{x}^*A\mathbf{x}, \qquad b = \mathbf{x}^*B\mathbf{x}$$

then $a^2 + b^2 \leq r$ or $r^2 \leq r$. This implies $r \leq 1$ and $r^2 \leq 1$ from which (3) follows.
(ii) An example is $A = \sigma_1$, $B = \sigma_2$ since $\sigma_1^2 = I_2$, $\sigma_2^2 = I_2$ and $\sigma_1\sigma_2 + \sigma_2\sigma_1 = 0_2$, where σ_1, σ_2, σ_3 are the Pauli spin matrices.

Problem 7. Let A, B be skew-hermitian matrices over $\mathbb{C}$, i.e. $A^* = -A$, $B^* = -B$. Is the commutator of A and B again skew-hermitian?

Solution 7. The answer is yes. We have

$$([A, B])^* = (AB - BA)^* = B^*A^* - A^*B^* = BA - AB = -([A, B]).$$

Problem 8. Let A, B be 2×2 skew-symmetric matrices over $\mathbb{R}$. Find the commutator $[A, B]$.

Solution 8. Since

$$A = \begin{pmatrix} 0 & a_{12} \\ -a_{12} & 0 \end{pmatrix}, \qquad B = \begin{pmatrix} 0 & b_{12} \\ -b_{12} & 0 \end{pmatrix}$$

we find $[A, B] = 0_2$.

Problem 9. (i) Let A, B be $n \times n$ matrices over $\mathbb{C}$. Let S be an invertible $n \times n$ matrix over $\mathbb{C}$ with $\tilde{A} = S^{-1}AS$, $\tilde{B} = S^{-1}BS$. Show that $[\tilde{A}, \tilde{B}] = S^{-1}[A, B]S$.
(ii) Let A, B be $n \times n$ matrices over $\mathbb{C}$. Assume that $[A, B] = 0_n$. Let U be a unitary matrix. Calculate $[U^*AU, U^*BU]$.
(iii) Let T be an invertible matrix. Show that $T^{-1}AT \equiv A + T^{-1}[A, T]$.

Solution 9. (i) Since $SS^{-1} = I_n$ we have

$$\begin{aligned}
[\tilde{A}, \tilde{B}] &= [S^{-1}AS, S^{-1}BS] = S^{-1}ASS^{-1}BS - S^{-1}BSS^{-1}AS \\
&= S^{-1}ABS - S^{-1}BAS = S^{-1}(AB - BA)S \\
&= S^{-1}[A, B]S.
\end{aligned}$$

(ii) Since $U^*U = UU^* = I_n$ we have

$$\begin{aligned}
[U^*AU, U^*BU] &= U^*AUU^*BU - U^*BUU^*AU = U^*ABU - U^*BAU \\
&= U^*([A, B])U = U^*0_nU \\
&= 0_n.
\end{aligned}$$

(iii) We have

$$A + T^{-1}[A, T] = A + T^{-1}(AT - TA) = A + T^{-1}AT - A = T^{-1}AT.$$

Problem 10. Can we find $n \times n$ matrices A, B over $\mathbb{C}$ such that

$$[A, B] = I_n \tag{1}$$

where I_n denotes the identity matrix?

Solution 10. Since $\operatorname{tr}(XY) = \operatorname{tr}(YX)$ for any $n \times n$ matrices X, Y we obtain $\operatorname{tr}([A, B]) = 0$. However for the right-hand side of (1) we find $\operatorname{tr}(I_n) = n$. Thus we have a contradiction and no such matrices exist.

We can find unbounded infinite-dimensional matrices X, Y with $[X, Y] = I$, where I is the infinite-dimensional unit matrix.

Problem 11. Show that any two 2×2 matrices which commute with the matrix

$$\begin{pmatrix} 0 & 1 \\ -1 & 0 \end{pmatrix}$$

commute with each other.

Solution 11. Any 2×2 matrix commuting with this skew-symmetric matrix takes the form

$$\begin{pmatrix} a & b \\ -b & a \end{pmatrix}.$$

Then we have for the commutator

$$\left[\begin{pmatrix} a & b \\ -b & a \end{pmatrix}, \begin{pmatrix} c & d \\ -d & c \end{pmatrix} \right] = \begin{pmatrix} 0 & 0 \\ 0 & 0 \end{pmatrix}.$$

Problem 12. Can we find 2×2 matrices A and B of the form

$$A = \begin{pmatrix} 0 & a_{12} \\ a_{21} & 0 \end{pmatrix}, \qquad B = \begin{pmatrix} 0 & b_{12} \\ b_{21} & 0 \end{pmatrix}$$

and singular (i.e. $\det(A) = 0$ and $\det(B) = 0$) such that $[A, B]_+ = I_2$?

Solution 12. We have

$$AB + BA = \begin{pmatrix} a_{12}b_{21} + a_{21}b_{12} & 0 \\ 0 & a_{12}b_{21} + a_{21}b_{12} \end{pmatrix} = \begin{pmatrix} 1 & 0 \\ 0 & 1 \end{pmatrix}.$$

Thus we obtain the equation $a_{12}b_{21} + a_{21}b_{12} = 1$. Since A and B are singular we obtain the two solutions $a_{12} = b_{21} = 1$, $a_{21} = b_{12} = 0$ and $a_{12} = b_{21} = 0$, $a_{21} = b_{12} = 1$.

Problem 13. Let A be an $n \times n$ hermitian matrix over $\mathbb{C}$. Assume that the eigenvalues of A, λ_1, ..., λ_n are nondegenerate and that the normalized eigenvectors $\mathbf{v}_j$ ($j = 1, \ldots, n$) of A form an orthonormal basis in $\mathbb{C}^n$. Let B be an $n \times n$ matrix over $\mathbb{C}$. Assume that $[A, B] = 0_n$. Show that

$$\mathbf{v}_k^* B \mathbf{v}_j = 0 \quad \text{for} \quad k \neq j. \tag{1}$$

Solution 13.   From $AB = BA$ it follows that $\mathbf{v}_k^*(AB\mathbf{v}_j) = \mathbf{v}_k^*(BA\mathbf{v}_j)$. The eigenvalues of a hermitian matrix are real. Since $A\mathbf{v}_j = \lambda_j \mathbf{v}_j$ and $\mathbf{v}_k^* A = \lambda_k \mathbf{v}_k^*$ we obtain

$$\lambda_k(\mathbf{v}_k^* B\mathbf{v}_j) = \lambda_j(\mathbf{v}_k^* B\mathbf{v}_j).$$

Consequently $(\lambda_k - \lambda_j)(\mathbf{v}_k^* B\mathbf{v}_j) = 0$. Since $\lambda_k \neq \lambda_j$ equation (1) follows.

Problem 14.   Let A, B be hermitian $n \times n$ matrices. Assume they have the same set of eigenvectors

$$A\mathbf{v}_j = \lambda_j \mathbf{v}_j, \quad B\mathbf{v}_j = \mu_j \mathbf{v}_j, \quad j = 1, \ldots, n$$

and that the normalized eigenvectors form an orthonormal basis in $\mathbb{C}^n$. Show that $[A, B] = 0_n$.

Solution 14.   Any vector $\mathbf{v}$ in $\mathbb{C}^n$ can be written as

$$\mathbf{v} = \sum_{j=1}^n (\mathbf{v}^* \mathbf{v}_j)\mathbf{v}_j.$$

It follows that

$$[A, B]\mathbf{v} = (AB - BA) \sum_{j=1}^n (\mathbf{v}^* \mathbf{v}_j)\mathbf{v}_j = \sum_{j=1}^n (\mathbf{v}^* \mathbf{v}_j)AB\mathbf{v}_j - \sum_{j=1}^n (\mathbf{v}^* \mathbf{v}_j)BA\mathbf{v}_j$$

$$= \sum_{j=1}^n (\mathbf{v}^* \mathbf{v}_j)\lambda_j \mu_j \mathbf{v}_j - \sum_{j=1}^n (\mathbf{v}^* \mathbf{v}_j)\mu_j \lambda_j \mathbf{v}_j$$

$$= 0.$$

Since this is true for an arbitrary vector $\mathbf{v}$ in $\mathbb{C}^n$ it follows that $[A, B] = 0_n$.

Problem 15.   Let A, B be $n \times n$ matrices. Then we have the expansion

$$e^A B e^{-A} = B + [A, B] + \frac{1}{2!}[A, [A, B]] + \frac{1}{3!}[A, [A, [A, B]]] + \cdots .$$

(i) Assume that $[A, B] = A$. Calculate $e^A B e^{-A}$.
(ii) Assume that $[A, B] = B$. Calculate $e^A B e^{-A}$.

Solution 15.   (i) From $[A, B] = A$ we obtain $[A, [A, B]] = [A, A] = 0_n$. Thus $e^A B e^{-A} = B + A$.
(ii) From $[A, B] = B$ we obtain $[A, [A, B]] = B$, $[A, [A, [A, B]]] = B$ etc. Thus

$$e^A B e^{-A} = B + B + \frac{1}{2!}B + \frac{1}{3!}B + \cdots = B\left(1 + 1 + \frac{1}{2!} + \frac{1}{3!} + \cdots\right) = eB.$$

Problem 16.   Let A be an arbitrary $n \times n$ matrix over $\mathbb{C}$ with $\mathrm{tr}(A) = 0$. Show that A can be written as commutator, i.e. there are $n \times n$ matrices X and Y such that $A = [X, Y]$.

Solution 16. The statement is valid if A is 1×1 or a larger $n \times n$ zero matrix. Assume that A is a nonzero $n \times n$ matrix of dimension larger than 1. We use induction. We assume the desired inference valid for all matrices of dimensions smaller than A's with trace zero. Since $\text{tr}(A) = 0$ the matrix A cannot be a nonzero scalar multiple of I_n. Thus there is some invertible matrix R such that

$$R^{-1}AR = \begin{pmatrix} 0 & d^T \\ c & B \end{pmatrix}$$

with $\text{tr}(B) = 0$. The induction hypothesis implies that B is a commutator. Thus $R^{-1}AR = XY - YX$ is also a commutator for some $n \times n$ matrices X and Y. It follows that

$$A = (RXR^{-1})(RYR^{-1}) - (RYR^{-1})(RXR^{-1})$$

must also be a commutator.

Problem 17. Let A, B be 2×2 symmetric matrices over $\mathbb{R}$. Assume that $AA^T = I_2$ and $BB^T = I_2$. Is $[A, B] = 0_2$? Prove or disprove.

Solution 17. Since $[AA^T, BB^T] = [I_2, I_2] = 0_2$ we obtain $[A, B] = 0_2$.

Problem 18. Let A, B, X, Y be $n \times n$ matrices over $\mathbb{C}$. Assume that $AX - XB = Y$. Let $z \in \mathbb{C}$. Show that $(A - zI_n)X - X(B - zI_n) = Y$.

Solution 18. We have

$$(A - zI_n)X - X(B - zI_n) = AX - zX - XB + zX = AX - XB = Y.$$

Problem 19. (i) Can we find nonzero symmetric 2×2 matrices H and A over $\mathbb{R}$ such that

$$[H, A] = \mu A$$

where $\mu \in \mathbb{R}$ and $\mu \neq 0$?
(ii) Let K be a nonzero $n \times n$ hermitian matrix. Let B be a nonzero $n \times n$ matrix. Assume that

$$[K, B] = aB$$

where $a \in \mathbb{R}$ and $a \neq 0$. Show that B cannot be hermitian.

Solution 19. (i) Since $\text{tr}([H, A]) = 0$ and $\mu \neq 0$ we find that $\text{tr}(A) = 0$. Thus A can be written as

$$A = \begin{pmatrix} a_{11} & a_{12} \\ a_{12} & -a_{11} \end{pmatrix}.$$

Now

$$[H, A] = \begin{pmatrix} 0 & a_{12}(h_{11} - h_{22}) \\ a_{12}(h_{22} - h_{11}) & 0 \end{pmatrix} = (h_{11} - h_{22}) \begin{pmatrix} 0 & a_{12} \\ -a_{12} & 0 \end{pmatrix}.$$

It follows that such a pair does not exist since A should be symmetric over $\mathbb{R}$ and nonzero.

(ii) Assume that B is hermitian. Then from $[K, B] = aB$ we obtain by taking the transpose and complex conjugate of this equation

$$[B, K] = aB.$$

Adding the two equations $[K, B] = aB$, $[B, K] = aB$ we obtain $2aB = 0_n$. Since B is a nonzero matrix we have $a = 0$. Thus we have a contradiction and B cannot be hermitian.

Problem 20. A *truncated Bose annihilation operator* is defined as the $n \times n$ $(n \geq 2)$ matrix

$$B_n = \begin{pmatrix} 0 & \sqrt{1} & 0 & 0 & 0 & \cdots & 0 \\ 0 & 0 & \sqrt{2} & 0 & 0 & \cdots & 0 \\ 0 & 0 & 0 & \sqrt{3} & 0 & \cdots & 0 \\ \vdots & \vdots & \vdots & \ddots & \vdots & \vdots & \vdots \\ 0 & 0 & 0 & 0 & \cdots & 0 & \sqrt{n-1} \\ 0 & 0 & 0 & 0 & \cdots & 0 & 0 \end{pmatrix}.$$

(i) Calculate $B_n^* B_n$.

(ii) Calculate the commutator $[B_n, B_n^*]$.

Solution 20. (i) We find the diagonal matrix $B_n^* B_n = \mathrm{diag}(0, 1, \ldots, n-1)$.

(ii) We obtain the diagonal matrix

$$[B_n, B_n^*] = B_n B_n^* - B_n^* B_n = \mathrm{diag}(1, 1, \ldots, 1, -(n-1)).$$

Problem 21. Find nonzero 2×2 matrices A, B such that $[A, B] \neq 0_2$, but

$$[A, [A, B]] = 0_2, \qquad [B, [A, B]] = 0_2.$$

Solution 21. For example if $[A, B] = A$, then $[A, [A, B]] = 0_2$ and $[B, [A, B]] = 0_2$. An example is

$$A = \begin{pmatrix} 0 & 1 \\ 0 & 0 \end{pmatrix}, \qquad B = \begin{pmatrix} 0 & 0 \\ 0 & 1 \end{pmatrix}.$$

Problem 22. Let

$$C = \begin{pmatrix} 0 & 1 \\ 1 & 0 \end{pmatrix}.$$

Find all 2×2 matrices A such that $[A, C] = 0_2$.

Solution 22. The commutator is the zero matrix if and only if $a_{11} = a_{22}$ and $a_{12} = a_{21}$.

Problem 23. Let A and B be positive semidefinite matrices. Can we conclude that $[A, B]_+$ is positive semidefinite?

Solution 23. No. Consider

$$A = \begin{pmatrix} 1 & 1 \\ 1 & 1 \end{pmatrix}, \quad B = \begin{pmatrix} 1 & 0 \\ 0 & 0 \end{pmatrix} \Rightarrow AB + BA = \begin{pmatrix} 2 & 1 \\ 1 & 0 \end{pmatrix}.$$

Both matrices A and B are positive semidefinite. However $[A, B]_+$ admits the eigenvalues $\lambda_1 = 1 + \sqrt{2}$ and $\lambda_2 = 1 - \sqrt{2}$. Thus $AB + BA$ is not positive semidefinite.

Problem 24. Let A, B be $n \times n$ matrices. Given the expression

$$A^2B + AB^2 + B^2A + BA^2 - 2ABA - 2BAB.$$

Write the expression in a more compact form using commutators.

Solution 24. We have

$$A^2B + AB^2 + B^2A + BA^2 - 2ABA - 2BAB = [A, [A, B]] + [B, [B, A]].$$

Problem 25. (i) Let A, B be $n \times n$ matrices over $\mathbb{C}$. Assume that $A^2 = I_n$ and $B^2 = I_n$. Find the commutators $[AB + BA, A]$, $[AB + BA, B]$. Give an example of such matrices for $n = 2$ and $A \neq B$.
(ii) Let A, B be $n \times n$ matrices over $\mathbb{C}$ such that $A^2 = I_n$, $B^2 = I_n$. Now assume that $[A, B]_+ = 0_n$. Show that there is no solution for A and B if n is odd.

Solution 25. (i) We find

$$[AB + BA, A] = ABA + BA^2 - A^2B - ABA = BA^2 - A^2B = B - B = 0_n.$$

Analogously we find $[AB + BA, B] = 0_n$. As an example consider the Pauli spin matrices $A = \sigma_1$, $B = \sigma_3$.
(ii) The proof follows from $\det(A)\det(B) = (-1)^n \det(A)\det(B)$ and $\det(A) \neq 0$, $\det(B) \neq 0$.

Problem 26. Let A_1, A_2, A_3 be $n \times n$ matrices over $\mathbb{C}$. The *ternary commutator* $[A_1, A_2, A_3]$ (also called the *ternutator*) is defined as

$$[A_1, A_2, A_3] := \sum_{\pi \in S_3} \text{sgn}(\pi) A_{\pi(1)} A_{\pi(2)} A_{\pi(3)}$$

$$\equiv A_1A_2A_3 + A_2A_3A_1 + A_3A_1A_2 - A_1A_3A_2 - A_2A_1A_3 - A_3A_2A_1.$$

Let $n = 2$ and consider the Pauli spin matrices σ_1, σ_2, σ_3. Calculate the ternutator $[\sigma_1, \sigma_2, \sigma_3]$.

Solution 26. We obtain

$$[\sigma_1, \sigma_2, \sigma_3] = \begin{pmatrix} 6i & 0 \\ 0 & 6i \end{pmatrix}.$$

Note that $\text{tr}([\sigma_1, \sigma_2, \sigma_3]) = 12i$.

Problem 27.   Let $\oplus$ be the *direct sum*.
(i) Let A, B, H be $n \times n$ matrices such that $[H, A] = 0_n$, $[H, B] = 0_n$. Show that

$$[H \oplus I_n + I_n \oplus H, A \oplus B] = 0_{2n}.$$

(ii) Let A_1, A_2 be $m \times m$ matrices over $\mathbb{C}$. Let B_1, B_2 be $n \times n$ matrices over $\mathbb{C}$. Show that

$$[A_1 \oplus B_1, A_2 \oplus B_2] = ([A_1, A_2]) \oplus ([B_1, B_2])$$

Solution 27.   (i) We have

$$
\begin{aligned}
[H \oplus I_n + I_n \oplus H, A \oplus B] &= [H \oplus I_n, A \oplus B] + [I_n \oplus H, A \oplus B] \\
&= (HA) \oplus B - (AH) \oplus B + A \oplus (HB) - A \oplus (BH) \\
&= [H, A] \oplus B + A \oplus [H, B] \\
&= 0_{2n}.
\end{aligned}
$$

(ii) We have

$$
\begin{aligned}
[A_1 \oplus B_1, A_2 \oplus B_2] &= (A_1 A_2) \oplus (B_1 B_2) - (A_2 A_1) \oplus (B_2 B_1) \\
&= (A_1 A_2 - A_2 A_1) \oplus (B_1 B_2 - B_2 B_1) \\
&= ([A_1, A_2]) \oplus ([B_1, B_2]).
\end{aligned}
$$

Problem 28.   (i) Can one find non-invertible 2×2 matrices A and B such the commutator $[A, B]$ is invertible?
(ii) Consider the 3×3 matrices

$$
A = \begin{pmatrix} a_1 & 0 & 0 \\ 0 & a_2 & 0 \\ 0 & 0 & a_3 \end{pmatrix}, \quad
B = \begin{pmatrix} 0 & 0 & b_1 \\ 0 & b_2 & 0 \\ b_3 & 0 & 0 \end{pmatrix}.
$$

Can we find a_j, b_j ($j = 1, 2, 3$) such that the commutator $[A, B]$ is invertible?

Solution 28.   (i) Yes. An example is

$$
A = \begin{pmatrix} 1 & 0 \\ 0 & 0 \end{pmatrix}, \quad
B = \begin{pmatrix} 1 & 1 \\ 1 & 1 \end{pmatrix} \quad \Rightarrow \quad
[A, B] = \begin{pmatrix} 0 & 1 \\ -1 & 0 \end{pmatrix}.
$$

(ii) We obtain

$$
[A, B] = \begin{pmatrix} 0 & 0 & b_1(a_1 - a_3) \\ 0 & 0 & 0 \\ b_3(a_3 - a_1) & 0 & 0 \end{pmatrix}.
$$

Thus $[A, B]$ is not invertible.

Problem 29. Let A, B be $n \times n$ matrices over $\mathbb{C}$. Assume that B is invertible. Show that

$$[A, B^{-1}] \equiv -B^{-1}[A, B]B^{-1}.$$

Solution 29. We have

$$-B^{-1}[A, B]B^{-1} = -B^{-1}(AB - BA)B^{-1} = -B^{-1}A + AB^{-1} = [A, B^{-1}].$$

Problem 30. Consider the 3×3 matrices over $\mathbb{C}$

$$A = \begin{pmatrix} a_{11} & 0 & 0 \\ 0 & a_{22} & 0 \\ 0 & 0 & a_{33} \end{pmatrix}, \qquad B = \begin{pmatrix} 0 & b_{12} & b_{13} \\ b_{21} & 0 & b_{23} \\ b_{31} & b_{32} & 0 \end{pmatrix}.$$

(i) Calculate the commutator $[A, B]$ and $\det([A, B])$.
(ii) Set $a_{11} = e^{i\phi_1}$, $a_{22} = e^{i\phi_2}$, $a_{33} = e^{i\phi_3}$. Find the condition on ϕ_1, ϕ_2, ϕ_3, b_{12}, b_{13}, b_{21}, b_{23}, b_{31}, b_{32} such that $[A, B]$ is unitary.

Solution 30. (i) We obtain

$$[A, B] = \begin{pmatrix} 0 & b_{12}(a_{11} - a_{22}) & b_{13}(a_{11} - a_{33}) \\ b_{21}(a_{22} - a_{11}) & 0 & b_{23}(a_{22} - a_{33}) \\ b_{31}(a_{33} - a_{11}) & b_{32}(a_{33} - a_{22}) & 0 \end{pmatrix}.$$

Thus

$$\det([A, B]) = (b_{12}b_{23}b_{31} - b_{13}b_{21}b_{32})(a_{11} - a_{22})(a_{22} - a_{33})(a_{11} - a_{33}).$$

(ii) Setting $\phi_0 = 0$, $\phi_1 = 2\pi/3$, $\phi_2 = 4\pi/3$ with

$$\cos(2\pi/3) = -\frac{1}{2}, \ \sin(2\pi/3) = \frac{1}{2}\sqrt{3}, \ \cos(4\pi/3) = -\frac{1}{2}, \ \sin(4\pi/3) = -\frac{1}{2}\sqrt{3}$$

we have $a_{11} + a_{22} + a_{33} = 0$. Thus

$$(a_{11} - a_{22})(a_{22} - a_{33})(a_{11} - a_{33}) = \frac{1}{2}(3 - i\sqrt{3})i\sqrt{3}\frac{1}{2}(3 + i\sqrt{3}).$$

Problem 31. Consider the set of six 3×3 matrices

$$A_1 = \begin{pmatrix} 1 & 0 & 0 \\ 0 & 0 & 0 \\ 0 & 0 & 0 \end{pmatrix}, \quad A_2 = \begin{pmatrix} 0 & 0 & 0 \\ 0 & 1 & 0 \\ 0 & 0 & 0 \end{pmatrix}, \quad A_3 = \begin{pmatrix} 0 & 0 & 0 \\ 0 & 0 & 0 \\ 0 & 0 & 1 \end{pmatrix},$$

$$A_{12} = \begin{pmatrix} 0 & 1 & 0 \\ 1 & 0 & 0 \\ 0 & 0 & 0 \end{pmatrix}, \quad A_{23} = \begin{pmatrix} 0 & 0 & 0 \\ 0 & 0 & 1 \\ 0 & 1 & 0 \end{pmatrix}, \quad A_{13} = \begin{pmatrix} 0 & 0 & 1 \\ 0 & 0 & 0 \\ 1 & 0 & 0 \end{pmatrix}.$$

Calculate the anticommutator and thus show that we have a basis of a *Jordan algebra*.

Solution 31. We obtain

$$[A_1, A_2]_+ = [A_1, A_3]_+ = [A_2, A_3]_+ = 0_3$$

$$[A_1, A_1]_+ = 2A_1, \quad [A_2, A_2]_+ = 2A_2, \quad [A_3, A_3]_+ = 2A_3$$

$$[A_{12}, A_{23}]_+ = A_{13}, \quad [A_{12}, A_{23}]_+ = A_{23}, \quad [A_{23}, A_{13}]_+ = A_{12}$$

$$[A_1, A_{12}]_+ = A_{12}, \quad [A_1, A_{23}]_+ = 0_3, \quad [A_1, A_{13}]_+ = A_{13}$$

$$[A_2, A_{12}]_+ = A_{12}, \quad [A_2, A_{23}]_+ = A_{23}, \quad [A_2, A_{13}]_+ = 0_3$$

$$[A_3, A_{12}]_+ = 0_3, \quad [A_3, A_{23}]_+ = A_{23}, \quad [A_3, A_{13}]_+ = A_{13}.$$

Problem 32. Let A, B be hermitian matrices, i.e. $A^* = A$ and $B^* = B$. Then in general $A + iB$ is nonnormal. What are the conditions on A and B such that $A + iB$ is normal?

Solution 32. From $(A + iB)^*(A + iB) = (A + iB)(A + iB)^*$ we find that the commutator of A and B must vanish, i.e. $[A, B] = 0_n$.

Problem 33. Show that one can find a 3×3 matrix over $\mathbb{R}$ such that

$$A^2 A^T + A^T A^2 = 2A, \quad AA^T A = 2A, \quad A^3 = 0_3$$

and

$$\frac{1}{2}[A^T, A] = \begin{pmatrix} 1 & 0 & 0 \\ 0 & 0 & 0 \\ 0 & 0 & -1 \end{pmatrix}.$$

Solution 33. We obtain the nonnormal matrix

$$A = \sqrt{2} \begin{pmatrix} 0 & 0 & 0 \\ 1 & 0 & 0 \\ 0 & 1 & 0 \end{pmatrix} \Rightarrow A^T = \sqrt{2} \begin{pmatrix} 0 & 1 & 0 \\ 0 & 0 & 1 \\ 0 & 0 & 0 \end{pmatrix}.$$

Problem 34. Consider the 3×3 matrices

$$D = \begin{pmatrix} d_1 & 0 & 0 \\ 0 & d_2 & 0 \\ 0 & 0 & d_3 \end{pmatrix}, \quad M = \begin{pmatrix} 0 & a & 0 \\ a & 0 & b \\ 0 & b & 0 \end{pmatrix}.$$

Find the conditions on d_1, d_2, d_3, a, b such that

$$[D, M] = \begin{pmatrix} 0 & a & 0 \\ -a & 0 & b \\ 0 & -b & 0 \end{pmatrix}.$$

Solution 34. We have

$$[D, M] = \begin{pmatrix} 0 & a(d_1 - d_2) & 0 \\ a(d_2 - d_1) & 0 & b(d_2 - d_3) \\ 0 & b(d_3 - d_2) & 0 \end{pmatrix}.$$

Hence $d_1 - d_2 = 1$ and $d_2 - d_3 = 1$.

Problem 35. (i) Consider the Pauli spin matrix σ_1 with eigenvalues $+1$ and -1 and corresponding normalized eigenvectors

$$\frac{1}{\sqrt{2}} \begin{pmatrix} 1 \\ 1 \end{pmatrix}, \quad \frac{1}{\sqrt{2}} \begin{pmatrix} 1 \\ -1 \end{pmatrix}.$$

Find all 2×2 matrices A such that $[\sigma_1, A] = 0_2$. Find the eigenvalues and eigenvectors of A. Discuss.
(ii) Consider the counter-diagonal unit matrix

$$J_3 = \begin{pmatrix} 0 & 0 & 1 \\ 0 & 1 & 0 \\ 1 & 0 & 0 \end{pmatrix}$$

with eigenvalues $+1$ (twice) and -1 and the corresponding normalized eigenvectors

$$\frac{1}{\sqrt{2}} \begin{pmatrix} 1 \\ 0 \\ 1 \end{pmatrix}, \quad \begin{pmatrix} 0 \\ 1 \\ 0 \end{pmatrix}, \quad \frac{1}{\sqrt{2}} \begin{pmatrix} 1 \\ 0 \\ -1 \end{pmatrix}.$$

Find all 3×3 matrices A such that $[J_3, A] = 0_3$. Find the eigenvalues and normalized eigenvectors of A and compare with the eigenvectors of J_3. Discuss.
(iii) Consider the 4×4 matrix (counter-diagonal unit matrix)

$$J_4 = \begin{pmatrix} 0 & 0 & 0 & 1 \\ 0 & 0 & 1 & 0 \\ 0 & 1 & 0 & 0 \\ 1 & 0 & 0 & 0 \end{pmatrix} = \begin{pmatrix} 0 & 1 \\ 1 & 0 \end{pmatrix} \otimes \begin{pmatrix} 0 & 1 \\ 1 & 0 \end{pmatrix} = \sigma_1 \otimes \sigma_1$$

with the eigenvalues $+1$ (twice) and -1 (twice) and the corresponding non-entangled eigenvectors

$$\frac{1}{\sqrt{2}} \begin{pmatrix} 1 \\ 1 \end{pmatrix} \otimes \frac{1}{\sqrt{2}} \begin{pmatrix} 1 \\ 1 \end{pmatrix}, \quad \frac{1}{\sqrt{2}} \begin{pmatrix} 1 \\ -1 \end{pmatrix} \otimes \frac{1}{\sqrt{2}} \begin{pmatrix} 1 \\ -1 \end{pmatrix},$$

$$\frac{1}{\sqrt{2}} \begin{pmatrix} 1 \\ 1 \end{pmatrix} \otimes \frac{1}{\sqrt{2}} \begin{pmatrix} 1 \\ -1 \end{pmatrix}, \quad \frac{1}{\sqrt{2}} \begin{pmatrix} 1 \\ -1 \end{pmatrix} \otimes \frac{1}{\sqrt{2}} \begin{pmatrix} 1 \\ 1 \end{pmatrix}.$$

Find entangled eigenvectors. Find all 4×4 matrices A such that $[A, J_4] \equiv AJ_4 - J_4A = 0_4$.
(iv) Let J_n be the $n \times n$ counter unit matrix. Let A be an $n \times n$ matrix with $[J_n, A] = 0_n$. Show that $[J_n \otimes J_n, A \otimes A] = 0_{n^2}$.

Solution 35. (i) From $[\sigma_1, A] = 0_2$ we obtain $a_{11} = a_{22}$ and $a_{12} = a_{21}$, i.e.

$$A = \begin{pmatrix} a_{11} & a_{12} \\ a_{12} & a_{11} \end{pmatrix} = a_{11}I_2 + a_{12}J_2.$$

The eigenvalues are $a_{11} + a_{12}$ and $a_{11} - a_{12}$ with the corresponding normalized eigenvectors are given by σ_1.
(ii) We find the matrix

$$A = \begin{pmatrix} a_{13} & a_{12} & a_{11} \\ a_{21} & a_{22} & a_{21} \\ a_{11} & a_{12} & a_{13} \end{pmatrix}$$

with $a_{11}, a_{12}, a_{13}, a_{21}$ arbitrary. The three eigenvalues are $a_{13} - a_{11}$

$$\frac{1}{2}\left(\sqrt{8a_{12}a_{21} + a_{13}^2 + 2a_{11}a_{12} + a_{11}^2} + a_{13} + a_{11}\right),$$

$$-\frac{1}{2}\left(\sqrt{8a_{12}a_{21} + a_{13}^2 + 2a_{11}a_{12} + a_{11}^2} - a_{13} - a_{11}\right).$$

The matrix A and J_3 have the eigenvector

$$\frac{1}{\sqrt{2}}\begin{pmatrix} 1 \\ 0 \\ -1 \end{pmatrix}$$

in common.
(iii) By linear combinations we obtain the *Bell basis* as eigenvectors

$$\frac{1}{\sqrt{2}}\begin{pmatrix} 1 \\ 0 \\ 0 \\ 1 \end{pmatrix}, \quad \frac{1}{\sqrt{2}}\begin{pmatrix} 0 \\ 1 \\ 1 \\ 0 \end{pmatrix}, \quad \frac{1}{\sqrt{2}}\begin{pmatrix} 1 \\ 0 \\ 0 \\ -1 \end{pmatrix}, \quad \frac{1}{\sqrt{2}}\begin{pmatrix} 0 \\ 1 \\ -1 \\ 0 \end{pmatrix}.$$

From $[A, J_4] = 0_4$ we obtain the matrix

$$\begin{pmatrix} a_{11} & a_{12} & a_{13} & a_{14} \\ a_{21} & a_{22} & a_{23} & a_{24} \\ a_{24} & a_{23} & a_{22} & a_{21} \\ a_{14} & a_{13} & a_{12} & a_{11} \end{pmatrix}$$

with $a_{11}, a_{12}, a_{13}, a_{14}, a_{21}, a_{22}, a_{23}, a_{24}$ arbitrary. What common eigenvectors do J_4 and A have? Find the corresponding eigenvalue.
(iv) We have

$$[J_n \otimes J_n, A \otimes A] = (J_n \otimes J_n)(A \otimes A) - (A \otimes A)(J_n \otimes J_n)$$
$$= (J_nA) \otimes (J_nA) - (AJ_n) \otimes (AJ_n)$$
$$= (AJ_n) \otimes (AJ_n) - (AJ_n) \otimes (AJ_n)$$
$$= 0_{2^n}.$$

Problem 36. (i) Let M, A, B be $n \times n$ matrices over $\mathbb{C}$. Assume that

$$[M, A] = 0_n, \qquad [M, B] = 0_n.$$

Calculate the commutator $[M \otimes I_n + I_n \otimes M, A \otimes B]$.

(ii) Let A, B be $n \times n$ matrices over $\mathbb{C}$. Calculate the commutator $[A \otimes I_n + I_n \otimes A, B \otimes B]$. Assume that $[A, B] = 0_n$.

(iii) Find the commutator

$$[A \otimes B + B \otimes A, A \otimes A - B \otimes B].$$

Simplify the result for $[A, B] = 0_n$. Simplify the result for $A^2 = B^2 = 0_n$. Simplify the result for $A^2 = B^2 = I_n$.

Solution 36. (i) We have

$$[M \otimes I_n + I_n \otimes M, A \otimes B] = [M, A] \otimes B + A \otimes [M, B] = 0_n.$$

(iI) We have

$$[A \otimes I_n + I_n \otimes A, B \otimes B] = [A, B] \otimes B + B \otimes [A, B].$$

It follows that $[A \otimes I_n + I_n \otimes A, B \otimes B] = 0_{n^2}$.

(iii) We have

$$[A \otimes B + B \otimes A, A \otimes A - B \otimes B] = (A^2 + B^2) \otimes [B, A] + [B, A] \otimes (A^2 + B^2).$$

If $[A, B] = 0_n$ the commutator is the $n^2 \times n^2$ zero matrix. If $A^2 = B^2 = 0_n$ the commutator is the $n^2 \times n^2$ zero matrix. We find

$$2I_n \otimes [B, A] + 2[B, A] \otimes I_n.$$

Problem 37. The four *Dirac matrices* which play a role in the Dirac equation are given by

$$\gamma_1 = \begin{pmatrix} 0 & 0 & 0 & -i \\ 0 & 0 & -i & 0 \\ 0 & i & 0 & 0 \\ i & 0 & 0 & 0 \end{pmatrix}, \quad \gamma_2 = \begin{pmatrix} 0 & 0 & 0 & -1 \\ 0 & 0 & 1 & 0 \\ 0 & 1 & 0 & 0 \\ -1 & 0 & 0 & 0 \end{pmatrix},$$

$$\gamma_3 = \begin{pmatrix} 0 & 0 & -i & 0 \\ 0 & 0 & 0 & i \\ i & 0 & 0 & 0 \\ 0 & -i & 0 & 0 \end{pmatrix}, \quad \gamma_4 = \begin{pmatrix} 1 & 0 & 0 & 0 \\ 0 & 1 & 0 & 0 \\ 0 & 0 & -1 & 0 \\ 0 & 0 & 0 & -1 \end{pmatrix}.$$

(i) Show that the matrices can be written as Kronecker product of the Pauli spin matrices σ_1, σ_2, σ_3 and the 2×2 identity matrix I_2.

(ii) Calculate the anticommutators $[\gamma_4, \gamma_1]_+$, $[\gamma_4, \gamma_2]_+$, $[\gamma_4, \gamma_3]_+$.

Solution 37. (i) We find $\gamma_1 = \sigma_2 \otimes \sigma_1$, $\gamma_2 = \sigma_2 \otimes \sigma_2$, $\gamma_3 = \sigma_2 \otimes \sigma_3$, $\gamma_4 = \sigma_3 \otimes I_2$. Find the eigenvalues of γ_1, γ_2, γ_3, γ_4 using this result.

(ii) We obtain for the anticommutators

$$[\gamma_4, \gamma_1]_+ = 0_4, \qquad [\gamma_4, \gamma_2]_+ = 0_4, \qquad [\gamma_4, \gamma_3]_+ = 0_4.$$

The matrix γ_4 also satisfies $\gamma_4^2 = I_4$. The matrix γ_4 is called a *chirality matrix*.

Problem 38.   Let A, B be $n \times n$ matrices over $\mathbb{C}$. What is the condition on A, B such that

$$[A \otimes B, B \otimes A] = 0_{n^2}.$$

Solution 38.   We have

$$[A \otimes B, B \otimes A] = (AB) \otimes (BA) - (BA) \otimes (AB).$$

Thus if $AB = BA$ (i.e. A and B commute) we obtain the zero matrix. This is sufficient, but is it also necessary?

Problem 39.   Let A, B be $n \times n$ matrices over $\mathbb{C}$ with commutator $[A, B]$. Let R be an $n \times n$ matrix over $\mathbb{C}$ with $R^2 = I_n$. Find the commutator $[A \otimes R, B \otimes R]$.

Solution 39.   We have

$$\begin{aligned}
[A \otimes R, B \otimes R] &= (AB) \otimes I_n - (BA) \otimes I_n = (AB - BA) \otimes I_n \\
&= [A, B] \otimes I_n.
\end{aligned}$$

Problem 40.   Let A_1, A_2, A_3 be $m \times m$ matrices. Let B_1, B_2, B_3 be $n \times n$ matrices. Consider the $(n \cdot m) \times (n \cdot m)$ matrices

$$X_1 = A_1 \otimes I_n + I_m \otimes B_1, \quad X_2 = A_2 \otimes I_n + I_m \otimes B_2, \quad X_3 = A_3 \otimes I_n + I_m \otimes B_3.$$

(i) Calculate the commutators and show that the results can again be expressed using commutators.
(ii) Assume that

$$[A_1, A_2] = A_3, \quad [A_2, A_3] = A_1, \quad [A_3, A_1] = A_2$$

and

$$[B_1, B_2] = B_3, \quad [B_2, B_3] = B_1, \quad [B_3, B_1] = B_2.$$

Use these commutation relations to simplify the results of (i).

Solution 40.   (i) We have

$$\begin{aligned}
[X_1, X_2] &= (A_1 \otimes I_n + I_m \otimes B_1)(A_2 \otimes I_n + I_m \otimes B_2) \\
&= (A_1 A_2) \otimes I_n - (A_2 A_1) \otimes I_n + I_m \otimes (B_1 B_2) - I_m \otimes (B_2 B_1) \\
&= [A_1, A_2] \otimes I_n + I_m \otimes [B_1, B_2].
\end{aligned}$$

Analogously we have

$$[X_2, X_3] = [A_2, A_3] \otimes I_n + I_m \otimes [B_2, B_3]$$

$$[X_3, X_1] = [A_3, A_1] \otimes I_n + I_m \otimes [B_3, B_1].$$

(ii) Using the commutation relations we obtain

$$[X_1, X_2] = A_3 \otimes I_n + I_m \otimes B_3 = X_3$$
$$[X_2, X_3] = A_1 \otimes I_n + I_m \otimes B_1 = X_1$$
$$[X_3, X_1] = A_2 \otimes I_n + I_m \otimes B_2 = X_2.$$

Problem 41. Let A, B be $n \times n$ matrices over $\mathbb{C}$. Assume that $[A, B] = 0_n$. Can we conclude that $A \otimes B - B \otimes A = 0_{n^2}$?

Solution 41. Obviously not in general. For example let

$$A = \begin{pmatrix} 1 & 0 \\ 0 & 1 \end{pmatrix}, \qquad B = \begin{pmatrix} 1 & 2 \\ 3 & 4 \end{pmatrix}.$$

Then $A \otimes B \neq B \otimes A$.

Problem 42. Let A be a 2×2 matrix with $A^2 = I_2$. Find the commutator and anticommutator of

$$X = A \otimes \begin{pmatrix} 0 & 1 \\ 0 & 0 \end{pmatrix}, \quad Y = A \otimes \begin{pmatrix} 0 & 0 \\ 1 & 0 \end{pmatrix}.$$

Solution 42. We obtain

$$[X, Y] = I_2 \otimes \sigma_3, \qquad [X, Y]_+ = A^2 \otimes \begin{pmatrix} 1 & 0 \\ 0 & 1 \end{pmatrix} = I_2 \otimes I_2.$$

Problem 43. Let A, B be $n \times n$ matrices over $\mathbb{C}$. We define the quasi-multiplication

$$A \circ B := \frac{1}{2}(AB + BA).$$

Obviously $A \circ B = B \circ A$. Show that $(A^2 \circ B) \circ A = A^2 \circ (B \circ A)$. This is called the *Jordan identity*.

Solution 43. We have

$$4(A^2 \circ B) \circ A = (A^2 B + BA^2)A + A(A^2 B + BA^2)$$
$$= A^2 BA + BA^3 + A^3 B + ABA^2$$
$$= A^2(BA + AB) + (BA + AB)A^2$$
$$= 4A^2 \circ (B \circ A).$$

Supplementary Problems

Problem 1. Consider the 2×2 matrices

$$A(\alpha) = \begin{pmatrix} \cos(\alpha) & -\sin(\alpha) \\ \sin(\alpha) & \cos(\alpha) \end{pmatrix}, \quad B(\beta) = \begin{pmatrix} \cosh(\beta) & \sinh(\beta) \\ \sinh(\beta) & \cosh(\beta) \end{pmatrix}$$

where $\alpha, \beta \in \mathbb{R}$. Calculate the commutator $[A(\alpha), B(\beta)]$. What is the condition on α, β such that $[A(\alpha), B(\beta)] = 0_2$?

Problem 2. Let A be an arbitrary 2×2 matrix. Show that

$$[A \otimes \sigma_3, I_2 \otimes A] = A \otimes ([\sigma_3, A]).$$

Problem 3. (i) Find all nonzero 2×2 matrices A, B such that $[A, B] = A + B$. Since $\operatorname{tr}([A, B]) = 0$ we have $\operatorname{tr}(A + B) = 0$.
(ii) Find all 2×2 matrices A and B such that $[A, B] = A - B$.
(iii) Find all 2×2 matrices A and B such that $[A \otimes A, B \otimes B] = A \otimes A - B \otimes B$.

Problem 4. Find all 2×2 matrices A over $\mathbb{C}$ such that

$$[A, A^*]_+ = I_2, \quad [A, A^*] = \sigma_3 = \begin{pmatrix} 1 & 0 \\ 0 & -1 \end{pmatrix}.$$

From the two conditions we find

$$AA^* + A^*A = I_2, \quad AA^* - A^*A = I_2.$$

(ii) Find all 2×2 matrices B over $\mathbb{C}$ such that $[B, B^*]_+ = I_2$. Start with

$$B = \begin{pmatrix} r_{11}e^{i\phi_{11}} & r_{12}e^{i\phi_{12}} \\ r_{21}e^{i\phi_{21}} & r_{22}e^{i\phi_{22}} \end{pmatrix} \Rightarrow B^* = \begin{pmatrix} r_{11}e^{-i\phi_{11}} & r_{21}e^{-i\phi_{21}} \\ r_{12}e^{-i\phi_{21}} & r_{22}e^{-i\phi_{22}} \end{pmatrix}.$$

(iii) Find all 2×2 matrices C over $\mathbb{C}$ such that $[C, C^*]_+ = I_2$, $[C, C^*] = \sigma_3$, where σ_3 is the third Pauli spin matrix.

Problem 5. (i) Find all nonzero 2×2 matrices K_+, K_-, K_3 such that

$$[K_3, K_+] = K_+, \quad [K_3, K_-] = -K_-, \quad [K_+, K_-] = -2K_3$$

where $(K_+)^* = K_-$. Since $\operatorname{tr}([A, B]) = 0$ for any $n \times n$ matrices A, B we have

$$\operatorname{tr}(K_+) = \operatorname{tr}(K_-) = \operatorname{tr}(K_3) = 0.$$

(ii) Find all nonzero 2×2 matrices A_1, A_2, A_3 such that

$$[A_1, A_2] = 0_2, \quad [A_1, A_3] = A_1, \quad [A_2, A_3] = A_2.$$

(iii) Find all 2×2 matrices A, B, C such that $[A, B] \neq 0_2$, $[A, C] \neq 0_2$, $[B, C] \neq 0_2$ and

$$[A, [B, C]] = 0_2.$$

Problem 6. Can one find a 2×2 matrix A over $\mathbb{R}$ such that

$$[A^T, A] = \begin{pmatrix} 0 & 1 \\ -1 & 0 \end{pmatrix}?$$

Problem 7. Let $\alpha, \beta \in \mathbb{C}$.
(i) Show that the 2×2 matrices

$$A = \begin{pmatrix} 0 & 1 \\ 0 & 0 \end{pmatrix}, \qquad B = \begin{pmatrix} \beta & -\beta^2 \\ 1 & -\beta \end{pmatrix}$$

satisfy the conditions $[A, B]_+ = I_2$, $[A, A]_+ = 0_2$, $[B, B]_+ = 0_2$.
(ii) Show that the 2×2 matrices

$$A = \begin{pmatrix} \alpha & 1 \\ -\alpha^2 & -\alpha \end{pmatrix}, \qquad B = \begin{pmatrix} 0 & 0 \\ 1 & 0 \end{pmatrix}$$

satisfy the conditions $[A, B]_+ = I_2$, $[A, A]_+ = 0_2$, $[B, B]_+ = 0_2$.

Problem 8. (i) Find all nonzero 2×2 matrices A, B such that

$$[A, B]_+ = 0_2, \qquad \text{tr}(AB^*) = 0.$$

(ii) Find all nonzero 2×2 matrices A and B such that $[A, [A, B]] = 0_2$.

Problem 9. (i) Let A, B, C 3×3 nonzero matrices with the commutation relation

$$[A, B] = C, \quad [B, C] = A, \quad [C, A] = B.$$

Consider the six 27×27 matrices

$$A \otimes B \otimes C, \ A \otimes C \otimes B, \ B \otimes A \otimes C, \ B \otimes C \otimes A, \ C \otimes A \otimes B, \ C \otimes B \otimes A.$$

Find all the commutators for these six matrices. Discuss.
(ii) Let A_1, A_2, A_3 be nonzero 3×3 matrices which satisfy the commutation relations

$$[A_1, A_2] = A_3, \quad [A_2, A_3] = A_1, \quad [A_3, A_1] = A_2.$$

Let

$$X := \sum_{k=1}^{3} c_k A_k.$$

Find the commutator

$$[\sum_{j=1}^{3} (A_j \otimes A_j), X \otimes I_3 + I_3 \otimes X].$$

Problem 10. (i) Let A, B be $n \times n$ matrices over $\mathbb{C}$. Show that if A and B commute and if A is normal, then A^* and B commute.

(ii) Let A be an $n \times n$ matrix over $\mathbb{C}$. It can be written as $A = HU$, where H is a non-negative definite hermitian matrix and U is unitary. Show that the matrices H and A commutate if and only if A is normal.

Problem 11. Let σ_j $(j = 0, 1, 2, 3)$ be the Pauli spin matrices, where $\sigma_0 = I_2$. Form the four 4×4 *Dirac matrices*

$$\gamma_k = \begin{pmatrix} 0_2 & \sigma_k \\ -\sigma_k & 0_2 \end{pmatrix}, \quad k = 0, 1, 2, 3.$$

(i) Are the matrices γ_k linearly independent?
(ii) Find the eigenvalues and eigenvectors of the γ_k's.
(iii) Are the matrices γ_k invertible. Use the result from (ii). If so, find the inverse.
(iv) Find the commutators $[\gamma_k, \gamma_\ell]$ for $k, \ell = 0, 1, 2, 3$. Find the anticommutators $[\gamma_k, \gamma_\ell]_+$ for $k, \ell = 0, 1, 2, 3$.

Problem 12. Let A, B be $n \times n$ matrices over $\mathbb{C}$. Assume that $[A, B] \neq 0_n$. Can we conclude that

$$[e^A, e^B] \neq 0_n?$$

Problem 13. Let A, B be invertible $n \times n$ matrices over $\mathbb{C}$. Assume that $[A, B] = 0_n$. Can we conclude that $[A^{-1}, B^{-1}] = 0_n$?

Problem 14. (i) Let σ_1, σ_2, σ_3 be the Pauli spin matrices. Consider the three nonnormal matrices

$$A = \sigma_1 + i\sigma_2, \quad B = \sigma_2 + i\sigma_3, \quad C = \sigma_3 + i\sigma_1.$$

Find the commutators and anticommutators. Discuss.
(ii) Consider the three nonnormal matrices

$$X = \sigma_1 \otimes \sigma_1 + i\sigma_2 \otimes \sigma_2, \quad Y = \sigma_2 \otimes \sigma_2 + i\sigma_3 \otimes \sigma_3, \quad Z = \sigma_3 \otimes \sigma_3 + i\sigma_1 \otimes \sigma_1.$$

Find the commutators and anticommutators. Discuss.

Problem 15. (i) Let σ_1, σ_2, σ_3 be the Pauli spin matrices. Consider the nonnormal matrices

$$A = \begin{pmatrix} 1 & 1 \\ 0 & 1 \end{pmatrix}, \quad B = \frac{1}{2}\begin{pmatrix} 3 & -1 \\ 1 & 1 \end{pmatrix}, \quad C = \frac{1}{2}\begin{pmatrix} 3 & i \\ i & 1 \end{pmatrix}$$

where $\det(A) = \det(B) = \det(C) = 1$, i.e. A, B, C are elements of the Lie group $SL(2, \mathbb{C})$. Show that $[A, A^*] = \sigma_3$, $[B, B^*] = \sigma_1$, $[C, C^*] = \sigma_2$.
(ii) Consider the unitary matrices

$$U = \frac{1}{\sqrt{2}}\begin{pmatrix} 1 & 1 \\ 1 & -1 \end{pmatrix}, \quad V = \begin{pmatrix} 1 & 0 \\ 0 & i \end{pmatrix}.$$

Show that $B = UAU^*$, $C = VBV^*$.

(iii) Consider the nonnormal and noninvertible matrices

$$X = \begin{pmatrix} 0 & 1 \\ 0 & 0 \end{pmatrix}, \quad Y = \frac{1}{2}\begin{pmatrix} 1 & -1 \\ 1 & -1 \end{pmatrix}, \quad Z = \frac{1}{2}\begin{pmatrix} 1 & i \\ i & -1 \end{pmatrix}.$$

All have trace zero and thus are elements of the Lie algebra $s\ell(2, \mathbb{C})$. Show that

$$[X, X^*] = \sigma_3, \quad [Y, Y^*] = \sigma_1, \quad [Z, Z^*] = \sigma_2.$$

Hint. Obviously we have $X = A - I_2$, $Y = B - I_2$, $Z = C - I_2$.
(iv) Show that $Y = UXU^*$, $Z = VYV^*$.
(v) Study the commutators

$$[X \otimes X, X^* \otimes X^*], \qquad [X \otimes X^*, X^* \otimes X],$$

$$[Y \otimes Y, Y^* \otimes Y^*], \qquad [Y \otimes Y^*, Y^* \otimes Y],$$

$$[Z \otimes Z, Z^* \otimes Z^*], \qquad [Z \otimes Z^*, Z^* \otimes Z].$$

Problem 16. (i) Can one find $n \times n$ matrices A and B over $\mathbb{C}$ such that the following conditions are satisfied $[A, B] = 0_n$, $[A, B]_+ = 0_n$ and both A and B are invertible?
(ii) Can one find $n \times n$ matrices A and B over $\mathbb{C}$ such that the following conditions are satisfied

$$[A, B] = 0_n, \qquad [A, B]_+ = I_n$$

and both A and B are invertible?

Problem 17. Consider the hermitian 3×3 matrices

$$L_1 = \begin{pmatrix} 0 & 0 & 0 \\ 0 & 0 & -i \\ 0 & i & 0 \end{pmatrix}, \quad L_2 = \begin{pmatrix} 0 & 0 & i \\ 0 & 0 & 0 \\ -i & 0 & 0 \end{pmatrix}, \quad L_3 = \begin{pmatrix} 0 & -i & 0 \\ i & 0 & 0 \\ 0 & 0 & 0 \end{pmatrix}.$$

Find the commutators $[L_1, L_2]$, $[L_2, L_3]$, $[L_3, L_1]$. Find the 3×3 matrix

$$A = c_{23}[L_2, L_3]_+ + c_{13}[L_3, L_1]_+ + c_{12}[L_1, L_2]_+.$$

Problem 18. (i) Find 4×4 matrices C and 2×2 matrices A such that

$$[C, A \otimes I_2 + I_2 \otimes A] = 0_4.$$

(ii) Find the conditions on the two 2×2 hermitian matrices B, C such that

$$[B \otimes C, P] = 0_4$$

where P is the permutation matrix

$$P = \begin{pmatrix} 1 & 0 & 0 & 0 \\ 0 & 0 & 1 & 0 \\ 0 & 1 & 0 & 0 \\ 0 & 0 & 0 & 1 \end{pmatrix}.$$

(iii) Let E, F be arbitrary 2×2 matrices. Then

$$[E \otimes I_2, I_2 \otimes F] = 0_4.$$

Let X, Y be 4×4 matrices with $[X, Y] = 0_4$. Can X, Y be written as $X = E \otimes I_2$, $Y = I_2 \otimes F$ or $Y = E \otimes I_2$, $X = I_2 \otimes F$?

Problem 19.   (i) Consider the two 2×2 matrices (counter diagonal matrices)

$$A = \begin{pmatrix} 0 & a_{12} \\ a_{21} & 0 \end{pmatrix} \qquad B = \begin{pmatrix} 0 & b_{12} \\ b_{21} & 0 \end{pmatrix}.$$

Find the condition on A and B such that $[A, B] = 0_2$.
(ii) Consider the two 3×3 matrices (counter diagonal matrices)

$$A = \begin{pmatrix} 0 & 0 & a_{13} \\ 0 & a_{22} & 0 \\ a_{31} & 0 & 0 \end{pmatrix} \qquad B = \begin{pmatrix} 0 & 0 & b_{13} \\ 0 & b_{22} & 0 \\ b_{31} & 0 & 0 \end{pmatrix}.$$

Find the condition on A and B such that $[A, B] = 0_3$.
(iii) Extend to n dimensions.

Problem 20.   Consider $(m + n) \times (m + n)$ matrices of the form

$$\begin{pmatrix} m \times m & m \times n \\ n \times m & n \times n \end{pmatrix}.$$

Let

$$B = \begin{pmatrix} B_1 & 0 \\ 0 & B_2 \end{pmatrix}, \quad \tilde{B} = \begin{pmatrix} \tilde{B}_1 & 0 \\ 0 & \tilde{B}_2 \end{pmatrix}, \quad F = \begin{pmatrix} 0 & F_1 \\ F_2 & 0 \end{pmatrix}, \quad \tilde{F} = \begin{pmatrix} 0 & \tilde{F}_1 \\ \tilde{F}_2 & 0 \end{pmatrix}.$$

Show that

$$[B, \tilde{B}] = \begin{pmatrix} B_1 \tilde{B}_1 - \tilde{B}_1 B_1 & 0 \\ 0 & B_2 \tilde{B}_2 - \tilde{B}_2 B_2 \end{pmatrix},$$

$$[B, F] = \begin{pmatrix} 0 & B_1 F_1 - F_1 B_2 \\ B_2 F_2 - F_2 B_1 & 0 \end{pmatrix},$$

$$[F, \tilde{F}]_+ = \begin{pmatrix} F_1 \tilde{F}_2 + \tilde{F}_1 F_2 & 0 \\ 0 & F_2 \tilde{F}_1 + \tilde{F}_2 F_1 \end{pmatrix}.$$

Problem 21.   Let $n \geq 1$ and A, B be $n \times n$ matrices over $\mathbb{C}$. Show that

$$[A, B^n] = \sum_{j=0}^{n-1} A^j [A, B] B^{n-1-j}.$$

Problem 22.   Let A_1, A_2, B_1, B_2 be $n \times n$ matrices over $\mathbb{C}$ with $A_1^2 = A_2^2 = B_1^2 = B_2^2 = I_n$. Let

$$T = A_1 \otimes (B_1 + B_2) + A_2 \otimes (B_1 - B_2).$$

Show that $T^2 = 4I_n \otimes I_n - [A_1, A_2] \otimes [B_1, B_2]$.

Problem 23. Let A, B be 2×2 matrices.
(i) Solve the equation $A \otimes B - B \otimes A = [A, B] \otimes I_2 + I_2 \otimes [A, B]$.
(ii) Solve the equation $A \otimes B - B \otimes A = [A \otimes A, B \otimes B]$.

Problem 24. Let A be an $n \times n$ matrix. Find the commutator

$$\left[\begin{pmatrix} 0_n & A \\ A & 0_n \end{pmatrix}, \begin{pmatrix} 0_n & A \\ -A & 0_n \end{pmatrix} \right]$$

and the anticommutator

Problem 25. Let A, B be $n \times n$ matrices and T a (fixed) invertible $n \times n$ matrix. We define the bracket

$$[A, B]_T := ATB - BTA.$$

Let

$$T = \begin{pmatrix} 0 & 1 \\ 1 & 0 \end{pmatrix}, \quad X = \begin{pmatrix} 0 & 1 \\ 0 & 0 \end{pmatrix}, \quad Y = \begin{pmatrix} 0 & 0 \\ 1 & 0 \end{pmatrix}, \quad H = \begin{pmatrix} 1 & 0 \\ 0 & -1 \end{pmatrix}.$$

Find $[X, Y]_T$, $[X, H]_T$, $[Y, H]_T$.

Problem 26. Can we find 3×3 matrices A and B such that $[A, B]_+ = 0_3$ and $A^2 = B^2 = I_3$?

Problem 27. (i) Let X, Y be $n \times n$ matrices over $\mathbb{C}$. Assume that $[X, Y] = 0_n$. Do $X \otimes X$ and $Y \otimes Y$ commutate? Do $X \otimes X$ and $Y \otimes I_n + I_n \otimes X$ commutate?
(ii) Let X, Y be $n \times n$ matrices over $\mathbb{C}$. Assume that $[X, Y]_+ = 0_n$. Do $X \otimes X$ and $Y \otimes Y$ anticommutate? Do $X \otimes X$ and $Y \otimes I_n + I_n \otimes X$ anticommutate?

Problem 28. Find all 2×2 matrices over $\mathbb{C}$ such that the commutator is an invertible diagonal matrix D, i.e. $d_{11} \neq 0$ and $d_{22} \neq 0$. An example is

$$A = \begin{pmatrix} 0 & -i \\ i & 0 \end{pmatrix}, \quad B = \begin{pmatrix} 0 & i \\ i & 0 \end{pmatrix} \Rightarrow D = \begin{pmatrix} 2 & 0 \\ 0 & -2 \end{pmatrix}.$$

Problem 29. Consider the skew-symmetric 3×3 matrices

$$A_1 = \begin{pmatrix} 0 & 0 & 0 \\ 0 & 0 & -1 \\ 0 & 1 & 0 \end{pmatrix}, \quad A_2 = \begin{pmatrix} 0 & 0 & 1 \\ 0 & 0 & 0 \\ -1 & 0 & 0 \end{pmatrix}, \quad A_3 = \begin{pmatrix} 0 & -1 & 0 \\ 1 & 0 & 0 \\ 0 & 0 & 0 \end{pmatrix}$$

with $[A_1, A_2] = A_3$, $[A_2, A_3] = A_1$, $[A_3, A_1] = A_2$. Find the commutators for the 9×9 matrices

$$A_1 \otimes I_3 + I_3 \otimes A_1, \quad A_2 \otimes I_3 + I_3 \otimes A_2, \quad A_3 \otimes I_3 + I_3 \otimes A_3.$$

Problem 30.   Let A, B, C be $n \times n$ matrices. Show that

$$\text{tr}([A, B]C) \equiv \text{tr}(A[B, C])$$

applying cyclic invariance.

Problem 31.   Show that $A = i\sigma_1/2$, $B = -i\sigma_2$ satisfy

$$-A = [B, [B, A]], \qquad -B = [A, [A, B]].$$

Problem 32.   Consider the linear operators J_+, J_-, J_z satisfying the commutation relations $[J_z, J_+] = J_+$, $[J_z, J_-] = -J_-$, $[J_+, J_-] = 2J_z$, where $(J_+)^* = J_-$. Show that the 2×2 matrices

$$J_+ = \begin{pmatrix} 0 & 1 \\ 0 & 0 \end{pmatrix}, \quad J_- = \begin{pmatrix} 0 & 0 \\ 1 & 0 \end{pmatrix}, \quad J_z = \frac{1}{2}\begin{pmatrix} 1 & 0 \\ 0 & -1 \end{pmatrix}$$

satisfy these commutations relations.

Problem 33.   (i) Consider the 3×3 matrices

$$P = \begin{pmatrix} 0 & 1 & 0 \\ 0 & 0 & 1 \\ 1 & 0 & 0 \end{pmatrix}, \qquad B = \begin{pmatrix} b_{11} & b_{12} & b_{13} \\ b_{13} & b_{11} & b_{12} \\ b_{12} & b_{13} & b_{11} \end{pmatrix}.$$

Both matrices are circulant matrices. Find the commutator $[P, B]$. Discuss.
(ii) Consider the invertible 3×3 matrices

$$P = \begin{pmatrix} 0 & 1 & 0 \\ 0 & 0 & 1 \\ 1 & 0 & 0 \end{pmatrix}, \qquad A = \begin{pmatrix} 1 & 0 & 0 \\ 0 & e^{2i\pi/3} & 0 \\ 0 & 0 & e^{-2i\pi/3} \end{pmatrix}.$$

Is the matrix $[A, P]$ invertible?

Problem 34.   Let $c \in \mathbb{R}$ and A be an 2×2 matrix over $\mathbb{R}$. Find the commutator of the 3×3 matrices $c \oplus A$ and $A \oplus c$, where $\oplus$ denotes the direct sum.

Problem 35.   Let U, V be $n \times n$ unitary matrices. Show that

$$[V^{-1} \otimes U, V \otimes U^{-1}] = 0_n.$$

Problem 36.   Let A, B be $n \times n$ matrices over $\mathbb{C}$. We define (*Jordan product*)

$$A \circ B = \frac{1}{2}(AB + BA).$$

Show that $A \circ B$ is commutative and satisfies $A \circ (A^2 \circ B) = A^2 \circ (A \circ B)$, but is not associative in general, i.e. $(A \circ B) \circ C \neq A \circ (B \circ C)$ in general.

Chapter 8

Decomposition of Matrices

A matrix decomposition (or matrix factorization) is the right-hand side matrix product

$$A = F_1 F_2 \cdots F_n$$

for an input matrix A. The number of factor matrices $F_1, F_2, \ldots, F_n$ depends on the chosen decomposition. In most cases $n = 2$ or $n = 3$.

The most common decompositions are:

1) *LU-decomposition*. A square matrix A is factorized into a product of a lower triangular matrix, L, and an upper triangular matrix U, i.e. $A = LU$.

2) *QR-decomposition*. An $n \times m$ matrix with linearly independent columns is factorized as $A = QR$, where Q is an $n \times m$ matrix with orthonormal columns and R is an invertible $m \times m$ upper triangular matrix.

3) The *polar decomposition* of $A \in \mathbb{C}^{n \times n}$ factors A as the product $A = UH$, where U is unitary and H is hermitian positive semi-definite. The hermitian factor H is always unique and can be expressed as $(A^*A)^{1/2}$, and the unitary factor is unique if A is nonsingular.

4) In the *singular value decomposition* an $m \times n$ matrix can be written as

$$A = U \Sigma V^T$$

where U is an $m \times m$ orthogonal matrix, V is an $n \times n$ orthogonal matrix, Σ is an $m \times n$ diagonal matrix with nonnegative entries and the superscript T denotes the transpose.

Other important decompositions are the cosine-sine decomposition for unitary matrices, Iwasawa decomposition (for Lie groups), Schur decomposition, Cholesky decomposition (for positive semi-definite matrices) and the Jordan decomposition.

Problem 1.   Find the LU-decomposition of the 3×3 matrix

$$A = \begin{pmatrix} 3 & 6 & -9 \\ 2 & 5 & -3 \\ -4 & 1 & 10 \end{pmatrix}.$$

The triangular matrices L and U are not uniquely determined by the matrix equation $A = LU$. These two matrices together contain $n^2 + n$ unknown elements. Thus when comparing elements on the left- and right-hand side of $A = LU$ we have n^2 equations and $n^2 + n$ unknowns. We require a further n conditions to uniquely determine the matrices. There are three additional sets of n conditions that are commonly used. These are *Doolittle's method* with $\ell_{jj} = 1$, $j = 1, \dots, n$; *Choleski's method* with $\ell_{jj} = u_{jj}$, $j = 1, \dots, n$; *Crout's method* with $u_{jj} = 1$, $j = 1, \dots, n$. Apply Crout's method.

Solution 1.   From

$$L = \begin{pmatrix} \ell_{11} & 0 & 0 \\ \ell_{21} & \ell_{22} & 0 \\ \ell_{31} & \ell_{32} & \ell_{33} \end{pmatrix}, \qquad U = \begin{pmatrix} 1 & u_{12} & u_{13} \\ 0 & 1 & u_{23} \\ 0 & 0 & 1 \end{pmatrix}$$

and $A = LU$ we obtain the 9 equations

$$\ell_{11} = 3$$
$$\ell_{11} u_{12} = 6$$
$$\ell_{11} u_{13} = -9$$
$$\ell_{21} = 2$$
$$\ell_{21} u_{12} + \ell_{22} = 5$$
$$\ell_{21} u_{13} + \ell_{22} u_{23} = -3$$
$$\ell_{31} = -4$$
$$\ell_{31} u_{12} + \ell_{32} = 1$$
$$\ell_{31} u_{13} + \ell_{32} u_{23} + \ell_{33} = 10$$

with the solution $\ell_{11} = 3$, $\ell_{21} = 2$, $\ell_{22} = 1$, $\ell_{31} = -4$, $\ell_{32} = 9$, $\ell_{33} = -29$,

$$u_{12} = 2, \quad u_{13} = -3, \quad u_{23} = 3.$$

Thus we obtain

$$L = \begin{pmatrix} 3 & 0 & 0 \\ 2 & 1 & 0 \\ -4 & 9 & -29 \end{pmatrix}, \qquad U = \begin{pmatrix} 1 & 2 & -3 \\ 0 & 1 & 3 \\ 0 & 0 & 1 \end{pmatrix}.$$

Problem 2.   Let A be an $n \times n$ matrix over $\mathbb{R}$. Consider the LU-decomposition $A = LU$, where L is a unit lower triangular matrix and U is an upper triangular matrix. The LDU-decomposition is defined as $A = LDU$, where L is unit lower triangular, D is diagonal and U is unit upper triangular. Let

$$A = \begin{pmatrix} 2 & 4 & -2 \\ 4 & 9 & -3 \\ -2 & -3 & 7 \end{pmatrix}.$$

Find the *LDU*-decomposition via the *LU*-decomposition.

Solution 2. We have the *LU*-decomposition

$$L = \begin{pmatrix} 1 & 0 & 0 \\ 2 & 1 & 0 \\ -1 & 1 & 1 \end{pmatrix}, \qquad U = \begin{pmatrix} 2 & 4 & -2 \\ 0 & 1 & 1 \\ 0 & 0 & 4 \end{pmatrix}.$$

Then the *LDU*-decomposition follows as

$$A = \begin{pmatrix} 1 & 0 & 0 \\ 2 & 1 & 0 \\ -1 & 1 & 1 \end{pmatrix} \begin{pmatrix} 2 & 0 & 0 \\ 0 & 1 & 0 \\ 0 & 0 & 4 \end{pmatrix} \begin{pmatrix} 1 & 2 & -1 \\ 0 & 1 & 1 \\ 0 & 0 & 1 \end{pmatrix}.$$

Problem 3. Find the *QR*-decomposition of the 3×3 matrix

$$A = \begin{pmatrix} 2 & 1 & 3 \\ -1 & 0 & 7 \\ 0 & -1 & -1 \end{pmatrix}.$$

Solution 3. The columns of A are

$$\mathbf{c}_1 = \begin{pmatrix} 2 \\ -1 \\ 0 \end{pmatrix}, \quad \mathbf{c}_2 = \begin{pmatrix} -1 \\ 0 \\ -1 \end{pmatrix}, \quad \mathbf{c}_3 = \begin{pmatrix} 3 \\ 7 \\ -1 \end{pmatrix}.$$

They are linearly independent. Applying the *Gram-Schmidt orthonormalization process* we obtain

$$\mathbf{v}_1 = \begin{pmatrix} 2/\sqrt{5} \\ -1/\sqrt{5} \\ 0 \end{pmatrix}, \quad \mathbf{v}_2 = \begin{pmatrix} 1/\sqrt{30} \\ 2/\sqrt{30} \\ -5/\sqrt{30} \end{pmatrix}, \quad \mathbf{v}_3 = \begin{pmatrix} 1/\sqrt{6} \\ 2/\sqrt{6} \\ 1/\sqrt{6} \end{pmatrix}.$$

Thus we obtain

$$Q = \begin{pmatrix} 2/\sqrt{5} & 1/\sqrt{30} & 1/\sqrt{6} \\ -1/\sqrt{5} & 2/\sqrt{30} & 2/\sqrt{6} \\ 0 & -5/\sqrt{30} & 1/\sqrt{6} \end{pmatrix}, \quad R = \begin{pmatrix} \sqrt{5} & 2/\sqrt{5} & -1/\sqrt{5} \\ 0 & 6/\sqrt{30} & 22/\sqrt{30} \\ 0 & 0 & 16/\sqrt{6} \end{pmatrix}.$$

Problem 4. We consider 3×3 matrices over $\mathbb{R}$. An orthogonal matrix Q such that $\det(Q) = 1$ is called a *rotation matrix*. Let $1 \le p < r \le 3$ and ϕ be a real number. An orthogonal 3×3 matrix $Q_{pr}(\phi) = (q_{ij})_{1 \le i,j \le 3}$ given by

$$q_{pp} = q_{rr} = \cos(\phi)$$
$$q_{ii} = 1 \quad \text{if } i \ne p, r$$
$$q_{pr} = -q_{rp} = -\sin(\phi)$$
$$q_{ip} = q_{pi} = q_{ir} = q_{ri} = 0 \quad i \ne p, r$$
$$q_{ij} = 0 \quad \text{if } i \ne p, r \text{ and } j \ne p, r$$

will be called a plane rotation through ϕ in the plane span (e_p, e_r). Let $Q = (q_{ij})_{1 \le i,j \le 3}$ be a rotation matrix. Show that there exist angles $\phi \in [0, \pi)$, $\theta, \psi \in (-\pi, \pi]$ called the *Euler angles* of Q such that

$$Q = Q_{12}(\phi)Q_{23}(\theta)Q_{12}(\psi). \tag{1}$$

Solution 4.   To prove it we use the QR factorization of the rotation matrix. We set

$$Q_1 = Q, \quad Q_2 = Q_{12}(-\phi)Q_1, \quad Q_3 = Q_{23}(-\theta)Q_2, \quad Q_4 = Q_{12}(-\psi)Q_3$$

where $Q_k = (q_{ij}^k)_{1 \le i,j \le 3}$. We then choose ϕ, θ, ψ to be the numbers that subsequently annihilate q_{13}^2, q_{23}^3, q_{12}^4, that is, such that

$$\cot(\phi) = -q_{23}^1/q_{13}^1, \quad \cot(\theta) = -q_{33}^2/q_{23}^2, \quad \cot(\psi) = -q_{22}^3/q_{12}^3.$$

The matrix Q_4 is a rotation lower triangular matrix and therefore it is the 3×3 identity matrix. Since $Q_{pr}(-\psi) = Q_{pr}(\psi)^{-1}$ we obtain (1).

Problem 5.   Consider a square non-singular square matrix A over $\mathbb{C}$, i.e. A^{-1} exists. The *polar decomposition theorem* states that A can be written as $A = UH$, where U is a unitary matrix and H is a hermitian positive definite matrix. Show that A has a unique polar decomposition.

Solution 5.   Since A is invertible, so are A^* and A^*A. The positive square root H of A^*A is also invertible. Set $U := AH^{-1}$. Then U is invertible and

$$U^*U = H^{-1}A^*AH^{-1} = H^{-1}H^2H^{-1} = I$$

so that U is unitary. Since H is invertible, it is obvious that AH^{-1} is the only possible choice for U.

Problem 6.   For any $n \times n$ matrix A over $\mathbb{C}$, there exists a positive semi-definite matrix H and a unitary matrix such that $A = HU$. If A is nonsingular, then H is positive definite and U and H are unique.
(i) Find the polar decomposition for the invertible 3×3 matrix

$$A = \begin{pmatrix} 1 & 0 & -4 \\ 0 & 5 & 4 \\ -4 & 4 & 3 \end{pmatrix}.$$

The positive definite matrix H is the unique square root of the positive definite matrix A^*A and then U is defined by $U = AH^{-1}$.
(ii) Apply the polar decomposition to the nonnormal matrix A

$$A = \begin{pmatrix} 1 & 1 \\ 0 & 1 \end{pmatrix} \quad \Rightarrow \quad A^* = \begin{pmatrix} 1 & 0 \\ 1 & 1 \end{pmatrix}$$

which is an element of the Lie group $SL(2, \mathbb{R})$.

Solution 6. (i) The unique decomposition is

$$H = \frac{1}{3} \begin{pmatrix} 11 & -4 & -4 \\ -4 & 17 & 8 \\ -4 & 8 & 17 \end{pmatrix}, \qquad U = \frac{1}{9} \begin{pmatrix} 1 & 4 & -8 \\ 4 & 7 & 4 \\ -8 & 4 & 1 \end{pmatrix}.$$

(ii) We have

$$A^*A = \begin{pmatrix} 1 & 1 \\ 1 & 2 \end{pmatrix}$$

with the positive eigenvalues $\lambda_\pm = 3/2 \pm \sqrt{5}/2$. The positive definite square root of this matrix is

$$P = \frac{1}{\sqrt{5}} \begin{pmatrix} 2 & 1 \\ 1 & 3 \end{pmatrix} \quad \Rightarrow \quad U = AP^{-1} = \frac{1}{\sqrt{5}} \begin{pmatrix} 2 & 1 \\ -1 & 2 \end{pmatrix}.$$

Problem 7. Let A be an arbitrary $m \times n$ matrix over $\mathbb{R}$, i.e. $A \in \mathbb{R}^{m \times n}$. Then A can be written as

$$A = U \Sigma V^T$$

where U is an $m \times m$ orthogonal matrix, V is an $n \times n$ orthogonal matrix, Σ is an $m \times n$ diagonal matrix with nonnegative entries and the superscript ^{T} denotes the transpose. This is called the *singular value decomposition*. An algorithm to find the singular value decomposition is as follows.

1) Find the eigenvalues λ_j $(j = 1, \ldots, n)$ of the $n \times n$ matrix $A^T A$. Arrange the eigenvalues $\lambda_1, \ldots, \lambda_n$ in descending order.
2) Find the number of nonzero eigenvalues of the matrix $A^T A$. We call this number r.
3) Find the orthogonal eigenvectors $\mathbf{v}_j$ of the matrix $A^T A$ corresponding to the obtained eigenvalues, and arrange them in the same order to form the column-vectors of the $n \times n$ matrix V.
4) Form an $m \times n$ diagonal matrix Σ placing on the leading diagonal of it the square root $\sigma_j := \sqrt{\lambda_j}$ of $p = \min(m, n)$ first eigenvalues of the matrix $A^T A$ found in 1) in descending order.
5) Find the first r column vectors of the $m \times m$ matrix U

$$\mathbf{u}_j = \frac{1}{\sigma_j} A \mathbf{v}_j, \quad j = 1, 2, \ldots, r.$$

6) Add to the matrix U the rest of the $m - r$ vectors using the Gram-Schmidt orthogonalization process.

We have $A\mathbf{v}_j = \sigma_j \mathbf{u}_j$, $A^T \mathbf{u}_j = \sigma_j \mathbf{v}_j$ and therefore $A^T A \mathbf{v}_j = \sigma_j^2 \mathbf{v}_j$, $AA^T \mathbf{u}_j = \sigma_j^2 \mathbf{u}_j$. Apply the algorithm to the matrix

$$A = \begin{pmatrix} 0.96 & 1.72 \\ 2.28 & 0.96 \end{pmatrix}.$$

Solution 7. 1) We find

$$A^T A = \begin{pmatrix} 6.12 & 3.84 \\ 3.84 & 3.88 \end{pmatrix}.$$

The eigenvalues are (arranged in descending order) $\lambda_1 = 9$ and $\lambda_2 = 1$.
2) The number of nonzero eigenvalues is $r = 2$.
3) The orthonormal normalized eigenvectors of the matrix $A^T A$, corresponding to the eigenvalues λ_1 and λ_2 are given by

$$\mathbf{v}_1 = \begin{pmatrix} 0.8 \\ 0.6 \end{pmatrix}, \qquad \mathbf{v}_2 = \begin{pmatrix} 0.6 \\ -0.8 \end{pmatrix}.$$

We obtain the 2×2 matrix V (V^T follows by taking the transpose)

$$V = (\mathbf{v}_1 \ \mathbf{v}_2) = \begin{pmatrix} 0.8 & 0.6 \\ 0.6 & -0.8 \end{pmatrix}.$$

4) From the eigenvalues we find the singular matrix taking the square roots of the eigenvalues

$$\Sigma = \begin{pmatrix} 3 & 0 \\ 0 & 1 \end{pmatrix}.$$

5) Next we find two column vectors of the 2×2 matrix U. Using the equation given above we find ($\sigma_1 = 3, \sigma_2 = 1$)

$$\mathbf{u}_1 = \frac{1}{\sigma_1} A \mathbf{v}_1 = \begin{pmatrix} 0.6 \\ 0.8 \end{pmatrix}, \qquad \mathbf{u}_2 = \frac{1}{\sigma_2} A \mathbf{v}_2 = \begin{pmatrix} -0.8 \\ 0.6 \end{pmatrix}.$$

It follows that

$$U = (\mathbf{u}_1 \ \mathbf{u}_2) = \begin{pmatrix} 0.6 & -0.8 \\ 0.8 & 0.6 \end{pmatrix}.$$

Thus we have found the singular value decomposition of the matrix A

$$A = U \Sigma V^T = \begin{pmatrix} 0.6 & -0.8 \\ 0.8 & 0.6 \end{pmatrix} \begin{pmatrix} 3 & 0 \\ 0 & 1 \end{pmatrix} \begin{pmatrix} 0.8 & 0.6 \\ 0.6 & -0.8 \end{pmatrix}^T.$$

Problem 8. Find the singular value decomposition $A = U \Sigma V^T$ of the matrix (row vector) $A = (2 \ 1 \ -2)$.

Solution 8. First we find the eigenvalues of the 3×3 matrix $A^T A$, where ^{T} denotes transpose. Note that $AA^T = 9$. Thus the eigenvalues of

$$A^T A = \begin{pmatrix} 4 & 2 & -4 \\ 2 & 1 & -2 \\ -4 & -2 & 4 \end{pmatrix}$$

are $\lambda_1 = 9$, $\lambda_2 = 0$, $\lambda_3 = 0$. Thus the eigenvalue 0 is degenerate. Next we find the number r of nonzero eigenvalues which is $r = 1$. Now we calculate the normalized eigenvectors. For the eigenvalue $\lambda_1 = 9$ we find

$$\mathbf{v}_1 = (-2/3 \ -1/3 \ 2/3)^T.$$

To find the two normalized eigenvectors for the eigenvalue 0 we have to apply the Gram-Schmidt orthogonalization process. Then we find

$$\mathbf{v}_2 = \begin{pmatrix} -\sqrt{5}/5 \\ 2\sqrt{5}/5 \\ 0 \end{pmatrix}, \qquad \mathbf{v}_3 = \begin{pmatrix} 4\sqrt{5}/15 \\ 2\sqrt{5}/15 \\ 5\sqrt{5}/15 \end{pmatrix}.$$

This provides the orthonormal matrix

$$V = \begin{pmatrix} -2/3 & -\sqrt{5}/5 & 4\sqrt{5}/15 \\ -1/3 & 2\sqrt{5}/5 & 2\sqrt{5}/15 \\ 2/3 & 0 & \sqrt{5}/3 \end{pmatrix}.$$

Next we form the singular value matrix $\Sigma = (3\ 0\ 0)$. Finally we calculate the unique column-vector of the matrix U

$$\mathbf{u}_1 = \frac{1}{3} A \mathbf{v}_1 = \frac{1}{3}(2\ 1\ -2) \begin{pmatrix} -2/3 \\ -1/3 \\ 2/3 \end{pmatrix} = (-1).$$

Hence the singular value decomposition of the matrix A is

$$A = U \Sigma V^T = (-1)(3\ 0\ 0) \begin{pmatrix} -2/3 & -1/3 & 2/3 \\ \sqrt{5}/5 & 2\sqrt{5}/5 & 0 \\ 4\sqrt{5}/15 & 2\sqrt{5}/15 & \sqrt{5}/3 \end{pmatrix}.$$

Problem 9. (i) Let $n \geq 2$ and $n = 2k$. Let A be an $n \times k$ matrix and $A^*A = I_k$. Find the $n \times n$ matrix AA^* using the singular value decomposition. Calculate $\operatorname{tr}(AA^*)$.
(ii) Let $n \geq 2$ and $n = 2k$. Let A be an $n \times k$ matrix and $A^*A = I_k$. Let S be a positive definite $n \times n$ matrix. Show that

$$1 \leq \frac{\operatorname{tr}(A^* S^2 A)}{\operatorname{tr}((A^* S A)^2)}.$$

Solution 9. (i) Using the singular value decomposition we have $A = U\Sigma V^*$, $A^* = V\Sigma^* U^*$, where U is a unitary $n \times n$ matrix, V is a unitary $k \times k$ matrix and Σ is an $n \times k$ matrix diagonal matrix with nonnegative entries. Thus

$$A^*A = (V\Sigma^* U^*)(U\Sigma V^*) = V\Sigma^*\Sigma V^* = I_k.$$

It follows that $\Sigma^*\Sigma = I_k$. On the other hand we have

$$AA^* = (U\Sigma V^*)(V\Sigma^* U^*) = U\Sigma\Sigma^* U^* = U \begin{pmatrix} I_k & 0_k \\ 0_k & 0_k \end{pmatrix} U^* = U(I_k \oplus 0_k)U^*$$

where 0_k is the $k \times k$ zero matrix and $\oplus$ the direct sum. Obviously

$$\operatorname{tr}(A^*A) = \operatorname{tr}(AA^*).$$

(ii) Using the result from the previous problem we have

$$(A^*SA)^2 = (A^*SA)(A^*SA) = A^*SU \begin{pmatrix} I_k & 0_k \\ 0_k & 0_k \end{pmatrix} U^*SA.$$

We also have

$$A^*S^2A = A^*SU \begin{pmatrix} I_k & 0_k \\ 0_k & 0_k \end{pmatrix} U^*SA + A^*SU \begin{pmatrix} 0_k & 0_k \\ 0_k & I_k \end{pmatrix} U^*SA.$$

Thus

$$\frac{\text{tr}(A^*S^2A)}{\text{tr}((A^*SA)^2)} = 1 + \frac{\text{tr}(A^*SU(0_k \oplus I_k)U^*SA)}{\text{tr}((A^*SA)^2)} \geq 1.$$

Problem 10. Let A be an $n \times n$ matrix over $\mathbb{R}$. Assume that A^{-1} exists. Given the singular value decomposition of A, i.e. $A = UWV^T$. Find the singular value decomposition for A^{-1}.

Solution 10. We have $A^{-1} = (UWV^T)^{-1} = (V^T)^{-1}W^{-1}U^{-1} = VW^{-1}U^T$.

Problem 11. Find the *Moore-Penrose pseudo inverses* of

$$\begin{pmatrix} 1 & 0 \\ 0 & 1 \\ -1 & 0 \end{pmatrix}, \quad \begin{pmatrix} 1 \\ 0 \end{pmatrix}, \quad (1 \ 1)$$

applying the singular value decomposition.

Solution 11. The singular value decompositions are

$$\begin{pmatrix} 1 & 0 \\ 0 & 1 \\ -1 & 0 \end{pmatrix} = \begin{pmatrix} 1/\sqrt{2} & 0 & 1/\sqrt{2} \\ 0 & 1 & 0 \\ -1/\sqrt{2} & 0 & 1/\sqrt{2} \end{pmatrix} \begin{pmatrix} \sqrt{2} & 0 \\ 0 & 1 \\ 0 & 0 \end{pmatrix} \begin{pmatrix} 1 & 0 \\ 0 & 1 \end{pmatrix}^*$$

$$\begin{pmatrix} 1 \\ 0 \end{pmatrix} = \begin{pmatrix} 1 & 0 \\ 0 & 1 \end{pmatrix} \begin{pmatrix} 1 \\ 0 \end{pmatrix} (1)^*$$

$$(1 \ 1) = (1)(\sqrt{2} \ 0)\frac{1}{\sqrt{2}} \begin{pmatrix} 1 & 1 \\ 1 & -1 \end{pmatrix}^*.$$

Thus the Moore-Penrose pseudo inverses are

$$\begin{pmatrix} 1 & 0 \\ 0 & 1 \\ -1 & 0 \end{pmatrix}^- = \begin{pmatrix} 1 & 0 \\ 0 & 1 \end{pmatrix} \begin{pmatrix} 1/\sqrt{2} & 0 & 0 \\ 0 & 1 & 0 \end{pmatrix} \begin{pmatrix} 1/\sqrt{2} & 0 & -1/\sqrt{2} \\ 0 & 1 & 0 \\ 1/\sqrt{2} & 0 & 1/\sqrt{2} \end{pmatrix} = \begin{pmatrix} \frac{1}{2} & 0 & -\frac{1}{2} \\ 0 & 1 & 0 \end{pmatrix},$$

$$\begin{pmatrix} 1 \\ 0 \end{pmatrix}^- = (1)(1 \ 0) \begin{pmatrix} 1 & 0 \\ 0 & 1 \end{pmatrix}^* = (1 \ 0),$$

$$(1 \ 1)^- = \frac{1}{\sqrt{2}} \begin{pmatrix} 1 & 1 \\ 1 & -1 \end{pmatrix} \begin{pmatrix} 1/\sqrt{2} \\ 0 \end{pmatrix} (1)^* = \frac{1}{2} \begin{pmatrix} 1 \\ 1 \end{pmatrix}.$$

Problem 12. Any unitary $2^n \times 2^n$ matrix U can be decomposed as

$$U = \begin{pmatrix} U_1 & 0 \\ 0 & U_2 \end{pmatrix} \begin{pmatrix} C & S \\ -S & C \end{pmatrix} \begin{pmatrix} U_3 & 0 \\ 0 & U_4 \end{pmatrix}$$

where U_1, U_2, U_3, U_4 are $2^{n-1} \times 2^{n-1}$ unitary matrices and C and S are the $2^{n-1} \times 2^{n-1}$ diagonal matrices

$$C = \text{diag}(\cos(\alpha_1), \cos(\alpha_2), \dots, \cos(\alpha_{2^n}/2)),$$
$$S = \text{diag}(\sin(\alpha_1), \sin(\alpha_2), \dots, \sin(\alpha_{2^n}/2))$$

where $\alpha_j \in \mathbb{R}$. This decomposition is called *cosine-sine decomposition*.
(i) Consider the unitary 2×2 matrix

$$U = \begin{pmatrix} 0 & i \\ -i & 0 \end{pmatrix}.$$

Show that U can be written as

$$U = \begin{pmatrix} u_1 & 0 \\ 0 & u_2 \end{pmatrix} \begin{pmatrix} \cos(\alpha) & \sin(\alpha) \\ -\sin(\alpha) & \cos(\alpha) \end{pmatrix} \begin{pmatrix} u_3 & 0 \\ 0 & u_4 \end{pmatrix}$$

where $\alpha \in \mathbb{R}$ and $u_1, u_2, u_3, u_4 \in U(1)$ (i.e. u_1, u_2, u_3, u_4 are complex numbers with length 1). Find α, u_1, u_2, u_3, u_4.
(ii) Find the cosine-sine decomposition of the unitary matrix

$$U = \frac{1}{\sqrt{2}} \begin{pmatrix} 1 & 1 \\ 1 & -1 \end{pmatrix}.$$

Solution 12. (i) Matrix multiplication yields

$$\begin{pmatrix} 0 & i \\ -i & 0 \end{pmatrix} = \begin{pmatrix} u_1 u_3 \cos(\alpha) & u_1 u_4 \sin(\alpha) \\ -u_2 u_3 \sin(\alpha) & u_2 u_4 \cos(\alpha) \end{pmatrix}.$$

Since $u_1, u_2, u_3, u_4 \neq 0$ we obtain $\cos(\alpha) = 0$. We select the solution $\alpha = \pi/2$. Since $\sin(\pi/2) = 1$ it follows that $u_1 u_4 = i$, $u_2 u_3 = i$. Thus we can select the solution $u_1 = u_2 = 1$, $u_3 = u_4 = i$.
(ii) We have ($\alpha \in \mathbb{R}$)

$$\frac{1}{\sqrt{2}} \begin{pmatrix} 1 & 1 \\ 1 & -1 \end{pmatrix} = \begin{pmatrix} u_1 & 0 \\ 0 & u_2 \end{pmatrix} \begin{pmatrix} \cos(\alpha) & \sin(\alpha) \\ -\sin(\alpha) & \cos(\alpha) \end{pmatrix} \begin{pmatrix} u_3 & 0 \\ 0 & u_4 \end{pmatrix}$$

where $u_1, u_2, u_3, u_4 \in \mathbb{C}$ with $|u_1| = |u_2| = |u_3| = |u_4| = 1$. Matrix multiplication yields

$$\frac{1}{\sqrt{2}} \begin{pmatrix} 1 & 1 \\ 1 & -1 \end{pmatrix} = \begin{pmatrix} u_1 u_3 \cos(\alpha) & u_1 u_4 \sin(\alpha) \\ -u_2 u_3 \sin(\alpha) & u_2 u_4 \cos(\alpha) \end{pmatrix}.$$

We obtain the four equations

$$\frac{1}{\sqrt{2}} = u_1 u_3 \cos(\alpha), \qquad \frac{1}{\sqrt{2}} = -u_2 u_4 \cos(\alpha),$$

$$\frac{1}{\sqrt{2}} = u_1 u_4 \sin(\alpha), \qquad \frac{1}{\sqrt{2}} = -u_2 u_3 \sin(\alpha)$$

with a solution $\alpha = \pi/4$ and $u_1 = u_3 = u_4 = 1$, $u_2 = -1$. Hence the decomposition is

$$\frac{1}{\sqrt{2}} \begin{pmatrix} 1 & 1 \\ 1 & -1 \end{pmatrix} = \begin{pmatrix} 1 & 0 \\ 0 & -1 \end{pmatrix} \begin{pmatrix} 1/\sqrt{2} & 1/\sqrt{2} \\ -1/\sqrt{2} & 1/\sqrt{2} \end{pmatrix} \begin{pmatrix} 1 & 0 \\ 0 & 1 \end{pmatrix}.$$

Problem 13.   Find the cosine-sine decomposition of the 4×4 unitary matrix (*Bell matrix*)

$$\frac{1}{\sqrt{2}} \begin{pmatrix} 1 & 0 & 0 & 1 \\ 0 & 1 & 1 & 0 \\ 0 & 1 & -1 & 0 \\ 1 & 0 & 0 & -1 \end{pmatrix}.$$

Solution 13.   We have to solve

$$\frac{1}{\sqrt{2}} \begin{pmatrix} 1 & 0 & 0 & 1 \\ 0 & 1 & 1 & 0 \\ 0 & 1 & -1 & 0 \\ 1 & 0 & 0 & -1 \end{pmatrix} = \begin{pmatrix} U_1 & 0_2 \\ 0_2 & U_2 \end{pmatrix} \begin{pmatrix} C & S \\ -S & C \end{pmatrix} \begin{pmatrix} U_3 & 0_2 \\ 0_2 & U_4 \end{pmatrix}$$

where

$$C = \begin{pmatrix} \cos(\alpha_1) & 0 \\ 0 & \cos(\alpha_2) \end{pmatrix}, \qquad S = \begin{pmatrix} \sin(\alpha_1) & 0 \\ 0 & \sin(\alpha_2) \end{pmatrix}.$$

U_1, U_2, U_3, U_4 are 2×2 unitary matrices and 0_2 is the 2×2 zero matrix. Then from the condition we obtain the four matrix equations

$$U_1 C U_3 = \frac{1}{\sqrt{2}} \begin{pmatrix} 1 & 0 \\ 0 & 1 \end{pmatrix}, \qquad U_2 C U_4 = -\frac{1}{\sqrt{2}} \begin{pmatrix} 1 & 0 \\ 0 & 1 \end{pmatrix},$$

$$U_2 S U_3 = -\frac{1}{\sqrt{2}} \begin{pmatrix} 0 & 1 \\ 1 & 0 \end{pmatrix}, \qquad U_1 S U_4 = \frac{1}{\sqrt{2}} \begin{pmatrix} 0 & 1 \\ 1 & 0 \end{pmatrix}.$$

The solution is

$$C = \frac{1}{\sqrt{2}} \begin{pmatrix} 1 & 0 \\ 0 & 1 \end{pmatrix}, \qquad S = \frac{1}{\sqrt{2}} \begin{pmatrix} 1 & 0 \\ 0 & 1 \end{pmatrix}$$

and

$$U_1 = U_3 = \begin{pmatrix} 0 & 1 \\ 1 & 0 \end{pmatrix}, \qquad U_2 = \begin{pmatrix} -1 & 0 \\ 0 & -1 \end{pmatrix}, \qquad U_4 = \begin{pmatrix} 1 & 0 \\ 0 & 1 \end{pmatrix}.$$

Problem 14.   For any $n \times n$ matrix A there exists an $n \times n$ unitary matrix ($U^* = U^{-1}$) such that

$$U^* A U = T \tag{1}$$

where T is an $n \times n$ matrix in upper triangular form. Equation (1) is called a *Schur decomposition*. The diagonal elements of T are the eigenvalues of A. Note

that such a decomposition is not unique. An iterative algorithm to find a Schur decomposition for an $n \times n$ matrix is as follows.
It generates at each step matrices U_k and T_k ($k = 1, \ldots, n-1$) with the properties: each U_k is unitary, and each T_k has only zeros below its main diagonal in its first k columns. T_{n-1} is in upper triangular form, and $U = U_1 U_2 \cdots U_{n-1}$ is the unitary matrix that transforms A into T_{n-1}. We set $T_0 = A$. The k-th step in the iteration is as follows.
Step 1. Denote as A_k the $(n - k + 1) \times (n - k + 1)$ submatrix in the lower right portion of T_{k-1}.
Step 2. Determine an eigenvalue and the corresponding normalized eigenvector for A_k.
Step 3. Construct a unitary matrix N_k which has as its first column the normalized eigenvector found in step 2.
Step 4. For $k = 1$, set $U_1 = N_1$, for $k > 1$, set

$$U_k = \begin{pmatrix} I_{k-1} & 0 \\ 0 & N_k \end{pmatrix}$$

where I_{k-1} is the $(k - 1) \times (k - 1)$ identity matrix.
Step 5. Calculate $T_k = U_k^* T_{k-1} U_k$.
Apply the algorithm to the 3×3 symmetric matrix

$$A = \begin{pmatrix} 1 & 0 & 1 \\ 0 & 1 & 0 \\ 1 & 0 & 1 \end{pmatrix}.$$

Solution 14. Since $\operatorname{rank}(A) = 2$ one eigenvalue is 0. The corresponding normalized eigenvector of the eigenvalue 0 is

$$\frac{1}{\sqrt{2}}(1 \ 0 \ -1)^T.$$

Thus we can construct the unitary matrix

$$N_1 = U_1 = \frac{1}{\sqrt{2}} \begin{pmatrix} 1 & 0 & 1 \\ 0 & \sqrt{2} & 0 \\ -1 & 0 & 1 \end{pmatrix}.$$

The two other column vectors in the matrix we construct from the fact that they must be normalized and orthogonal to the first column vector and orthogonal to each other. Now

$$T_1 = U_1^* A U_1 = \begin{pmatrix} 0 & 0 & 0 \\ 0 & 1 & 0 \\ 0 & 0 & 2 \end{pmatrix}$$

which is already the solution of the problem. Thus the eigenvalues of A are 0, 1, 2.

Problem 15. Let A be an $n \times n$ matrix over $\mathbb{C}$. Then there exists an $n \times n$ unitary matrix Q, such that

$$Q^* A Q = D + N$$

where $D = \text{diag}(\lambda_1, \lambda_2, \ldots, \lambda_n)$ is the diagonal matrix composed of the eigenvalues of A and N is a strictly upper triangular matrix (i.e. N has zero entries on the diagonal). The matrix Q is said to provide a *Schur decomposition* of A. Let

$$A = \begin{pmatrix} 3 & 8 \\ -2 & 3 \end{pmatrix}, \qquad Q = \frac{1}{\sqrt{5}} \begin{pmatrix} 2i & 1 \\ -1 & -2i \end{pmatrix}.$$

Show that Q provides a Schur decomposition of A.

Solution 15.   Obviously, $Q^*Q = QQ^* = I_2$. Now

$$Q^*AQ = \frac{1}{5} \begin{pmatrix} -2i & -1 \\ 1 & 2i \end{pmatrix} \begin{pmatrix} 3 & 8 \\ -2 & 3 \end{pmatrix} \begin{pmatrix} 2i & 1 \\ -1 & -2i \end{pmatrix} = \begin{pmatrix} 3+4i & -6 \\ 0 & 3-4i \end{pmatrix}$$

$$= \begin{pmatrix} 3+4i & 0 \\ 0 & 3-4i \end{pmatrix} + \begin{pmatrix} 0 & -6 \\ 0 & 0 \end{pmatrix}.$$

Consequently, we obtained a Schur decomposition of the matrix A with the given Q.

Problem 16.   We say that a matrix is upper triangular if all their entries below the main diagonal are 0, and that it is strictly upper triangular if in addition all the entries on the main diagonal are equal to 1. Any invertible real $n \times n$ matrix A can be written as the product of three real $n \times n$ matrices

$$A = ODN$$

where N is strictly upper triangular, D is diagonal with positive entries, and O is orthogonal. This is known as the *Iwasawa decomposition* of the matrix A. The decomposition is unique. In other words, that if $A = O'D'N'$, where O', D' and N' are orthogonal, diagonal with positive entries and strictly upper triangular, respectively, then $O' = O$, $D = D'$ and $N' = N$.
(i) Find the Iwasawa decomposition of the matrix

$$A = \begin{pmatrix} 0 & 1 \\ 1 & 2 \end{pmatrix}.$$

(ii) Consider the 2×2 matrix

$$M = \begin{pmatrix} a & b \\ c & d \end{pmatrix}$$

where $a, b, c, d \in \mathbb{C}$ and $ad - bc = 1$. Thus M is an element of the Lie group $SL(2, \mathbb{C})$. The Iwasawa decomposition is given by

$$\begin{pmatrix} a & b \\ c & d \end{pmatrix} = \begin{pmatrix} \alpha & \beta \\ -\bar{\beta} & \bar{\alpha} \end{pmatrix} \begin{pmatrix} \delta^{-1/2} & 0 \\ 0 & \delta^{1/2} \end{pmatrix} \begin{pmatrix} 1 & \eta \\ 0 & 1 \end{pmatrix}$$

where $\alpha, \beta, \eta \in \mathbb{C}$ and $\delta \in \mathbb{R}^+$. Find α, β, δ and η.

Solution 16.   (i) We obtain

$$O = \begin{pmatrix} 0 & 1 \\ 1 & 0 \end{pmatrix}, \qquad D = \begin{pmatrix} 1 & 0 \\ 0 & 1 \end{pmatrix}, \qquad N = \begin{pmatrix} 1 & 2 \\ 0 & 1 \end{pmatrix}.$$

(ii) Matrix multiplication yields the four equations

$$a = \alpha\delta^{-1/2}, \quad b = \alpha\eta\delta^{-1/2} + \beta\delta^{1/2}, \quad c = -\bar{\beta}\delta^{-1/2}, \quad d = -\bar{\beta}\eta\delta^{-1/2} + \bar{\alpha}\delta^{1/2}$$

where $\alpha\bar{\alpha} + \beta\bar{\beta} = 1$. Solving these four equations yields

$$\delta = (|a|^2 + |b|^2)^{-1}, \quad \alpha = a\delta^{1/2}, \quad \beta = -\bar{c}\delta^{1/2}, \quad \eta = (\bar{a}b + \bar{c}d)\delta.$$

Problem 17. Let A be a unitary $n \times n$ matrix. Let P be an invertible $n \times n$ matrix. Let $B := AP$. Show that PB^{-1} is unitary.

Solution 17. Since A and P are invertible, the matrix B is also invertible. We have

$$\begin{aligned}
PB^{-1}(PB^{-1})^* &= PB^{-1}B^{-1*}P^* = P(AP)^{-1}(AP)^{-1*}P^* \\
&= P(P^{-1}A^{-1})(P^{-1}A^{-1})^*P^* = P(P^{-1}A^{-1})(A^{-1*}P^{-1*})P^* \\
&= P(P^{-1}A^{-1}AP^{-1*})P^* = PP^{-1}P^{-1*}P^* \\
&= I_n.
\end{aligned}$$

Problem 18. Show that every 2×2 matrix A of determinant 1 is the product of three elementary matrices. This means that matrix A can be written as

$$\begin{pmatrix} a_{11} & a_{12} \\ a_{21} & a_{22} \end{pmatrix} = \begin{pmatrix} 1 & x \\ 0 & 1 \end{pmatrix}\begin{pmatrix} 1 & 0 \\ y & 1 \end{pmatrix}\begin{pmatrix} 1 & z \\ 0 & 1 \end{pmatrix}. \tag{1}$$

Solution 18. We find the four equations

$$a_{11} = 1 + xy, \quad a_{12} = z(1 + xy), \quad a_{21} = y, \quad a_{22} = yz + 1. \tag{2}$$

Case $a_{21} \neq 0$. Then we solve the system of equations (2) for y, x, z in that order using all but the second equation. We obtain

$$y = a_{21}, \quad x = \frac{a_{11} - 1}{a_{21}}, \quad z = \frac{a_{22} - 1}{a_{21}}.$$

For these choices of x, y, z the second equation of (2) is also satisfied since

$$z(1 + xy) + x = a_{21}^{-1}(a_{11}a_{22} - 1) = a_{12}$$

where we used that $\det(A) = 1$.
Case $a_{21} = 0$. We have $y = 0$. Thus we have the representation

$$\begin{pmatrix} a_{11} & a_{12} \\ 0 & a_{11}^{-1} \end{pmatrix} = \begin{pmatrix} a_{11} & 0 \\ 0 & 1 \end{pmatrix}\begin{pmatrix} 1 & a_{12} \\ 0 & 1 \end{pmatrix}\begin{pmatrix} 1 & 0 \\ 0 & a_{11}^{-1} \end{pmatrix}.$$

Problem 19. Almost any 2×2 matrix A can be factored (*Gaussian decomposition*) as

$$\begin{pmatrix} a_{11} & a_{12} \\ a_{21} & a_{22} \end{pmatrix} = \begin{pmatrix} 1 & \alpha \\ 0 & 1 \end{pmatrix}\begin{pmatrix} \lambda & 0 \\ 0 & \mu \end{pmatrix}\begin{pmatrix} 1 & 0 \\ \beta & 1 \end{pmatrix}.$$

Find the decomposition of the matrix

$$A = \begin{pmatrix} 1 & 1 \\ 1 & 1 \end{pmatrix}.$$

Solution 19.   Matrix multiplication yields

$$\begin{pmatrix} 1 & 1 \\ 1 & 1 \end{pmatrix} = \begin{pmatrix} \lambda + \mu\alpha\beta & \alpha\mu \\ \beta\mu & \mu \end{pmatrix}.$$

We obtain the four equations $\lambda + \alpha\beta\mu = 1$, $\alpha\mu = 1$, $\beta\mu = 1$, $\mu = 1$. This leads to the unique solution $\mu = \alpha = \beta = 1$, $\lambda = 0$. Thus we have the decomposition

$$\begin{pmatrix} 1 & 1 \\ 1 & 1 \end{pmatrix} = \begin{pmatrix} 1 & 1 \\ 0 & 1 \end{pmatrix} \begin{pmatrix} 0 & 0 \\ 0 & 1 \end{pmatrix} \begin{pmatrix} 1 & 0 \\ 1 & 1 \end{pmatrix}.$$

Problem 20.   Consider the symmetric 3×3 matrix A over $\mathbb{R}$ and the orthogonal matrix O, respectively

$$A = \begin{pmatrix} 2 & -1 & 0 \\ -1 & 2 & -1 \\ 0 & -1 & 2 \end{pmatrix}, \qquad O = \begin{pmatrix} \cos(\phi) & -\sin(\phi) & 0 \\ \sin(\phi) & \cos(\phi) & 0 \\ 0 & 0 & 1 \end{pmatrix}.$$

Calculate $\tilde{A} = O^{-1}AO$. Can we find an angle ϕ such that $\tilde{a}_{12} = \tilde{a}_{21} = 0$?

Solution 20.   From the orthogonal matrix O we obtain the inverse via $\phi \to -\phi$

$$O^{-1} = \begin{pmatrix} \cos(\phi) & \sin(\phi) & 0 \\ -\sin(\phi) & \cos(\phi) & 0 \\ 0 & 0 & 1 \end{pmatrix}.$$

Thus

$$\tilde{A} = O^{-1}AO = \begin{pmatrix} 2 - 2\sin(\phi)\cos(\phi) & \sin^2(\phi) - \cos^2(\phi) & -\sin(\phi) \\ \sin^2(\phi) - \cos^2(\phi) & 2 + 2\sin(\phi)\cos(\phi) & -\cos(\phi) \\ -\sin(\phi) & -\cos(\phi) & 2 \end{pmatrix}.$$

Thus the condition is $\tilde{a}_{12} = \tilde{a}_{21} = \sin^2(\phi) - \cos^2(\phi) = 0$. This leads to the solution $\phi = \pi/4 + k\pi/2$ ($k \in \mathbb{Z}$) and the matrix $\tilde{A}$ takes the form

$$\tilde{A} = O^{-1}AO = \begin{pmatrix} 1 & 0 & -1/\sqrt{2} \\ 0 & 3 & -1/\sqrt{2} \\ -1/\sqrt{2} & -1/\sqrt{2} & 2 \end{pmatrix}$$

since $\sin(\pi/4) = \cos(\pi/4) = 1/\sqrt{2}$.

Problem 21.   Let A be an $m \times m$ matrix. Let B be an $n \times n$ matrix. Let X be an $m \times n$ matrix such that

$$AX = XB. \tag{1}$$

We can find non-singular matrices V and W such that $V^{-1}AV = J_A$, $W^{-1}BW = J_B$, where J_A, J_B are the Jordan canonical form of A and B, respectively. Show that from (1) it follows that $J_A Y = Y J_B$, where $Y := V^{-1}XW$.

Solution 21. From $AX = XB$ we have

$$V J_A V^{-1} X = X W J_B W^{-1}$$
$$\Rightarrow V J_A V^{-1} X W = X W J_B$$
$$\Rightarrow V J_A Y = X W J_B$$
$$\Rightarrow J_A Y = V^{-1} X W J_B$$
$$\Rightarrow J_A Y = Y J_B.$$

Problem 22. (i) Let A be a square nonsingular matrix of order n_A with LU factorization $A = P_A^T L_A U_A$ and B be a square nonsingular matrix of order n_B with LU factorization $B = P_B^T L_B U_B$. Find the LU factorization for $A \otimes B$.
(ii) Let A be a positive definite matrix of order n_A with Cholesky factor G_A and B be a positive definite matrix of order n_B with Cholesky factor G_B. Find the Cholesky factorization for $A \otimes B$.
(iii) Let A be an $m_A \times n_A$ matrix with linear independent columns and QR factorization $A = Q_A R_A$, where Q_A is an $m_A \times n_A$ matrix with orthonormal columns and R_A is an $n_A \times n_A$ upper triangular matrix. B is similar defined with $B = Q_B R_B$ as its QR factorization. Find the QR factorization for $A \otimes B$.
(iv) Let A be a square matrix of order n_A with Schur decomposition $A = U_A T_A U_A^T$, where U_A is unitary and T_A is upper triangular. Let B be a square matrix of order n_B with Schur decomposition $B = U_B T_B U_B^T$, where U_B is unitary and T_B is upper triangular. Find the Schur decomposition for $A \otimes B$.
(v) Let A be an $m_A \times n_A$ matrix with singular value decomposition $A = U_A \Sigma_A V_A^T$ and B be an $m_B \times n_B$ matrix with singular value decomposition $B = U_B \Sigma_B V_B^T$. Find the singular value decomposition for $A \otimes B$.

Solution 22. (i) We find

$$A \otimes B = (P_A^T L_A U_A) \otimes (P_B^T L_B U_B) = (P_A \otimes P_B)^T (L_A \otimes L_B)(U_A \otimes U_B).$$

(ii) We find

$$A \otimes B = (G_A^T G_A) \otimes (G_B^T G_B) = (G_A \otimes G_B)^T (G_A \otimes G_B).$$

(iii) We find

$$A \otimes B = (Q_A R_A) \otimes (Q_B R_B) = (Q_A \otimes Q_B)(R_A \otimes R_B).$$

(iv) We find

$$A \otimes B = (U_A T_A U_A^T) \otimes (U_B T_B U_B^T) = (U_A \otimes U_B)(T_A \otimes T_B)(U_A \otimes U_B)^T.$$

(v) We find

$$A \otimes B = (U_A \Sigma_A V_A^T) \otimes (U_B \Sigma_B V_B^T) = (U_A \otimes U_B)(\Sigma_A \otimes \Sigma_B)(V_A \otimes V_B)^T.$$

Problem 23.   Write the matrix

$$H = \begin{pmatrix} 1 & 1 & 1 & 1 \\ 1 & -i & -1 & i \\ 1 & -1 & 1 & -1 \\ 1 & i & -1 & -i \end{pmatrix}$$

as the product of two 4×4 matrices A and B such that each of these matrices has precisely two non-zero entries in each row.

Solution 23.   We have

$$A = \begin{pmatrix} 1 & 1 & 0 & 0 \\ 0 & 0 & 1 & 1 \\ 1 & -1 & 0 & 0 \\ 0 & 0 & 1 & -1 \end{pmatrix}, \quad B = \begin{pmatrix} 1 & 0 & 1 & 0 \\ 0 & 1 & 0 & 1 \\ 1 & 0 & -1 & 0 \\ 0 & -i & 0 & i \end{pmatrix}$$

with $H = AB$.

Problem 24.   An $n \times n$ matrix A is *persymmetric* if $J_n A J_n = A^T$ and *skew persymmetric* if $J_n A J_n = A^T$, where J_n is the $n \times n$ *exchange matrix*, i.e. the matrix with 1's on the counter diagonal and 0's otherwise. Hence $J_n^2 = I_n$. Let B be an $n \times n$ symmetric matrix.
(i) Show that $B + J_n B J_n$ is persymmetric.
(ii) Show that $B - J_n B J_n$ is skew persymmetric.
(iii) Show that B can be written as a sum of a persymmetric matrix and a skew persymmetric matrix.

Solution 24.   (i) We have that

$$J_n (B + J_n B J_n) J_n = J_n B J_n + J_n^2 B J_n^2 = B + J_n B J_n = (B + J_n B J_n)^T$$

since $B^T = B$ and $J_n^T = J_n$.
(ii) We have that

$$J_n (B - J_n B J_n) J_n = J_n B J_n - J_n^2 B J_n^2 = -B + J_n B J_n = -(B - J_n B J_n)^T$$

since $B^T = B$ and $J_n^T = J_n$.
(iii) Clearly

$$B = \frac{1}{2}(B + J_n B J_n) + \frac{1}{2}(B - J_n B J_n)$$

where $\frac{1}{2}(B + J_n B J_n)$ is persymmetric and $\frac{1}{2}(B - J_n B J_n)$ is skew persymmetric.

Problem 25.   Decompose the 3×3 matrix

$$\begin{pmatrix} 1 & 0 & 1 \\ 0 & 0 & 0 \\ 1 & 0 & -1 \end{pmatrix}$$

as the sum of a persymmetric and a skew persymmetric matrix.

Solution 25. The matrix

$$\frac{1}{2}\begin{pmatrix} 1 & 0 & 1 \\ 0 & 0 & 0 \\ 1 & 0 & -1 \end{pmatrix} + \frac{1}{2}\begin{pmatrix} 0 & 0 & 1 \\ 0 & 1 & 0 \\ 1 & 0 & 0 \end{pmatrix}\begin{pmatrix} 1 & 0 & 1 \\ 0 & 0 & 0 \\ 1 & 0 & -1 \end{pmatrix}\begin{pmatrix} 0 & 0 & 1 \\ 0 & 1 & 0 \\ 1 & 0 & 0 \end{pmatrix} = \begin{pmatrix} 0 & 0 & 1 \\ 0 & 0 & 0 \\ 1 & 0 & 0 \end{pmatrix}$$

is persymmetric, and the matrix

$$\frac{1}{2}\begin{pmatrix} 1 & 0 & 1 \\ 0 & 0 & 0 \\ 1 & 0 & -1 \end{pmatrix} - \frac{1}{2}\begin{pmatrix} 0 & 0 & 1 \\ 0 & 1 & 0 \\ 1 & 0 & 0 \end{pmatrix}\begin{pmatrix} 1 & 0 & 1 \\ 0 & 0 & 0 \\ 1 & 0 & -1 \end{pmatrix}\begin{pmatrix} 0 & 0 & 1 \\ 0 & 1 & 0 \\ 1 & 0 & 0 \end{pmatrix} = \begin{pmatrix} 1 & 0 & 0 \\ 0 & 0 & 0 \\ 0 & 0 & -1 \end{pmatrix}$$

is skew persymmetric. We have

$$\begin{pmatrix} 1 & 0 & 1 \\ 0 & 0 & 0 \\ 1 & 0 & -1 \end{pmatrix} = \begin{pmatrix} 0 & 0 & 1 \\ 0 & 0 & 0 \\ 1 & 0 & 0 \end{pmatrix} + \begin{pmatrix} 1 & 0 & 0 \\ 0 & 0 & 0 \\ 0 & 0 & -1 \end{pmatrix}.$$

Supplementary Problems

Problem 1. If $A \in \mathbb{R}^{n \times n}$, then there exists an orthogonal $Q \in \mathbb{R}^{n \times n}$ such that

$$Q^T A Q = \begin{pmatrix} R_{11} & R_{12} & \cdots & R_{1m} \\ 0 & R_{22} & \cdots & R_{2m} \\ \vdots & \vdots & \ddots & \vdots \\ 0 & 0 & \cdots & R_{mn} \end{pmatrix}$$

where each R_{ii} is either a 1×1 matrix or a 2×2 matrix having complex conjugate eigenvalues. Find Q for the 3×3 matrix

$$A = \begin{pmatrix} 0 & 1 & 0 \\ 2 & 0 & 3 \\ 0 & 4 & 0 \end{pmatrix}.$$

Then calculate $Q^T A Q$.

Problem 2. Find a cosine-sine decomposition of the matrices

$$\frac{1}{\sqrt{2}}\begin{pmatrix} i & i \\ i & -i \end{pmatrix}, \quad \frac{1}{\sqrt{2}}\begin{pmatrix} i & i \\ i & -i \end{pmatrix} \otimes \frac{1}{\sqrt{2}}\begin{pmatrix} i & i \\ i & -i \end{pmatrix}.$$

Problem 3. (i) Consider the 4×4 matrices

$$\Omega = \begin{pmatrix} 0 & 1 & 0 & 0 \\ -1 & 0 & 1 & 0 \\ 0 & -1 & 0 & 1 \\ 0 & 0 & -1 & 0 \end{pmatrix}, \quad \tilde{\Omega} = \begin{pmatrix} 0 & 0 & 1 & 0 \\ 0 & 0 & 0 & 1 \\ -1 & 0 & 0 & 0 \\ 0 & -1 & 0 & 0 \end{pmatrix}.$$

Can one find 4×4 permutation matrices P, Q such that $\Omega = P\tilde{\Omega}Q$?
(ii) Consider the $2n \times 2n$ matrices

$$
\Omega = \begin{pmatrix}
0 & 1 & 0 & \cdots & 0 & 0 & 0 \\
-1 & 0 & 1 & \cdots & 0 & 0 & \\
0 & -1 & 0 & \cdots & 0 & 0 & 0 \\
 & & & \ddots & & & \\
0 & 0 & 0 & \cdots & -1 & 0 & 1 \\
0 & 0 & 0 & \cdots & 0 & -1 & 0
\end{pmatrix}, \qquad
\tilde{\Omega} = \begin{pmatrix} 0_n & I_n \\ -I_n & 0_n \end{pmatrix}.
$$

Can one find $2n \times 2n$ permutation matrices P, Q such that $\Omega = P\tilde{\Omega}Q$?

Problem 4.   (i) Given the 3×2 matrix

$$
A = \begin{pmatrix} 0 & 1 \\ 1 & 1 \\ 1 & 0 \end{pmatrix}.
$$

Find the singular value decomposition of A.
(ii) Let $\alpha \in \mathbb{R}$. Find the singular value decomposition of the 2×3 matrix

$$
X(\alpha) = \begin{pmatrix} \cos(\alpha) & \sin(\alpha) & 0 \\ 0 & \cos(\alpha) & \sin(\alpha) \end{pmatrix}.
$$

Problem 5.   The Cholesky decomposition of the a positive semi-definite matrix M is the decomposition into a product of a lower triangular matrix L and its conjugate transpose L^*, i.e. $M = LL^*$. Find the Cholesky decomposition for the *density matrix* (pure state)

$$
\rho = \frac{1}{\sqrt{3}} \begin{pmatrix} 1 \\ -i \\ i \end{pmatrix} \frac{1}{\sqrt{3}} \begin{pmatrix} 1 & i & -i \end{pmatrix} \equiv \frac{1}{3} \begin{pmatrix} 1 & i & -i \\ -i & 1 & -1 \\ i & -1 & 1 \end{pmatrix}.
$$

Problem 6.   Show that (cosine-sine decomposition)

$$
\begin{pmatrix} 0 & 1 \\ 1 & 0 \end{pmatrix} = \begin{pmatrix} 1 & 0 \\ 0 & i \end{pmatrix} \begin{pmatrix} 0 & 1 \\ -1 & 0 \end{pmatrix} \begin{pmatrix} i & 0 \\ 0 & 1 \end{pmatrix}.
$$

Problem 7.   Let U be an $n \times n$ unitary matrix. The matrix U can always be diagonalized by a unitary matrix V such that

$$
U = V \begin{pmatrix} e^{i\theta_1} & \cdots & 0 \\ \vdots & \ddots & \vdots \\ 0 & \cdots & e^{i\theta_n} \end{pmatrix} V^*
$$

where $e^{i\theta_j}$, $\theta_j \in [0, 2\pi)$ are the eigenvalues of U. Let $n = 2$ and

$$
U = \begin{pmatrix} 0 & 1 \\ 1 & 0 \end{pmatrix}.
$$

Thus the eigenvalues of U are 1 and -1. Show that

$$U = V \begin{pmatrix} 1 & 0 \\ 0 & -1 \end{pmatrix} V^* \quad \text{with} \quad V = V^* = \frac{1}{\sqrt{2}} \begin{pmatrix} 1 & 1 \\ 1 & -1 \end{pmatrix}.$$

Problem 8. If $A \in SL(2, \mathbb{R})$, then it can be uniquely be written in the form

$$\begin{pmatrix} \cos(\phi) & \sin(\phi) \\ -\sin(\phi) & \cos(\phi) \end{pmatrix} \exp \begin{pmatrix} a & b \\ b & -a \end{pmatrix}.$$

Find this decomposition for the 2×2 nonnormal matrix

$$A = \begin{pmatrix} 1 & 1 \\ 0 & 1 \end{pmatrix}.$$

Chapter 9

Functions of Matrices

Let p be a polynomial

$$p(x) = \sum_{j=0}^{n} c_j x^j$$

then the corresponding matrix function of an $n \times n$ matrix A over $\mathbb{C}$ can be defined by

$$p(A) = \sum_{j=0}^{n} c_j A^j$$

with the convention $A^0 = I_n$. If a function f of a complex variable z has a *MacLaurin series expansion* $f(z) = \sum_{j=0}^{\infty} c_j z^j$ which converges for $|z| < R$ $(R > 0)$, then the matrix series

$$f(A) = \sum_{j=0}^{\infty} c_j A^j$$

converges, provided A is square and each of its eigenvalues has absolute value less than R. The most used matrix function is $\exp(A)$ defined by

$$\exp(A) := \sum_{j=0}^{\infty} \frac{A^j}{j!} \equiv \lim_{n \to \infty} \left(I_n + \frac{A}{n} \right)^n.$$

For the analytic functions $\sinh(z)$ and $\cosh(z)$ we have

$$\sinh(A) := \sum_{j=0}^{\infty} \frac{A^{2j+1}}{(2j+1)!}, \qquad \cosh(A) := \sum_{j=0}^{\infty} \frac{A^{2j}}{(2j)!}.$$

Let M be an $n \times n$ matrix over $\mathbb{C}$. If there exists a matrix B such that $B^2 = M$, then B is called a square root of M.

Problem 1. Consider the 2×2 nonnormal matrix

$$A = \begin{pmatrix} 3 & -4 \\ 1 & -1 \end{pmatrix}.$$

Calculate A^n, where $n \in \mathbb{N}$.

Solution 1. Note that the determinant of A is $+1$. Thus $\det(A^n) = +1$ for all n. We have

$$A^2 = \begin{pmatrix} 5 & -8 \\ 2 & -3 \end{pmatrix}, \qquad A^3 = \begin{pmatrix} 7 & -12 \\ 3 & -5 \end{pmatrix}, \qquad A^4 = \begin{pmatrix} 9 & -16 \\ 4 & -7 \end{pmatrix}.$$

This suggests the ansatz

$$A^n = \begin{pmatrix} 1+2n & -4n \\ n & 1-2n \end{pmatrix}. \tag{1}$$

We use *proof by induction*. We see that $n = 1$ in (1) gives A. Suppose that (1) is valid for $n = k$. Then we have

$$A^{k+1} = A^k A = \begin{pmatrix} 1+2k & -4k \\ k & 1-2k \end{pmatrix} \begin{pmatrix} 3 & -4 \\ 1 & -1 \end{pmatrix} = \begin{pmatrix} 3+2k & -4-4k \\ k+1 & -1-2k \end{pmatrix}$$

$$= \begin{pmatrix} 1+2(k+1) & -4(k+1) \\ k+1 & 1-2(k+1) \end{pmatrix}$$

showing that (1) is correct for $n = k+1$. Thus (1) is proved by induction.

Problem 2. Consider the 2×2 matrix $A = (a_{jk})$ over the complex numbers. Find $\exp(A)$.

Solution 2. If

$$(a_{11} - a_{22})^2 + 4a_{12}a_{21} \equiv (\operatorname{tr}(A))^2 - 4\det(A) = 0$$

then

$$e^A = e^{(a_{11}+a_{22})/2} \begin{pmatrix} 1 + (a_{11} - a_{22})/2 & a_{12} \\ a_{21} & 1 - (a_{11} - a_{22})/2 \end{pmatrix}.$$

If $(a_{11} - a_{22})^2 + 4a_{12}a_{21} \equiv (\operatorname{tr}(A))^2 - 4\det(A) \neq 0$, then

$$e^A = e^{(a_{11}+a_{22})/2} \begin{pmatrix} \cosh(\Delta) + \frac{a_{11}-a_{22}}{2} \frac{\sinh(\Delta)}{\Delta} & a_{12} \frac{\sinh(\Delta)}{\Delta} \\ a_{21} \frac{\sinh(\Delta)}{\Delta} & \cosh(\Delta) - \frac{a_{11}-a_{22}}{2} \frac{\sinh(\Delta)}{\Delta} \end{pmatrix}$$

where $\Delta := \sqrt{(a_{11} - a_{22})^2 + 4a_{12}a_{21}}$.

Problem 3. (i) Consider the 2×2 matrix

$$A = \begin{pmatrix} a_{11} & a_{12} \\ 0 & a_{22} \end{pmatrix}.$$

Find $\exp(tA)$.

(ii) Let $\alpha, \beta \in \mathbb{R}$ and

$$M(\alpha, \beta) = \begin{pmatrix} \alpha & -\beta \\ \beta & \alpha \end{pmatrix}.$$

Calculate $\exp(M(\alpha, \beta))$.

Solution 3. (i) If $a_{11} \neq a_{22}$ we have

$$\exp(tA) = \begin{pmatrix} e^{a_{11}t} & a_{12} \frac{e^{a_{11}t} - e^{a_{22}t}}{a_{11} - a_{22}} \\ 0 & e^{a_{22}t} \end{pmatrix}.$$

If $a_{11} = a_{22}$ we obtain

$$e^{tA} = \begin{pmatrix} e^{a_{11}t} & a_{12}te^{a_{11}t} \\ 0 & e^{a_{22}t} \end{pmatrix}.$$

(ii) Since M can be written as

$$M(\alpha, \beta) = \alpha I_2 + \beta \begin{pmatrix} 0 & -1 \\ 1 & 0 \end{pmatrix}$$

and the two matrices on the right-hand side commute we have

$$\exp(M(\alpha, \beta)) = \exp\left(\alpha \begin{pmatrix} 1 & 0 \\ 0 & 1 \end{pmatrix} \right) \exp\left(\beta \begin{pmatrix} 0 & -1 \\ 1 & 0 \end{pmatrix} \right)$$
$$= \begin{pmatrix} e^{\alpha} \cos(\beta) & -e^{\alpha} \sin(\beta) \\ e^{\alpha} \sin(\beta) & e^{\alpha} \cos(\beta) \end{pmatrix}.$$

Problem 4. Consider the 2×2 unitary matrix

$$U = \begin{pmatrix} 0 & 1 \\ 1 & 0 \end{pmatrix}.$$

Can we find an $\alpha \in \mathbb{R}$ such that $U = \exp(\alpha A)$, where

$$A = \begin{pmatrix} 0 & 1 \\ -1 & 0 \end{pmatrix}?$$

Solution 4. Since $A^2 = -I_2$, $A^3 = -A$, $A^4 = I_2$ etc. we obtain

$$e^{\alpha A} = I_2 \cos(\alpha) + A \sin(\alpha).$$

This leads to the matrix equation

$$\begin{pmatrix} \cos(\alpha) & \sin(\alpha) \\ -\sin(\alpha) & \cos(\alpha) \end{pmatrix} = \begin{pmatrix} 0 & 1 \\ 1 & 0 \end{pmatrix}$$

which has no solution.

Problem 5.   Let H be a hermitian matrix, i.e. $H = H^*$. Now $U := e^{iH}$ is a unitary matrix. Let

$$H = \begin{pmatrix} a & b \\ \bar{b} & a \end{pmatrix}, \qquad a \in \mathbb{R}, \ b \in \mathbb{C}$$

with $b \neq 0$.
(i) Calculate e^{iH} using the normalized eigenvectors of H to construct a unitary matrix V such that $V^* H V$ is a diagonal matrix.
(ii) Specify a, b such that we find the unitary matrix

$$U = \begin{pmatrix} 0 & 1 \\ 1 & 0 \end{pmatrix}.$$

Solution 5.   (i) The eigenvalues of H are given by $\lambda_1 = a + \sqrt{b\bar{b}}$, $\lambda_2 = a - \sqrt{b\bar{b}}$. The corresponding normalized eigenvectors are

$$\mathbf{v}_1 = \frac{1}{\sqrt{2}} \begin{pmatrix} 1 \\ \sqrt{b\bar{b}}/b \end{pmatrix}, \qquad \mathbf{v}_2 = \frac{1}{\sqrt{2}} \begin{pmatrix} 1 \\ -\sqrt{b\bar{b}}/b \end{pmatrix}.$$

Thus the unitary matrices V, V^* which diagonalize H are

$$V = \frac{1}{\sqrt{2}} \begin{pmatrix} 1 & 1 \\ \sqrt{b\bar{b}}/b & -\sqrt{b\bar{b}}/b \end{pmatrix}, \qquad V^* = \frac{1}{\sqrt{2}} \begin{pmatrix} 1 & b/\sqrt{b\bar{b}} \\ 1 & -b/\sqrt{b\bar{b}} \end{pmatrix}$$

with

$$D := V^* H V = \begin{pmatrix} a + \sqrt{b\bar{b}} & 0 \\ 0 & a - \sqrt{b\bar{b}} \end{pmatrix}.$$

From $U = e^{iH}$ it follows that $V^* U V = V^* e^{iH} V = e^{iV^* H V} = e^{iD}$. Thus

$$e^{iD} = \begin{pmatrix} e^{i(a+\sqrt{b\bar{b}})} & 0 \\ 0 & e^{i(a-\sqrt{b\bar{b}})} \end{pmatrix}$$

and since $V^* = V^{-1}$ the unitary matrix U is given by $U = V e^{iD} V^*$. We obtain

$$U = e^{ia} \begin{pmatrix} \cos(\sqrt{b\bar{b}}) & b/\sqrt{b\bar{b}} \sin(\sqrt{b\bar{b}}) \\ \sqrt{b\bar{b}}/b \sin(\sqrt{b\bar{b}}) & \cos(\sqrt{b\bar{b}}) \end{pmatrix}.$$

(ii) If $a = \pi/2$ and $b = -\pi/2$ we find the unitary matrix, where we used that $\cos(\pi/2) = 0$.

Problem 6.   Let $z \in \mathbb{C}$. Let A, B be $n \times n$ matrices over $\mathbb{C}$. We say that B is *invariant* with respect to A if

$$e^{zA} B e^{-zA} = B.$$

Obviously e^{-zA} is the inverse of e^{zA}. Show that, if this condition is satisfied, one has $[A, B] = 0_n$. If e^{zA} would be unitary we have $U B U^* = B$.

Solution 6. We set

$$f(z) := e^{zA}Be^{-zA}.$$

Then

$$\frac{df(z)}{dz} = e^{zA}ABe^{-zA} - e^{zA}BAe^{-zA} = e^{zA}([A,B])e^{-zA}.$$

Since $dB/dz = 0_n$ and z is arbitrary we find $[A,B] = 0_n$.

Problem 7. Let $z \in \mathbb{C}$.
(i) Consider the 2×2 matrices

$$A = \begin{pmatrix} 0 & 1 \\ 1 & 0 \end{pmatrix}, \quad B = \begin{pmatrix} b_{11} & b_{12} \\ b_{12} & b_{11} \end{pmatrix}.$$

Calculate $\exp(zA)$, $\exp(-zA)$ and $\exp(zA)B\exp(-zA)$. Calculate the commutator $[A,B]$.
(ii) Consider the 2×2 matrices

$$C = \begin{pmatrix} 0 & -i \\ i & 0 \end{pmatrix}, \quad D = \begin{pmatrix} d_{11} & d_{12} \\ -d_{12} & d_{11} \end{pmatrix}.$$

Calculate $\exp(zC)$, $\exp(-zC)$ and $\exp(zC)D\exp(-zC)$. Calculate the commutator $[C,D]$.
(iii) Consider the 2×2 matrices

$$E = \begin{pmatrix} 0 & 1 \\ 0 & 0 \end{pmatrix}, \quad F = \begin{pmatrix} f_{11} & f_{12} \\ 0 & f_{11} \end{pmatrix}.$$

Calculate $\exp(zE)$, $\exp(-zE)$ and $\exp(zE)F\exp(-zE)$. Calculate the commutator $[E,F]$.

Solution 7. (i) We find

$$e^{zA} = \begin{pmatrix} \cosh(z) & \sinh(z) \\ \sinh(z) & \cosh(z) \end{pmatrix} \Rightarrow e^{-zA} = \begin{pmatrix} \cosh(z) & -\sinh(z) \\ -\sinh(z) & \cosh(z) \end{pmatrix}$$

and

$$\exp(zA)\begin{pmatrix} b_{11} & b_{12} \\ b_{12} & b_{11} \end{pmatrix}\exp(-zA) = \begin{pmatrix} b_{11} & b_{12} \\ b_{12} & b_{11} \end{pmatrix} = B.$$

We obtain $[A,B] = 0_2$.
(ii) We find

$$e^{zC} = \begin{pmatrix} \cosh(z) & -i\sinh(z) \\ i\sinh(z) & \cosh(z) \end{pmatrix} \Rightarrow e^{-zC} = \begin{pmatrix} \cosh(z) & i\sinh(z) \\ -i\sinh(z) & \cosh(z) \end{pmatrix}$$

and

$$\exp(zC)D\exp(-zC) = D.$$

We obtain $[C,D] = 0_2$.
(iii) We find

$$e^{zE} = \begin{pmatrix} 1 & z \\ 0 & 1 \end{pmatrix} \Rightarrow e^{-zE} = \begin{pmatrix} 1 & -z \\ 0 & 1 \end{pmatrix}$$

and
$$\exp(zE) \begin{pmatrix} f_{11} & f_{12} \\ 0 & f_{11} \end{pmatrix} \exp(-zE) = \begin{pmatrix} f_{11} & f_{12} \\ 0 & f_{11} \end{pmatrix} = F.$$

We obtain $[E, F] = 0_2$.

Problem 8. Let σ_1, σ_2, σ_3 be the Pauli spin matrices. Consider the 2×2 matrix
$$U(\alpha, \beta, \gamma) = e^{-i\alpha\sigma_3/2} e^{-i\beta\sigma_2/2} e^{-i\gamma\sigma_3/2}$$

where α, β, γ are the three *Euler angles* with the range $0 \le \alpha < 2\pi$, $0 \le \beta \le \pi$ and $0 \le \gamma < 2\pi$. Show that
$$U(\alpha, \beta, \gamma) = \begin{pmatrix} e^{-i\alpha/2}\cos(\beta/2)e^{-i\gamma/2} & -e^{-i\alpha/2}\sin(\beta/2)e^{i\gamma/2} \\ e^{-i\alpha/2}\sin(\beta/2)e^{-i\gamma/2} & e^{i\alpha/2}\cos(\beta/2)e^{i\gamma/2} \end{pmatrix}. \qquad (1)$$

Solution 8. Since $\sigma_3^2 = I_2$ and $\sigma_2^2 = I_2$ we have
$$e^{-i\alpha\sigma_3/2} = I_2 \cosh(i\alpha/2) - \sigma_3 \sinh(i\alpha/2)$$
$$e^{-i\beta\sigma_2/2} = I_2 \cosh(i\beta/2) - \sigma_2 \sinh(i\beta/2)$$
$$e^{-i\gamma\sigma_3/2} = I_2 \cosh(i\gamma/2) - \sigma_3 \sinh(i\gamma/2)$$

where we used $\cosh(-z) = \cosh(z)$ and $\sinh(-z) = -\sinh(z)$. Using $(x \in \mathbb{R})$
$$\cosh(ix) \equiv \cos(x), \qquad \sinh(ix) \equiv i\sin(x)$$

we arrive at
$$U(\alpha, \beta, \gamma) = \begin{pmatrix} e^{-i\alpha/2} & 0 \\ 0 & e^{i\alpha/2} \end{pmatrix} \begin{pmatrix} \cos(\beta/2) & \sin(\beta/2) \\ -\sin(\beta/2) & \cos(\beta/2) \end{pmatrix} \begin{pmatrix} e^{-i\gamma/2} & 0 \\ 0 & e^{i\gamma/2} \end{pmatrix}.$$

Applying matrix multiplication equation (1) follows.

Problem 9. (i) Let $\alpha \in \mathbb{R}$. Consider the 3×3 symmetric matrix
$$A(\alpha) = \begin{pmatrix} 0 & 0 & \alpha \\ 0 & 0 & 0 \\ \alpha & 0 & 0 \end{pmatrix}.$$

Find $\exp(A(\alpha))$.
(ii) Consider the 3×3 matrix
$$B = \begin{pmatrix} 0 & -1 & 0 \\ 1 & 0 & 1 \\ 0 & 1 & 0 \end{pmatrix}.$$

Let $\alpha \in \mathbb{R}$. Find $\exp(\alpha B)$.
(iii) Let $a_{12}, a_{13}, a_{23} \in \mathbb{R}$. Consider the skew-symmetric 3×3 matrix
$$A(a_{12}, a_{13}, a_{23}) = \begin{pmatrix} 0 & a_{12} & a_{13} \\ -a_{12} & 0 & a_{23} \\ -a_{13} & -a_{23} & 0 \end{pmatrix}.$$

Find $\exp(A)$.

Solution 9.   (i) We obtain

$$e^{A(\alpha)} = \begin{pmatrix} \cosh(\alpha) & 0 & \sinh(\alpha) \\ 0 & 1 & 0 \\ \sinh(\alpha) & 0 & \cosh(\alpha) \end{pmatrix}.$$

(ii) We obtain

$$\exp(\alpha B) = \begin{pmatrix} 1 - \alpha^2/2 & -\alpha & -\alpha^2/2 \\ \alpha & 1 & \alpha \\ \alpha^2/2 & \alpha & \alpha^2/2 \end{pmatrix}.$$

(iii) Let $\Delta := \sqrt{a_{12}^2 + a_{13}^2 + a_{23}^2}$. Then $A^3 + \Delta^2 A = 0$ and we find

$$\exp(A) = I_3 + \frac{\sin(\Delta)}{\Delta} A + \frac{(1 - \cos(\Delta))}{\Delta^2} A^2.$$

Problem 10.   Let A be an $n \times n$ matrix over $\mathbb{C}$ with $A^2 = rA$, where $r \in \mathbb{C}$ and $r \neq 0$.
(i) Calculate e^{zA}, where $z \in \mathbb{C}$.
(ii) Let $U(z) = e^{zA}$. Let $z' \in \mathbb{C}$. Calculate $U(z)U(z')$.

Solution 10.   (i) From $A^2 = rA$, we obtain $A^3 = r^2 A$, $A^4 = r^3 A$, $\ldots$, $A^n = r^{n-1} A$. Thus we obtain

$$e^{zA} = \sum_{j=0}^{\infty} \frac{(zA)^j}{j!} = I_n + \frac{z}{1!} A + \frac{rz^2}{2!} A + \frac{r^2 z^3}{3!} A + \cdots + \frac{r^{n-1} z^n}{n!} A + \cdots$$

$$= I_n + \frac{A}{r} \left(rz + \frac{(rz)^2}{2!} + \frac{(rz)^3}{3!} + \cdots + \frac{(rz)^n}{n!} + \cdots \right)$$

$$= I_n + \frac{A}{r} \left(1 + rz + \frac{(rz)^2}{2!} + \frac{(rz)^3}{3!} + \cdots + \frac{(rz)^n}{n!} + \cdots \right) - \frac{A}{r}$$

$$= I_n + \frac{A}{r} (e^{rz} - 1).$$

(ii) Straightforward calculation yields

$$U(z)U(z') = I_n + \frac{A}{r} (e^{r(z+z')} - 1) = U(z + z').$$

This result is obvious since $[A, A] = 0_n$ and therefore $e^{zA} e^{z'A} = e^{(z+z')A}$.

Problem 11.   Can we find a 2×2 matrix B over the real numbers $\mathbb{R}$ such that

$$\sin(B) = \begin{pmatrix} 1 & 4 \\ 0 & 1 \end{pmatrix}?$$

$$(1)$$

Solution 11. Owing to the right-hand side of equation (1) the 2×2 matrix B can only be of the form

$$B = \begin{pmatrix} x & y \\ 0 & x \end{pmatrix}$$

with $x \neq 0$. Straightforward calculation yields

$$\sin \begin{pmatrix} x & y \\ 0 & x \end{pmatrix} = \sum_{k=0}^{\infty} \frac{(-1)^k}{(2k+1)!} \begin{pmatrix} x & y \\ 0 & x \end{pmatrix}^{2k+1} = \sum_{k=0}^{\infty} \frac{(-1)^k x^{2k+1}}{(2k+1)!} \begin{pmatrix} 1 & y/x \\ 0 & 1 \end{pmatrix}^{2k+1}$$

$$= \sum_{k=0}^{\infty} \frac{(-1)^k x^{2k+1}}{(2k+1)!} \begin{pmatrix} 1 & (2k+1)y/x \\ 0 & 1 \end{pmatrix}$$

$$= \begin{pmatrix} \sin(x) & y\cos(x) \\ 0 & \sin(x) \end{pmatrix}.$$

Thus $\sin(x) = 1$ and $y\cos(x) = 4$. From the first condition we find that $\cos(x) = 0$ and therefore the second condition $y\cos(x) = 4$ cannot be satisfied. Thus there is no such matrix.

Problem 12. Let A be an $n \times n$ matrix over $\mathbb{C}$. Assume that $A^2 = cI_n$, where $c \in \mathbb{R}$.

(i) Calculate $\exp(A)$.
(ii) Apply the result to the 2×2 matrix $(z \neq 0)$

$$B = \begin{pmatrix} 0 & z \\ -\bar{z} & 0 \end{pmatrix}.$$

Thus B is skew-hermitian, i.e. $\bar{B}^T = -B$.

Solution 12. (i) We have $A^3 = cA$, $A^4 = c^2 I_n$, $A^5 = c^2 A$, $A^6 = c^3 I_n$. Thus, in general

$$A^m = \begin{cases} c^{m/2} I_n & m \text{ even} \\ c^{(m-1)/2} A & m \text{ odd}. \end{cases}$$

It follows that

$$e^A = I_n + A + \frac{c}{2!}I_n + \frac{c}{3!}A + \frac{c^2}{4!}I_n + \frac{c^2}{5!}A + \frac{c^3}{6!}I_n + \cdots$$

$$= I_n\left(1 + \frac{c}{2!} + \frac{c^2}{4!} + \frac{c^3}{6!} + \cdots\right) + A\left(1 + \frac{c}{3!} + \frac{c^2}{5!} + \frac{c^3}{7!} + \cdots\right)$$

$$= I_n\left(1 + \frac{c}{2!} + \frac{c^2}{4!} + \frac{c^3}{6!} + \cdots\right) + \frac{A}{\sqrt{c}}\left(\sqrt{c} + \frac{c\sqrt{c}}{3!} + \frac{c^2\sqrt{c}}{5!} + \cdots\right)$$

$$= I_n\cosh(\sqrt{c}) + \frac{A}{\sqrt{c}}\sinh(\sqrt{c}).$$

(ii) We have $B^2 = -rI_2$, $r = z^*z$, $r > 0$. Thus the condition $B^2 = cI_2$ is satisfied with $c = -r$. It follows that

$$e^B = I_2\cosh(\sqrt{-r}) + \frac{B}{\sqrt{-r}}\sinh(\sqrt{-r}).$$

With $\sqrt{-r} = i\sqrt{r}$, $\cosh(i\sqrt{r}) = \cos(\sqrt{r})$, $\sinh(i\sqrt{r}) = i\sin(\sqrt{r})$ we obtain

$$e^B = I_2 \cos(\sqrt{r}) + \frac{B}{\sqrt{r}} \sin(\sqrt{r}).$$

Problem 13. For any $n \times n$ matrix A we have the identity

$$\det(e^A) \equiv \exp(\mathrm{tr}(A)).$$

(i) Any $n \times n$ unitary matrix U can be written as $U = \exp(iK)$, where K is a hermitian matrix. Assume that $\det(U) = -1$. What can be said about the trace of K?

(ii) Let

$$A = \begin{pmatrix} 2 & 3 \\ 7 & -2 \end{pmatrix}.$$

Calculate $\det(e^A)$.

Solution 13. (i) It follows that

$$-1 = \det(U) = \det(e^{iK}) = \exp(i\mathrm{tr}(K)).$$

Thus $\mathrm{tr}(K) = \pi$ $(\pm n2\pi)$ with $n \in \mathbb{N}$.

(ii) To calculate $\det(e^A)$ would be quite clumsy. We use the identity given above. Since $\mathrm{tr}(A) = 0$ we obtain $\det(\exp(A)) = e^0 = 1$.

Problem 14. Let A be an $n \times n$ matrix. Then $\exp(A)$ can also be calculated as

$$e^A = \lim_{m \to \infty} \left(I_n + \frac{A}{m} \right)^m.$$

Use this definition to show that $\det(e^A) \equiv \exp(\mathrm{tr}(A))$.

Solution 14. We have

$$\det(\exp(A)) = \det(\lim_{m \to \infty} (I_n + A/m)^m) = \lim_{m \to \infty} \det((I_n + A/m)^m)$$
$$= \lim_{m \to \infty} (\det(I_n + A/m))^m = \lim_{m \to \infty} (1 + (\mathrm{tr}(A))/m + O(m^{-2}))^m$$
$$= \exp(\mathrm{tr}(A)).$$

Problem 15. Let A, B be $n \times n$ matrices. Then we have the identity

$$\det(e^A e^B e^{-A} e^{-B}) \equiv \exp(\mathrm{tr}([A, B])).$$

Show that $\det(e^A e^B e^{-A} e^{-B}) = 1$.

Solution 15. Since $\mathrm{tr}([A, B]) = 0$ and $e^0 = 1$ we have $\det(e^A e^B e^{-A} e^{-B}) = 1$.

Problem 16. The *MacLaurin series* for $\arctan(z)$ is defined as

$$\arctan(z) = z - \frac{z^3}{3} + \frac{z^5}{5} - \frac{z^7}{7} + \cdots = \sum_{j=0}^{\infty} \frac{(-1)^j z^{2j+1}}{2j+1}$$

which converges for all complex values of z having absolute value less than 1, i.e. $|z| < 1$. Let A be an $n \times n$ matrix. Thus the series expansion

$$\arctan(A) = A - \frac{A^3}{3} + \frac{A^5}{5} - \frac{A^7}{7} + \cdots = \sum_{j=0}^{\infty} \frac{(-1)^j A^{2j+1}}{2j+1}$$

is well-defined for A if all eigenvalues λ of A satisfy $|\lambda| < 1$. Let

$$A = \frac{1}{2} \begin{pmatrix} 1 & 1 \\ 1 & 1 \end{pmatrix}.$$

Does $\arctan(A)$ exist?

Solution 16. The eigenvalues of A are 0 and 1. Thus $\arctan(A)$ does not exist.

Problem 17. Let A, B be $n \times n$ matrices over $\mathbb{C}$. Assume that

$$[A, [A, B]] = [B, [A, B]] = 0_n. \tag{1}$$

Show that

$$e^{A+B} = e^A e^B e^{-\frac{1}{2}[A,B]} \tag{2a}$$

$$e^{A+B} = e^B e^A e^{+\frac{1}{2}[A,B]}. \tag{2b}$$

Use the *technique of parameter differentiation*, i.e. consider the matrix-valued function

$$f(\epsilon) := e^{\epsilon A} e^{\epsilon B}$$

where ϵ is a real parameter. Then take the derivative of f with respect to ϵ.

Solution 17. If we differentiate f with respect to ϵ we find

$$\frac{df(\epsilon)}{d\epsilon} = A e^{\epsilon A} e^{\epsilon B} + e^{\epsilon A} e^{\epsilon B} B = (A + e^{\epsilon A} B e^{-\epsilon A}) f(\epsilon)$$

since $e^{\epsilon A} e^{-\epsilon A} = I_n$. We have $e^{\epsilon A} B e^{-\epsilon A} = B + \epsilon[A, B]$, where we have taken (1) into account. Thus we obtain the linear matrix-valued differential equation

$$\frac{df(\epsilon)}{d\epsilon} = ((A + B) + \epsilon[A, B]) f(\epsilon).$$

Since the matrix $A + B$ commutes with $[A, B]$ we may treat $A + B$ and $[A, B]$ as ordinary commuting variables and integrate this linear differential equation with the initial condition $f(0) = I_n$. We find

$$f(\epsilon) = e^{\epsilon(A+B)+(\epsilon^2/2)[A,B]} = e^{\epsilon(A+B)} e^{(\epsilon^2/2)[A,B]}$$

where the last form follows since $A + B$ commutes with $[A, B]$. If we set $\epsilon = 1$ and multiply both sides by $e^{-[A,B]/2}$ then (2a) follows. Likewise we can prove the second form of the identity (2b).

Problem 18.   Let A, B, C_2, ..., C_m, ... be $n \times n$ matrices over $\mathbb{C}$. The *Zassenhaus formula* is given by

$$\exp(A + B) = \exp(A)\exp(B)\exp(C_2)\cdots\exp(C_m)\cdots.$$

The left-hand side is called the *disentangled form* and the right-hand side is called the *undisentangled form*. Find $C_2, C_3, \ldots$, using the *comparison method*. In the comparison method the disentangled and undisentangled forms are expanded in terms of an ordering scalar α and matrix coefficients of equal powers of α are compared. From

$$\exp(\alpha(A + B)) = \exp(\alpha A)\exp(\alpha B)\exp(\alpha^2 C_2)\exp(\alpha^3 C_3)\cdots$$

we obtain

$$\sum_{k=0}^{\infty} \frac{\alpha^k}{k!}(A + B)^k = \sum_{r_0,r_1,r_2,r_3,\ldots=0}^{\infty} \frac{\alpha^{r_0+r_1+2r_2+3r_3+\ldots}}{r_0!r_1!r_2!r_3!\cdots}A^{r_0}B^{r_1}C_2^{r_2}C_3^{r_3}\cdots.$$

(i) Find C_2 and C_3.
(ii) Assume that $[A, [A, B]] = 0_n$ and $[B, [A, B]] = 0_n$. What conclusion can we draw for the Zassenhaus formula?

Solution 18.   (i) For α^2 we have the decompositions $(r_0, r_1, r_2) = (2, 0, 0)$, $(1, 1, 0)$, $(0, 2, 0)$, $(0, 0, 1)$. Thus we obtain

$$(A + B)^2 = A^2 + 2AB + B^2 + 2C_2.$$

Thus it follows that $C_2 = -\frac{1}{2}[A, B]$. For α^3 we obtain

$$(A + B)^3 = A^3 + 3A^2B + 3AB^2 + B^3 + 6AC_2 + 6BC_2 + 6C_3.$$

Using C_2 given above we obtain

$$C_3 = \frac{1}{3}[B, [A, B]] + \frac{1}{6}[A, [A, B]].$$

(ii) Since $[B, [A, B]] = 0_n$, $[A, [A, B]] = 0_n$ we find that $C_3 = C_4 = \cdots = 0_n$. Thus

$$\exp(\alpha(A + B)) = \exp(\alpha A)\exp(\alpha B)\exp(-\alpha^2[A, B]/2).$$

Problem 19.   Let A, B be $n \times n$ matrices and $t \in \mathbb{R}$. Show that

$$e^{t(A+B)} - e^{tA}e^{tB} = \frac{t^2}{2}(BA - AB) + \text{higher order terms in } t. \tag{1}$$

Solution 19. We have

$$e^{tA}e^{tB} = (I_n + tA + \frac{t^2}{2!}A^2 + \cdots)(I_n + tB + \frac{t^2}{2!}B^2 + \cdots)$$

$$= I_n + t(A+B) + \frac{t^2}{2!}(A^2 + B^2 + 2AB) + \cdots$$

and

$$e^{t(A+B)} = I_n + t(A+B) + \frac{t^2}{2!}(A+B)^2 + \cdots$$

$$= I_n + t(A+B) + \frac{t^2}{2!}(A^2 + B^2 + AB + BA) + \cdots.$$

Subtracting the two equations yields (1).

Problem 20. Let A be an $n \times n$ matrix with $A^3 = -A$ and $\alpha \in \mathbb{R}$. Calculate $\exp(\alpha A)$.

Solution 20. From $A^3 = -A$ it follows that $A^{2k+1} = (-1)^k A$, $k = 1, 2, \ldots$. Thus

$$\exp(\alpha A) = I_n + \sum_{k=1}^{\infty} \frac{\alpha^k A^k}{k!} = I_n + \sum_{k=0}^{\infty} \frac{\alpha^{2k+1} A^{2k+1}}{(2k+1)!} + \sum_{k=1}^{\infty} \frac{\alpha^{2k} A^{2k}}{(2k)!}$$

$$= I_n + \sum_{k=0}^{\infty} \frac{\alpha^{2k+1}(-1)^k A}{(2k+1)!} - \sum_{k=1}^{\infty} \frac{\alpha^{2k}(-1)^k A^2}{(2k)!}$$

$$= I_n + A\sin(\alpha) + A^2(1 - \cos(\alpha)).$$

Problem 21. Let X be an $n \times n$ matrix over $\mathbb{C}$. Assume that $X^2 = I_n$. Let Y be an arbitrary $n \times n$ matrix over $\mathbb{C}$. Let $z \in \mathbb{C}$.
(i) Calculate $\exp(zX)Y\exp(-zX)$ using the *Baker-Campbell-Hausdorff formula*

$$e^{zX}Ye^{-zX} = Y + z[X,Y] + \frac{z^2}{2!}[X,[X,Y]] + \frac{z^3}{3!}[X,[X,[X,Y]]] + \cdots.$$

(ii) Calculate $\exp(zX)Y\exp(-zX)$ by first calculating $\exp(zX)$ and $\exp(-zX)$ and then doing the matrix multiplication. Compare the two methods.

Solution 21. (i) Using $X^2 = I_n$ we find for the first three commutators

$$[X,[X,Y]] = [X, XY - YX] = 2(Y - XYX)$$
$$[X,[X,[X,Y]]] = 2^2[X,Y]$$
$$[X,[X,[X,[X,Y]]]] = 2^3(Y - XYX).$$

If the number of X's in the commutator is even (say, m, $m \geq 2$) we have

$$[X,[X,\ldots[X,Y]\ldots]] = 2^{m-1}(Y - XYX).$$

If the number of X's in the commutator is odd (say m, $m \geq 3$) we have

$$[X, [X, \ldots [X, Y] \ldots]] = 2^{m-1}[X, Y].$$

Thus

$$e^{zX}Ye^{-zX} = Y\left(1 + \frac{2^1 z^2}{2!} + \frac{2^3 z^4}{4!} + \cdots\right) + [X, Y]\left(z + \frac{2^2 z^3}{3!} + \cdots\right)$$
$$-XYX\left(\frac{2^1 z^2}{2!} + \frac{2^3 z^4}{4!} + \cdots\right).$$

Consequently

$$e^{zX}Ye^{-zX} = Y\cosh^2(z) + [X, Y]\sinh(z)\cosh(z) - XYX\sinh^2(z).$$

(ii) Using the expansion

$$e^{zX} = \sum_{j=0}^{\infty} \frac{(zX)^j}{j!}$$

and $X^2 = I_n$ we have

$$e^{zX} = I_n \cosh(z) + X\sinh(z), \quad e^{-zX} = I_n\cosh(z) - X\sinh(z).$$

Matrix multiplication yields

$$e^{zX}Ye^{-zX} = Y\cosh^2(z) + [X, Y]\sinh(z)\cosh(z) - XYX\sinh^2(z).$$

Problem 22. Let K be a hermitian matrix. Then $U := \exp(iK)$ is a unitary matrix. A method to find the hermitian matrix K from the unitary matrix U is to consider the principal logarithm of a matrix $A \in \mathbb{C}^{n \times n}$ with no eigenvalues on $\mathbb{R}^-$ (the closed negative real axis). This logarithm is denoted by $\log(A)$ and is the unique matrix B such that $\exp(B) = A$ and the eigenvalues of B have imaginary parts lying strictly between $-\pi$ and π. For $A \in \mathbb{C}^{n \times n}$ with no eigenvalues on $\mathbb{R}^-$ we have the following integral representation

$$\log(s(A - I_n) + I_n) = \int_0^s (A - I_n)(t(A - I_n) + I_n)^{-1}dt.$$

Thus with $s = 1$ we obtain

$$\log(A) = \int_0^1 (A - I_n)(t(A - I_n) + I_n)^{-1}dt.$$

Find $\log(U)$ of the unitary matrix

$$U = \frac{1}{\sqrt{2}}\begin{pmatrix} 1 & -1 \\ 1 & 1 \end{pmatrix}.$$

First test whether the method can be applied.

Solution 22. We calculate $\log(U)$ to find iK given by $U = \exp(iK)$. We set $B = iK$ in the following. The eigenvalues of U are given by

$$\lambda_1 = \frac{1}{\sqrt{2}}(1 + i), \qquad \lambda_2 = \frac{1}{\sqrt{2}}(1 - i).$$

Thus the condition to apply the integral representation is satisfied. We consider first the general case $U = (u_{jk})$ and then simplify to $u_{11} = u_{22} = 1/\sqrt{2}$ and $u_{21} = -u_{12} = 1/\sqrt{2}$. We obtain

$$t(U - I_2) + I_2 = \begin{pmatrix} 1 + t(u_{11} - 1) & tu_{12} \\ tu_{21} & 1 + t(u_{22} - 1) \end{pmatrix}$$

and

$$d(t) := \det(t(U - I_2) + I_2) = 1 + t(-2 + \mathrm{tr}(U)) + t^2(1 - \mathrm{tr}(U) + \det(U)).$$

Let $X \equiv \det(U) - \mathrm{tr}(U) + 1$. Then

$$(U - I_2)(t(U - I_2) + I_2)^{-1} = \frac{1}{d(t)} \begin{pmatrix} tX + u_{11} - 1 & u_{12} \\ u_{21} & tX + u_{22} - 1 \end{pmatrix}.$$

With $u_{11} = u_{22} = 1/\sqrt{2}$, $u_{21} = -u_{12} = 1/\sqrt{2}$ we obtain

$$d(t) = 1 + t(-2 + \sqrt{2}) + t^2(2 - \sqrt{2})$$

and $X = 2 - \sqrt{2}$. Thus the matrix takes the form

$$\frac{1}{d(t)} \begin{pmatrix} t(2 - \sqrt{2}) + 1/\sqrt{2} - 1 & -1/\sqrt{2} \\ 1/\sqrt{2} & t(2 - \sqrt{2}) + 1/\sqrt{2} - 1 \end{pmatrix}.$$

Since

$$\int_0^1 \frac{1}{d(t)} dt = \sqrt{2}\frac{\pi}{4}, \qquad \int_0^1 \frac{t}{d(t)} dt = \frac{1}{\sqrt{2}}\frac{\pi}{4}$$

we obtain

$$K = \begin{pmatrix} 0 & i\pi/4 \\ -i\pi/4 & 0 \end{pmatrix}.$$

Problem 23. Consider the matrix ($\zeta \in \mathbb{R}$)

$$S(\zeta) = \begin{pmatrix} \cosh(\zeta) & 0 & 0 & \sinh(\zeta) \\ 0 & \cosh(\zeta) & \sinh(\zeta) & 0 \\ 0 & \sinh(\zeta) & \cosh(\zeta) & 0 \\ \sinh(\zeta) & 0 & 0 & \cosh(\zeta) \end{pmatrix}.$$

(i) Show that the matrix is invertible, i.e. find the determinant.
(ii) Calculate the inverse of $S(\zeta)$.
(iii) Calculate

$$A := \frac{d}{d\zeta} S(\zeta) \Big|_{\zeta=0}$$

and then calculate $\exp(\zeta A)$.

(iv) Do the matrices $S(\zeta)$ form a group under matrix multiplication?

Solution 23.   (i) We find $\det(S(\zeta)) = (\cosh^2(\zeta) - \sinh^2(\zeta))^2 = 1$.

(ii) The inverse is obtained by setting $\zeta \to -\zeta$, i.e.

$$
S^{-1}(\zeta) = S(-\zeta) = \begin{pmatrix} \cosh(\zeta) & 0 & 0 & -\sinh(\zeta) \\ 0 & \cosh(\zeta) & -\sinh(\zeta) & 0 \\ 0 & -\sinh(\zeta) & \cosh(\zeta) & 0 \\ -\sinh(\zeta) & 0 & 0 & \cosh(\zeta) \end{pmatrix}.
$$

(iii) We have

$$
A = \begin{pmatrix} 0 & 0 & 0 & 1 \\ 0 & 0 & 1 & 0 \\ 0 & 1 & 0 & 0 \\ 1 & 0 & 0 & 0 \end{pmatrix}.
$$

Since $A^2 = I_4$ it follows that $\exp(\zeta A) = S(\zeta)$.

(iv) Yes. For $\zeta = 0$ we have the identity matrix I_4.

Problem 24.   Let M be an $n \times n$ matrix with $m_{jk} = 1$ for all $j, k = 1, \ldots, n$. Let $z \in \mathbb{C}$. Find $\exp(zM)$. Then consider the special case $zn = i\pi$

Solution 24.   Since

$$
M^k = \begin{pmatrix} n^{k-1} & \cdots & n^{k-1} \\ \vdots & \ddots & \vdots \\ n^{k-1} & \cdots & n^{k-1} \end{pmatrix}
$$

we obtain $e^{zM} = I_n + \frac{1}{n}(e^{zn} - 1)M$. If $zn = i\pi$ we obtain

$$
e^{zM} = I_n - \frac{2}{n}M.
$$

Problem 25.   Let X, Y be $n \times n$ matrices. Show that

$$
[e^X, Y] = \sum_{k=1}^{\infty} \frac{[X^k, Y]}{k!}.
$$

Solution 25.   We have

$$
[e^X, Y] = e^X Y - Y e^X = \left(I_n + X + \frac{X^2}{2!} + \cdots \right) Y - Y \left(I_n + X + \frac{X^2}{2!} + \cdots \right)
$$

$$
= [X, Y] + \frac{1}{2!}[X^2, Y] + \frac{1}{3!}[X^3, Y] + \cdots
$$

$$
= \sum_{k=1}^{\infty} \frac{[X^k, Y]}{k!}.
$$

Problem 26. Let A, B be $n \times n$ matrices over $\mathbb{C}$ and $\alpha \in \mathbb{C}$. The *Baker-Campbell-Hausdorff formula* states that

$$e^{\alpha A} B e^{-\alpha A} = B + \alpha [A, B] + \frac{\alpha^2}{2!} [A, [A, B]] + \cdots = \sum_{j=0}^{\infty} \frac{\alpha^j}{j!} \{A^j, B\} = \tilde{B}(\alpha)$$

where $[A, B] := AB - BA$ and $\{A^j, B\} := [A, \{A^{j-1}, B\}]$ is the *repeated commutator*.

(i) Extend the formula to $e^{\alpha A} B^k e^{-\alpha A}$, where $k \geq 1$.
(ii) Extend the formula to $e^{\alpha A} e^B e^{-\alpha A}$.

Solution 26. (i) Using that $e^{-\alpha A} e^{\alpha A} = I_n$ we have

$$e^{\alpha A} B^k e^{-\alpha A} = e^{\alpha A} B e^{-\alpha A} e^{\alpha A} B^{k-1} e^{-\alpha A} = \tilde{B}(\alpha) e^{\alpha A} B^{k-1} e^{-\alpha A} = (\tilde{B}(\alpha))^k.$$

(ii) Using the result from (i) we obtain $e^{\alpha A} e^B e^{-\alpha A} = \exp(\tilde{B}(\alpha))$.

Problem 27. Consider the $n \times n$ matrix $(n \geq 2)$

$$A = \begin{pmatrix} 0 & 1 & 0 & \cdots & 0 \\ 0 & 0 & 1 & \cdots & 0 \\ \vdots & \vdots & \vdots & \ddots & \vdots \\ 0 & 0 & 0 & \cdots & 1 \\ 0 & 0 & 0 & \cdots & 0 \end{pmatrix}.$$

Let $f : \mathbb{R} \to \mathbb{R}$ be an analytic function. Calculate

$$f(0) I_n + \frac{f'(0)}{1!} A + \frac{f''(0)}{2!} A^2 + \cdots + \frac{f^{n-1}(0)}{(n-1)!} A^{n-1}$$

where $'$ denotes differentiation. Discuss.

Solution 27. Since A^n is the zero matrix the expansion given above is $f(A)$. We obtain

$$f(A) = \begin{pmatrix} f(0) & f'(0)/1! & \cdots & f^{(n-1)}(0)/(n-1)! \\ 0 & f(0) & \cdots & f^{(n-2)}(0)/(n-2)! \\ \vdots & \vdots & \ddots & \vdots \\ 0 & 0 & \cdots & f(0) \end{pmatrix}.$$

Problem 28. Let A, B be $n \times n$ matrices with $A^2 = I_n$ and $B^2 = I_n$. Assume that $[A, B] = 0_n$. Let $a, b \in \mathbb{C}$. Calculate $e^{aA + bB}$.

Solution 28. Since $[A, B] = 0_n$ we have $e^{aA + bB} = e^{aA} e^{bB}$. Now

$$e^{aA} = I_n + aA + \frac{1}{2!} a^2 A^2 + \frac{1}{3!} a^3 A^3 + \cdots = I_n + aA + \frac{1}{2!} a^2 I_n + \frac{1}{3!} a^3 A + \cdots$$

$$= I_n \left(1 + \frac{1}{2!} a^2 + \cdots \right) + A \left(a + \frac{1}{3!} a^3 + \cdots \right)$$

$$= I_n \cosh(a) + A \sinh(a).$$

Analogously $e^{bB} = I_n \cosh(b) + B \sinh(b)$. Thus

$$e^{aA+bB} = I_n \cosh(a)\cosh(b) + A\sinh(a)\cosh(b) + B\sinh(b)\cosh(a)$$
$$+ AB\sinh(a)\sinh(b).$$

Problem 29.   Let A, B be $n \times n$ matrices with $A^2 = I_n$ and $B^2 = I_n$. Assume that $[A, B]_+ = 0_n$. Let $a, b \in \mathbb{C}$. Calculate e^{aA+bB}.

Solution 29.   We have

$$e^{aA+bB} = I_n + (aA + bB) + \frac{1}{2!}(aA + bB)^2 + \frac{1}{3!}(aA + bB)^3 + \cdots.$$

Since

$$(aA + bB)^2 = a^2 A^2 + b^2 B^2 + ab(AB + BA) = (a^2 + b^2)I_n$$
$$(aA + bB)^3 = (a^2 + b^2)(aA + bB)$$

we have in general

$$(aA + bB)^n = (a^2 + b^2)^{n/2} I_n \quad n \text{ even}$$
$$(aA + bB)^n = (a^2 + b^2)^{(n-1)/2}(aA + bB) \quad n \text{ odd}.$$

Thus

$$e^{aA+bB} = I_n \cosh(\sqrt{a^2 + b^2}) + \frac{aA + bB}{\sqrt{a^2 + b^2}} \sinh(\sqrt{a^2 + b^2}).$$

Problem 30.   (i) Let A, B be $n \times n$ matrices over $\mathbb{C}$. Assume that $A^2 = I_n$ and $B^2 = I_n$. Find $\exp(\alpha A \otimes \beta B)$, where $\alpha, \beta \in \mathbb{C}$.
(ii) Let A, B, C be $n \times n$ matrices over $\mathbb{C}$ such that $A^2 = I_n$, $B^2 = I_n$ and $C^2 = I_n$. Furthermore assume that

$$[A, B]_+ = 0_n, \quad [B, C]_+ = 0_n, \quad [C, A]_+ = 0_n.$$

Let $\alpha, \beta, \gamma \in \mathbb{C}$. Calculate $e^{\alpha A + \beta B + \gamma C}$ using

$$e^{\alpha A + \beta B + \gamma C} = \sum_{j=0}^{\infty} \frac{(\alpha A + \beta B + \gamma C)^j}{j!}.$$

Solution 30.   (i) Since $(\alpha A \otimes \beta B)^2 = (\alpha^2 \beta^2) I_n \otimes I_n$ we find

$$\exp(\alpha A \otimes \beta B) = (I_n \otimes I_n) \cosh(\alpha\beta) + (A \otimes B) \sinh(\alpha\beta).$$

(ii) Since the anticommutators vanish we have

$$(\alpha A + \beta B + \gamma C)^2 = (\alpha^2 + \beta^2 + \gamma^2)^2 I_n$$
$$(\alpha A + \beta B + \gamma C)^3 = (\alpha^2 + \beta^2 + \gamma^2)(\alpha A + \beta B + \gamma C).$$

In general we have for positive n

$$(\alpha A + \beta B + \gamma C)^n = (\alpha^2 + \beta^2 + \gamma^2)^{n/2} I_n \quad \text{for} \quad n \text{ even}$$

and

$$(\alpha A + \beta B + \gamma C)^n = (\alpha^2 + \beta^2 + \gamma^2)^{n/2-1} \quad \text{for} \quad n \text{ odd.}$$

Thus we have the expansion ($\delta^2 := \alpha^2 + \beta^2 + \gamma^2$)

$$e^{\alpha A + \beta B + \gamma C} = I_n (1 + \frac{1}{2!}\delta^2 + \frac{1}{4!}(\delta^2)^2 + \frac{1}{6!}(\delta^2)^3 + \cdots)$$

$$+ (\alpha A + \beta B + \gamma C)(1 + \frac{1}{3!}(\delta^2) + \frac{1}{5!}(\delta^2)^2 + \cdots).$$

This can be summed up to

$$e^{\alpha A + \beta B + \gamma C} = I_n \cosh(\sqrt{\delta^2}) + \frac{\alpha A + \beta B + \gamma C}{\sqrt{\delta}} \sinh(\sqrt{\delta^2}).$$

Problem 31. Let A be an $n \times n$ matrix over $\mathbb{C}$. Assume that all eigenvalues $\lambda_1, \ldots, \lambda_n$ are pairwise distinct.
(i) Then e^{tA} can be calculated as follows (*Lagrange interpolation*)

$$e^{tA} = \sum_{j=1}^{n} e^{\lambda_j t} \prod_{\substack{k=1 \\ k \neq j}}^{n} \frac{(A - \lambda_k I_n)}{(\lambda_j - \lambda_k)}.$$

Consider the Pauli spin matrix σ_1. Calculate e^{tA} using this method.
(ii) e^{tA} can also be calculated as follows (*Newton interpolation*)

$$e^{tA} = e^{\lambda_1 t} I_n + \sum_{j=2}^{n} [\lambda_1, \ldots, \lambda_j] \prod_{k=1}^{j-1} (A - \lambda_k I_n).$$

The divided differences $[\lambda_1, \ldots, \lambda_j]$ depend on t and are defined recursively by

$$[\lambda_1, \lambda_2] := \frac{e^{\lambda_1 t} - e^{\lambda_2 t}}{\lambda_1 - \lambda_2}$$

$$[\lambda_1, \ldots, \lambda_{k+1}] := \frac{[\lambda_1, \ldots, \lambda_k] - [\lambda_2, \ldots, \lambda_{k+1}]}{\lambda_1 - \lambda_{k+1}}, \quad k \geq 2.$$

Calculate $e^{t\sigma_1}$ using this method.

Solution 31. (i) The eigenvalues of A are $\lambda_1 = 1$ and $\lambda_2 = -1$. Thus

$$e^{tA} = e^t \frac{A - \lambda_2 I_2}{\lambda_1 - \lambda_2} + e^{-t} \frac{A - \lambda_1 I_2}{\lambda_2 - \lambda_1}.$$

It follows that

$$e^{tA} = \frac{1}{2} e^t \left(\begin{pmatrix} 0 & 1 \\ 1 & 0 \end{pmatrix} + \begin{pmatrix} 1 & 0 \\ 0 & 1 \end{pmatrix} \right) - \frac{1}{2} e^{-t} \left(\begin{pmatrix} 0 & 1 \\ 1 & 0 \end{pmatrix} - \begin{pmatrix} 1 & 0 \\ 0 & 1 \end{pmatrix} \right).$$

Finally

$$e^{tA} = \frac{1}{2} \begin{pmatrix} e^t + e^{-t} & e^t - e^{-t} \\ e^t - e^{-t} & e^t + e^{-t} \end{pmatrix} = \begin{pmatrix} \cosh(t) & \sinh(t) \\ \sinh(t) & \cosh(t) \end{pmatrix}.$$

(ii) We have $n = 2$ and the eigenvalues of A are $\lambda_1 = 1$ and $\lambda_2 = -1$. Thus

$$e^{tA} = e^{\lambda_1 t} I_2 + [\lambda_1, \lambda_2] \prod_{k=1}^{1} (A - \lambda_k I_2) = e^t I_2 + [\lambda_1, \lambda_2](A - I_2).$$

Since

$$[\lambda_1, \lambda_2] = \frac{e^{\lambda_1 t} - e^{\lambda_2 t}}{\lambda_1 - \lambda_2} = \frac{e^t - e^{-t}}{2}$$

it follows that

$$e^{tA} = \frac{1}{2} \begin{pmatrix} e^t + e^{-t} & e^t - e^{-t} \\ e^t - e^{-t} & e^t + e^{-t} \end{pmatrix} = \begin{pmatrix} \cosh(t) & \sinh(t) \\ \sinh(t) & \cosh(t) \end{pmatrix}.$$

Problem 32.   Let A be an $n \times n$ matrix. The *characteristic polynomial*

$$\det(\lambda I_n - A) = \lambda^n + a_1 \lambda^{n-1} + \cdots + a_n = p(\lambda)$$

is closely related the *resolvent* $(\lambda I_n - A)^{-1}$ trough the formula

$$(\lambda I_n - A)^{-1} = \frac{N_1 \lambda^{n-1} + N_2 \lambda^{n-2} + \cdots + N_n}{\lambda^n + a_1 \lambda^{n-1} + \cdots + a_n} = \frac{N(\lambda)}{p(\lambda)}$$

where the adjugate matrix $N(\lambda)$ is a polynomial in λ of degree $n - 1$ with constant $n \times n$ coefficient matrices $N_1, \ldots, N_n$. The *Laplace transform* of the matrix exponential is the resolvent

$$\mathcal{L}(e^{tA}) = (\lambda I_n - A)^{-1}.$$

The N_k matrices and a_k coefficients may be computed recursively as follows

$$N_1 = I_n, \qquad a_1 = -\frac{1}{1} \mathrm{tr}(AN_1)$$

$$N_2 = AN_1 + a_1 I_n, \qquad a_2 = -\frac{1}{2} \mathrm{tr}(AN_2)$$

$$\vdots$$

$$N_n = AN_{n-1} + a_{n-1} I_n, \qquad a_n = -\frac{1}{n} \mathrm{tr}(AN_n)$$

$$0_n = AN_n + a_n I_n.$$

Consider the 2×2 unitary matrix

$$U = \frac{1}{\sqrt{2}} \begin{pmatrix} 1 & 1 \\ 1 & -1 \end{pmatrix}.$$

Find the N_k matrices and the coefficients a_k and thus calculate the resolvent.

Solution 32. We have $N_1 = I_2$, $a_1 = 0$, $N_2 = U$, $a_2 = -1$. Thus the resolvent is

$$(\lambda I_2 - U)^{-1} = \frac{1}{\lambda^2 - 1}(\lambda I_2 + U).$$

Problem 33. (i) Let A, B be $n \times n$ matrices over $\mathbb{C}$. Calculate $e^A B e^A$. Set $f(\epsilon) = e^{\epsilon A} B e^{\epsilon A}$, where ϵ is a real parameter. Then differentiate with respect to ϵ. For $\epsilon = 1$ we have $e^A B e^A$.
(ii) Let $\epsilon \in \mathbb{R}$. Calculate the matrix-valued function

$$f(\epsilon) = e^{-\epsilon \sigma_2} \sigma_3 e^{\epsilon \sigma_2}.$$

Differentiate the matrix-valued function f with respect to ϵ and solve the initial value problem of the resulting ordinary differential equation.

Solution 33. (i) From $f(\epsilon) = e^{\epsilon A} B e^{\epsilon A}$ we obtain

$$\frac{df(\epsilon)}{d\epsilon} = e^{\epsilon A} A B e^{\epsilon A} + e^{\epsilon A} B A e^{\epsilon A} = e^{\epsilon A}[A, B]_+ e^{\epsilon A}.$$

The second derivative yields

$$\frac{d^2 f(\epsilon)}{d\epsilon^2} = e^{\epsilon A} A[A, B]_+ e^{\epsilon A} + e^{\epsilon A}[A, B]_+ A e^{\epsilon A} = e^{\epsilon A}[A, [A, B]_+]_+ e^{\epsilon A}$$

etc. Using the *Taylor expansion* we obtain

$$f(\epsilon) = B + \frac{1}{1!}\frac{df}{d\epsilon} + \frac{1}{2!}\frac{d^2 f}{d\epsilon^2} + \frac{1}{3!}\frac{d^3 f}{d\epsilon^3} + \cdots$$

and therefore

$$e^A B e^A = B + \frac{1}{1!}[A, B]_+ + \frac{1}{2!}[A, [A, B]_+]_+ + \frac{1}{3!}[A, [A, [A, B]_+]_+]_+ + \cdots.$$

Since $e^A B e^{-A} \equiv e^A B e^A e^{-2A}$ we also have

$$e^A B e^{-A} = (B + \frac{1}{1!}[A, B]_+ + \frac{1}{2!}[A, [A, B]_+]_+ + \cdots)e^{-2A}.$$

(ii) We obtain

$$\frac{df(\epsilon)}{d\epsilon} = -e^{\epsilon \sigma_2}[\sigma_2, \sigma_3]e^{\epsilon \sigma_2} = -2ie^{-\epsilon \sigma_2}\sigma_1 e^{-\epsilon \sigma_2}$$

and

$$\frac{d^2 f(\epsilon)}{d\epsilon^2} = -2ie^{-\epsilon \sigma_2}2i\sigma_3 e^{\epsilon \sigma_2} = 4f(\epsilon).$$

Thus we have to solve the linear matrix differential equation

$$\frac{d^2 f(\epsilon)}{d\epsilon^2} = 4f(\epsilon)$$

with the initial conditions $f(0) = \sigma_3$ and $df(0)/d\epsilon = -2i\sigma_1$. We obtain the solution

$$f(\epsilon) = \frac{1}{2}(\sigma_3 - i\sigma_1)e^{2\epsilon} + \frac{1}{2}(\sigma_3 + i\sigma_1)e^{-2\epsilon}.$$

Problem 34. Let A be an $n \times n$ matrix over $\mathbb{C}$. Let T be a nilpotent matrix over $\mathbb{C}$ satisfying $T^*A + AT = 0_n$. Show that $(e^T)^*Ae^T = A$.

Solution 34. Since T is nilpotent we have $T^k = 0_n$ for some non-negative integer k. Thus

$$e^T = \sum_{j=0}^{k-1} \frac{1}{j!}T^j.$$

Thus

$$(e^T)^*A = e^{T^*}A = \sum_{j=0}^{k-1} \frac{1}{j!}(T^*)^jA = \sum_{j=0}^{k-1} \frac{(-1)^j}{j!}AT^j = Ae^{-T}.$$

It follows that $(e^T)^*Ae^T = A$.

Problem 35. (i) Let Π be an $n \times n$ projection matrix. Let $\epsilon \in \mathbb{R}$. Calculate $\exp(\epsilon\Pi)$.
(ii) Let $\Pi_1, \Pi_2, \ldots, \Pi_n$ be $n \times n$ projection matrices. Assume that $\Pi_j\Pi_k = 0_n$ ($j \neq k$) for all $j, k = 1, \ldots, n$. Let $\epsilon_j \in \mathbb{R}$ with $j = 1, \ldots, n$. Calculate

$$\exp(\epsilon_1\Pi_1 + \epsilon_2\Pi_2 + \cdots + \epsilon_n\Pi_n).$$

Assume additionally that $\Pi_1 + \Pi_2 + \cdots + \Pi_n = I_n$. Simplify the result from (i) using this condition.

Solution 35. (i) Using $\Pi^2 = \Pi$ we find

$$\exp(\epsilon\Pi) = I_n + \epsilon\Pi + \frac{1}{2!}\epsilon^2\Pi^2 + \frac{1}{3!}\epsilon^3\Pi^3 + \cdots$$
$$= I_n + \epsilon\Pi + \frac{1}{2!}\epsilon^2\Pi + \frac{1}{3!}\epsilon^3\Pi + \cdots$$
$$= I_n + (e^\epsilon - 1)\Pi.$$

(ii) Using $\Pi_j^2 = \Pi_j$ and $\Pi_j\Pi_k = 0_n$ for $j \neq k$ we find

$$e^{\epsilon_1\Pi_1 + \cdots + \epsilon_n\Pi_n} = (I_n + (e^{\epsilon_1} - 1)\Pi_1)(I_n + (e^{\epsilon_2} - 1)\Pi_2) \cdots (e^{\epsilon_n} - 1)\Pi_n)$$
$$= I_n + (e^{\epsilon_1} - 1)\Pi_1 + (e^{\epsilon_2} - 1)\Pi_2 + \cdots + (e^{\epsilon_n} - 1)\Pi_n.$$

Using $\Pi_1 + \Pi_2 + \cdots + \Pi_n = I_n$ the result simplifies to

$$e^{\epsilon_1\Pi_1 + \cdots + \epsilon_n\Pi_n} = \sum_{j=1}^{n} e^{\epsilon_j}\Pi_j.$$

Problem 36.   Let A, B be $n \times n$ matrices over $\mathbb{C}$. Assume that A and B commute with the commutator $[A, B]$. Then

$$\exp(A + B) = \exp(A)\exp(B)\exp\left(-\frac{1}{2}[A, B]\right).$$

Can this formula be applied to the matrices

$$C = \begin{pmatrix} 1 & 0 \\ 0 & 0 \end{pmatrix}, \qquad D = \begin{pmatrix} 0 & 1 \\ 0 & 0 \end{pmatrix}.$$

Solution 36.   We have

$$[C, D] = \begin{pmatrix} 0 & 1 \\ 0 & 0 \end{pmatrix} = D.$$

Now D does not commute with C. Thus the formula cannot be applied.

Problem 37.   Let $\epsilon \in \mathbb{R}$. Let A, B be $n \times n$ matrices over $\mathbb{C}$. Expand $e^{\epsilon A}e^{\epsilon B}e^{-\epsilon A}e^{-\epsilon B}$ up to second order in ϵ.

Solution 37.   There are no first order terms. We find

$$e^{\epsilon A}e^{\epsilon B}e^{-\epsilon A}e^{-\epsilon B} \approx I_n - \epsilon^2[A, B] + \epsilon^2(A^2 + B^2).$$

Problem 38.   Let $\alpha \in \mathbb{R}$. Consider the 2×2 matrix

$$A(\alpha) = \begin{pmatrix} \cos(\alpha) & \sin(\alpha) \\ \sin(\alpha) & -\cos(\alpha) \end{pmatrix}.$$

(i) Show that the matrix is orthogonal.
(ii) Find the determinant of $A(\alpha)$. Is the matrix an element of $SO(2, \mathbb{R})$?
(iii) Do these matrices form a group under matrix multiplication?
(iv) Calculate

$$X = \frac{d}{d\alpha}A(\alpha)\bigg|_{\alpha=0}.$$

Calculate $\exp(\alpha X)$ and compare this matrix with $A(\alpha)$. Discuss.
(v) Let $\beta \in \mathbb{R}$ and

$$B(\beta) = \begin{pmatrix} \cos(\beta) & \sin(\beta) \\ \sin(\beta) & -\cos(\beta) \end{pmatrix}.$$

Is the matrix $A(\alpha) \otimes B(\beta)$ orthogonal? Find the determinant of $A(\alpha) \otimes B(\alpha)$. Is this matrix an element of $SO(4, \mathbb{R})$?

Solution 38.   (i) We have $A(\alpha) = A^T(\alpha)$ and thus

$$A(\alpha)A^T(\alpha) = \begin{pmatrix} 1 & 0 \\ 0 & 1 \end{pmatrix}.$$

(ii) We obtain $\det(A(\alpha)) = -1$. Thus $A(\alpha)$ is not an element of $SO(2, \mathbb{R})$.

(iii) No since $\det(A(\alpha)) = -1$. No neutral element (identity matrix) exists.

(iv) We have

$$\frac{d}{dx}A(\alpha)\Big|_{\alpha=0} = \begin{pmatrix} 0 & 1 \\ 1 & 0 \end{pmatrix}, \quad \exp(\alpha \begin{pmatrix} 0 & 1 \\ 1 & 0 \end{pmatrix}) = \begin{pmatrix} \cosh(\alpha) & \sinh(\alpha) \\ \sinh(\alpha) & \cosh(\alpha) \end{pmatrix}.$$

(v) Yes. We have

$$(A(\alpha) \otimes B(\beta))^T = A^T(\alpha) \otimes B^T(\beta).$$

We find $\det(A(\alpha) \otimes B(\beta)) = (-1) \cdot (-1) = +1$. Yes $A(\alpha) \otimes B(\beta)$ is an element of $SO(4, \mathbb{R})$.

Problem 39. Let A be a normal matrix with eigenvalues $\lambda_1, \ldots, \lambda_n$ and corresponding normalized pairwise orthogonal eigenvectors $\mathbf{u}_1, \ldots, \mathbf{u}_n$ (column vectors). Let $\mathbf{w}, \mathbf{v} \in \mathbb{C}^n$. Find $\mathbf{w}^* \exp(A)\mathbf{v}$ by expanding $\mathbf{w}$ and $\mathbf{v}$ with respect to the basis $\mathbf{u}_j$ $(j = 1, \ldots, n)$.

Solution 39. Using the expansion

$$\mathbf{w} = \sum_{j=1}^{n} c_j \mathbf{u}_j, \qquad \mathbf{v} = \sum_{k=1}^{n} d_k \mathbf{u}_k$$

and $e^A \mathbf{u}_k = e^{\lambda_k} \mathbf{u}_k$ we obtain

$$\mathbf{w}^* e^A \mathbf{v} = \sum_{j=1}^{n}\sum_{j=1}^{n} c_j^* d_k e^{\lambda_k} \mathbf{u}_j^* \mathbf{u}_k = \sum_{j=1}^{n}\sum_{k=1}^{n} c_j^* d_k e^{\lambda_k} \delta_{jk} = \sum_{k=1}^{n} c_k^* d_k e^{\lambda_k}.$$

Problem 40. Let B be an $n \times n$ matrix with $B^2 = I_n$. Show that

$$\exp\left(-\frac{1}{2}i\pi(B - I_n)\right) \equiv B.$$

Solution 40. Let $z \in \mathbb{C}$. Then we have

$$e^{zB} \equiv \cosh(z)I_n + \sinh(z)B.$$

Setting $z = -i\pi/2$ we obtain the identity utilizing that $\sin(\pi/2) = 1$ and $\cos(\pi/2) = 0$.

Problem 41. Consider the 2×2 nonnormal matrices

$$A = \begin{pmatrix} 0 & 1 \\ 0 & 0 \end{pmatrix}, \qquad B = \begin{pmatrix} 0 & 0 \\ 1 & 0 \end{pmatrix} = A^*.$$

Are the matrices

$$\exp(A), \qquad \exp(B)$$

normal? Are the matrices $\sin(A)$, $\sin(B)$, $\cos(A)$, $\cos(B)$ normal?

Solution 41. We have

$$\exp(A) = \begin{pmatrix} 1 & 1 \\ 0 & 1 \end{pmatrix}, \qquad \exp(B) = \begin{pmatrix} 1 & 0 \\ 1 & 1 \end{pmatrix}.$$

Both matrices are nonnormal. We have

$$\sin(A) = A, \quad \sin(B) = B, \quad \cos(A) = I_2, \quad \cos(B) = I_2.$$

Thus $\sin(A)$, $\sin(B)$ are nonnormal and $\cos(A)$, $\cos(B)$ are normal.

Problem 42. Let $z \in \mathbb{C}$. Let A be an $n \times n$ matrix over $\mathbb{C}$. Assume that $A^3 = zA$. Find $\exp(A)$.

Solution 42. If $z = 0$ we have $\exp(A) = I_n + A + A^2/2$. If $z \neq 0$, then

$$\exp(A) = I_n + \frac{\sinh(\sqrt{z})}{\sqrt{z}} A + \frac{\cosh(\sqrt{z}) - 1}{z} A^2.$$

Problem 43. Let $\mathbf{v}_1$, $\mathbf{v}_2$ be an orthonormal set in $\mathbb{C}^2$. Consider the 2×2 matrix

$$A = -i\mathbf{v}_1\mathbf{v}_1^* + i\mathbf{v}_2\mathbf{v}_2^*.$$

Thus $I_2 = \mathbf{v}_1\mathbf{v}_1^* + \mathbf{v}_2\mathbf{v}_2^*$. Find K such that $\exp(K) = A$.

Solution 43. Since $(\mathbf{v}_j\mathbf{v}_j^*)^2 = \mathbf{v}_j\mathbf{v}_j^*$ $(j = 1, 2)$ we obtain

$$K = \left(-i\frac{\pi}{2} + 2ik_1\pi\right) \mathbf{v}_1\mathbf{v}_1^* + \left(i\frac{\pi}{2} + 2ik_2\pi\right) \mathbf{v}_2\mathbf{v}_2^*$$

where $k_1, k_2 \in \mathbb{Z}$. Thus

$$K = \pi \left(\frac{1}{2} - (k_1 - k_2)\right) A + i\pi(k_1 + k_2)I_2.$$

Problem 44. Any 2×2 matrix A can be written as a linear combination of the Pauli spin matrices and the 2×2 identity matrix I_2

$$A = aI_2 + b\sigma_1 + c\sigma_2 + d\sigma_3$$

where $a, b, c, d \in \mathbb{C}$.
(i) Find A^2 and A^3.
(ii) Use the result from (i) to find all matrices A such that $A^3 = \sigma_1$, i.e. calculate the third root $\sqrt[3]{\sigma_1}$.

Solution 44.   (i) We obtain

$$A^2 = (b^2 + c^2 + d^2 - a^2)I_2 + 2aA$$
$$A^3 = a(a^2 + 3b^2 + 3c^2 + 3d^2)I_2 + b(3a^2 + b^2 + c^2 + d^2)\sigma_1$$
$$+ c(3a^2 + b^2 + c^2 + d^2)\sigma_2 + d(3a^2 + b^2 + c^2 + d^2)\sigma_3.$$

(ii) The Pauli spin matrices σ_1, σ_2, σ_3 together with I_2 form a basis for the vector space of the 2×2 matrices. Setting $A^3 = \sigma_1$ yields the four equations

$$a(a^2 + 3b^2 + 3c^2 + 3d^2) = 0$$
$$b(3a^2 + b^2 + c^2 + d^2) = 1$$
$$c(3a^2 + b^2 + c^2 + d^2) = 0$$
$$d(3a^2 + b^2 + c^2 + d^2) = 0.$$

Thus $b \neq 0$ and $c = d = 0$. Case 1. If $a = 0$, then $b^3 = 1$. Therefore $A = b\sigma_1$, where $b \in \{1, e^{2\pi i/3}, e^{4\pi i/3}\}$. Case 2. If $a \neq 0$, then

$$a^2 + 3b^2 = 0, \qquad b(3a^2 + b^2) = 1.$$

It follows that $b^3 = -1/8$, i.e. $b = -1/2$ and $a = \pm\sqrt{3}i/2$. Thus

$$A = \frac{1}{2}\begin{pmatrix} \pm i\sqrt{3} & -1 \\ -1 & \pm i\sqrt{3} \end{pmatrix}.$$

Problem 45.   Let A, B, H, K be $n \times n$ matrices over $\mathbb{C}$. Assume that $A^2 = B^2 = 0_n$. Let $z \in \mathbb{C}$. Calculate

$$e^{z(A \otimes B)}(H \otimes K)e^{-z(A \otimes B)}$$

using the *Baker-Campbell-Hausdorff formula*.

Solution 45.   We have

$$[A \otimes B, H \otimes K] = (AH) \otimes (BK) - (HA) \otimes (KB) \qquad (1)$$

and using (1)

$$[A \otimes B, [A \otimes B, H \otimes K]] = -2(AHA) \otimes (BKB).$$

Since

$$[A \otimes B, (AHA) \otimes (BKB)] = (A^2HA) \otimes (B^2KB) - (AHA^2) \otimes (BKB^2)$$

we obtain

$$e^{z(A \otimes B)}(H \otimes K)e^{-z(A \otimes B)} = H \otimes K + z((AH) \otimes (BK) - (HA) \otimes (KB))$$
$$- z^2(AHA) \otimes (BKB).$$

Problem 46. (i) Let A, B be $n \times n$ matrices over $\mathbb{C}$, $z \in \mathbb{C}$ and $\mathbf{u}, \mathbf{v} \in \mathbb{C}^n$. Calculate

$$\exp(z(A \otimes B))(\mathbf{u} \otimes \mathbf{v}).$$

(ii) Simplify the result from (i) if $A^2 = I_n$ and $B^2 = I_n$.

Solution 46. (i) Since $(A \otimes B)(\mathbf{u} \otimes \mathbf{v}) = (A\mathbf{u}) \otimes (B\mathbf{v})$ we have

$$\exp(z(A \otimes B))(\mathbf{u} \otimes \mathbf{v}) = (\mathbf{u} \otimes \mathbf{v}) + \sum_{j=1}^{\infty} \frac{z^j}{j!} (A^j\mathbf{u}) \otimes (B^j\mathbf{v}).$$

(ii) If $A^2 = I_n$ and $B^2 = I_n$ we obtain

$$\exp(z(A \otimes B))(\mathbf{u} \otimes \mathbf{v}) = (\mathbf{u} \otimes \mathbf{v}) \cosh(z) + (A\mathbf{u}) \otimes (B\mathbf{v}) \sinh(z).$$

Problem 47. Let A be an $n \times n$ matrix. Calculate

$$\exp \begin{pmatrix} 0_n & A \\ A & 0_n \end{pmatrix}$$

using the identity

$$\begin{pmatrix} 0 & 1 \\ 1 & 0 \end{pmatrix} \otimes A \equiv \begin{pmatrix} 0_n & A \\ A & 0_n \end{pmatrix}.$$

Solution 47. We have

$$\exp \begin{pmatrix} 0_n & A \\ A & 0_n \end{pmatrix} = \sum_{j=0}^{\infty} \frac{1}{j!} \begin{pmatrix} 0_n & A \\ A & 0_n \end{pmatrix}^j$$

$$= I_{2n} + \begin{pmatrix} 0_n & A \\ A & 0_n \end{pmatrix} + \frac{1}{2!} \begin{pmatrix} 0_n & A \\ A & 0_n \end{pmatrix}^2 + \frac{1}{3!} \begin{pmatrix} 0_n & A \\ A & 0_n \end{pmatrix}^3 + \cdots$$

$$= I_{2n} + \begin{pmatrix} 0 & 1 \\ 1 & 0 \end{pmatrix} \otimes A + \frac{1}{2!} \begin{pmatrix} 0 & 1 \\ 1 & 0 \end{pmatrix}^2 \otimes A^2 + \cdots$$

$$= \begin{pmatrix} \cosh(A) & \sinh(A) \\ \sinh(A) & \cosh(A) \end{pmatrix}.$$

Problem 48. (i) Let $A_1, A_2, \ldots, A_m$ be $n \times n$ matrices with $A_j^2 = I_n$ for $j = 1, \ldots, m$. Let $z \in \mathbb{C}$. Calculate

$$\exp(z(A_1 \otimes A_2 \otimes \cdots \otimes A_m)).$$

(ii) Use the result from (i) to calculate $\exp(-i\theta(\sigma_1 \otimes \sigma_2 \otimes \sigma_3))$, where $\theta \in \mathbb{R}$.

Solution 48. (i) Since $(A_1 \otimes A_2 \otimes \cdots \otimes A_m)^2 = I_n \otimes I_n \otimes \cdots \otimes I_n$ we obtain

$$\exp(z(A_1 \otimes A_2 \otimes \cdots \otimes A_m)) = (I_n \otimes \cdots \otimes I_n) \cosh(z) + (A_1 \otimes \cdots \otimes A_m) \sinh(z).$$

(ii) Since $\cosh(-i\theta) = \cos(\theta)$ and $\sinh(-i\theta) = -i\sin(\theta)$ we have

$$\exp(-i\theta(\sigma_1 \otimes \sigma_2 \otimes \sigma_3)) = \cos(\theta)(I_2 \otimes I_2 \otimes I_2) - (i\sin(\theta))(\sigma_1 \otimes \sigma_2 \otimes \sigma_3).$$

Problem 49.   (i) Let A be an $n \times n$ matrix over $\mathbb{C}$ and Π be an $m \times m$ projection matrix. Let $z \in \mathbb{C}$. Calculate

$$\exp(z(A \otimes \Pi)).$$

(ii) Let A_1, A_2 be $n \times n$ matrices over $\mathbb{C}$. Let Π_1, Π_2 be $m \times m$ projection matrices with $\Pi_1\Pi_2 = 0_n$. Calculate

$$\exp(z(A_1 \otimes \Pi_1 + A_2 \otimes \Pi_2)).$$

(iii) Use the result from (ii) to find the unitary matrix

$$U(t) = \exp(-i\hat{H}t/\hbar)$$

where $\hat{H} = \hbar\omega(A_1 \otimes \Pi_1 + A_2 \otimes \Pi_2)$ and we assume that A_1 and A_2 are hermitian matrices.

Solution 49.   (i) We find $\exp(z(A \otimes \Pi)) = I_n \otimes I_m + (e^{zA} - I_n) \otimes \Pi$.
(ii) We obtain

$$\exp(z(A_1 \otimes \Pi_1 + A_2 \otimes \Pi_2)) = I_n \otimes I_m + (e^{zA_1} - I_n) \otimes \Pi_1 + (e^{zA_2} - I_n) \otimes \Pi_2.$$

(iii) Since

$$U(t) = \exp(-i\omega t(A_1 \otimes \Pi_1 + A_2 \otimes \Pi_2))$$

with $z = -i\omega t$ we obtain the unitary matrix

$$U(t) = I_n \otimes I_m + (e^{-i\omega tA_1} - I_n) \otimes \Pi_1 + (e^{-i\omega tA_2} - I_n) \otimes \Pi_2.$$

Problem 50.   Let A, B be $n \times n$ matrices over $\mathbb{C}$. Let $\mathbf{u}$ be an eigenvector of A and $\mathbf{v}$ be an eigenvector of B, respectively, i.e. $A\mathbf{u} = \lambda\mathbf{u}$, $B\mathbf{v} = \mu\mathbf{v}$.
(i) Find $e^{A \otimes B}(\mathbf{u} \otimes \mathbf{v})$.
(ii) Find $e^{A \otimes I_n + I_n \otimes B}(\mathbf{u} \otimes \mathbf{v})$.
(iii) Find $e^{A \otimes I_n + I_n \otimes B + A \otimes B}(\mathbf{u} \otimes \mathbf{v})$.

Solution 50.   (i) We have

$$e^{A \otimes B}(\mathbf{u} \otimes \mathbf{v}) = (I_n \otimes I_n + A \otimes B + \frac{1}{2!}A^2 \otimes B^2 + \frac{1}{3!}A^3 \otimes B^3 + \cdots)(\mathbf{u} \otimes \mathbf{v})$$

$$= \mathbf{u} \otimes \mathbf{v} + A\mathbf{u} \otimes B\mathbf{v} + \frac{1}{2!}(A^2\mathbf{u}) \otimes (B^2\mathbf{v}) + \cdots$$

$$= \mathbf{u} \otimes \mathbf{v} + \lambda\mathbf{u} \otimes \mu\mathbf{v} + \frac{1}{2!}\lambda^2\mathbf{u} \otimes \mu^2\mathbf{v} + \cdots$$

$$= (1 + \lambda\mu + \frac{1}{2!}\lambda^2\mu^2 + \cdots)(\mathbf{u} \otimes \mathbf{v})$$

$$= e^{\lambda\mu}(\mathbf{u} \otimes \mathbf{v}).$$

(ii) Since $e^{A \otimes I_n + I_n \otimes B} = e^A \otimes e^B$ we obtain

$$e^{A \otimes I_n + I_n \otimes B}(\mathbf{u} \otimes \mathbf{v}) = e^{\lambda + \mu}(\mathbf{u} \otimes \mathbf{v}).$$

(iii) Using $[A \otimes I_n + I_n \otimes B, A \otimes B] = 0_{n^2}$ we find

$$e^{A \otimes I_n + I_n \otimes B + A \otimes B}(\mathbf{u} \otimes \mathbf{v}) = e^{A \otimes I_n + I_n \otimes B} e^{A \otimes B}(\mathbf{u} \otimes \mathbf{v})$$
$$= (e^A \otimes e^B)(e^{\lambda \mu} \mathbf{u} \otimes \mathbf{v})$$
$$= e^{\lambda \mu} e^{\lambda} e^{\mu}(\mathbf{u} \otimes \mathbf{v})$$
$$= e^{\lambda + \mu + \lambda \mu}(\mathbf{u} \otimes \mathbf{v}).$$

Obviously this is an eigenvalue equation again.

Problem 51. For every positive definite matrix A, there is a unique positive definite matrix Q such that $Q^2 = A$. The matrix Q is called the *square root* of A. Can we find the square root of the matrix

$$B = \frac{1}{2}\begin{pmatrix} 5 & 3 \\ 3 & 5 \end{pmatrix}?$$

Solution 51. First we see that B is symmetric over $\mathbb{R}$ and therefore hermitian. The matrix is also positive definite since the eigenvalues are 1 and 4. Thus we can find an orthogonal matrix U such that $U^T B U = \text{diag}(4, 1)$ with U given by

$$U = \frac{1}{\sqrt{2}}\begin{pmatrix} 1 & -1 \\ 1 & 1 \end{pmatrix}.$$

Obviously the square root of the diagonal matrix $\text{diag}(4, 1)$ is $D = \text{diag}(2, 1)$. Then the square root of B is

$$Q = UDU^T = \frac{1}{2}\begin{pmatrix} 3 & 1 \\ 1 & 3 \end{pmatrix}.$$

Problem 52. Consider the 2×2 matrix

$$A = \begin{pmatrix} 1 & 1 \\ 1 & 2 \end{pmatrix}.$$

The matrix is symmetric over $\mathbb{R}$ and the eigenvalues are positive. Thus the matrix is positive definite. Show that

$$X = \frac{1}{\sqrt{5}}\begin{pmatrix} 2 & 1 \\ 1 & 3 \end{pmatrix}, \qquad Y = \begin{pmatrix} 0 & 1 \\ 1 & 1 \end{pmatrix}$$

are *square roots* of A. Discuss.

Solution 52. Direct calculation yields $X^2 = A$ and $Y^2 = A$. The matrix X is positive definite again, whereas the matrix Y is not positive definite (one negative eigenvalue). Note that $-X$ and $-Y$ are also square roots.

Problem 53.   Find the *square root* of the Pauli matrix σ_1 applying the *spectral theorem*.

Solution 53.   The eigenvalues of σ_1 are $\lambda_1 = +1$ and $\lambda_2 = -1$ with the corresponding normalized eigenvectors

$$\mathbf{v}_1 = \frac{1}{\sqrt{2}}\begin{pmatrix} 1 \\ 1 \end{pmatrix}, \quad \mathbf{v}_2 = \frac{1}{\sqrt{2}}\begin{pmatrix} 1 \\ -1 \end{pmatrix}.$$

Then the spectral representation of σ_1 is $\sigma_1 = \lambda_1\mathbf{v}_1\mathbf{v}_1^* + \lambda_2\mathbf{v}_2\mathbf{v}_2^*$. Now $\sqrt{1} = \pm 1$ and $\sqrt{i} = \pm i$. Thus we find the four roots of σ_1

$$R_1 = 1\mathbf{v}_1\mathbf{v}_1^* + i\mathbf{v}_2\mathbf{v}_2^*, \quad R_2 = 1\mathbf{v}_1\mathbf{v}_1^* - i\mathbf{v}_2\mathbf{v}_2^*$$
$$R_3 = -1\mathbf{v}_1\mathbf{v}_1^* + i\mathbf{v}_2\mathbf{v}_2^*, \quad R_4 = -1\mathbf{v}_1\mathbf{v}_1^* - i\mathbf{v}_2\mathbf{v}_2^*.$$

Note that $-R_1, -R_2, -R_3, -R_4$ are also square roots. Extend the result to $U \otimes U$.

Problem 54.   Let A be an $n \times n$ matrix over $\mathbb{C}$. The $n \times n$ matrix B over $\mathbb{C}$ is a square root of A iff $B^2 = A$. The number of square roots of a given matrix A may be zero, finite or infinite. Does the matrix

$$A = \begin{pmatrix} 0 & 1 \\ 0 & 0 \end{pmatrix}$$

admit a square root?

Solution 54.   From $B^2 = A$ we find the four conditions

$$b_{11}^2 + b_{12}b_{21} = 0, \quad b_{12}(b_{11} + b_{22}) = 1, \quad b_{21}(b_{11} + b_{22}) = 0, \quad b_{12}b_{21} + b_{22}^2 = 0.$$

From the second equation we find that $b_{11} + b_{22} \neq 0$. Thus from the third equation we have $b_{21} = 0$. Thus we arrive at

$$b_{11}^2 = 0, \quad b_{12}(b_{11} + b_{22}) = 1, \quad b_{22}^2 = 0.$$

Thus $b_{11} = b_{22} = 0$ and we find a contradiction since $b_{12}(b_{11} + b_{22}) = 1$. Hence A does not admit a square root.

Problem 55.   Find a square root of the nonnormal 3×3 matrix

$$A = \begin{pmatrix} 0 & 1 & 0 \\ 0 & 0 & 0 \\ 0 & 0 & 0 \end{pmatrix}$$

i.e. find the matrices B such that $B^2 = A$.

Solution 55.   Let $\beta \neq 0$. We find

$$B = \begin{pmatrix} 0 & \alpha & \beta \\ 0 & 0 & 0 \\ 0 & \beta^{-1} & 0 \end{pmatrix}$$

where α is arbitrary.

Problem 56. Consider the two-dimensional *rotation matrix*

$$R(\alpha) = \begin{pmatrix} \cos(\alpha) & -\sin(\alpha) \\ \sin(\alpha) & \cos(\alpha) \end{pmatrix}$$

with $0 \leq \alpha \leq \pi$. Find a square root $R^{1/2}(\alpha)$ of $R(\alpha)$, i.e. a matrix S such that $S^2 = R(\alpha)$.

Solution 56. Obviously $R^{1/2}$ is the rotation matrix with rotation angle $\alpha/2$

$$R^{1/2}(\alpha) = \begin{pmatrix} \cos(\alpha/2) & -\sin(\alpha/2) \\ \sin(\alpha/2) & \cos(\alpha/2) \end{pmatrix}.$$

Problem 57. Let V_1 be a hermitian $n \times n$ matrix. Let V_2 be a positive semidefinite $n \times n$ matrix. Let k be a positive integer. Show that

$$\mathrm{tr}((V_2 V_1)^k)$$

can be written as $\mathrm{tr}(V^k)$, where $V := V_2^{1/2} V_1 V_2^{1/2}$.

Solution 57. Since V_2 is positive semidefinite we calculate the square root. Thus applying cyclic invariance of the trace we have

$$\mathrm{tr}((V_2 V_1)^k) = \mathrm{tr}(V_2^{1/2} V_1 V_2^{1/2} V_2^{1/2} V_1 \cdots V_2^{1/2} V_1 V_2^{1/2}) = \mathrm{tr}(V^k).$$

Problem 58. Let A, B be $n \times n$ hermitian matrices. Then (*Golden-Thompson inequality*)

$$\mathrm{tr}(e^{A+B}) \leq \mathrm{tr}(e^A e^B).$$

Let $A = \sigma_3$ and $B = \sigma_1$. Calculate the left and right-hand side of the inequality.

Solution 58. We have $(z \in \mathbb{C})$

$$e^{z\sigma_1} = I_2 \cosh(z) + \sigma_1 \sinh(z), \qquad e^{z\sigma_3} = I_2 \cosh(z) + \sigma_3 \sinh(z).$$

Thus $\mathrm{tr}(e^{z\sigma_3} e^{z\sigma_1}) = 2\cosh^2(z)$. On the other hand since $(\sigma_1 + \sigma_3)^2 = 2I_2$ we have

$$e^{z(\sigma_1+\sigma_3)} = I_2 \cosh(\sqrt{2}z) + (\sigma_1 + \sigma_3) \sinh(\sqrt{2}z).$$

Thus

$$\mathrm{tr}(e^{z(\sigma_1+\sigma_3)}) = 2\cosh(\sqrt{2}z).$$

With $z = 1$ we have the inequality $\cosh(\sqrt{2}) < \cosh^2(1)$.

Problem 59. Let A be an $n \times n$ matrix over $\mathbb{R}$. The (p,q) *Padé approximation* to $\exp(A)$ is defined by

$$R_{pq}(A) := (D_{pq}(A))^{-1} N_{pq}(A)$$

where

$$N_{pq}(A) = \sum_{j=0}^{p} \frac{(p+q-j)!p!}{(p+q)!j!(p-j)!} A^j, \quad D_{pq}(A) = \sum_{j=0}^{q} \frac{(p+q-j)!q!}{(p+q)!j!(q-j)!} (-A)^j.$$

Nonsingularity of $D_{pq}(A)$ is assured if p and q are large enough or if the eigenvalues of A are negative. Find the Padé approximation for the matrix

$$A = \begin{pmatrix} 0 & 1 \\ 1 & 0 \end{pmatrix}$$

and $p = q = 2$. Compare with the exact solution.

Solution 59.   Since $A^2 = I_2$ we find

$$N_{22}(A) = \begin{pmatrix} 13/12 & 1/2 \\ 1/2 & 13/12 \end{pmatrix}, \quad D_{22}(A) = \begin{pmatrix} 13/12 & -1/2 \\ -1/2 & 13/12 \end{pmatrix}$$

with the inverse

$$(D_{22}(A))^{-1} = \frac{144}{133} \begin{pmatrix} 13/12 & 1/2 \\ 1/2 & 13/12 \end{pmatrix}.$$

Thus

$$R_{22}(A) = (D_{22}(A))^{-1} N_{22}(A) = \frac{1}{133} \begin{pmatrix} 205 & 156 \\ 156 & 205 \end{pmatrix}$$

which compares well with the exact solution

$$e^A = \begin{pmatrix} \cosh(1) & \sinh(1) \\ \sinh(1) & \cosh(1) \end{pmatrix}.$$

Problem 60.   Let A be an $n \times n$ matrix. We define the $j - k$ approximant of $\exp(A)$ by

$$f_{j,k}(A) := \left(\sum_{\ell=0}^{k} \frac{1}{\ell!} \left(\frac{A}{j} \right)^{\ell} \right)^j. \tag{1}$$

We have the inequality

$$\| e^A - f_{j,k}(A) \| \leq \frac{1}{j^k (k+1)!} \|A\|^{k+1} e^{\|A\|} \tag{2}$$

and $f_{j,k}(A)$ converges to e^A, i.e.

$$\lim_{j \to \infty} f_{j,k}(A) = \lim_{k \to \infty} f_{j,k}(A) = e^A.$$

Let

$$A = \begin{pmatrix} 0 & 1 \\ 0 & 0 \end{pmatrix}.$$

Find $f_{2,2}(A)$ and e^A. Calculate the right-hand side of the inequality (2).

Solution 60. We have

$$f_{2,2}(A) = \left(\sum_{\ell=0}^{2} \frac{1}{\ell!}\left(\frac{A}{2}\right)^{\ell}\right)^2 = \left(I_2 + \frac{1}{2}A + \frac{1}{2! \cdot 4}A^2\right)^2.$$

Since $A^2 = 0_2$ we obtain

$$f_{2,2}(A) = \left(I_2 + \frac{1}{2}A\right)^2 = \begin{pmatrix} 1 & 1 \\ 0 & 1 \end{pmatrix}.$$

For $\exp(A)$ we have

$$e^A = \begin{pmatrix} 1 & 1 \\ 0 & 1 \end{pmatrix}.$$

Thus the approximant gives the correct result for $\exp(A)$.

Problem 61. The *Denman-Beavers* iteration for the *square root* of an $n \times n$ matrix A with no eigenvalues on $\mathbb{R}^-$ is

$$Y_{k+1} = \frac{1}{2}(Y_k + Z_k^{-1}), \quad Z_{k+1} = \frac{1}{2}(Z_k + Y_k^{-1})$$

with $k = 0, 1, 2, \ldots$ and $Z_0 = I_n$ and $Y_0 = A$. The iteration has the properties that

$$\lim_{k \to \infty} Y_k = A^{1/2}, \qquad \lim_{k \to \infty} Z_k = A^{-1/2}$$

and, for all k,

$$Y_k = A Z_k, \quad Y_k Z_k = Z_k Y_k, \quad Y_{k+1} = \frac{1}{2}(Y_k + A Y_k^{-1}).$$

(i) Show that the Denman-Beavers iteration be applied to the matrix

$$A = \begin{pmatrix} 1 & 1 \\ 1 & 2 \end{pmatrix}.$$

(ii) Find Y_1 and Z_1.

Solution 61. (i) For the eigenvalues of A we find

$$\lambda_{1,2} = \frac{1}{2}(3 \pm \sqrt{5}).$$

Thus there is no eigenvalue on $\mathbb{R}^-$ and the Denman-Beavers iteration can be applied.
(ii) We have $\det(A) = 1$. For the inverse of A we find

$$A^{-1} = \begin{pmatrix} 2 & -1 \\ -1 & 1 \end{pmatrix}$$

and therefore

$$Y_1 = \begin{pmatrix} 1 & 1/2 \\ 1/2 & 3/2 \end{pmatrix}, \quad Z_1 = \begin{pmatrix} 3/2 & -1/2 \\ -1/2 & 1/2 \end{pmatrix}.$$

For $n \to \infty$, Y_n converges to

$$\begin{pmatrix} 0.89443 & 0.44721 \\ 0.44721 & 1.34164 \end{pmatrix}.$$

Supplementary Problems

Problem 1.   Let $z_1, z_2, z_3 \in \mathbb{C}$. Consider the 2×2 matrix

$$A(z_1, z_2, z_3) = \begin{pmatrix} z_1 & z_2 \\ z_3 & 0 \end{pmatrix}.$$

Calculate $\exp(A(z_1, z_2, z_3))$.

Problem 2.   Let $\alpha \in \mathbb{R}$. Calculate the vector in $\mathbb{R}^2$

$$\exp\left(\alpha \begin{pmatrix} 0 & 1 \\ 0 & 0 \end{pmatrix}\right) \begin{pmatrix} 1 \\ 0 \end{pmatrix}$$

in two different ways. Compare and discuss.

Problem 3.   Let A, B be $n \times n$ matrices.
(i) Show that

$$[e^A, e^B] = e^A e^B - e^B e^A = \sum_{j,k=1}^{\infty} \frac{1}{j!k!} [A^j, B^k].$$

(ii) Let A, B be $n \times n$ matrices. Show that

$$e^{A+B} - e^A \equiv \int_0^1 e^{(1-t)A} B e^{t(A+B)} dt.$$

(ii) Assume that for all eigenvalues λ we have $\Re(\lambda) < 0$. Let B be an arbitrary $n \times n$ matrix over $\mathbb{C}$. Let

$$R(t) := \int_0^\infty e^{tA^*} B e^{tA} dt.$$

Show that the matrix R satisfies $R(t)A + A^*R(t) = -B$.

Problem 4.   Let A, B be $n \times n$ matrices over $\mathbb{C}$. Is $\text{tr}(e^A \otimes e^B) = \text{tr}(e^{A \otimes B})$? Prove or disprove.

Problem 5.   (i) Let $A_1, A_2, \ldots, A_p$ be $n \times n$ matrices over $\mathbb{C}$. The generalized *Trotter formula* is given by

$$\exp\left(\sum_{j=1}^n A_j\right) = \lim_{n \to \infty} f_n(\{A_j\}) \tag{1}$$

where the n-th approximant $f_n(\{A_j\})$ is defined by

$$f_n(\{A_j\}) := \left(\exp\left(\frac{1}{n}A_1\right)\exp\left(\frac{1}{n}A_2\right)\cdots\exp\left(\frac{1}{n}A_p\right)\right)^n.$$

Let $p = 2$ and

$$A_1 = \begin{pmatrix} 1 & 0 \\ 0 & 1 \end{pmatrix}, \qquad A_2 = \begin{pmatrix} 0 & 1 \\ 1 & 0 \end{pmatrix}.$$

Calculate the left and right-hand side of (1).
(ii) Let A, B be $n \times n$ matrices over $\mathbb{C}$ with $A^2 = B^2 = I_n$ and $[A, B]_+ = 0_n$.
Let $z \in \mathbb{C}$. The *Lie-Trotter formula* is given by

$$\exp(z(A + B)) = \lim_{p \to \infty} \left(e^{zA/p}e^{zB/p}\right)^p.$$

Calculate $e^{z(A+B)}$ using the right-hand side. Note that

$$e^{zA/p} = I_n \cosh(z/p) + A \sinh(z/p), \quad e^{zB/p} = I_n \cosh(z/p) + B \sinh(z/p).$$

Problem 6. (i) Let A, B be $n \times n$ matrices with $AB = BA$. Show that

$$\cos(A + B) \equiv \cos(A)\cos(B) - \sin(A)\sin(B).$$

(ii) Let $x, y \in \mathbb{R}$. We know that $\sin(x + y) \equiv \sin(x)\cos(y) + \cos(x)\sin(y)$. Let

$$C = \begin{pmatrix} 1 & 1 \\ 1 & 1 \end{pmatrix}, \qquad D = \begin{pmatrix} 0 & 1 \\ 1 & 0 \end{pmatrix}.$$

Is $\sin(C + D) = \sin(C)\cos(D) + \cos(C)\sin(D)$? Prove or disprove.

Problem 7. Let A, B be $n \times n$ matrices over $\mathbb{C}$. The *Laplace transform* $\mathcal{L}$ of the matrix exponential is the *resolvent*

$$\mathcal{L}(e^{tA}) = (\lambda I_n - A)^{-1}.$$

(i) Calculate $\mathcal{L}(e^{tA} \otimes e^{tB})$.
(ii) Let $\mathcal{L}^{-1}$ be the inverse Laplace transform. Find

$$\mathcal{L}^{-1}((\lambda I_n - A)^{-1} \otimes (\lambda I_n - B)^{-1}).$$

Problem 8. Let A be an $n \times n$ matrix. The *characteristic polynomial*

$$\det(\lambda I_n - A) = \lambda^n + a_1\lambda^{n-1} + \cdots + a_n = p(\lambda)$$

is closely related the *resolvent* $(\lambda I_n - A)^{-1}$ trough the formula

$$(\lambda I_n - A)^{-1} = \frac{N_1\lambda^{n-1} + N_2\lambda^{n-2} + \cdots + N_n}{\lambda^n + a_1\lambda^{n-1} + \cdots + a_n} = \frac{N(\lambda)}{p(\lambda)}$$

where the adjugate matrix $N(\lambda)$ is a polynomial in λ of degree $n-1$ with constant $n \times n$ coefficient matrices $N_1, \ldots, N_n$. The *Laplace transform* of the matrix exponential is the resolvent

$$\mathcal{L}(e^{tA}) = (\lambda I_n - A)^{-1}.$$

The N_k matrices and a_k coefficients may be computed recursively as follows

$$N_1 = I_n, \qquad a_1 = -\frac{1}{1}\mathrm{tr}(AN_1)$$

$$N_2 = AN_1 + a_1 I_n, \qquad a_2 = -\frac{1}{2}\mathrm{tr}(AN_2)$$

$$\vdots$$

$$N_n = AN_{n-1} + a_{n-1}I_n, \qquad a_n = -\frac{1}{n}\mathrm{tr}(AN_n)$$

$$0_n = AN_n + a_n I_n.$$

Show that

$$\mathrm{tr}(\mathcal{L}(e^{tA})) = \frac{1}{p(\lambda)}\frac{dp(\lambda)}{d\lambda}.$$

Problem 9.   Let A be an arbitrary $n \times n$ matrix. Can we conclude that $\exp(A^*) = (\exp(A))^*$?

Problem 10.   Let A be an $n \times n$ matrix over $\mathbb{C}$. Assume that A is hermitian, i.e. $A^* = A$. Thus A has only real eigenvalues. Assume that $A^5 + A^3 + A = 3I_n$. Show that $A = I_n$.

Problem 11.   Let A, B be $n \times n$ hermitian matrices. There exists $n \times n$ unitary matrices U and V (depending on A and B) such that

$$\exp(iA)\exp(iB) = \exp(iUAU^{-1} + iVBV^{-1}).$$

Consider $n = 2$ and

$$A = \begin{pmatrix} 0 & 1 \\ 1 & 0 \end{pmatrix}, \quad B = \frac{1}{\sqrt{2}}\begin{pmatrix} 1 & 1 \\ 1 & -1 \end{pmatrix} \Rightarrow [A, B] = \sqrt{2}\begin{pmatrix} 0 & -1 \\ 1 & 0 \end{pmatrix}.$$

Find U and V. Note that A and B are also unitary.

Problem 12.   Let A, B be $n \times n$ matrices over $\mathbb{C}$ such that $[A, B] = A$. What can be said about the commutator $[e^A, e^B]$?

Problem 13.   (i) Let A be an $n \times n$ nilpotent matrix. Is the matrix $\cos(A)$ invertible?
(ii) Let N be a *nilpotent* $n \times n$ matrix with $N^j = 0_n$ with $j \geq 1$. Is

$$\ln(I_n - N) = -\left(N + \frac{1}{2}N + \cdots + \frac{1}{j-1}N^{j-1}\right)$$

and $\exp(\ln(I_n - N)) = I_n - N$.

Problem 14. Let $H = H_1 + H_2 + H_3$ and H_1, H_2, H_3 be $n \times n$ hermitian matrices. Show that

$$e^{-\beta H} = e^{-\beta H_1/2} e^{-\beta H_2/2} e^{-\beta H_3} e^{-\beta H_2/2} e^{-\beta H_1/2} + \left(-\frac{1}{2}\beta\right)^3 S$$

where

$$S = \frac{1}{6}([[H_2 + H_3, H_1], H_1 + 2H_2 + 2H_3] + [[H_3, H_2], H_2 + 2H_3]).$$

Problem 15. (i) Let r, s be integers with $r \geq 2$ and $s \geq 2$. Let A, B be $n \times n$ matrices with $A^r = 0_n$ and $B^s = 0_n$. Calculate $\exp(A \otimes B)$.
(ii) Let A, B be the basis (given by $n \times n$ matrices) of the non-commutative two-dimensional Lie algebra with $[A, B] = A$. Calculate $\exp(A \otimes I_n + I_n \otimes A + B \otimes B)$.
(iii) Let X_1, X_2, X_3 be the basis (given by $n \times n$ matrices) of a Lie algebra with the commutation relations

$$[X_1, X_2] = 0_n, \quad [X_1, X_3] = X_1, \quad [X_2, X_3] = 0_n.$$

Calculate $\exp(X_1 + X_2 + X_3)$ and

$$\exp(X_1 \otimes I_n \otimes I_n + I_n \otimes X_2 \otimes I_n + I_n \otimes I_n \otimes X_3 + X_1 \otimes X_2 \otimes X_3).$$

Problem 16. Consider the 3×3 matrix

$$A = \begin{pmatrix} 1 & 0 & 1 \\ 0 & 1 & 0 \\ 1 & 0 & 1 \end{pmatrix} = I_3 + B, \qquad B = \begin{pmatrix} 0 & 0 & 1 \\ 0 & 0 & 0 \\ 1 & 0 & 0 \end{pmatrix}.$$

Calculate e^A using $e^A = e^{I_3} e^B$.

Problem 17. Let $z \in \mathbb{C}$. Construct all 2×2 matrices A and B over $\mathbb{C}$ such that

$$\exp(zA) B \exp(-zA) = e^{-z} B.$$

Problem 18. Let A be an $n \times n$ matrix over $\mathbb{C}$. Consider the *Taylor series*

$$(I_n + A)^{1/2} = I_n + \frac{1}{2}A - \frac{1}{2 \cdot 4}A^2 + \frac{1 \cdot 3}{2 \cdot 4 \cdot 6}A^3 - \cdots$$

and

$$(I_n + A)^{-1/2} = I_n - \frac{1}{2}A + \frac{1 \cdot 3}{2 \cdot 4}A^2 - \frac{1 \cdot 3 \cdot 5}{2 \cdot 4 \cdot 6}A^3 + \cdots$$

What is the condition (the norm) on A such that the Taylor series exist? Can it be applied to the matrix

$$A = \begin{pmatrix} 1 & 1 \\ 1 & 1 \end{pmatrix} ?$$

Note that for $n = 1$ we have the condition $-1 < A \le +1$.

Problem 19.   Let U be an $n \times n$ unitary matrix. Let $H = U + U^*$. Calculate $\exp(zH)$.

Problem 20.   Let A, B be $n \times n$ matrices over $\mathbb{C}$ and $\exp(A)\exp(B) = \exp(C)$. Then the matrix C can be given as an infinite series of commutators of A and B. Let $z \in \mathbb{C}$. We write

$$\exp(zA)\exp(zB) = \exp(C(zA, zB))$$

where

$$C(zA, zB) = \sum_{j=1}^{\infty} c_j(A, B)z^j.$$

Show that the expansion up to fourth order is given by

$$c_1(A, B) = A + B, \qquad c_2(A, B) = \frac{1}{2}[A, B],$$

$$c_3(A, B) = \frac{1}{12}[A, [A, B]] - \frac{1}{12}[B, [A, B]], \quad c_4(A, B) = -\frac{1}{24}[A, [B, [A, B]].$$

Problem 21.   Let $A_1, \ldots, A_m$ be $n \times n$ matrices. Show that

$$e^{A_1} \cdots e^{A_m} = \exp\left(\frac{1}{2} \sum_{j<k}^{m} [A_j, A_k]\right) \exp(A_1 + \cdots + A_m)$$

if the matrices A_j $(j = 1, \ldots, m)$ satisfy

$$[[A_j, A_k], A_\ell] = 0_n \quad \text{for all } j, k, \ell \in \{1, \ldots, m\}.$$

Problem 22.   Let A, B, C be $n \times n$ positive semidefinite matrices. We define

$$A \circ B := A^{1/2} B A^{1/2}$$

where $A^{1/2}$ denotes the unique positive square root of A. Is $A \circ (B \circ C) = (A \circ B) \circ C$? Prove or disprove.

Problem 23.   (i) Let σ_1 be the Pauli spin matrix. Show that

$$\exp\left(-\frac{1}{2}i\pi(\sigma_1 - I_2)\right) = \sigma_1.$$

(ii) Find all 2×2 matrices A and $c \in \mathbb{C}$ such that $\exp(c(A - I_2)) = A$. Note that

$$\exp(c(A - I_2)) = \exp(cA)\exp(-cI_2) = \exp(cA)\begin{pmatrix} e^{-c} & 0 \\ 0 & e^{-c} \end{pmatrix}.$$

Problem 24.   Let A, B be real symmetric positive definite matrices. Show that

$$\frac{1}{\det(A)}\det(A+B) \le \exp(\operatorname{tr}(A^{-1}B)).$$

Problem 25.   Let $\epsilon \in \mathbb{R}$. Let $I_n - \epsilon A$ be a positive definite matrix. Show that

$$\exp(\operatorname{tr}(\ln(I_n - \epsilon A))) = \det(I_n - \epsilon A)$$

using the identity $\det(e^M) \equiv \exp(\operatorname{tr}(M))$.

Problem 26.   Let $\mathbf{v}$ be a normalized (column) vector in $\mathbb{C}^n$ and let A be an $n \times n$ hermitian matrix. Is $\mathbf{v}^* \exp(A)\mathbf{v} \ge \exp(\mathbf{v}^* A\mathbf{v})$ for all normalized $\mathbf{v}$? Prove or disprove.

Problem 27.   (i) Let $\epsilon \in \mathbb{R}$. Let A be an $n \times n$ matrix over $\mathbb{C}$. Find

$$\lim_{\epsilon \to 0} \frac{\sinh(2\epsilon A)}{\sinh(\epsilon)}.$$

(ii) Assume that $A^2 = I_n$. Calculate

$$\frac{\sinh(2\epsilon A)}{\sinh(\epsilon)}.$$

(iii) Assume that $A^2 = 0_n$. Calculate

$$\frac{\sinh(2\epsilon A)}{\sinh(\epsilon)}.$$

Utilize the expansion

$$\sinh(2\epsilon A) = 2\epsilon A + \frac{(2\epsilon)^3}{3!}A^3 + \cdots.$$

Problem 28.   Let $\alpha > 0$. Consider the 4×4 matrix

$$A(\alpha) = \begin{pmatrix} 0 & 0 & 0 & i/\alpha \\ 0 & 0 & i & 0 \\ 0 & i & 0 & 0 \\ \alpha i & 0 & 0 & 0 \end{pmatrix}.$$

Show that $\exp(\pi A(\alpha)) = -I_4$.

Problem 29.   Let V be the 2×2 matrix $V = v_0 I_2 + v_1\sigma_1 + v_2\sigma_2 + v_3\sigma_3$, where $v_0, v_1, v_2, v_3 \in \mathbb{R}$. Consider the equation

$$\exp(i\epsilon V) = (I_2 - iW)(I_2 + iW)^{-1}$$

where ϵ is real. Show that W as a function of V is given by

$$W = -\tan\left(\frac{1}{2}\epsilon(v_0 I_2 + v_1\sigma_1 + v_2\sigma_2 + v_3\sigma_3)\right).$$

Problem 30. Let A, B be $n \times n$ matrices. Show that

$$e^{A\otimes I_n + I_n \otimes A}e^{B\otimes I_n + I_n \otimes B} = (e^A \otimes e^B)(e^B \otimes e^B) = (e^A e^B) \otimes (e^A e^B).$$

Problem 31. Let $\alpha \in \mathbb{R}$ and Π be an $n \times n$ projection matrix. Let $Q = \Pi^{1/2}$ be a square root of Π, i.e. $Q^2 = \Pi$. Show that

$$U(\alpha) = \exp(i\alpha\Pi^{1/2}) = I_n + i\alpha\Pi^{1/2} + (\cos(\alpha) - 1)\Pi + i(\sin(\alpha) - \alpha)\Pi\Pi^{1/2}.$$

Problem 32. Let A be an $n \times n$ matrix over $\mathbb{R}$. Then we have the *Taylor expansion*

$$\sin(A) := \sum_{k=0}^{\infty} \frac{(-1)^k}{(2k+1)!} A^{2k+1}, \qquad \cos(A) := \sum_{k=0}^{\infty} \frac{(-1)^k}{(2k)!} A^{2k}.$$

To calculate $\sin(A)$ and $\cos(A)$ from a truncated Taylor series approximation is only worthwhile near the origin. We can use the repeated application of the *double angle formula*

$$\cos(2A) \equiv 2\cos^2(A) - I_n, \qquad \sin(2A) \equiv 2\sin(A)\cos(A).$$

We can find $\sin(A)$ and $\cos(A)$ of a matrix A from a suitably truncated Taylor series approximates as follows

$S_0 = $ Taylor approximate to $\sin(A/2^k)$, $C_0 = $ Taylor approximate to $\cos(A/2^k)$

and the recursion

$$S_j = 2S_{j-1}C_{j-1}, \qquad C_j = 2C_{j-1}^2 - I_n$$

where $j = 1, 2, \ldots$. Here k is a positive integer chosen so that, say $\|A\|_\infty \approx 2^k$. Apply this recursion to calculate sine and cosine of the 2×2 matrix

$$A = \begin{pmatrix} 2 & 1 \\ 1 & 2 \end{pmatrix}.$$

Use $k = 2$.

Problem 33. Let T be the 2×2 matrix

$$T = \frac{1}{3}\begin{pmatrix} 1 & 1 \\ 1 & 1 \end{pmatrix}.$$

Calculate $\ln(\det(I_2 - T))$ using the right-hand side of the identity

$$\ln(\det(I_2 - T)) \equiv -\sum_{k=1}^{\infty} \frac{1}{k}\operatorname{tr}(T^k).$$

Chapter 10

Cayley-Hamilton Theorem

Consider an $n \times n$ matrix A over $\mathbb{C}$ and the polynomial

$$p(\lambda) = \det(A - \lambda I_n)$$

with the characteristic equation

$$p(\lambda) = 0.$$

The Cayley-Hamilton theorem states that substituting the matrix A in the characteristic polynomial results in the $n \times n$ zero matrix, i.e.

$$p(A) = 0_n.$$

For example, let

$$A = \begin{pmatrix} 0 & -i \\ i & 0 \end{pmatrix}.$$

Then $\det(A - \lambda I_2) = \lambda^2 - 1 = 0$. Consequently

$$A^2 - I_2 = 0_2 \quad \Rightarrow \quad A^2 = I_2.$$

The Cayley-Hamilton theorem can be utilized to calculate A^2 from A and A^{-1} (if it exists) from A. The Cayley-Hamilton theorem can also be used to calculate $\exp(A)$ and other entire functions for an $n \times n$ matrix. Let A be an $n \times n$ matrix over $\mathbb{C}$. Let f be an *entire function*, i.e. an analytic function on the whole complex plane, for example $\exp(z)$, $\sin(z)$, $\cos(z)$. An infinite series expansion for $f(A)$ is an alternative for computing $f(A)$.

Problem 1. Let A be an $n \times n$ matrix over $\mathbb{C}$ and

$$p(\lambda) = \det(A - \lambda I_n).$$

Show that $p(A) = 0_n$, i.e. the matrix A satisfies its *characteristic equation* (*Cayley-Hamilton theorem*).

Solution 1. There exists an invertible $n \times n$ matrix X over $\mathbb{C}$ such that

$$X^{-1}AX = J = \operatorname{diag}(J_1, J_2, \ldots, J_t)$$

where

$$J_i = \begin{pmatrix} \lambda_i & 1 & 0 & \cdots & 0 \\ 0 & \lambda_i & 1 & \ddots & \vdots \\ \vdots & \ddots & \ddots & \ddots & \vdots \\ \vdots & \ddots & \ddots & \ddots & 1 \\ 0 & \cdots & \cdots & 0 & \lambda_i \end{pmatrix}$$

is an upper bidiagonal $m_i \times m_i$ matrix (*Jordan block*) that has on its main diagonal the eigenvalue λ_i of the matrix A (at least m_i-multiple eigenvalue of the matrix A since to this eigenvalue may correspond some more Jordan blocks) and $m_1 + \cdots + m_t = n$. Since $J_i - \lambda I_n = (\delta_{k,j-1})$, then $(\delta_{k,j-1})^{m_i} = 0$ and $(J_i - \lambda_i I_n)^{m_i} = 0$. If $p(\lambda)$ is the *characteristic polynomial* of the matrix A and the zeros of this polynomial are $\lambda_1, \ldots, \lambda_t$, then

$$p(\lambda) = (-1)^n (\lambda - \lambda_1)^{m_1} (\lambda - \lambda_2)^{m_2} \cdots (\lambda - \lambda_t)^{m_t}$$

and

$$p(J) = (-1)^n (J - \lambda_1 I_n)^{m_1} (J - \lambda_2 I_n)^{m_2} \cdots (J - \lambda_t I_n)^{m_t}.$$

We show that $p(J) = 0$. Let the matrix J have the block form

$$J = \begin{pmatrix} J_1 & 0 & 0 & \cdots & 0 \\ 0 & J_2 & 0 & \cdots & 0 \\ 0 & 0 & J_3 & \cdots & 0 \\ \vdots & \vdots & \vdots & \ddots & \vdots \\ 0 & 0 & 0 & \cdots & J_t \end{pmatrix}.$$

Then we obtain

$$p(J) = (-1)^n (J - \lambda_1 I_n)^{m_1} (J - \lambda_2 I_n)^{m_2} \cdots (J - \lambda_t I_n)^{m_t}$$

$$= (-1)^n \prod_{k=1}^{t} (J - \lambda_k I_n)^{m_k}$$

$$= (-1)^n \prod_{k=1}^{t} \begin{pmatrix} J_1 - \lambda_k I_{m_k} & 0 & \cdots & 0 \\ 0 & J_2 - \lambda_k I_{m_k} & \cdots & 0 \\ \vdots & \vdots & \ddots & \vdots \\ 0 & 0 & \cdots & J_t - \lambda_k I_{m_k} \end{pmatrix}^{m_k}$$

$$= (-1)^n \prod_{k=1}^{t} \begin{pmatrix} (J_1 - \lambda_k I_{m_k})^{m_k} & 0 & \cdots & 0 \\ 0 & (J_2 - \lambda_k I_{m_k})^{m_k} & \cdots & 0 \\ \vdots & \vdots & \ddots & \vdots \\ 0 & 0 & \cdots & (J_t - \lambda_k I_{m_k})^{m_k} \end{pmatrix}$$

$$= 0_n.$$

From the $X^{-1}AX = J$ it follows that $A = XJX^{-1}$. Thus

$$\begin{aligned} p(A) &= p(XJX^{-1}) \\ &= (-1)^n (XJX^{-1} - X\lambda_1 I_n X^{-1}) \cdots (XJX^{-1} - X\lambda_n I_n X^{-1}) \\ &= (-1)^n X(J - \lambda_1 I_n)X^{-1} X(J - \lambda_2 I_n)X^{-1} \cdots X(J - \lambda_n I_n)X^{-1} \\ &= Xp(J)X^{-1} \\ &= 0_n \end{aligned}$$

since $p(J) = 0_n$.

Problem 2. Let A be an arbitrary 2×2 matrix. Show that

$$A^2 - A\mathrm{tr}(A) + I_2 \det(A) = 0$$

and therefore $(\mathrm{tr}(A))^2 = \mathrm{tr}(A^2) + 2\det(A)$.

Solution 2. The characteristic polynomial of A is

$$\lambda^2 - (a_{11} + a_{22})\lambda + a_{11}a_{22} - a_{12}a_{21} = 0.$$

Applying the Cayley-Hamilton theorem we have

$$A^2 - (a_{11} + a_{22})A + (a_{11}a_{22} - a_{12}a_{21})I_2 = 0_2.$$

Since $\mathrm{tr}(A) = a_{11} + a_{22}$ and $\det(A) = a_{11}a_{22} - a_{12}a_{21}$ we obtain the result.

Problem 3. Consider the 2×2 nonnormal matrix

$$A = \begin{pmatrix} 1 & 1 \\ 0 & 1 \end{pmatrix}.$$

Calculate $f(A) = A^3$ using the Cayley-Hamilton theorem.

Solution 3. The characteristic equation $\det(A - \lambda I_2) = 0$ of A is given by $\lambda^2 - 2\lambda + 1 = 0$. Hence $A^2 = 2A - I_2$. Using this expression it follows that

$$A^3 = AA^2 = A(2A - I_2) = 2A^2 - A = 3A - 2I_2.$$

Consequently

$$A^3 = \begin{pmatrix} 3 & 3 \\ 0 & 3 \end{pmatrix} - \begin{pmatrix} 2 & 0 \\ 0 & 2 \end{pmatrix} = \begin{pmatrix} 1 & 3 \\ 0 & 1 \end{pmatrix}.$$

Problem 4. Apply the Cayley-Hamilton theorem to the 3×3 matrix $A = (a_{jk})$ and express the result using the trace and determinant of A.

Solution 4.   Let

$$S := a_{11}a_{22} + a_{11}a_{33} + a_{22}a_{33} - a_{13}a_{31} - a_{23}a_{32} - a_{12}a_{21}.$$

From $\det(A - \lambda I_3) = 0$ we obtain

$$\lambda^3 - \operatorname{tr}(A)\lambda^2 + S\lambda - \det(A) = 0.$$

Thus we have

$$A^3 - \operatorname{tr}(A)A^2 + SA - \det(A)I_3 = 0_3.$$

Problem 5.   Let A be an $n \times n$ matrix. Assume that the inverse matrix of A exists. The inverse matrix can be calculated as follows (*Csanky's algorithm*). Let

$$p(x) := \det(xI_n - A). \tag{1}$$

The roots are, by definition, the eigenvalues $\lambda_1, \lambda_2, \ldots, \lambda_n$ of A. We write

$$p(x) = x^n + c_1 x^{n-1} + \cdots + c_{n-1}x + c_n \tag{2}$$

where $c_n = (-1)^n \det(A)$. Since A is nonsingular we have $c_n \neq 0$ and vice versa. The Cayley-Hamilton theorem states that

$$p(A) = A^n + c_1 A^{n-1} + \cdots + c_{n-1}A + c_n I_n = 0_n. \tag{3}$$

Multiplying this equation with A^{-1} we obtain

$$A^{-1} = \frac{1}{-c_n}(A^{n-1} + c_1 A^{n-2} + \cdots + c_{n-1}I_n). \tag{4}$$

If we have the coefficients c_j we can calculate the inverse matrix A. Let

$$s_k := \sum_{j=1}^{n} \lambda_j^k.$$

Then the s_j and c_j satisfy the following $n \times n$ lower triangular system of linear equations

$$\begin{pmatrix} 1 & 0 & 0 & \cdots & 0 \\ s_1 & 2 & 0 & \cdots & 0 \\ s_2 & s_1 & 3 & \cdots & 0 \\ \vdots & \vdots & \vdots & \ddots & \vdots \\ s_{n-1} & s_{n-2} & \cdots & s_1 & n \end{pmatrix} \begin{pmatrix} c_1 \\ c_2 \\ c_3 \\ \vdots \\ c_n \end{pmatrix} = \begin{pmatrix} -s_1 \\ -s_2 \\ -s_3 \\ \vdots \\ -s_n \end{pmatrix}.$$

Since

$$\operatorname{tr}(A^k) = \lambda_1^k + \lambda_2^k + \cdots + \lambda_n^k = s_k$$

we find s_k for $k = 1, \ldots, n$. Thus we can solve the linear equation for c_j. Finally, using (4) we obtain the inverse matrix of A. Apply Csanky's algorithm to the 3×3 matrix

$$A = \begin{pmatrix} 1 & 0 & 1 \\ 1 & 0 & 0 \\ 0 & 1 & 1 \end{pmatrix}.$$

Solution 5. Since

$$A^2 = \begin{pmatrix} 1 & 1 & 2 \\ 1 & 0 & 1 \\ 1 & 1 & 1 \end{pmatrix}, \quad A^3 = \begin{pmatrix} 2 & 2 & 3 \\ 1 & 1 & 2 \\ 2 & 1 & 2 \end{pmatrix}$$

we find $\text{tr}(A) = 2 = s_1$, $\text{tr}(A^2) = 2 = s_2$, $\text{tr}(A^3) = 5 = s_3$. We obtain the system of linear equations

$$\begin{pmatrix} 1 & 0 & 0 \\ 2 & 2 & 0 \\ 2 & 2 & 3 \end{pmatrix} \begin{pmatrix} c_1 \\ c_2 \\ c_3 \end{pmatrix} = \begin{pmatrix} -2 \\ -2 \\ -5 \end{pmatrix}$$

with the solution $c_1 = -2$, $c_2 = 1$, $c_3 = -1$. Since $c_3 = -1$ the inverse exists and $\det(A) = 1$. Hence

$$A^{-1} = \frac{1}{-c_3}(A^2 + c_1 A + c_2 I_3) = \begin{pmatrix} 0 & 1 & 0 \\ -1 & 1 & 1 \\ 1 & -1 & 0 \end{pmatrix}.$$

Problem 6. Let A be an $n \times n$ matrix. Let

$$c(z) := \det(z I_n - A) = z^n - \sum_{k=0}^{n-1} c_k z^k$$

be the characteristic polynomial of A. Apply the Cayley-Hamilton theorem $c(A) = 0_n$ to calculate $\exp(A)$.

Solution 6. From the Cayley-Hamilton theorem we have

$$A^n = c_0 I_n + c_1 A + \cdots + c_{n-1} A^{n-1}.$$

It follows that any power n of A can be expressed in terms of $I_n, A, \ldots, A^{n-1}$

$$A^k = \sum_{j=0}^{n-1} \beta_{kj} A^j.$$

This implies that e^{tA} is a polynomial in A with analytic coefficients in t

$$e^{tA} = \sum_{k=0}^{\infty} \frac{t^k A^k}{k!} = \sum_{k=0}^{\infty} \frac{t^k}{k!} \left(\sum_{j=0}^{n-1} \beta_{kj} A^j \right) = \sum_{j=0}^{n-1} \left(\sum_{k=0}^{\infty} \beta_{kj} \frac{t^k}{k!} \right) A^j.$$

The coefficients β_{jk} can be found as follows

$$\beta_{jk} = \begin{cases} \delta_{kj} & k < n \\ c_j & k = n \\ c_0 \beta_{k-1,n-1} & k > n, \, j = 0 \\ c_j \beta_{k-1,n-1} + \beta_{k-1,j-1} & k > n, \, j > 0. \end{cases}$$

Problem 7.   (i) Let A be an $n \times n$ matrix with $A^3 = I_n$. Calculate $\exp(A)$ using

$$\exp(A) = \sum_{j=0}^{\infty} \frac{A^j}{j!}.$$

(ii) Let

$$B = \begin{pmatrix} 0 & 1 & 0 \\ 0 & 0 & 1 \\ 1 & 0 & 0 \end{pmatrix}$$

with $B^3 = I_3$. Calculate $\exp(B)$ applying the Cayley-Hamilton theorem.

Solution 7.   (i) We have for e^A utilizing $A^3 = I_n$

$$I_n \left(1 + \frac{1}{3!} + \frac{1}{6!} + \cdots \right) + A \left(1 + \frac{1}{4!} + \frac{1}{7!} + \cdots \right) + A^2 \left(\frac{1}{2!} + \frac{1}{5!} + \frac{1}{8!} + \cdots \right).$$

(ii) Applying the Cayley-Hamilton theorem we start from

$$e^B = b_2 B^2 + b_1 B + b_0 I_3 = \begin{pmatrix} b_0 & b_1 & b_2 \\ b_2 & b_0 & b_1 \\ b_1 & b_2 & b_0 \end{pmatrix}.$$

The eigenvalues of B are

$$\lambda_1 = 1, \qquad \lambda_2 = -\frac{1}{2} + i\frac{\sqrt{3}}{2}, \qquad \lambda_3 = -\frac{1}{2} - i\frac{\sqrt{3}}{2}.$$

Thus we obtain the three equations for the three eigenvalues

$$e^{\lambda_1} = e^1 = b_2 \lambda_1^2 + b_1 \lambda_1 + b_0 = b_2 + b_1 + b_0$$

$$e^{\lambda_2} = e^{-1/2} e^{i\sqrt{3}/2} = b_2 \lambda_2^2 + b_1 \lambda_2 + b_0 = -\frac{b_2}{2}(1 + i\sqrt{3}) + \frac{b_1}{2}(-1 + i\sqrt{3}) + b_0$$

$$e^{\lambda_3} = e^{-1/2} e^{-i\sqrt{3}/2} = b_2 \lambda_3^2 + b_1 \lambda_3 + b_0 = -\frac{b_2}{2}(1 - i\sqrt{3}) - \frac{b_1}{2}(1 + i\sqrt{3}) + b_0.$$

Adding and subtracting the second and third equations we arrive at

$$e = b_2 + b_1 + b_0$$

$$2e^{-1/2} \cos(\sqrt{3}/2) = -b_2 - b_1 + 2b_0$$

$$2\frac{e^{-1/2} \sin(\sqrt{3}/2)}{\sqrt{3}} = -b_2 + b_1.$$

The solution of these linear equations is

$$\begin{pmatrix} b_0 \\ b_1 \\ b_2 \end{pmatrix} = \begin{pmatrix} 1/3 & 1/3 & 0 \\ 1/3 & -1/6 & 1/2 \\ 1/3 & -1/6 & -1/2 \end{pmatrix} \begin{pmatrix} e \\ 2e^{-1/2} \cos(\sqrt{3}/2) \\ 2e^{-1/2} \sin(\sqrt{3}/2)/\sqrt{3} \end{pmatrix}.$$

Thus

$$b_0 = \frac{e}{3} + \frac{2}{3}e^{-1/2}\cos(\sqrt{3}/2)$$

$$b_1 = \frac{e}{3} - \frac{1}{3}e^{-1/2}\cos(\sqrt{3}/2) + \frac{1}{\sqrt{3}}e^{-1/2}\sin(\sqrt{3}/2)$$

$$b_2 = \frac{e}{3} - \frac{1}{3}e^{-1/2}\cos(\sqrt{3}/2) - \frac{1}{\sqrt{3}}e^{-1/2}\sin(\sqrt{3}/2).$$

Problem 8. Let A be an $n \times n$ matrix over $\mathbb{C}$. Let f be an *entire function*, i.e. an analytic function on the whole complex plane, for example $\exp(z)$, $\sin(z)$, $\cos(z)$. Using the Cayley-Hamilton theorem we can write

$$f(A) = a_{n-1}A^{n-1} + a_{n-2}A^{n-2} + \cdots + a_2A^2 + a_1A + a_0I_n \qquad (1)$$

where the complex numbers $a_0, a_1, \ldots, a_{n-1}$ are determined as follows: Let

$$r(\lambda) := a_{n-1}\lambda^{n-1} + a_{n-2}\lambda^{n-2} + \cdots + a_2\lambda^2 + a_1\lambda + a_0$$

which is the right-hand side of (1) with A^j replaced by λ^j, where $j = 0, 1, \ldots, n-1$ of each distinct eigenvalue λ_j of the matrix A, we consider the equation

$$f(\lambda_j) = r(\lambda_j). \qquad (2)$$

If λ_j is an eigenvalue of multiplicity k, for $k > 1$, then we consider also the following equations

$$f'(\lambda)|_{\lambda=\lambda_j} = r'(\lambda)|_{\lambda=\lambda_j}$$

$$f''(\lambda)|_{\lambda=\lambda_j} = r''(\lambda)|_{\lambda=\lambda_j}$$

$$\cdots = \cdots$$

$$f^{(k-1)}(\lambda)\Big|_{\lambda=\lambda_j} = r^{(k-1)}(\lambda)\Big|_{\lambda=\lambda_j}.$$

Apply this technique to find $\exp(A)$ with

$$A = \begin{pmatrix} c & c \\ c & c \end{pmatrix}, \qquad c \in \mathbb{R}, \quad c \neq 0.$$

Solution 8. (i) We have

$$e^A = a_1A + a_0I_2 = c\begin{pmatrix} a_1 & a_1 \\ a_1 & a_1 \end{pmatrix} + \begin{pmatrix} a_0 & 0 \\ 0 & a_0 \end{pmatrix} = \begin{pmatrix} a_0 + ca_1 & ca_1 \\ ca_1 & a_0 + ca_1 \end{pmatrix}.$$

The eigenvalues of A are 0 and $2c$. Thus we obtain the two linear equations

$$e^0 = 0a_1 + a_0 = a_0, \qquad e^{2c} = 2ca_1 + a_0.$$

Solving these two linear equations yields $a_0 = 1$, $a_1 = (e^{2c} - 1)/(2c)$. It follows that

$$e^A = \begin{pmatrix} a_0 + ca_1 & ca_1 \\ ca_1 & ca_1 + a_0 \end{pmatrix} = \begin{pmatrix} (e^{2c} + 1)/2 & (e^{2c} - 1)/2 \\ (e^{2c} - 1)/2 & (e^{2c} + 1)/2 \end{pmatrix}.$$

Problem 9. Let $z_1, z_2, z_3 \in \mathbb{C}$. Assume that at least two of the three complex numbers are nonzero. Consider the 2×2 matrix

$$A = \begin{pmatrix} z_1 & z_2 \\ z_3 & 0 \end{pmatrix}.$$

Calculate $\exp(A)$ applying the Cayley-Hamilton theorem. The characteristic equation for A is given by $\lambda^2 - z_1\lambda - z_2 z_3 = 0$ with the eigenvalues

$$\lambda_+ = \frac{1}{2}z_1 + \sqrt{z_1^2 + 4z_2 z_3}, \quad \lambda_- = \frac{1}{2}z_1 - \sqrt{z_1^2 + 4z_2 z_3}.$$

Cayley-Hamilton theorem then states that $A^2 - z_1 A - z_1 z_3 I_2 = 0_2$.

Solution 9. We can write $\exp(A) = a_1 A + a_0 I_2$ and a_0, a_1 are given by the system of linear equations

$$e^{\lambda_+} = a_1 \lambda_+ + a_0, \qquad e^{\lambda_-} = a_1 \lambda_- + a_0.$$

The solution is

$$a_0 = \frac{e^{\lambda_-}\lambda_+ - e^{\lambda_+}\lambda_-}{\lambda_+ - \lambda_-}, \qquad a_1 = \frac{e^{\lambda_+} - e^{\lambda_-}}{\lambda_+ - \lambda_-}$$

with $\exp(A) = a_1 A + a_0 I_2$ and $\lambda_+ - \lambda_- = \sqrt{z_1^2 + 4z_2 z_3}$.

Problem 10. Let $\alpha \in \mathbb{R}$ and $\alpha \neq 0$. Consider the 2×2 matrix

$$A(\alpha) = \begin{pmatrix} 0 & \alpha \\ \alpha & 0 \end{pmatrix}.$$

Calculate $\sin(A(\alpha))$ applying the Cayley-Hamilton theorem.

Solution 10. The eigenvalues of $A(\alpha)$ are $\lambda_+(\alpha) = \alpha$ and $\lambda_-(\alpha) = -\alpha$. Now

$$\sin(A(\alpha)) = b_1 A + b_0 I_2 = \begin{pmatrix} b_0 & b_1\alpha \\ b_1\alpha & b_0 \end{pmatrix}$$

and

$$\sin(\lambda_+) = \sin(\alpha) = b_1\lambda_+ + b_0, \quad \sin(\lambda_-) = \sin(-\alpha) = b_1\lambda_- + b_0.$$

Note that $\sin(-\alpha) = -\sin(\alpha)$. The solution of the two equations

$$\sin(\alpha) = b_1\alpha + b_0, \qquad -\sin(\alpha) = -b_1\alpha + b_0$$

provides $b_0 = 0$ and $b_1 = \sin(\alpha)/\alpha$. Thus

$$\sin(A(\alpha)) = \begin{pmatrix} 0 & \sin(\alpha) \\ \sin(\alpha) & 0 \end{pmatrix}.$$

Problem 11. Calculate

$$
\sec \begin{pmatrix} \frac{\pi}{\sqrt{2}} & 0 & 0 & \frac{\pi}{\sqrt{2}} \\ 0 & \frac{\pi}{\sqrt{2}} & \frac{\pi}{\sqrt{2}} & 0 \\ 0 & \frac{\pi}{\sqrt{2}} & -\frac{\pi}{\sqrt{2}} & 0 \\ \frac{\pi}{\sqrt{2}} & 0 & 0 & -\frac{\pi}{\sqrt{2}} \end{pmatrix} .
$$

Solution 11. Consider the 4×4 matrix

$$
A = \begin{pmatrix} 1 & 0 & 0 & 1 \\ 0 & 1 & 1 & 0 \\ 0 & 1 & -1 & 0 \\ 1 & 0 & 0 & -1 \end{pmatrix} .
$$

The eigenvalues $\sqrt{2}, -\sqrt{2}, \sqrt{2}, -\sqrt{2}$. Thus the eigenvalues of $\frac{\pi}{\sqrt{2}} A$ are π, π, $-\pi$, $-\pi$. We apply the Cayley-Hamilton theorem

$$
\sec(\pi) = -1 = c_0 + c_1\pi + c_2\pi^2 + c_3\pi^3
$$
$$
\sec(\pi)\tan(\pi) = 0 = c_1 + 2c_2\pi + 3c_3\pi^2
$$
$$
\sec(-\pi) = -1 = c_0 - c_1\pi + c_2\pi^2 - c_3\pi^3
$$
$$
\sec(-\pi)\tan(-\pi) = 0 = c_1 - 2c_2\pi + 3c_3\pi^2 .
$$

Subtracting the fourth equation from the second yields $0 = 4c_2\pi$ so that $c_2 = 0$. Adding the first and third equations yields $-1 = c_0 + c_2\pi^2 = c_0$. Adding the second and fourth equations yields $0 = 2c_1 + 6c_3\pi^2$ and subtracting the third equation from the first yields $0 = 2c_1\pi + 2c_3\pi^3$ so that $2c_1 = -2c_3\pi^2$, $c_3 = 0$ and $c_1 = 0$. Consequently

$$
\sec \begin{pmatrix} \frac{\pi}{\sqrt{2}} & 0 & 0 & \frac{\pi}{\sqrt{2}} \\ 0 & \frac{\pi}{\sqrt{2}} & \frac{\pi}{\sqrt{2}} & 0 \\ 0 & \frac{\pi}{\sqrt{2}} & -\frac{\pi}{\sqrt{2}} & 0 \\ \frac{\pi}{\sqrt{2}} & 0 & 0 & -\frac{\pi}{\sqrt{2}} \end{pmatrix} = - \begin{pmatrix} 1 & 0 & 0 & 0 \\ 0 & 1 & 0 & 0 \\ 0 & 0 & 1 & 0 \\ 0 & 0 & 0 & 1 \end{pmatrix} .
$$

Problem 12. Let A, B be $n \times n$ matrices such that $ABAB = 0_n$. Can we conclude that $BABA = 0_n$?

Solution 12. There are no 1×1 or 2×2 counterexamples. For 2×2 matrices $ABAB = 0_n$ implies $B(ABAB)A = 0_n$. Therefore the matrix BA is *nilpotent*, i.e. $(BA)^3 = 0_n$. If a 2×2 matrix C is nilpotent, its characteristic polynomial is λ^2 and therefore $C^2 = 0_n$ by the *Cayley-Hamilton theorem*. Thus $BABA = 0_n$. For $n = 3$ we find the counterexample

$$
A = \begin{pmatrix} 0 & 0 & 1 \\ 0 & 0 & 0 \\ 0 & 1 & 0 \end{pmatrix} , \quad B = \begin{pmatrix} 0 & 0 & 1 \\ 1 & 0 & 0 \\ 0 & 0 & 0 \end{pmatrix} \quad \Rightarrow \quad BABA = \begin{pmatrix} 0 & 0 & 1 \\ 0 & 0 & 0 \\ 0 & 0 & 0 \end{pmatrix} .
$$

Supplementary Problems

Problem 1. Consider the nonnormal invertible 2×2 matrix

$$A = \begin{pmatrix} 1 & 1 \\ 0 & -1 \end{pmatrix}.$$

Find A^2, A^3 and A^{-1} utilizing the Cayley-Hamilton theorem.

Problem 2. Let $\epsilon \in \mathbb{R}$ and $\epsilon \neq 0$. Let

$$A(\epsilon) = \begin{pmatrix} 0 & \epsilon \\ \epsilon & 0 \end{pmatrix}.$$

(i) Calculate $\cos(A(\epsilon))$ and $\cos(A(\epsilon) \otimes A(\epsilon))$ applying the Cayley-Hamilton theorem.
(ii) Calculate $\tanh(A(\epsilon))$ applying the Cayley-Hamilton theorem.

Problem 3. Let $\epsilon \in \mathbb{R}$. Consider the 4×4 matrix

$$A(\epsilon) = \begin{pmatrix} 0 & \epsilon \\ \epsilon & \epsilon \end{pmatrix} \otimes \begin{pmatrix} 0 & \epsilon \\ \epsilon & \epsilon \end{pmatrix}.$$

Find $\exp(A(\epsilon))$ applying the Cayley-Hamilton theorem.

Problem 4. Consider the *Bell matrix*

$$B = \frac{1}{\sqrt{2}} \begin{pmatrix} 1 & 0 & 0 & 1 \\ 0 & 1 & 1 & 0 \\ 0 & 1 & -1 & 0 \\ 1 & 0 & 0 & -1 \end{pmatrix}$$

which is a unitary matrix. Apply the Cayley-Hamilton theorem to find the skew-hermitian matrix K such that $B = e^K$.

Chapter 11

Hadamard Product

Suppose $A = (a_{ij})$ and $B = (b_{ij})$ are two $n \times m$ matrices with entries in some fields. Then the *Hadamard product* (sometimes called the *Schur product*) is the *entrywise product* of A and B, that is, the $m \times n$ matrix $A \bullet B$ whose (i,j) entry is $a_{ij}b_{ij}$, i.e.

$$A \bullet B := (a_{ij}b_{ij}).$$

We have the following properties. Suppose A, B, C are matrices of the same size and c is a scalar. Then

$$A \bullet (B \bullet C) = (A \bullet B) \bullet C$$
$$A \bullet B = B \bullet A$$
$$A \bullet (B + C) = A \bullet B + A \bullet C$$
$$A \bullet (cB) = c(A \bullet B).$$

Let J be the $n \times m$ matrix with entries equal to one. Then $J \bullet A = A \bullet J = A$. If A, B are $n \times n$ diagonal matrices, then $A \bullet B = AB$ is again a diagonal matrix. If A, B are $n \times n$ positive definite matrices and (a_{jj}) are the diagonal entries of A, then

$$\det(A \bullet B) \geq \det(B) \prod_{j=1}^{n} a_{jj}$$

with equality if and only if A is a diagonal matrix.

If A and B are binary matrices, then $A \bullet B$ is a binary matrix. If A and B are matrices with entries ± 1, then $A \bullet B$ is again a matrix with entries ± 1.

Problem 1.   Let

$$A = \begin{pmatrix} 0 & 1 \\ 1 & 0 \end{pmatrix}, \quad B = \begin{pmatrix} 3 & 4 \\ 7 & 1 \end{pmatrix}, \quad J = \begin{pmatrix} 1 & 1 \\ 1 & 1 \end{pmatrix}.$$

Calculate $A \bullet B$, $A \bullet A^{-1}$, $A \bullet J$, $J \bullet B$.

Solution 1.   Entrywise multiplication of the matrices yields

$$A \bullet B = \begin{pmatrix} 0 & 4 \\ 7 & 0 \end{pmatrix}$$

and $A \bullet A^{-1} = A$, $A \bullet J = A$, $J \bullet B = B$.

Problem 2.   (i) Consider the 2×2 matrices

$$A = \begin{pmatrix} 1 & 1 \\ 1 & -1 \end{pmatrix}, \quad B = \begin{pmatrix} 1 & -1 \\ 1 & 1 \end{pmatrix}$$

which are invertible. Is $A \bullet B$ invertible?
(ii) Let U be an $n \times n$ unitary matrix. Can we conclude that $U \bullet U^*$ is a unitary matrix?

Solution 2.   (i) We find the non-invertible matrix

$$A \bullet B = \begin{pmatrix} 1 & -1 \\ 1 & -1 \end{pmatrix}.$$

(ii) In general we cannot conclude that $U \bullet U^*$ is a unitary matrix. For example, let

$$U = \frac{1}{\sqrt{2}} \begin{pmatrix} 1 & 1 \\ 1 & -1 \end{pmatrix}.$$

Then $U^* = U = U^{-1}$ and

$$U \bullet U^* = \frac{1}{2} \begin{pmatrix} 1 & 1 \\ 1 & 1 \end{pmatrix}.$$

Note that $U \bullet U^*$ is a projection matrix.

Problem 3.   Let $n \in \mathbb{N}$ and

$$A = \begin{pmatrix} 1 & n \\ 0 & 1 \end{pmatrix} \Rightarrow A^{-1} = \begin{pmatrix} 1 & -n \\ 0 & 1 \end{pmatrix}.$$

Thus $\det(A) = \det(A^{-1}) = 1$. Find $\det(A \bullet A^{-1})$.

Solution 3.   Obviously we find $\det(A \bullet A^{-1}) = 1$.

Problem 4.   Let C, D be $m \times n$ matrices. Show that $\operatorname{rank}(A \bullet B) \leq (\operatorname{rank}(A))(\operatorname{rank}(B))$.

Solution 4. Any matrix of rank r can be written as a sum of r rank one matrices, each of which is an outer product of two vectors (column vector times row vector). Thus, if $\mathrm{rank}(A) = r_1$ and $\mathrm{rank}(B) = r_2$, we have

$$A = \sum_{i=1}^{r_1} \mathbf{x}_i \mathbf{y}_i^*, \qquad B = \sum_{j=1}^{r_2} \mathbf{u}_j \mathbf{v}_j^*$$

where $\mathbf{x}_i, \mathbf{u}_j \in \mathbb{C}^m$, $\mathbf{y}_i, \mathbf{v}_j \in \mathbb{C}^n$, $i = 1, \ldots, r_1$ and $j = 1, \ldots, r_2$. Then

$$A \bullet B = \sum_{i=1}^{r_1} \sum_{j=1}^{r_2} (\mathbf{x}_i \bullet \mathbf{u}_j)(\mathbf{y}_i \bullet \mathbf{v}_j)^*.$$

This shows that $A \bullet B$ is a sum of at most $r_1 r_2$ rank one matrices. Thus $\mathrm{rank}(A \bullet B) \le r_1 r_2 = (\mathrm{rank}(A))(\mathrm{rank}(B))$.

Problem 5. Let

$$A = \begin{pmatrix} a_1 & a_2 \\ a_2 & a_3 \end{pmatrix}, \qquad B = \begin{pmatrix} b_1 & b_2 \\ b_2 & b_3 \end{pmatrix}$$

be symmetric matrices over $\mathbb{R}$. Assume that A and B are positive definite. Show that $A \bullet B$ is positive definite using the trace and determinant.

Solution 5. We have

$$A \quad \text{positive definite} \quad \Leftrightarrow \quad \mathrm{tr}(A) > 0 \quad \det(A) > 0$$

$$B \quad \text{positive definite} \quad \Leftrightarrow \quad \mathrm{tr}(B) > 0 \quad \det(B) > 0.$$

Thus we have $a_1 + a_3 > 0$, $a_1 a_3 - a_2^2 > 0$ (or $a_1 a_3 > a_2^2$) and $b_1 + b_3 > 0$, $b_1 b_3 - b_2^2 > 0$ (or $b_1 b_3 > b_2^2$). Now using the conditions $a_1 a_3 > a_2^2$ and $b_1 b_3 > b_2^2$ we have

$$\det(A \bullet B) = (a_1 a_3)(b_1 b_3) - (a_2^2)(b_2^2) > (a_2^2)(b_2^2) - (a_2^2)(b_2^2) = 0.$$

Thus $\det(A \bullet B) > 0$. For the trace we have $\mathrm{tr}(A \bullet B) = a_1 b_1 + a_3 b_3$. Now using $a_1 a_3 > a_2^2$ and $b_1 b_3 > b_2^2$ again we have

$$(a_1 + a_3)(a_1 b_1 + a_3 b_3)(b_1 + b_3) > 0.$$

Since $a_1 + a_3 > 0$ and $b_1 + b_3 > 0$ it follows that $a_1 b_1 + a_3 b_3 > 0$ and therefore $\mathrm{tr}(A \bullet B) > 0$. Thus since $\det(A \bullet B) > 0$ and $\mathrm{tr}(A \bullet B) > 0$ it follows that $A \bullet B$ is positive definite.

Problem 6. Let A be an $n \times n$ matrix over $\mathbb{C}$. The *spectral radius* $\rho(A)$ is the radius of the smallest circle in the complex plane that contains all its eigenvalues. Every characteristic polynomial has at least one root. Let A, B be nonnegative matrices. Then $\rho(A \bullet B) \le \rho(A)\rho(B)$. Apply this inequality to the nonnegative matrices

$$A = \begin{pmatrix} 1/4 & 0 \\ 0 & 3/4 \end{pmatrix}, \qquad B = \begin{pmatrix} 1 & 1 \\ 1 & 1 \end{pmatrix}.$$

Solution 6. The eigenvalues of A are $1/4$ and $3/4$. Thus the spectral radius is $3/4$. The eigenvalues of B are 0 and 2. Thus the spectral radius is 2. Now $A \bullet B = \text{diag}(1/4 \ 3/4)$. The eigenvalues of $A \bullet B$ are $1/4$ and $3/4$ with the spectral radius is $3/4$. Thus $\rho(A \bullet B) = 3/4 \leq 3/4 \cdot 2 = 3/2$.

Problem 7. There exists an $n^2 \times n$ *selection matrix* J such that $A \bullet B = J^T(A \otimes B)J$, where J^T is defined as the $n \times n^2$ matrix $[E_{11} \ E_{22} \ \cdots \ E_{nn}]$ with E_{ii} the $n \times n$ matrix of zeros except for a 1 in the (i,i)-th position (*elementary matrices*). Prove this identity for the special case $n = 2$.

Solution 7. From the definition for the selection matrix we have

$$
J^T = \begin{pmatrix} 1 & 0 & 0 & 0 \\ 0 & 0 & 0 & 1 \end{pmatrix} \Rightarrow J = \begin{pmatrix} 1 & 0 \\ 0 & 0 \\ 0 & 0 \\ 0 & 1 \end{pmatrix}.
$$

We have

$$
A \bullet B = \begin{pmatrix} a_{11}b_{11} & a_{12}b_{12} \\ a_{21}b_{21} & a_{22}b_{22} \end{pmatrix}.
$$

Now

$$
J^T(A \otimes B)J = \begin{pmatrix} 1 & 0 & 0 & 0 \\ 0 & 0 & 0 & 1 \end{pmatrix} \begin{pmatrix} a_{11}b_{11} & a_{11}b_{12} & a_{12}b_{11} & a_{12}b_{12} \\ a_{11}b_{21} & a_{11}b_{22} & a_{12}b_{21} & a_{12}b_{22} \\ a_{21}b_{11} & a_{21}b_{12} & a_{22}b_{11} & a_{22}b_{12} \\ a_{21}b_{21} & a_{21}b_{22} & a_{22}b_{21} & a_{22}b_{22} \end{pmatrix} \begin{pmatrix} 1 & 0 \\ 0 & 0 \\ 0 & 0 \\ 0 & 1 \end{pmatrix}
$$

$$
= A \bullet B.
$$

Problem 8. If A, B are $n \times n$ positive definite matrices and (a_{jj}) are the diagonal entries of A, then

$$
\det(A \bullet B) \geq \det(B) \prod_{j=1}^{n} a_{jj}
$$

with equality if and only if A is a diagonal matrix. Let

$$
A = \begin{pmatrix} 5 & 1 \\ 1 & 1 \end{pmatrix}, \qquad B = \begin{pmatrix} 13 & 4 \\ 4 & 4 \end{pmatrix}.
$$

First show that A and B are positive definite and then calculate the left- and right-hand side of (1).

Solution 8. The matrix A is symmetric over $\mathbb{R}$ and the eigenvalues are positive. The matrix B is symmetric over $\mathbb{R}$ and the eigenvalues are positive. Thus both A and B are positive definite. We obtain

$$
\det(A \bullet B) = \begin{pmatrix} 65 & 4 \\ 4 & 4 \end{pmatrix} = 244, \qquad \det(B) \prod_{j=1}^{2} a_{jj} = 180.
$$

Problem 9. Consider the symmetric 4×4 matrices

$$A = \begin{pmatrix} 1 & 0 & 0 & 1 \\ 0 & 0 & 0 & 0 \\ 0 & 0 & 0 & 0 \\ 1 & 0 & 0 & 1 \end{pmatrix}, \quad B = \begin{pmatrix} 1 & 0 & 0 & 0 \\ 0 & 0 & 1 & 0 \\ 0 & 1 & 0 & 0 \\ 0 & 0 & 0 & 1 \end{pmatrix}.$$

Calculate the Hadamard product $A \bullet B$. Show that $\|A \bullet B\| \leq \|A^*A\| \, \|B^*B\|$, where the norm is given by the Hilbert-Schmidt norm.

Solution 9. We have

$$A \bullet B = \begin{pmatrix} 1 & 0 & 0 & 0 \\ 0 & 0 & 0 & 0 \\ 0 & 0 & 0 & 0 \\ 0 & 0 & 0 & 1 \end{pmatrix}.$$

Thus $(A \bullet B)(A \bullet B)^T = A \bullet B$ and $\mathrm{tr}((A \bullet B)(A \bullet B)^T) = 2$. Now

$$A^T A = \begin{pmatrix} 2 & 0 & 0 & 2 \\ 0 & 0 & 0 & 0 \\ 0 & 0 & 0 & 0 \\ 2 & 0 & 0 & 2 \end{pmatrix}, \quad B^T B = \begin{pmatrix} 1 & 0 & 0 & 0 \\ 0 & 1 & 0 & 0 \\ 0 & 0 & 1 & 0 \\ 0 & 0 & 0 & 1 \end{pmatrix}.$$

Thus the inequality given in the problem is true for any two matrices A and B.

Problem 10. Let A, B, C and D^T be $n \times n$ matrices over $\mathbb{R}$. Show that

$$\mathrm{tr}((A \bullet B)(C^T \bullet D)) = \mathrm{tr}((A \bullet B \bullet C)D).$$

Solution 10. We define $A \bullet B = E = (e_{ij})$ and $C \bullet D^T = F = (f_{ij})$. Then

$$\mathrm{tr}((A \bullet B)(C^T \bullet D)) = \mathrm{tr}(EF^T) = \sum_{i=1}^{n} \sum_{j=1}^{n} e_{ij} f_{ij} = \sum_{i=1}^{n} \sum_{j=1}^{n} (a_{ij} b_{ij} c_{ij}) d_{ji}$$

$$= \mathrm{tr}((A \bullet B \bullet C)D).$$

Problem 11. Let V, W be two matrices of the same order. If all entries of V are nonzero, then we say that X is Schur invertible and define its *Schur inverse*, $V^{(-)}$, by $V^{(-)} \bullet V = J$, where J is the matrix with all 1's.

The vector space $M_n(\mathbb{F})$ of $n \times n$ matrices acts on itself in three distinct ways: if $C \in M_n(\mathbb{F})$ we can define endomorphisms X_C, Δ_C and Y_C by

$$X_C M := CM, \qquad \Delta_C M := C \bullet M, \qquad Y_C := MC^T.$$

Let A, B be $n \times n$ matrices. Assume that X_A is invertible and Δ_B is invertible in the sense of Schur. Note that X_A is invertible if and only if A is, and Δ_B is invertible if and only if the Schur inverse $B^{(-)}$ is defined. We say that (A, B) is a *one-sided Jones pair* if

$$X_A \Delta_B X_A = \Delta_B X_A \Delta_B.$$

We call this the *braid relation*. Give an example for a one-sided Jones pair.

Solution 11.   A trivial example is the pair (I_n, J_n), where J_n is the $n \times n$ matrix with all ones. Note that for the left-hand side we have

$$(X_A \Delta_B X_A)(M) = X_A \Delta_B (AM) = X_A(B \bullet (AM)) = A(B \bullet (AM)).$$

Thus with $A = I_n$ and $B = J_n$ we obtain M. For the right-hand side we have

$$(\Delta_B X_A \Delta_B)M = \Delta_B X_A(B \bullet M) = \Delta_B(A(B \bullet M)) = B \bullet (A(B \bullet M)).$$

Thus with $A = I_n$ and $B = J_n$ we also obtain M.

Problem 12.   Let A, B be $n \times n$ matrices. Let $e_1, \ldots, e_n$ be the standard basis vectors in $\mathbb{C}^n$. We form the n^2 column vectors $(Ae_j) \bullet (Be_k)$, $j, k = 1, \ldots, n$. If A is invertible and B is Schur invertible, then for any j

$$\{ (Ae_1) \bullet (Be_j), \ (Ae_2) \bullet (Be_j), \ \ldots, \ (Ae_n) \bullet (Be_j) \}$$

is a basis of the vector space $\mathbb{C}^n$. Let

$$A = \begin{pmatrix} 1 & 1 & 1 \\ 0 & 1 & 1 \\ 0 & 0 & 1 \end{pmatrix}, \qquad B = \begin{pmatrix} 1 & -1 & 1 \\ -1 & 1 & -1 \\ 1 & -1 & 1 \end{pmatrix}.$$

Find these bases for these matrices.

Solution 12.   For $j = 1$ we have

$$(Ae_1) \bullet (Be_1) = \begin{pmatrix} 1 \\ 0 \\ 0 \end{pmatrix}, \ (Ae_2) \bullet (Be_1) = \begin{pmatrix} 1 \\ -1 \\ 0 \end{pmatrix}, \ (Ae_3) \bullet (Be_1) = \begin{pmatrix} 1 \\ -1 \\ 1 \end{pmatrix}.$$

The three vectors are linearly independent and thus form a basis in $\mathbb{C}^3$. For $j = 2$ we obtain

$$(Ae_1) \bullet (Be_2) = \begin{pmatrix} -1 \\ 0 \\ 0 \end{pmatrix}, \ (Ae_2) \bullet (Be_2) = \begin{pmatrix} -1 \\ 1 \\ 0 \end{pmatrix}, \ (Ae_3) \bullet (Be_2) = \begin{pmatrix} -1 \\ 1 \\ -1 \end{pmatrix}.$$

For $j = 3$ we obtain

$$(Ae_1) \bullet (Be_3) = \begin{pmatrix} 1 \\ 0 \\ 0 \end{pmatrix}, \ (Ae_2) \bullet (Be_3) = \begin{pmatrix} 1 \\ -1 \\ 0 \end{pmatrix}, \ (Ae_3) \bullet (Be_3) = \begin{pmatrix} 1 \\ -1 \\ 1 \end{pmatrix}.$$

Problem 13. Let $B = (b_{jk})$ be a diagonalizable $n \times n$ matrix with eigenvalues $\lambda_1, \lambda_2, \ldots, \lambda_n$. Thus, there is nonsingular $n \times n$ matrix A such that

$$B = A(\text{diag}(\lambda_1, \lambda_2, \ldots, \lambda_n))A^{-1}.$$

Show that

$$\begin{pmatrix} b_{11} \\ b_{22} \\ \vdots \\ b_{nn} \end{pmatrix} = (A \bullet (A^{-1})^T) \begin{pmatrix} \lambda_1 \\ \lambda_2 \\ \vdots \\ \lambda_n \end{pmatrix}.$$

Thus the vector of eigenvalues of B is transformed to the vector of its diagonal entries by the coefficient matrix $A \bullet (A^{-1})^T$.

Solution 13. Let $(A)_{ij}$ denote the entry in the i-th row and j-th column of A. Then

$$(\text{diag}(\lambda_1, \lambda_2, \ldots, \lambda_n)A^{-1})_{ij} = \lambda_i (A^{-1})_{ij}.$$

It follows that

$$b_{jj} = (A(\text{diag}(\lambda_1, \lambda_2, \ldots, \lambda_n))A^{-1})_{jj} = \sum_{k=1}^{n} (A)_{jk} (\text{diag}(\lambda_1, \lambda_2, \ldots, \lambda_n)A^{-1})_{kj}$$

$$= \sum_{k=1}^{n} (A)_{jk}(A^{-1})_{kj}\lambda_k = \sum_{k=1}^{n} (A)_{jk} ((A^{-1})^T)_{jk} \lambda_k = \sum_{k=1}^{n} (A \bullet (A^{-1})^T)_{jk} \lambda_k.$$

Problem 14. Let A, B, C, D be $n \times n$ matrices over $\mathbb{R}$. Let $\mathbf{s}^T = (1 \ldots 1)$ be a row vector in $\mathbb{R}^n$. Show that

$$\mathbf{s}^T (A \bullet B)(C^T \bullet D)\mathbf{s} = \text{tr}(C\Gamma D)$$

where $\Gamma = (\gamma_{ij})$ is a diagonal matrix with $\gamma_{jj} = \sum_{i=1}^{n} a_{ij}b_{ij}$ and $j = 1, \ldots, n$.

Solution 14. We have

$$\mathbf{s}^T (A \bullet B)(C^T \bullet D)\mathbf{s} = \mathbf{s}^T E F^T \mathbf{s} = \sum_{i=1}^{n} \sum_{j=1}^{n} \sum_{k=1}^{n} e_{ij}f_{kj}$$

$$= \sum_{j=1}^{n} \sum_{k=1}^{n} \left(\sum_{i=1}^{n} a_{ij}b_{ij} \right) c_{kj}d_{jk} = \sum_{j=1}^{n} \sum_{k=1}^{n} \gamma_{jj}c_{kj}d_{jk}$$

$$= \text{tr}(C\Gamma D).$$

Problem 15. Given two matrices A and B of the same size. If all entries of A are nonzero, then we say that A is *Schur invertible* and define its Schur inverse, $A^{(-)}$ by $A_{ij}^{(-)} := 1/A_{ij}$. Equivalently, we have $A^{(-)} \bullet A = J$, where J is the matrix with all ones. An $n \times n$ matrix W is a *type-II matrix* if

$$WW^{(-)T} = nI_n$$

where I_n is the $n \times n$ identity matrix. Find such a matrix for $n = 2$.

Solution 15. Such a matrix is

$$W = \begin{pmatrix} 1 & 1 \\ 1 & -1 \end{pmatrix}$$

with $W = W^{(-)}$.

Problem 16. Let A, B, C, D be 2×2 matrices and $\otimes$ be the Kronecker product. Is

$$(A \otimes B) \bullet (C \otimes D) = (A \bullet C) \otimes (B \bullet D)?$$

Solution 16. Yes. A Maxima program to prove it is

```
/* Note that * is the Hadamard product and . is the matrix product */
A: matrix([a11,a12],[a21,a22]); B: matrix([b11,b12],[b21,b22]);
C: matrix([c11,c12],[c21,c22]); D: matrix([d11,d12],[d21,d22]);
T1: kronecker_product(A,B); T2: kronecker_product(C,D);
T3: A*C; T4: B*D;
R1: T1*T2;
R2: kronecker_product(T3,T4);
R3: R1-R2; print("R3=",R3);
```

Supplementary Problems

Problem 1. Show that the Hadamard product is linear. Show that the Hadamard product is associative.

Problem 2. Let A, B be $n \times n$ positive semidefinite matrices. Show that $A \bullet B$ is also positive semidefinite.

Problem 3. (i) Let A, B be $n \times n$ symmetric matrices over $\mathbb{R}$. Is $A \bullet B$ a symmetric matrix over $\mathbb{R}$?
(ii) Let C, D be $n \times n$ hermitian matrices. Is $C \bullet D$ a hermitian matrix?

Problem 4. Let A be a positive semidefinite $n \times n$ matrix. Let B be an $n \times n$ matrix with $\|B\| \leq 1$, where $\| \, . \, \|$ denotes the spectral norm. Show that

$$\max\{ \|A \bullet B\| \, : \, \|B\| \leq 1 \} = \max(a_{jj})$$

where $\| \, \|$ denotes the spectral norm and $j = 1, \dots, n$.

Problem 5. Let A, B be $m \times n$ matrices and D and E be $m \times m$ and $n \times n$ diagonal matrices, respectively. Show that

$$D(A \bullet B)E = A \bullet (DBE).$$

Problem 6. Let A be an $n \times n$ diagonalizable matrix over $\mathbb{C}$ with eigenvalues $\lambda_1, \ldots, \lambda_n$ and diagonalization $A = TDT^{-1}$, where D is a diagonal matrix such that $D_{jj} = \lambda_j$ for all $j = 1, \ldots, n$. Show that

$$\begin{pmatrix} a_{11} \\ a_{22} \\ \vdots \\ a_{nn} \end{pmatrix} = (T \bullet (T^{-1})^T) \begin{pmatrix} \lambda_1 \\ \lambda_2 \\ \vdots \\ \lambda_n \end{pmatrix}.$$

Problem 7. Show that $\text{tr}(A(B \bullet C)) \equiv (\text{vec}(A^T \bullet B))^T \text{vec}(C)$. The left-hand side yields

$$\text{tr}(A(B \bullet C)) = \sum_{j=1}^{n} \sum_{r=1}^{n} a_{jr} b_{rj} c_{rj}.$$

Problem 8. Consider the $n \times n$ matrices with entries ± 1. If A, B such matrices, then $A \bullet B$ is again such a matrix. Do these matrices form a group under the Hadamard product? The neutral element would be the matrix J with all entries 1.

Problem 9. Consider the $n \times n$ matrices with entries 0 and 1. Let $\oplus$ be the XOR operation and we define $A \bullet B = (a_{ij} \oplus b_{ij})$. Then $A \bullet B$ is again a matrix with entries 0 and 1. Do these matrices form a group?

Chapter 12

Norms and Scalar Products

A linear space V is called a *normed space*, if for every $\mathbf{v} \in V$ there is associated a real number $\|\mathbf{v}\|$, the norm of the vector $\mathbf{v}$, such that

$$\|\mathbf{v}\| \geq 0, \qquad \|\mathbf{v}\| = 0 \text{ iff } \mathbf{v} = 0$$
$$\|c\mathbf{v}\| = |c| \, \|\mathbf{v}\| \quad \text{where } c \in \mathbb{C}$$
$$\|\mathbf{v} + \mathbf{w}\| \leq \|\mathbf{v}\| + \|\mathbf{w}\| \quad \text{for all } \mathbf{v}, \mathbf{w} \in V.$$

Consider $\mathbb{C}^n$ and $\mathbf{v} = (v_1, v_2, \dots, v_n)^T$. The most common *vector norms* are

(i) The *Euclidean norm*: $\|\mathbf{v}\|_2 := \sqrt{\mathbf{v}^*\mathbf{v}}$
(ii) The ℓ_1 norm: $\|\mathbf{v}\|_1 := |v_1| + |v_2| + \cdots + |v_n|$
(iii) The ℓ_∞ norm: $\|\mathbf{v}\|_\infty := \max(|v_1|, |v_2|, \dots, |v_n|)$
(iv) The ℓ_p norm ($p \geq 1$): $\|\mathbf{v}\|_p := (|v_1|^p + |v_2|^p + \cdots + |v_n|^p)^{1/p}$

A scalar product of two vectors $\mathbf{v}$ and $\mathbf{u}$ in $\mathbb{C}^n$ is given by $\mathbf{v}^*\mathbf{u}$. This implies the norm $\|\mathbf{v}\|^2 = \mathbf{v}^*\mathbf{v}$.

Let A be an $n \times n$ matrix over $\mathbb{C}$ and $\mathbf{x} \in \mathbb{C}^n$. Each vector norm induces the *matrix norm*

$$\|A\| := \max_{\|\mathbf{x}\|=1} \|A\mathbf{x}\|.$$

Another important matrix norm is

$$\|A\| := \sqrt{\mathrm{tr}(AA^*)}.$$

This is called the *Frobenius norm* (also called *Hilbert-Schmidt norm*). This relates to the scalar product $\langle A, B \rangle := \mathrm{tr}(AB^*)$ of two $n \times n$ matrices A and B.

318

Problem 1. Consider the vectors in $\mathbb{C}^4$

$$\mathbf{v} = \begin{pmatrix} i \\ 1 \\ -1 \\ -i \end{pmatrix}, \quad \mathbf{w} = \begin{pmatrix} i \\ 0 \\ 0 \\ i \end{pmatrix}.$$

Find the Euclidean norm and then normalize the vectors. Are the vectors orthogonal to each other?

Solution 1. We have

$$\mathbf{v}^*\mathbf{v} = (-i \ 1 \ -1 \ i) \begin{pmatrix} i \\ 1 \\ -1 \\ -i \end{pmatrix} = 4, \quad \mathbf{w}^*\mathbf{w} = (-i \ 0 \ 0 \ -i) \begin{pmatrix} i \\ 0 \\ 0 \\ i \end{pmatrix} = 2.$$

Thus $\|\mathbf{v}\| = 2$ and $\|\mathbf{w}\| = \sqrt{2}$ and the normalized vectors are

$$\frac{1}{2} \begin{pmatrix} i \\ 1 \\ -1 \\ -i \end{pmatrix}, \quad \frac{1}{\sqrt{2}} \begin{pmatrix} i \\ 0 \\ 0 \\ i \end{pmatrix}.$$

Yes we have $\mathbf{v}^*\mathbf{w} = 0$.

Problem 2. Consider the vectors in $\mathbb{R}^3$

$$\mathbf{v}_1 = \begin{pmatrix} 1 \\ 1 \\ 1 \end{pmatrix}, \quad \mathbf{v}_2 = \begin{pmatrix} 1 \\ -1 \\ 1 \end{pmatrix}, \quad \mathbf{v}_3 = \begin{pmatrix} 1 \\ -1 \\ -1 \end{pmatrix}.$$

(i) Show that the vectors are linearly independent.
(ii) Apply the *Gram-Schmidt orthonormalization process* to these vectors.

Solution 2. (i) The equation $a\mathbf{v}_1 + b\mathbf{v}_2 + c\mathbf{v}_3 = \mathbf{0}$ only admits the solution $a = b = c = 0$.
(ii) Since $\mathbf{v}_1^T\mathbf{v}_1 = 3$ we set $\mathbf{w}_1 = \mathbf{v}_1/\sqrt{3}$. Since $\alpha_1 := \mathbf{w}_1^T\mathbf{v}_2 = 1/\sqrt{3}$ we set

$$\mathbf{z}_2 = \mathbf{v}_2 - \alpha_1\mathbf{w}_1 = \begin{pmatrix} 2/3 \\ -4/3 \\ 2/3 \end{pmatrix}.$$

Next we normalize the vector $\mathbf{z}_2$ and set

$$\mathbf{w}_2 = \frac{1}{\sqrt{8/3}}\mathbf{z}_2 = \begin{pmatrix} 1/\sqrt{6} \\ -\sqrt{2}/\sqrt{3} \\ 1/\sqrt{6} \end{pmatrix}.$$

Finally we find that $\mathbf{w}_1^T\mathbf{v}_3 = -1/\sqrt{3}$, $\mathbf{w}_2^T\mathbf{v}_3 = \sqrt{2}/\sqrt{3}$ and therefore

$$\mathbf{z}_3 = \mathbf{v}_3 - (\mathbf{w}_1^T\mathbf{v}_3)\mathbf{w}_1 - (\mathbf{w}_1^T\mathbf{v}_3)\mathbf{w}_2 = \begin{pmatrix} 1 \\ 0 \\ -1 \end{pmatrix}.$$

Normalizing yields

$$\mathbf{w}_3 = \frac{1}{\sqrt{2}} \begin{pmatrix} 1/\sqrt{2} \\ 0 \\ -1/\sqrt{2} \end{pmatrix}.$$

Thus we find the orthonormal basis

$$\mathbf{w}_1 = \begin{pmatrix} 1/\sqrt{3} \\ 1/\sqrt{3} \\ 1/\sqrt{3} \end{pmatrix}, \quad \mathbf{w}_2 = \begin{pmatrix} 1/\sqrt{6} \\ -\sqrt{2}/\sqrt{3} \\ 1/\sqrt{6} \end{pmatrix}, \quad \mathbf{w}_3 = \begin{pmatrix} 1/\sqrt{2} \\ 0 \\ -1/\sqrt{2} \end{pmatrix}.$$

Problem 3.   Let A be an $n \times n$ matrix over $\mathbb{R}$. The *spectral norm* is

$$\|A\|_2 := \max_{\mathbf{x} \neq 0} \frac{\|A\mathbf{x}\|_2}{\|\mathbf{x}\|_2}.$$

It can be shown that $\|A\|_2$ can also be calculated as

$$\|A\|_2 = \sqrt{\text{largest eigenvalue of } A^T A}.$$

The eigenvalues of $A^T A$ are real and nonnegative. Consider the 2×2 matrix

$$A = \begin{pmatrix} 2 & 5 \\ 1 & 3 \end{pmatrix}.$$

Calculate $\|A\|_2$ using this method.

Solution 3.   We have

$$A^T A = \begin{pmatrix} 5 & 13 \\ 13 & 34 \end{pmatrix}.$$

The eigenvalues of $A^T A$ are 0.0257 and 38.9743. Thus $\|A\|_2 = \sqrt{38.9743} = 6.2429$.

Problem 4.   Let $\{ \mathbf{v}_j : j = 1, \ldots, r \}$ be an orthogonal set of vectors in $\mathbb{R}^n$ with $r \leq n$. Show that

$$\| \sum_{j=1}^{r} \mathbf{v}_j \|^2 = \sum_{j=1}^{r} \|\mathbf{v}_j\|^2.$$

Solution 4.   We have

$$\| \sum_{j=1}^{r} \mathbf{v}_j \|^2 = \langle \sum_{j=1}^{r} \mathbf{v}_j, \sum_{k=1}^{r} \mathbf{v}_k \rangle = \sum_{j=1}^{r} \sum_{k=1}^{r} \langle \mathbf{v}_j, \mathbf{v}_k \rangle = \sum_{j=1}^{k} \|\mathbf{v}_j\|^2$$

where we used that for the scalar product we have $\langle \mathbf{v}_j, \mathbf{v}_k \rangle \equiv \mathbf{v}_j^* \mathbf{v}_k = 0$ for $j \neq k$.

Problem 5.   Consider the 2×2 matrix A over $\mathbb{C}$ Find the norm of A implied by the scalar product $\langle A, A \rangle = \sqrt{\text{tr}(AA^*)}$.

Solution 5. Since

$$A^* = \begin{pmatrix} a_{11}^* & a_{21}^* \\ a_{12}^* & a_{22}^* \end{pmatrix}$$

we obtain

$$AA^* = \begin{pmatrix} a_{11}a_{11}^* + a_{12}a_{12}^* & a_{11}a_{21}^* + a_{12}a_{22}^* \\ a_{21}a_{11}^* + a_{22}a_{12}^* & a_{21}a_{21}^* + a_{22}a_{22}^* \end{pmatrix}$$

where a_{ij}^* is the conjugate complex of a_{ij}. Thus

$$\mathrm{tr}(A^*A) = a_{11}a_{11}^* + a_{12}a_{12}^* + a_{21}a_{21}^* + a_{22}a_{22}^*.$$

Therefore the norm of A is $\|A\| = \sqrt{a_{11}a_{11}^* + a_{12}a_{12}^* + a_{21}a_{21}^* + a_{22}a_{22}^*}$.

Problem 6. Let A, B be $n \times n$ matrices over $\mathbb{C}$. A *scalar product* can be defined as

$$\langle A, B \rangle := \mathrm{tr}(AB^*).$$

The scalar product implies a *norm* $\|A\|^2 = \langle A, A \rangle = \mathrm{tr}(AA^*)$. This norm is called the *Hilbert-Schmidt norm*.
(i) Consider the two *Dirac matrices*

$$\gamma_0 := \begin{pmatrix} 1 & 0 & 0 & 0 \\ 0 & 1 & 0 & 0 \\ 0 & 0 & -1 & 0 \\ 0 & 0 & 0 & -1 \end{pmatrix}, \qquad \gamma_1 := \begin{pmatrix} 0 & 0 & 0 & 1 \\ 0 & 0 & 1 & 0 \\ 0 & -1 & 0 & 0 \\ -1 & 0 & 0 & 0 \end{pmatrix}.$$

Calculate $\langle \gamma_0, \gamma_1 \rangle$.
(ii) Let U be a unitary $n \times n$ matrix. Find $\langle UA, UB \rangle$.
(iii) Let C, D be $m \times m$ matrices over $\mathbb{C}$. Find $\langle A \otimes C, B \otimes D \rangle$.
(iv) Let U be a unitary matrix. Calculate $\langle U, U \rangle$. Then find the norm implied by the scalar product.
(v) Calculate $\|U\| := \max_{\|\mathbf{v}\|=1} \|U\mathbf{v}\|$.

Solution 6. (i) We find $\langle \gamma_0, \gamma_1 \rangle = \mathrm{tr}(\gamma_0 \gamma_1^*) = 0$.
(ii) Since

$$\mathrm{tr}(UA(UB)^*) = \mathrm{tr}(UAB^*U^*) = \mathrm{tr}(U^*UAB^*) = \mathrm{tr}(AB)$$

and using *cyclic invariance* we find that $\langle UA, UB \rangle = \langle A, B \rangle$. Thus the scalar product is invariant under the unitary transformation.
(iii) Since

$$\mathrm{tr}((A \otimes C)(B \otimes D)^*) = \mathrm{tr}((AB^*) \otimes (CD^*)) = \mathrm{tr}(AB^*)\mathrm{tr}(CD^*)$$

we find $\langle A \otimes C, B \otimes D \rangle = \langle A, B \rangle \langle C, D \rangle$.
(iv) We have $\langle U, U \rangle = \mathrm{tr}(UU^*) = \mathrm{tr}(I_n) = n$, where we used that $U^{-1} = U^*$. Thus the norm is given by $\|U\| = \sqrt{n}$.
(v) With $\|\mathbf{v}\| = 1$ and $UU^* = I_n$ we find $\|U\|^2 = \mathbf{v}^*U^*U\mathbf{v} = \mathbf{v}^*\mathbf{v} = 1$.

Problem 7. (i) Let $\{ \mathbf{x}_j : j = 1, \ldots, n \}$, $\{ \mathbf{y}_j : j = 1, \ldots, n \}$ be orthonormal bases in $\mathbb{C}^n$. Show that

$$(U_{jk}) := (\mathbf{x}_j^*\mathbf{y}_k)$$

is a unitary matrix, where $\mathbf{x}_j^* \mathbf{y}_k$ is the scalar product of the vectors $\mathbf{x}_j$ and $\mathbf{y}_k$. This means showing that $UU^* = I_n$.

(ii) Consider the two orthonormal bases in $\mathbb{C}^2$

$$\mathbf{x}_1 = \frac{1}{\sqrt{2}} \begin{pmatrix} 1 \\ 1 \end{pmatrix}, \quad \mathbf{x}_2 = \frac{1}{\sqrt{2}} \begin{pmatrix} 1 \\ -1 \end{pmatrix}; \qquad \mathbf{y}_1 = \frac{1}{\sqrt{2}} \begin{pmatrix} 1 \\ i \end{pmatrix}, \quad \mathbf{y}_2 = \frac{1}{\sqrt{2}} \begin{pmatrix} 1 \\ -i \end{pmatrix}.$$

Use these bases to construct the corresponding 2×2 unitary matrix.

Solution 7.   (i) Using the *completeness relation*

$$\sum_{\ell=1}^{n} \mathbf{y}_\ell \mathbf{y}_\ell^* = I_n$$

and the property that matrix multiplication is associative we have

$$\sum_{\ell=1}^{n} (\mathbf{x}_j^* \mathbf{y}_\ell)(\mathbf{y}_\ell^* \mathbf{x}_k) = \sum_{\ell=1}^{n} \mathbf{x}_j^* (\mathbf{y}_\ell \mathbf{y}_\ell^*) \mathbf{x}_k = \mathbf{x}_j^* (\sum_{\ell=1}^{n} \mathbf{y}_\ell \mathbf{y}_\ell^*) \mathbf{x}_k = \mathbf{x}_j^* I_n \mathbf{x}_k = \mathbf{x}_j^* \mathbf{x}_k$$
$$= \delta_{jk}.$$

(ii) We have

$$u_{11} = \mathbf{x}_1^* \mathbf{y}_1 = \frac{1}{2}(1+i), \qquad u_{12} = \mathbf{x}_1^* \mathbf{y}_2 = \frac{1}{2}(1-i),$$

$$u_{21} = \mathbf{x}_2^* \mathbf{y}_1 = \frac{1}{2}(1-i), \qquad u_{22} = \mathbf{x}_2^* \mathbf{y}_2 = \frac{1}{2}(1+i).$$

Thus

$$U = \frac{1}{2} \begin{pmatrix} 1+i & 1-i \\ 1-i & 1+i \end{pmatrix} \quad \Rightarrow \quad U^* = U^{-1} = \frac{1}{2} \begin{pmatrix} 1-i & 1+i \\ 1+i & 1-i \end{pmatrix}.$$

Problem 8.   Find the norm $\|A\| = \sqrt{\operatorname{tr}(A^*A)}$ of the skew-hermitian matrix

$$A = \begin{pmatrix} i & 2+i \\ -2+i & 3i \end{pmatrix}$$

without calculating A^*.

Solution 8.   Since $A^* = -A$ for a skew-hermitian matrix we obtain $\|A\| = \sqrt{-\operatorname{tr}(A^2)}$. Since

$$A^2 = \begin{pmatrix} -6 & * \\ * & -14 \end{pmatrix}$$

we obtain $\|A\| = \sqrt{20}$.

Problem 9.   Let A and B be 2×2 diagonal matrices over $\mathbb{R}$. Assume that

$$\operatorname{tr}(AA^T) = \operatorname{tr}(BB^T)$$

and

$$\max_{\|\mathbf{x}\|=1} \|A\mathbf{x}\| = \max_{\|\mathbf{x}\|=1} \|B\mathbf{x}\|.$$

Can we conclude that $A = B$?

Solution 9. No we cannot conclude that $A = B$, for example the two 2×2 matrices

$$A = \begin{pmatrix} 1 & 0 \\ 0 & 3 \end{pmatrix}, \qquad B = \begin{pmatrix} 3 & 0 \\ 0 & 1 \end{pmatrix}$$

satisfy both conditions.

Problem 10. Let A be an $n \times n$ matrix over $\mathbb{C}$. Let $\|.\|$ be a subordinate matrix norm for which $\|I_n\| = 1$. Assume that $\|A\| < 1$.
(i) Show that the matrix $(I_n - A)$ is nonsingular.
(ii) Show that $\|(I_n - A)^{-1}\| \leq (1 - \|A\|)^{-1}$.

Solution 10. (i) Let $\lambda_1, \ldots, \lambda_n$ be the eigenvalues of A. Then the eigenvalues of $I_n - A$ are $1 - \lambda_1, \ldots, 1 - \lambda_n$. Since $\|A\| < 1$, we have $|\lambda_j| < 1$ for each j. Thus, none of the numbers $1 - \lambda_1, \ldots, 1 - \lambda_n$ is equal to 0. This proves that $I_n - A$ is nonsingular.
(ii) Since $\|A\| < 1$, we can write the expansion

$$(I_n - A)^{-1} = I_n + A + A^2 + \cdots = \sum_{j=0}^{\infty} A^j. \tag{1}$$

Now we have $\|A^j\| \leq \|A\|^j$. Taking the norm on both sides of (1) we have

$$\|(I_n - A)^{-1}\| \leq \sum_{j=0}^{\infty} \|A\|^j = (1 - \|A\|)^{-1}$$

since $\|I_n\| = 1$. Note that the infinite series $1 + x + x^2 + \cdots$ converges to $1/(1-x)$ if and only if $|x| < 1$.

Problem 11. Let A be an $n \times n$ matrix. Assume that $\|A\| < 1$. Show that

$$\|(I_n - A)^{-1} - I_n\| \leq \frac{\|A\|}{1 - \|A\|}.$$

Solution 11. For any two $n \times n$ nonsingular matrices X, Y we have the identity

$$X^{-1} - Y^{-1} \equiv X^{-1}(Y - X)Y^{-1}.$$

If $Y = I_n$ and $X = I_n - A$, then $(I_n - A)^{-1} - I_n = (I_n - A)^{-1}A$. Taking the norm on both sides yields

$$\|(I_n - A)^{-1} - I_n\| \leq \|I_n - A\|^{-1}\|A\|.$$

Using $\|I_n - A\|^{-1} \le (1 - \|A\|)^{-1}$ we obtain the result.

Problem 12.   Let A be an $n \times n$ nonsingular matrix and B an $n \times n$ matrix. Assume that $\|A^{-1}B\| < 1$.
(i) Show that $A - B$ is nonsingular.
(ii) Show that

$$\frac{\|A^{-1} - (A - B)^{-1}\|}{\|A^{-1}\|} \le \frac{\|A^{-1}B\|}{1 - \|A^{-1}B\|}.$$

Solution 12.   (i) We have $A - B \equiv A(I_n - A^{-1}B)$. Since $\|A^{-1}B\| < 1$ we have that $I_n - A^{-1}B$ is nonsingular. Therefore, $A - B$ which is the product of the nonsingular matrices A and $I_n - A^{-1}B$ is also nonsingular.
(ii) Consider the identity for invertible matrices X and Y

$$X^{-1} - Y^{-1} \equiv X^{-1}(Y - X)Y^{-1}.$$

Substituting $X = A$ and $Y = A - B$, we obtain $A^{-1} - (A - B)^{-1} = -A^{-1}B(A - B)^{-1}$. Taking the norm on both sides yields

$$\|A^{-1} - (A - B)^{-1}\| \le \|A^{-1}B\|\,\|(A - B)^{-1}\|.$$

For nonsingular matrices X and Y we have the identities

$$Y \equiv X - (X - Y) \equiv X(I_n - X^{-1}(X - Y))$$

and therefore $Y^{-1} \equiv (I_n - X^{-1}(X - Y))^{-1}X^{-1}$. If we substitute $Y = A - B$ and $X = A$ we have

$$(A - B)^{-1} = (I_n - A^{-1}B)^{-1}A^{-1}.$$

Taking norms yields

$$\|(A - B)^{-1}\| \le \|A^{-1}\|\,\|(I_n - A^{-1}B)^{-1}\|.$$

We know that

$$\|(I_n - A^{-1}B)^{-1}\| \le (1 - \|A^{-1}B\|)^{-1}$$

and therefore

$$\|(A - B)^{-1}\| \le \frac{\|A^{-1}\|}{1 - \|A^{-1}B\|}.$$

It follows that

$$\|A^{-1} - (A - B)^{-1}\| \le \frac{\|A^{-1}B\| \cdot \|A^{-1}\|}{1 - \|A^{-1}B\|} \quad \text{or} \quad \frac{\|A^{-1} - (A - B)^{-1}\|}{\|A^{-1}\|} \le \frac{\|A^{-1}B\|}{1 - \|A^{-1}B\|}.$$

Problem 13.   Let A, B be $n \times n$ matrices over $\mathbb{R}$ and $t \in \mathbb{R}$. Let $\|\ \|$ be a matrix norm. Show that

$$\|e^{tA}e^{tB} - I_n\| \le \exp(|t|(\|A\| + \|B\|)) - 1.$$

Solution 13. We have

$$
\begin{aligned}
\|e^{tA}e^{tB} - I_n\| &= \|(e^{tA} - I_n) + (e^{tB} - I_n) + (e^{tA} - I_n)(e^{tB} - I_n)\| \\
&\leq \|e^{tA} - I_n\| + \|e^{tB} - I_n\| + \|e^{tA} - I_n\| \cdot \|e^{tB} - I_n\| \\
&\leq (\exp(|t| \, \|A\|) - 1) + (\exp(|t| \, \|B\|) - 1) \\
&\quad + (\exp(|t| \, \|A\|) - 1)(\exp(|t| \, \|B\|) - 1) \\
&= \exp(|t|(\|A\| + \|B\|)) - 1.
\end{aligned}
$$

Problem 14. Let $A_1, A_2, \ldots, A_p$ be $m \times m$ matrices over $\mathbb{C}$. Then we have the inequality

$$
\| \exp(\sum_{j=1}^{p} A_j) - (e^{A_1/n} \cdots e^{A_p/n})^n \| \leq \frac{2}{n} \left(\sum_{j=1}^{p} \|A_j\| \right)^2 \exp\left(\frac{n+2}{n} \sum_{j=1}^{p} \|A_j\| \right)
$$

and (*Trotter formula*)

$$
\lim_{n \to \infty} (e^{A_1/n} e^{A_2/n} \cdots e^{A_p/n})^n = \exp(\sum_{j=1}^{p} A_j).
$$

Let $p = 2$. Find the estimate for the 2×2 matrices

$$
A_1 = \begin{pmatrix} 0 & 1 \\ 1 & 0 \end{pmatrix}, \qquad A_2 = \begin{pmatrix} 1 & 0 \\ 0 & 2 \end{pmatrix}.
$$

Solution 14. We have $\|A_1\| = 1$, $\|A_2\| = 2$. Thus for the right-hand side we find

$$
\frac{2}{n} (\sum_{j=1}^{2} \|A_j\|)^2 \exp\left(\frac{n+2}{n} \sum_{j=1}^{2} \|A_j\| \right) = \frac{18}{n} \exp\left(\frac{3(n+2)}{n} \right).
$$

Problem 15. Let U_1, U_2 be unitary $n \times n$ matrices. Let $\mathbf{v}$ be a normalized vector in $\mathbb{C}^n$. Consider the norm of a $k \times k$ matrix M

$$
\|M\| = \max_{\|\mathbf{v}\|=1} \|M\mathbf{v}\|
$$

where $\|\mathbf{v}\|$ denotes the Euclidean norm. Show that if $\|U_1 - U_2\| \leq \epsilon$ then $\|U_1\mathbf{v} - U_2\mathbf{v}\| \leq \epsilon$.

Solution 15. We have

$$
\|U_1\mathbf{v} - U_2\mathbf{v}\| = \|(U_1 - U_2)\mathbf{v}\| \leq \max_{\|\mathbf{y}\|=1} \|(U_1 - U_2)\mathbf{y}\| = \|U_1 - U_2\| \leq \epsilon.
$$

Problem 16. Given the 2×2 matrix (*rotation matrix*)

$$
R(\alpha) = \begin{pmatrix} \cos(\alpha) & -\sin(\alpha) \\ \sin(\alpha) & \cos(\alpha) \end{pmatrix}.
$$

Calculate $\|R(\alpha)\| = \sup_{\|\mathbf{x}\|=1} \|R(\alpha)\mathbf{x}\|$.

Solution 16.   Since $R^T(\alpha) = R^{-1}(\alpha)$ we have $R^T(\alpha)R(\alpha) = I_2$. The largest eigenvalue of I_2 is obviously 1 and the positive square root is 1. Thus $\|R(\alpha)\| = 1$ and the norm is independent of α.

Problem 17.   Let A be an $n \times n$ matrix. Let $\rho(A)$ be the *spectral radius* of A. Then we have

$$\rho(A) \leq \min\{\max_{1 \leq i \leq n} \sum_{j=1}^{n} |a_{ij}|, \max_{1 \leq j \leq n} \sum_{i=1}^{n} |a_{ij}|\}.$$

Let

$$A = \begin{pmatrix} 0 & 1 & 2 \\ 3 & 4 & 5 \\ 6 & 7 & 8 \end{pmatrix}.$$

Calculate $\rho(A)$ and the right-hand side of the inequality.

Solution 17.   For the right-hand side we have

$$\min\{\max\{3, 12, 21\}, \max\{9, 12, 15\}\} = 15.$$

For the left-hand side we find

$$\rho(A) = \max_{1 \leq j \leq 3} |\lambda_j| = \max\{6 - 3\sqrt{6}, 6 + 3\sqrt{6}, 0\} = 6 + 3\sqrt{6} \approx 13.3485.$$

Problem 18.   Let $t \in \mathbb{R}$. Consider the symmetric 3×3 matrix over $\mathbb{R}$

$$A(t) = \begin{pmatrix} t & 1 & 0 \\ 1 & t & 1 \\ 0 & 1 & t \end{pmatrix}.$$

Find the condition on t such that $\rho(A(t)) < 1$, where $\rho(A(t))$ denotes the spectral radius of $A(t)$.

Solution 18.   We have

$$\rho(A) = \max_{1 \leq j \leq 3} |\lambda_j|.$$

The eigenvalues depending on t are $\{t, t - \sqrt{2}, t + \sqrt{2}\}$ with the maximum $t + \sqrt{2}$. Thus we have the condition $t < 1 - \sqrt{2}$.

Problem 19.   (i) Let A be an $n \times n$ positive semidefinite matrix. Show that $(I_n + A)^{-1}$ exists.
(ii) Let B be an arbitrary $n \times n$ matrix. Show that the inverse of $I_n + B^*B$ exists.

Solution 19. Consider the linear equation $(I_n + A)\mathbf{u} = \mathbf{0}$. Then $-\mathbf{u} = A\mathbf{u}$ and since A is positive semidefinite we have

$$0 \le (A\mathbf{u})^*\mathbf{u} = \mathbf{u}^*A\mathbf{u} = -\mathbf{u}^*\mathbf{u} = -\|\mathbf{u}\| \le 0.$$

Thus $\|\mathbf{u}\| = 0$ and therefore $\mathbf{u} = \mathbf{0}$. Then $(I_n + A)^{-1}$ exists. We could also prove it as follows. Since A is a positive semidefinite matrix there is an $n \times n$ unitary matrix U such that UAU^* is a diagonal matrix and the diagonal elements are nonnegative. Since $U(I_n + A)U^* = I_n + UAU^*$ the inverse of $U(I_n + A)U^*$ exists and therefore the inverse of $I_n + A$ exists.
(ii) Since B^*B is a positive semidefinite matrix we can apply the result from (i) and therefore the inverse of $I_n + B^*B$ exists.

Problem 20. Let A be an $n \times n$ matrix. One approach to calculate $\exp(A)$ is to compute an eigenvalue decomposition $A = XBX^{-1}$ and then apply the formula $e^A = Xe^BX^{-1}$. We have using the *Schur decomposition*

$$U^*AU = \operatorname{diag}(\lambda_1, \ldots, \lambda_n) + N$$

where U is unitary, the matrix $N = (n_{jk})$ is a strictly upper triangular ($n_{jk} = 0, j \ge k$) and $\lambda(A) = \{\lambda_1, \ldots, \lambda_n\}$ is the spectrum of A. Using the *Padé approximation* to calculate e^A we have

$$R_{pq} = (D_{pq}(A))^{-1}N_{pq}(A)$$

where

$$N_{pq}(A) := \sum_{j=0}^{p} \frac{(p+q-j)!p!}{(p+q)!j!(p-j)!}A^j, \quad D_{pq}(A) := \sum_{j=0}^{q} \frac{(p+q-j)!q!}{(p+q)!j!(q-j)!}(-A)^j.$$

Consider the nonnormal matrix

$$A = \begin{pmatrix} 0 & 6 & 0 & 0 \\ 0 & 0 & 6 & 0 \\ 0 & 0 & 0 & 6 \\ 0 & 0 & 0 & 0 \end{pmatrix}.$$

Calculate $\|R_{11} - e^A\|$, where $\|\cdot\|$ denotes the 2-norm.

Solution 20. We have $p = q = 1$, $N_{11} = I_4 + \frac{1}{2}A$, $D_{11} = I_4 - \frac{1}{2}A$

$$N_{11}(A) = \begin{pmatrix} 1 & 3 & 0 & 0 \\ 0 & 1 & 3 & 0 \\ 0 & 0 & 1 & 3 \\ 0 & 0 & 0 & 1 \end{pmatrix}, \quad D_{11}(A) = \begin{pmatrix} 1 & -3 & 0 & 0 \\ 0 & 1 & -3 & 0 \\ 0 & 0 & 1 & -3 \\ 0 & 0 & 0 & 1 \end{pmatrix}$$

and

$$D_{11}(A)^{-1} = \begin{pmatrix} 1 & 3 & 9 & 27 \\ 0 & 1 & 3 & 9 \\ 0 & 0 & 1 & 3 \\ 0 & 0 & 0 & 1 \end{pmatrix}, \quad R_{11}(A) = \begin{pmatrix} 1 & 6 & 18 & 54 \\ 0 & 1 & 6 & 18 \\ 0 & 0 & 1 & 6 \\ 0 & 0 & 0 & 1 \end{pmatrix}.$$

We also have

$$A^2 = 36 \begin{pmatrix} 0 & 0 & 1 & 0 \\ 0 & 0 & 0 & 1 \\ 0 & 0 & 0 & 0 \\ 0 & 0 & 0 & 0 \end{pmatrix}, \qquad A^3 = 216 \begin{pmatrix} 0 & 0 & 0 & 1 \\ 0 & 0 & 0 & 0 \\ 0 & 0 & 0 & 0 \\ 0 & 0 & 0 & 0 \end{pmatrix}$$

and A^4 is the 4×4 zero matrix. Thus

$$e^A = I_4 + A + \frac{1}{2}A^2 + \frac{1}{6}A^3 = \begin{pmatrix} 1 & 6 & 18 & 36 \\ 0 & 1 & 6 & 18 \\ 0 & 0 & 1 & 6 \\ 0 & 0 & 0 & 1 \end{pmatrix}.$$

Thus

$$R_{11}(A) - e^A = \begin{pmatrix} 0 & 0 & 0 & 18 \\ 0 & 0 & 0 & 0 \\ 0 & 0 & 0 & 0 \\ 0 & 0 & 0 & 0 \end{pmatrix}.$$

It follows that $\|R_{11}(A) - e^A\| = 18$. In general, the bounds on the accuracy of an approximation $f(A)$ to e^A deteriorate with loss of normality of the matrix A.

Problem 21. Let A be an $n \times n$ matrix with $\|A\| < 1$. Then $\ln(I_n + A)$ exists. Show that

$$\| \ln(I_n + A) \| \leq \frac{\|A\|}{1 - \|A\|}.$$

Solution 21. Since $\|A\| < 1$ we have

$$\ln(I_n + A) = \sum_{k=1}^{\infty} (-1)^{k+1} \frac{A^k}{k}.$$

Thus

$$\| \ln(I_n + A) \| \leq \sum_{k=1}^{\infty} \frac{\|A\|^k}{k} \leq \|A\| \sum_{k=0}^{\infty} \|A\|^k = \frac{\|A\|}{1 - \|A\|}.$$

Problem 22. Let A, B be $n \times n$ matrices over $\mathbb{C}$. Show that $\|[A, B]\| \leq 2\|A\|\,\|B\|$.

Solution 22. We have

$$\|[A, B]\| = \|AB - BA\| \leq \|AB\| + \|BA\| = \|A\|\,\|B\| + \|B\|\,\|A\| = 2\|A\|\,\|B\|.$$

Problem 23. Let A, B be $n \times n$ matrices over $\mathbb{C}$. Then

$$\|[A, B]\| \leq \|AB\| + \|BA\| \leq 2\|A\|\,\|B\|.$$

Let

$$A = \begin{pmatrix} 0 & 1 \\ 1 & 0 \end{pmatrix}, \qquad B = \frac{1}{\sqrt{2}} \begin{pmatrix} 1 & 1 \\ 1 & -1 \end{pmatrix}.$$

Calculate $\|[A, B]\|$, $\|AB\|$, $\|BA\|$, $\|A\|$, $\|B\|$ and thus verify the inequality for these matrices. The norm is given by $\|C\| = \sqrt{\mathrm{tr}(CC^*)}$.

Solution 23. We have $\|A\| = \sqrt{2}$, $\|B\| = \sqrt{2}$ and $\|AB\| = \sqrt{2}$, $\|BA\| = \sqrt{2}$. The commutator of A and B is given by

$$[A, B] = AB - BA = \begin{pmatrix} 0 & -\sqrt{2} \\ \sqrt{2} & 0 \end{pmatrix}.$$

Thus $\|[A, B]\| = 2$.

Problem 24. Denote by $\|\cdot\|_{HS}$ the Hilbert-Schmidt norm and by $\|\cdot\|_O$ the operator norm, i.e.

$$\|A\|_{HS} := \sqrt{\mathrm{tr}(AA^*)}, \qquad \|A\|_O := \sup_{\|\mathbf{x}\|=1} \|A\mathbf{x}\| = \sup_{\substack{\lambda \in \mathbb{C}, \, \|\mathbf{x}\|=1 \\ (A^*A)\mathbf{x}=\lambda\mathbf{x}}} \sqrt{\lambda}$$

where A is an $m \times n$ matrix over $\mathbb{C}$, $\mathbf{x} \in \mathbb{C}^n$ and $\|\mathbf{x}\|$ is the Euclidean norm.
(i) Let

$$M = \frac{1}{2} \begin{pmatrix} 1 & 0 & 1 \\ 1 & 0 & 1 \end{pmatrix}.$$

Calculate $\|M\|_{HS}$ and $\|M\|_O$.
(ii) Let A be an $m \times n$ matrix over $\mathbb{C}$. Find the conditions on A such that $\|A\|_{HS} = \|A\|_O$.

Solution 24. The first calculation is straightforward

$$\|M\|_{HS} = \sqrt{\mathrm{tr}(MM^T)} = 1.$$

To calculate $\|M\|_O$ we need to find the largest eigenvalue of

$$M^T M = \frac{1}{2} \begin{pmatrix} 1 & 0 & 1 \\ 0 & 0 & 0 \\ 1 & 0 & 1 \end{pmatrix}.$$

The eigenvalues are 0, 0 and 1. Thus $\|M\|_O = \sqrt{1} = 1$.
(ii) Let $\lambda_1, \ldots, \lambda_m$ be the nonnegative eigenvalues of AA^*, then

$$\|A\|_{HS} = \sqrt{\lambda_1 + \lambda_2 + \cdots + \lambda_m}.$$

Let $\mu_1, \mu_2, \ldots, \mu_n$ be the non-negative eigenvalues of A^*A, then from $\mathrm{tr}(A^*A) = \mathrm{tr}(AA^*)$

$$\|A\|_{HS} = \sqrt{\mu_1 + \mu_2 + \cdots + \mu_n}.$$

Now

$$\|A\|_O = \sqrt{\max\{\mu_1, \mu_2, \ldots, \mu_n\}}.$$

Setting

$$\|A\|_{HS} = \sqrt{\mu_1 + \mu_2 + \cdots + \mu_n} = \sqrt{\max\{\mu_1, \mu_2, \ldots, \mu_n\}} = \|A\|_O$$

we find

$$\mu_1 + \mu_2 + \cdots + \mu_n = \max\{\mu_1, \mu_2, \ldots, \mu_n\}.$$

Since the values μ_j are non-negative all the eigenvalues, except perhaps for one, must be zero. In other words

$$\|A\|_{HS} = \|A\|_O$$

if and only if A^*A has at most one non-zero eigenvalue (or equivalently AA^* has at most one non-zero eigenvalue).

Problem 25.   Let A be an $n \times n$ matrix. The *logarithmic norm* is defined by

$$\mu[A] := \lim_{h \to 0^+} \frac{\|I_n + hA\| - 1}{h}.$$

Let $\|A\| := \sup_{x=1} |Ax\|$. Let A be the $n \times n$ identity matrix I_n. Find $\mu[I_n]$.

Solution 25.   Since $\|I_n\| = 1$ we have

$$\mu[A] = \lim_{h \to 0^+} \frac{\|I_n + hI_n\| - 1}{h} = \lim_{h \to 0^+} \frac{\|(1+h)I_n\| - 1}{h} = \frac{(1+h)\|I_n\| - 1}{h} = 1.$$

Problem 26.   Consider the Hilbert space $\mathbb{R}^n$ and $x, y \in \mathbb{R}^n$. The scalar product $\langle x, y \rangle$ is given by

$$\langle x, y \rangle := x^T y = \sum_{j=1}^{n} x_j y_j.$$

Thus the norm is given by $\|x\| = \sqrt{\langle x, x \rangle}$. Show that $|x^T y| \le \|x\| \cdot \|y\|$.

Solution 26.   If $y = 0$, then the left- and right-hand side of the inequality are equal to 0. Now assume that $y \ne 0$. We set $(\epsilon \in \mathbb{R})$

$$f(\epsilon) := \langle x - \epsilon y, x - \epsilon y \rangle.$$

Obviously $f(\epsilon) \ge 0$. Now we have $f(\epsilon) = \langle x, x \rangle - \langle \epsilon y, x \rangle - \langle x, \epsilon y \rangle + \langle \epsilon y, \epsilon y \rangle$ or since $\langle x, y \rangle = \langle y, x \rangle$

$$f(\epsilon) = \langle x, x \rangle - 2\epsilon \langle x, y \rangle + \epsilon^2 \langle y, y \rangle.$$

Differentiation of f with respect to ϵ yields

$$\frac{df(\epsilon)}{d\epsilon} = -2\langle x, y \rangle + 2\epsilon \langle y, y \rangle, \qquad \frac{d^2 f(\epsilon)}{d\epsilon^2} = 2\langle y, y \rangle \ge 0.$$

Therefore we find that the condition $df/d\epsilon = 0$ leads to the minimum

$$\langle \mathbf{x}, \mathbf{y} \rangle = \epsilon \langle \mathbf{y}, \mathbf{y} \rangle \quad \Rightarrow \quad \epsilon = \frac{\langle \mathbf{x}, \mathbf{y} \rangle}{\langle \mathbf{y}, \mathbf{y} \rangle}.$$

Inserting ϵ into the function $f(\epsilon)$ yields

$$f\left(\epsilon = \frac{\langle \mathbf{x}, \mathbf{y} \rangle}{\langle \mathbf{y}, \mathbf{y} \rangle} \right) = \langle \mathbf{x}, \mathbf{x} \rangle - \frac{\langle \mathbf{x}, \mathbf{y} \rangle^2}{\langle \mathbf{y}, \mathbf{y} \rangle} \geq 0.$$

Therefore $\langle \mathbf{x}, \mathbf{x} \rangle \langle \mathbf{y}, \mathbf{y} \rangle - \langle \mathbf{x}, \mathbf{y} \rangle^2 \geq 0$. Thus $\langle \mathbf{x}, \mathbf{x} \rangle \langle \mathbf{y}, \mathbf{y} \rangle \geq \langle \mathbf{x}, \mathbf{y} \rangle^2$ and the inequality follows.

Problem 27. Let A be an $n \times n$ hermitian matrix. Let $\mathbf{u}, \mathbf{v} \in \mathbb{C}^n$ and $\lambda \in \mathbb{C}$. Consider the equation

$$A\mathbf{u} - \lambda\mathbf{u} = \mathbf{v}.$$

(i) Show that for λ nonreal (i.e. it has an imaginary part) the vector $\mathbf{v}$ cannot vanish unless $\mathbf{u}$ vanishes.
(ii) Show that for λ nonreal we have

$$\| (A - \lambda I_n)^{-1} \mathbf{v} \| \leq \frac{1}{|\Im(\lambda)|} \| \mathbf{v} \|.$$

Solution 27. (i) From the equation by taking the scalar product with $\mathbf{u}$ we obtain

$$\mathbf{u}^* A \mathbf{u} - \lambda \mathbf{u}^* \mathbf{u} = \mathbf{u}^* \mathbf{v}.$$

Taking the imaginary part of this equation and taking into account that $\mathbf{u}^* A \mathbf{u}$ is real yields

$$-\Im(\lambda) \| \mathbf{u} \|^2 = \Im(\mathbf{u}^* \mathbf{v}).$$

This shows that for λ nonreal, $\mathbf{v}$ cannot vanish unless $\mathbf{u}$ vanishes. Thus λ is not an eigenvalue of A, and $(A - \lambda I_n)$ has an inverse.
(ii) By the *Schwarz inequality* we have

$$|\Im(\lambda)| \| \mathbf{u} \|^2 = \| \Im(\mathbf{u}^* \mathbf{v}) \| \leq |\mathbf{u}^* \mathbf{v}| \leq \| \mathbf{u} \| \, \| \mathbf{v} \|.$$

However $\mathbf{u} = (A - \lambda I_n)^{-1} \mathbf{v}$. Therefore

$$\| (A - \lambda I_n)^{-1} \mathbf{v} \| \leq \frac{1}{|\Im(\lambda)|} \| \mathbf{v} \|.$$

Problem 28. Let M be an $m \times n$ matrix over $\mathbb{C}$. The Frobenius norm of M is given by

$$\| M \|_F := \sqrt{\operatorname{tr}(M^* M)} = \sqrt{\operatorname{tr}(M M^*)}.$$

Let U_m be $m \times m$ unitary matrix and U_n be an $n \times n$ unitary matrix. Show that

$$\| U_m M \|_F = \| M U_n \|_F = \| M \|.$$

Show that $\|M\|_F$ is the square root of the sum of the squares of the singular values of M.

Solution 28. Using the property that $\text{tr}(AB) = \text{tr}(BA)$ we find

$$\|U_m M\|_F = \sqrt{\text{tr}((U_m M)^*(U_m M))} = \sqrt{\text{tr}(M^* M)} = \|M\|_F$$

and

$$\|MU_n\|_F = \sqrt{\text{tr}((MU_n)^*(MU_n))} = \sqrt{\text{tr}(M^* M)} = \|M\|_F.$$

Let $M = U\Sigma V^*$ be a singular value decomposition of M, then

$$\|M\|_F = \|U^* M\|_F = \|U^* MV\|_F = \|\Sigma\|_F = \sqrt{\text{tr}(\Sigma^* \Sigma)}.$$

Σ is an $m \times n$ matrix with

$$(\Sigma)_{ij} = \begin{cases} \sigma_i & i = j \\ 0 & i \neq j \end{cases}$$

where the σ_j are the singular values (with the convention that $\sigma_j = 0$ when $j > m$ and $j > n$). It follows that $\Sigma^* = \Sigma^T$ and

$$(\Sigma^* \Sigma)_{ij} = \sum_{k=1}^{m} (\Sigma)_{ki}(\Sigma)_{kj} = \sum_{k=1}^{m} \sigma_k^2 \delta_{ik}\delta_{jk} = \sigma_i^2 \delta_{ij}.$$

Finally

$$\|M\|_f = \sqrt{\text{tr}(\Sigma^* \Sigma)} = \sqrt{\sum_{i=1}^{n} (\Sigma^* \Sigma)_{ii}} = \sqrt{\sum_{i=1}^{n} \sigma_j^2}.$$

Problem 29. Let M be an $m \times n$ matrix over $\mathbb{C}$. Find the rank-1 $m \times n$ matrix A over $\mathbb{C}$ which minimizes

$$\|M - A\|_F.$$

Hint. Find the singular value decomposition of $M = U\Sigma V^*$ and find A' with rank 1 which minimizes

$$\|\Sigma - A'\|_F.$$

Apply the method to the 3×2 matrix

$$M = \begin{pmatrix} 0 & 1 \\ 1 & 0 \\ 0 & 1 \end{pmatrix}.$$

Solution 29. Using the singular value decomposition $M = U\Sigma V^*$ of M and the properties in 1. We obtain

$$\|M - A\|_F = \|U^*(M - A)V\|_F = \|\Sigma - A'\|_F$$

where $A' := U^*AV$. Since

$$\|\Sigma - A'\|_F = \sqrt{\sum_{i=1}^{m}\sum_{j=1}^{n}|\Sigma_{ij} - (A')_{ij}|^2} = \sqrt{\sum_{i=1}^{m}\sum_{j=1}^{n}|\delta_{ij}\sigma_i - (A')_{ij}|^2}$$

we find that minimizing $\|M - A\|$ implies that A' must be "diagonal"

$$\|\Sigma - A'\|_F = \sqrt{\sum_{i=1}^{\min\{m,n\}}|\sigma_i - (A')_{ii}|^2}.$$

Since A must have rank 1, A' must have rank 1 (since unitary operators are rank preserving). Since A' is "diagonal" only one entry can be non-zero. Thus to minimize $\|M - A\|$ we set

$$A' = \sigma_1 \mathbf{e}_{1,m}\mathbf{e}_{1,n}^T$$

where σ_1 is the largest singular value of M by convention. Finally $A = UA'V^*$. For the given M we find the singular value decomposition

$$M = \begin{pmatrix} 0 & 1 \\ 1 & 0 \\ 0 & 1 \end{pmatrix} = \begin{pmatrix} 1/\sqrt{2} & 0 & 1/\sqrt{2} \\ 0 & 1 & 0 \\ 1/\sqrt{2} & 0 & -1/\sqrt{2} \end{pmatrix} \begin{pmatrix} \sqrt{2} & 0 \\ 0 & 1 \\ 0 & 0 \end{pmatrix} \begin{pmatrix} 0 & 1 \\ 1 & 0 \end{pmatrix}.$$

Thus we obtain

$$A = \begin{pmatrix} 1/\sqrt{2} & 0 & 1/\sqrt{2} \\ 0 & 1 & 0 \\ 1/\sqrt{2} & 0 & -1/\sqrt{2} \end{pmatrix} \begin{pmatrix} \sqrt{2} & 0 \\ 0 & 0 \\ 0 & 0 \end{pmatrix} \begin{pmatrix} 0 & 1 \\ 1 & 0 \end{pmatrix} = \begin{pmatrix} 0 & 1 \\ 0 & 0 \\ 0 & 1 \end{pmatrix}.$$

/

Problem 30. Let A be an $n \times n$ matrix over $\mathbb{C}$. The *spectral radius* of the matrix A is the nonnegative number $\rho(A)$ defined by

$$\rho(A) := \max\{ |\lambda_j(A)| : 1 \le j \le n \}$$

where $\lambda_j(A)$ $(j = 1, \ldots, n)$ are the eigenvalues of A. We define the norm of A as

$$\|A\| := \sup_{\|\mathbf{x}\|=1} \|A\mathbf{x}\|$$

where $\|A\mathbf{x}\|$ denotes the Euclidean norm. Is $\rho(A) \le \|A\|$? Prove or disprove.

Solution 30. From the eigenvalue equation $A\mathbf{x} = \lambda\mathbf{x}$ ($\mathbf{x} \ne \mathbf{0}$) we obtain

$$|\lambda|\|\mathbf{x}\| = \|\lambda\mathbf{x}\| = \|A\mathbf{x}\| \le \|A\|\|\mathbf{x}\|.$$

Therefore since $\|\mathbf{x}\| \ne 0$ we obtain $|\lambda| \le \|A\|$ and $\rho(A) \le \|A\|$.

Problem 31. Let A be an $n \times n$ matrix over $\mathbb{C}$ and I_m be the $m \times m$ unit matrix. Consider the norm of a $k \times k$ matrix M

$$\|M\| = \max_{\|\mathbf{x}\|=1} \|M\mathbf{x}\|$$

where $\|\mathbf{x}\|$ denotes the Euclidean norm. Show that $\|A \otimes I_m\| = \|A\|$.

Solution 31. The norm of $\|A\|$ is the square root of the largest eigenvalue of the positive semi-definite matrix AA^*. Now

$$(A \otimes I_m)(A^* \otimes I_m) \equiv (AA^*) \otimes I_m.$$

Now the largest eigenvalue of the positive semidefinite matrix $(AA^*) \otimes I_m$ is equal to the largest eigenvalue of AA^* since the eigenvalues of I_m are 1. Thus $\|A\| = \|A \otimes I_m\|$ follows.

Problem 32. Find 2×2 matrices A and B over $\mathbb{C}$ which minimize

$$\left\| \begin{pmatrix} 0 & 0 & 0 & 0 \\ 0 & 1 & 1 & 0 \\ 0 & 1 & 1 & 0 \\ 0 & 0 & 0 & 0 \end{pmatrix} - A \otimes B \right\|_F$$

and determine the minimum.

Solution 32. Consider the 2×2 matrices

$$A = \begin{pmatrix} a_1 & a_2 \\ a_3 & a_4 \end{pmatrix}, \qquad B = \begin{pmatrix} b_1 & b_2 \\ b_3 & b_4 \end{pmatrix}.$$

The reshaping operator $R(A \otimes B)$ is given by

$$R \begin{pmatrix} a_1 b_1 & a_1 b_2 & a_2 b_1 & a_2 b_2 \\ a_1 b_3 & a_1 b_4 & a_2 b_3 & a_2 b_4 \\ a_3 b_1 & a_3 b_2 & a_4 b_1 & a_4 b_2 \\ a_3 b_3 & a_3 b_4 & a_4 b_3 & a_4 b_4 \end{pmatrix} = \mathrm{vec}(A)(\mathrm{vec}(B))^T.$$

Thus

$$R \begin{pmatrix} 0 & 0 & 0 & 0 \\ 0 & 1 & 1 & 0 \\ 0 & 1 & 1 & 0 \\ 0 & 0 & 0 & 0 \end{pmatrix} = \begin{pmatrix} 0 & 0 & 0 & 1 \\ 0 & 0 & 1 & 0 \\ 0 & 1 & 0 & 0 \\ 1 & 0 & 0 & 0 \end{pmatrix}$$

which has four (of many) singular value decompositions

$$\begin{pmatrix} 0 & 0 & 0 & 1 \\ 0 & 0 & 1 & 0 \\ 0 & 1 & 0 & 0 \\ 1 & 0 & 0 & 0 \end{pmatrix} = I_4 I_4 \begin{pmatrix} 0 & 0 & 0 & 1 \\ 0 & 0 & 1 & 0 \\ 0 & 1 & 0 & 0 \\ 1 & 0 & 0 & 0 \end{pmatrix}^* = \begin{pmatrix} 0 & 0 & 0 & 1 \\ 0 & 0 & 1 & 0 \\ 0 & 1 & 0 & 0 \\ 1 & 0 & 0 & 0 \end{pmatrix} I_4 I_4^*$$

$$= \begin{pmatrix} 0 & 0 & 0 & 1 \\ 0 & 0 & 1 & 0 \\ 1 & 0 & 0 & 0 \\ 0 & 1 & 0 & 0 \end{pmatrix} I_4 \begin{pmatrix} 0 & 1 & 0 & 0 \\ 1 & 0 & 0 & 0 \\ 0 & 0 & 1 & 0 \\ 0 & 0 & 0 & 1 \end{pmatrix}^*$$

$$= \begin{pmatrix} 0 & 0 & 0 & 1 \\ 1 & 0 & 0 & 0 \\ 0 & 1 & 0 & 0 \\ 0 & 0 & 1 & 0 \end{pmatrix} I_4 \begin{pmatrix} 0 & 0 & 1 & 0 \\ 0 & 1 & 0 & 0 \\ 1 & 0 & 0 & 0 \\ 0 & 0 & 0 & 1 \end{pmatrix}^*.$$

The first decomposition yields

$$\text{vec}(A) = 1 \cdot I_4 \begin{pmatrix} 1 \\ 0 \\ 0 \\ 0 \end{pmatrix} = \begin{pmatrix} 1 \\ 0 \\ 0 \\ 0 \end{pmatrix}, \qquad A = \begin{pmatrix} 1 & 0 \\ 0 & 0 \end{pmatrix}$$

$$\text{vec}(B) = \begin{pmatrix} 0 & 0 & 0 & 1 \\ 0 & 0 & 1 & 0 \\ 0 & 1 & 0 & 0 \\ 1 & 0 & 0 & 0 \end{pmatrix} \begin{pmatrix} 1 \\ 0 \\ 0 \\ 0 \end{pmatrix} = \begin{pmatrix} 0 \\ 0 \\ 0 \\ 1 \end{pmatrix}, \qquad B = \begin{pmatrix} 0 & 0 \\ 0 & 1 \end{pmatrix}.$$

Thus $\|M - A \otimes B\|_F = \sqrt{3}$. The second decomposition yields

$$A = \begin{pmatrix} 0 & 0 \\ 0 & 1 \end{pmatrix}, \quad B = \begin{pmatrix} 1 & 0 \\ 0 & 0 \end{pmatrix}, \quad \|M - A \otimes B\|_F = \sqrt{3}.$$

The third decomposition yields

$$A = \begin{pmatrix} 0 & 1 \\ 0 & 0 \end{pmatrix}, \quad B = \begin{pmatrix} 0 & 0 \\ 1 & 0 \end{pmatrix}, \quad \|M - A \otimes B\|_F = \sqrt{3}.$$

The fourth decomposition yields

$$A = \begin{pmatrix} 0 & 0 \\ 1 & 0 \end{pmatrix}, \quad B = \begin{pmatrix} 0 & 1 \\ 0 & 0 \end{pmatrix}, \quad \|M - A \otimes B\|_F = \sqrt{3}.$$

Problem 33. Consider the 4×4 matrix (Hamilton operator)

$$\hat{H} = \frac{\hbar\omega}{2}(\sigma_1 \otimes \sigma_1 - \sigma_2 \otimes \sigma_2)$$

where ω is the frequency and $\hbar$ is the Planck constant divided by 2π. Find the *norm* of $\hat{H}$, i.e.

$$\|\hat{H}\| := \max_{\|\mathbf{x}\|=1} \|\hat{H}\mathbf{x}\|, \qquad \mathbf{x} \in \mathbb{C}^4$$

applying two different methods. In the first method apply the *Lagrange multiplier method*, where the constraint is $\|\mathbf{x}\| = 1$. In the second method we calculate $\hat{H}^* \hat{H}$ and find the square root of the largest eigenvalue. This is then $\|\hat{H}\|$. Note that $\hat{H}^* \hat{H}$ is positive semi-definite.

Solution 33. In the first method we use the *Lagrange multiplier method*, where the constraint $\|\mathbf{x}\| = 1$ can be written as $x_1^2 + x_2^2 + x_3^2 + x_4^2 = 1$. Since

$$\sigma_1 \otimes \sigma_1 = \begin{pmatrix} 0 & 0 & 0 & 1 \\ 0 & 0 & 1 & 0 \\ 0 & 1 & 0 & 0 \\ 1 & 0 & 0 & 0 \end{pmatrix}, \quad \sigma_2 \otimes \sigma_2 = \begin{pmatrix} 0 & 0 & 0 & -1 \\ 0 & 0 & 1 & 0 \\ 0 & 1 & 0 & 0 \\ -1 & 0 & 0 & 0 \end{pmatrix}$$

we have

$$\hat{H} = \hbar\omega \begin{pmatrix} 0 & 0 & 0 & 1 \\ 0 & 0 & 0 & 0 \\ 0 & 0 & 0 & 0 \\ 1 & 0 & 0 & 0 \end{pmatrix}.$$

Let $\mathbf{x} = (x_1, x_2, x_3, x_4)^T \in \mathbb{C}^4$. We maximize

$$f(\mathbf{x}) := \|\hat{H}\mathbf{x}\|^2 - \lambda(x_1^2 + x_2^2 + x_3^2 + x_4^2 - 1)$$

where λ is the *Lagrange multiplier*. To find the extrema we solve the equations

$$\frac{\partial f}{\partial x_1} = 2\hbar^2\omega^2 x_1 - 2\lambda x_1 = 0$$

$$\frac{\partial f}{\partial x_2} = -2\lambda x_2 = 0$$

$$\frac{\partial f}{\partial x_3} = -2\lambda x_3 = 0$$

$$\frac{\partial f}{\partial x_4} = 2\hbar^2\omega^2 x_4 - 2\lambda x_4 = 0$$

together with the constraint $x_1^2 + x_2^2 + x_3^2 + x_4^2 = 1$. These equations can be written in the matrix form

$$\begin{pmatrix} \hbar^2\omega^2 - \lambda & 0 & 0 & 0 \\ 0 & -\lambda & 0 & 0 \\ 0 & 0 & -\lambda & 0 \\ 0 & 0 & 0 & \hbar^2\omega^2 - \lambda \end{pmatrix} \begin{pmatrix} x_1 \\ x_2 \\ x_3 \\ x_4 \end{pmatrix} = \begin{pmatrix} 0 \\ 0 \\ 0 \\ 0 \end{pmatrix}.$$

If $\lambda = 0$ then $x_1 = x_4 = 0$ and $\|\hat{H}\mathbf{x}\| = 0$, which is a minimum. If $\lambda \neq 0$ then $x_2 = x_3 = 0$ and $x_1^2 + x_4^2 = 1$ so that $\|\hat{H}\mathbf{x}\| = \hbar\omega$, which is the maximum. Thus we find $\|\hat{H}\| = \hbar\omega$.

In the second method we calculate $\hat{H}^*\hat{H}$ and find the square root of the largest eigenvalue. Since $H^* = H$ we find

$$\hat{H}^*\hat{H} = \hbar^2\omega^2 \begin{pmatrix} 1 & 0 & 0 & 0 \\ 0 & 0 & 0 & 0 \\ 0 & 0 & 0 & 0 \\ 0 & 0 & 0 & 1 \end{pmatrix}.$$

Thus the maximum eigenvalue is $\hbar^2\omega^2$ (twice degenerate) and $\|\hat{H}\| = \hbar\omega$.

Problem 34. Let A be an invertible $n \times n$ matrix over $\mathbb{R}$. Consider the linear system $A\mathbf{x} = \mathbf{b}$. The *condition number* of A is defined as

$$\mathrm{Cond}(A) := \|A\| \cdot \|A^{-1}\|.$$

Find the condition number for the matrix

$$A = \begin{pmatrix} 1 & 0.9999 \\ 0.9999 & 1 \end{pmatrix}$$

for the infinity norm, 1-norm and 2-norm.

Solution 34. The inverse matrix of A is given by

$$A^{-1} = 10^3 \begin{pmatrix} 5.000250013 & -4.999749987 \\ -4.999749987 & 5.000250013 \end{pmatrix}.$$

Thus for the infinity norm we obtain

$$\|A\|_\infty = 1.9999, \qquad \|A^{-1}\|_\infty = 10^4.$$

Therefore $\text{Cond}_\infty(A) = 1.9999 \cdot 10^4$. For the 1-norm we obtain

$$\|A\|_1 = 1.9999, \qquad \|A^{-1}\|_1 = 10^4.$$

Therefore $\text{Cond}_1(A) = 1.9999 \cdot 10^4$. Thus for the 2-norm we obtain

$$\|A\|_2 = 1.9999, \qquad \|A^{-1}\|_2 = 10^4.$$

Therefore $\text{Cond}_2(A) = 1.9999 \cdot 10^4$. In the given problem the condition number is the same with respect to the norms. Note that this is not always the case.

Problem 35. Let A, B be nonsingular $n \times n$ matrices. Show that

$$\text{cond}(A \otimes B) = \text{cond}(A)\text{cond}(B)$$

for all matrix norms, where cond denotes the condition number.

Solution 35. We have

$$\begin{aligned}
\text{cond}(A \otimes B) &= \|A \otimes B\| \, \|(A \otimes B)^{-1}\| = \|A \otimes B\| \, \|A^{-1} \otimes B^{-1}\| \\
&= \|A\| \, \|B\| \, \|A^{-1}\| \, \|B^{-1}\| \\
&= \text{cond}(A)\text{cond}(B).
\end{aligned}$$

Supplementary Problems

Problem 1. Let A, B be $n \times n$ matrices over $\mathbb{C}$. Given the distance $\|A - B\|$. Find the distances $\|A \otimes I_n - B \otimes I_n\|$ and $\|A \otimes I_n - I_n \otimes B\|$.

Problem 2. Let A be an $n \times n$ positive semidefinite matrix. Show that

$$|\mathbf{x}^* A \mathbf{y}| \le \sqrt{\mathbf{x}^* A \mathbf{x}} \sqrt{\mathbf{y}^* A \mathbf{y}}$$

for all $\mathbf{x}, \mathbf{y} \in \mathbb{C}$.

Problem 3. (i) Let A and C be invertible $n \times n$ matrices over $\mathbb{R}$. Let B be an $n \times n$ matrix over $\mathbb{R}$. Assume that $\|A\| \le \|B\| \le \|C\|$. Is B invertible?
(ii) Let A, B, C be invertible $n \times n$ matrices over $\mathbb{R}$ with $\|A\| \le \|B\| \le \|C\|$. Is

$$\|A^{-1}\| \ge \|B^{-1}\| \ge \|C^{-1}\|?$$

Problem 4. Let A be an $n \times n$ matrix over $\mathbb{R}$. Assume that $\|A\| < 1$, where

$$\|A\| := \sup_{\|\mathbf{x}\|=1} \|A\mathbf{x}\|.$$

Show that the matrix $B = I_n + A$ is invertible, i.e. $B \in GL(n.\mathbb{R})$. To show that the expansion

$$I_n - A + A^2 - A^3 + \cdots$$

converges apply

$$\|A^m - A^{m+1} + A^{m+2} - \cdots \pm A^{m+k-1}\| \le \|A^m\| \cdot \|1 + \|A\| + \cdots + \|A\|^{k-1}\|$$

$$= \|A\|^m \frac{1 - \|A\|^k}{1 - \|A\|}.$$

Then calculate $(I_n + A)(I_n - A + A^2 - A^3 + \cdots)$.

Problem 5. Let A, B be 2×2 matrices over $\mathbb{R}$. Find A, B such that

$$\min \|[A, B] - I_2\|.$$

For the norm $\|\ \|$ consider the Frobenius norm and max-norm.

Problem 6. Let A be an $n \times n$ matrix over $\mathbb{C}$. Assume that $\|A\| < 1$. Then $I_n + A$ is invertible. Can we conclude that $I_n \otimes I_n + A \otimes A$ is invertible?

Problem 7. Let A be an $n \times n$ positive semidefinite (and thus hermitian) matrix. Is $\|A^{1/2}\| = \|A\|^{1/2}$?

Problem 8. Let A be a given symmetric 4×4 matrix over $\mathbb{R}$. Find symmetric 2×2 matrices B, C over $\mathbb{R}$ such that

$$f(B, C) = \|A - B \otimes C\|$$

is a minimum. The norm $\|\ \|$ is the Hilbert-Schmidt norm over $\mathbb{R}$ given by

$$\|X - Y\|^2 = \operatorname{tr}((X - Y)(X - Y)^T)$$

where X, Y are $n \times n$ matrices over $\mathbb{R}$. Such a problem is called the *nearest Kronecker product problem*. Note that since $A^T = A$, $B^T = B$, $C^T = C$, $\operatorname{tr}(B^2 \otimes C^2) = \operatorname{tr}(B^2)\operatorname{tr}(C^2)$ and $\operatorname{tr}(A(B \otimes C)) = \operatorname{tr}((B \otimes C)A)$ we have

$$
\begin{aligned}
\|A - B \otimes C\|^2 &= \operatorname{tr}((A - B \otimes C)(A - B \otimes C)) \\
&= \operatorname{tr}(A^2 - A(B \otimes C) - (B \otimes C)A + B^2 \otimes C^2) \\
&= \operatorname{tr}(A^2) - \operatorname{tr}(A(B \otimes C)) - \operatorname{tr}((B \otimes C)A) + \operatorname{tr}(B^2)\operatorname{tr}(C^2) \\
&= \operatorname{tr}(A^2) - 2\operatorname{tr}(A(B \otimes C)) + \operatorname{tr}(B^2)\operatorname{tr}(C^2).
\end{aligned}
$$

Since

$$
B \otimes C = \begin{pmatrix} b_{11} & b_{12} \\ b_{12} & b_{22} \end{pmatrix} \otimes \begin{pmatrix} c_{11} & c_{12} \\ c_{12} & c_{22} \end{pmatrix} = \begin{pmatrix} b_{11}c_{11} & b_{11}c_{12} & b_{12}c_{11} & b_{12}c_{12} \\ b_{11}c_{12} & b_{11}c_{22} & b_{12}c_{12} & b_{12}c_{22} \\ b_{12}c_{11} & b_{12}c_{12} & b_{22}c_{11} & b_{22}c_{12} \\ b_{12}c_{12} & b_{12}c_{22} & b_{22}c_{12} & b_{22}c_{22} \end{pmatrix}
$$

we have

$$\text{tr}(A(B \otimes C)) = a_{11}b_{11}c_{11} + a_{22}b_{11}c_{22} + a_{33}b_{22}c_{11} + a_{44}b_{22}c_{22}$$
$$+ 2(a_{12}b_{11}c_{12} + a_{13}b_{12}c_{11} + a_{14}b_{12}c_{12}$$
$$+ a_{23}b_{12}c_{12} + a_{24}b_{12}c_{22} + a_{34}b_{22}c_{12}).$$

We also have $\text{tr}(B^2) = b_{11}^2 + 2b_{12}^2 + b_{22}^2$, $\text{tr}(C^2) = c_{11}^2 + 2c_{12}^2 + c_{22}^2$.

Problem 9. Let A be an $n \times n$ matrix over $\mathbb{C}$. Is $\|I_n + A\| \leq 1 + \|A\|$?

Problem 10. (i) Let A, B be positive definite $n \times n$ matrices. Show that $\|A + B\|_F \geq \|A\|_F$, where $\|.\|_F$ denotes the Frobenius norm.
(ii) Let A be an $n \times n$ matrix over $\mathbb{C}$ and $\mathbf{v}, \mathbf{u} \in \mathbb{C}^n$. Is $\|A\mathbf{v} - A\mathbf{u}\| \leq \|A\| \cdot \|\mathbf{v} - \mathbf{u}\|$?

Problem 11. Let $\{\mathbf{v}_1, \ldots, \mathbf{v}_m\}$ be a linearly independent set of vectors in the Euclidean space $\mathbb{R}^n$ with $m \leq n$.
(i) Show that there is a number $c > 0$ such that for every choice of real numbers $c_1, \ldots, c_m$ we have

$$\|c_1\mathbf{v}_1 + \cdots + c_m\mathbf{v}_m\| \geq c(|c_1| + \cdots + |c_m|).$$

(ii) Consider $\mathbb{R}^2$ and the linearly independent vectors

$$\mathbf{v}_1 = \begin{pmatrix} 1 \\ 0 \end{pmatrix}, \qquad \mathbf{v}_2 = \begin{pmatrix} 1 \\ 1 \end{pmatrix}.$$

Show that
$$\|c_1\mathbf{v}_1 + c_2\mathbf{v}_2\| = \sqrt{(c_1 + c_2)^2 + c_2^2} \geq c(|c_1| + |c_2|)$$

and $c = 1/2$ will provide the answer.

Chapter 13

vec Operator

Let A be an $m \times n$ matrix over $\mathbb{C}$. One defines the *vec operator* (vectorization) of A as

$$\text{vec}(A) := (a_{11}, a_{21}, \ldots, a_{m1}, a_{12}, a_{22}, \ldots, a_{m2}, \ldots, a_{1n}, a_{2n}, \ldots, a_{mn})^T$$

where T denotes transpose. Thus $\text{vec}(A)$ is a column vector in $\mathbb{C}^{mn}$. Thus the vec operator stacks the columns of a matrix on top of each other to form a column vector, for example

$$\text{vec} \begin{pmatrix} 1 & 2 & 3 \\ 4 & 5 & 6 \end{pmatrix} = \begin{pmatrix} 1 \\ 4 \\ 2 \\ 5 \\ 3 \\ 6 \end{pmatrix}.$$

One defines the *reshaping operator* of A as

$$\text{res}(A) = (a_{11}, a_{12}, \ldots, a_{1n}, a_{21}, a_{22}, \ldots, a_{2n}, \ldots, a_{m1}, a_{m2}, \ldots, a_{mn})^T$$

where T denotes transpose. Thus $\text{res}(A)$ is a column vector in $\mathbb{C}^{mn}$. One has $\text{res}(A) = \text{vec}(A^T)$. For $n \times n$ matrices over $\mathbb{C}$ we have

$$\text{vec}(ABC) \equiv (C^T \otimes A)\text{vec}(B)$$
$$\text{res}(ABC) \equiv (A \otimes C^T)\text{res}(B)$$
$$\text{vec}(AB) \equiv (I_n \otimes A)\text{vec}(B) \equiv (B^T \otimes I_n)\text{vec}(A)$$
$$\text{vec}(A \bullet B) \equiv \text{vec}(A) \bullet \text{vec}(B)$$
$$\text{tr}(A^*B) \equiv (\text{vec}(A))^*\text{vec}(B) \equiv (\text{res}(A))^*\text{res}(B)$$

where $\bullet$ denotes the Hadamard product.

The vec operator can be considered as a bijective operator between the $m \times n$ matrices and $\mathbb{C}^{mn}$. However, we need to know m and n. For example the vec operator is a bijection between the 3×2 matrices and $\mathbb{C}^6$ and is also a bijection between the 2×3 matrices and $\mathbb{C}^6$.

Let I_n be the $n \times n$ identity matrix. Let e_j $(j = 1, \ldots, n)$ be the standard basis in $\mathbb{R}^n$. Then we have

$$\mathrm{vec}(I_n) = \sum_{j=1}^{n} e_j \otimes e_j, \qquad \mathrm{vec}(A) = (I_n \otimes A)\mathrm{vec}(I_n).$$

This is a straightforward consequence of the identity

$$\mathrm{vec}(ABC) = (C^T \otimes A)\mathrm{vec}(B).$$

We define the permutation matrix $P_{m,n}$ (*perfect shuffle*) by

$$\mathrm{vec}(A) = P_{m,n}\mathrm{vec}(A^T)$$

for all $m \times n$ matrices A. Thus

$$P_{m,n}^{-1} = P_{m,n}^T = P_{n,m}.$$

Let $\mathbf{x} \in \mathbb{C}^m$ and $\mathbf{y} \in \mathbb{C}^n$. It follows that

$$P_{m,n}(\mathbf{x} \otimes \mathbf{y}) = \mathbf{y} \otimes \mathbf{x}.$$

The permutation matrices $P_{n,t}$ and $P_{s,m}$ provide the relation for commuting the Kronecker product. Let A be and $m \times n$ matrix and B be an $s \times t$ matrix, Then

$$A \otimes B = P_{s,m}(B \otimes A)P_{n,t}.$$

Using the vec-operator we can cast matrix equations into vector equations.

1) *Sylvester equation*

$$FX + XG^T = C \iff (I_n \otimes F + G \otimes I_m)\mathrm{vec}(X) = \mathrm{vec}(C).$$

2) *Generalized Sylvester equation*

$$FXH^T + KXG^T = C \iff (H \otimes F + G \otimes K)\mathrm{vec}(X) = \mathrm{vec}(C).$$

3) *Lyapunov equation*

$$FX + XF^T = C \iff (I_n \otimes F + F \otimes I_n)\mathrm{vec}(X) = \mathrm{vec}(C).$$

Problem 1. Consider the 2×3 matrix

$$A = \begin{pmatrix} a_{11} & a_{12} & a_{13} \\ a_{21} & a_{22} & a_{23} \end{pmatrix}.$$

Let $B = A^T$. Thus B is a 3×2 matrix. Find the 6×6 permutation matrix P such that

$$\text{vec}(B) = P\text{vec}(A).$$

Solution 1. We obtain

$$\begin{pmatrix} a_{11} \\ a_{12} \\ a_{13} \\ a_{21} \\ a_{22} \\ a_{23} \end{pmatrix} = \begin{pmatrix} 1 & 0 & 0 & 0 & 0 & 0 \\ 0 & 0 & 1 & 0 & 0 & 0 \\ 0 & 0 & 0 & 0 & 1 & 0 \\ 0 & 1 & 0 & 0 & 0 & 0 \\ 0 & 0 & 0 & 1 & 0 & 0 \\ 0 & 0 & 0 & 0 & 0 & 1 \end{pmatrix} \begin{pmatrix} a_{11} \\ a_{21} \\ a_{12} \\ a_{22} \\ a_{13} \\ a_{23} \end{pmatrix}.$$

Problem 2. Let A be an $m \times n$ matrix over $\mathbb{C}$ and B be a $s \times t$ matrix over $\mathbb{C}$. Find the permutation matrix P such that

$$\text{vec}(A \otimes B) = P(\text{vec}(A) \otimes \text{vec}(B)).$$

Solution 2. We find

$$\text{vec}_{m \times n}(A) = (I_n \otimes A) \sum_{j=1}^{n} \mathbf{e}_{j,n} \otimes \mathbf{e}_{j,n} = (A^T \otimes I_m) \sum_{j=1}^{m} \mathbf{e}_{j,m} \otimes \mathbf{e}_{j,m}.$$

Since

$$\text{vec}_{ms \times nt}(A \otimes B) = \sum_{j=1}^{n} \sum_{k=1}^{t} \mathbf{e}_{j,n} \otimes \mathbf{e}_{k,t} \otimes (A\mathbf{e}_{j,n}) \otimes (B\mathbf{e}_{k,t})$$

$$= \sum_{j=1}^{m} \sum_{k=1}^{t} (A^T \mathbf{e}_{j,m}) \otimes \mathbf{e}_{k,t} \otimes \mathbf{e}_{j,m} \otimes (B\mathbf{e}_{k,t})$$

and

$$\text{vec}_{m \times n}(A) \otimes \text{vec}_{s \times t}(B) = \sum_{j=1}^{n} \sum_{k=1}^{t} \mathbf{e}_{j,n} \otimes (A\mathbf{e}_{j,n}) \otimes \mathbf{e}_{k,t} \otimes (B\mathbf{e}_{k,t})$$

$$= \sum_{j=1}^{m} \sum_{k=1}^{t} (A^T \mathbf{e}_{j,m}) \otimes \mathbf{e}_{j,m} \otimes \mathbf{e}_{k,t} \otimes (B\mathbf{e}_{k,t})$$

there is an $(mnst) \times (mnst)$ permutation matrix P such that

$$\text{vec}_{ms \times nt}(A \otimes B) = P(\text{vec}_{m \times n}(A) \otimes \text{vec}_{s \times t}(B)).$$

This permutation matrix P is given by

$$P = I_n \otimes \left(\sum_{j=1}^{m} \sum_{k=1}^{t} (\mathbf{e}_{k,t} \otimes \mathbf{e}_{j,m})(\mathbf{e}_{j,m}^* \otimes \mathbf{e}_{k,t}^*) \right) \otimes I_s.$$

Problem 3. (i) Let A, B be two $n \times n$ matrices. Is

$$\text{vec}(A \otimes B) = (\text{vec}(A)) \otimes (\text{vec}(B))?$$

(ii) Let A be an arbitrary 2×2 matrix. Find the condition on A such that

$$\text{vec}(A) \otimes \text{vec}(A) = \text{vec}(A \otimes A).$$

What can we conclude about the rank of A?

Solution 3. (i) This is not true in general. Consider, for example, with $n = 2$

$$A = \begin{pmatrix} 1 & 2 \\ 3 & 4 \end{pmatrix}, \qquad B = \begin{pmatrix} 5 & 6 \\ 7 & 8 \end{pmatrix}.$$

However there is an $(n^2 \times n^2) \times (n^2 \times n^2)$ permutation matrix P such that

$$\text{vec}(A \otimes B) = P(\text{vec}(A) \otimes \text{vec}(B)).$$

Find P for the example given above.
(ii) We obtain the four conditions

$$a_{11}(a_{12} - a_{21}) = 0, \ a_{22}(a_{12} - a_{21}) = 0, \ a_{11}a_{22} - a_{12}^2 = 0, \ a_{11}a_{22} - a_{21}^2 = 0.$$

If $a_{11} = 0$, then $a_{12} = a_{21} = 0$ and a_{22} is arbitrary. If $a_{22} = 0$, then $a_{12} = a_{21} = 0$ and a_{11} is arbitrary. If $a_{12} = a_{21}$, then $a_{11}a_{22} = a_{12}^2$. Thus we can conclude that the rank of A must be 1 or 0.

Problem 4. Let $\mathbf{v}$ be a column vector in $\mathbb{R}^n$. Show that $\text{vec}(\mathbf{v}\mathbf{v}^T) = \mathbf{v} \otimes \mathbf{v}$.

Solution 4. Since

$$\mathbf{v}\mathbf{v}^T = \begin{pmatrix} v_1 v_1 & v_1 v_2 & \cdots & v_1 v_n \\ v_2 v_1 & v_2 v_2 & \cdots & v_2 v_n \\ \vdots & \vdots & \ddots & \vdots \\ v_n v_1 & v_n v_2 & \cdots & v_n v_n \end{pmatrix}$$

we obtain the identity.

Problem 5. Let A be an $m \times n$ matrix over $\mathbb{C}$. Using

$$\text{vec}_{m \times n}(A) := \sum_{j=1}^{n} \mathbf{e}_{j,n} \otimes (A\mathbf{e}_{j,n}) = (I_n \otimes A) \sum_{j=1}^{n} \mathbf{e}_{j,n} \otimes \mathbf{e}_{j,n}$$

and

$$\text{vec}_{m \times n}^{-1}(\mathbf{x}) = \sum_{i=1}^{m} \sum_{j=1}^{n} \left((\mathbf{e}_{j,n} \otimes \mathbf{e}_{i,m})^* \mathbf{x} \right) \mathbf{e}_{i,m} \otimes \mathbf{e}_{j,n}^*.$$

Show that $\text{vec}_{m \times n}^{-1}(\text{vec}_{m \times n}(A)) = A$.

Solution 5.   We have

$$
\begin{aligned}
\text{vec}_{m \times n}^{-1}(\text{vec}_{m \times n}(A)) &= \sum_{i=1}^{m} \sum_{j=1}^{n} \left((\mathbf{e}_{j,n} \otimes \mathbf{e}_{i,m})^* \sum_{k=1}^{n} \mathbf{e}_{k,n} \otimes (A\mathbf{e}_{k,n}) \right) \mathbf{e}_{i,m} \otimes \mathbf{e}_{j,n}^* \\
&= \sum_{i=1}^{m} \sum_{j=1}^{n} \left(\sum_{k=1}^{n} (\mathbf{e}_{j,n}^* \mathbf{e}_{k,n}) \otimes (\mathbf{e}_{i,m}^* A\mathbf{e}_{k,n}) \right) \mathbf{e}_{i,m} \otimes \mathbf{e}_{j,n}^* \\
&= \sum_{i=1}^{m} \sum_{j=1}^{n} \left(\sum_{k=1}^{n} \delta_{j,k} (\mathbf{e}_{i,m}^* A\mathbf{e}_{k,n}) \right) \mathbf{e}_{i,m} \otimes \mathbf{e}_{j,n}^* \\
&= \sum_{i=1}^{m} \sum_{j=1}^{n} (\mathbf{e}_{i,m}^* A\mathbf{e}_{j,n}) \mathbf{e}_{i,m} \mathbf{e}_{j,n}^* \\
&= A.
\end{aligned}
$$

Problem 6.   (i) Let $AX + XB = C$, where C is an $m \times n$ matrix over $\mathbb{R}$. What are the dimensions of A, B, and X?
(ii) Solve the equation

$$
\begin{pmatrix} 0 & 1 \\ 1 & 0 \end{pmatrix} X + X \begin{pmatrix} 1 & 1 & 1 \\ 1 & 1 & 1 \\ 1 & 1 & 1 \end{pmatrix} = \begin{pmatrix} 1 & 0 & -1 \\ 0 & 1 & 0 \end{pmatrix}
$$

for the real-valued 2×3 matrix X.

Solution 6.   (i) From $AX + XB = C$ we find A has m rows, X has n columns, X has m rows and B has n columns. Since X has m rows, A must have m columns. Since X has n columns, B must have n rows. Thus A is $m \times m$, B is $n \times n$ and X is $m \times n$.
(ii) The equation

$$
\begin{pmatrix} 0 & 1 \\ 1 & 0 \end{pmatrix} X + X \begin{pmatrix} 1 & 1 & 1 \\ 1 & 1 & 1 \\ 1 & 1 & 1 \end{pmatrix} = \begin{pmatrix} 1 & 0 & -1 \\ 0 & 1 & 0 \end{pmatrix}
$$

can be written as

$$
\left(I_3 \otimes \begin{pmatrix} 0 & 1 \\ 1 & 0 \end{pmatrix} + \begin{pmatrix} 1 & 1 & 1 \\ 1 & 1 & 1 \\ 1 & 1 & 1 \end{pmatrix}^T \otimes I_2 \right) \text{vec}(X) = \begin{pmatrix} 1 \\ 0 \\ 0 \\ 1 \\ -1 \\ 0 \end{pmatrix}
$$

with the solution

$$X = \begin{pmatrix} -3/8 & 5/8 & -3/8 \\ 9/8 & 1/8 & -7/8 \end{pmatrix}.$$

Problem 7. Let A be an $m \times n$ matrix over $\mathbb{C}$ and B be an $s \times t$ matrix over $\mathbb{C}$. For the rearrangement operator R by

$$R(A \otimes B) := \text{vec}(A)(\text{vec}(B))^T$$

and linear extension. Find

$$R\left(\begin{pmatrix} 0 & 1 \\ 1 & 0 \end{pmatrix} \otimes \begin{pmatrix} 1 \\ 2 \\ 3 \end{pmatrix} \right).$$

Solution 7. We have

$$R\left(\begin{pmatrix} 0 & 1 \\ 1 & 0 \end{pmatrix} \otimes \begin{pmatrix} 1 \\ 2 \\ 3 \end{pmatrix} \right) = \begin{pmatrix} 0 \\ 1 \\ 1 \\ 0 \end{pmatrix} (1 \quad 2 \quad 3) = \begin{pmatrix} 0 & 0 & 0 \\ 1 & 2 & 3 \\ 1 & 2 & 3 \\ 0 & 0 & 0 \end{pmatrix}.$$

Problem 8. (i) Let K be a given $n \times n$ nonnormal matrix over $\mathbb{C}$. We want to find all unitary $n \times n$ matrices such that $UKU^* = K^*$. Now we can write $KU^* = U^*K^*$ and therefore

$$KU^* - U^*K^* = 0_n$$

where 0_n is the $n \times n$ zero matrix. This is a matrix equation (a special case of Sylvester) and using the vec-operator and Kronecker product we can cast the matrix equation into a vector equation. Write down this linear equation.
(ii) Apply it to the case $n = 2$ with

$$K = \begin{pmatrix} -i & i \\ 0 & i \end{pmatrix}.$$

Solution 8. (i) The vector equation is

$$(I_n \otimes K - (K^*)^T \otimes I_n)\text{vec}(U^*) = \mathbf{0}$$

where we still have to take into account that U is unitary.
(ii) For the given 2×2 matrix we have

$$K = \begin{pmatrix} -i & i \\ 0 & i \end{pmatrix} \Rightarrow K^* = \begin{pmatrix} i & 0 \\ -i & -i \end{pmatrix} \Rightarrow (K^*)^T = \begin{pmatrix} i & -i \\ 0 & -i \end{pmatrix}.$$

Thus we arrive at the system of linear equations

$$\begin{pmatrix} -2i & i & i & 0 \\ 0 & 0 & 0 & i \\ 0 & 0 & 0 & i \\ 0 & 0 & 0 & 2i \end{pmatrix} \begin{pmatrix} \bar{u}_{11} \\ \bar{u}_{12} \\ \bar{u}_{21} \\ \bar{u}_{22} \end{pmatrix} = \begin{pmatrix} 0 \\ 0 \\ 0 \\ 0 \end{pmatrix}.$$

Note that the determinant on the right-hand side is 0 so we obtain a non-trivial solution. We find $u_{22} = 0$ and the equation $-2u_{11} + u_{12} + u_{21} = 0$. Taking into account that U is unitary provides the two solutions $u_{11} = 0$, $u_{12} = 1$, $u_{21} = -1$ and $u_{11} = 0$, $u_{12} = -1$, $u_{21} = 1$.

Problem 9. Consider Lyapunov $FX + XF^T = C$ and

$$F = \begin{pmatrix} 0 & 1 \\ 1 & 0 \end{pmatrix}, \qquad C = \begin{pmatrix} 0 & 1 \\ 1 & 0 \end{pmatrix}.$$

Find X.

Solution 9. We have $F^T = F$ and

$$(I_2 \otimes F + F \otimes I_2)\mathrm{vec}(X) = \begin{pmatrix} 0 & 1 & 1 & 0 \\ 1 & 0 & 0 & 1 \\ 1 & 0 & 0 & 1 \\ 0 & 1 & 1 & 0 \end{pmatrix} \begin{pmatrix} x_{11} \\ x_{21} \\ x_{12} \\ x_{22} \end{pmatrix} = \begin{pmatrix} 0 \\ 1 \\ 1 \\ 0 \end{pmatrix}.$$

It follows that $x_{21} + x_{12} = 0$, $x_{11} + x_{22} = 1$. Thus

$$X = \begin{pmatrix} x_{11} & -x_{12} \\ x_{12} & 1 - x_{11} \end{pmatrix}$$

with x_{11}, x_{12} arbitrary.

Problem 10. Consider the matrices

$$A = \begin{pmatrix} 0 & 1 & 0 \\ 2 & 3 & 1 \\ -1 & -2 & 0 \end{pmatrix}, \qquad B = \begin{pmatrix} 1 & 0 \\ -2 & 1 \end{pmatrix}, \qquad C = \begin{pmatrix} 1 & 2 \\ 0 & 1 \\ 1 & -2 \end{pmatrix}.$$

Find the solution of the equation $AX + XB = C$, where X is the 3×2 matrix

$$X = \begin{pmatrix} x_{11} & x_{12} \\ x_{21} & x_{22} \\ x_{31} & x_{32} \end{pmatrix}.$$

Hint. Using the vec-operation we can write the equation as linear equation

$$(I_2 \otimes A + B^T \otimes I_3)\mathrm{vec}(X) = \mathrm{vec}(C).$$

Solution 10. We obtain

$$I_2 \otimes A + B^T \otimes I_3 = \begin{pmatrix} 1 & 1 & 0 & -2 & 0 & 0 \\ 2 & 4 & 1 & 0 & -2 & 0 \\ -1 & -2 & 1 & 0 & 0 & -2 \\ 0 & 0 & 0 & 1 & 1 & 0 \\ 0 & 0 & 0 & 2 & 4 & 1 \\ 0 & 0 & 0 & -1 & -2 & 1 \end{pmatrix}.$$

The determinant of $I_2 \otimes A + B^T \otimes I_3$ is nonzero and thus the inverse matrix exists. Note that there is a unique solution for X iff no eigenvalue of A is the negative of the eigenvalue of B. This is the case here and therefore the determinant of $I_2 \otimes A + B^T \otimes I_3$ is nonzero. The solution is

$$X = \begin{pmatrix} 43/3 & 3 \\ -22/3 & -1 \\ -4/3 & -1 \end{pmatrix}.$$

Problem 11. Let A, B, X be $n \times n$ matrices, where the matrices A, B are given. Assume that $AX = B$. Find X using the vec-operator and the Kronecker product. Then consider the case that A is invertible.

Solution 11. Applying the vec operator to the left- and right-hand side of $AX = B$ and using the Kronecker product we have

$$\text{vec}(AX) \equiv (I_n \otimes A)\text{vec}(X) = \text{vec}(B).$$

If A is invertible we have $\text{vec}(X) = (I_n \otimes A^{-1})\text{vec}(B)$.

Problem 12. Let A, B, C, X be $n \times n$ matrices, where the matrices A, B, C are given. Assume that $AXB = C$. Find X using the vec operator and the Kronecker product. Then consider the case that A and B are invertible.

Solution 12. Applying the vec operator to the left- and right-hand side of $AXB = C$ and using the Kronecker product we have

$$(B^T \otimes A)\text{vec}(X) = \text{vec}(C).$$

If A and B are invertible we have

$$\text{vec}(X) = ((B^T)^{-1} \otimes A^{-1})\text{vec}(C).$$

Problem 13. Let A, B, C, X Y be $n \times n$ matrices, where the matrices A, B, C are given. Assume that $AX + YB = C$. Apply the vec operator to this equation and then express the left-hand side using the Kronecker product.

Solution 13. Applying the vec operator to the left- and right-hand side of $AX + YB = C$ and using the Kronecker product we have

$$(I_n \otimes A)\text{vec}(X) + (B^T \otimes I_n)\text{vec}(Y) = \text{vec}(C).$$

Problem 14. Let U be an $n \times n$ unitary matrix. Show that the column vector $\text{vec}(U)$ cannot be written as a Kronecker product of two vectors in $\mathbb{C}^n$.

Solution 14. We have

$$\text{vec}(U) = \sum_{j=1}^{n} \mathbf{e}_j \otimes (U\mathbf{e}_j)$$

where $\{\, e_1, \ldots, e_n \,\}$ is the standard basis in $\mathbb{C}^n$. Let $\mathbf{a}, \mathbf{b} \in \mathbb{C}^n$. Suppose

$$\text{vec}(U) = \mathbf{a} \otimes \mathbf{b}.$$

Since

$$\mathbf{a} = \sum_{j=1}^{n} (e_j^T \mathbf{a}) e_j$$

we find

$$\sum_{j=1}^{n} e_j \otimes (U e_j) = \sum_{j=1}^{n} (e_j^T \mathbf{a}) e_j \otimes \mathbf{b}.$$

Equating yields $U e_j = (e_j^T \mathbf{a}) \mathbf{b}$ i.e. the columns of U are linearly dependent. However, this contradicts the fact that U is unitary. Consequently the vector $\text{vec}(U)$ is entangled in $\mathbb{C}^n \otimes \mathbb{C}^n$.

A generalization is: Let A be an $n \times n$ matrix with rank greater than 1. Then $\text{vec}(A)$ is entangled in $\mathbb{C}^n \otimes \mathbb{C}^n$.

Problem 15.   Let A be an arbitrary $n \times n$ matrix. Let U be an $n \times n$ unitary matrix $U^{-1} = U^*$. Can we conclude that

$$\text{vec}(U^* A U) = \text{vec}(A)\,? \tag{1}$$

Solution 15.   If (1) would be true we would have

$$(U^T \otimes U^*)\text{vec}(A) = \text{vec}(A). \tag{2}$$

This is not true in general. A solution of (2) would be $U = e^{i\phi} I_n$, $\phi \in \mathbb{R}$.

Problem 16.   Consider the *Frobenius norm*. Let M be an $ms \times nt$ matrix over $\mathbb{C}$. Find the $m \times n$ matrices A over $\mathbb{C}$ and the $s \times t$ matrices B over $\mathbb{C}$ which minimize

$$\|M - A \otimes B\|_F = \|R(M) - \text{vec}(A)(\text{vec}(B))^T\|_F.$$

Apply the method for $m = 2, n = 1, s = 2, t = 2$ to the 4×2 matrix

$$M = \begin{pmatrix} 0 & 1 \\ 1 & 0 \\ 0 & 1 \\ 1 & 0 \end{pmatrix}.$$

Solution 16.   We have

$$\|M - A \otimes B\|_F = \|R(M - A \otimes B)\|_F = \|R(M) - R(A \otimes B)\|_F$$
$$= \|R(M) - \text{vec}(A)(\text{vec}(B))^T\|_F.$$

Notice that $R(M)$ is $mn \times st$ while M is $ms \times nt$. Clearly $\text{vec}(A)(\text{vec}(B))^T$ has rank 1. Let $R(M) = U\Sigma V^*$ be a *singular value decomposition* of $R(M)$, then the minimum is found when

$$\text{vec}(A)(\text{vec}(B))^T = U(\sigma_1 \mathbf{e}_{1,mn}\mathbf{e}_{1,st}^T)V^* = (\sigma_1 U\mathbf{e}_{1,mn})(V\mathbf{e}_{1,st})^*.$$

Thus we may set (amongst other choices)

$$\text{vec}(A) = \sigma_1 U\mathbf{e}_{1,mn} \quad \text{and} \quad \text{vec}(B) = V\mathbf{e}_{1,st}.$$

Since

$$M = \begin{pmatrix} 1 \\ 0 \end{pmatrix} \otimes \begin{pmatrix} 0 & 1 \\ 1 & 0 \end{pmatrix} + \begin{pmatrix} 0 \\ 1 \end{pmatrix} \otimes \begin{pmatrix} 0 & 1 \\ 1 & 0 \end{pmatrix}$$

and applying the rearrangement operator to M yields

$$R(M) = \begin{pmatrix} 0 & 1 & 1 & 0 \\ 0 & 1 & 1 & 0 \end{pmatrix}.$$

The singular value decomposition of $R(M)$ is

$$\begin{pmatrix} 0 & 1 & 1 & 0 \\ 0 & 1 & 1 & 0 \end{pmatrix} = \frac{1}{\sqrt{2}}\begin{pmatrix} 1 & 1 \\ 1 & -1 \end{pmatrix}\begin{pmatrix} 2 & 0 & 0 & 0 \\ 0 & 0 & 0 & 0 \end{pmatrix}\begin{pmatrix} 0 & 1 & 0 & 0 \\ \frac{1}{\sqrt{2}} & 0 & 0 & \frac{1}{\sqrt{2}} \\ \frac{1}{\sqrt{2}} & 0 & 0 & -\frac{1}{\sqrt{2}} \\ 0 & 0 & 1 & 0 \end{pmatrix}^*.$$

Thus we set

$$\text{vec}(A) = \frac{2}{\sqrt{2}}\begin{pmatrix} 1 & 1 \\ 1 & -1 \end{pmatrix}\begin{pmatrix} 1 \\ 0 \end{pmatrix} = \sqrt{2}\begin{pmatrix} 1 \\ 1 \end{pmatrix}, \quad A = \sqrt{2}\begin{pmatrix} 1 \\ 1 \end{pmatrix}$$

$$\text{vec}(B) = \begin{pmatrix} 0 & 1 & 0 & 0 \\ \frac{1}{\sqrt{2}} & 0 & 0 & \frac{1}{\sqrt{2}} \\ \frac{1}{\sqrt{2}} & 0 & 0 & -\frac{1}{\sqrt{2}} \\ 0 & 0 & 1 & 0 \end{pmatrix}\begin{pmatrix} 1 \\ 0 \\ 0 \\ 0 \end{pmatrix} = \frac{1}{\sqrt{2}}\begin{pmatrix} 0 \\ 1 \\ 1 \\ 0 \end{pmatrix}, \quad B = \frac{1}{\sqrt{2}}\begin{pmatrix} 0 & 1 \\ 1 & 0 \end{pmatrix}.$$

In this case M is exactly the Kronecker product $A \otimes B$.

Problem 17.   Let P be an $n \times n$ permutation matrix and H be an $n \times n$ hermitian matrix.
(i) Given the hermitian matrix H we want to find the permutation matrices P such that $HP = PH$. The equation can be written as $HP - PH = 0_n$. This is a special case of the Sylvester equation. Applying the vec-operation and the Kronecker product we can write the equation as

$$(I_n \otimes H - H^T \otimes I_n)\text{vec}(P) = \mathbf{0}.$$

Consider $H = \sigma_1$. Find all P.
(ii) Given the permutation matrix P we want to find the hermitian matrices H such that $PH - HP = 0_n$. This is a special case of the Sylvester equation. Applying the vec-operation and the Kronecker product we can write the equation as

$$(I_n \otimes P - P^T \otimes I_n)\text{vec}(H) = \mathbf{0}.$$

Consider $P = \sigma_1$. Find all H.

Solution 17.   (i) Since

$$
I_n \otimes H - H^T \otimes I_n = \begin{pmatrix} 0 & 1 & -1 & 0 \\ 1 & 0 & 0 & -1 \\ -1 & 0 & 0 & 1 \\ 0 & -1 & 1 & 0 \end{pmatrix}
$$

we find the two equations $p_{11} - p_{22} = 0$, $p_{12} - p_{21} = 0$ with the condition that P is a permutation matrix. Thus we find two solutions, namely $p_{11} = 1$, $p_{12} = 0$, $p_{21} = 0$, $p_{22} = 1$ (identity matrix) and $p_{11} = 0$, $p_{12} = 1$, $p_{21} = 1$, $p_{22} = 0$.
(ii) Since

$$
I_n \otimes P - P^T \otimes I_n = \begin{pmatrix} 0 & 1 & -1 & 0 \\ 1 & 0 & 0 & -1 \\ -1 & 0 & 0 & 1 \\ 0 & -1 & 1 & 0 \end{pmatrix}
$$

we find the two equations $h_{11} - h_{22} = 0$, $h_{12} - h_{21} = 0$ with the condition that H is a hermitian matrix. Thus

$$
H = \begin{pmatrix} h_{11} & h_{12} \\ h_{12} & h_{11} \end{pmatrix}
$$

with $H = H^*$. This means that H must be real.

Problem 18.   Consider the Pauli spin matrices σ_1, σ_3. The unitary and hermitian normal matrices σ_1 and σ_3 have the same eigenvalues, namely $+1$ and -1. Find all invertible 2×2 matrices X such that

$$
X^{-1}\sigma_1 X = \sigma_3.
$$

Proceed as follows: Multiply the equation with X and after rearrangements we have

$$
\sigma_1 X - X\sigma_3 = 0_2.
$$

This is a special case of the Sylvester equation. It can be written as

$$
(I_2 \otimes \sigma_1 - \sigma_3^T \otimes I_2)\mathrm{vec}(X) = \begin{pmatrix} 0 \\ 0 \\ 0 \\ 0 \end{pmatrix}, \quad \mathrm{vec}(X) := \begin{pmatrix} x_{11} \\ x_{21} \\ x_{12} \\ x_{22} \end{pmatrix}.
$$

Solve this linear equation to find the matrix X, but one still has to take into account that the matrix X is invertible.

Solution 18.   We have

$$
I_2 \otimes \begin{pmatrix} 0 & 1 \\ 1 & 0 \end{pmatrix} + \begin{pmatrix} -1 & 0 \\ 0 & 1 \end{pmatrix} \otimes I_2 = \begin{pmatrix} -1 & 1 & 0 & 0 \\ 1 & -1 & 0 & 0 \\ 0 & 0 & 1 & 1 \\ 0 & 0 & 1 & 1 \end{pmatrix}.
$$

Thus we have the system of linear equations

$$
\begin{pmatrix}
-1 & 1 & 0 & 0 \\
1 & -1 & 0 & 0 \\
0 & 0 & 1 & 1 \\
0 & 0 & 1 & 1
\end{pmatrix}
\begin{pmatrix}
x_{11} \\
x_{21} \\
x_{12} \\
x_{22}
\end{pmatrix}
=
\begin{pmatrix}
0 \\
0 \\
0 \\
0
\end{pmatrix}
$$

with the solution $x_{21} = x_{11}$, $x_{12} = -x_{22}$. Therefore the matrix X is

$$
X = \begin{pmatrix} x_{11} & -x_{22} \\ x_{11} & x_{22} \end{pmatrix}
$$

with determinant $\det(X) = 2x_{11}x_{22}$. This implies that $x_{11} \neq 0$ and $x_{22} \neq 0$. Then the inverse of X is

$$
X^{-1} = \frac{1}{2x_{11}x_{22}} \begin{pmatrix} x_{22} & x_{22} \\ -x_{11} & x_{11} \end{pmatrix}.
$$

Problem 19. Let $\mathbf{A} \in \mathbb{C}^{n^m} \equiv \bigotimes^m \mathbb{C}^n$ and $\mathbf{x} \in \mathbb{C}^n$. Define the product

$$
\mathbf{A}\mathbf{x}^{m-1} := [I_n \otimes \underbrace{\mathbf{x} \otimes \cdots \otimes \mathbf{x}}_{(m-1)\,\text{times}}]^T \mathbf{A}.
$$

Let A be an $n \times n$ matrix over $\mathbb{C}$. Find $\mathrm{vec}\,(A^T)\,\mathbf{x}^1$. What is the relationship between this product and the matrix product $A\mathbf{x}$?

Solution 19. We have

$$
\mathrm{vec}(A^T)\mathbf{x}^1 = (I_n \otimes \mathbf{x}^T)\mathrm{vec}(A^T) = \sum_{j=1}^{n} \mathbf{e}_j \otimes (\mathbf{x}^T A^T \mathbf{e}_j) = \sum_{j=1}^{n} (\mathbf{e}_j^T A\mathbf{x})\mathbf{e}_j = A\mathbf{x}
$$

where $\{\mathbf{e}_1, \ldots, \mathbf{e}_n\}$ is the standard basis in $\mathbb{C}^n$.

Problem 20. Let $\mathbf{A} \in \mathbb{C}^{n^m} \equiv \bigotimes^m \mathbb{C}^n$. If $\lambda \in \mathbb{C}$ and $\mathbf{x} \in \mathbb{C}^n$, $\mathbf{x} \neq \mathbf{0}$, satisfies $\mathbf{A}\mathbf{x}^{m-1} = \lambda\mathbf{x}$ then $(\lambda, \mathbf{x})$ is said to be an *eigenpair* of $\mathbf{A}$, where λ is the *eigenvalue* of $\mathbf{A}$ and $\mathbf{x}$ is a corresponding *eigenvector*. Two eigenpairs $(\lambda, \mathbf{x})$ and $(\lambda', \mathbf{x}')$ are equivalent when $t^{m-2}\lambda = \lambda'$ and $t\mathbf{x} = \mathbf{x}'$ for some non-zero complex number t.
(i) Find the eigenpairs for the tensor $\mathbf{A} = (0\ \ 1\ \ 1\ \ 0)^T$ and $n = 2$.
(ii) Find the eigenpairs for the tensor $\mathbf{A} = (1\ \ 1\ \ 1\ \ 1\ \ 1\ \ 1\ \ 1\ \ 1)^T$ and $n = 2$.

Solution 20. (i) Here we have $m = 2$. Since

$$
\mathrm{vec}\left(\begin{pmatrix} 0 & 1 \\ 1 & 0 \end{pmatrix}^T\right) = \mathbf{A}
$$

the eigenvalue equation becomes

$$
\mathbf{A}\mathbf{x}^{m-1} = \lambda\mathbf{x} \quad \Leftrightarrow \quad \begin{pmatrix} 0 & 1 \\ 1 & 0 \end{pmatrix}\mathbf{x} = \lambda\mathbf{x}
$$

i.e. we solve the matrix eigenvalue problem for $\begin{pmatrix} 0 & 1 \\ 1 & 0 \end{pmatrix}$ to find the eigenpairs

$$\left(1, \begin{pmatrix} t \\ t \end{pmatrix}\right), \ t \in \mathbb{C} \setminus \{0\} \qquad \left(-1, \begin{pmatrix} t \\ -t \end{pmatrix}\right), \ t \in \mathbb{C} \setminus \{0\}.$$

(ii) Here we have $m = 3$. Note that

$$\mathbf{A} = \begin{pmatrix} 1 \\ 1 \end{pmatrix} \otimes \begin{pmatrix} 1 \\ 1 \end{pmatrix} \otimes \begin{pmatrix} 1 \\ 1 \end{pmatrix}.$$

Let $\mathbf{x} = \begin{pmatrix} a & b \end{pmatrix}^T$. The eigenvalue problem is

$$\left[\begin{pmatrix} 1 & 0 \\ 0 & 1 \end{pmatrix} \otimes \begin{pmatrix} a & b \end{pmatrix} \otimes \begin{pmatrix} a & b \end{pmatrix} \right] \begin{pmatrix} 1 \\ 1 \end{pmatrix} \otimes \begin{pmatrix} 1 \\ 1 \end{pmatrix} \otimes \begin{pmatrix} 1 \\ 1 \end{pmatrix} = \lambda \begin{pmatrix} a \\ b \end{pmatrix}.$$

Thus

$$(a + b)^2 \begin{pmatrix} 1 \\ 1 \end{pmatrix} = \lambda \begin{pmatrix} a \\ b \end{pmatrix}.$$

This equation has the solution $\lambda = 0$ and $a = -b$. If $\lambda \neq 0$, then $a = b$ so that $\lambda x = 4a^2$. Since $\mathbf{x} \neq \mathbf{0}$, we find $\lambda = 4a$. Thus we have the eigenpairs

$$\left(0, \begin{pmatrix} t \\ -t \end{pmatrix}\right), \ t \in \mathbb{C} \setminus \{0\}, \qquad \left(4t, \begin{pmatrix} t \\ t \end{pmatrix}\right), \ t \in \mathbb{C} \setminus \{0\}.$$

Since the sets of eigenpairs are equivalent, we need only consider the representative pairs

$$\left(0, \begin{pmatrix} 1 \\ -1 \end{pmatrix}\right), \qquad \left(4, \begin{pmatrix} 1 \\ 1 \end{pmatrix}\right).$$

Supplementary Problems

Problem 1.   Let A, B be $n \times n$ matrices over $\mathbb{C}$.
(i) Consider the maps

$$f_1(A, B) = A \otimes B$$
$$f_2(A, B) = \text{vec}(A)(\text{vec}(B))^T$$
$$f_3(A, B) = \text{vec}(A)(\text{vec}(B^T))^T$$
$$f_4(A, B) = \text{vec}(B)(\text{vec}(A^T))^T.$$

Is

$$\text{tr}(f_1(A, B)) = \text{tr}(f_2(A, B)),$$
$$\text{tr}(f_2(A, B)) = \text{tr}(f_3(A, B)),$$
$$\text{tr}(f_3(A, B)) = \text{tr}(f_4(A, B))?$$

(ii) Show that

$$\text{tr}(\text{vec}(B)(\text{vec}(A))^T) \equiv \text{tr}(A^T B).$$

Problem 2. Let A be an 2×2 matrix.
(i) Find $\mathrm{vec}(A \otimes I_2 + I_2 \otimes A)$ and $\mathrm{vec}(A \otimes A)$.
(ii) Find the condition on A such that $\mathrm{vec}(A \otimes A) = \mathrm{vec}(A \otimes I_2 + I_2 \otimes A)$.

Problem 3. Let A, B be $n \times n$ matrices. Show that

$$\mathrm{vec}(A \bullet B) = \mathrm{vec}(A) \bullet \mathrm{vec}(B)$$

where $\bullet$ denotes the Hadamard product.

Problem 4. Find an invertible 3×3 matrix X which satisfies the following three matrix equations simultaneously

$$X^{-1} \begin{pmatrix} 0 & -i & 0 \\ i & 0 & 0 \\ 0 & 0 & 0 \end{pmatrix} X = \begin{pmatrix} 1 & 0 & 0 \\ 0 & 0 & 0 \\ 0 & 0 & -1 \end{pmatrix}$$

$$X^{-1} \begin{pmatrix} 0 & 0 & 0 \\ 0 & 0 & -i \\ 0 & i & 0 \end{pmatrix} X = \frac{1}{\sqrt{2}} \begin{pmatrix} 0 & 1 & 0 \\ 1 & 0 & 1 \\ 0 & 1 & 0 \end{pmatrix}$$

$$X^{-1} \begin{pmatrix} 0 & 0 & i \\ 0 & 0 & 0 \\ -i & 0 & 0 \end{pmatrix} X = \frac{i}{\sqrt{2}} \begin{pmatrix} 0 & -1 & 0 \\ 1 & 0 & -1 \\ 0 & 1 & 0 \end{pmatrix}.$$

Cast the equations into a *Sylvester equation* and then solve the equations. The first equation can be written as

$$(I_3 \otimes \begin{pmatrix} 0 & -i & 0 \\ i & 0 & 0 \\ 0 & 0 & 0 \end{pmatrix} - \begin{pmatrix} 1 & 0 & 0 \\ 0 & 0 & 0 \\ 0 & 0 & -1 \end{pmatrix}^T \otimes I_3) \mathrm{vec}(X) = \mathbf{0}$$

where $\mathbf{0}$ is the zero column vector.

Problem 5. Consider the Pauli spin matrices σ_1, σ_2 and

$$R := \sigma_1 \otimes \sigma_2 \equiv \begin{pmatrix} 0 & 1 \\ 1 & 0 \end{pmatrix} \otimes \begin{pmatrix} 0 & -i \\ i & 0 \end{pmatrix} \equiv \begin{pmatrix} 0 & 0 & 0 & -i \\ 0 & 0 & i & 0 \\ 0 & -i & 0 & 0 \\ i & 0 & 0 & 0 \end{pmatrix}.$$

Apply the vec-operator on R, i.e. $\mathrm{vec}(R)$. Calculate the discrete Fourier transform of the vector in $\mathbb{C}^4$. Then apply vec^{-1} to this vector. Compare this 4×4 matrix with the 4×4 matrix R. Discuss.

Problem 6. Let A be an $n \times n$ matrix over $\mathbb{C}$. Find a matrix S_n such that $S_n \mathrm{vec}(A) = \mathrm{tr}(A)$. Prove the equivalence. Illustrate the answer with

$$A = \begin{pmatrix} 1 & 2 & 3 \\ 3 & 1 & 2 \\ 2 & 3 & 1 \end{pmatrix}$$

where

$$S_n = \text{vec}(I_3) = (1 \quad 0 \quad 0 \quad 0 \quad 1 \quad 0 \quad 0 \quad 0 \quad 1)^T$$

and

$$S_n \text{vec}(A) = (1 \quad 0 \quad 0 \quad 0 \quad 1 \quad 0 \quad 0 \quad 0 \quad 1) \, \text{vec}(A) = 3.$$

Problem 7.   (i) Let A, B be $n \times n$ matrices over $\mathbb{C}$. Show that

$$\text{tr}(\text{vec}(B)(\text{vec}(A))^T) = \text{tr}(A^T B)$$

applying $\text{tr}(\text{vec}(B)(\text{vec}(A))^T) = \text{tr}((\text{vec}(I_n))^T (I \otimes A^T B)\text{vec}(I))$.
(ii) Let A be an $m \times n$ matrix, B be an $p \times p$ matrix, X be a $p \times n$ matrix and C be an $n \times n$ matrix. The underlying field is $\mathbb{R}$. Show that

$$\text{tr}(AX^T BXC) \equiv (\text{vec}(X))^T ((CA) \otimes B^T)\text{vec}(X).$$

Chapter 14

Nonnormal Matrices

A square matrix M over $\mathbb{C}$ is called normal if

$$MM^* = M^*M.$$

A square matrix M over $\mathbb{C}$ is called nonnormal if

$$MM^* \neq M^*M.$$

Examples of normal matrices are: hermitian, skew-hermitian, unitary, projection matrices. Examples for nonnormal matrices are

$$\begin{pmatrix} 0 & 1 \\ 0 & 0 \end{pmatrix}, \quad \begin{pmatrix} 0 & 1 & 1 \\ 0 & 0 & 1 \\ 0 & 0 & 0 \end{pmatrix}.$$

If M is a nonnormal matrix, then M^* and M^T are nonnormal matrices. If M is nonnormal and invertible, then M^{-1} is nonnormal. If A and B are nonnormal, then $A \oplus B$ is nonnormal.

If M is nonnormal, then $M + M^*$, MM^*, the commutator $[M, M^*]$ and the anti-commutator $[M, M^*]_+$ are normal matrices.

The Kronecker product of two normal matrices is normal again. The direct sum of two normal matrices is normal again. The star product of two normal matrices is normal again.

Any non-diagonalizeable matrix M is nonnormal. Every normal matrix is diagonalizable by the spectral decomposition. There are nonnormal matrices which are diagonalizable.

Problem 1.   Let A be a nonnormal $n \times n$ matrix over $\mathbb{C}$.
(i) Show that $A + A^*$ is a normal matrix.
(ii) Show that AA^* is a normal matrix.

Solution 1.   (i) Let $M = A + A^*$. Then $M^* = A^* + A = M$. Hence $MM^* = M^*M$.
(ii) Let $M = AA^*$. Then $M^* = AA^*$ and $M^*M = MM^*$.

Problem 2.   Let A be a nonnormal matrix. Show that $[A, A^*]$ is a normal matrix. Show that $[A, A^*]_+$ is a normal matrix.

Solution 2.   Let $M = [A, A^*] = AA^* - A^*A$. Then $M^* = AA^* - A^*A = M$ and $[A, A^*]$ is normal. Let $N = [A, A^*]_+ = AA^* + A^*A$. Then $N^* = AA^* + A^*A = N$ and $[A, A^*]_+$ is normal.

Problem 3.   Let $\alpha, \beta \in \mathbb{C}$. What are the conditions on α, β such that the matrix

$$M(\alpha, \beta) = \begin{pmatrix} \alpha & \beta \\ 0 & 1 \end{pmatrix} \Rightarrow M^*(\alpha, \beta) = \begin{pmatrix} \bar{\alpha} & 0 \\ \bar{\beta} & 1 \end{pmatrix}$$

is normal?

Solution 3.   We have

$$M^*(\alpha, \beta)M(\alpha, \beta) = \begin{pmatrix} \alpha\bar{\alpha} & \bar{\alpha}\beta \\ \alpha\bar{\beta} & \beta\bar{\beta} + 1 \end{pmatrix}, \quad M(\alpha, \beta)M^*(\alpha, \beta) = \begin{pmatrix} \alpha\bar{\alpha} + \beta\bar{\beta} & \beta \\ \bar{\beta} & 1 \end{pmatrix}.$$

Thus the condition is $\beta\bar{\beta} = 0$ which implies that $\beta = 0$.

Problem 4.   Let $\theta \in [0, 2\pi)$. What are the conditions on θ such that the matrix $A(\theta)$

$$A(\theta) = \begin{pmatrix} 0 & \sin(\theta) \\ \cos(\theta) & 0 \end{pmatrix} \Rightarrow A^*(\theta) = \begin{pmatrix} 0 & \cos(\theta) \\ \sin(\theta) & 0 \end{pmatrix}$$

is normal?

Solution 4.   We have

$$A(\theta)A^*(\theta) = \begin{pmatrix} \sin^2(\theta) & 0 \\ 0 & \cos^2(\theta) \end{pmatrix}, \quad A^*(\theta)A(\theta) = \begin{pmatrix} \cos^2(\theta) & 0 \\ 0 & \sin^2(\theta) \end{pmatrix}.$$

Thus the condition for the matrix to be normal is $\cos^2(\theta) = \sin^2(\theta)$ with solutions $\theta = \pi/4$, $\theta = 3\pi/4$, $\theta = 5\pi/4$, $\theta = 7\pi/4$.

Problem 5.   Let $\theta \in \mathbb{R}$. What are the conditions on θ such that the matrix $A(\theta)$

$$A(\theta) = \begin{pmatrix} 0 & \sinh(\theta) \\ \cosh(\theta) & 0 \end{pmatrix} \Rightarrow A^*(\theta) = \begin{pmatrix} 0 & \cosh(\theta) \\ \sinh(\theta) & 0 \end{pmatrix}$$

is normal?

Solution 5. We have

$$A(\theta)A^*(\theta) = \begin{pmatrix} \sinh^2(\theta) & 0 \\ 0 & \cosh^2(\theta) \end{pmatrix}, \quad A^*(\theta)A(\theta) = \begin{pmatrix} \cosh^2(\theta) & 0 \\ 0 & \sinh^2(\theta) \end{pmatrix}.$$

Thus the condition for the matrix to be normal is $\cosh^2(\theta) = \sinh^2(\theta)$. There are no solutions to this equation. Thus $A(\theta)$ is nonnormal for all θ.

Problem 6. Let $z \in \mathbb{C}$ and $z \neq 0$. Show that the 2×2 matrix

$$A = \begin{pmatrix} 1 & z \\ 0 & -1 \end{pmatrix}$$

is nonnormal. Give the eigenvalues.

Solution 6. We have

$$AA^* = \begin{pmatrix} 1 + z\bar{z} & * \\ * & * \end{pmatrix}, \quad A^*A = \begin{pmatrix} 1 & * \\ * & * \end{pmatrix}.$$

Thus the matrix is nonnormal if $z \neq 0$. The eigenvalues are $+1$ and -1.

Problem 7. (i) Can we conclude that an invertible matrix A is normal? (ii) Let B be an $n \times n$ matrix over $\mathbb{C}$ with $B^2 = I_n$. Can we conclude that B is normal?

Solution 7. (i) No. Consider the example

$$A = \begin{pmatrix} 1 & 1 \\ 0 & 1 \end{pmatrix} \Rightarrow A^{-1} = \begin{pmatrix} 1 & -1 \\ 0 & 1 \end{pmatrix}.$$

Both are nonnormal. Is the matrix $A \otimes A^{-1}$ normal?
(ii) This is not true in general, for example

$$B = \begin{pmatrix} 1 & 1 & 1 \\ 0 & 0 & -1 \\ 0 & -1 & 0 \end{pmatrix}$$

with $B^2 = I_3$ and $B^*B \neq BB^*$.

Problem 8. Show that not all nonnormal matrices are non-diagonalizable.

Solution 8. Consider the 2×2 matrix

$$A = \begin{pmatrix} 1 & 2 \\ 1 & 1 \end{pmatrix}$$

which is nonnormal and the invertible matrix

$$B = \begin{pmatrix} 1 & 1 \\ 1/\sqrt{2} & -1/\sqrt{2} \end{pmatrix} \Rightarrow B^{-1} = \begin{pmatrix} 1/2 & 1/\sqrt{2} \\ 1/2 & -1/\sqrt{2} \end{pmatrix}.$$

Then the matrix $B^{-1}AB$ is diagonal. Is B nonnormal?

Problem 9. Find all 2×2 matrices over $\mathbb{C}$ which are nonnormal but diagonalizable.

Solution 9. The matrix must have distinct eigenvalues (otherwise it is a multiple of the identity matrix). Thus

$$(\operatorname{tr}(A))^2 \neq 4 \det(A).$$

Let

$$A = \begin{pmatrix} a & b \\ c & d \end{pmatrix}.$$

Then the condition $(\operatorname{tr}(A))^2 \neq 4 \det(A)$ becomes $(a - d)^2 \neq -4bc$. If A is nonnormal, then the two equations

$$|b| = |c|, \qquad a\bar{c} + b\bar{d} = \bar{a}b + \bar{c}d$$

have no solution. So we have three classes of solutions

$$(a - d)^2 \neq -4bc, \quad |b| \neq |c| \quad \text{and} \quad a\bar{c} + b\bar{d} = \bar{a}b + \bar{c}d$$
$$(a - d)^2 \neq -4bc, \quad a\bar{c} + b\bar{d} \neq \bar{a}b + \bar{c}d \quad \text{and} \quad |b| = |c|$$
$$(a - d)^2 \neq -4bc, \quad |b| \neq |c| \quad \text{and} \quad a\bar{c} + b\bar{d} \neq \bar{a}b + \bar{c}d.$$

Problem 10. Let A be a normal $n \times n$ matrix and let S be an invertible nonnormal matrix. Is SAS^{-1} a normal matrix? Study the case

$$A = \frac{1}{\sqrt{2}} \begin{pmatrix} 1 & 1 \\ 1 & -1 \end{pmatrix}, \quad S = \begin{pmatrix} 1 & -1 \\ 0 & 1 \end{pmatrix}.$$

Of course the eigenvalues of A and SAS^{-1} are the same.

Solution 10. We have

$$S^{-1} = \begin{pmatrix} 1 & 1 \\ 0 & 1 \end{pmatrix} \quad \Rightarrow \quad SAS^{-1} = \frac{1}{\sqrt{2}} \begin{pmatrix} 2 & -2 \\ 1 & -2 \end{pmatrix}.$$

The matrix SAS^{-1} is nonnormal.

Problem 11. Consider the nonnormal 2×2 matrix

$$A = \begin{pmatrix} 0 & 1 \\ 0 & 0 \end{pmatrix}.$$

Find $\sinh(A)$ and $\tanh^{-1}(A)$.

Solution 11. Since

$$\sinh(A) = A + \frac{1}{3!}A^3 + \frac{1}{5!}A^5 + \cdots$$

and $A^3 = 0_2$ we find $\sinh(A) = A$. Since

$$\tanh^{-1}(A) = A + \frac{1}{3}A^3 + \frac{1}{5}A^5 + \cdots$$

and $A^3 = 0_2$ we have $\tanh^{-1}(A) = A$.

Problem 12. (i) Consider the nonnormal 2×2 matrix

$$A = \begin{pmatrix} 0 & i \\ 0 & 0 \end{pmatrix} \Rightarrow A^* = \begin{pmatrix} 0 & 0 \\ -i & 0 \end{pmatrix}.$$

Find a unitary matrix U such that $UAU^* = A^*$.
(ii) Consider the nonnormal 2×2 matrix

$$K = \begin{pmatrix} -i & i \\ 0 & i \end{pmatrix}.$$

Find a unitary matrix V such $VKV^* = K^*$. Find the eigenvalues and normalized eigenvectors of K and K^*. Discuss.

Solution 12. (i) We find

$$U = \begin{pmatrix} 0 & 1 \\ -1 & 0 \end{pmatrix}.$$

(ii) We find

$$U = \begin{pmatrix} 0 & 1 \\ -1 & 0 \end{pmatrix}, \qquad K^* = \begin{pmatrix} i & 0 \\ -i & -i \end{pmatrix}.$$

The eigenvalues of K are $-i$ and i with the corresponding normalized eigenvectors

$$\begin{pmatrix} 1 \\ 0 \end{pmatrix}, \qquad \frac{2}{\sqrt{5}}\begin{pmatrix} 1/2 \\ 1 \end{pmatrix}.$$

The eigenvectors are not orthonormal to each other, but linearly independent. The eigenvalues of K^* are also $-i$ and i with the corresponding normalized eigenvectors

$$\begin{pmatrix} 0 \\ 1 \end{pmatrix}, \qquad \frac{2}{\sqrt{5}}\begin{pmatrix} 1 \\ -1/2 \end{pmatrix}.$$

The eigenvectors are not orthogonal to each other, but linearly independent. If we compare the eigenvectors of K and K^*, then the eigenvectors for the eigenvalue $-i$ are orthogonal to each other and the eigenvectors for the eigenvalue i are orthogonal to each other.

Problem 13. Let A, B be normal matrices. Can we conclude that AB is a normal matrix?

Solution 13. We have $AA^* = A^*A$ and $BB^* = B^*B$. We have to show that

$$(AB)(AB)^* = (AB)^*(AB).$$

It follows that $ABB^*A^* = B^*A^*AB$. This equation can only be satisfied if $AB = BA$, i.e. A and B commute. For example, let

$$A = \begin{pmatrix} 1 & 0 \\ 0 & 0 \end{pmatrix}, \quad B = \begin{pmatrix} 1 & 1 \\ 1 & 1 \end{pmatrix}.$$

Then A and B do not commute and we find the nonnormal matrices

$$AB = \begin{pmatrix} 1 & 1 \\ 0 & 0 \end{pmatrix}, \quad BA = \begin{pmatrix} 1 & 0 \\ 1 & 0 \end{pmatrix}.$$

Problem 14.   Consider the nonnormal 3×3 matrix

$$B = \begin{pmatrix} 0 & 1 & 1 \\ 0 & 1 & 1 \\ 0 & 0 & 0 \end{pmatrix}.$$

Can we find a 3×3 matrix Y such that $Y^2 = B$, i.e. Y would be the square root of B.

Solution 14.   A solution is

$$Y = B = \begin{pmatrix} 0 & 1 & 1 \\ 0 & 1 & 1 \\ 0 & 0 & 0 \end{pmatrix}.$$

Problem 15.   Let A be an normal $n \times n$ matrix. Is the matrix $A - iI_n$ normal?

Solution 15.   Yes. We have

$$(A - iI_n)(A - iI_n)^* = (A - iI_n)(A^* + iI_n) = AA^* + i(A - A^*) + I_n$$
$$(A - iI_n)^*(A - iI_n) = (A^* + iI_n)(A - iI_n) = A^*A + i(A - A^*) + I_n.$$

Problem 16.   A *measure of nonnormality* is given by

$$m(A) := \|A^*A - AA^*\|$$

where $\| \cdot \|$ denotes some matrix norm. Let S_3 and S_1 be real-valued and symmetric $n \times n$ matrices (for example, spin matrices). Calculate $m(S_3 + \exp(i\phi)S_1)$.

Solution 16.   The matrices S_1 and S_3 are symmetric and real-valued. So for

$$A := S_3 + \exp(i\phi)S_1$$

we have

$$AA^* - A^*A = 2i\sin(\phi)[S_1, S_3]$$

so that

$$\|AA^* - A^*A\| = 2|\sin(\phi)| \, \|[S_1, S_3]\|.$$

Thus A is normal when $\sin(\phi) = 0$ and nonnormal otherwise. The commutator $[S_1, S_3]$ determines the extent to which A is nonnormal under the measure m.

Problem 17. Consider the nonnormal matrix

$$A = \begin{pmatrix} i\pi & 1 \\ 0 & -i\pi \end{pmatrix}.$$

Is $e^{A+A^*} = e^A e^{A^*}$?

Solution 17. No. We have

$$e^{A+A^*} = \begin{pmatrix} 0 & e \\ e & 0 \end{pmatrix}, \qquad e^A e^{A^*} = \begin{pmatrix} 1 & 0 \\ 0 & 1 \end{pmatrix} = I_2.$$

Problem 18. Prove or disprove the following statements.
(i) If the $n \times n$ matrix A is nonnormal, then there exists no matrix B such that $B^2 = A$.
(ii) Let A, B be nonzero $n \times n$ matrices with $B^2 = A$. Then the matrix A is normal.

Solution 18. (i) Let B be the 2×2 nonnormal matrix

$$B = \begin{pmatrix} 1 & 1/2 \\ 0 & 1 \end{pmatrix} \Rightarrow B^2 = \begin{pmatrix} 1 & 1 \\ 0 & 1 \end{pmatrix} = A.$$

Then $AA^* \neq A^*A$, i.e. A is nonnormal. Thus in general the statement is incorrect.
(ii) We can take the example form (i).

Problem 19. We know that all real symmetric matrices are diagonalizable. Are all complex symmetric matrices are diagonalizable?

Solution 19. This is not true in general, for example the matrix

$$A = \begin{pmatrix} -1 & -2i \\ -2i & 3 \end{pmatrix}$$

is not diagonalizable. The matrix A is also nonnormal.

Problem 20. Let A be a nonnormal $n \times n$ matrix and let U be a unitary $n \times n$ matrix. Is $\tilde{A} = UAU^{-1}$ nonnormal for all U?

Solution 20. We have to prove that $\tilde{A}\tilde{A}^* \neq \tilde{A}^*\tilde{A}$ if $A^*A \neq A^*A$. Now we have

$$\tilde{A}\tilde{A}^* = UAU^{-1}UA^*U^{-1} = UAA^*U^{-1}$$
$$\tilde{A}^*\tilde{A} = UA^*U^{-1}UAU^{-1} = UA^*AU^{-1}.$$

Thus if $AA^* \neq A^*A$ it follows that $\tilde{A}\tilde{A}^* \neq \tilde{A}^*A$.

Problem 21.   Let A be a nonnormal matrix. Show that A^*A and AA^* are linearly independent.

Solution 21.   Consider the equation $c_1A^*A + c_2AA^* = 0_n$. This equation yields

$$\text{tr}(c_1A^*A + c_2AA^*) = (c_1 + c_2)\text{tr}(A^*A) = 0.$$

Since A^*A is nonzero and positive semidefinite, $\text{tr}(A^*A) \neq 0$ and $c_1 = -c_2$. Thus $c_1A^*A = c_1AA^*$. Since A is nonnormal we must have $c_1 = c_2 = 0$.

Problem 22.   (i) Show that if A and B are nonnormal, then $A \otimes B$ is nonnormal.

(ii) Show that if A and B are nonnormal, then $A \oplus B$ is nonnormal, where $\oplus$ denotes the direct sum.

Solution 22.   (i) The matrices A^*A and AA^* are linearly independent and the matrices B^*B and BB^* are linearly independent. Now

$$(A \otimes B)(A \otimes B)^* - (A \otimes B)(A \otimes B)^* = (AA^*) \otimes (BB^*) - (A^*A) \otimes (B^*B) \neq 0_{n^2}$$

by the linear independence of AA^* and A^*A and the linear independence of BB^* and B^*B.

(ii) Since $(A \oplus B)(A \oplus B)^* = (A \oplus B)(A^* \oplus B^*) = (AA^*) \oplus (BB^*)$, $A \oplus B$ is normal if and only if both A and B are normal.

Supplementary Problems

Problem 1.   Are all nonzero nilpotent matrices nonnormal?

Problem 2.   (i) Let $a_{11}, a_{22}, \epsilon \in \mathbb{R}$. What is the condition on a_{11}, a_{22}, ϵ such that the 2×2 matrix

$$A = \begin{pmatrix} a_{11} & e^\epsilon \\ e^{-\epsilon} & a_{22} \end{pmatrix}$$

is normal?

(ii) Let $a > 0$, $b \geq 0$ and $\phi \in [0, \pi]$. What are the conditions a, b, ϕ such that

$$B(a, b, \phi) = \begin{pmatrix} 0 & a \\ e^{i\phi}b & 0 \end{pmatrix}$$

is a normal matrix?

(iii) Let $z \in \mathbb{C}$. Is the 2×2 matrix

$$\begin{pmatrix} 1 & 0 \\ 1 - e^z & e^{-z} \end{pmatrix}$$

nonnormal?

Problem 3. (i) Let $x_1, x_2, x_3 \in \mathbb{R}$. What is the condition such that the 3×3 matrix

$$A(x_1, x_2, x_3) = \begin{pmatrix} 0 & x_1 & 0 \\ 0 & 0 & x_2 \\ x_3 & 0 & 0 \end{pmatrix}$$

is normal? Show that $\lambda^3 = x_1 x_2 x_3$ for $\det(A - \lambda I_3) = 0$. Find the eigenvalues and normalized eigenvectors of A.

(ii) Let $x \in \mathbb{R}$. For which values of x is the 3×3 matrix

$$A(x) = \begin{pmatrix} -1 + ix & ix & ix \\ ix & ix & ix \\ ix & ix & 1 + ix \end{pmatrix}$$

nonnormal.

Problem 4. Show that if A is nonnormal, then $A \bullet A^*$ is normal, where $\bullet$ denotes the Hadamard product. Notice that $A \bullet B = B \bullet A$ and $(A \bullet B)^* = A^* \bullet B^*$.

Problem 5. Find all nonnormal 2×2 matrices A such that $AA^* + A^*A = I_2$. An example is

$$A = \begin{pmatrix} 0 & 1 \\ 0 & 0 \end{pmatrix} \Rightarrow A^* = \begin{pmatrix} 0 & 0 \\ 1 & 0 \end{pmatrix}.$$

Problem 6. (i) Can one find a nonnormal matrix A such that $e^A e^{A^*} = e^{A^*} e^A$?
(ii) Can one find a nonnormal matrix A such that $e^A e^{A^*} = e^{A+A^*}$?

Problem 7. Let M be an invertible $n \times n$ matrix. Then we can form

$$M_+ = \frac{1}{2}(M + M^{-1}), \qquad M_- = \frac{1}{2}(M - M^{-1}).$$

Consider the nonnormal matrix

$$M = \begin{pmatrix} 1 & 1 \\ 0 & 1 \end{pmatrix}.$$

Find M_+ and M_-. Are the matrices M_+ and M_- normal?

Problem 8. Let $s_j := 2\sin(2\pi j/5)$ with $j = 1, \ldots, 5$. Consider the 5×5 matrix

$$M = \begin{pmatrix} s_1 & 1 & 0 & 0 & -1 \\ -1 & s_2 & 1 & 0 & 0 \\ 0 & -1 & s_3 & 1 & 0 \\ 0 & 0 & -1 & s_4 & 1 \\ 1 & 0 & 0 & -1 & s_5 \end{pmatrix}.$$

Show that the matrix is nonnormal. Find the eigenvalues and eigenvectors. Is the matrix diagonalizable?

Problem 9.   Let A, B be nonzero $n \times n$ hermitian matrices. We form the matrix K as $K := A + iB$. This matrix is nonnormal if $[A, B] \neq 0_n$ and normal if $AB = BA$. In the following we assume that K is nonnormal. Assume that we can find a unitary $n \times n$ matrix U such that $UKU^* = K^*$. What can be said about the eigenvalues of K? In physics such a unitary matrix is called a *quasi-symmetry operator*.

Problem 10.   Let Q be a nonnormal invertible $n \times n$ matrix. Is $Q \otimes Q^{-1}$ nonnormal?

Problem 11.   Let M be an $n \times n$ nonnormal matrix. Are the $2n \times 2n$ matrices

$$A = \begin{pmatrix} 0_n & M^* \\ M & 0_n \end{pmatrix}, \qquad B = \begin{pmatrix} 0_n & -M^* \\ M & 0_n \end{pmatrix}$$

nonnormal?

Problem 12.   Let A be a nonnormal invertible matrix. How do we construct a matrix B such that $A = \exp(B)$? Study first the two examples

$$A_1 = \begin{pmatrix} 1 & 1 \\ 0 & 1 \end{pmatrix}, \qquad A_2 = \begin{pmatrix} 2 & 1 \\ 0 & 1/2 \end{pmatrix}.$$

Both are elements of the Lie group $SL(2, \mathbb{R})$. The matrix A_1 admits only one linearly independent eigenvector, namely $(1, 0)^T$. The matrix

$$B_1 = \begin{pmatrix} 0 & 1 \\ 0 & 0 \end{pmatrix}$$

generates A_1.

Problem 13.   Let A be an $n \times n$ nonnormal matrix and $Q \in SL(n, \mathbb{C})$. Can one find Q such that $\tilde{A} = QAQ^{-1}$ is normal?

Chapter 15

Binary Matrices

An $m \times n$ matrix A is a *binary matrix* if $a_{jk} \in \{0,1\}$ for $j = 1, \ldots, m$ and $k = 1, \ldots, n$. A binary matrix is a matrix over the two-element field $GF(2) = \{0,1\}$. In $GF(2) = \{0,1\}$ we have that $-1 = 1$, i.e. $1 + 1 = 0$. Thus $A + A = 0$ for all binary matrices A, where 0 is the zero matrix. The special linear group and general linear group coincide over $GF(2)$, $GL_n(GF(2)) = SL_n(GF(2))$. The set of invertible 2×2 binary matrices $SL_2(GF(2))$ is

$$\left\{ \begin{pmatrix} 1 & 0 \\ 0 & 1 \end{pmatrix}, \begin{pmatrix} 1 & 1 \\ 0 & 1 \end{pmatrix}, \begin{pmatrix} 1 & 0 \\ 1 & 1 \end{pmatrix}, \begin{pmatrix} 0 & 1 \\ 1 & 0 \end{pmatrix}, \begin{pmatrix} 1 & 1 \\ 1 & 0 \end{pmatrix}, \begin{pmatrix} 0 & 1 \\ 1 & 1 \end{pmatrix} \right\}.$$

Over the complex numbers, each of the above matrices can be expressed as a group commutator $ABA^{-1}B^{-1}$ for some 2×2 complex matrices A and B. Over $GF(2)$, this is not the case. For example

$$\begin{pmatrix} 1 & 1 \\ 0 & 1 \end{pmatrix} = \begin{pmatrix} 1 & t \\ 0 & -1 \end{pmatrix} \begin{pmatrix} 1 & -\frac{1}{2} \\ 0 & 1 \end{pmatrix} \begin{pmatrix} 1 & t \\ 0 & -1 \end{pmatrix}^{-1} \begin{pmatrix} 1 & -\frac{1}{2} \\ 0 & 1 \end{pmatrix}^{-1}$$

over the complex numbers, but

$$\begin{pmatrix} 1 & 1 \\ 0 & 1 \end{pmatrix} \neq ABA^{-1}B^{-1}$$

for all $A, B \in SL_2(GF(2))$.

Permutation matrices are binary matrices with the underlying field $\mathbb{R}$.

The Kronecker product, the direct sum and the star product of binary matrices are again binary matrices.

Problem 1. Consider the four 2×2 binary matrices

$$A = \begin{pmatrix} 1 & 1 \\ 0 & 0 \end{pmatrix}, \quad B = \begin{pmatrix} 0 & 1 \\ 0 & 1 \end{pmatrix}, \quad C = \begin{pmatrix} 1 & 0 \\ 1 & 0 \end{pmatrix}, \quad D = \begin{pmatrix} 0 & 0 \\ 1 & 1 \end{pmatrix}$$

with the underlying field $\mathbb{F}$, $\text{char}(\mathbb{F}) = 2$. Find $A + B + C + D$ and $ABCD$.

Solution 1. We have

$$A + B + C + D = 0_2, \qquad ABCD = 0_2.$$

Problem 2. For a 2×2 binary matrix

$$A = \begin{pmatrix} a_{11} & a_{12} \\ a_{21} & a_{22} \end{pmatrix}, \quad a_{jk} \in \{0, 1\}$$

we define the determinant as

$$\det(A) = (a_{11} \cdot a_{22}) \oplus (a_{12} \cdot a_{21})$$

where $\cdot$ is the AND-operation and $\oplus$ is the XOR-operation.
(i) Find the determinant for the following 2×2 matrices

$$\begin{pmatrix} 1 & 0 \\ 0 & 1 \end{pmatrix}, \quad \begin{pmatrix} 0 & 1 \\ 1 & 0 \end{pmatrix}, \quad \begin{pmatrix} 1 & 1 \\ 0 & 1 \end{pmatrix}, \quad \begin{pmatrix} 0 & 1 \\ 1 & 1 \end{pmatrix}, \quad \begin{pmatrix} 1 & 0 \\ 1 & 1 \end{pmatrix}, \quad \begin{pmatrix} 1 & 1 \\ 1 & 0 \end{pmatrix}.$$

(ii) Find the determinant for the following 2×2 matrices

$$\begin{pmatrix} 0 & 0 \\ 0 & 0 \end{pmatrix}, \quad \begin{pmatrix} 1 & 0 \\ 0 & 0 \end{pmatrix}, \quad \begin{pmatrix} 0 & 1 \\ 0 & 0 \end{pmatrix}, \quad \begin{pmatrix} 1 & 0 \\ 0 & 0 \end{pmatrix}, \quad \begin{pmatrix} 0 & 0 \\ 0 & 1 \end{pmatrix}$$

$$\begin{pmatrix} 1 & 1 \\ 0 & 0 \end{pmatrix}, \quad \begin{pmatrix} 1 & 0 \\ 1 & 0 \end{pmatrix}, \quad \begin{pmatrix} 0 & 0 \\ 1 & 1 \end{pmatrix}, \quad \begin{pmatrix} 0 & 1 \\ 0 & 1 \end{pmatrix}, \quad \begin{pmatrix} 1 & 1 \\ 1 & 1 \end{pmatrix}.$$

Solution 2. (i) Since $0 \cdot 0 = 0$, $0 \cdot 1 = 0$, $1 \cdot 0 = 0$, $1 \cdot 1 = 1$ and $0 \oplus 0 = 0$, $0 \oplus 1 = 1$, $1 \oplus 0 = 1$, $1 \oplus 1 = 0$ we find that all 6 matrices have determinant equal to 1.
(ii) Since $0 \cdot 0 = 0$, $0 \cdot 1 = 0$, $1 \cdot 0 = 0$, $1 \cdot 1 = 1$ and $0 \oplus 0 = 0$, $0 \oplus 1 = 1$, $1 \oplus 0 = 1$, $1 \oplus 1 = 0$ we find that all 10 matrices have determinant equal to 0.

Problem 3. The determinant of a 3×3 matrix $A = (a_{jk})$ is given by

$$a_{11}a_{22}a_{33} + a_{12}a_{23}a_{31} + a_{13}a_{21}a_{32} - a_{13}a_{22}a_{31} - a_{11}a_{23}a_{32} - a_{12}a_{21}a_{33}.$$

For a binary matrix B we replace this expression by

$$\det(B) = (b_{11} \cdot b_{22} \cdot b_{33}) \oplus (b_{12} \cdot b_{23} \cdot b_{31}) \oplus (b_{13} \cdot b_{21} \cdot b_{32})$$
$$\oplus (b_{13} \cdot b_{22} \cdot b_{31}) \oplus (b_{11} \cdot b_{23} \cdot b_{32}) \oplus (b_{12} \cdot b_{21} \cdot b_{33}).$$

(i) Calculate the determinant for the binary matrices

$$\begin{pmatrix} 1 & 0 & 0 \\ 0 & 1 & 0 \\ 0 & 0 & 1 \end{pmatrix}, \qquad \begin{pmatrix} 1 & 1 & 1 \\ 0 & 1 & 1 \\ 0 & 0 & 1 \end{pmatrix}.$$

(ii) Calculate the determinant for the binary matrices

$$\begin{pmatrix} 1 & 1 & 0 \\ 1 & 1 & 0 \\ 0 & 0 & 0 \end{pmatrix}, \qquad \begin{pmatrix} 1 & 0 & 0 \\ 1 & 0 & 0 \\ 1 & 0 & 0 \end{pmatrix}, \qquad \begin{pmatrix} 1 & 0 & 1 \\ 0 & 1 & 0 \\ 1 & 0 & 1 \end{pmatrix}.$$

Solution 3. (i) For both matrices the determinant is 1.
(ii) For all three matrices the determinant is 0.

Problem 4. The finite field $GF(2)$ consists of the elements 0 and 1 (bits) which satisfies the following addition (XOR) and multiplication (AND) tables

$\oplus$	0	1
0	0	1
1	1	0

$\cdot$	0	1
0	0	0
1	0	1

Find the determinant of the binary matrices

$$A = \begin{pmatrix} 1 & 0 & 1 \\ 0 & 1 & 0 \\ 1 & 0 & 1 \end{pmatrix}, \qquad B = \begin{pmatrix} 1 & 1 & 1 \\ 0 & 1 & 1 \\ 0 & 0 & 1 \end{pmatrix}.$$

Solution 4. Applying the *rule of Sarrus* provides

$$\det(A) = (1 \cdot 1 \cdot 1) \oplus (0 \cdot 0 \cdot 1) \oplus (1 \cdot 0 \cdot 0) \oplus (1 \cdot 1 \cdot 1) \oplus (0 \cdot 0 \cdot 1) \oplus (1 \cdot 0 \cdot 0) = 0$$

$$\det(B) = (1 \cdot 1 \cdot 1) \oplus (1 \cdot 1 \cdot 0) \oplus (1 \cdot 0 \cdot 0) \oplus (0 \cdot 1 \cdot 1) \oplus (0 \cdot 1 \cdot 1) \oplus (1 \cdot 0 \cdot 1) = 1.$$

Problem 5. A *boolean function* $f : \{0, 1\}^n \to \{0, 1\}$ can be transformed from the domain $\{0, 1\}$ into the spectral domain by a linear transformation

$$T\mathbf{y} = \mathbf{s}$$

where T is a $2^n \times 2^n$ orthogonal matrix, $\mathbf{y} = (y_0, y_1, \dots, y_{2^n-1})^T$, is the two valued ($\{+1, -1\}$ with $0 \leftrightarrow 1, 1 \leftrightarrow -1$) truth table vector of the boolean function and s_j ($j = 0, 1, \dots, 7$) are the spectral coefficients ($\mathbf{s} = (s_0, s_1, \dots, s_{2^n-1})^T$). Since T is invertible we have

$$T^{-1}\mathbf{s} = \mathbf{y}.$$

For T we select the Hadamard matrix. The $2^n \times 2^n$ *Hadamard matrix* $H(n)$ is recursively defined as

$$H(n) = \begin{pmatrix} H(n-1) & H(n-1) \\ H(n-1) & -H(n-1) \end{pmatrix}, \qquad n = 1, 2, \dots$$

with $H(0) = (1)$ (1×1 matrix). The inverse of $H(n)$ is given by

$$H^{-1}(n) = \frac{1}{2^n} H(n).$$

Now any boolean function $f(x_1, \ldots, x_n)$ can be expanded as the arithmetical polynomial

$$\frac{1}{2^{n+1}} \left(2^n - s_0 - s_1(-1)^{x_n} - s_2(-1)^{x_{n-1}} - \cdots - s_{2^n-1}(-1)^{x_1 \oplus x_2 \oplus \cdots \oplus x_n} \right)$$

where $\oplus$ denotes the modulo-2 addition.

Consider the boolean function $f : \{0,1\}^3 \to \{0,1\}$ given by

$$f(x_1, x_2, x_3) = \bar{x}_1 \cdot \bar{x}_2 \cdot \bar{x}_3 + \bar{x}_1 \cdot x_2 \cdot \bar{x}_3 + x_1 \cdot x_2 \cdot \bar{x}_3.$$

Find the truth table, the vector $\mathbf{y}$ and then calculate, using $H(3)$, the spectral coefficients s_j, $(j = 0, 1, \ldots, 7)$.

Solution 5. The truth table with the vector $\mathbf{y}$ is given by

(x_1, x_2, x_3)	$f(x_1, x_2, x_3)$	$\mathbf{y}$
0 0 0	1	-1
0 0 1	0	1
0 1 0	1	-1
0 1 1	0	1
1 0 0	0	1
1 0 1	0	1
1 1 0	1	-1
1 1 1	0	1

where we used the map $0 \leftrightarrow 1$, $1 \leftrightarrow -1$ to find the vector $\mathbf{y}$. Now the Hadamard matrix $H(3)$ is given by

$$H(3) = \begin{pmatrix} H(2) & H(2) \\ H(2) & -H(2) \end{pmatrix} = \begin{pmatrix} H(1) & H(1) & H(1) & H(1) \\ H(1) & -H(1) & H(1) & -H(1) \\ H(1) & H(1) & -H(1) & -H(1) \\ H(1) & -H(1) & -H(1) & H(1) \end{pmatrix}.$$

Thus

$$H(3) = \begin{pmatrix} 1 & 1 & 1 & 1 & 1 & 1 & 1 & 1 \\ 1 & -1 & 1 & -1 & 1 & -1 & 1 & -1 \\ 1 & 1 & -1 & -1 & 1 & 1 & -1 & -1 \\ 1 & -1 & -1 & 1 & 1 & -1 & -1 & 1 \\ 1 & 1 & 1 & 1 & -1 & -1 & -1 & -1 \\ 1 & -1 & 1 & -1 & -1 & 1 & -1 & 1 \\ 1 & 1 & -1 & -1 & -1 & -1 & 1 & 1 \\ 1 & -1 & -1 & 1 & -1 & 1 & 1 & -1 \end{pmatrix}.$$

From $T\mathbf{y} = \mathbf{s}$ we find $\mathbf{s} = (2, -6, 2, 2, -2, -2, -2, -2)^T$ and the expansion is

$$f(x_1, x_2, x_3) = \frac{1}{16} \left(8 - 2 + 6(-1)^{x_3} - 2(-1)^{x_2} - 2(-1)^{x_2 \oplus x_3} \right.$$
$$\left. + 2(-1)^{x_1} + 2(-1)^{x_1 \oplus x_3} + 2(-1)^{x_1 \oplus x_2} + 2(-1)^{x_1 \oplus x_2 \oplus x_3} \right).$$

Problem 6. Consider the binary matrices

$$A = \begin{pmatrix} 0 & 1 \\ 1 & 0 \end{pmatrix}, \qquad B = \begin{pmatrix} 1 & 1 \\ 1 & 1 \end{pmatrix}.$$

Calculate the Hadamard product $A \bullet B$ and the star product $A \star B$.

Solution 6. We find

$$A \bullet B = \begin{pmatrix} 0 & 1 \\ 1 & 0 \end{pmatrix}, \qquad A \star B = \begin{pmatrix} 1 & 0 & 0 & 1 \\ 0 & 0 & 1 & 0 \\ 0 & 1 & 0 & 0 \\ 1 & 0 & 0 & 1 \end{pmatrix}.$$

Both are binary matrices again.

Problem 7. A *binary Hadamard matrix* is an $n \times n$ matrix M (where n is even) with entries in $\{0, 1\}$ such that any two distinct rows or columns of M have *Hamming distance* $n/2$. The Hamming distance between two vectors is the number of entries at which they differ. Find a 4×4 binary Hadamard matrix.

Solution 7. Hadamard matrices are in *bijection* with binary Hadamard matrices with the mapping $1 \to 1$ and $-1 \to 0$. Thus using the result from the previous problem we obtain

$$M = \begin{pmatrix} 1 & 1 & 0 & 1 \\ 1 & 1 & 1 & 0 \\ 1 & 0 & 1 & 1 \\ 1 & 0 & 0 & 0 \end{pmatrix}.$$

Problem 8. How many 2×2 binary matrices can one form which contain two 1's?

Solution 8. There are six such matrices

$$\begin{pmatrix} 1 & 1 \\ 0 & 0 \end{pmatrix}, \begin{pmatrix} 1 & 0 \\ 1 & 0 \end{pmatrix}, \begin{pmatrix} 1 & 0 \\ 0 & 1 \end{pmatrix}, \begin{pmatrix} 0 & 1 \\ 1 & 0 \end{pmatrix}, \begin{pmatrix} 0 & 1 \\ 0 & 1 \end{pmatrix}, \begin{pmatrix} 0 & 0 \\ 1 & 1 \end{pmatrix}.$$

Find the commutator between these matrices.

Supplementary Problems

Problem 1. How many 3×3 binary matrices can one form which contain three 1's? Write down these matrices. Which of them are invertible?

Problem 2.   Consider the four 8×8 binary matrices

$$
S_0 = \begin{pmatrix}
0 & 0 & 0 & 0 & 0 & 0 & 0 & 0 \\
0 & 0 & 0 & 0 & 0 & 0 & 0 & 0 \\
0 & 0 & 0 & 0 & 0 & 0 & 0 & 0 \\
0 & 0 & 0 & 1 & 1 & 0 & 0 & 0 \\
0 & 0 & 0 & 1 & 1 & 0 & 0 & 0 \\
0 & 0 & 0 & 0 & 0 & 0 & 0 & 0 \\
0 & 0 & 0 & 0 & 0 & 0 & 0 & 0 \\
0 & 0 & 0 & 0 & 0 & 0 & 0 & 0
\end{pmatrix}, \quad
S_1 = \begin{pmatrix}
0 & 0 & 0 & 0 & 0 & 0 & 0 & 0 \\
0 & 0 & 0 & 0 & 0 & 0 & 0 & 0 \\
0 & 0 & 1 & 0 & 0 & 1 & 0 & 0 \\
0 & 0 & 0 & 0 & 0 & 0 & 0 & 0 \\
0 & 0 & 0 & 0 & 0 & 0 & 0 & 0 \\
0 & 0 & 1 & 0 & 0 & 1 & 0 & 0 \\
0 & 0 & 0 & 0 & 0 & 0 & 0 & 0 \\
0 & 0 & 0 & 0 & 0 & 0 & 0 & 0
\end{pmatrix},
$$

$$
S_2 = \begin{pmatrix}
0 & 0 & 0 & 0 & 0 & 0 & 0 & 0 \\
0 & 1 & 0 & 0 & 0 & 0 & 1 & 0 \\
0 & 0 & 0 & 0 & 0 & 0 & 0 & 0 \\
0 & 0 & 0 & 0 & 0 & 0 & 0 & 0 \\
0 & 0 & 0 & 0 & 0 & 0 & 0 & 0 \\
0 & 0 & 0 & 0 & 0 & 0 & 0 & 0 \\
0 & 1 & 0 & 0 & 0 & 0 & 1 & 0 \\
0 & 0 & 0 & 0 & 0 & 0 & 0 & 0
\end{pmatrix}, \quad
S_3 = \begin{pmatrix}
1 & 0 & 0 & 0 & 0 & 0 & 0 & 1 \\
0 & 0 & 0 & 0 & 0 & 0 & 0 & 0 \\
0 & 0 & 0 & 0 & 0 & 0 & 0 & 0 \\
0 & 0 & 0 & 0 & 0 & 0 & 0 & 0 \\
0 & 0 & 0 & 0 & 0 & 0 & 0 & 0 \\
0 & 0 & 0 & 0 & 0 & 0 & 0 & 0 \\
0 & 0 & 0 & 0 & 0 & 0 & 0 & 0 \\
1 & 0 & 0 & 0 & 0 & 0 & 0 & 1
\end{pmatrix}.
$$

Find the 8×8 matrices Q_0, Q_1, Q_2, Q_3 such that

$$
Q_0 S_0 Q_0^T = S_1, \quad Q_1 S_1 Q_1^T = S_2, \quad Q_2 S_2 Q_2^T = S_3, \quad Q_3 S_3 Q_3^T = S_0.
$$

Problem 3.   Consider the four 2×2 binary matrices A and B with the underlying field $\mathbb{F}$, $\text{char}(\mathbb{F}) = 2$. Find the condition on A and B such that $A + B = AB$.

Problem 4.   Consider the two permutation matrices (NOT-gate and XOR-gate)

$$
N = \begin{pmatrix}
0 & 0 & 0 & 1 \\
0 & 0 & 1 & 0 \\
0 & 1 & 0 & 0 \\
1 & 0 & 0 & 0
\end{pmatrix}, \quad
X = \begin{pmatrix}
1 & 0 & 0 & 0 \\
0 & 1 & 0 & 0 \\
0 & 0 & 0 & 1 \\
0 & 0 & 1 & 0
\end{pmatrix}.
$$

Can we generate all other permutation matrices from these two permutation matrices? Note that $N^2 = I_4$, $X^2 = I_4$ and

$$
NX = \begin{pmatrix}
0 & 0 & 1 & 0 \\
0 & 0 & 0 & 1 \\
0 & 1 & 0 & 0 \\
1 & 0 & 0 & 0
\end{pmatrix}, \quad
XN = \begin{pmatrix}
0 & 0 & 0 & 1 \\
0 & 0 & 1 & 0 \\
1 & 0 & 0 & 0 \\
0 & 1 & 0 & 0
\end{pmatrix}.
$$

Chapter 16

Star Product

Consider the 2×2 matrices

$$A = \begin{pmatrix} a_{11} & a_{12} \\ a_{21} & a_{22} \end{pmatrix}, \qquad B = \begin{pmatrix} b_{11} & b_{12} \\ b_{21} & b_{22} \end{pmatrix}.$$

We define the composition $A \star B$ (star product) and $\widehat{A \star B}$ as

$$A \star B := \begin{pmatrix} b_{11} & 0 & 0 & b_{12} \\ 0 & a_{11} & a_{12} & 0 \\ 0 & a_{21} & a_{22} & 0 \\ b_{21} & 0 & 0 & b_{22} \end{pmatrix}, \quad \widehat{A \star B} := \begin{pmatrix} 0 & a_{11} & a_{12} & 0 \\ b_{11} & 0 & 0 & b_{12} \\ b_{21} & 0 & 0 & b_{22} \\ 0 & a_{21} & a_{22} & 0 \end{pmatrix}.$$

Note that

$$I_2 \star I_2 = I_4, \qquad 0_2 \star 0_2 = 0_4.$$

This definition can be extended to higher dimensions. For example the extension to 4×4 matrices $A = (a_{jk})$ and $B = (b_{jk})$ is the 8×8 matrix for $A \star B$

$$\begin{pmatrix} b_{11} & b_{12} & 0 & 0 & 0 & 0 & b_{13} & b_{14} \\ b_{21} & b_{22} & 0 & 0 & 0 & 0 & b_{23} & b_{24} \\ 0 & 0 & a_{11} & a_{12} & a_{13} & a_{14} & 0 & 0 \\ 0 & 0 & a_{21} & a_{22} & a_{23} & a_{24} & 0 & 0 \\ 0 & 0 & a_{31} & a_{32} & a_{33} & a_{34} & 0 & 0 \\ 0 & 0 & a_{41} & a_{42} & a_{43} & a_{44} & 0 & 0 \\ b_{31} & b_{32} & 0 & 0 & 0 & 0 & b_{33} & b_{34} \\ b_{41} & b_{42} & 0 & 0 & 0 & 0 & b_{43} & b_{44} \end{pmatrix}$$

and analogously we define $\widehat{A \star B}$.

Problem 1.   Consider the unitary and hermitian 2×2 matrix

$$U = \frac{1}{\sqrt{2}} \begin{pmatrix} 1 & 1 \\ 1 & -1 \end{pmatrix}.$$

(i) Find $U \star U$. Is $U \star U$ unitary? Is $U \star U$ hermitian?
(ii) Find $\widehat{U \star U}$. Is $\widehat{U \star U}$ unitary? Is $\widehat{U \star U}$ hermitian?

Solution 1.   (i) We obtain the *Bell matrix*

$$U \star U = \frac{1}{\sqrt{2}} \begin{pmatrix} 1 & 0 & 0 & 1 \\ 0 & 1 & 1 & 0 \\ 0 & 1 & -1 & 0 \\ 1 & 0 & 0 & -1 \end{pmatrix}.$$

Obviously $U \star U$ is unitary and hermitian.
(ii) We obtain the hermitian and unitary matrix

$$\widehat{U \star U} = \frac{1}{\sqrt{2}} \begin{pmatrix} 0 & 1 & 1 & 0 \\ 1 & 0 & 0 & 1 \\ 1 & 0 & 0 & -1 \\ 0 & 1 & -1 & 0 \end{pmatrix}.$$

Problem 2.   Consider the nonnormal 2×2 matrices

$$A = \begin{pmatrix} 0 & 1 \\ 0 & 0 \end{pmatrix}, \qquad B = \begin{pmatrix} 0 & 0 \\ 1 & 0 \end{pmatrix} = A^T$$

with the eigenvalues 0 ($2\times$).
(i) Find $A \star B$ and $B \star A$ and the eigenvalues.
(ii) Is $A \star B$ nonnormal? Is $B \star A$ nonnormal?

Solution 2.   (i) We have

$$A \star B = \begin{pmatrix} 0 & 0 & 0 & 0 \\ 0 & 0 & 1 & 0 \\ 0 & 0 & 0 & 0 \\ 1 & 0 & 0 & 0 \end{pmatrix}, \qquad B \star A = \begin{pmatrix} 0 & 0 & 0 & 1 \\ 0 & 0 & 0 & 0 \\ 0 & 1 & 0 & 0 \\ 0 & 0 & 0 & 0 \end{pmatrix} = (A \star B)^T.$$

For both matrices the eigenvalues are 0 ($4\times$).
(ii) Yes. Both $A \star B$ and $B \star A$ are nonnormal. Note that $(A \star B)(B \star A)$ is normal.

Problem 3.   (i) What can be said about the trace of $A \star B$? What can be said about the determinant of $A \star B$?
(i) What can be said about the trace of $\widehat{A \star B}$? What can be said about the determinant of $\widehat{A \star B}$?

Solution 3.   (i) We have

$$\text{tr}(A \star B) = \text{tr}(A) + \text{tr}(B), \qquad \det(A \star B) = \det(A) \det(B).$$

(ii) We have $\text{tr}(\widehat{A \star B}) = 0$ and $\det(\widehat{A \star B}) = \det(A)\det(B)$.

Problem 4.   Let A, B be 2×2 matrices. Can one find a 4×4 permutation matrix P such that $P(A \star B)P^T = B \oplus A$? Here $\oplus$ denotes the direct sum.

Solution 4.   We find the permutation matrix

$$P = \begin{pmatrix} 1 & 0 & 0 & 0 \\ 0 & 0 & 0 & 1 \\ 0 & 1 & 0 & 0 \\ 0 & 0 & 1 & 0 \end{pmatrix}.$$

Problem 5.   Let A, B be 2×2 matrices. Let $A \star B$ be the star product. Show that one can find a 4×4 permutation matrix P such that

$$P(A \star B)P^T = \begin{pmatrix} a_{11} & a_{12} & 0 & 0 \\ a_{21} & a_{22} & 0 & 0 \\ 0 & 0 & b_{11} & b_{12} \\ 0 & 0 & b_{21} & b_{22} \end{pmatrix}.$$

Solution 5.   Owing to the structure of the problem one can set $p_{11} = 1$. The permutation matrix is

$$P = \begin{pmatrix} 1 & 0 & 0 & 0 \\ 0 & 0 & 0 & 1 \\ 0 & 1 & 0 & 0 \\ 0 & 0 & 1 & 0 \end{pmatrix} \Rightarrow P^T = \begin{pmatrix} 1 & 0 & 0 & 0 \\ 0 & 0 & 1 & 0 \\ 0 & 0 & 0 & 1 \\ 0 & 1 & 0 & 0 \end{pmatrix}.$$

Note that

$$P(A \star B) = \begin{pmatrix} a_{11} & 0 & 0 & a_{12} \\ a_{21} & 0 & 0 & a_{22} \\ 0 & b_{11} & b_{12} & 0 \\ 0 & b_{21} & b_{22} & 0 \end{pmatrix}.$$

Problem 6.   The 2×2 matrices

$$I = \begin{pmatrix} 1 & 0 \\ 0 & 1 \end{pmatrix}, \qquad J = \begin{pmatrix} 0 & 1 \\ 1 & 0 \end{pmatrix}$$

form a group under matrix multiplication. Do the four 4×4 matrices $I \star I$, $I \star J$, $J \star I$, $J \star J$ form a group under matrix multiplication?

Solution 6.   Yes. Note that the four matrices are permutation matrices and form a subgroup of all 4×4 permutation matrices. The neutral element is the 4×4 matrix $I \star I$. Furthermore

$$(J \star J)(J \star J) = I \star I, \quad (I \star J)(I \star J) = I \star I, \quad (J \star I)(I \star J) = J \star J.$$

Problem 7.   Given the eigenvalues and eigenvectors of the 2×2 matrices A and B. What can be said about the eigenvalues and eigenvectors of $A \star B$?

Solution 7.   From

$$
\det \begin{pmatrix} b_{11} - \lambda & 0 & 0 & b_{12} \\ 0 & a_{11} - \lambda & a_{12} & 0 \\ 0 & a_{21} & a_{11} - \lambda & 0 \\ b_{21} & 0 & 0 & b_{22} - \lambda \end{pmatrix} = 0
$$

we obtain

$$
\det(A - \lambda I_2) \det(B - \lambda I_2) = 0.
$$

Thus if λ_1, λ_2 are the eigenvalues of A and μ_1, μ_2 are the eigenvalues of B, then the eigenvalues of $A \star B$ are λ_1, λ_2, μ_1, μ_2. The eigenvectors have the form

$$
\begin{pmatrix} 0 \\ v_{11} \\ v_{21} \\ 0 \end{pmatrix}, \begin{pmatrix} 0 \\ v_{21} \\ v_{22} \\ 0 \end{pmatrix}, \begin{pmatrix} u_{11} \\ 0 \\ 0 \\ u_{12} \end{pmatrix}, \begin{pmatrix} u_{21} \\ 0 \\ 0 \\ u_{22} \end{pmatrix}.
$$

Problem 8.   The 2×2 matrix

$$
A(\alpha) = \begin{pmatrix} \cos(\alpha) & -\sin(\alpha) \\ \sin(\alpha) & \cos(\alpha) \end{pmatrix}
$$

admits the eigenvalues $\lambda_+(\alpha) = e^{i\alpha}$ and $\lambda_-(\alpha) = e^{-i\alpha}$ with the corresponding normalized eigenvectors

$$
\frac{1}{\sqrt{2}} \begin{pmatrix} 1 \\ -i \end{pmatrix}, \quad \frac{1}{\sqrt{2}} \begin{pmatrix} 1 \\ i \end{pmatrix}.
$$

Let

$$
A(\alpha) \star A(\alpha) = \begin{pmatrix} \cos(\alpha) & 0 & 0 & -\sin(\alpha) \\ 0 & \cos(\alpha) & -\sin(\alpha) & 0 \\ 0 & \sin(\alpha) & \cos(\alpha) & 0 \\ \sin(\alpha) & 0 & 0 & \cos(\alpha) \end{pmatrix}.
$$

Find the eigenvalues and normalized eigenvectors of $A(\alpha) \star A(\alpha)$.

Solution 8.   For the eigenvalues we find $e^{i\alpha}$ (twice), $e^{-i\alpha}$ (twice) with the corresponding normalized eigenvectors

$$
\frac{1}{\sqrt{2}} \begin{pmatrix} 1 \\ 0 \\ 0 \\ -i \end{pmatrix}, \frac{1}{\sqrt{2}} \begin{pmatrix} 0 \\ 1 \\ -i \\ 0 \end{pmatrix}, \frac{1}{\sqrt{2}} \begin{pmatrix} 1 \\ 0 \\ 0 \\ i \end{pmatrix}, \frac{1}{\sqrt{2}} \begin{pmatrix} 0 \\ 1 \\ i \\ 0 \end{pmatrix}.
$$

Problem 9.   Let A, B be two 2×2 matrices and $A \star B$ be the star product. Can one find a 4×4 permutation matrix P such that

$$
P(A \star B) = \widehat{A \star B}?
$$

Solution 9.   We obtain

$$P = \begin{pmatrix} 0 & 1 & 0 & 0 \\ 1 & 0 & 0 & 0 \\ 0 & 0 & 0 & 1 \\ 0 & 0 & 1 & 0 \end{pmatrix} = P^{-1}.$$

Problem 10.   Let A, B be 2×2 matrices. Find $A \star B - B \star A$.

Solution 10.   We obtain

$$A \star B - B \star A = \begin{pmatrix} c_{11} & c_{12} \\ c_{21} & c_{22} \end{pmatrix} \star \begin{pmatrix} -c_{11} & -c_{12} \\ -c_{21} & -c_{22} \end{pmatrix}$$

where

$$c_{11} = a_{11} - b_{11}, \quad c_{12} = a_{12} - b_{12}, \quad c_{21} = a_{21} - b_{21}, \quad c_{22} = a_{22} - b_{22}.$$

Thus $A \star B - B \star A$ can be written as a star product.

Supplementary Problems

Problem 1.   Consider the star product $A \star B$ of two 2×2 matrices A and B and the product

$$A \diamond B := \begin{pmatrix} b_{11} & 0 & 0 & b_{12} \\ b_{21} & 0 & 0 & b_{22} \\ 0 & a_{11} & a_{12} & 0 \\ 0 & a_{21} & a_{22} & 0 \end{pmatrix}.$$

Show that there is a 4×4 permutation matrix P such that $P(A \star B) = A \diamond B$.

Problem 2.   Consider the 2×2 matrices A, B over $\mathbb{C}$. Let $A \star B$ be the star product.
(i) Let A and B be normal matrices. Show that $A \star B$ normal.
(ii) Let A and B be unitary matrices. Show that $A \star B$ a unitary matrix.
(iii) Let A and B be nilpotent matrices. Show that $A \star B$ a nilpotent matrix.

Problem 3.   Let A, B be invertible 2×2 matrices. Let $A \star B$ be the star product. Show that

$$(A \star B)^{-1} = A^{-1} \star B^{-1}.$$

Problem 4.   Let A be a 3×3 matrix and B be a 2×2 matrix. We define

$$A \star B := \begin{pmatrix} b_{11} & 0 & 0 & 0 & b_{12} \\ 0 & a_{11} & a_{12} & a_{13} & 0 \\ 0 & a_{21} & a_{22} & a_{23} & 0 \\ 0 & a_{31} & a_{32} & a_{33} & 0 \\ b_{21} & 0 & 0 & 0 & b_{22} \end{pmatrix}.$$

(i) Find $\det(A \star B)$ and $\operatorname{tr}(A \star B)$.
(ii) Let A, B be normal matrices. Is $A \star B$ normal?
(iii) Let A, B be unitary. Is $A \star B$ unitary?
(iv) Let

$$A = \frac{1}{\sqrt{2}} \begin{pmatrix} 1 & 1 \\ 1 & -1 \end{pmatrix}, \qquad B = I_3.$$

Find $A \star B$ and calculate the eigenvalues and normalized eigenvectors of $A \star I_3$.

Problem 5.   Let Π_1 and Π_2 be 2×2 projection matrices. Is the 4×4 matrix $\Pi_1 \star \Pi_2$ a projection matrix? Apply it to $\Pi_1 \star \Pi_1$ where

$$\Pi_1 = \frac{1}{2} \begin{pmatrix} 1 & 1 \\ 1 & 1 \end{pmatrix}.$$

Problem 6.   (i) Let U and V be elements of the Lie group $SU(2)$. Is $U \star V$ an element of the Lie group $SU(4)$?
(ii) Let X and Y be elements of $SL(2, \mathbb{R})$. Is $X \star Y$ an element of $SL(4, \mathbb{R})$?

Problem 7.   Let A, B be normal 2×2 matrices with eigenvalues λ_1, λ_2 and μ_1, μ_2, respectively. What can be said about the eigenvalues of $A \star B - B \star A$?

Problem 8.   Let A, B be 2×2 matrices.
(i) What are the conditions on A and B such that $A \star B = A \otimes B$?
(ii) What are the conditions on A and B such that $\widehat{A \star B} = A \otimes B$?
(iii) What are the conditions on A and B such that $A \star B = A \oplus B$?
(i) What are the conditions on A and B such that $\widehat{A \star B} = A \oplus B$?

Chapter 17

Unitary Matrices

An $n \times n$ matrix U over $\mathbb{C}$ is called a unitary matrix if

$$U^* = U^{-1}.$$

Thus $UU^* = I_n$ for a unitary matrix. If U and V are unitary $n \times n$ matrices, then UV is an $n \times n$ unitary matrix. The columns in a unitary matrix are pairwise orthonormal. Thus the column vectors in a unitary matrix form an orthonormal basis in $\mathbb{C}^n$. If $\mathbf{v}$ is a normalized (column) vector in $\mathbb{C}^n$, then $U\mathbf{v}$ is also a normalized vector. The $n \times n$ unitary matrices form a group under matrix multiplication. Let K be a skew-hermitian matrix, then $\exp(K)$ is a unitary matrix. For any unitary matrix U we can find a skew-hermitian matrix K such that $U = \exp(K)$. Let H be an $n \times n$ hermitian matrix and $U := \exp(iK)$ is a unitary matrix. The eigenvalues of a unitary matrix are of the form $e^{i\phi}$ with $\phi \in \mathbb{R}$. Let A be an arbitrary $n \times n$ matrix over $\mathbb{C}$. Then under the map $A \mapsto UAU^*$ the eigenvalues are preserved and thus the trace and the determinant of A.

If U is an $n \times n$ unitary matrix and V is an $m \times m$ unitary matrix, then $U \otimes V$ is a unitary matrix, where $\otimes$ denotes the Kronecker product. If U is an $n \times n$ unitary matrix and V is an $m \times m$ unitary matrix, then $U \oplus V$ is a unitary matrix, where $\oplus$ denotes the direct sum product. If U is a 2×2 unitary matrix and V is an 2×2 unitary matrix, then $U \star V$ is a unitary matrix, where $\star$ denotes the star operation. The Fourier matrices and the permutation matrices are important special cases of unitary matrices. The square roots of a unitary matrix are unitary again. Any unitary $n \times n$ matrix is conjugate to a diagonal matrix of the form

$$\text{diag}(\, e^{i\phi_1} \ e^{i\phi_2} \ \ldots \ e^{i\phi_n} \,).$$

Problem 1.   (i) Let $\phi_1, \phi_2 \in \mathbb{R}$. Are the 2×2 matrices

$$U_1 = \begin{pmatrix} e^{i\phi_1} & 0 \\ 0 & e^{i\phi_2} \end{pmatrix}, \qquad U_2 = \begin{pmatrix} 0 & e^{i\phi_1} \\ e^{i\phi_2} & 0 \end{pmatrix}$$

unitary?
(ii) Are the 2×2 matrices

$$V_1 = \begin{pmatrix} 1/2 & \sqrt{3}/2 \\ \sqrt{3}/2 & -1/2 \end{pmatrix}, \qquad V_2 = \begin{pmatrix} 1/2 & -\sqrt{3}/2 \\ -\sqrt{3}/2 & -1/2 \end{pmatrix}$$

unitary?
(iii) Let $\theta, \phi \in \mathbb{C}$. Is the 2×2 matrix

$$U(\theta, \phi) = \begin{pmatrix} \cos(\theta) & e^{i\phi} \sin(\theta) \\ e^{-i\phi} \sin(\theta) & -\cos(\theta) \end{pmatrix}$$

unitary? If so find the inverse.

Solution 1.   (i) Yes. We find $U_1 U_1^* = I_2$ and $U_2 U_2^* = I_2$.
(ii) Yes. We have $V_1 V_1^* = I_2$ and $V_2 V_2^* = I_2$.
(iii) We have

$$U^*(\theta, \phi) = \begin{pmatrix} \cos(\theta) & e^{i\phi} \sin(\theta) \\ e^{-i\phi} \sin(\theta) & -\cos(\theta) \end{pmatrix}$$

and $U(\theta, \phi)U^*(\theta, \phi) = I_2$. Thus $U^*(\theta, \phi) = U^{-1}(\theta, \phi)$ and $U(\theta, \phi)$ is unitary.

Problem 2.   (i) Can one find a 2×2 unitary matrix such that

$$U \begin{pmatrix} -1 & 0 \\ 0 & 1 \end{pmatrix} U^{-1} = \begin{pmatrix} 0 & -1 \\ -1 & 0 \end{pmatrix}?$$

(ii) Find a 2×2 unitary matrix V such that

$$V^* \begin{pmatrix} 0 & 1 \\ 0 & 0 \end{pmatrix} V = \begin{pmatrix} 0 & 0 \\ 1 & 0 \end{pmatrix}.$$

(iii) Can one find a 2×2 unitary matrix W such that

$$W \begin{pmatrix} 1 & 0 \\ 1 & 0 \end{pmatrix} W^{-1} = \begin{pmatrix} 0 & 1 \\ 0 & 1 \end{pmatrix}?$$

Solution 2.   (i) Note that both matrices are hermitian and have the same eigenvalues, namely $+1, -1$. We find the *Hadamard matrix*

$$U = U^{-1} = \frac{1}{\sqrt{2}} \begin{pmatrix} 1 & 1 \\ 1 & -1 \end{pmatrix}.$$

(ii) Let $\phi \in \mathbb{R}$. We find

$$V = e^{i\phi} \begin{pmatrix} 0 & 1 \\ 1 & 0 \end{pmatrix}.$$

(iii) One obtains

$$W = W^{-1} = \begin{pmatrix} 0 & 1 \\ 1 & 0 \end{pmatrix}.$$

Problem 3. Let $0 \le r_{jk} \le 1$ and $\phi_{jk} \in \mathbb{R}$ ($j, k = 1, 2$). Find the conditions on r_{jk}, ϕ_{jk} such that the 2×2 matrix

$$U(r_{jk}, \phi_{jk}) = \begin{pmatrix} r_{11}e^{i\phi_{11}} & r_{12}e^{i\phi_{12}} \\ r_{21}e^{i\phi_{21}} & r_{22}e^{i\phi_{22}} \end{pmatrix}$$

is unitary. Then simplify to the special case $\phi_{jk} = 0$ for $j, k = 1, 2$.

Solution 3. We have

$$U^*(r_{jk}, \phi_{jk}) = \begin{pmatrix} r_{11}e^{-i\phi_{11}} & r_{21}e^{-i\phi_{21}} \\ r_{12}e^{-i\phi_{12}} & r_{22}e^{-i\phi_{22}} \end{pmatrix}.$$

Then from $U(r_{jk}, \phi_{jk})U^*(r_{jk}, \phi_{jk}) = I_2$ we obtain the four conditions

$$r_{11}^2 + r_{12}^2 = 1, \qquad r_{21}^2 + r_{22}^2 = 1,$$

$$r_{11}r_{21}e^{i(\phi_{11}-\phi_{21})} + r_{12}r_{22}e^{i(\phi_{12}-\phi_{22})} = 0,$$

$$r_{11}r_{21}e^{i(\phi_{21}-\phi_{11})} + r_{12}r_{22}e^{i(\phi_{22}-\phi_{12})} = 0.$$

Using addition and subtraction the last two equation can be cast into

$$r_{11}r_{21}\cos(\phi_{11} - \phi_{21}) + r_{12}r_{22}\cos(\phi_{12} - \phi_{22}) = 0$$
$$r_{11}r_{21}\sin(\phi_{11} - \phi_{21}) + r_{12}r_{22}\sin(\phi_{12} - \phi_{22}) = 0.$$

If $\phi_{11} = \phi_{12} = \phi_{21} = \phi_{22} = 0$ we arrive at the three equations

$$r_{11}^2 + r_{12}^2 = 1, \quad r_{21}^2 + r_{22}^2 = 1, \quad r_{11}r_{21} + r_{12}r_{22} = 0.$$

Problem 4. (i) Is the 3×3 matrix

$$U_1 = \begin{pmatrix} -1/\sqrt{2} & 0 & 1/\sqrt{2} \\ -i/\sqrt{2} & 0 & -i/\sqrt{2} \\ 0 & 1 & 0 \end{pmatrix}$$

unitary?

(ii) Is the 3×3 matrix

$$U_2 = \begin{pmatrix} 1/\sqrt{3} & 1/\sqrt{3} & 1/\sqrt{3} \\ 1/\sqrt{6} & 1/\sqrt{6} & -\sqrt{2}/\sqrt{3} \\ 1/\sqrt{2} & -1/\sqrt{2} & 0 \end{pmatrix}$$

a unitary matrix?

(iii) Consider the 3×3 unitary matrix

$$U = \begin{pmatrix} 1/\sqrt{2} & 0 & -1/\sqrt{2} \\ 1/\sqrt{2} & 0 & 1/\sqrt{2} \\ 0 & -1 & 0 \end{pmatrix}$$

and the normalized vector in $\mathbb{C}^3$

$$\mathbf{v} = \frac{1}{\sqrt{3}} \begin{pmatrix} 1 \\ 1 \\ 1 \end{pmatrix}.$$

Find the vectors $U\mathbf{v}$ and $U^2\mathbf{v}$.

Solution 4.   (i) We have $\det(U_1) = i$ and the rows are pairwise orthonormal to each other. Thus the matrix is unitary.
(ii) Yes. We have $U_2U_2^* = I_3$.
(iii) We find the normalized vectors

$$U\mathbf{v} = \frac{1}{\sqrt{3}} \begin{pmatrix} 0 \\ \sqrt{2} \\ -1 \end{pmatrix}, \quad U^2\mathbf{v} = \begin{pmatrix} 1/\sqrt{6} \\ -1/\sqrt{6} \\ -\sqrt{2}/\sqrt{3} \end{pmatrix}.$$

Study $U^n\mathbf{v}$ with $n \to \infty$.

Problem 5.   Consider the orthonormal basis in $\mathbb{C}^3$

$$\mathbf{v}_1 = \frac{1}{\sqrt{2}} \begin{pmatrix} i \\ 0 \\ i \end{pmatrix}, \quad \mathbf{v}_2 = \begin{pmatrix} 0 \\ 1 \\ 0 \end{pmatrix}, \quad \mathbf{v}_3 = \frac{1}{\sqrt{2}} \begin{pmatrix} i \\ 0 \\ -i \end{pmatrix}.$$

(i) Is $U = (\begin{array}{ccc} \mathbf{v}_1 & \mathbf{v}_2 & \mathbf{v}_3 \end{array})$ a unitary matrix?
(ii) Let $\phi_1, \phi_2, \phi_3 \in \mathbb{R}$. Is $V(\phi_1, \phi_2, \phi_3) = e^{i\phi_1}\mathbf{v}_1\mathbf{v}_1^* + e^{i\phi_2}\mathbf{v}_2\mathbf{v}_2^* + e^{i\phi_3}\mathbf{v}_3\mathbf{v}_3^*$ a unitary matrix?

Solution 5.   (i) Yes. We have $UU^* = I_3$.
(ii) Yes. The matrix is given by

$$V(\phi_1, \phi_2, \phi_3) = \begin{pmatrix} e^{i\phi_1}/2 + e^{i\phi_3}/2 & 0 & e^{i\phi_1}/2 - e^{i\phi_3}/2 \\ 0 & e^{i\phi_2} & 0 \\ e^{i\phi_1}/2 - e^{i\phi_3}/2 & 0 & e^{i\phi_1}/2 + e^{i\phi_3}/2 \end{pmatrix}.$$

This is an application of the *spectral theorem*.

Problem 6.   (i) Let A, B be $n \times n$ matrices over $\mathbb{R}$. Show that one can find a $2n \times 2n$ unitary matrix U such that

$$U \begin{pmatrix} A & B \\ -B & A \end{pmatrix} U^* = \begin{pmatrix} A + iB & 0_n \\ 0_n & A - iB \end{pmatrix}.$$

Here 0_n denotes the $n \times n$ zero matrix.

(ii) Use the result from (i) to show that

$$\det \begin{pmatrix} A & B \\ -B & A \end{pmatrix} = \det(A + iB)\overline{\det(A + iB)} \geq 0.$$

Solution 6. (i) Let I_n be the $n \times n$ identity matrix. We find

$$U = \frac{1}{\sqrt{2}} \begin{pmatrix} I_n & -iI_n \\ -iI_n & I_n \end{pmatrix}.$$

(ii) We have

$$\det \begin{pmatrix} A & B \\ -B & A \end{pmatrix} = \det(U \begin{pmatrix} A & B \\ -B & A \end{pmatrix} U^*) = \det \begin{pmatrix} A + iB & 0_n \\ 0_n & A - iB \end{pmatrix}$$
$$= \det(A + iB)\det(A - iB) = \det(A + iB)\overline{\det(A + iB)}$$
$$\geq 0.$$

Note that $A - iB = \overline{A + iB}$ since A and B are real matrices.

Problem 7. Let $\mathbf{v}$ be a column vector in $\mathbb{C}^n$ with $\mathbf{v}^*\mathbf{v} = 1$, i.e. the vector is normalized. Consider the matrix $U = I_n - 2\mathbf{v}\mathbf{v}^*$.
(i) Show that U is hermitian.
(ii) Show that U is unitary.
(iii) Let $n = 2$ and $\mathbf{v}^* = \frac{1}{\sqrt{2}}(1\ 1)$. Find U.

Solution 7. (i) Since $(\mathbf{v}^*)^* = \mathbf{v}$ we have

$$U^* = (I_n - 2\mathbf{v}\mathbf{v}^*)^* = I_n - 2\mathbf{v}\mathbf{v}^* = U.$$

(ii) Since $U = U^*$ and $\mathbf{v}^*\mathbf{v} = 1$ we have

$$UU^* = UU = (I_n - 2\mathbf{v}\mathbf{v}^*)(I_n - 2\mathbf{v}\mathbf{v}^*) = I_n - 2\mathbf{v}\mathbf{v}^* - 2\mathbf{v}\mathbf{v}^* + 4\mathbf{v}\mathbf{v}^* = I_n.$$

(iii) We find

$$U = \begin{pmatrix} 1 & 0 \\ 0 & 1 \end{pmatrix} - \begin{pmatrix} 1 & 1 \\ 1 & 1 \end{pmatrix} = \begin{pmatrix} 0 & -1 \\ -1 & 0 \end{pmatrix}.$$

Problem 8. (i) Let U be a unitary matrix with $U^2 = I_n$. Can we conclude that U is hermitian?
(ii) Let $\alpha \in \mathbb{R}$. Calculate $\exp(\alpha U)$.
(iii) Let $\alpha \in \mathbb{R}$. Consider the 2×2 unitary matrix

$$U_1 = \begin{pmatrix} 0 & -i \\ i & 0 \end{pmatrix}.$$

Calculate $\exp(i\alpha U_1)$.

(iv) Let $\alpha \in \mathbb{R}$. Consider the 2×2 unitary matrix

$$U_2 = \frac{1}{\sqrt{2}} \begin{pmatrix} 1 & 1 \\ 1 & -1 \end{pmatrix}.$$

Calculate $\exp(\alpha U_2)$.

(v) Let σ_1, σ_2 be the Pauli spin matrices. We define the unitary 2×2 matrix

$$\sigma_\phi := \cos(\phi)\sigma_1 + \sin(\phi)\sigma_2 = \begin{pmatrix} 0 & \cos(\phi) - i\sin(\phi) \\ \cos(\phi) + i\sin(\phi) & 0 \end{pmatrix}.$$

Calculate $\exp(-i\theta\sigma_\phi/2)$.

Solution 8. (i) We have $UU^* = UU$. Thus U is hermitian.
(ii) With $U^2 = I_n$ we obtain $\exp(\alpha U) = I_n \cos(\alpha) + U \sin(\alpha)$.
(iii) Using the result from (ii) and $i^2 = -1$, $i(-i) = 1$ we arrive at the *rotation matrix*

$$\exp(i\alpha U_1) = \begin{pmatrix} \cos(\alpha) & \sin(\alpha) \\ -\sin(\alpha) & \cos(\alpha) \end{pmatrix}.$$

(iv) Using the result from (ii) we obtain $\exp(\alpha U_2) = I_2 \cos(\alpha) + U_2 \sin(\alpha)$.
(v) We find

$$\exp(-i\theta\sigma_\phi/2) = \cos(\theta/2)I_2 - i\sin(\theta/2)\sigma_\phi.$$

Problem 9. Consider the two 2×2 unitary matrices

$$U_1 = \begin{pmatrix} 1 & 0 \\ 0 & 1 \end{pmatrix}, \qquad U_2 = \begin{pmatrix} 0 & 1 \\ 1 & 0 \end{pmatrix}.$$

Can one find a unitary 2×2 matrix V such that $U_1 = VU_2V^*$?

Solution 9. Assume that $U_1 = VU_2V^*$ is true. Taking the trace of the left-hand side yields $\text{tr}(U_1) = 2$. However, taking the trace of the right-hand side yields

$$\text{tr}(VU_2V^*) = \text{tr}(U_2) = 0.$$

Thus we have a contradiction. Therefore no such V can be found.

Problem 10. Let U be an $n \times n$ unitary matrix.
(i) Is $U + U^*$ invertible?
(ii) Is $U + U^*$ hermitian?

Solution 10. (i) In general $U + U^*$ is not invertible. For example

$$U = \begin{pmatrix} i & 0 \\ 0 & i \end{pmatrix}$$

yields $U + U^* = 0_2$.
(ii) Obviously $U + U^*$ is hermitian since $(U^*)^* = U$.

Problem 11. Find the condition on the $n \times n$ matrix A over $\mathbb{C}$ such that $I_n + A$ is a unitary matrix.

Solution 11. From $(I_n + A)(I_n + A^*) = I_n$ we obtain the condition $A + A^* + AA^* = 0_n$.

Problem 12. Let $z_1, z_2, w_1, w_2 \in \mathbb{C}$. Consider the 2×2 matrices

$$U = \begin{pmatrix} 0 & z_1 \\ z_2 & 0 \end{pmatrix}, \qquad V = \begin{pmatrix} 0 & w_1 \\ w_2 & 0 \end{pmatrix}$$

where $z_1 \bar{z}_1 = 1$, $z_2 \bar{z}_2 = 1$, $w_1 \bar{w}_1 = 1$, $w_2 \bar{w}_2 = 1$. This means the matrices U, V are unitary. Find the condition on z_1, z_2, w_1, w_2 such that the commutator $[U, V]$ is again a unitary matrix.

Solution 12. We have

$$[U, V] = \begin{pmatrix} 0 & z_1 \\ z_2 & 0 \end{pmatrix} \begin{pmatrix} 0 & w_1 \\ w_2 & 0 \end{pmatrix} = \begin{pmatrix} z_1 w_2 - w_1 z_2 & 0 \\ 0 & w_1 z_2 - w_2 z_1 \end{pmatrix}.$$

Thus the conditions for this matrix to be unitary are

$$z_1 \bar{z}_1 w_2 \bar{w}_2 - z_1 \bar{z}_2 \bar{w}_1 w_2 - \bar{z}_1 z_2 w_1 \bar{w}_2 + z_2 \bar{z}_2 w_1 \bar{w}_1 = 1$$

$$z_2 \bar{z}_2 w_1 \bar{w}_1 - z_2 \bar{z}_1 w_1 \bar{w}_2 - z_1 \bar{z}_2 \bar{w}_1 w_2 - z_1 \bar{z}_1 w_2 \bar{w}_2 = 1.$$

Thus

$$z_1 \bar{z}_2 \bar{w}_1 w_2 + \bar{z}_1 z_2 w_1 \bar{w}_2 = 1, \qquad z_2 \bar{z}_1 w_1 \bar{w}_2 + z_1 \bar{z}_2 \bar{w}_1 w_2 = 1.$$

Using $z_1 = r_1 e^{i\phi}$, $z_2 = r_2 e^{i\phi}$, $w_1 = s_1 e^{i\theta_1}$, $w_2 = s_2 e^{i\theta_2}$ we obtain

$$r_1 r_2 s_1 s_2 \cos(\phi_1 - \phi_2 - \theta_1 + \theta_2) = \frac{1}{2}.$$

Problem 13. Find a unitary 3×3 matrix U with $\det(U) = -1$ such that

$$U \begin{pmatrix} 0 \\ 1 \\ 0 \end{pmatrix} = \frac{1}{\sqrt{2}} \begin{pmatrix} 1 \\ 0 \\ 1 \end{pmatrix}.$$

Solution 13. An example is

$$U = \begin{pmatrix} 1/\sqrt{2} & 1/\sqrt{2} & 0 \\ 0 & 0 & 1 \\ -1/\sqrt{2} & 1/\sqrt{2} & 0 \end{pmatrix}.$$

Problem 14. Show that the two matrices

$$A = \begin{pmatrix} e^{i\theta} & 0 \\ 0 & e^{-i\theta} \end{pmatrix}, \qquad B = \begin{pmatrix} \cos(\theta) & \sin(\theta) \\ -\sin(\theta) & \cos(\theta) \end{pmatrix}$$

are conjugate in the Lie group $SU(2)$.

Solution 14.   Let $U \in SU(2)$ given by

$$U = \frac{1}{\sqrt{2}} \begin{pmatrix} 1 & i \\ i & 1 \end{pmatrix} \Leftrightarrow U^{-1} = \frac{1}{\sqrt{2}} \begin{pmatrix} 1 & -i \\ -i & 1 \end{pmatrix}.$$

Using $e^{i\theta} = \cos(\theta) + i\sin(\theta)$ straightforward calculation yields $UAU^{-1} = B$.

Problem 15.   Let U be an $n \times n$ unitary matrix. Let V be an $n \times n$ unitary matrix such that $V^{-1}UV = D$ is a diagonal matrix D. Is $V^{-1}U^*V$ a diagonal matrix?

Solution 15.   Note that $U^{-1} = U^*$. Then $(V^{-1}UV)^{-1} = V^{-1}U^{-1}V = D^{-1}$. Thus $V^{-1}U^*V$ is a diagonal matrix.

Problem 16.   Let $n \geq 2$ and even. Let U be a unitary antisymmetric $n \times n$ matrix. Show that there exists a unitary matrix V such that

$$V^T U V = \begin{pmatrix} 0 & 1 \\ -1 & 0 \end{pmatrix} \oplus \cdots \oplus \begin{pmatrix} 0 & 1 \\ -1 & 0 \end{pmatrix}$$

where $\oplus$ denotes the direct sum.

Solution 16.   The unitary matrix U can be written as $U = S_1 + iS_2$ with real and antisymmetric S_1 and S_2. Therefore

$$I_n = U^*U = -S_1^2 - S_2^2 - i[S_1, S_2].$$

It follows that $[S_1, S_2] = 0_n$ and there is an orthogonal $n \times n$ matrix O such that

$$O^T S_j O = \begin{pmatrix} 0 & a_1^{(j)} \\ -a_1^{(j)} & 0 \end{pmatrix} \oplus \cdots \oplus \begin{pmatrix} 0 & a_{n/2}^{(j)} \\ -a_{n/2}^{(j)} & 0 \end{pmatrix}$$

for $j = 1, 2$. Since U is unitary we find $|a_k^{(1)} + ia_k^{(2)}| = 1$, $k = 1, \ldots, n/2$. Hence there exists a diagonal phase matrix P such that $V = OP$.

Problem 17.   Let U be a unitary and symmetric matrix. Show that there exists a unitary and symmetric matrix V such that $U = V^2$.

Solution 17.   We have $U = S_1 + iS_2$, where S_1 and S_2 are real and symmetric. Now S_1 and S_2 commute and therefore can be diagonalized simultaneously by an orthogonal matrix O. Then $P := O^T U O$ is a diagonal matrix and V is given by $V = O\sqrt{P}O^T$.

Problem 18.   Let U be an $n \times n$ unitary matrix. Show that $|\det(U)| = 1$.

Solution 18.   We have

$$1 = \det(I_n) = \det(UU^*) = \det(U)\det(U^*) = \det(U)\overline{\det(U)} = |\det(U)|^2.$$

Since $|\det(U)| \geq 0$ we have $|\det(U)| = 1$.

Problem 19. Let A be an $n \times n$ matrix over $\mathbb{R}$. Show that if λ is an eigenvalue of A, then $\bar{\lambda}$ is also an eigenvalue of A. Give an example for a 2×2 matrix, where the eigenvalues are complex.

Solution 19. If A is real, then the characteristic polynomial $p_A(t)$ of A has real coefficients. Thus $\overline{p_A(t)} = p_A(\bar{t})$. If λ is an eigenvalue of A, then from $p_A(\lambda) = 0$ it follows that $\overline{p_A(\lambda)} = 0$ and then $p_A(\bar{\lambda}) = 0$. Hence $\bar{\lambda}$ is also an eigenvalue of A. An example is the 2×2 matrix

$$A = \begin{pmatrix} 0 & 1 \\ -1 & 0 \end{pmatrix}$$

with the eigenvalues i and $-i$.

Problem 20. Let V be an $n \times n$ normal matrix over $\mathbb{C}$. Assume that all its eigenvalues have absolute value of 1, i.e. they are of the form $e^{i\phi}$. Show that V is unitary.

Solution 20. All the eigenvalues of a unitary matrix have absolute value 1. Thus we only need to show the converse of the statement. Since V is normal it is unitarily diagonalizable. This means there is a unitary matrix U such that $U^*VU = \text{diag}(\lambda_1, \ldots, \lambda_n)$, where by assumption $|\lambda_j| = 1$ for $j = 1, \ldots, n$. This implies that

$$U^*VU(U^*VU)^* = I_n.$$

Thus U^*VU is a unitary matrix and it follows that V is a unitary matrix.

Problem 21. Find *square roots* of the Pauli spin matrices

$$\sigma_1 = \begin{pmatrix} 0 & 1 \\ 1 & 0 \end{pmatrix}, \quad \sigma_2 = \begin{pmatrix} 0 & -i \\ i & 0 \end{pmatrix}, \quad \sigma_3 = \begin{pmatrix} 1 & 0 \\ 0 & -1 \end{pmatrix}.$$

Solution 21. First we note that the square root of 1 is $1 = e^0$ and $-1 = e^{i\pi}$. The square root of -1 is $i = e^{i\pi/2}$ and $-i = e^{i3\pi/2} = e^{-i\pi/2}$. For the Pauli spin matrix σ_3 we find

$$R_{30} = \begin{pmatrix} 1 & 0 \\ 0 & i \end{pmatrix}, \quad R_{31} = \begin{pmatrix} -1 & 0 \\ 0 & i \end{pmatrix}, \quad R_{32} = \begin{pmatrix} 1 & 0 \\ 0 & -i \end{pmatrix}, \quad R_{33} = \begin{pmatrix} -1 & 0 \\ 0 & -i \end{pmatrix}.$$

For the square roots of σ_1 we apply the spectral decomposition theorem. For the pairs $(1, i)$, $(1, -i)$, $(-1, i)$, $(-1, -i)$, respectively, we find

$$R_{10} = \frac{1}{2} \begin{pmatrix} 1+i & 1-i \\ 1-i & 1+i \end{pmatrix}, \quad R_{11} = \frac{1}{2} \begin{pmatrix} 1-i & 1+i \\ 1+i & 1-i \end{pmatrix},$$

$$R_{12} = \frac{1}{2} \begin{pmatrix} -1+i & -1-i \\ -1-i & -1+i \end{pmatrix}, \quad R_{13} = \frac{1}{2} \begin{pmatrix} -1-i & -1+i \\ -1+i & -1-i \end{pmatrix}.$$

For the square roots of σ_2 we apply the spectral decomposition theorem. For the pairs $(1, i)$, $(1, -i)$, $(-1, i)$, $(-1, -i)$, respectively, we find

$$R_{20} = \frac{1}{2}\begin{pmatrix} 1+i & -1-i \\ 1+i & 1+i \end{pmatrix}, \quad R_{21} = \frac{1}{2}\begin{pmatrix} 1-i & 1-i \\ -1+i & 1-i \end{pmatrix},$$

$$R_{22} = \frac{1}{2}\begin{pmatrix} -1+i & -1+i \\ 1-i & -1+i \end{pmatrix}, \quad R_{23} = \frac{1}{2}\begin{pmatrix} -1-i & 1+i \\ -1-i & -1-i \end{pmatrix}.$$

Problem 22. Can one find a unitary matrix U with $\det(U) = 1$, i.e. U is an element of $SU(2)$, such that

$$\begin{pmatrix} 1 \\ 0 \end{pmatrix} = U\begin{pmatrix} 0 \\ 1 \end{pmatrix}?$$

Solution 22. The condition

$$\begin{pmatrix} 1 \\ 0 \end{pmatrix} = \begin{pmatrix} u_{11} & u_{12} \\ u_{21} & u_{22} \end{pmatrix}\begin{pmatrix} 0 \\ 1 \end{pmatrix}$$

yields $u_{22} = 0$ and $u_{12} = 1$. Then the conditions that U is unitary and $\det(U) = 1$ provides $u_{11} = 0$ and $u_{21} = -1$.

Problem 23. Can one 2×2 find unitary matrices such that the rows and columns add up to one? Of course the 2×2 identity matrix is one of them.

Solution 23. Another example is

$$U = \frac{1}{2}\begin{pmatrix} 1+i & 1-i \\ 1-i & 1+i \end{pmatrix}.$$

Problem 24. Let U be an $n \times n$ unitary matrix. Then

$$\det(U - \lambda I_n) = \prod_{j=1}^{N}(e^{i\phi_j} - \lambda)$$

where $\phi_j \in [0, 2\pi)$. Let $n = 2$. Find λ for the matrix

$$U = \begin{pmatrix} 0 & -i \\ i & 0 \end{pmatrix}.$$

Then calculate $|e^{i\phi_1} - e^{i\phi_2}|$.

Solution 24. We have

$$\lambda^2 - 1 = \lambda^2 - \lambda(e^{i\phi_1} - e^{i\phi_2}) + e^{i(\phi_1 + \phi_2)}.$$

Thus $e^{i\phi_1} + e^{i\phi_2} = 0$, $e^{i(\phi_1+\phi_2)} = -1$. Consequently $\phi_1 = 0$, $\phi_2 = \pi$ or $\phi_1 = \pi$, $\phi_2 = 0$. Using this result we obtain $|e^{i\phi_1} - e^{i\phi_2}| = 2$.

Problem 25. Find a unitary 2×2 matrix such that $\sigma_2 U - U\sigma_3 = 0_2$.

Solution 25. An example is

$$U = \frac{1}{\sqrt{2}} \begin{pmatrix} i & 1 \\ -1 & -i \end{pmatrix}.$$

Problem 26. The Pauli spin matrices σ_1, σ_2, σ_3 are hermitian and unitary. Together with the 2×2 identity matrix $\sigma_0 = I_2$ they form an orthogonal basis in Hilbert space of the 2×2 matrices over $\mathbb{C}$ with the scalar product $\text{tr}(AB^*)$. Let X be an $n \times n$ hermitian matrix. Then $(X + iI_n)^{-1}$ exists and

$$U = (X - iI_n)(X + iI_n)^{-1}$$

is unitary. This is the *Cayley transform* of X. Find the Cayley transform of the Pauli spin matrix σ_2.

Solution 26. We have

$$\sigma_2 - iI_2 = \begin{pmatrix} -i & -i \\ i & -i \end{pmatrix}, \quad \sigma_2 + iI_2 = \begin{pmatrix} i & -i \\ i & i \end{pmatrix}, \quad (\sigma_2 + iI_2)^{-1} = \frac{1}{2}\begin{pmatrix} -i & -i \\ i & -i \end{pmatrix}.$$

Thus

$$U_2 = \begin{pmatrix} 0 & -1 \\ 1 & 0 \end{pmatrix}.$$

We see that U_2 is not hermitian anymore, but skew-hermitian. For the identity matrix we find

$$U_0 = \begin{pmatrix} -i & 0 \\ 0 & -i \end{pmatrix} = -iI_2.$$

Problem 27. Let A be an $n \times n$ hermitian matrix. Then $(A + iI_n)^{-1}$ exists and

$$U = (A - iI_n)(A + iI_n)^{-1} \tag{1}$$

is a unitary matrix (*Cayley transform* of A).
(i) Show that $+1$ cannot be an eigenvalue of U.
(ii) Show that $A = i(U + I_n)(I_n - U)^{-1}$.

Solution 27. (i) Assume that $+1$ is an eigenvalue, i.e. $U\mathbf{x} = \mathbf{x}$, where $\mathbf{x}$ is the eigenvector. Then

$$(A - iI_n)(A + iI_n)^{-1}\mathbf{x} = \mathbf{x} = I_n\mathbf{x} = (A + iI_n)(A + iI_n)^{-1}\mathbf{x}.$$

Thus

$$2iI_n(A + iI_n)^{-1}\mathbf{x} = 0$$

or $(A + iI_n)^{-1}\mathbf{x} = \mathbf{0}$. Therefore $\mathbf{x} = \mathbf{0}$ (contradiction).

(ii) We set $\mathbf{y} = (A + iI_n)^{-1}\mathbf{x}$, where $\mathbf{x} \in \mathbb{C}^n$ is arbitrary. Then $\mathbf{x} = (A + iI_n)\mathbf{y}$. From (1) we obtain

$$U\mathbf{x} = (A - iI_n)(A + iI_n)^{-1}\mathbf{x}$$

and therefore $U\mathbf{x} = (A - iI_n)\mathbf{y}$. Adding and subtracting the equation $\mathbf{x} = (A + iI_n)\mathbf{y}$ yields the two equations

$$(U + I_n)\mathbf{x} = 2A\mathbf{y}, \qquad (I_n - U)\mathbf{x} = 2iI_n\mathbf{y}.$$

Thus $(U + I_n)\mathbf{x} = -iA(I_n - U)\mathbf{x}$. It follows that

$$(U + I_n) = -iA(I_n - U) \Rightarrow (U + I_n)(I_n - U)^{-1} = -iA.$$

Therefore $A = i(U + I_n)(I_n - U)^{-1}$.

Problem 28. Find the *Cayley transform* of the hermitian matrix

$$H = \begin{pmatrix} h_{11} & h_{12} \\ \bar{h}_{12} & h_{22} \end{pmatrix}, \qquad h_{11}, h_{22} \in \mathbb{R}, \quad h_{12} \in \mathbb{C}.$$

Solution 28. We have

$$H - iI_2 = \begin{pmatrix} h_{11} - i & h_{12} \\ \bar{h}_{12} & h_{22} - i \end{pmatrix}, \qquad H + iI_2 = \begin{pmatrix} h_{11} + i & h_{12} \\ \bar{h}_{12} & h_{22} + i \end{pmatrix},$$

$$(H + iI_2)^{-1} = \frac{1}{D} \begin{pmatrix} h_{22} + i & -h_{12} \\ -\bar{h}_{12} & h_{11} + i \end{pmatrix}$$

where

$$D \equiv \det(H + iI_2) = (h_{11} + i)(h_{22} + i) - h_{12}\bar{h}_{12}.$$

Consequently the Cayley transform $(H - iI_2)(H + iI_2)^{-1}$ of H is

$$\frac{1}{D} \begin{pmatrix} (h_{11} - i)(h_{22} + i) - h_{12}\bar{h}_{12} & 2ih_{12} \\ 2i\bar{h}_{12} & (h_{11} + i)(h_{22} - i) - h_{12}\bar{h}_{12} \end{pmatrix}.$$

Problem 29. Let $\delta_j, \eta_j \in \mathbb{R}$ with $j = 1, 2, 3$. Any 3×3 unitary symmetric matrix U can be written in the product form

$$U = \begin{pmatrix} e^{i\delta_1} & 0 & 0 \\ 0 & e^{i\delta_2} & 0 \\ 0 & 0 & e^{i\delta_3} \end{pmatrix} \begin{pmatrix} \eta_1 & \gamma_{12} & \gamma_{13} \\ \gamma_{12} & \eta_2 & \gamma_{23} \\ \gamma_{13} & \gamma_{23} & \eta_3 \end{pmatrix} \begin{pmatrix} e^{i\delta_1} & 0 & 0 \\ 0 & e^{i\delta_2} & 0 \\ 0 & 0 & e^{i\delta_3} \end{pmatrix}$$

where $\gamma_{jk} = N_{jk} \exp(i\beta_{jk})$ with $N_{jk}, \beta_{jk} \in \mathbb{R}$. It follows that

$$U_{jj} = \eta_j \exp(2i\delta_j), \qquad U_{jk} = N_{jk} \exp(i(\delta_j + \delta_k + \beta_{jk})).$$

The unitary condition $UU^* = I_3$ provides

$$\sum_{k \neq j}^{3} N_{jk}^2 + \eta_j = 1, \quad j = 1, 2, 3$$

and

$$N_{12}(\eta_1 \exp(i\beta_{12}) + \eta_2 \exp(-i\beta_{12})) = N_{13}N_{23} \exp(i(\pi + \beta_{23} - \beta_{13}))$$

and cyclic $(1 \to 2 \to 3 \to 1)$. Write the unitary symmetric matrix

$$W = \begin{pmatrix} 0 & 0 & i \\ 0 & i & 0 \\ i & 0 & 0 \end{pmatrix}$$

in this form.

Solution 29. Owing to the form of W we can start from the ansatz

$$W = \begin{pmatrix} 1 & 0 & 0 \\ 0 & e^{i\alpha} & 0 \\ 0 & 0 & 1 \end{pmatrix} \begin{pmatrix} 0 & 0 & i \\ 0 & 1 & 0 \\ i & 0 & 0 \end{pmatrix} \begin{pmatrix} 1 & 0 & 0 \\ 0 & e^{i\alpha} & 0 \\ 0 & 0 & 1 \end{pmatrix}$$

and determine $\alpha \in [0, 2\pi)$. We obtain the equation $\cos(2\alpha) + i\sin(2\alpha) = i$ with the solution $\alpha = \pi/4$.

Problem 30. (i) Consider the matrix

$$R = \begin{pmatrix} 1/2 & 1/2 & 1/\sqrt{2} \\ 1/2 & 1/2 & -1/\sqrt{2} \\ 1/\sqrt{2} & -1/\sqrt{2} & 0 \end{pmatrix}.$$

Show that $R^{-1} = R^* = R$. Use these properties and $\operatorname{tr}(R)$ to find all the eigenvalues of R. Find the eigenvectors.
(ii) Let

$$A_1 = \begin{pmatrix} 0 & 1 & 0 \\ 1 & 0 & 0 \\ 0 & 0 & 1 \end{pmatrix}, \qquad A_2 = \begin{pmatrix} 0 & 0 & 1 \\ 0 & 0 & 1 \\ 1 & 1 & 0 \end{pmatrix}.$$

Calculate RA_1R^{-1} and RA_2R^{-1}. Discuss.

Solution 30. (i) The matrix R is hermitian and unitary. Thus the eigenvalues can only the $+1$ and -1. Together with $\operatorname{tr}(R) = 1$ we obtain the eigenvalues $+1$ (twice) and -1.
(ii) We obtain

$$RA_1R^{-1} = \begin{pmatrix} 1 & 0 & 0 \\ 0 & 1 & 0 \\ 0 & 0 & -1 \end{pmatrix}, \qquad RA_2R^{-1} = \begin{pmatrix} \sqrt{2} & 0 & 0 \\ 0 & -\sqrt{2} & 0 \\ 0 & 0 & 0 \end{pmatrix}.$$

Supplementary Problems

Problem 1. Let U be a unitary 2×2 matrix. Can the matrix be reconstructed from $\operatorname{tr}(U)$, $\operatorname{tr}(U^2)$, $\operatorname{tr}(U^3)$?

Problem 2.   Consider the Hadamard matrix

$$U = \frac{1}{\sqrt{2}} \begin{pmatrix} 1 & 1 \\ 1 & -1 \end{pmatrix}$$

which is a unitary and hermitian matrix with eigenvalues $+1$ and -1.   We multiply each column with a phase factor $e^{i\phi_j}$ and obtain the matrix

$$V(\phi_1, \phi_2) = \frac{1}{\sqrt{2}} \begin{pmatrix} e^{i\phi_1} & e^{i\phi_2} \\ e^{i\phi_1} & -e^{i\phi_2} \end{pmatrix}.$$

Is the matrix still unitary? If not find the conditions on ϕ_1 and ϕ_2 such that $V(\phi_1, \phi_2)$ is unitary.

Problem 3.   Can one find a θ such that

$$\begin{pmatrix} \cos(\theta)e^{-i\theta} & i\sin(\theta)e^{-i\theta} \\ i\sin(\theta)e^{-i\theta} & \cos(\theta)e^{-i\theta} \end{pmatrix} = \frac{1}{2} \begin{pmatrix} 1+i & 1-i \\ 1-i & 1+i \end{pmatrix}?$$

Problem 4.   (i) Is the 2×2 matrix

$$U = \begin{pmatrix} u_1 + iu_2 & u_3 + iu_4 \\ -u_3 + iu_4 & u_1 - iu_2 \end{pmatrix}$$

with $u_1 = \cos(\alpha)$, $u_2 = \sin(\alpha)\cos(\beta)$ and

$$u_3 = \sin(\alpha)\sin(\beta)\cos(\gamma), \quad u_4 = \sin(\alpha)\sin(\beta)\sin(\gamma)$$

unitary?

(ii) Is the 3×3 matrix

$$U = \frac{1}{\sqrt{3}} \begin{pmatrix} 1 & 1 & 1 \\ i & i & -i \\ i & -i & -i \end{pmatrix}$$

unitary?

(iii) Show that the 3×3 matrix (*Fourier matrix*)

$$V = \frac{1}{\sqrt{3}} \begin{pmatrix} 1 & 1 & 1 \\ 1 & \exp(i2\pi/3) & \exp(i4\pi/3) \\ 1 & \exp(i4\pi/3) & \exp(i2\pi/3) \end{pmatrix}$$

is unitary.

(iv) Are the 4×4 matrices

$$A = \frac{1}{2} \begin{pmatrix} 1 & 1 & 1 & 1 \\ -1 & 1 & -1 & 1 \\ -1 & -1 & 1 & 1 \\ 1 & -1 & -1 & 1 \end{pmatrix}, \quad B = \frac{1}{2} \begin{pmatrix} 1 & 1 & 1 & 1 \\ 1 & -1 & 1 & -1 \\ 1 & 1 & -1 & -1 \\ -1 & 1 & 1 & -1 \end{pmatrix}$$

unitary?

(v) Show that the 4×4 matrices

$$X = \frac{1}{\sqrt{2}} \begin{pmatrix} 1 & 0 & 0 & -i \\ 0 & 1 & i & 0 \\ 0 & -i & -1 & 0 \\ i & 0 & 0 & -1 \end{pmatrix}, \quad Y = \frac{1}{\sqrt{2}} \begin{pmatrix} 0 & -1 & 0 & 1 \\ 1 & 0 & -1 & 0 \\ 0 & 1 & 0 & 1 \\ 1 & 0 & 1 & 0 \end{pmatrix}.$$

are unitary.

(vi) Let

$$\sigma := \frac{1}{2}(1 - \sqrt{5}), \qquad \tau := \frac{1}{2}(1 + \sqrt{5})$$

with τ the *golden mean number*. Are the 4×4 matrices

$$U_1 = -\frac{1}{2} \begin{pmatrix} 1 & -\tau & -\sigma & 0 \\ \tau & 1 & 0 & \sigma \\ \sigma & .0 & 1 & -\tau \\ 0 & -\sigma & \tau & 1 \end{pmatrix}, \quad U_2 = -\frac{1}{2} \begin{pmatrix} 1 & -\tau & -\sigma & 0 \\ \tau & 1 & 0 & -\sigma \\ \sigma & 0 & 1 & \tau \\ 0 & \sigma & -\tau & 1 \end{pmatrix}$$

unitary? Prove or disprove.

(vii) Show that the $2n \times 2n$ matrix

$$V = \frac{1}{\sqrt{2}} \begin{pmatrix} I_n & iI_n \\ I_n & -iI_n \end{pmatrix}$$

is unitary.

Problem 5. Let A be an $n \times n$ hermitian matrix with $A^2 = I_n$.
(i) Show that $U = \frac{1}{\sqrt{2}}(I_n + iA)$ is unitary.
(ii) Show that $U = \frac{1}{\sqrt{2}}(I_n - iA)$ is unitary.
(iii) Let $\phi \in [0, 2\pi)$. Show that the condition on ϕ such that

$$V(\phi) = \frac{1}{\sqrt{2}}(e^{i\phi} I_n + e^{-i\phi} A)$$

is unitary is given by $\phi = \pi/2$ and $\phi = 3\pi/2$.

Problem 6. Let I_n be the $n \times n$ identity matrix and J_n be the $n \times n$ *backward identity matrix* i.e. the entries are 1 at the counter diagonal and 0 otherwise. Are the matrices

$$U_1 = \frac{1}{\sqrt{2}}(I_n + iJ_n), \quad U_2 = \frac{1}{\sqrt{2}}(I_n - iJ_n), \quad U_3 = \frac{1}{\sqrt{2}}((1+i)I_n + (1-i)J_n)$$

unitary?

Problem 7. Let U be a unitary 4×4 matrix. Assume that each column (vector) can be written as the Kronecker product of two vectors in $\mathbb{C}^2$. Can we conclude that the eigenvectors of such a unitary matrix can also be written a Kronecker product of two vectors in $\mathbb{C}^2$?

Problem 8.   Find all 2×2 hermitian and unitary matrices A, B such that $AB = e^{i\pi} BA$.

Problem 9.   (i) Find the conditions on $u_{11}, u_{12}, u_{21}, u_{22}, u_{23}, u_{32}, u_{33} \in \mathbb{C}$ such that the 3×3 matrix

$$U = \begin{pmatrix} u_{11} & u_{12} & 0 \\ u_{21} & u_{22} & u_{23} \\ 0 & u_{32} & u_{33} \end{pmatrix}$$

is unitary.

(ii) Let $1 \geq r_{jk} \geq 0$, $\phi_{jk} \in \mathbb{R}$ $(j, k = 1, 2, 3)$. Find the conditions on r_{jk} and ϕ_{jk} such that

$$\begin{pmatrix} r_{11}e^{i\phi_{11}} & r_{12}e^{i\phi_{12}} & r_{13}e^{i\phi_{13}} \\ r_{21}e^{i\phi_{21}} & r_{22}e^{i\phi_{22}} & r_{23}e^{i\phi_{23}} \\ r_{31}e^{i\phi_{31}} & r_{32}e^{i\phi_{32}} & r_{33}e^{i\phi_{33}} \end{pmatrix}$$

is a unitary matrix.

Problem 10.   Find all $(n+1) \times (n+1)$ matrices A such that $A^*UA = U$, where U is the unitary matrix

$$U = \begin{pmatrix} 0 & 0 & i \\ 0_{n-1} & I_{n-1} & 0_{n-1} \\ -i & 0 & 0 \end{pmatrix}$$

and $\det(A) = 1$. Consider first the case $n = 2$. For $n = 2$ we have

$$A^* \begin{pmatrix} 0 & 0 & i \\ 0 & 1 & 0 \\ -i & 0 & 0 \end{pmatrix} A = \begin{pmatrix} 0 & 0 & i \\ 0 & 1 & 0 \\ -i & 0 & 0 \end{pmatrix}$$

and find nine equations.

Problem 11.   Consider $n \times n$ unitary matrices. A scalar product of two $n \times n$ matrices U, V can be defined as

$$\langle U, V \rangle := \frac{1}{n} \mathrm{tr}(UV^*).$$

Find two 2×2 unitary matrices U, V such that $\langle U, V \rangle = \frac{1}{2}$.

Problem 12.   Let U be a unitary matrix. Then the determinant of U must be of the form $e^{i\phi}$. Find the determinant of $U + U^*$.

Problem 13.   (i) Consider the unitary matrix

$$Y = \begin{pmatrix} 0 & 1 \\ -1 & 0 \end{pmatrix} \equiv i\sigma_2.$$

Find all 2×2 matrices S such that $YSY^{-1} = S^T$, $S = S^*$.

(ii) Can one find a 2×2 unitary matrix U such that

$$\begin{pmatrix} \cos(\theta) & e^{i\phi}\sin(\theta) \\ e^{-i\phi}\sin(\theta) & -\cos(\theta) \end{pmatrix} = U \begin{pmatrix} 1 & 0 \\ 0 & -1 \end{pmatrix} U^* \ ?$$

Problem 14. Let U be an $n \times n$ unitary matrix and A an arbitrary $n \times n$ matrix. Then we know that $Ue^A U^{-1} = e^{UAU^{-1}}$. Calculate $Ue^A U$ with $U \neq U^{-1}$.

Problem 15. Consider the unit vectors in $\mathbb{C}^3$

$$\mathbf{z} = \begin{pmatrix} z_1 \\ z_2 \\ z_3 \end{pmatrix}, \qquad \mathbf{w} = \begin{pmatrix} w_1 \\ w_2 \\ w_3 \end{pmatrix}$$

i.e. $|z_1|^2 + |z_2|^2 + |z_3|^2 = 1$, $|w_1|^2 + |w_2|^2 + |w_3|^2 = 1$. Assume that (complex unit cone) $z_1 w_1 + z_2 w_2 + z_3 w_3 = 0$. Show that $U \in SU(3)$ can be written as

$$U = \begin{pmatrix} \bar{z}_1 & \bar{z}_2 & \bar{z}_3 \\ w_1 & w_2 & w_3 \\ u_1 & u_2 & u_3 \end{pmatrix}$$

where $u_j := \sum_{k,\ell=1}^{3} \epsilon_{jk\ell} z_k \bar{w}_\ell$.

Problem 16. (i) Let $\phi_1, \phi_2 \in \mathbb{R}$. Show that

$$U(\phi_1, \phi_2) = \frac{1}{\sqrt{2}} \begin{pmatrix} e^{i\phi_1} & -e^{-i\phi_2} \\ e^{i\phi_2} & e^{-i\phi_1} \end{pmatrix}$$

is unitary. Is $U(\phi_1, \phi_2)$ an element of $SU(2)$? Find the eigenvalues and eigenvectors of $U(\phi_1, \phi_2)$. Do the eigenvalues and eigenvectors depend on the ϕ's? Show that

$$V(\phi_1, \phi_2) = \frac{1}{\sqrt{2}} \begin{pmatrix} e^{i\phi_1} & e^{-i\phi_2} \\ e^{i\phi_2} & -e^{-i\phi_1} \end{pmatrix}$$

is unitary. Is $V(\phi_1, \phi_2)$ an element of $SU(2)$? Find the eigenvalues and eigenvectors of $V(\phi_1, \phi_2)$. Do the eigenvalues and eigenvectors depend on the ϕ's?
(ii) Let $\phi_1, \phi_2, \phi_3, \phi_4 \in \mathbb{R}$. Show that

$$U(\phi_1, \phi_2, \phi_3, \phi_4) = \frac{1}{\sqrt{2}} \begin{pmatrix} e^{i\phi_1} & 0 & 0 & -e^{-i\phi_2} \\ 0 & e^{i\phi_3} & -e^{-i\phi_4} & 0 \\ 0 & e^{i\phi_4} & e^{-i\phi_3} & 0 \\ e^{i\phi_2} & 0 & 0 & e^{-i\phi_1} \end{pmatrix}$$

is unitary. Is $U(\phi_1, \phi_2, \phi_3, \phi_4)$ an element of $SU(4)$? Find the eigenvalues and eigenvectors of $U(\phi_1, \phi_2, \phi_3, \phi_4)$. Do the eigenvalues and eigenvectors depend on the ϕ's?
(iii) Let $\phi_1, \phi_2 \in \mathbb{R}$. Show that

$$U(\phi_1, \phi_2) = \frac{1}{\sqrt{2}} \begin{pmatrix} e^{i\phi_1} & 0 & -e^{-i\phi_2} \\ 0 & \sqrt{2} & 0 \\ e^{i\phi_2} & 0 & e^{-i\phi_1} \end{pmatrix}$$

is unitary. Is $U(\phi_1, \phi_2)$ an element of $SU(2)$? Find the eigenvalues and eigenvectors of $U(\phi_1, \phi_2)$. Do the eigenvalues and eigenvectors depend on the ϕ's?

Problem 17. (i) What can be said about the eigenvalues of a matrix which is unitary and hermitian?
(ii) What can be said about the eigenvalues of a matrix which is unitary and skew-hermitian?
(iii) What can be said about the eigenvalues of a matrix which is unitary and $U^T = U$?
(iv) What can be said about the eigenvalues of a matrix which is unitary and $U^3 = U$?

Problem 18. Let n be a prime number and $j, k = 1, \ldots, n$. Show that the matrices

$$(U_1)_{jk} = \frac{1}{\sqrt{n}} \exp((2\pi i/n)(j+k-1)^2)$$

$$\vdots$$

$$(U_r)_{jk} = \frac{1}{\sqrt{n}} \exp((2\pi i/n)r(j+k-1)^2)$$

$$\vdots$$

$$(U_{n-1})_{jk} = \frac{1}{\sqrt{n}} \exp(((2\pi i/n)(n-1)(j+k-1)^2)$$

$$(U_n)_{jk} = \frac{1}{\sqrt{n}} \exp((2\pi i/n)jk)$$

$$(U_{n+1})_{jk} = \delta_{jk}$$

are unitary.

Problem 19. In the Hilbert space $\mathbb{C}^4$ consider the *Bell states*

$$\frac{1}{\sqrt{2}} \begin{pmatrix} 1 \\ 0 \\ 0 \\ 1 \end{pmatrix}, \quad \frac{1}{\sqrt{2}} \begin{pmatrix} 0 \\ 1 \\ 1 \\ 0 \end{pmatrix}, \quad \frac{1}{\sqrt{2}} \begin{pmatrix} 0 \\ 1 \\ -1 \\ 0 \end{pmatrix}, \quad \frac{1}{\sqrt{2}} \begin{pmatrix} 1 \\ 0 \\ 0 \\ -1 \end{pmatrix}.$$

(i) Let $\omega := e^{2\pi i/4}$. Apply the *Fourier transformation*

$$U_F = \frac{1}{2} \begin{pmatrix} 1 & 1 & 1 & 1 \\ 1 & \omega & \omega^2 & \omega^3 \\ 1 & \omega^2 & 1 & \omega^2 \\ 1 & \omega^3 & \omega^2 & \omega \end{pmatrix}$$

to the Bell states and study the entanglement of these states.
(ii) Apply the *Haar wavelet transform*

$$U_H = \frac{1}{2} \begin{pmatrix} 1 & 1 & 1 & 1 \\ 1 & 1 & -1 & -1 \\ \sqrt{2} & -\sqrt{2} & 0 & 0 \\ 0 & 0 & \sqrt{2} & -\sqrt{2} \end{pmatrix}$$

to the Bell states and study the entanglement of these states.
(iii) Apply the *Walsh-Hadamard transform*

$$U_W = \frac{1}{2} \begin{pmatrix} 1 & 1 & 1 & 1 \\ -1 & -1 & 1 & 1 \\ -1 & 1 & 1 & -1 \\ 1 & -1 & 1 & -1 \end{pmatrix}$$

to the Bell states and study the entanglement of these states.
Extend to the Hilbert space $\mathbb{C}^{2^n}$ with the first Bell state given by

$$\frac{1}{\sqrt{2}} \begin{pmatrix} 1 & 0 & \cdots & 0 & 1 \end{pmatrix}^T.$$

Problem 20. Let U be an $n \times n$ unitary matrix. Then $U + U^*$ is a hermitian matrix. Can any hermitian matrix represented in this form? For example for the $n \times n$ zero matrix we have $U = \text{diag}(i, i, \ldots, i)$.

Problem 21. Let $\{ |a_0\rangle, |a_1\rangle, \ldots, |a_{n-1}\rangle \}$ be an orthonormal basis in the Hilbert space $\mathbb{C}^n$. The *discrete Fourier transform*

$$|b_j\rangle = \frac{1}{\sqrt{n}} \sum_{k=0}^{n-1} \omega^{jk} |a_k\rangle, \qquad j = 0, 1, \ldots, n-1$$

where $\omega := \exp(2\pi i/n)$ is the primitive n-th root of unity.
(i) Apply the discrete Fourier transform to the standard basis in $\mathbb{C}^4$

$$\begin{pmatrix} 1 \\ 0 \\ 0 \\ 0 \end{pmatrix}, \quad \begin{pmatrix} 0 \\ 1 \\ 0 \\ 0 \end{pmatrix}, \quad \begin{pmatrix} 0 \\ 0 \\ 1 \\ 0 \end{pmatrix}, \quad \begin{pmatrix} 0 \\ 0 \\ 0 \\ 1 \end{pmatrix}.$$

(ii) Apply the discrete Fourier transform to the Bell basis in $\mathbb{C}^4$

$$\frac{1}{\sqrt{2}} \begin{pmatrix} 1 \\ 0 \\ 0 \\ 1 \end{pmatrix}, \quad \frac{1}{\sqrt{2}} \begin{pmatrix} 1 \\ 0 \\ 0 \\ -1 \end{pmatrix}, \quad \frac{1}{\sqrt{2}} \begin{pmatrix} 0 \\ 1 \\ 1 \\ 0 \end{pmatrix}, \quad \frac{1}{\sqrt{2}} \begin{pmatrix} 0 \\ 1 \\ -1 \\ 0 \end{pmatrix}.$$

Problem 22. Consider the 3×3 unitary matrices

$$U_1(\phi_1, \phi_2, \phi_3) = \begin{pmatrix} e^{i\phi_1} & 0 & 0 \\ 0 & e^{i\phi_2} & 0 \\ 0 & 0 & e^{i\phi_3} \end{pmatrix}, \quad U_2(\phi_4, \phi_5, \phi_6) = \begin{pmatrix} 0 & 0 & e^{i\phi_4} \\ 0 & e^{i\phi_5} & 0 \\ e^{i\phi_6} & 0 & 0 \end{pmatrix}.$$

What is the condition on $\phi_1, \ldots, \phi_6$ such that

$$[U_1(\phi_1, \phi_2; \phi_3), U_2(\phi_1, \phi_2, \phi_3)] = 0_3 \, ?$$

Problem 23.   Let U be a unitary matrix with $U = U^T$. Show that U can be written as $U = V^T V$, where V is unitary. Note that $(V^T V)^T = V^T V$.

Problem 24.   Let $n \geq 2$. Consider the $n \times n$ matrix (counting from $(0,0)$)

$$(a_{jk}) = \frac{1}{\sqrt{n}} (e^{i\pi(j-k)})$$

where $j, k = 0, 1, \ldots, n-1$. Is the matrix unitary? Study first the cases $n = 2$ and $n = 3$.

Problem 25.   Consider the *rotation matrix*

$$R(\alpha) = \begin{pmatrix} \cos(\alpha) & -\sin(\alpha) \\ \sin(\alpha) & \cos(\alpha) \end{pmatrix}.$$

Find all 2×2 matrices

$$M(\beta) = \begin{pmatrix} f_{11}(\beta) & f_{12}(\beta) \\ f_{21}(\beta) & f_{22}(\beta) \end{pmatrix}$$

such that $R(\alpha)M(\beta)R^{-1}(\alpha) = M(\beta + \alpha)$.

Problem 26.   (i) Consider an $n \times n$ unitary matrix $U = (u_{jk})$ with $j, k = 1, \ldots, n$. Show that

$$S = (s_{jk}) = (u_{jk}\bar{u}_{jk})$$

is a double stochastic matrix.
(ii) Given a double stochastic $n \times n$ matrix S. Can we construct the unitary matrix U which generates the double stochastic matrix as described in (i).

Problem 27.   Let σ_1, σ_2, σ_3 be the Pauli spin matrices and $\alpha_j \in \mathbb{R}$ for $j = 1, 2, 3$.
(i) Find the 2×2 matrices

$$U_k := (I_2 - i\alpha_k\sigma_k)(I_2 + i\alpha_k\sigma_k)^{-1}$$

where $k = 1, 2, 3$. Are the matrices unitary?
(ii) Find the 4×4 matrices

$$V_k := (I_4 - i\alpha_k\sigma_k \otimes \sigma_k)(I_4 + i\alpha_k\sigma_k \otimes \sigma_k)^{-1}$$

where $k = 1, 2, 3$. Are the matrices unitary?

Problem 28.   Let U be a unitary matrix. Calculate $\exp(\epsilon(U + U^*))$, where $\epsilon \in \mathbb{R}$. Note that

$$(U + U^*)^2 = U^2 + (U^*)^2 + 2I_n, \quad (U + U^*)^3 = U^3 + (U^*)^3 + 3U + 3U^*.$$

Problem 29.   Let U, V be $n \times n$ unitary matrices. Is the $2n \times 2n$ matrix

$$\begin{pmatrix} 0_n & U \\ V & 0_n \end{pmatrix}$$

unitary?

Problem 30. Let $\phi \in \mathbb{R}$ and U be an 2×2 unitary matrix. Show that the 4×4 matrix

$$V = \begin{pmatrix} 0 & 0 \\ 0 & 1 \end{pmatrix} \otimes U + \begin{pmatrix} e^{i\phi} & 0 \\ 0 & 0 \end{pmatrix} \otimes I_2$$

is unitary. Note that $V = e^{i\phi} I_2 \oplus U$.

Chapter 18

Groups, Lie Groups and Matrices

A *group* G is a set of objects $\{a, b, c, \ldots\}$ (not necessarily countable) together with a binary operation which associates with any ordered pair of elements a, b in G a third element ab in G (closure). The binary operation (called group multiplication) is subject to the following requirements:

1) There exists an element e in G called the *identity element* such that $eg = ge = g$ for all $g \in G$.
2) For every $g \in G$ there exists an *inverse element* g^{-1} in G such that $gg^{-1} = g^{-1}g = e$.
3) *Associative law.* The identity $(ab)c = a(bc)$ is satisfied for all $a, b, c \in G$.

If $ab = ba$ for all $a, b \in G$ we call the group *commutative*.

If G has a finite number of elements it has *finite order* $n(G)$, where $n(G)$ is the number of elements. Otherwise, G has infinite order.

Groups have matrix representations with invertible $n \times n$ matrices and matrix multiplication as group multiplication. The identity element is the identity matrix. The inverse element is the inverse matrix. An important subgroup is the set of unitary matrices, where $U^* = U^{-1}$.

Let $(G_1, *)$ and $(G_2, \circ)$ be groups. A function $f : G_1 \to G_2$ with

$$f(a * b) = f(a) \circ f(b), \qquad \text{for all } a, b \in G_1$$

is called a *homomorphism*.

Problem 1. Let $\phi \in \mathbb{R}$. Do the 2×2 matrices

$$I_2 = \begin{pmatrix} 1 & 0 \\ 0 & 1 \end{pmatrix}, \qquad U(\phi) = \begin{pmatrix} 0 & e^{i\phi} \\ e^{-i\phi} & 0 \end{pmatrix}$$

form a group under matrix multiplication?

Solution 1. Since $U(\phi)U(\phi) = I_2$ we have a commutative group with two elements. Note that $U(\phi) = U^*(\phi) = U^{-1}(\phi)$.

Problem 2. (i) Show that the 2×2 matrix

$$U = \begin{pmatrix} 0 & 1 \\ i & 0 \end{pmatrix}$$

is unitary.
(ii) Find the eigenvalues and normalized eigenvectors of U.
(iii) Find the group generated by U.
(iv) Find the group generated by $U \otimes U$.

Solution 2. (i) We have

$$UU^* = \begin{pmatrix} 0 & 1 \\ i & 0 \end{pmatrix} \begin{pmatrix} 0 & -i \\ 1 & 0 \end{pmatrix} = I_2.$$

(ii) The eigenvalues of U are $e^{i\pi/4}$, $e^{i\pi 5/4} \equiv -e^{i\pi/4}$ with the corresponding normalized eigenvectors are

$$\mathbf{v}_1 = \frac{1}{\sqrt{2}} \begin{pmatrix} 1 \\ e^{i\pi/4} \end{pmatrix}, \qquad \mathbf{v}_2 = \frac{1}{\sqrt{2}} \begin{pmatrix} 1 \\ e^{i\pi 5/4} \end{pmatrix}.$$

(iii) We obtain $U^8 = I_2$ with

$$U^2 = iI_2, \quad U^3 = \begin{pmatrix} 0 & i \\ -1 & 0 \end{pmatrix} = -U^*, \quad U^4 = -I_2,$$

$$U^5 = \begin{pmatrix} 0 & -1 \\ -i & 0 \end{pmatrix}, \quad U^6 = -iI_2, \quad U^7 = \begin{pmatrix} 0 & -i \\ 1 & 0 \end{pmatrix}.$$

Hence the (commutative) group consists of 8 elements.
(iv) We have

$$(U \otimes U)(U \otimes U) = (U^2 \otimes U^2) = iI_2 \otimes iI_2 = -I_4.$$

Thus $(U \otimes U)^4 = I_4$. Hence the group consists of the four elements $U \otimes U$, $(U \otimes U)^2$, $(U \otimes U)^3$, $(U \otimes U)^4 = I_4$.

Problem 3. Find the group generated by the 3×3 matrices

$$G_1 = \begin{pmatrix} 1 & 0 & 0 \\ 0 & -1 & 0 \\ 0 & 0 & -1 \end{pmatrix}, \qquad G_2 = \begin{pmatrix} 0 & 0 & 1 \\ 1 & 0 & 0 \\ 0 & 1 & 0 \end{pmatrix}.$$

Set $G_0 = G_1^2 = I_3$.

Solution 3.   Note that $G_2^3 = I_3$ and $G_1^2 = I_3$. We find the *tetrahedral group* of order 12.

Problem 4.   (i) Consider the permutation group S_3 which consists of 6 elements. The matrix representation of the permutations (12) and (13) are

$$(12) \mapsto \begin{pmatrix} 0 & 1 & 0 \\ 1 & 0 & 0 \\ 0 & 0 & 1 \end{pmatrix}, \qquad (13) \mapsto \begin{pmatrix} 0 & 0 & 1 \\ 0 & 1 & 0 \\ 1 & 0 & 0 \end{pmatrix}.$$

Can the other permutations of S_3 be constructed from matrix products of these two matrices?

(ii) Consider the two 3×3 permutation matrices

$$C_1 = \begin{pmatrix} 0 & 1 & 0 \\ 0 & 0 & 1 \\ 1 & 0 & 0 \end{pmatrix}, \qquad A = \begin{pmatrix} 0 & 0 & 1 \\ 0 & 1 & 0 \\ 1 & 0 & 0 \end{pmatrix}.$$

Can the remaining four 3×3 matrices be generated from C_1 and A using matrix multiplication?

Solution 4.   (i) Yes. All six permutation matrices can be generated from the two matrices.

(ii) Yes. We have $A^2 = I_3$ and

$$C_1^2 = \begin{pmatrix} 0 & 0 & 1 \\ 1 & 0 & 0 \\ 0 & 1 & 0 \end{pmatrix}, \qquad C_1^3 = \begin{pmatrix} 1 & 0 & 0 \\ 0 & 1 & 0 \\ 0 & 0 & 1 \end{pmatrix},$$

$$AC_1 = \begin{pmatrix} 1 & 0 & 0 \\ 0 & 0 & 1 \\ 0 & 1 & 0 \end{pmatrix}, \qquad AC_2 = \begin{pmatrix} 0 & 1 & 0 \\ 1 & 0 & 0 \\ 0 & 0 & 1 \end{pmatrix}.$$

Note that C_1, C_1^2, C_1^3 forms a subgroup under matrix multiplication.

Problem 5.   Find the group generated by the two 2×2 matrices

$$\sigma_3 = \begin{pmatrix} 1 & 0 \\ 0 & -1 \end{pmatrix}, \qquad R = \frac{1}{2}\begin{pmatrix} -1 & -\sqrt{3} \\ \sqrt{3} & -1 \end{pmatrix}$$

under matrix multiplication. Is the group commutative?

Solution 5.   With $\sigma_3^2 = I_2$ we find the group elements $\sigma_3, R, R^2, RS, SR, I_2$. The group is not commutative. Note that $R\sigma_3 \neq \sigma_3 R$.

Problem 6.   (i) Show that the set of matrices

$$E = \begin{pmatrix} 1 & 0 \\ 0 & 1 \end{pmatrix}, \quad C_3 = \begin{pmatrix} -1/2 & -\sqrt{3}/2 \\ \sqrt{3}/2 & -1/2 \end{pmatrix}, \quad C_3^{-1} = \begin{pmatrix} -1/2 & \sqrt{3}/2 \\ -\sqrt{3}/2 & -1/2 \end{pmatrix}$$

$$\sigma_1 = \begin{pmatrix} 1 & 0 \\ 0 & -1 \end{pmatrix}, \quad \sigma_2 = \begin{pmatrix} -1/2 & -\sqrt{3}/2 \\ -\sqrt{3}/2 & 1/2 \end{pmatrix}, \quad \sigma_3 = \begin{pmatrix} -1/2 & \sqrt{3}/2 \\ \sqrt{3}/2 & 1/2 \end{pmatrix}$$

form a group G under matrix multiplication, where C_3^{-1} is the inverse matrix of C_3.

(ii) Find the determinant of all these matrices. Does the set of numbers

$$\{\, \det(E), \ \det(C_3), \ \det(C_3^{-1}), \ \det(\sigma_1), \ \det(\sigma_2), \ \det(\sigma_3) \,\}$$

form a group under multiplication?

(iii) Find two proper subgroups.

(iv) Find the right coset decomposition. Find the left coset decomposition. We obtain the right *coset decomposition* as follows: Let G be a finite group of order g having a proper subgroup $\mathcal{H}$ of order h. Take some element g_2 of G which does not belong to the subgroup $\mathcal{H}$, and make a right coset $\mathcal{H}g_2$. If $\mathcal{H}$ and $\mathcal{H}g_2$ do not exhaust the group G, take some element g_3 of G which is not an element of $\mathcal{H}$ and $\mathcal{H}g_2$, and make a right coset $\mathcal{H}g_3$. Continue making right cosets $\mathcal{H}g_j$ in this way. If G is a finite group, all elements of G will be exhausted in a finite number of steps and we obtain the right coset decomposition.

Solution 6. (i) Matrix multiplication reveals that the set is closed under matrix multiplication. The neutral element is the 2×2 identity matrix. Each element has exactly one inverse. The associative law holds for $n \times n$ matrices. The *group table* is given by

$\cdot$	E	C_3	C_3^{-1}	σ_1	σ_2	σ_3
E	E	C_3	C_3^{-1}	σ_1	σ_2	σ_3
C_3	C_3	C_3^{-1}	E	σ_3	σ_1	σ_2
C_3^{-1}	C_3^{-1}	E	C_3	σ_2	σ_3	σ_1
σ_1	σ_1	σ_2	σ_3	E	C_3	C_3^{-1}
σ_2	σ_2	σ_3	σ_1	C_3^{-1}	E	C_3
σ_3	σ_3	σ_1	σ_2	C_3	C_3^{-1}	E

(ii) For the determinants we find

$$\det(E) = \det(C_3) = \det(C_3^{-1}) = 1, \quad \det(\sigma_1) = \det(\sigma_2) = \det(\sigma_3) = -1.$$

The set $\{\, +1, -1 \,\}$ forms a commutative group under multiplication.

(iii) From the group table we see that $\{\, E, \sigma_1 \,\}$ is a proper subgroup. Another proper subgroup is $\{\, E, C_3, C_3^{-1} \,\}$.

(iv) We start from the proper subgroup $\mathcal{H} = \{\, E, \sigma_1 \,\}$. If we multiply the elements of $\mathcal{H}$ with σ_2 on the right we obtain the set

$$\mathcal{H}\sigma_2 = \{\, \sigma_2, \sigma_1\sigma_2 \,\} = \{\, \sigma_2, C_3 \,\}.$$

Similarly, we have

$$\mathcal{H}\sigma_3 = \{\, \sigma_3, \sigma_1\sigma_3 \,\} = \{\, \sigma_3, C_3^{-1} \,\}.$$

We see that the six elements of the group G are just exhausted by the three right cosets so that

$$G = \mathcal{H} + \mathcal{H}\sigma_2 + \mathcal{H}\sigma_3, \qquad \mathcal{H} = \{\, E, \sigma_1 \,\}.$$

The left coset decomposition is given by

$$G = \mathcal{H} + \sigma_2\mathcal{H} + \sigma_3\mathcal{H}, \quad \mathcal{H} = \{E, \sigma_1\}$$

with

$$\sigma_2\mathcal{H} = \{\sigma_2, C_3^{-1}\}, \quad \sigma_3\mathcal{H} = \{\sigma_3, C_3\}.$$

Problem 7. We know that the set of 2×2 matrices

$$E = \begin{pmatrix} 1 & 0 \\ 0 & 1 \end{pmatrix}, \quad C_3 = \begin{pmatrix} -1/2 & -\sqrt{3}/2 \\ \sqrt{3}/2 & -1/2 \end{pmatrix}, \quad C_3^{-1} = \begin{pmatrix} -1/2 & \sqrt{3}/2 \\ -\sqrt{3}/2 & -1/2 \end{pmatrix}$$

$$\sigma_1 = \begin{pmatrix} 1 & 0 \\ 0 & -1 \end{pmatrix}, \quad \sigma_2 = \begin{pmatrix} -1/2 & -\sqrt{3}/2 \\ -\sqrt{3}/2 & 1/2 \end{pmatrix}, \quad \sigma_3 = \begin{pmatrix} -1/2 & \sqrt{3}/2 \\ \sqrt{3}/2 & 1/2 \end{pmatrix}$$

forms a group G under matrix multiplication, where C_3^{-1} is the inverse matrix of C_3. The set of matrices (3×3 *permutation matrices*)

$$I = P_0 = \begin{pmatrix} 1 & 0 & 0 \\ 0 & 1 & 0 \\ 0 & 0 & 1 \end{pmatrix}, \quad P_1 = \begin{pmatrix} 0 & 0 & 1 \\ 1 & 0 & 0 \\ 0 & 1 & 0 \end{pmatrix},$$

$$P_2 = \begin{pmatrix} 0 & 1 & 0 \\ 0 & 0 & 1 \\ 1 & 0 & 0 \end{pmatrix}, \quad P_3 = \begin{pmatrix} 1 & 0 & 0 \\ 0 & 0 & 1 \\ 0 & 1 & 0 \end{pmatrix},$$

$$P_4 = \begin{pmatrix} 0 & 0 & 1 \\ 0 & 1 & 0 \\ 1 & 0 & 0 \end{pmatrix}, \quad P_5 = \begin{pmatrix} 0 & 1 & 0 \\ 1 & 0 & 0 \\ 0 & 0 & 1 \end{pmatrix}$$

also forms a group G under matrix multiplication. Are the two groups isomorphic? A homomorphism which is $1 - 1$ and onto is an *isomorphism*.

Solution 7. Using the map

$$E \to I = P_0, \quad C_3 \to P_1, \quad C_3^{-1} \to P_2, \quad \sigma_1 \to P_3, \quad \sigma_2 \to P_4, \quad \sigma_3 \to P_5$$

we see that the two groups are isomorphic.

Problem 8. (i) Show that the 4×4 matrices

$$A = \begin{pmatrix} 1 & 0 & 0 \\ 0 & 1 & 0 \\ 0 & 0 & 1 \end{pmatrix}, \quad B = \begin{pmatrix} 0 & 0 & 1 \\ 0 & 1 & 0 \\ 1 & 0 & 0 \end{pmatrix},$$

$$C = \begin{pmatrix} -1 & 0 & 0 \\ 0 & -1 & 0 \\ 0 & 0 & -1 \end{pmatrix}, \quad D = \begin{pmatrix} 0 & 0 & -1 \\ 0 & -1 & 0 \\ -1 & 0 & 0 \end{pmatrix}$$

form a group under matrix multiplication.

(ii) Show that the 4×4 matrices

$$X = \begin{pmatrix} 1 & 0 & 0 & 0 \\ 0 & 1 & 0 & 0 \\ 0 & 0 & 1 & 0 \\ 0 & 0 & 0 & 1 \end{pmatrix}, \quad Y = \begin{pmatrix} 0 & 0 & 1 & 0 \\ 0 & 1 & 0 & 0 \\ 1 & 0 & 0 & 0 \\ 0 & 0 & 0 & 1 \end{pmatrix},$$

$$V = \begin{pmatrix} -1 & 0 & 0 & 0 \\ 0 & -1 & 0 & 0 \\ 0 & 0 & -1 & 0 \\ 0 & 0 & 0 & -1 \end{pmatrix}, \quad W = \begin{pmatrix} 0 & 0 & -1 & 0 \\ 0 & -1 & 0 & 0 \\ -1 & 0 & 0 & 0 \\ 0 & 0 & 0 & -1 \end{pmatrix}$$

form a group under matrix multiplication.
(iii) Show that the two groups (so-called *Vierergruppe*) are isomorphic.

Solution 8. (i) The group table is

·	A	B	C	D
A	A	B	C	D
B	B	A	D	C
C	C	D	A	B
D	D	C	B	A

The neutral element is A. Each element is its own inverse $A^{-1} = A$, $B^{-1} = B$, $C^{-1} = C$, $D^{-1} = D$.
(ii) The group table is

·	X	Y	V	W
X	X	Y	V	W
Y	Y	X	W	V
V	V	W	X	Y
W	W	V	Y	X

The neutral element is X. Each element is its own inverse $X^{-1} = X$, $Y^{-1} = Y$, $V^{-1} = V$, $W^{-1} = W$.
(iii) The map $A \leftrightarrow X$, $B \leftrightarrow Y$, $C \leftrightarrow V$, $D \leftrightarrow W$ provides the isomorphism.

Problem 9. (i) Let $x \in \mathbb{R}$. Show that the 2×2 matrices

$$A(x) = \begin{pmatrix} 1 & x \\ 0 & 1 \end{pmatrix}$$

form a group under matrix multiplication.
(ii) Is the group commutative?
(iii) Find a group that is isomorphic to this group.

Solution 9. (i) We have (closure)

$$A(x)A(y) = \begin{pmatrix} 1 & x \\ 0 & 1 \end{pmatrix} \begin{pmatrix} 1 & y \\ 0 & 1 \end{pmatrix} = \begin{pmatrix} 1 & x+y \\ 0 & 1 \end{pmatrix} = A(x+y).$$

The neutral element is the identity matrix I_2. The inverse element is given by

$$A(-x) = \begin{pmatrix} 1 & -x \\ 0 & 1 \end{pmatrix}.$$

Matrix multiplication is associative. Thus we have a group.
(ii) Since $A(x)A(y) = A(y)A(x)$ the group is commutative.
(iii) The group is isomorphic to the group $(\mathbb{R}, +)$.

Problem 10.   Let $a, b, c, d \in \mathbb{Z}$. Show that the 2×2 matrices

$$A = \begin{pmatrix} a & b \\ c & d \end{pmatrix}$$

with $ad - bc = 1$ form a group under matrix multiplication.

Solution 10.   We have

$$\begin{pmatrix} a_1 & b_1 \\ c_1 & d_1 \end{pmatrix} \begin{pmatrix} a_2 & b_2 \\ c_2 & d_2 \end{pmatrix} = \begin{pmatrix} a_1a_2 + b_1c_2 & a_1b_2 + b_1d_2 \\ c_1a_2 + d_1c_2 & c_1b_2 + d_1d_2 \end{pmatrix}.$$

Since $a_1, b_1, c_1, d_1 \in \mathbb{Z}$ and $a_2, b_2, c_2, d_2 \in \mathbb{Z}$ the entries of the matrix on the right-hand side are again elements in $\mathbb{Z}$. Since $\det(A_1A_2) = \det(A_1)\det(A_2)$ and $\det(A_1) = 1$, $\det(A_2) = 1$ we have $\det(A_1A_2) = 1$. The inverse element of A is given by

$$A^{-1} = \begin{pmatrix} d & -b \\ -c & a \end{pmatrix}.$$

Thus the entries of A^{-1} are again elements in the set of integers $\mathbb{Z}$.

Problem 11.   (i) Let $c \in \mathbb{R}$ and $c \neq 0$. Do the 4×4 matrices

$$A(c) = \begin{pmatrix} c & c & c & c \\ c & c & c & c \\ c & c & c & c \\ c & c & c & c \end{pmatrix}$$

form a group under matrix multiplication?
(ii) Find the eigenvalues of $A(c)$.

Solution 11.   (i) The neutral element and inverse element are, respectively

$$\begin{pmatrix} 1/4 & 1/4 & 1/4 & 1/4 \\ 1/4 & 1/4 & 1/4 & 1/4 \\ 1/4 & 1/4 & 1/4 & 1/4 \\ 1/4 & 1/4 & 1/4 & 1/4 \end{pmatrix}, \quad \begin{pmatrix} 1/(16c) & 1/(16c) & 1/(16c) & 1/(16c) \\ 1/(16c) & 1/(16c) & 1/(16c) & 1/(16c) \\ 1/(16c) & 1/(16c) & 1/(16c) & 1/(16c) \\ 1/(16c) & 1/(16c) & 1/(16c) & 1/(16c) \end{pmatrix}.$$

(ii) The rank of the matrix is 1 and $\text{tr}(A(c)) = 4c$. Thus the eigenvalues are 0 (3 times) and $4c$.

Problem 12.   The *Heisenberg group* is the set of upper 3×3 matrices of the form

$$H = \begin{pmatrix} 1 & a & c \\ 0 & 1 & b \\ 0 & 0 & 1 \end{pmatrix}$$

where a, b, c can be taken from some (arbitrary) commutative ring.

(i) Find the inverse of H.

(ii) Given two elements x, y of a group G, we define the *commutator* of x and y, denoted by $[x, y]$ to be the element $x^{-1}y^{-1}xy$. If a, b, c are integers (in the ring $\mathbb{Z}$ of the integers) we obtain the discrete Heisenberg group H_3. It has two generators

$$x = \begin{pmatrix} 1 & 1 & 0 \\ 0 & 1 & 0 \\ 0 & 0 & 1 \end{pmatrix}, \qquad y = \begin{pmatrix} 1 & 0 & 0 \\ 0 & 1 & 1 \\ 0 & 0 & 1 \end{pmatrix}.$$

Find $z = xyx^{-1}y^{-1}$. Show that $xz = zx$ and $yz = zy$, i.e. z is the generator of the center of H_3.

(iii) The *derived subgroup* (or commutator subgroup) of a group G is the subgroup $[G, G]$ generated by the set of commutators of every pair of elements of G. Find $[G, G]$ for the Heisenberg group.

(iv) Let

$$A = \begin{pmatrix} 0 & a & c \\ 0 & 0 & b \\ 0 & 0 & 0 \end{pmatrix}$$

and $a, b, c \in \mathbb{R}$. Find $\exp(A)$.

(v) The Heisenberg group is a simple connected Lie group whose Lie algebra consists of matrices

$$L = \begin{pmatrix} 0 & a & c \\ 0 & 0 & b \\ 0 & 0 & 0 \end{pmatrix}.$$

Find the commutators $[L, L']$ and $[[L, L'], L']$, where $[L, L'] := LL' - L'L$.

Solution 12.   (i) An inverse exists since $\det(H) = 1$. Multiplying two upper triangular matrices yields again an upper triangular matrix. From the condition

$$\begin{pmatrix} 1 & a & c \\ 0 & 1 & b \\ 0 & 0 & 1 \end{pmatrix} \begin{pmatrix} 1 & e & g \\ 0 & 1 & f \\ 0 & 0 & 1 \end{pmatrix} = \begin{pmatrix} 1 & 0 & 0 \\ 0 & 1 & 0 \\ 0 & 0 & 1 \end{pmatrix}$$

we obtain $e + a = 0$, $g + af + c = 0$, $f + b = 0$. Thus $e = -a$, $f = -b$, $g = ab - c$. Consequently

$$H^{-1} = \begin{pmatrix} 1 & -a & ab - c \\ 0 & 1 & -b \\ 0 & 0 & 1 \end{pmatrix}.$$

(ii) Since

$$x^{-1} = \begin{pmatrix} 1 & -1 & 0 \\ 0 & 1 & 0 \\ 0 & 0 & 1 \end{pmatrix}, \qquad y^{-1} = \begin{pmatrix} 1 & 0 & 0 \\ 0 & 1 & -1 \\ 0 & 0 & 1 \end{pmatrix}$$

we obtain

$$z = xyx^{-1}y^{-1} = \begin{pmatrix} 1 & 0 & 1 \\ 0 & 1 & 0 \\ 0 & 0 & 1 \end{pmatrix}$$

and $xz = zx$, $yz = zy$. Thus z is the generator of the center of H_3.

(iii) We have $xyx^{-1}y^{-1} = z$, $xzx^{-1}z^{-1} = I$, $yzy^{-1}z^{-1} = I$, where I is the neutral element (3×3 identity matrix) of the group. Thus the derived subgroup has only one generator z.

(iv) We use the expansion

$$e^A = \sum_{k=0}^{\infty} \frac{A^k}{k!}.$$

Since

$$A^2 = \begin{pmatrix} 0 & 0 & ab \\ 0 & 0 & 0 \\ 0 & 0 & 0 \end{pmatrix}, \qquad A^3 = \begin{pmatrix} 0 & 0 & 0 \\ 0 & 0 & 0 \\ 0 & 0 & 0 \end{pmatrix}$$

(the matrix A is nilpotent) we obtain

$$e^A = \begin{pmatrix} 1 & a & c + ab/2 \\ 0 & 1 & b \\ 0 & 0 & 1 \end{pmatrix}.$$

(v) We obtain

$$[L, L'] = \begin{pmatrix} 0 & 0 & ab' - a'b \\ 0 & 0 & 0 \\ 0 & 0 & 0 \end{pmatrix}.$$

Using this result we find the zero matrix $[[L, L'], L'] = 0_3$.

Problem 13.   Define

$$M : \mathbb{R}^3 \to V := \{\, \mathbf{a} \cdot \boldsymbol{\sigma} : \mathbf{a} \in \mathbb{R}^3 \,\} \subset \{\, 2 \times 2 \text{ complex matrices} \,\}$$
$$\mathbf{a} \to M(\mathbf{a}) = \mathbf{a} \cdot \boldsymbol{\sigma} = a_1\sigma_1 + a_2\sigma_2 + a_3\sigma_3.$$

This is a linear *bijection* between $\mathbb{R}^3$ and V. Each $U \in SU(2)$ determines a linear map $S(U)$ on $\mathbb{R}^3$ by

$$M(S(U)\mathbf{a}) = U^{-1}M(\mathbf{a})U.$$

The right-hand side is clearly linear in $\mathbf{a}$. Show that $U^{-1}M(\mathbf{a})U$ is in V, that is, of the form $M(\mathbf{b})$.

Solution 13.   Let $U = x_0 I_2 + i\mathbf{x} \cdot \boldsymbol{\sigma}$ with $(x_0, \mathbf{x}) \in \mathbb{R}^4$ obeying $\|x_0\|^2 + \|\mathbf{x}\|^2 = 1$ and compute $U^{-1}M(\mathbf{a})U$ explicitly, where $U^{-1} = U^*$. We have

$$U^{-1}M(\mathbf{a})U = (x_0 I_2 - i\mathbf{x} \cdot \boldsymbol{\sigma})(\mathbf{a} \cdot \boldsymbol{\sigma})(x_0 I_2 + i\mathbf{x} \cdot \boldsymbol{\sigma})$$
$$= (x_0 I_2 - i\mathbf{x} \cdot \boldsymbol{\sigma})(x_0 \mathbf{a} \cdot \boldsymbol{\sigma} + i(\mathbf{a} \cdot \mathbf{x})I_2 - (\mathbf{a} \times \mathbf{x}) \cdot \boldsymbol{\sigma})$$
$$= x_0^2 \mathbf{a} \cdot \boldsymbol{\sigma} - 2x_0(\mathbf{a} \times \mathbf{x}) \cdot \boldsymbol{\sigma} + (\mathbf{a} \cdot \mathbf{x})(\mathbf{x} \cdot \boldsymbol{\sigma}) - (\mathbf{x} \times (\mathbf{a} \times \mathbf{x})) \cdot \boldsymbol{\sigma}$$

since $\mathbf{x}$ is perpendicular to $\mathbf{a} \times \mathbf{x}$, where $\times$ denotes the vector product. Using the identity

$$\mathbf{c} \cdot (\mathbf{a} \times \mathbf{b}) \equiv (\mathbf{b} \cdot \mathbf{c})\mathbf{a} - (\mathbf{a} \cdot \mathbf{c})\mathbf{b}$$

we obtain

$$U^{-1}M(\mathbf{a})U = (x_0^2 - \|\mathbf{x}\|^2)\mathbf{a} \cdot \boldsymbol{\sigma} - 2x_0(\mathbf{a} \times \mathbf{x}) \cdot \boldsymbol{\sigma} + 2(\mathbf{a} \cdot \mathbf{x})(\mathbf{x} \cdot \boldsymbol{\sigma}).$$

This shows, not only that $U^{-1}M(\mathbf{a})U \in V$, but also that, for $U = x_0 I_2 + i\mathbf{x} \cdot \boldsymbol{\sigma}$ we have

$$S(U)\mathbf{a} = (x_0^2 - \|\mathbf{x}\|^2)\mathbf{a} + 2x_0\mathbf{x} \times \mathbf{a} + 2(\mathbf{a} \cdot \mathbf{x})\mathbf{x}.$$

Problem 14. Show that the nonnormal 2×2 matrices

$$\begin{pmatrix} 1 & 1 \\ 0 & 1 \end{pmatrix}, \qquad \begin{pmatrix} 1 & -1 \\ 0 & 1 \end{pmatrix}$$

are *conjugate* in $SL(2, \mathbb{C})$ but not in $SL(2, \mathbb{R})$ (the real matrices in $SL(2, \mathbb{C})$).

Solution 14. Let

$$S = \begin{pmatrix} s_{11} & s_{12} \\ s_{21} & s_{22} \end{pmatrix}$$

with $\det(S) = 1$, i.e. $S \in SL(2, \mathbb{C})$. Now

$$S \begin{pmatrix} 1 & 1 \\ 0 & 1 \end{pmatrix} S^{-1} = \begin{pmatrix} s_{11} & s_{12} \\ s_{21} & s_{22} \end{pmatrix} \begin{pmatrix} 1 & 1 \\ 0 & 1 \end{pmatrix} \begin{pmatrix} s_{22} & -s_{12} \\ -s_{21} & s_{11} \end{pmatrix}$$

$$= \begin{pmatrix} 1 - s_{11}s_{21} & s_{11}^2 \\ -s_{21}^2 & 1 + s_{11}s_{21} \end{pmatrix} = \begin{pmatrix} 1 & -1 \\ 0 & 1 \end{pmatrix}.$$

We obtain the four conditions $1 - s_{11}s_{21} = 1$, $s_{11}^2 = -1$, $-s_{21}^2 = 0$, $1 + s_{11}s_{21} = 1$. It follows that $s_{21} = 0$, $s_{11} = i$. The entry s_{22} follows from $\det(S) = 1$. We have $s_{11}s_{22} = 1$ and therefore $s_{22} = -i$. The entry s_{12} is arbitrary. Thus

$$S = \begin{pmatrix} i & s_{12} \\ 0 & -i \end{pmatrix} \Rightarrow S^{-1} = \begin{pmatrix} -i & -s_{12} \\ 0 & i \end{pmatrix}.$$

Problem 15. (i) Let G be a finite set of real $n \times n$ matrices $\{A_j\}$, $1 \leq j \leq r$, which forms a group under matrix multiplication. Suppose that

$$\text{tr}\left(\sum_{j=1}^{r} A_j\right) \equiv \sum_{j=1}^{r} \text{tr}(A_j) = 0$$

where tr denotes the trace. Show that

$$\sum_{j=1}^{r} A_j = 0_n.$$

(ii) Let $\omega := \exp(2\pi i/3)$. Show that the 2×2 matrices

$$B_1 = \begin{pmatrix} 1 & 0 \\ 0 & 1 \end{pmatrix}, \quad B_2 = \begin{pmatrix} \omega & 0 \\ 0 & \omega^2 \end{pmatrix}, \quad B_3 = \begin{pmatrix} \omega^2 & 0 \\ 0 & \omega \end{pmatrix}$$

$$B_4 = \begin{pmatrix} 0 & 1 \\ 1 & 0 \end{pmatrix}, \quad B_5 = \begin{pmatrix} 0 & \omega \\ \omega^2 & 0 \end{pmatrix}, \quad B_6 = \begin{pmatrix} 0 & \omega^2 \\ \omega & 0 \end{pmatrix}$$

form a group under matrix multiplication. Show that

$$\sum_{j=1}^{6} \operatorname{tr}(B_j) = 0.$$

Solution 15.   (i) Let

$$S := \sum_{i=1}^{r} A_i.$$

For any j, the sequence of matrices $A_j A_1, A_j A_2, \ldots, A_j A_r$ is a permutation of the elements of G, and summing yields $A_j S = S$. Thus

$$\sum_{j=1}^{r} A_j S = S \sum_{j=1}^{r} A_j = \sum_{j=1}^{r} S \Rightarrow S^2 = rS.$$

Therefore the minimal polynomial of S divides $x^2 - rx$, and every eigenvalue of S is either 0 or r since $x^2 - rx \equiv x(x - r)$. However the eigenvalues counted with multiplicity sum to $\operatorname{tr}(S) = 0$. Thus they are all 0. Now every eigenvalue of $S - rI_n$ is $-r \neq 0$, so the matrix $S - rI_n$ is invertible. Therefore from $S(S - rI_n) = 0_n$ we obtain $S = 0_n$.

(ii) The group table is

·	B_1	B_2	B_3	B_4	B_5	B_6
B_1	B_1	B_2	B_3	B_4	B_5	B_6
B_2	B_2	B_3	B_1	B_5	B_6	B_4
B_3	B_3	B_1	B_2	B_6	B_4	B_5
B_4	B_4	B_6	B_5	B_1	B_3	B_2
B_5	B_5	B_4	B_6	B_2	B_1	B_3
B_6	B_6	B_5	B_4	B_3	B_2	B_1

We have

$$\sum_{j=1}^{6} \operatorname{tr}(B_j) = 2 + 2(\omega + \omega^2) = 0$$

where we used that $\omega + \omega^2 = -1$. The matrices B_2, B_3, B_5, B_6 contain complex numbers as entries. This is a special case of

$$\sum_{j=1}^{6} B_j = 0_2.$$

Problem 16. Consider the unitary 2×2 matrix

$$J := \begin{pmatrix} 0 & 1 \\ -1 & 0 \end{pmatrix}.$$

(i) Find all 2×2 matrices A over $\mathbb{R}$ such that $A^T J A = J$.
(ii) Do these 2×2 matrices form a group under matrix multiplication?

Solution 16. (i) From the condition

$$\begin{pmatrix} a_{11} & a_{21} \\ a_{12} & a_{22} \end{pmatrix} \begin{pmatrix} 0 & 1 \\ -1 & 0 \end{pmatrix} \begin{pmatrix} a_{11} & a_{12} \\ a_{21} & a_{22} \end{pmatrix} = \begin{pmatrix} 0 & 1 \\ -1 & 0 \end{pmatrix}$$

we obtain

$$\begin{pmatrix} 0 & a_{11}a_{22} - a_{12}a_{21} \\ -a_{11}a_{22} + a_{12}a_{21} & 0 \end{pmatrix} = \begin{pmatrix} 0 & 1 \\ -1 & 0 \end{pmatrix}.$$

Thus $\det(A) = 1$.
(ii) All $n \times n$ matrices A with $\det(A) = 1$ form a subgroup of the general linear group $GL(n, \mathbb{R})$. Note that $\det(AB) \equiv \det(A)\det(B)$. Thus if $\det(A) = 1$ and $\det(B) = 1$ we have $\det(AB) = 1$.

Problem 17. Let J be the $2n \times 2n$ matrix

$$J := \begin{pmatrix} 0_n & I_n \\ -I_n & 0_n \end{pmatrix}.$$

Show that the $2n \times 2n$ matrices A satisfying $A^T J A = J$ form a group under matrix multiplication. This group is called the *symplectic group* $Sp(2n)$.

Solution 17. Let $A^T J A = J$, $B^T J B = J$. Then

$$(AB)^T J(AB) = B^T(A^T J A)B = B^T J B = J.$$

Obviously, $I_{2n} J I_{2n} = J$ and $\det(J) \neq 0$. From $A^T J A = J$ it follows that

$$\det(A^T J A) = \det(A^T)\det(J)\det(A) = \det(J).$$

Thus $\det(A^T)\det(A) = 1$. Since $\det(A^T) = \det(A)$ we have $(\det(A))^2 = 1$. From $A^T J A = J$ we obtain $(A^T J A)^{-1} = J^{-1}$. Since $J^T = J^{-1}$ we arrive at

$$A^{-1}J^{-1}(A^T)^{-1} = A^{-1}J^T(A^{-1})^T = J^T.$$

Applying the transpose we find $A^{-1}J(A^T)^{-1} = J$. Let $B = (A^{-1})^T$. Then $B^T = A^{-1}$ and $B^T J B = J$.

Problem 18. The Pauli spin matrix σ_1 is not only hermitian, unitary and its own inverse, but also a permutation matrix. Find all 2×2 matrices A such that $\sigma_1^{-1} A \sigma_1 = A$.

Solution 18.   From

$$\begin{pmatrix} 0 & 1 \\ 1 & 0 \end{pmatrix} \begin{pmatrix} a_{11} & a_{12} \\ a_{21} & a_{22} \end{pmatrix} \begin{pmatrix} 0 & 1 \\ 1 & 0 \end{pmatrix} = \begin{pmatrix} a_{11} & a_{12} \\ a_{21} & a_{22} \end{pmatrix}$$

we find the solution $a_{22} = a_{11}$, $a_{21} = a_{12}$.

Problem 19.   For the vector space of the $n \times n$ matrices over $\mathbb{R}$ we can introduce a scalar product via

$$\langle A, B \rangle := \text{tr}(AB^T).$$

Consider the Lie group $SL(2, \mathbb{R})$ of the 2×2 matrices with determinant 1. Find $X, Y \in SL(2, \mathbb{R})$ such that

$$\langle X, Y \rangle = 0$$

where neither X nor Y can be 2×2 identity matrix.

Solution 19.   The 2×2 matrices

$$X = \begin{pmatrix} 1 & 1 \\ 1 & 2 \end{pmatrix}, \qquad Y = \begin{pmatrix} 0 & -1 \\ 1 & 0 \end{pmatrix}$$

satisfy the equations, i.e. $\det(X) = \det(Y) = 1$ and $\text{tr}(XY^T) = 0$.

Problem 20.   The numbers 1, i, -1, $-i$ form a group under multiplication. Do the vectors

$$\mathbf{v}_1 = \frac{1}{2} \begin{pmatrix} 1 \\ i \\ -1 \\ -i \end{pmatrix}, \quad \mathbf{v}_2 = \frac{1}{2} \begin{pmatrix} i \\ -1 \\ -i \\ 1 \end{pmatrix}, \quad \mathbf{v}_3 = \frac{1}{2} \begin{pmatrix} -1 \\ -i \\ 1 \\ i \end{pmatrix}, \quad \mathbf{v}_4 = \frac{1}{2} \begin{pmatrix} -i \\ 1 \\ i \\ -1 \end{pmatrix}$$

form an orthonormal basis in $\mathbb{C}^4$?

Solution 20.   No. The equation $a_1\mathbf{v}_1 + a_2\mathbf{v}_2 + a_3\mathbf{v}_3 + a_4\mathbf{v}_4 = \mathbf{0}$ admits the solution $a_1 - a_3 = 0$, $a_2 - a_4 = 0$.

Problem 21.   (i) Consider the group G of all 4×4 permutation matrices. Show that

$$\frac{1}{|G|} \sum_{g \in G} g$$

is a *projection matrix*. Here $|G|$ denotes the number of elements in the group.
(ii) Consider the subgroup given by the matrices

$$\begin{pmatrix} 1 & 0 & 0 & 0 \\ 0 & 1 & 0 & 0 \\ 0 & 0 & 1 & 0 \\ 0 & 0 & 0 & 1 \end{pmatrix}, \qquad \begin{pmatrix} 0 & 0 & 0 & 1 \\ 0 & 0 & 1 & 0 \\ 0 & 1 & 0 & 0 \\ 1 & 0 & 0 & 0 \end{pmatrix}.$$

Show that

$$\frac{1}{|G|} \sum_{g \in G} g$$

is a projection matrix.

Solution 21. (i) We have $|G| = 24$. Thus

$$\Pi = \frac{1}{|G|} \sum_{g \in G} g = \frac{1}{4} \begin{pmatrix} 1 & 1 & 1 & 1 \\ 1 & 1 & 1 & 1 \\ 1 & 1 & 1 & 1 \\ 1 & 1 & 1 & 1 \end{pmatrix}.$$

We find $\Pi = \Pi^*$ (i.e. Π is hermitian) and $\Pi^2 = \Pi$. Hence Π is a projection matrix.

(ii) We have $|G| = 2$. Thus

$$\frac{1}{|G|} \sum_{g \in G} g = \frac{1}{2} \begin{pmatrix} 1 & 0 & 0 & 1 \\ 0 & 1 & 1 & 0 \\ 0 & 1 & 1 & 0 \\ 1 & 0 & 0 & 1 \end{pmatrix}.$$

This symmetric matrix over $\mathbb{R}$ is a projection matrix.

Problem 22. Let A be the 2×2 matrix

$$A = \begin{pmatrix} r_{11} e^{i\phi_{11}} & r_{12} e^{i\phi_{12}} \\ r_{21} e^{i\phi_{21}} & r_{22} e^{i\phi_{22}} \end{pmatrix}$$

where $r_{jk} \in \mathbb{R}$, $r_{jk} > 0$ for $j, k = 1, 2$ and $r_{12} = r_{21}$. We also have $\phi_{jk} \in \mathbb{R}$ for $j, k = 1, 2$ and impose $\phi_{12} = \phi_{21}$. What are the conditions on r_{jk} and ϕ_{jk} such that $I_2 + iA$ is a unitary matrix?

Solution 22. Let $U = I_2 + A$. Then $U^* = I_2 - iA^*$. Thus from $UU^* = I_2$ we find

$$i(A - A^*) + AA^* = 0_2.$$

A solution is $r_{11} = r_{22} = 0$, $r_{12} = r_{21} = 1$ and

$$\phi_{11} = \phi_{22} = \frac{1}{2}\pi, \qquad \phi_{12} \text{ arbitrary.}$$

Problem 23. Show that the four 2×2 matrices

$$A = \begin{pmatrix} 1 & 0 \\ 0 & 1 \end{pmatrix}, \quad B = \begin{pmatrix} -1 & 0 \\ 0 & -1 \end{pmatrix}, \quad C = \begin{pmatrix} 0 & 1 \\ 1 & 0 \end{pmatrix}, \quad D = \begin{pmatrix} 0 & -1 \\ -1 & 0 \end{pmatrix}$$

from a group under matrix multiplication. Is the group abelian?

Solution 23. The group is abelian. We have

$$AA = A, \quad AB = BA = B, \quad AC = CA = C, \quad AD = DA = D$$

$$BB = A, \quad BC = CB = D, \quad BD = DB = C, \quad CD = DC = B.$$

Problem 24. (i) Consider the symmetric 2×2 matrix

$$A = \begin{pmatrix} a_{11} & a_{12} \\ a_{12} & a_{11} \end{pmatrix}, \quad a_{11}, a_{12} \in \mathbb{R}.$$

Find all invertible 2×2 matrices S over $\mathbb{R}$ such that $SAS^{-1} = A$. Obviously the identity matrix I_2 would be such as matrix.
(ii) Do the matrices S form a group under matrix multiplication? Prove or disprove.
(iii) Use the result form (i) to calculate $(S \otimes S)(A \otimes A)(S \otimes S)^{-1}$. Discuss.

Solution 24. (i) With

$$S = \begin{pmatrix} s_{11} & s_{12} \\ s_{21} & s_{22} \end{pmatrix} \Rightarrow S^{-1} = \frac{1}{\det(S)} \begin{pmatrix} s_{22} & -s_{12} \\ -s_{21} & s_{11} \end{pmatrix}$$

we obtain for $\det(S)SAS^{-1}$

$$\begin{pmatrix} a_{11}\det(S) + a_{12}(s_{12}s_{22} - s_{11}s_{21}) & a_{12}(s_{11}^2 - s_{12}^2) \\ a_{12}(s_{22}^2 - s_{21}^2) & a_{11}\det(S) + a_{12}(-s_{12}s_{22} - s_{11}s_{21}) \end{pmatrix}.$$

Since a_{11} and a_{12} are arbitrary we find the conditions

$$s_{12}s_{22} - s_{11}s_{21} = 0, \quad s_{11}^2 - s_{12}^2 = \det(S), \quad s_{22}^2 - s_{21}^2 = \det(S).$$

A solution is $s_{11} = s_{22} = 0$, $s_{12} = s_{21} = 1$.
(ii) Yes. The neutral element is the identity matrix. Let $S_1 A S_1^{-1} = A$, $S_2 A S_2^{-1} = A$ we have

$$(S_1 S_2)A(S_1 S_2)^{-1} = (S_1 S_2)A(S_2 S_1^{-1}) = S_1 A S_1^{-1} = A.$$

(iii) We find

$$(S \otimes S)(A \otimes A)(S \otimes S)^{-1} = (SAS^{-1}) \otimes (SAS^{-1}) = A \otimes A.$$

Problem 25. Find all 2×2 matrices S over $\mathbb{C}$ with determinant 1 (i.e. they are elements of $SL(2, \mathbb{C})$) such that

$$S^{-1} \begin{pmatrix} 0 & 1 \\ 1 & 0 \end{pmatrix} S = \begin{pmatrix} 0 & 1 \\ 1 & 0 \end{pmatrix}.$$

Obviously, the 2×2 identity matrix is such an element.

Solution 25. Let

$$S = \begin{pmatrix} s_{11} & s_{12} \\ s_{21} & s_{22} \end{pmatrix}, \quad S^{-1} = \begin{pmatrix} s_{22} & -s_{12} \\ -s_{21} & s_{11} \end{pmatrix}$$

with $\det(S) = s_{11}s_{22} - s_{12}s_{21} = 1$. Then

$$S^{-1} \begin{pmatrix} 0 & 1 \\ 1 & 0 \end{pmatrix} S = \begin{pmatrix} -s_{11}s_{12} + s_{21}s_{22} & -s_{12}^2 + s_{22}^2 \\ s_{11}^2 - s_{21}^2 & s_{11}s_{12} - s_{21}s_{22} \end{pmatrix} = \begin{pmatrix} 0 & 1 \\ 1 & 0 \end{pmatrix}.$$

Thus we have to solve the four equations

$$s_{11}s_{12} - s_{21}s_{22} = 0, \ \ s_{11}^2 - s_{21}^2 = 1, \ \ -s_{12}^2 + s_{22} = 1, \ \ s_{11}s_{22} - s_{12}s_{21} = 0.$$

We have to do a case study.
Case 1. $s_{11} = 0$. Then $s_{21}^2 = -1$, $s_{22} = 0$, $s_{12}^2 = -1$.
Case 2. $s_{12} = 0$. Then $s_{22}^2 = 1$, $s_{21} = 0$, $s_{11}^2 = 1$.
Case 3. $s_{21} = 0$. Then $s_{11}^2 = 1$, $s_{12} = 0$, $s_{22}^2 = 1$.
Case 4. $s_{11} \neq 0$. Then multiplying $s_{11}s_{12} = -s_{21}s_{22}$ by s_{11} and inserting $s_{11}^2 = 1 + s_{21}^2$, $s_{11}s_{22} = 1 + s_{12}s_{21}$ yields $s_{21} = 1$. Thus $s_{11}^2 = 2$ and $s_{11}s_{12} = s_{22}$, $-s_{12}^2 + s_{22}^2 = 1$, $s_{11}s_{22} - s_{12} = 1$. From $s_{11}s_{12} = s_{22}$ we obtain $s_{11}^2 s_{12} = s_{11}s_{22}$ and therefore $s_{12} = 1$. Finally $s_{22}^2 = 1$.

Problem 26. (i) Show that the 2×2 matrices

$$A = \begin{pmatrix} 0 & 1 \\ 0 & 1 \end{pmatrix}, \qquad B = \begin{pmatrix} 1 & 0 \\ 1 & 0 \end{pmatrix}$$

are *similar*. This means find an invertible 2×2 matrix S, i.e. $S \in GL(2, \mathbb{R})$, such that $SAS^{-1} = B$.
(ii) Is there an invertible 2×2 matrix S such that

$$(S \otimes S)(A \otimes A)(S^{-1} \otimes S^{-1}) = B \otimes B?$$

Solution 26. We start from

$$S = \begin{pmatrix} s_{11} & s_{12} \\ s_{21} & s_{22} \end{pmatrix} \quad \Rightarrow \quad S^{-1} = \frac{1}{\det(S)} \begin{pmatrix} s_{22} & -s_{12} \\ -s_{21} & s_{11} \end{pmatrix}.$$

Thus

$$SAS^{-1} = \frac{1}{\det(S)} \begin{pmatrix} -(s_{11} + s_{12})s_{21} & (s_{11} + s_{12})s_{11} \\ -(s_{21} + s_{22})s_{21} & (s_{21} + s_{22})s_{11} \end{pmatrix} = \begin{pmatrix} 1 & 0 \\ 1 & 0 \end{pmatrix}.$$

Thus we obtain four conditions

$$(s_{11} + s_{12})s_{11} = 0, \qquad (s_{21} + s_{22})s_{11} = 0,$$

$$-(s_{11} + s_{12})s_{21} = \det(S), \qquad -(s_{21} + s_{22})s_{21} = \det(S).$$

Case: $s_{11} \neq 0$. Then $s_{11} = -s_{12}$ and $s_{21} = -s_{22}$ from the first two equations. This contradicts the last two equations ($\det(S) \neq 0$).
Case: $s_{11} = 0$. Then we arrive at

$$-s_{12}s_{21} = -s_{12}s_{21}, \qquad -(s_{21} + s_{22})s_{21} = -s_{12}s_{21}$$

with $s_{12}, s_{21} \neq 0$. Thus $s_{21}(-s_{21} - s_{22} + s_{12}) = 0$ or $s_{22} = s_{12} - s_{21}$. It follows that $\det(S) = -s_{12}s_{21}$. Therefore

$$S = \begin{pmatrix} 0 & s_{12} \\ s_{21} & s_{12} - s_{21} \end{pmatrix} \Rightarrow S^{-1} = \frac{1}{s_{12}s_{21}} \begin{pmatrix} -s_{12} + s_{21} & s_{12} \\ s_{21} & 0 \end{pmatrix}.$$

The simplest special case would be $s_{12} = s_{21} = 1$.

(ii) We have

$$(S \otimes S)(A \otimes A)(S^{-1} \otimes S^{-1}) = (SAS^{-1}) \otimes (SAS^{-1}).$$

Thus the S constructed in (i) can be utilized.

Problem 27.   Find all $M \in GL(2, \mathbb{C})$ such that

$$M^{-1} \begin{pmatrix} 0 & -i \\ i & 0 \end{pmatrix} M = \begin{pmatrix} 0 & -i \\ i & 0 \end{pmatrix}.$$

Thus we consider the invariance of the Pauli spin matrix σ_2. Show that these matrices form a group under matrix multiplication.

Solution 27.   Let $M_1, M_2 \in GL(2, \mathbb{C})$ with $M_1^{-1}\sigma_2 M_1 = \sigma_2$ and $M_2^{-1}\sigma_2 M_2 = \sigma_2$. Then

$$(M_1 M_2)^{-1}\sigma_2(M_1 M_2) = M_2^{-1}(M_1^{-1}\sigma_2 M_1)M_2 = M_2^{-1}\sigma_2 M_2 = \begin{pmatrix} 0 & -i \\ i & 0 \end{pmatrix}.$$

If $M^{-1}\sigma_2 M = \sigma_2$, then $MM^{-1}\sigma_2 MM^{-1} = M\sigma_2 M^{-1}$ and thus $\sigma_2 = M\sigma_2 M^{-1}$. The neutral element is the 2×2 identity matrix.

Problem 28.   Consider the cyclic 3×3 matrix

$$C = \begin{pmatrix} 0 & 1 & 0 \\ 0 & 0 & 1 \\ 1 & 0 & 0 \end{pmatrix}.$$

(i) Show that the matrices C, C^2, C^3 form a group under matrix multiplication. Is the group commutative?

(ii) Find the eigenvalues of C and show that the form a group under multiplication.

(iii) Find the normalized eigenvalues of C. Show that the form a orthonormal basis in $\mathbb{C}^3$.

(iv) Use the eigenvalues and normalized eigenvectors to write down the spectral decomposition of C.

Solution 28.   (i) We have

$$C^2 = \begin{pmatrix} 0 & 0 & 1 \\ 0 & 1 & 0 \\ 1 & 0 & 0 \end{pmatrix}, \quad C^3 = I_3 = \begin{pmatrix} 1 & 0 & 0 \\ 0 & 1 & 0 \\ 0 & 0 & 1 \end{pmatrix}.$$

The inverse of C is C^2. The group is commutative.

(ii) The eigenvalues are 1, $e^{i2\pi/3}$, $e^{i4\pi/3}$. Obviously they form a group under multiplication. Note that $e^{i4\pi/3} = e^{-i2\pi/3}$.

(iii) The normalized eigenvectors for the eigenvalues $\lambda_1 = 1$, $\lambda_2 = e^{i2\pi/3}$, $\lambda_3 = e^{i4\pi/3}$ are

$$\mathbf{v}_1 = \frac{1}{\sqrt{3}} \begin{pmatrix} 1 \\ 1 \\ 1 \end{pmatrix}, \quad \mathbf{v}_2 = \frac{1}{\sqrt{3}} \begin{pmatrix} 1 \\ e^{i2\pi/3} \\ e^{i4\pi/3} \end{pmatrix}, \quad \mathbf{v}_3 = \frac{1}{\sqrt{3}} \begin{pmatrix} 1 \\ e^{i4\pi/3} \\ e^{i2\pi/3} \end{pmatrix}.$$

These vectors form an orthonormal basis in $\mathbb{C}^3$.

(iv) Hence the spectral decomposition is $C = \lambda_1 \mathbf{v}_1 \mathbf{v}_1^* + \lambda_2 \mathbf{v}_2 \mathbf{v}_2^* + \lambda_3 \mathbf{v}_3 \mathbf{v}_3^*$.

Problem 29. (i) Find the group generated by the 2×2 matrix

$$A = \begin{pmatrix} 0 & 1 \\ -1 & 0 \end{pmatrix}$$

under matrix multiplication. Is the group commutative? Find the determinant of all these matrices. Do these (real) numbers form a group under multiplication?

(ii) Find the group generated by the 4×4 matrix

$$A \otimes A = \begin{pmatrix} 0 & 1 \\ -1 & 0 \end{pmatrix} \otimes \begin{pmatrix} 0 & 1 \\ -1 & 0 \end{pmatrix}$$

under matrix multiplication.

Solution 29. (i) We have

$$A^2 = \begin{pmatrix} -1 & 0 \\ 0 & -1 \end{pmatrix} = B = -I_2.$$

Now $B^2 = I_2$, $BA = A^T$ and $AA^T = A^T A = I_2$. Thus the group consists of four elements A, A^T, $B = -I_2$ and I_2. Of course I_2 is the neutral element. The group is commutative. The determinant of all these matrices is $+1$.

(ii) We have

$$(A \otimes A)(A \otimes A) = A^2 \otimes A^2 = B \otimes B = (-I_2) \otimes (-I_2) = I_4.$$

The group consists of the two elements $I_4 = I_2 \otimes I_2$ and $A \otimes A$.

Problem 30. Let $n \geq 2$. An invertible integer matrix, $A \in GL_n(\mathbb{Z})$, generates a toral automorphism $f : \mathbb{T}^n \to \mathbb{T}^n$ via the formula

$$f \circ \pi = \pi \circ A, \qquad \pi : \mathbb{R}^n \to \mathbb{T}^n := \mathbb{R}^n / \mathbb{Z}^n.$$

The set of fixed points of f is given by

$$\#\mathrm{Fix}(f) := \{ x^* \in \mathbb{T}^n : f(x^*) = x^* \}.$$

Now we have: if $\det(I_n - A) \neq 0$, then

$$\#\text{Fix}(f) = |\det(I_n - A)|.$$

Let $n = 2$ and

$$A = \begin{pmatrix} 2 & 1 \\ 1 & 1 \end{pmatrix}.$$

Show that $\det(I_2 - A) \neq 0$ and find $\#\text{Fix}(f)$.

Solution 30.   We have $\det(I_2 - A) = -1$. Thus $|\det(I_2 - A)| = 1$.

Problem 31.   Let A be an $n \times n$ matrix over $\mathbb{R}$. Consider the $2n \times 2n$ matrices

$$S = \begin{pmatrix} I_n & I_n \\ A & I_n + A \end{pmatrix}, \quad \tilde{S} = \begin{pmatrix} A + 2I_n & -I_n \\ I_n & 0_n \end{pmatrix}.$$

Can we find an invertible $2n \times 2n$ matrix T such that $\tilde{S} = T^{-1}ST$?

Solution 31.   We find

$$T = \begin{pmatrix} 0_n & I_n \\ I_n & -I_n \end{pmatrix} = T^{-1}.$$

Problem 32.   Let $R \in \mathbb{C}^{m \times m}$ and $S \in \mathbb{C}^{n \times n}$ be nontrivial *involutions*. This means that $R = R^{-1} \neq \pm I_m$ and $S = S^{-1} \neq I_n$. A matrix $A \in \mathbb{C}^{m \times n}$ is called (R, S)-symmetric if $RAS = A$. Consider the case $m = n = 2$ and the Pauli spin matrices

$$R = \begin{pmatrix} 0 & 1 \\ 1 & 0 \end{pmatrix}, \quad S = \begin{pmatrix} 0 & -i \\ i & 0 \end{pmatrix}.$$

Find all 2×2 matrices A over $\mathbb{C}$ such that $RAS = A$.

Solution 32.   From

$$\begin{pmatrix} 0 & 1 \\ 1 & 0 \end{pmatrix} \begin{pmatrix} a_{11} & a_{12} \\ a_{21} & a_{22} \end{pmatrix} \begin{pmatrix} 0 & -i \\ i & 0 \end{pmatrix} = \begin{pmatrix} a_{11} & a_{12} \\ a_{21} & a_{22} \end{pmatrix}$$

we obtain

$$A = \begin{pmatrix} a_{11} & a_{12} \\ ia_{12} & -ia_{11} \end{pmatrix}.$$

Problem 33.   Find all 2×2 matrices A over $\mathbb{R}$ such that $\det(A) = 1$ (i.e. A is an element of the Lie group $SL(2, \mathbb{R})$) and

$$A \frac{1}{\sqrt{2}} \begin{pmatrix} 1 \\ -1 \end{pmatrix} = \frac{1}{\sqrt{2}} \begin{pmatrix} 1 \\ -1 \end{pmatrix}.$$

Do these matrices form a group under matrix multiplication?

Solution 33. The condition provides the two equations $a_{11} - a_{12} = 1$, $a_{21} - a_{22} = -1$. Thus $a_{12} = a_{11} - 1$ and $a_{21} = a_{22} - 1$. Inserting this into $\det(A) = 1$ yields $a_{22} = 2 - a_{11}$. Thus we end up with the matrix

$$A = \begin{pmatrix} a_{11} & a_{11} - 1 \\ 1 - a_{11} & 2 - a_{11} \end{pmatrix}$$

with $a_{11} \in \mathbb{R}$ and $\det(A) = 1$. We set $a_{11} = \alpha$ and

$$A(\alpha) = \begin{pmatrix} \alpha & \alpha - 1 \\ 1 - \alpha & 2 - \alpha \end{pmatrix}.$$

For $\alpha = 1$ we obtain the 2×2 identity matrix. Now

$$A(\alpha)A(\beta) = \begin{pmatrix} \alpha + \beta - 1 & \alpha + \beta - 2 \\ -\alpha - \beta + 2 & -\alpha - \beta + 3 \end{pmatrix}.$$

Setting $\gamma = \alpha + \beta - 1$ we obtain

$$A(\gamma) = \begin{pmatrix} \gamma & \gamma - 1 \\ 1 - \gamma & 2 - \gamma \end{pmatrix}.$$

The inverse of $A(\alpha)$ is given by

$$A^{-1}(\alpha) = \begin{pmatrix} 2 - \alpha & 1 - \alpha \\ \alpha - 1 & \alpha \end{pmatrix}.$$

Thus the matrices $A(\alpha)$ form a group under matrix multiplication.

Problem 34. Let $\alpha \in \mathbb{R}$. Consider the hermitian matrix which is an element of the noncompact Lie group $SO(1,1)$

$$A(\alpha) = \begin{pmatrix} \cosh(\alpha) & \sinh(\alpha) \\ \sinh(\alpha) & \cosh(\alpha) \end{pmatrix}.$$

Find the *Cayley transform*

$$B = (A - iI_2)(A + iI_2)^{-1}.$$

Note that B is a unitary matrix and therefore an element of the compact Lie group $U(n)$. Find $B(\alpha \to \infty)$.

Solution 34. We have

$$A - iI_2 = \begin{pmatrix} \cosh(\alpha) - i & \sinh(\alpha) \\ \sinh(\alpha) & \cosh(\alpha) - i \end{pmatrix}, \quad A + iI_2 = \begin{pmatrix} \cosh(\alpha) + i & \sinh(\alpha) \\ \sinh(\alpha) & \cosh(\alpha) + i \end{pmatrix}.$$

Thus

$$(A + iI_2)^{-1} = \frac{1}{2i\cosh(\alpha)} \begin{pmatrix} \cosh(\alpha) + i & -\sinh(\alpha) \\ -\sinh(\alpha) & \cosh(\alpha) + i \end{pmatrix}$$

and

$$B = (A - iI_2)(A + iI_2)^{-1} = \frac{1}{\cosh(\alpha)} \begin{pmatrix} -i & \sinh(\alpha) \\ \sinh(\alpha) & -i \end{pmatrix}$$

with $\det(U) = -1$. If $\alpha \to \infty$ we obtain

$$B(\alpha \to \infty) = \begin{pmatrix} 0 & 1 \\ 1 & 0 \end{pmatrix}.$$

Problem 35.   (i) Consider the Lie group $SL(n, \mathbb{C})$, i.e. the $n \times n$ matrices over $\mathbb{C}$ with determinant 1. Can we find $A, B \in SL(n, \mathbb{C})$ such that $[A, B]$ is an element of $SL(n, \mathbb{C})$?
(ii) Consider the 2×2 matrices

$$A(z) = \begin{pmatrix} 1 & 0 \\ z & 1 \end{pmatrix}, \qquad B = \begin{pmatrix} 0 & 1 \\ -1 & 0 \end{pmatrix}.$$

Both are elements of the non-compact Lie group $SL(2, \mathbb{C})$. Can one finds $z \in \mathbb{C}$ such that the commutator $[A(z), B]$ is again an element of $SL(2, \mathbb{C})$?

Solution 35.   (i) Consider the case $n = 2$. Let $\epsilon \in \mathbb{R}$ and $\epsilon \neq 0$. Then an example is

$$A = \begin{pmatrix} 1 & \epsilon \\ 0 & 1 \end{pmatrix}, \quad B = \begin{pmatrix} 0 & i\epsilon \\ i/\epsilon & 0 \end{pmatrix} \Rightarrow [A, B] = \begin{pmatrix} i & 0 \\ 0 & -i \end{pmatrix}.$$

Note that $\det(A \otimes B - B \otimes A) = 0$. Another example is

$$C = \begin{pmatrix} 1 & i \\ 0 & 1 \end{pmatrix}, \quad D = \begin{pmatrix} 0 & 1 \\ -1 & 0 \end{pmatrix} \Rightarrow [C, D] = \begin{pmatrix} -i & 0 \\ 0 & i \end{pmatrix}.$$

(ii) We find

$$[A(z), B] = \begin{pmatrix} -z & 0 \\ 0 & z \end{pmatrix}$$

with the solutions $z = i$ and $z = -i$.

Problem 36.   The Lie group $SU(2)$ is defined by

$$SU(2) := \{\, U \ 2 \times 2 \ \text{matrix} : UU^* = I_2, \ \det(U) = 1 \,\}.$$

Let (3-sphere)

$$S^3 := \{\, (x_1, x_2, x_3, x_4) \in \mathbb{R}^4 \ : \ x_1^2 + x_2^2 + x_3^2 + x_4^2 = 1 \,\}.$$

Show that $SU(2)$ can be identified as a real manifold with the 3-sphere S^3.

Solution 36.   Let $a, b, c, d \in \mathbb{C}$ and

$$U = \begin{pmatrix} a & b \\ c & d \end{pmatrix}.$$

Imposing the conditions $UU^* = I_2$ and $\det(U) = 1$ we find that U has the form

$$U = \begin{pmatrix} a & b \\ -\bar{b} & \bar{a} \end{pmatrix}$$

where $a\bar{a} + b\bar{b} = 1$. Now we embed $SU(2)$ as a subset of $\mathbb{R}^4$ by writing out the complex numbers a and b in terms of their real and imaginary parts

$$a = x_1 + ix_2, \quad b = x_3 + ix_4, \qquad x_1, x_2, x_3, x_4 \in \mathbb{R}$$

whence

$$SU(2) \longrightarrow \mathbb{R}^4 : (a, b) \to (x_1, x_2, x_3, x_4).$$

The image of this map are points such that $a\bar{a} + b\bar{b} = 1$, that is $x_1^2 + x_2^2 + x_3^2 + x_4^2 = 1$. Given a and b we find

$$x_1 = \frac{a + \bar{a}}{2}, \quad x_2 = \frac{a - \bar{a}}{2i}, \quad x_3 = \frac{b + \bar{b}}{2}, \quad x_4 = \frac{b - \bar{b}}{2i}.$$

Problem 37. Consider the matrix ($\alpha \in \mathbb{R}$)

$$A(\alpha) = \begin{pmatrix} \cos(\alpha) & \sin(\alpha) \\ \sin(\alpha) & -\cos(\alpha) \end{pmatrix}.$$

(i) Is $A(\alpha)$ an element of $SO(2)$?

(ii) Consider the transformation

$$\begin{pmatrix} x_1' \\ x_2' \end{pmatrix} = A(\alpha) \begin{pmatrix} x_1 \\ x_2 \end{pmatrix}.$$

Is $(x_1')^2 + (x_2')^2 = x_1^2 + x_2^2$? Prove or disprove.

(iii) We define the star product

$$A(\alpha) \star A(\alpha) = \begin{pmatrix} \cos(\alpha) & 0 & 0 & \sin(\alpha) \\ 0 & \cos(\alpha) & \sin(\alpha) & 0 \\ 0 & \sin(\alpha) & -\cos(\alpha) & 0 \\ \sin(\alpha) & 0 & 0 & -\cos(\alpha) \end{pmatrix}.$$

Is $A(\alpha) \star A(\alpha)$ an element of $SO(4)$?

(iv) Is $A(\alpha) \otimes A(\alpha)$ an element of $SO(4)$?

(v) Find

$$X = \left.\frac{dA(\alpha)}{d\alpha}\right|_{\alpha=0}$$

and then $B(\alpha) = \exp(\alpha X)$. Compare $B(\alpha)$ and $A(\alpha)$ and discuss.

Solution 37. (i) We have $A(\alpha)^T = A(\alpha)$ and $A(\alpha)^T A(\alpha) = I_2$. Thus $A(\alpha)$ is an element of $O(2)$. Now $\det(A(\alpha)) = -1$ and thus $A(\alpha)$ is not an element of $SO(2)$.

(ii) Yes we have $x_1^2 + x_2^2 = (x_1')^2 + (x_2')^2$.

(iii) The eigenvalues are $+1$ (2 times) and -1 (2 times) Thus $\det(A(\alpha) \star A(\alpha)) = 1$.

(iv) We have $\det(A(\alpha) \otimes A(\alpha)) = \det(A(\alpha)) \det(A(\alpha)) = 1$.

(v) We obtain

$$X = \begin{pmatrix} 0 & 1 \\ 1 & 0 \end{pmatrix} \Rightarrow B(\alpha) = \begin{pmatrix} \cosh(\alpha) & \sinh(\alpha) \\ \sinh(\alpha) & \cosh(\alpha) \end{pmatrix}.$$

Thus $A(\alpha) \neq B(\alpha)$. $B(\alpha)$ is not an element of $O(2)$.

Problem 38. Given two 2×2 matrices A and B in the Lie group $GL(2,\mathbb{C})$, i.e. $\det(A) \neq 0$, $\det(B) \neq 0$. The action of an $n \times n$ matrix $M = (m_{jk})$ on a polynomial $p(\mathbf{x}) \in \mathbf{R}_n = \mathbb{C}[x_1, x_2, \ldots, x_n]$ may be defined by setting

$$T_M p(\mathbf{x}) := p(\mathbf{x}M)$$

where $\mathbf{x}M$ is the multiplication of the row n-vector $\mathbf{x}$ by the $n \times n$ matrix M. Thus $\mathbf{x}M$ is a row vector again. We denote by

$$\mathbf{R}_4^{GL(2,\mathbb{C}) \otimes GL(2,\mathbb{C})}$$

the ring of polynomials in $\mathbf{R}_4$ that are invariant under the action of $A \otimes B$ for all pairs $A, B \in GL(2,\mathbb{C})$, i.e.

$$\mathbf{R}_4^{GL(2,\mathbb{C}) \otimes GL(2,\mathbb{C})} := \{\, p \in \mathbf{R}_4 : T_{A \otimes B} p(\mathbf{x}) = p(\mathbf{x}) \,\}.$$

The action preserves degree and homogeneity. Consider the polynomial

$$p(x_1, x_2, x_3, x_4) = x_1^2 + x_2^2 + x_3^2 + x_4^2 + x_1 x_2 + x_2 x_3 + x_3 x_4 + x_4 x_1$$

and

$$A = B = \begin{pmatrix} 0 & 1 \\ 1 & 0 \end{pmatrix}.$$

Show that $T_{A \otimes B} p(\mathbf{x}) = p(\mathbf{x})$.

Solution 38. Since

$$\begin{pmatrix} x_1 & x_2 & x_3 & x_4 \end{pmatrix} (A \otimes B) = \begin{pmatrix} x_4 & x_3 & x_2 & x_1 \end{pmatrix}$$

we have

$$T_{A \otimes B} p(\mathbf{x}) = p(\mathbf{x}(A \otimes B)) = p(x_4, x_3, x_2, x_1) = p(x_1, x_2, x_3, x_4).$$

Problem 39. Let $\alpha, \beta \in \mathbb{R}$. Then the matrices

$$A(\alpha) = \begin{pmatrix} \cos(\alpha) & \sin(\alpha) \\ -\sin(\alpha) & \cos(\alpha) \end{pmatrix}, \qquad B(\beta) = \begin{pmatrix} \cos(\beta) & \sin(\beta) \\ -\sin(\beta) & \cos(\beta) \end{pmatrix}$$

are elements of the Lie group $SO(2)$ and $A(\alpha) \otimes B(\beta)$ is an element of the Lie group $SO(4)$. Find the "infinitesimal generators" X and Y such that

$$\exp(\alpha X + \beta Y) = A(\alpha) \otimes B(\beta).$$

Solution 39. We have

$$X = \frac{\partial}{\partial \alpha}(A(\alpha) \otimes B(\beta))\Big|_{\alpha=0,\beta=0} = \begin{pmatrix} 0 & 1 \\ -1 & 0 \end{pmatrix} \otimes \begin{pmatrix} 1 & 0 \\ 0 & 1 \end{pmatrix}$$

and

$$Y = \frac{\partial}{\partial \beta}(A(\alpha) \otimes B(\beta))\Big|_{\alpha=0,\beta=0} = \begin{pmatrix} 1 & 0 \\ 0 & 1 \end{pmatrix} \otimes \begin{pmatrix} 0 & 1 \\ -1 & 0 \end{pmatrix}$$

with $\exp(\alpha X + \beta Y) = A(\alpha) \otimes B(\alpha)$. Note that $[X, Y] = 0_2$.

Problem 40. A *topological group* G is both a group and a topological space, the two structures are related by the requirement that the maps $x \mapsto x^{-1}$ (of G onto G) and $(x, y) \mapsto xy$ (of $G \times G$ onto G) are continuous. $G \times G$ is given by the product topology.
(i) Given a topological group G, define the maps

$$\phi(x) := xax^{-1}$$

and

$$\psi(x) := xax^{-1}a^{-1} \equiv [x, a].$$

How are the iterates of the maps ϕ and ψ related?
(ii) Consider $G = SO(2)$ and

$$x = \begin{pmatrix} \cos(\alpha) & -\sin(\alpha) \\ \sin(\alpha) & \cos(\alpha) \end{pmatrix}, \qquad a = \begin{pmatrix} 0 & 1 \\ -1 & 0 \end{pmatrix}$$

with $x, a \in SO(2)$. Calculate ϕ and ψ. Discuss.

Solution 40. (i) The iterates $\phi^{(n)}$ and $\psi^{(n)}$ are related by

$$\phi^{(n)}(x) = \psi^{(n)}(x)a$$

since $\phi(x) = \psi(x)a$ and by induction

$$\phi^{(n+1)}(x) = (\psi^{(n)}(x)a)a(\psi^{(n)}(x)a)^{-1} = \psi^{(n)}(x)a(\psi^{(n)}(x))^{-1} = \psi^{n+1}(x)a.$$

(ii) We find for $\phi(x) = xax^{-1}$

$$\begin{pmatrix} \cos(\alpha) & -\sin(\alpha) \\ \sin(\alpha) & \cos(\alpha) \end{pmatrix} \begin{pmatrix} 0 & 1 \\ -1 & 0 \end{pmatrix} \begin{pmatrix} \cos(\alpha) & \sin(\alpha) \\ -\sin(\alpha) & \cos(\alpha) \end{pmatrix} = \begin{pmatrix} 0 & 1 \\ -1 & 0 \end{pmatrix}.$$

Thus a is a *fixed point*, i.e. $\phi(a) = a$, of ϕ. Since $\psi(x) = xax^{-1}a^{-1}$ we obtain

$$xax^{-1}a^{-1} = \begin{pmatrix} 1 & 0 \\ 0 & 1 \end{pmatrix} = I_2.$$

Thus $\psi^{(n)}(x) = I_2$ for all n.

Problem 41. The unitary matrices are elements of the Lie group $U(n)$. The corresponding Lie algebra $u(n)$ is the set of matrices with the condition $X^* = -X$. An important subgroup of $U(n)$ is the Lie group $SU(n)$ with the condition that $\det(U) = 1$. The unitary matrices

$$\frac{1}{\sqrt{2}} \begin{pmatrix} 1 & 1 \\ 1 & -1 \end{pmatrix}, \qquad \begin{pmatrix} 0 & 1 \\ 1 & 0 \end{pmatrix}$$

are not elements of the Lie group $SU(2)$ since the determinants of these unitary matrices are -1. The corresponding Lie algebra $su(n)$ of the Lie group $SU(n)$ are the $n \times n$ matrices given by

$$X^* = -X, \qquad \text{tr}(X) = 0.$$

Let σ_1, σ_2, σ_3 be the Pauli spin matrices. Then any unitary matrix in $U(2)$ can be represented by

$$U(\alpha, \beta, \gamma, \delta) = e^{i\alpha I_2} e^{-i\beta\sigma_3/2} e^{-i\gamma\sigma_2/2} e^{-i\delta\sigma_3/2}$$

where $0 \leq \alpha < 2\pi$, $0 \leq \beta < 2\pi$, $0 \leq \gamma \leq \pi$ and $0 \leq \delta < 2\pi$. Calculate the right-hand side.

Solution 41.   We find

$$e^{i\alpha I_2} = \begin{pmatrix} e^{i\alpha} & 0 \\ 0 & e^{i\alpha} \end{pmatrix}$$

$$e^{-i\beta\sigma_3/2} = \begin{pmatrix} e^{-i\beta/2} & 0 \\ 0 & e^{i\beta/2} \end{pmatrix}$$

$$e^{-i\gamma\sigma_2/2} = \begin{pmatrix} \cos(\gamma/2) & -\sin(\gamma/2) \\ \sin(\gamma/2) & \cos(\gamma/2) \end{pmatrix}$$

$$e^{-i\delta\sigma_3/2} = \begin{pmatrix} e^{-i\delta/2} & 0 \\ 0 & e^{i\delta/2} \end{pmatrix}.$$

This is the *cosine-sine decomposition*. Each of the four matrices on the right-hand side is unitary and $e^{i\alpha}$ is unitary. Thus U is unitary and $\det(U) = e^{2i\alpha}$. We obtain the special case of the Lie group $SU(2)$ if $\alpha = 0$.

Problem 42.   We consider the following subgroups of the Lie group $SL(2, \mathbb{R})$. Let

$$K := \left\{ \begin{pmatrix} \cos(\theta) & -\sin(\theta) \\ \sin(\theta) & \cos(\theta) \end{pmatrix} : \theta \in [0, 2\pi) \right\}$$

$$A := \left\{ \begin{pmatrix} r^{1/2} & 0 \\ 0 & r^{-1/2} \end{pmatrix} : r > 0 \right\}$$

$$N := \left\{ \begin{pmatrix} 1 & t \\ 0 & 1 \end{pmatrix} : t \in \mathbb{R} \right\}.$$

Any matrix $m \in SL(2, \mathbb{R})$ can be written in a unique way as the product $m = kan$ with $k \in K$, $a \in A$ and $n \in N$. This decomposition is called *Iwasawa decomposition* and has a natural generalization to $SL(n, \mathbb{R})$, $n \geq 3$. The notation of the subgroups comes from the fact that K is a compact subgroup, A is an abelian subgroup and N is a nilpotent subgroup of $SL(2, \mathbb{R})$. Find the Iwasawa decomposition of the matrix

$$\begin{pmatrix} \sqrt{2} & 1 \\ 1 & \sqrt{2} \end{pmatrix}.$$

Solution 42.   From

$$\begin{pmatrix} \sqrt{2} & 1 \\ 1 & \sqrt{2} \end{pmatrix} = \begin{pmatrix} \cos(\theta) & -\sin(\theta) \\ \sin(\theta) & \cos(\theta) \end{pmatrix} \begin{pmatrix} r^{1/2} & 0 \\ 0 & r^{-1/2} \end{pmatrix} \begin{pmatrix} 1 & t \\ 0 & 1 \end{pmatrix}$$

we obtain

$$\begin{pmatrix} \sqrt{2} & 1 \\ 1 & \sqrt{2} \end{pmatrix} = \begin{pmatrix} r^{1/2}\cos(\theta) & tr^{1/2}\cos(\theta) - r^{-1/2}\sin(\theta) \\ r^{1/2}\sin(\theta) & tr^{1/2}\sin(\theta) + r^{-1/2}\cos(\theta) \end{pmatrix}.$$

Thus we have the four conditions

$$r^{1/2}\cos(\theta) = \sqrt{2}, \quad r^{1/2}t\cos(\theta) - r^{-1/2}\sin(\theta) = 1$$
$$r^{1/2}\sin(\theta) = 1, \quad r^{1/2}t\sin(\theta) + r^{-1/2}\cos(\theta) = \sqrt{2}$$

for the three unknowns r, t, θ. From the first and third conditions we obtain $\tan(\theta) = 1/\sqrt{2}$ and therefore

$$\sin(\theta) = \frac{1}{\sqrt{3}}, \qquad \cos(\theta) = \frac{\sqrt{2}}{\sqrt{3}}.$$

It also follows that $r^{1/2} = \sqrt{3}$. From the second and fourth conditions it finally follows that $t = 2\sqrt{2}/3$. Thus we have the decomposition

$$\begin{pmatrix} \sqrt{2} & 1 \\ 1 & \sqrt{2} \end{pmatrix} = \begin{pmatrix} \sqrt{2}/\sqrt{3} & -1/\sqrt{3} \\ 1/\sqrt{3} & \sqrt{2}/\sqrt{3} \end{pmatrix} \begin{pmatrix} \sqrt{3} & 0 \\ 0 & 1/\sqrt{3} \end{pmatrix} \begin{pmatrix} 1 & 2\sqrt{2}/3 \\ 0 & 1 \end{pmatrix}.$$

Problem 43.   Let $0 \le \alpha < \pi/4$. Consider the transformation

$$X(x, y, \alpha) = \frac{1}{\sqrt{\cos(2\alpha)}}(x\cos(\alpha) + iy\sin(\alpha)),$$

$$Y(x, y, \alpha) = \frac{1}{\sqrt{\cos(2\alpha)}}(-ix\sin(\alpha) + y\cos(\alpha)).$$

(i) Show that $X^2 + Y^2 = x^2 + y^2$.
(ii) Do the matrices

$$\frac{1}{\sqrt{\cos(2\alpha)}}\begin{pmatrix} \cos(\alpha) & i\sin(\alpha) \\ -i\sin(\alpha) & \cos(\alpha) \end{pmatrix}$$

form a group under matrix multiplication?

Solution 43.   (i) Using the identity $\cos^2(\alpha) - \sin^2(\alpha) \equiv \cos(2\alpha)$ we find $X^2 + Y^2 = x^2 + y^2$.
(ii) The determinant of the matrix is 1. Thus it is an element of the Lie group $SL(2, \mathbb{C})$. For $\alpha = 0$ we have the identity matrix.

Problem 44.   The *Schur-Weyl duality theorem* states that if the natural representation of the unitary group $U(N)$ (or of the general linear group $GL(n, \mathbb{C})$)

is tensored (Kronecker product) n times, then under the joint action of tbe symmetry group S_n (permutation matrices) and $GL(n, \mathbb{C})$ this tensor product is decomposed into a direct sum of multiplicity-free irreducible subrepresentations uniquelly labelled by Young diagrams. Let $n = 4$. Consider the permutation matrix (flip matrix)

$$P = \begin{pmatrix} 1 & 0 & 0 & 0 \\ 0 & 0 & 1 & 0 \\ 0 & 1 & 0 & 0 \\ 0 & 0 & 0 & 1 \end{pmatrix}.$$

and the *Hadamard matrix*

$$U_H = \frac{1}{\sqrt{2}} \begin{pmatrix} 1 & 1 \\ 1 & -1 \end{pmatrix}$$

which is an element of the Lie group $U(2)$. Then

$$U_H \otimes U_H = \frac{1}{2} \begin{pmatrix} 1 & 1 & 1 & 1 \\ 1 & -1 & 1 & -1 \\ 1 & 1 & -1 & -1 \\ 1 & -1 & -1 & 1 \end{pmatrix}$$

which is an element of the Lie group $U(4)$. Let

$$\mathbf{v}_1 = \begin{pmatrix} v_{1,1} \\ v_{1,2} \end{pmatrix}, \quad \mathbf{v}_2 = \begin{pmatrix} v_{2,1} \\ v_{2,2} \end{pmatrix}$$

be elements of $\mathbb{C}^2$.
(i) Find the commutator $[P, U_H \otimes U_H]$.
(ii) Find $P(U_H \otimes U_H)(\mathbf{v}_1 \otimes \mathbf{v}_2)$. Discuss.

Solution 44. (i) We find $[P, U_H \otimes U_H] = 0_4$.
(ii) We obtain

$$P(U_H \otimes U_H)(\mathbf{v}_1 \otimes \mathbf{v}_2) = \begin{pmatrix} v_{1,1}(v_{2,1} + v_{2,2}) + v_{1,2}(v_{2,1} + v_{2,2}) \\ v_{1,1}v_{2,1} - v_{1,2}v_{2,2} \\ v_{1,1}v_{2,1} - v_{1,2}v_{2,2} \\ v_{1,1}(v_{2,1} - v_{2,2}) - v_{1,2}(v_{2,1} - v_{2,2}) \end{pmatrix}$$

$$+ \begin{pmatrix} 0 \\ v_{1,1}v_{2,2} - v_{1,2}v_{2,1} \\ -v_{1,1}v_{2,2} + v_{1,2}v_{2,1} \\ 0 \end{pmatrix}.$$

The two vectors on the right-hand side are orthogonal to each other. The eigenvalues of P are $+1$ (three times) and -1 (one times) with the corresponding eigenspaces

$$\left\{ \begin{pmatrix} a \\ b \\ b \\ d \end{pmatrix} : a, b, d \in \mathbb{C} \right\} \quad \left\{ \begin{pmatrix} 0 \\ x \\ -x \\ 0 \end{pmatrix} : x \in \mathbb{C} \setminus 0 \right\}.$$

Problem 45. The *octonion algebra* $\mathcal{O}$ is an 8-dimensional non-associative algebra. It is defined in terms of the basis elements e_μ ($\mu = 0, 1, \ldots, 7$) and their multiplication table. e_0 is the unit element. We use Greek indices ($\mu, \nu, \ldots$) to include the 0 and latin indices ($i, j, k, \ldots$) when we exclude the 0. We define

$$\hat{e}_k := e_{4+k} \quad \text{for} \quad k = 1, 2, 3.$$

The multiplication rules among the basis elements of octonions e_μ are given by

$$e_i e_j = -\delta_{ij} e_0 + \sum_{k=1}^{3} \epsilon_{ijk} e_k, \quad i, j, k = 1, 2, 3 \tag{1}$$

and

$$-e_4 e_i = e_i e_4 = \hat{e}_i, \quad e_4 \hat{e}_i = -\hat{e}_i e_4 = e_i, \quad e_4 e_4 = -e_0$$

$$\hat{e}_i \hat{e}_j = -\delta_{ij} e_0 - \sum_{k=1}^{3} \epsilon_{ijk} e_k, \quad i, j, k = 1, 2, 3$$

$$-\hat{e}_j e_i = e_i \hat{e}_j = -\delta_{ij} e_4 - \sum_{k=1}^{3} \epsilon_{ijk} \hat{e}_k, \quad i, j, k = 1, 2, 3$$

where δ_{ij} is the Kronecker delta and ϵ_{ijk} is $+1$ if (ijk) is an even permutation of (123), -1 if (ijk) is an odd permutation of (123) and 0 otherwise. We can formally summarize the multiplications as

$$e_\mu e_\nu = g_{\mu\nu} e_0 + \sum_{k=1}^{7} \gamma_{\mu\nu}^k e_k$$

where

$$g_{\mu\nu} = \text{diag}(1, -1, -1, -1, -1, -1, -1, -1), \quad \gamma_{ij}^k = -\gamma_{ji}^k.$$

with $\mu, \nu = 0, 1, \ldots, 7$; and $i, j, k = 1, 2, \ldots, 7$.
(i) Show that the set $\{e_0, e_1, e_2, e_3\}$ is a closed associative subalgebra.
(ii) Show that the octonian algebra $\mathcal{O}$ is non-associative.

Solution 45. (i) Owing to the relation (1) we have a faithful representation using the *Pauli spin matrices*

$$e_0 \to \sigma_0 \equiv I_2, \quad e_j \to -i\sigma_j \quad (j = 1, 2, 3).$$

We find the isomorphism

$$e_i e_j \Leftrightarrow -\sigma_i \sigma_j = -(\delta_{ij} + i \sum_{k=1}^{3} \epsilon_{ijk} \sigma_k) \Leftrightarrow -\delta_{ij} + \sum_{k=1}^{3} \epsilon_{ijk} e_k.$$

(ii) The algebra is non-associative owing to the relation

$$\hat{e}_i \hat{e}_j = -\delta_{ij} e_0 - \sum_{k=1}^{3} \epsilon_{ijk} e_k, \quad i, j, k = 1, 2, 3.$$

Thus we cannot find a matrix representation.

Supplementary Problems

Problem 1. (i) Let I_2, 0_2 be the 2×2 identity and zero matrix, respectively. Find the group generated by the 4×4 matrix

$$M = \begin{pmatrix} 0_2 & -I_2 \\ I_2 & 0_2 \end{pmatrix}.$$

(ii) Find the group generated by the 4×4 matrix

$$X = \frac{1}{\sqrt{2}} \begin{pmatrix} 1 & 1 \\ 1 & -1 \end{pmatrix} \otimes \frac{1}{\sqrt{2}} \begin{pmatrix} 1 & 1 \\ 1 & -1 \end{pmatrix}.$$

Find the eigenvalues of the matrix.
(iii) Find the group generated by the 6×6 matrix

$$B = \begin{pmatrix} 0_2 & 0_2 & I_2 \\ I_2 & 0_2 & 0_2 \\ 0_2 & I_2 & 0_2 \end{pmatrix}$$

under matrix multiplication.
(iv) Find the group (matrix multiplication) generated by the 4×4 matrix

$$A = \begin{pmatrix} 0 & 0 & 0 & 1 \\ 0 & 1 & 0 & 0 \\ 0 & 0 & 1 & 0 \\ 1 & 0 & 0 & 0 \end{pmatrix}.$$

Is the group commutative? Find the determinant of all these matrices from (i). Do these numbers form a group under multiplication? Find all the eigenvalues of these matrices. Do these numbers form a group under multiplication? Let $\alpha \in \mathbb{R}$. Find $\exp(\alpha A)$. Is $\exp(\alpha A)$ an element of $SL(4, \mathbb{R})$?

Problem 2. (i) What group is generated by the two 3×3 permutation matrices

$$A = \begin{pmatrix} 1 & 0 & 0 \\ 0 & 0 & 1 \\ 0 & 1 & 0 \end{pmatrix}, \quad B = \begin{pmatrix} 0 & 1 & 0 \\ 1 & 0 & 0 \\ 0 & 0 & 1 \end{pmatrix}$$

under matrix multiplication? Note that $A^2 = B^2 = I_3$.
(ii) Find the group generated by the 4×4 permutation matrices

$$P_1 = \begin{pmatrix} 0 & 1 & 0 & 0 \\ 0 & 0 & 1 & 0 \\ 0 & 0 & 0 & 1 \\ 1 & 0 & 0 & 0 \end{pmatrix}, \quad P_2 = \begin{pmatrix} 0 & 0 & 0 & 1 \\ 0 & 0 & 1 & 0 \\ 0 & 1 & 0 & 0 \\ 1 & 0 & 0 & 0 \end{pmatrix}$$

under matrix multiplication. There are twenty-four 4×4 permutation matrices which form a group under matrix multiplication.

Problem 3. (i) Let $\alpha, \beta \in \mathbb{R}$ and $\beta \neq 0$. Do the 2×2 matrices

$$M(\alpha, \beta) = \begin{pmatrix} \cos(\alpha) & \beta^{-1} \sin(\alpha) \\ -\beta \sin(\alpha) & \cos(\alpha) \end{pmatrix}$$

form a group under matrix multiplication? Note that $\det(M(\alpha, \beta)) = 1$.
(ii) Let $\alpha, \beta, \phi \in \mathbb{R}$ and $\alpha, \beta \neq 0$. Consider the 2×2 matrices

$$A(\alpha, \beta, \phi) = \begin{pmatrix} \alpha \cos(\phi) & -\beta \sin(\phi) \\ \beta^{-1} \sin(\phi) & \alpha^{-1} \cos(\phi) \end{pmatrix}.$$

Do the matrices form a group under matrix multiplication?
(iii) Let $\alpha, \theta \in \mathbb{R}$. Do the matrices

$$\begin{pmatrix} \cosh(\alpha) & e^{i\theta} \sinh(\alpha) \\ e^{-i\theta} \sinh(\alpha) & \cosh(\alpha) \end{pmatrix}$$

form a group under matrix multiplication? Are the matrices unitary?
(iv) Let $\alpha, \beta, \tau \in \mathbb{R}$. Do the 2×2 matrices

$$\begin{pmatrix} e^{i(\alpha+\beta)} \cosh(\tau) & e^{i(\alpha-\beta)} \sinh(\tau) \\ e^{-i(\alpha-\beta)} \sinh(\tau) & e^{-i(\alpha+\beta)} \cosh(\tau) \end{pmatrix}$$

form a group under matrix multiplication?

Problem 4. Do the eight 2×2 matrices

$$\begin{pmatrix} 1 & 0 \\ 0 & 1 \end{pmatrix}, \quad \begin{pmatrix} 1 & 0 \\ 0 & -1 \end{pmatrix}, \quad \begin{pmatrix} 0 & 1 \\ 1 & 0 \end{pmatrix}, \quad \begin{pmatrix} 0 & 1 \\ -1 & 0 \end{pmatrix},$$

$$\frac{1}{\sqrt{2}} \begin{pmatrix} 1 & 1 \\ 1 & -1 \end{pmatrix}, \quad \frac{1}{\sqrt{2}} \begin{pmatrix} 1 & -1 \\ 1 & 1 \end{pmatrix}, \quad \frac{1}{\sqrt{2}} \begin{pmatrix} -1 & 1 \\ 1 & 1 \end{pmatrix}, \quad \frac{1}{\sqrt{2}} \begin{pmatrix} 1 & 1 \\ -1 & 1 \end{pmatrix}$$

form a group under matrix multiplication? If not add the matrices so that one has a group.

Problem 5. (i) Do the 4×4 matrices

$$g(\phi, \theta) = \begin{pmatrix} e^{i\phi} \cosh(\theta) & 0 & 0 & \sinh(\theta) \\ 0 & e^{-i\phi} \cosh(\theta) & \sinh(\theta) & 0 \\ 0 & \sinh(\theta) & e^{i\phi} \cosh(\theta) & 0 \\ \sinh(\theta) & 0 & 0 & e^{-i\phi} \cosh(\theta) \end{pmatrix}$$

form a group under matrix multiplication?
(ii) Do the 4×4 matrices

$$g(\phi, \theta) = \begin{pmatrix} e^{i\phi} \cos(\theta) & 0 & 0 & -\sin(\theta) \\ 0 & e^{-i\phi} \cos(\theta) & -\sin(\theta) & 0 \\ 0 & \sin(\theta) & e^{i\phi} \cos(\theta) & 0 \\ \sin(\theta) & 0 & 0 & e^{-i\phi} \cos(\theta) \end{pmatrix}$$

form a group under matrix multiplication?

Problem 6. The Lie group $O(2)$ is generated by a rotation R_1 and a reflection R_2

$$R_1(\theta) = \begin{pmatrix} \cos(\theta) & -\sin(\theta) \\ \sin(\theta) & \cos(\theta) \end{pmatrix}, \quad R_2(\theta) = \begin{pmatrix} \cos(\theta) & \sin(\theta) \\ \sin(\theta) & -\cos(\theta) \end{pmatrix}.$$

Show that

$$\mathrm{tr}(R_1(\theta)) = 2\cos(\theta), \quad \mathrm{tr}(R_2(\theta)) = 0, \quad \det(R_1(\theta)) = 1, \quad \det(R_2(\theta)) = -1$$

and $\det(R_1(\theta)R_2(\theta)) = -1$, $\mathrm{tr}(R_1(\theta)R_2(\theta)) = 0$.

Problem 7. (i) Find all 3×3 permutation matrices P such that

$$P^{-1} \begin{pmatrix} 1/\sqrt{2} & 0 & 1/\sqrt{2} \\ 0 & 1 & 0 \\ 1/\sqrt{2} & 0 & -1/\sqrt{2} \end{pmatrix} P = \begin{pmatrix} 1/\sqrt{2} & 0 & 1/\sqrt{2} \\ 0 & 1 & 0 \\ 1/\sqrt{2} & 0 & -1/\sqrt{2} \end{pmatrix}.$$

Show that these matrices form a group, i.e. a subgroup of the 3×3 permutation matrices.

(ii) Find all 2×2 invertible matrices B with $\det(B) = 1$ such that

$$B \begin{pmatrix} 0 & -i \\ i & 0 \end{pmatrix} B^{-1} = \begin{pmatrix} 0 & -i \\ i & 0 \end{pmatrix}.$$

Show that these matrices form a group under matrix multiplication.

Problem 8. Prove or disprove: For any permutation matrix their eigenvalues form a group under multiplication. Hint. Each permutation matrix admits the eigenvalue $+1$. Why?

Problem 9. Let σ_j $(j = 0, 1, 2, 3)$ be the Pauli spin matrices, where $\sigma_0 = I_2$. Do the 4×4 matrices

$$\begin{pmatrix} \sigma_j & 0_2 \\ 0_2 & \sigma_k \end{pmatrix}, \quad \begin{pmatrix} 0_2 & \sigma_j \\ \sigma_k & 0_2 \end{pmatrix}, \quad j, k = 0, 1, 2, 3$$

form a group under matrix multiplication? If not add the elements to find a group.

Problem 10. Let $\omega := \exp(2\pi i/4)$. Consider the 3×3 unitary matrices

$$\Sigma = \begin{pmatrix} \omega & 0 & 0 \\ 0 & \omega^2 & 0 \\ 0 & 0 & \omega^3 \end{pmatrix}, \quad C = \begin{pmatrix} 0 & 0 & 1 \\ 0 & 1 & 0 \\ 1 & 0 & 0 \end{pmatrix},$$

$$\Omega = \begin{pmatrix} -i/2 & (1+i)/2 & 1/2 \\ (1+i)/2 & 0 & (1-i)/2 \\ 1/2 & (1-i)/2 & i/2 \end{pmatrix}.$$

Do the matrices of the set

$$\Lambda := \{ \Sigma^j C^k \Omega^\ell : 0 \le j \le 3, 0 \le k \le 1, 0 \le \ell \le 2 \}$$

form a group under matrix multiplication?

Problem 11. Consider the set of 2×2 matrices

$$\left\{ e = \begin{pmatrix} 1 & 0 \\ 0 & 1 \end{pmatrix}, \quad a = \begin{pmatrix} 0 & 1 \\ 1 & 0 \end{pmatrix} \right\}.$$

Then under matrix multiplication we have a group. Consider the set

$$\{ e \otimes e, \quad e \otimes a, \quad a \otimes e, \quad a \otimes a \}.$$

Show that this set forms a group under matrix multiplication. For example we have $(a \otimes a)(a \otimes a) = e \otimes e$. The neutral element is $e \otimes e$.

Problem 12. Consider the $2n \times 2n$ matrix

$$J := \begin{pmatrix} 0_n & I_n \\ -I_n & 0_n \end{pmatrix}.$$

We define that the $2n \times 2n$ matrix H over $\mathbb{R}$ is *Hamiltonian* if $(JH)^T = JH$. We define that the $2n \times 2n$ matrix S over $\mathbb{R}$ is *symplectic* if $S^T J S = J$. Show that if H is Hamiltonian and S is symplectic, then the matrix $S^{-1}HS$ is Hamiltonian. Note that since S invertible S^T is invertible. We have $(JS^{-1}HS)^T = S^T H^T (S^T)^{-1} J^T$.

Problem 13. Let $\alpha, \beta, \gamma \in \mathbb{R}$. The *Heisenberg group* $H_3(\mathbb{R})$ consists of all 3×3 upper triangle matrices of the form

$$M(\alpha, \beta, \gamma) = \begin{pmatrix} 1 & \alpha & \gamma \\ 0 & 1 & \beta \\ 0 & 0 & 1 \end{pmatrix}$$

with matrix multiplication as composition. Let $t \in \mathbb{R}$. Consider the matrices

$$A(t) = \begin{pmatrix} 1 & t & 0 \\ 0 & 1 & 0 \\ 0 & 0 & 1 \end{pmatrix}, \quad B(t) = \begin{pmatrix} 1 & 0 & 0 \\ 0 & 1 & t \\ 0 & 0 & 1 \end{pmatrix}, \quad C(t) = \begin{pmatrix} 1 & 0 & t \\ 0 & 1 & 0 \\ 0 & 0 & 1 \end{pmatrix}.$$

(i) Show that $\{A(t) : t \in \mathbb{R}\}$, $\{B(t) : t \in \mathbb{R}\}$, $\{C(t) : t \in \mathbb{R}\}$ are one-parameter subgroups in $H_3(\mathbb{R})$.
(ii) Show that $\{C(t) : t \in \mathbb{R}\}$ is the center of $H_3(\mathbb{R})$.

Problem 14. Let $\omega^3 = 1$. What group is generated by the 3×3 matrices

$$A = \begin{pmatrix} 1 & 0 & 0 \\ 0 & 0 & \omega \\ 0 & \omega^2 & 0 \end{pmatrix}, \quad B = \begin{pmatrix} 0 & 0 & \omega \\ 0 & 1 & 0 \\ \omega^2 & 0 & 0 \end{pmatrix}, \quad C = \begin{pmatrix} 0 & 0 & \omega \\ 0 & 1 & 0 \\ \omega^2 & 0 & 0 \end{pmatrix}$$

under matrix multiplication? Is the matrix $\Pi = \frac{1}{3}(A + B + C)$ a projection matrix?

Problem 15. Consider the six 3×3 permutation matrices denoted by P_0, P_1, P_2, P_3, P_4, P_5, where P_0 is the 3×3 identity matrix. Find all triples $(Q_1, Q_2. Q_3)$ of these permutation matrices such that $Q_1 Q_2 Q_3 = Q_3 Q_2 Q_1$, where $Q_1 \neq Q_2$, $Q_2 \neq Q_3$, $Q_3 \neq Q_1$.

Problem 16. Let $n \geq 2$ and $\omega = \exp(2\pi i / n)$, where ω is the n-th primitive root of unity with $\omega^n = 1$. Consider the diagonal and permutation matrices, respectively

$$
D = \begin{pmatrix} 1 & 0 & 0 & \cdots & 0 \\ 0 & \omega & 0 & \cdots & 0 \\ 0 & 0 & \omega^2 & \cdots & 0 \\ \vdots & \vdots & \vdots & & \vdots \\ 0 & 0 & 0 & \cdots & \omega^{n-1} \end{pmatrix}, \qquad
P = \begin{pmatrix} 0 & 1 & 0 & \cdots & 0 \\ 0 & 0 & 1 & \cdots & 0 \\ \vdots & \vdots & \vdots & & \vdots \\ 0 & 0 & 0 & \cdots & 1 \\ 1 & 0 & 0 & \cdots & 0 \end{pmatrix}.
$$

(i) Show that $D^n = P^n = I_n$.

(ii) Show that the set of matrices $\{ D^j P^k : j, k = 0, 1, \ldots, n-1 \}$ form a basis of the vector space of $n \times n$ matrices.

(iii) Show that $PD = \omega DP$, $P^j D^k = \omega^{jk} D^k P^j$.

(iv) Find the matrix $X = \zeta P + \zeta^{-1} P^{-1} + \eta D + \eta^{-1} D^{-1}$ and calculate the eigenvalues.

(v) Let U be the unitary matrix

$$
U = \frac{1}{\sqrt{n}} (\omega^{jk})
$$

where $j, k = 0, 1, \ldots, n-1$. Show that $UPU^{-1} = D$.

(v) Let $R = P \otimes I_n$ and $S = D \otimes I_n$. Find RS. Let $X = P \otimes P$ and $Y = D \otimes D$. Find XY. Find the commutator $[X, Y]$.

Problem 17. Consider the diagonal matrix D and the permutation matrix P

$$
D = \begin{pmatrix} 1 & 0 & 0 \\ 0 & \exp(2i\pi/3) & 0 \\ 0 & 0 & \exp(-2i\pi/3) \end{pmatrix}, \qquad
P = \begin{pmatrix} 0 & 1 & 0 \\ 0 & 0 & 1 \\ 1 & 0 & 0 \end{pmatrix}.
$$

(i) What group is generated by D and P?

(ii) Calculate the commutator $[D, P]$. Discuss.

Problem 18. In the Lie group $U(N)$ of the $N \times N$ unitary matrices one can find two $N \times N$ matrices U and V such that

$$
UV = e^{2\pi i / N} VU.
$$

Any $N \times N$ hermitian matrix H can be written in the form

$$
H = \sum_{j=0}^{N-1} \sum_{k=0}^{N-1} h_{jk} U^j V^k.
$$

Using the expansion coefficients h_{jk} one can associate to the hermitian matrix H the function

$$h(q, q) = \sum_{j=0}^{N-1} \sum_{k=1}^{N-1} h_{jk} e^{2\pi i (jp + kq)}$$

where $p = 0, 1, \ldots, N - 1$ and $q = 0, 1, \ldots, N - 1$. Consider the case $N = 2$ and

$$U = \begin{pmatrix} 1 & 0 \\ 0 & e^{i\pi} \end{pmatrix}, \qquad V = \begin{pmatrix} 0 & 1 \\ 1 & 0 \end{pmatrix}.$$

(i) Consider the hermitian and unitary matrix

$$\begin{pmatrix} 0 & -i \\ i & 0 \end{pmatrix}.$$

Find $h(p, q)$.

(ii) Consider the hermitian and projection matrix

$$\frac{1}{2} \begin{pmatrix} 1 & 1 \\ 1 & 1 \end{pmatrix}.$$

Find $h(p, q)$.

Problem 19. (i) Consider the 2×2 matrix

$$J := -i\sigma_2 = \begin{pmatrix} 0 & 1 \\ -1 & 0 \end{pmatrix}.$$

Show that any 2×2 matrix $A \in SL(2, \mathbb{C})$ satisfies $A^T J A = J$.
(ii) Let A satisfying $A^T J A = J$.
Is $(A \otimes A)^T (J \otimes J)(A \otimes A) = J \otimes J$?
Is $(A \oplus A)^T (J \oplus J)(A \oplus A) = J \oplus J$?
Is $(A \star A)^T (J \star J)(A \star A) = J \star J$?

Problem 20. The group $SL(2, \mathbb{F}_3)$ consists of unimodular 2×2 matrices with integer entries taken modulo three. Let

$$A = \begin{pmatrix} 1 & 2 \\ 0 & 1 \end{pmatrix}$$

be an element of $SL(2, \mathbb{F}_3)$. Find the inverse of A.

Problem 21. Consider the semi-simple Lie group $SL(2, \mathbb{R})$. Let $A \in SL(2, \mathbb{R})$. Then $A^{-1} \in SL(2, \mathbb{R})$. Explain why. Show that A and A^{-1} have the same eigenvalues. Is this still true for $A \in SL(3, \mathbb{R})$?

Problem 22. Let $A, B \in SL(2, \mathbb{R})$.
(i) Is $\text{tr}(A) = \text{tr}(A^{-1})$? Prove or disprove.
(ii) Is $\text{tr}(AB) = \text{tr}(A) \text{tr}(B) - \text{tr}(AB^{-1})$? Prove or disprove.

Problem 23. Let $\alpha \in \mathbb{R}$. Consider the 2×2 matrix

$$A(\alpha) = \frac{1}{\sqrt{2}} \begin{pmatrix} e^\alpha & -e^\alpha \\ e^{-\alpha} & e^{-\alpha} \end{pmatrix}.$$

(i) Find the trace, determinant, eigenvalues and normalized eigenvectors of the matrix $A(\alpha)$.
(ii) Calculate

$$X := \left. \frac{dA(\alpha)}{d\alpha} \right|_{\alpha=0}.$$

Find $\exp(\alpha X)$ and compare with $A(\alpha)$, i.e. is $\exp(\alpha X) = A(\alpha)$? Discuss.

Problem 24. (i) Let $\alpha \in \mathbb{R}$. The 2×2 *rotation matrix*

$$R(\alpha) = \begin{pmatrix} \cos(\alpha) & -\sin(\alpha) \\ \sin(\alpha) & \cos(\alpha) \end{pmatrix}$$

is an element of the Lie group $SO(2, \mathbb{R})$. Find the spectral decomposition of $R(\alpha)$.
(ii) Let $\beta \in \mathbb{R}$. The matrix

$$B(\beta) = \begin{pmatrix} \cosh(\beta) & \sinh(\beta) \\ \sinh(\beta) & \cosh(\beta) \end{pmatrix}$$

is an element of the Lie group $SO(1, 1, \mathbb{R})$. Find the spectral decomposition of $B(\beta)$.
(iii) Find the spectral decomposition of $R(\alpha) \otimes B(\beta)$.

Problem 25. The Lie group $SU(1, 1)$ consists of all 2×2 matrices T over the complex numbers with

$$TMT^* = M, \quad M = \begin{pmatrix} 1 & 0 \\ 0 & -1 \end{pmatrix}, \quad \det(T) = 1.$$

Show that the conditions on $\xi_0, \xi_1, \xi_2, \xi_3 \in \mathbb{R}$ such that

$$T = \begin{pmatrix} \xi_0 + i\xi_3 & \xi_1 + i\xi_2 \\ \xi_1 - i\xi_2 & \xi_0 - i\xi_3 \end{pmatrix}$$

is an element of $SU(1, 1)$ is given by $\det(T) = \xi_0^2 + \xi_3^2 - \xi_1^2 - \xi_2^2 = 1$.

Problem 26. The *unit sphere*

$$S^3 := \{ (x_1, x_2, x_3, x_4) \in \mathbb{R}^4 : \sum_{j=1}^{4} x_j^2 = 1 \}$$

we identify with the Lie group $SU(2)$ via the map

$$(x_1, x_2, x_3, x_4) \mapsto \begin{pmatrix} x_1 + ix_2 & -x_3 + ix_4 \\ x_3 + ix_4 & x_1 - ix_2 \end{pmatrix}.$$

(i) Map the standard basis of $\mathbb{R}^4$ into $SU(2)$ and express these matrices using the Pauli spin matrices and the 2×2 identity matrix.

(ii) Map the *Bell basis*

$$\frac{1}{\sqrt{2}} \begin{pmatrix} 1 \\ 0 \\ 0 \\ 1 \end{pmatrix}, \quad \frac{1}{\sqrt{2}} \begin{pmatrix} 1 \\ 0 \\ 0 \\ -1 \end{pmatrix}, \quad \frac{1}{\sqrt{2}} \begin{pmatrix} 0 \\ 1 \\ 1 \\ 0 \end{pmatrix}, \quad \frac{1}{\sqrt{2}} \begin{pmatrix} 0 \\ 1 \\ -1 \\ 0 \end{pmatrix}$$

into $SU(2)$ and express these matrices using the Pauli spin matrices and the 2×2 identity matrix.

Problem 27. Let A be an $n \times n$ matrix over $\mathbb{C}$. Show that if $A^T = -A$, then $e^A \in O(n, \mathbb{C})$.

Problem 28. (i) Can any element of the Lie group $SU(1,1)$ be written as

$$\begin{pmatrix} e^{i\alpha/2} & 0 \\ 0 & e^{-i\alpha/2} \end{pmatrix} \begin{pmatrix} \cosh(\zeta/2) & \sinh(\zeta/2) \\ \sinh(\zeta/2) & \cosh(\zeta/2) \end{pmatrix} \begin{pmatrix} e^{i\beta/2} & 0 \\ 0 & e^{-i\beta/2} \end{pmatrix} ?$$

Each element in the this product is an element of the Lie group $SU(1,1)$.

(ii) Can any element of the compact Lie group $SU(2)$ be written as

$$\begin{pmatrix} \cos(\alpha/2) & -\sin(\alpha/2) \\ \sin(\alpha/2) & \cos(\alpha/2) \end{pmatrix} \begin{pmatrix} e^{i\gamma/2} & 0 \\ 0 & e^{-i\gamma/2} \end{pmatrix} \begin{pmatrix} \cos(\beta/2) & \sin(\beta/2) \\ -\sin(\beta/2) & \cos(\beta/2) \end{pmatrix} ?$$

Problem 29. Consider the Lie group $SL(2, \mathbb{R})$, i.e. the set of all real 2×2 matrices with determinant equal to 1. A dynamical system in $SL(2, \mathbb{R})$ can be defined by

$$M_{k+2} = M_k M_{k+1}, \qquad k = 0, 1, 2, \ldots$$

with the initial matrices $M_0, M_1 \in SL(2, \mathbb{R})$. Let $F_k := \text{tr}(M_k)$. Is

$$F_{k+3} = F_{k+2} F_{k+1} - F_k \qquad k = 0, 1, 2, \ldots ?$$

Prove or disprove. Use that property that for any 2×2 matrix A we have (Cayley-Hamilton theorem)

$$A^2 - A\text{tr}(A) + I_2 \det(A) = 0_2.$$

Problem 30. (i) Let $\alpha \in \mathbb{R}$. Do the matrices

$$A(\alpha) = \begin{pmatrix} \cos(\alpha) & i \sin(\alpha) \\ i \sin(\alpha) & \cos(\alpha) \end{pmatrix}$$

form a group under matrix multiplication? Note that $A(0) = I_2$. So we have the neutral element. Furthermore $\det(A(\alpha)) = 1$.

(ii) Let $\alpha \in \mathbb{R}$. Do the matrices

$$B(\alpha) = \begin{pmatrix} \cosh(\alpha) & i\sinh(\alpha) \\ -i\sinh(\alpha) & \cosh(\alpha) \end{pmatrix}$$

form a group under matrix multiplication? Note $B(0) = I_2$. So we have a neutral element. Furthermore $\det(B(\alpha)) = 1$.

Problem 31.   Consider the 3×3 matrix

$$A(\theta) = \begin{pmatrix} \cos(\theta) & 0 & \sin(\theta) \\ 0 & 1 & 0 \\ -\sin(\theta) & 0 & \cos(\theta) \end{pmatrix}.$$

(i) Show that $A(\theta)$ is an element of the compact Lie group $SO(3, \mathbb{R})$.
(ii) Show that the eigenvalues of $A(\theta)$ are $\lambda_1 = 1$, $\lambda_2 = e^{i\theta}$, $\lambda_3 = e^{-i\theta}$.
(iii) Show that the 3×3 matrix

$$O(\theta, \phi) = \begin{pmatrix} \sin(\theta)\cos(\phi) & -\sin(\phi) & -\cos(\theta)\cos(\phi) \\ \sin(\theta)\sin(\phi) & \cos(\phi) & \cos(\theta)\sin(\phi) \\ \cos(\theta) & 0 & \sin(\theta) \end{pmatrix}$$

is an element of the compact Lie group $SO(3)$.

Problem 32.   Let n be an integer with $n \geq 2$. Let p, q be integers with $p, q \geq 1$ and $n = p + q$. Let I_p be the $p \times p$ identity matrix and I_q be the $q \times q$ identity matrix. Let

$$I_{p,q} = \begin{pmatrix} I_p & 0 \\ 0 & -I_q \end{pmatrix} \equiv I_p \oplus (-I_q).$$

The Lie group $O(p, q)$ is the set of all $n \times n$ matrices defined by

$$O(p, q) := \{ A \in GL(n, \mathbb{R}) : A^T I_{p,q} A = I_{p,q} \}.$$

Show that this is the group of all invertible linear maps of $\mathbb{R}^n$ that preserves the quadratic form

$$\sum_{j=1}^{p} x_j y_j - \sum_{j=p+1}^{n} x_j y_j.$$

Problem 33.   The Lie group $SU(2,2)$ is defined as the group of transformation on the four dimensional complex space $\mathbb{C}^4$ leaving invariant the indefinite quadratic form

$$|z_1|^2 + |z_2|^2 - |z_3|^2 - |z_4|^2.$$

The Lie algebra $su(2,2)$ is defined as the 4×4 matrices X with trace 0 and $X^*L + LX = 0_4$, where L is the 4×4 matrix

$$L = \begin{pmatrix} -I_2 & 0_2 \\ 0_2 & I_2 \end{pmatrix}.$$

Is

$$X = \begin{pmatrix} 0 & 1 \\ 1 & 0 \end{pmatrix} \otimes \begin{pmatrix} 0 & 1 \\ 1 & 0 \end{pmatrix}$$

an element of the Lie algebra $su(2,2)$? Find $\exp(zX)$. Discuss.

Problem 34. Let $\theta \in \mathbb{R}$.
(i) Is the 2×2 matrix

$$M_1(\theta) = \begin{pmatrix} \cos(\theta) & \sin(\theta) \\ -\sin(\theta) & \cos(\theta) \end{pmatrix} \frac{1}{\sqrt{2}} \begin{pmatrix} 1 & -1 \\ 1 & 1 \end{pmatrix} \begin{pmatrix} 1 & 0 \\ 0 & -i \end{pmatrix}$$

unitary? Prove or disprove. If so is the matrix an element of the Lie group $SU(2)$?
(ii) Is the 8×8 matrix

$$M_2(\theta) = \begin{pmatrix} \cos(\theta) & \sin(\theta) \\ -\sin(\theta) & \cos(\theta) \end{pmatrix} \otimes \frac{1}{\sqrt{2}} \begin{pmatrix} 1 & -1 \\ 1 & 1 \end{pmatrix} \otimes \begin{pmatrix} 1 & 0 \\ 0 & -i \end{pmatrix}$$

unitary? Prove or disprove. If so is the matrix an element of the Lie group $SU(4)$?
(iii) Let $\oplus$ be the direct sum. Is the 6×6 matrix

$$M_3(\theta) = \begin{pmatrix} \cos(\theta) & \sin(\theta) \\ -\sin(\theta) & \cos(\theta) \end{pmatrix} \oplus \frac{1}{\sqrt{2}} \begin{pmatrix} 1 & -1 \\ 1 & 1 \end{pmatrix} \oplus \begin{pmatrix} 1 & 0 \\ 0 & -i \end{pmatrix}$$

unitary? Prove or disprove. If so is the matrix an element of the Lie group $SU(4)$?

Problem 35. Write down two 2×2 matrices A and B which are elements of the Lie group $O(2, \mathbb{R})$ but **not** elements of $SO(2, \mathbb{R})$.
(i) Is AB an element of the Lie group $SO(2, \mathbb{R})$?
(ii) Is $A \otimes B$ an element of the Lie group $SO(4, \mathbb{R})$?
(iii) Is $A \oplus B$ an element of $SO(4, \mathbb{R})$?

Problem 36. Let $z = x + iy$ with $x, y \in \mathbb{R}$. Consider the 2×2 matrix

$$\begin{pmatrix} \cos(z) & -\sin(z) \\ \sin(z) & \cos(z) \end{pmatrix}.$$

One has the identities

$$\cos(x + iy) \equiv \cos(x)\cos(iy) - \sin(x)\sin(iy) \equiv \cos(x)\cosh(y) - i\sin(x)\sinh(y)$$
$$\sin(x + iy) \equiv \sin(x)\cos(iy) + \cos(x)\sin(iy) \equiv \sin(x)\cosh(y) + i\cos(x)\sinh(y).$$

Let $x = 0$. Then we arrive at the matrix

$$M(y) = \begin{pmatrix} \cosh(y) & -i\sinh(y) \\ i\sinh(y) & \cosh(y) \end{pmatrix}.$$

(i) Do these matrices ($y \in \mathbb{R}$) form a group under matrix multiplication?

(ii) Calculate

$$X = \frac{d}{dy} M(y)\Big|_{y=0}$$

and then $\exp(yX)$ with $y \in \mathbb{R}$. Discuss.

Problem 37.   The maximal compact Lie subgroup of $SL(2, \mathbb{C})$ is $SU(2)$. Give a 2×2 matrix A which is an element of $SL(2, \mathbb{C})$, but not an element of $SU(2)$.

Problem 38.   Let $A_1, A_2, \ldots, A_n$ be $m \times m$ matrices. We define

$$\sigma(A_1, \ldots, A_n) := \frac{1}{n!} \sum_{s \in S_n} A_{s(1)} \otimes \cdots \otimes A_{s(n)}$$

where S_n is the permutation group. Let $n = 3$ and consider the matrices

$$A_1 = \begin{pmatrix} 0 & 1 \\ 1 & 0 \end{pmatrix}, \quad A_2 = \begin{pmatrix} 0 & -i \\ i & 0 \end{pmatrix}, \quad A_3 = \begin{pmatrix} 1 & 0 \\ 0 & -1 \end{pmatrix}.$$

Find

$$\sigma(A_1, A_2, A_3) = \frac{1}{3!} \sum_{s \in S_3} A_{s(1)} \otimes A_{s(2)} \otimes A_{s(3)}.$$

Problem 39.   Let $\alpha, \beta, \gamma \in \mathbb{R}$. Do the 3×3 matrices

$$M(\alpha, \beta, \gamma) = \begin{pmatrix} \cos(\alpha) & \sin(\alpha) & \beta \\ -\sin(\alpha) & \cos(\alpha) & \gamma \\ 0 & 0 & 1 \end{pmatrix}$$

form a group under matrix multiplication?

Problem 40.   (i) Consider the non-compact Lie group $SO(1, 1, \mathbb{R})$ with the element

$$A(\alpha) = \begin{pmatrix} \cosh(\alpha) & \sinh(\alpha) \\ \sinh(\alpha) & \cosh(\alpha) \end{pmatrix}.$$

Show that the inverse of $A(\alpha)$ is given by

$$A^{-1}(\alpha) = A(-\alpha) = \begin{pmatrix} \cosh(\alpha) & -\sinh(\alpha) \\ -\sinh(\alpha) & \cosh(\alpha) \end{pmatrix}.$$

Show that the eigenvalues of $A(\alpha)$ are given by e^{α} and $e^{-\alpha}$.
(ii) Let $\oplus$ be the direct sum. Find the determinant, eigenvalues and normalized eigenvectors of the 3×3 matrix $(1) \oplus A(\alpha)$.

Problem 41.   (i) Let A, B be elements of $SL(n, \mathbb{R})$. Show that $A \otimes B$ is an element of $SL(n^2, \mathbb{R})$.
(ii) Let A, B be elements of $SL(n, \mathbb{R})$. Show that $A \oplus B$ is an element of $SL(2n, \mathbb{R})$.

(iii) Let A, B be elements of $SL(2, \mathbb{R})$. Let $\star$ be the star product. Show that $A \star B$ an element of $SL(4, \mathbb{R})$.

(iv) Are the matrices

$$X = \begin{pmatrix} 1 & 1 & 1 \\ 1 & 0 & 1 \\ 1 & 1 & 0 \end{pmatrix}, \qquad Y = \begin{pmatrix} 1 & 1 & 1 & 1 & 1 \\ 1 & 1 & 0 & 1 & 1 \\ 1 & 0 & 0 & 1 & 1 \\ 1 & 1 & 1 & 0 & 0 \\ 1 & 1 & 1 & 0 & 1 \end{pmatrix}$$

elements of $SL(3, \mathbb{R})$ and $SL(5, \mathbb{R})$, respectively?

Problem 42. Let A be an $n \times n$ matrix over $\mathbb{C}$. Assume that $\|A\| < 1$. Then $I_n + A$ is invertible ($I_n + A \in GL(n, \mathbb{C})$). Can we conclude that $I_n \otimes I_n + A \otimes A$ is invertible?

Problem 43. Let σ_1, σ_2, σ_3 be the Pauli spin matrices. We define

$$\tau_1 = i\sigma_1, \qquad \tau_2 = i\sigma_2, \qquad \tau_3 = i\sigma_3$$

and $\tau_0 = I_2$. Thus τ_j ($j = 0, 1, 2, 3$) are elements of the Lie group $SU(2)$.
(i) We define the operator R as ($k \geq 2$)

$$R(\tau_{j_1} \otimes \tau_{j_2} \otimes \cdots \otimes \tau_{j_k}) := \frac{1}{2}(\tau_{j_1}\tau_{j_2}^* \otimes \tau_{j_2}\tau_{j_3}^* \otimes \cdots \otimes \tau_{j_k}\tau_{j_1}^*).$$

Find $R(\tau_1 \otimes \tau_2 \otimes \tau_3)$.
(ii) Consider the unitary matrix (constructed from the four Bell states)

$$U = \frac{1}{\sqrt{2}} \begin{pmatrix} 1 & 0 & 0 & -i \\ 0 & -i & -1 & 0 \\ 0 & -i & 1 & 0 \\ 1 & 0 & 0 & i \end{pmatrix} \Rightarrow U^* = \frac{1}{\sqrt{2}} \begin{pmatrix} 1 & 0 & 0 & 1 \\ 0 & i & i & 0 \\ 0 & -1 & 1 & 0 \\ i & 0 & 0 & -i \end{pmatrix}.$$

Show that $U(i\sigma_1 \otimes i\sigma_1)U$, $U(i\sigma_2 \otimes i\sigma_2)U$, $U(i\sigma_3 \otimes i\sigma_3)U$ are elements of the Lie group $SO(4)$.

Problem 44. Consider the 8×8 permutation matrix

$$P = (1) \oplus \begin{pmatrix} 0 & 0 & 0 & 1 & 0 & 0 \\ 1 & 0 & 0 & 0 & 0 & 0 \\ 0 & 0 & 0 & 0 & 1 & 0 \\ 0 & 1 & 0 & 0 & 0 & 0 \\ 0 & 0 & 0 & 0 & 0 & 1 \\ 0 & 0 & 1 & 0 & 0 & 0 \end{pmatrix} \oplus (1)$$

and

$$G = \begin{pmatrix} g_{11} & g_{12} \\ g_{21} & g_{22} \end{pmatrix}, \quad \mathbf{v}_1 = \begin{pmatrix} v_{1,1} \\ v_{1,2} \end{pmatrix}, \quad \mathbf{v}_2 = \begin{pmatrix} v_{2,1} \\ v_{2,2} \end{pmatrix}, \quad \mathbf{v}_3 = \begin{pmatrix} v_{3,1} \\ v_{3,2} \end{pmatrix}.$$

(i) Show that $[P, G \otimes G \otimes G] = 0_8$, where

$$G = \frac{1}{\sqrt{2}} \begin{pmatrix} 1 & 1 \\ 1 & -1 \end{pmatrix}.$$

(ii) Assume that G is the Hadamard matrix. Find the vector

$$(P(G \otimes G \otimes G))(\mathbf{v}_1 \otimes \mathbf{v}_2 \otimes \mathbf{v}_3).$$

Find the eigenvalues and normalized eigenvectors of P. Discuss.

Chapter 19

Lie Algebras and Matrices

A real Lie algebra L is a real vector space together with a bilinear map

$$[,] : L \times L \to L$$

called the Lie bracket, such that the following identities hold for all $a, b, c \in L$

$$[a, a] = 0$$

and the so-called *Jacobi identity*

$$[a, [b, c]] + [c, [a, b]] + [b, [c, a]] = 0.$$

It follows that $[b, a] = -[a, b]$. If $\mathcal{A}$ is an associative algebra over a field $\mathbb{F}$ (for example, the $n \times n$ matrices over $\mathbb{C}$ and matrix multiplication) with the definition

$$[a, b] := ab - ba, \quad a, b \in \mathcal{A}$$

then $\mathcal{A}$ acquires the structure of a Lie algebra.

If $X \subseteq L$ then $\langle X \rangle$ denotes the *Lie subalgebra* generated by X, that is, the smallest Lie subalgebra of L containing X. It consists of all elements obtainable from X by a finite sequence of vector space operations and Lie multiplications. A set of generators for L is a subset $X \subseteq L$ such that $L = \langle X \rangle$. If L has a finite set of generators we say that it is finitely generated.

Given two Lie algebras L_1 and L_2, a *homomorphism* of Lie algebras is a function, $f : L_1 \to L_2$, that is a linear map between vector spaces L_1 and L_2 and that preserves Lie brackets, i.e. $f([a, b]) = [f(a), f(b)]$ for all $a, b \in L_1$.

Problem 1.   Consider the $n \times n$ matrices E_{ij} having 1 in the (i, j) position and 0 elsewhere, where $i, j = 1, \ldots, n$ (*elementary matrices*). Calculate the commutator. Discuss.

Solution 1.   Since $E_{ij}E_{kl} = \delta_{jk}E_{il}$ it follows that

$$[E_{ij}, E_{kl}] = \delta_{ik}E_{il} - \delta_{li}E_{kj}.$$

The coefficients are all ± 1 or 0; in particular all of them lie in the field $\mathbb{C}$. Thus the E_{ij} are the standard basis in the vector space of all $n \times n$ matrices and thus the generators for the Lie algebra of all $n \times n$ matrices with the commutator $[A, B] := AB - BA$.

Problem 2.   Consider the vector space of the 2×2 matrices over $\mathbb{R}$ and the three 2×2 matrices with trace equal to 0

$$E = \begin{pmatrix} 0 & 1 \\ 0 & 0 \end{pmatrix}, \quad F = \begin{pmatrix} 0 & 0 \\ 1 & 0 \end{pmatrix}, \quad H = \begin{pmatrix} 1 & 0 \\ 0 & -1 \end{pmatrix}.$$

Show that we have a basis of a Lie algebra.

Solution 2.   We only have to calculate the commutators and show that the right-hand side can be represented as linear combination of the generators. We find $[E, E] = [F, F] = [H, H] = 0_2$ and

$$[E, H] = -2E, \quad [E, F] = H, \quad [H, F] = -2F.$$

Thus E, F, H are a basis of a Lie algebra.

Problem 3.   Consider the matrices

$$h_1 = \begin{pmatrix} 1 & 0 & 0 \\ 0 & 0 & 0 \\ 0 & 0 & 1 \end{pmatrix}, \quad h_2 = \begin{pmatrix} 0 & 0 & 0 \\ 0 & 1 & 0 \\ 0 & 0 & 0 \end{pmatrix}, \quad h_3 = \begin{pmatrix} 0 & 0 & 1 \\ 0 & 0 & 0 \\ 1 & 0 & 0 \end{pmatrix},$$

$$e = \begin{pmatrix} 0 & 1 & 0 \\ 0 & 0 & 0 \\ 0 & 1 & 0 \end{pmatrix}, \quad f = \begin{pmatrix} 0 & 0 & 0 \\ 1 & 0 & 1 \\ 0 & 0 & 0 \end{pmatrix}.$$

Show that the matrices form a basis of a Lie algebra.

Solution 3.   We know that all $n \times n$ matrices over $\mathbb{R}$ or $\mathbb{C}$ form a Lie algebra ($g\ell(n, \mathbb{R})$ and $g\ell(n, \mathbb{C})$) under the commutator. Thus we only have to calculate the commutators and prove that they can be written as linear combinations of h_1, h_2, h_3 and e, f. We find

$$[h_1, h_2] = [h_1, h_3] = [h_2, h_3] = 0_n, \quad [e, f] = h_1 + h_3 - 2h_2$$

$$[h_1, f] = [h_3, f] = -[h_2, f] = -f, \quad [h_1, e] = [h_3, e] = -[h_2, e] = e.$$

Thus we have a basis of a Lie algebra.

Problem 4. Let A, B be $n \times n$ matrices over $\mathbb{C}$. Calculate $\text{tr}([A, B])$. Discuss.

Solution 4. Since is the trace is linear and $\text{tr}(AB) = \text{tr}(BA)$ we have

$$\text{tr}([A, B]) = \text{tr}(AB - BA) = \text{tr}(AB) - \text{tr}(BA) = 0.$$

Thus the trace of the commutator of any two $n \times n$ matrices is 0.

Problem 5. An $n \times n$ matrix X over $\mathbb{C}$ is *skew-hermitian* if $X^* = -X$. Show that the commutator of two skew-hermitian matrices is again skew-hermitian. Discuss.

Solution 5. Let X and Y be $n \times n$ skew-hermitian matrices. Then

$$[X, Y]^* = (XY - YX)^* = (XY)^* - (YX)^*$$
$$= Y^* X^* - X^* Y^* = YX - XY$$
$$= -[X, Y].$$

Thus the skew-hermitian matrices form a Lie subalgebra of the Lie algebra $g\ell(n, \mathbb{C})$.

Problem 6. The Lie algebra $su(m)$ consists of all $m \times m$ matrices X over $\mathbb{C}$ with the conditions $X^* = -X$ (i.e. X is skew-hermitian) and $\text{tr}(X) = 0$. Note that $\exp(X)$ is a unitary matrix. Find a basis for $su(3)$.

Solution 6. Since we have the condition $\text{tr}(X) = 0$ the dimension of the Lie algebra is 8. A possible basis is

$$\begin{pmatrix} 0 & i & 0 \\ i & 0 & 0 \\ 0 & 0 & 0 \end{pmatrix}, \quad \begin{pmatrix} 0 & 1 & 0 \\ -1 & 0 & 0 \\ 0 & 0 & 0 \end{pmatrix}, \quad \begin{pmatrix} i & 0 & 0 \\ 0 & -i & 0 \\ 0 & 0 & 0 \end{pmatrix}, \quad \begin{pmatrix} 0 & 0 & i \\ 0 & 0 & 0 \\ i & 0 & 0 \end{pmatrix},$$

$$\begin{pmatrix} 0 & 0 & 1 \\ 0 & 0 & 0 \\ -1 & 0 & 0 \end{pmatrix}, \quad \begin{pmatrix} 0 & 0 & 0 \\ 0 & 0 & i \\ 0 & i & 0 \end{pmatrix}, \quad \begin{pmatrix} 0 & 0 & 0 \\ 0 & 0 & 1 \\ 0 & -1 & 0 \end{pmatrix}, \quad \frac{1}{\sqrt{3}}\begin{pmatrix} i & 0 & 0 \\ 0 & i & 0 \\ 0 & 0 & -2i \end{pmatrix}.$$

The eight matrices are linearly independent.

Problem 7. Any fixed element X of a Lie algebra L defines a linear transformation

$$\text{ad}(X) : Z \to [X, Z] \quad \text{for any} \quad Z \in L.$$

Show that for any $K \in L$ we have

$$[\text{ad}(Y), \text{ad}(Z)]K = \text{ad}([Y, Z])K.$$

The linear mapping ad gives a representation of the Lie algebra known as *adjoint representation*.

Solution 7. We have

$$[\mathrm{ad}(Y), \mathrm{ad}(Z)]K = \mathrm{ad}(Y)[Z, K] - \mathrm{ad}(Z)[Y, K]$$
$$= [Y, [Z, K]] - [Z, [Y, K]] = [[Y, Z], K]$$
$$= \mathrm{ad}([Y, Z])K$$

where we used the *Jacobi identity* $[Y, [Z, K]] + [K, [Y, Z]] + [Z, [K, Y]] = 0$.

Problem 8.   There is only one non-commutative Lie algebra L of dimension 2. If x, y are the generators (basis in L), then

$$[x, y] = x.$$

(i) Find the *adjoint representation* of this Lie algebra. Let v, w be two elements of a Lie algebra. Then we define

$$\mathrm{ad}v(w) := [v, w]$$

and $w\mathrm{ad}v := [v, w]$.
(ii) The *Killing form* is defined by

$$\kappa(x, y) := \mathrm{tr}(\mathrm{ad}x \, \mathrm{ad}y)$$

for all $x, y \in L$. Find the Killing form.

Solution 8.   (i) Since

$$\mathrm{ad}x(x) = [x, x] = 0, \ \mathrm{ad}x(y) = [x, y] = x$$

and

$$\mathrm{ad}y(x) = [y, x] = -x, \ \mathrm{ad}y(y) = [y, y] = 0$$

we obtain

$$(x \ \ y) \begin{pmatrix} \mathrm{ad}x_{11} & \mathrm{ad}x_{12} \\ \mathrm{ad}x_{21} & \mathrm{ad}x_{22} \end{pmatrix} = (0 \ \ x), \ (x \ \ y) \begin{pmatrix} \mathrm{ad}y_{11} & \mathrm{ad}y_{12} \\ \mathrm{ad}y_{21} & \mathrm{ad}y_{22} \end{pmatrix} = (-x \ \ 0).$$

Since x, y are basis elements in L we obtain

$$\mathrm{ad}x = \begin{pmatrix} 0 & 1 \\ 0 & 0 \end{pmatrix} \qquad \mathrm{ad}y = \begin{pmatrix} -1 & 0 \\ 0 & 0 \end{pmatrix}.$$

(ii) We find

$$\kappa(x, x) = \mathrm{tr}(\mathrm{ad}x \, \mathrm{ad}x) = 0, \quad \kappa(x, y) = \mathrm{tr}(\mathrm{ad}x \, \mathrm{ad}y) = 0,$$
$$\kappa(y, x) = \mathrm{tr}(\mathrm{ad}y \, \mathrm{ad}x) = 0, \quad \kappa(y, y) = \mathrm{tr}(\mathrm{ad}y \, \mathrm{ad}y) = 1.$$

This can be written in matrix form

$$\begin{pmatrix} 0 & 0 \\ 0 & 1 \end{pmatrix}.$$

Problem 9. Consider the Lie algebra $L = s\ell(2, \mathbb{F})$ with $\text{char}(\mathbb{F}) \neq 2$. Take as the standard basis for L the three matrices

$$E = \begin{pmatrix} 0 & 1 \\ 0 & 0 \end{pmatrix}, \quad F = \begin{pmatrix} 0 & 0 \\ 1 & 0 \end{pmatrix}, \quad H = \begin{pmatrix} 1 & 0 \\ 0 & -1 \end{pmatrix}.$$

(i) Find the commutators.
(ii) Find the adjoint representation of L with the ordered basis $\{\, E \; H \; F \,\}$.
(iii) Show that L is *simple*. If L has no ideals except itself and 0, and if moreover $[L, L] \neq 0$, we call L simple. A subspace I of a Lie algebra L is called an *ideal* of L if $x \in L$, $y \in I$ together imply $[x, y] \in I$. In other words show that the Lie algebra $s\ell(2, \mathbb{R})$ has no proper nontrivial ideals.

Solution 9. (i) We have $[E, F] = H$, $[H, E] = 2E$, $[H, F] = -2F$.
(ii) We find

$$\text{ad}(E) = \begin{pmatrix} 0 & -2 & 0 \\ 0 & 0 & 1 \\ 0 & 0 & 0 \end{pmatrix}, \quad \text{ad}(F) = \begin{pmatrix} 0 & 0 & 0 \\ -1 & 0 & 0 \\ 0 & 2 & 0 \end{pmatrix}, \quad \text{ad}(H) = \begin{pmatrix} 2 & 0 & 0 \\ 0 & 0 & 0 \\ 0 & 0 & -2 \end{pmatrix}.$$

The eigenvalues of $\text{ad}(H)$ are $2, -2, 0$. Since $\text{char}(\mathbb{F}) \neq 2$, these eigenvalues are distinct.
(iii) Assume that $J \subset s\ell(2, \mathbb{R})$ is an ideal. Let $A \in J$. Then

$$[E, A] = \begin{pmatrix} a_{21} & a_{22} - a_{11} \\ 0 & -a_{21} \end{pmatrix} \in J.$$

It follows that

$$[H, [E, A]] = 2 \begin{pmatrix} 0 & a_{11} - a_{22} \\ 0 & a_{21} \end{pmatrix} \in J$$

and

$$[E, [H, [E, A]]] = \begin{pmatrix} 0 & -2a_{21} \\ 0 & 0 \end{pmatrix}.$$

With $-2a_{21} = 1$ we obtain the element E. Thus $E \in J$. Consider

$$B = A - a_{11}E = \begin{pmatrix} a_{11} & 0 \\ a_{21} & a_{22} \end{pmatrix} \in J$$

since J is a subspace. Then we find the commutator

$$[H, B] = \begin{pmatrix} 0 & 0 \\ -2a_{21} & 0 \end{pmatrix}.$$

With $-2a_{21} = 1$ we obtain the element F. Thus $F \in J$. Finally $[E, F] = H$. Thus $H \in J$ and the result follows.

Problem 10. Consider the four-dimensional Lie algebra $g\ell(2, \mathbb{R})$. The elementary matrices

$$e_1 = \begin{pmatrix} 1 & 0 \\ 0 & 0 \end{pmatrix}, \quad e_2 = \begin{pmatrix} 0 & 1 \\ 0 & 0 \end{pmatrix}, \quad e_3 = \begin{pmatrix} 0 & 0 \\ 1 & 0 \end{pmatrix}, \quad e_4 = \begin{pmatrix} 0 & 0 \\ 0 & 1 \end{pmatrix}$$

form a basis of $g\ell(2, \mathbb{R})$. Find the *adjoint representation*.

Solution 10.   Since

$$\mathrm{ad}e_1(e_1) = [e_1, e_1] = 0_2, \qquad \mathrm{ad}e_1(e_2) = [e_1, e_2] = e_2,$$

$$\mathrm{ad}e_1(e_3) = [e_1, e_3] = -e_3, \qquad \mathrm{ad}e_1(e_4) = [e_1, e_4] = 0_2$$

we have $(e_1 \ e_2 \ e_3 \ e_4)\mathrm{ad}e_1 = (0 \ e_2 \ -e_3 \ 0)$. Therefore

$$\mathrm{ad}(e_1) = \begin{pmatrix} 0 & 0 & 0 & 0 \\ 0 & 1 & 0 & 0 \\ 0 & 0 & -1 & 0 \\ 0 & 0 & 0 & 0 \end{pmatrix}.$$

Analogously, we have

$$\mathrm{ad}(e_2) = \begin{pmatrix} 0 & 0 & 1 & 0 \\ -1 & 0 & 0 & 1 \\ 0 & 0 & 0 & 0 \\ 0 & 0 & -1 & 0 \end{pmatrix}, \qquad \mathrm{ad}(e_3) = \begin{pmatrix} 0 & -1 & 0 & 0 \\ 0 & 0 & 0 & 0 \\ 1 & 0 & 0 & -1 \\ 0 & 1 & 0 & 0 \end{pmatrix}$$

$$\mathrm{ad}(e_4) = \begin{pmatrix} 0 & 0 & 0 & 0 \\ 0 & -1 & 0 & 0 \\ 0 & 0 & 1 & 0 \\ 0 & 0 & 0 & 0 \end{pmatrix}.$$

Problem 11.   Show that the matrices

$$J_1 = \begin{pmatrix} 0 & 0 & 0 \\ 0 & 0 & -1 \\ 0 & 1 & 0 \end{pmatrix}, \quad J_2 = \begin{pmatrix} 0 & 0 & 1 \\ 0 & 0 & 0 \\ -1 & 0 & 0 \end{pmatrix}, \quad J_3 = \begin{pmatrix} 0 & -1 & 0 \\ 1 & 0 & 0 \\ 0 & 0 & 0 \end{pmatrix}$$

form generators of a Lie algebra. Is the Lie algebra simple? A Lie algebra is *simple* if it contains no ideals other than L and 0.

Solution 11.   We find the commutators $[J_1, J_2] = J_3$, $[J_2, J_3] = J_1$, $[J_3, J_1] = J_2$. The Lie algebra contains no ideals other than L and 0. Thus the given Lie algebra is simple.

Problem 12.   The *roots* of a semisimple Lie algebra are the Lie algebra weights occurring in its adjoint representation. The set of roots forms the root system, and is completely determined by the semisimple Lie algebra. Consider the semisimple Lie algebra $s\ell(2, \mathbb{R})$ with the generators

$$E = \begin{pmatrix} 0 & 1 \\ 0 & 0 \end{pmatrix}, \quad F = \begin{pmatrix} 0 & 0 \\ 1 & 0 \end{pmatrix}, \quad H = \begin{pmatrix} 1 & 0 \\ 0 & -1 \end{pmatrix}.$$

Find the roots.

Solution 12. We obtain $\mathrm{ad}H(E) = [H, E] = 2E$, $\mathrm{ad}H(F) = [H, F] = -2F$ and $[E, F] = H$. Thus there are two roots of $s\ell(2, \mathbb{R})$ given by $\alpha(H) = 2$ and $\alpha(H) = -2$. The Lie algebraic rank of $s\ell(2, \mathbb{R})$ is one, and it has one positive root.

Problem 13. The simple Lie algebra $s\ell(2, \mathbb{R})$ is spanned by the matrices

$$H = \begin{pmatrix} 1 & 0 \\ 0 & -1 \end{pmatrix}, \qquad E = \begin{pmatrix} 0 & 1 \\ 0 & 0 \end{pmatrix}, \qquad F = \begin{pmatrix} 0 & 0 \\ 1 & 0 \end{pmatrix}$$

with the commutators $[H, E] = 2E$, $[H, F] = -2F$ and $[E, F] = H$.
(i) Define

$$C := \frac{1}{2}H^2 + EF + FE.$$

Find C. Calculate the commutators $[C, H]$, $[C, E]$, $[C, F]$. Show that C can be written in the form

$$C = \frac{1}{2}H^2 + H + 2FE.$$

(ii) Consider the vector in $\mathbb{R}^2$

$$\mathbf{v} = \begin{pmatrix} 1 \\ 0 \end{pmatrix}.$$

Calculate the vectors $H\mathbf{v}$, $E\mathbf{v}$, $F\mathbf{v}$ and $C\mathbf{v}$. Give an interpretation.
(iii) The *universal enveloping algebra* $U(s\ell(2, \mathbb{R}))$ of the Lie algebra $s\ell(2, \mathbb{R})$ is the associative algebra with generators H, E, F and the relations

$$HE - EH = 2E, \qquad HF - FH = -2F, \qquad EF - FE = H.$$

Find a basis of the Lie algebra $s\ell(2, \mathbb{R})$ so that all matrices are invertible. Find the inverse matrices of these matrices. Give the commutation relations.
(iv) Which of the E, F, H matrices are nonnormal? Use linear combinations to find a basis where all elements are normal matrices.

Solution 13. (i) Straightforward calculation yields

$$C = \frac{1}{2}\begin{pmatrix} 1 & 0 \\ 0 & 1 \end{pmatrix} + \begin{pmatrix} 0 & 1 \\ 0 & 0 \end{pmatrix}\begin{pmatrix} 0 & 0 \\ 1 & 0 \end{pmatrix} + \begin{pmatrix} 0 & 0 \\ 1 & 0 \end{pmatrix}\begin{pmatrix} 0 & 1 \\ 0 & 0 \end{pmatrix}$$

$$= \frac{3}{2}\begin{pmatrix} 1 & 0 \\ 0 & 1 \end{pmatrix}.$$

Since C is the 2×2 identity matrix times $3/2$ we find $[C, E] = 0_2$, $[C, F] = 0_2$, $[C, H] = 0_2$. Since $EF = FE + [E, F] = FE + H$ we obtain

$$C = \frac{1}{2}H^2 + H + 2FE.$$

(ii) We find

$$H\mathbf{v} = \mathbf{v}, \qquad E\mathbf{v} = \begin{pmatrix} 0 \\ 0 \end{pmatrix}, \qquad F\mathbf{v} = \begin{pmatrix} 0 \\ 1 \end{pmatrix}, \qquad C\mathbf{v} = \frac{3}{2}\mathbf{v}.$$

Thus $H\mathbf{v} = \mathbf{v}$ is an eigenvalue equation with eigenvalue $+1$. $E\mathbf{v} = \mathbf{0}$ is also an eigenvalue equation with eigenvalue 0 and finally $C\mathbf{v} = \frac{3}{2}\mathbf{v}$ is an eigenvalue equation with eigenvalue $3/2$.

(iii) The matrix H is already invertible with $H^{-1} = H$. For the other two matrices we set

$$\widetilde{E} := E - F = \begin{pmatrix} 0 & 1 \\ -1 & 0 \end{pmatrix}, \qquad \widetilde{F} := E + F = \begin{pmatrix} 0 & 1 \\ 1 & 0 \end{pmatrix}$$

with $\widetilde{F}^{-1} = \widetilde{F}$ and $\widetilde{E} = \widetilde{E}^T$. The commutators are

$$[\widetilde{E}, H] = -2\widetilde{F}, \quad [\widetilde{F}, H] = -2\widetilde{E}, \quad [\widetilde{E}, \widetilde{F}] = 2H.$$

(iv) Obviously E and F are nonnormal, but H is normal. A basis with all elements are normal matrices is given by

$$E + F = \begin{pmatrix} 0 & 1 \\ 1 & 0 \end{pmatrix}, \quad E - F = \begin{pmatrix} 0 & 1 \\ -1 & 0 \end{pmatrix}, \quad H = \begin{pmatrix} 1 & 0 \\ 0 & -1 \end{pmatrix}.$$

Problem 14. Let $\{e, f\}$ with

$$e = \begin{pmatrix} 0 & 1 \\ 0 & 0 \end{pmatrix}, \qquad f = \begin{pmatrix} -1 & 0 \\ 0 & 0 \end{pmatrix}$$

a basis for a Lie algebra. We have $[e, f] = e$.
(i) Is $\{\, e \otimes I_2, f \otimes I_2 \,\}$ a basis of a Lie algebra?
(ii) Is $\{\, e \otimes e, e \otimes f, f \otimes e, f \otimes f \,\}$ a basis of a Lie algebra?

Solution 14. Let A, B, C, D be $n \times n$ matrices. Then we have the identity

$$[A \otimes B, C \otimes D] \equiv (AC) \otimes (BD) - (CA) \otimes (DB).$$

Using this identity we obtain

$$[e \otimes I_2, f \otimes I_2] = (ef) \otimes I_2 - (fe) \otimes I_2 = (ef - fe) \otimes I_2 = e \otimes I_2$$
$$[e \otimes I_2, e \otimes I_2] = (ee) \otimes I_2 - (ee) \otimes I_2 = (ee - ee) \otimes I_2 = 0_2 \otimes 0_2$$
$$[f \otimes I_2, f \otimes I_2] = (ff) \otimes I_2 - (ff) \otimes I_2 = (ff - ff) \otimes I_2 = 0_2 \otimes 0_2.$$

Thus $\{\, e \otimes I_2, f \otimes I_2 \,\}$ is a basis of a Lie algebra.
(ii) Straightforward calculation yields

$$[e \otimes e, e \otimes e] = 0_4$$
$$[e \otimes e, e \otimes f] = 0_4$$
$$[e \otimes e, f \otimes e] = 0_4$$
$$[e \otimes f, e \otimes f] = 0_4$$
$$[e \otimes f, f \otimes e] = 0_4$$
$$[e \otimes f, f \otimes f] = -e \otimes f$$
$$[f \otimes e, f \otimes e] = 0_4$$
$$[f \otimes e, f \otimes f] = -f \otimes e$$

$$[f \otimes f, f \otimes f] = 0_4$$

where 0_4 is the 4×4 zero matrix. Thus the set is a basis of a Lie algebra.

Problem 15. We know that

$$E = \begin{pmatrix} 0 & 1 \\ 0 & 0 \end{pmatrix}, \quad H = \begin{pmatrix} 1 & 0 \\ 0 & -1 \end{pmatrix}, \quad F = \begin{pmatrix} 0 & 0 \\ 1 & 0 \end{pmatrix}$$

is an ordered basis of the simple Lie algebra $s\ell(2, \mathbb{R})$ with

$$[E, H] = -2E, \quad [E, F] = H, \quad [F, H] = 2F.$$

Consider

$$\tilde{E} = \begin{pmatrix} 0 & 0 & 0 & 1 \\ 0 & 0 & 1 & 0 \\ 0 & 0 & 0 & 0 \\ 0 & 0 & 0 & 0 \end{pmatrix}, \quad \tilde{H} = \begin{pmatrix} 1 & 0 & 0 & 0 \\ 0 & 1 & 0 & 0 \\ 0 & 0 & -1 & 0 \\ 0 & 0 & 0 & -1 \end{pmatrix}, \quad \tilde{F} = \begin{pmatrix} 0 & 0 & 0 & 0 \\ 0 & 0 & 0 & 0 \\ 0 & 1 & 0 & 0 \\ 1 & 0 & 0 & 0 \end{pmatrix}.$$

Find the commutators $[\tilde{E}, \tilde{H}]$, $[\tilde{E}, \tilde{F}]$, $[\tilde{F}, \tilde{H}]$.

Solution 15. We find $[\tilde{E}, \tilde{H}] = -2\tilde{E}$, $[\tilde{E}, \tilde{F}] = \tilde{H}$, $[\tilde{F}, \tilde{H}] = 2\tilde{F}$.

Problem 16. The 2×2 matrices

$$E = \begin{pmatrix} 0 & 1 \\ 0 & 0 \end{pmatrix}, \quad F = \begin{pmatrix} 0 & 0 \\ 1 & 0 \end{pmatrix}, \quad H = \begin{pmatrix} 1 & 0 \\ 0 & -1 \end{pmatrix}$$

form a basis of the Lie algebra $s\ell(2, \mathbb{R})$ with the commutation relations $[E, H] = -2E$, $[E, F] = H$, $[F, H] = 2F$.
(i) Let $\star$ be the *star product*. Do the nine 4×4 matrices

$$E \star E, \quad E \star F, \quad E \star H, \quad F \star E, \quad F \star F, \quad F \star H, \quad H \star E, \quad H \star F, \quad H \star H$$

form a basis of a Lie algebra?
(ii) Define the 4×4 matrices

$$\tilde{H} := H \otimes I_2 + I_2 \otimes H, \quad \tilde{E} := E \otimes H^{-1} + H \otimes E, \quad \tilde{F} := F \otimes H^{-1} + H \otimes F.$$

Find the commutators $[\tilde{H}, \tilde{E}]$, $[\tilde{H}, \tilde{F}]$, $[\tilde{E}, \tilde{F}]$.
(iii) Consider

$$\tilde{H} = H \otimes I_2 + I_2 \otimes H, \quad \tilde{E} = E \otimes H + H \otimes E, \quad \tilde{F} = F \otimes H + H \otimes F.$$

Find the commutators $[\tilde{H}, \tilde{E}]$, $[\tilde{H}, \tilde{F}]$, $[\tilde{E}, \tilde{F}]$ and thus show that we have a basis of a Lie algebra. Is this Lie algebra isomorphic to $s\ell(2, \mathbb{R})$?

Solution 16. (i) Yes. We have

$$[E \star E, F \star F] = H \star H, \quad [E \star E, H \star H] = -2E \star E, \quad [F \star F, H \star H] = 2F.$$

(ii) Note that $H^{-1} = H$. Now we find

$$[\tilde{H}, \tilde{E}] = 2\tilde{E}, \quad [\tilde{H}, \tilde{F}] = -2\tilde{F}, \quad [\tilde{E}, \tilde{F}] = H.$$

(iii) We note that $H^2 = I_2$, $EH = -E$, $HE = E$, $FH = F$, $HF = -F$. Thus

$$[\tilde{E}, \tilde{F}] = \tilde{H}, \quad [\tilde{E}, \tilde{H}] = -2\tilde{E}, \quad [\tilde{F}, \tilde{H}] = 2\tilde{F}.$$

Thus the two Lie algebras are isomorphic.

Problem 17. A basis for the Lie algebra $su(N)$, for odd N, may be built from two unitary unimodular $N \times N$ matrices

$$g = \begin{pmatrix} 1 & 0 & 0 & \cdots & 0 \\ 0 & \omega & 0 & \cdots & 0 \\ 0 & 0 & \omega^2 & \cdots & 0 \\ \vdots & \vdots & \vdots & \ddots & \vdots \\ 0 & 0 & 0 & \cdots & \omega^{N-1} \end{pmatrix}, \quad h = \begin{pmatrix} 0 & 1 & 0 & \cdots & 0 \\ 0 & 0 & 1 & \cdots & 0 \\ \vdots & \vdots & \vdots & \ddots & \vdots \\ 0 & 0 & 0 & \cdots & 1 \\ 1 & 0 & 0 & \cdots & 0 \end{pmatrix}$$

where ω is a primitive N-th root of unity, i.e. with period not smaller than N, here taken to be $\exp(4\pi i/N)$. We obviously have

$$hg = \omega gh. \tag{1}$$

(i) Find g^N and h^N.
(ii) Find $\operatorname{tr}(g)$.
(iii) Let $\mathbf{m} = (m_1, m_2)$, $\mathbf{n} = (n_1, n_2)$ and define

$$\mathbf{m} \times \mathbf{n} := m_1 n_2 - m_2 n_1$$

where $m_1 = 0, 1, \ldots, N - 1$ and $m_2 = 0, 1, \ldots, N - 1$. The complete set of unitary unimodular $N \times N$ matrices

$$J_{m_1, m_2} := \omega^{m_1 m_2/2} g^{m_1} h^{m_2}$$

suffice to span the Lie algebra $su(N)$, where $J_{0,0} = I_N$. Find J^*.
(iv) Calculate $J_{\mathbf{m}} J_{\mathbf{n}}$.
(v) Find the commutator $[J_{\mathbf{m}}, J_{\mathbf{n}}]$.

Solution 17. (i) Since $\omega^N = 1$ we find $g^N = I_N$. We also obtain $h^N = I_N$.
(ii) Since

$$1 + \omega + \omega^2 + \cdots + \omega^{N-1} = 0$$

we find $\operatorname{tr}(g) = 0$.
(iii) Obviously we have $J^*_{(m_1, m_2)} = J_{(-m_1, -m_2)}$.
(iv) Using equation (1) we find $J_{\mathbf{m}} J_{\mathbf{n}} = \omega^{\mathbf{n} \times \mathbf{m}/2} J_{\mathbf{m}+\mathbf{n}}$.
(v) Using the result from (iv) we obtain

$$[J_{\mathbf{m}}, J_{\mathbf{n}}] = -2i \sin\left(\frac{2\pi}{N} \mathbf{m} \times \mathbf{n}\right) J_{\mathbf{m}+\mathbf{n}}.$$

Problem 18. Let L be a finite dimensional Lie algebra and $Z(L)$ the center of L. Show that ad $: L \to g\ell(L)$ is a homomorphism of the Lie algebra L with kernel $Z(L)$.

Solution 18. Using the bilinearity of the commutator we have

$$\mathrm{ad}(cx) = c\,\mathrm{ad}(x), \qquad \mathrm{ad}(x+y) = \mathrm{ad}(x) + \mathrm{ad}(y)$$

where $c \in \mathbb{F}$ and $x, y \in L$. Now let $x, y, z \in L$. Then we have

$$
\begin{aligned}
\mathrm{ad}([x,y])z &= [[x,y],z] = [z,[y,x]] \\
&= [x,[y,z]] - [y,[x,z]] = \mathrm{ad}(x)[y,z] - \mathrm{ad}(y)[x,z] \\
&= \mathrm{ad}(x)\mathrm{ad}(y)(z) - \mathrm{ad}(y)\mathrm{ad}(x)(z) \\
&= [\mathrm{ad}(x), \mathrm{ad}(y)](z).
\end{aligned}
$$

Consider $x \in \ker(\mathrm{ad})$ if and only if $0 = \mathrm{ad}(x)y = [x,y]$ for every $y \in L$, if and only if $x \in Z(L)$. Thus $\ker(\mathrm{ad}) = Z(L)$.

Problem 19. Consider the 2×2 matrices over $\mathbb{R}$

$$A = \begin{pmatrix} -1 & 0 \\ 1 & 0 \end{pmatrix}, \qquad B = \begin{pmatrix} 0 & 1 \\ 0 & -1 \end{pmatrix}.$$

Calculate the commutator $C = [A, B]$ and check whether C can be written as a linear combination of A and B. If so we have a basis of a Lie algebra.

Solution 19. We find

$$C = [A, B] = \begin{pmatrix} -1 & -1 \\ 1 & 1 \end{pmatrix} = A - B.$$

Problem 20. A *Chevalley basis* for the semisimple Lie algebra $s\ell(3, \mathbb{R})$ is given by

$$X_1 = \begin{pmatrix} 0 & 0 & 0 \\ 0 & 0 & 1 \\ 0 & 0 & 0 \end{pmatrix}, \quad X_2 = \begin{pmatrix} 0 & 1 & 0 \\ 0 & 0 & 0 \\ 0 & 0 & 0 \end{pmatrix}, \quad X_3 = \begin{pmatrix} 0 & 0 & 1 \\ 0 & 0 & 0 \\ 0 & 0 & 0 \end{pmatrix},$$

$$Y_1 = \begin{pmatrix} 0 & 0 & 0 \\ 0 & 0 & 0 \\ 0 & 1 & 0 \end{pmatrix}, \quad Y_2 = \begin{pmatrix} 0 & 0 & 0 \\ 1 & 0 & 0 \\ 0 & 0 & 0 \end{pmatrix}, \quad Y_3 = \begin{pmatrix} 0 & 0 & 0 \\ 0 & 0 & 0 \\ 1 & 0 & 0 \end{pmatrix},$$

$$H_1 = \begin{pmatrix} 0 & 0 & 0 \\ 0 & 1 & 0 \\ 0 & 0 & -1 \end{pmatrix}, \quad H_2 = \begin{pmatrix} 1 & 0 & 0 \\ 0 & -1 & 0 \\ 0 & 0 & 0 \end{pmatrix}$$

where $Y_j = X_j^T$ for $j = 1, 2, 3$. The Lie algebra has rank 2 owing to H_1, H_2 and $[H_1, H_2] = 0_3$. Another basis could be formed by looking at the linear combinations

$$U_j = X_j + Y_j, \qquad V_j = X_j - V_j.$$

(i) Find the table of the commutator.

(ii) Calculate the vectors of commutators

$$\begin{pmatrix} [H_1, X_1] \\ [H_2, X_1] \end{pmatrix}, \qquad \begin{pmatrix} [H_1, X_2] \\ [H_2, X_2] \end{pmatrix}, \qquad \begin{pmatrix} [H_1, X_3] \\ [H_2, X_3] \end{pmatrix}$$

and thus find the roots.

Solution 20.   (i) The commutators are summarized in the table

[,]	X_1	X_2	X_3	Y_1	Y_2	Y_3	H_1	H_2
X_1	0	$-X_3$	0	H_1	0	Y_2	$-2X_1$	X_1
X_2	.	0	0	0	H_2	$-Y_1$	X_2	$-2X_2$
X_3	.	.	.	X_2	$-X_1$	$H_1 + H_2$	$-X_3$	$-X_3$
Y_1	.	.	.	0	Y_3	0	$2Y_1$	$-Y_1$
Y_2	.	.	.	.	0	0	$-Y_2$	$2Y_2$
Y_3	.	.	.	.	.	0	Y_3	Y_3
H_1	.	.	.	.	.	.	0	0
H_2	.	.	.	.	.	.	.	0

(ii) We find

$$\begin{pmatrix} [H_1, X_1] \\ [H_2, X_1] \end{pmatrix} = \begin{pmatrix} 2X_1 \\ -X_1 \end{pmatrix} = \begin{pmatrix} 2 \\ -1 \end{pmatrix} X_1,$$

$$\begin{pmatrix} [H_1, X_2] \\ [H_2, X_2] \end{pmatrix} = \begin{pmatrix} -X_2 \\ 2X_2 \end{pmatrix} = \begin{pmatrix} -1 \\ 2 \end{pmatrix} X_2,$$

$$\begin{pmatrix} [H_1, X_3] \\ [H_2, X_3] \end{pmatrix} = \begin{pmatrix} X_3 \\ X_3 \end{pmatrix} = \begin{pmatrix} 1 \\ 1 \end{pmatrix} X_3$$

with the roots $(2, -1)^T$, $(-1, 2)^T$, $(1, 1)^T$.

Problem 21.   Consider the 3×3 matrices

$$h_1 = \begin{pmatrix} 1 & 0 & 0 \\ 0 & 0 & 0 \\ 0 & 0 & 1 \end{pmatrix}, \quad h_2 = \begin{pmatrix} 0 & 0 & 0 \\ 0 & 1 & 0 \\ 0 & 0 & 0 \end{pmatrix}, \quad h_3 = \begin{pmatrix} 0 & 0 & 1 \\ 0 & 0 & 0 \\ 1 & 0 & 0 \end{pmatrix},$$

$$e = \begin{pmatrix} 0 & 1 & 0 \\ 0 & 0 & 0 \\ 0 & 1 & 0 \end{pmatrix}, \quad f = \begin{pmatrix} 0 & 0 & 0 \\ 1 & 0 & 1 \\ 0 & 0 & 0 \end{pmatrix}.$$

Show that the matrices form a basis of a Lie algebra.

Solution 21.   We know that all $n \times n$ matrices over $\mathbb{R}$ or $\mathbb{C}$ form a Lie algebra $(g\ell(n, \mathbb{R})$ and $g\ell(n, \mathbb{C}))$ under the commutator. Thus we only have to calculate the commutators and prove that they can be written as linear combinations of h_1, h_2, h_3 and e, f. We find

$$[h_1, h_2] = [h_1, h_3] = [h_2, h_3] = 0_n, \qquad [e, f] = h_1 + h_3 - 2h_2$$

$$[h_1, f] = [h_3, f] = -[h_2, f] = -f, \qquad [h_1, e] = [h_3, e] = -[h_2, e] = e.$$

Thus we have a basis of a Lie algebra.

Problem 22. Consider the two 2×2 matrices

$$A = \begin{pmatrix} 1 & 0 \\ 0 & -1 \end{pmatrix}, \quad B = \begin{pmatrix} 0 & 1 \\ 1 & 0 \end{pmatrix}$$

where A is the Pauli spin matrix σ_3 and B the Pauli spin matrix σ_1. The two matrices A, B are linearly independent. Let A, B be the generators of a Lie algebra. Classify the Lie algebra generated.

Solution 22. We have

$$[A, B] = \begin{pmatrix} 0 & 2 \\ -2 & 0 \end{pmatrix} = C.$$

The matrices A, B, C are linearly independent. Now we set $\tilde{C} = \frac{1}{2}C$, where $\tilde{C} = i\sigma_2$. Then we find

$$[A, B] = 2\tilde{C}, \quad [A, \tilde{C}] = 2B, \quad [B, \tilde{C}] = -2A.$$

From the commutation relation we see that A, B, $\tilde{C}$ is a basis of a simple Lie algebra.

Problem 23. The $g\ell(1|1)$ superalgebra involves two even (denoted by H and Z) and two odd (denoted by E, F) generators. The following commutation and anti-commutation relations hold

$$[Z, E] = [Z, F] = [Z, H] = 0, \quad [H, E] = E, \quad [H, F] = -F, \quad [E, F]_+ = Z$$

and $E^2 = F^2 = 0$. Find a 2×2 matrix representation.

Solution 23. We find

$$E = \begin{pmatrix} 0 & 1 \\ 0 & 0 \end{pmatrix}, \quad F = \begin{pmatrix} 0 & 0 \\ 1 & 0 \end{pmatrix}, \quad Z = \begin{pmatrix} 1 & 0 \\ 0 & 1 \end{pmatrix}, \quad H = \frac{1}{2}\begin{pmatrix} 1 & 0 \\ 0 & -1 \end{pmatrix} = \frac{1}{2}\sigma_3.$$

Problem 24. A basis of the simple Lie algebra $s\ell(2, \mathbb{R})$ is given by the traceless 2×2 matrices

$$X_1 = \frac{1}{2}\begin{pmatrix} 0 & 1 \\ 1 & 0 \end{pmatrix}, \quad X_2 = \frac{1}{2}\begin{pmatrix} -1 & 0 \\ 0 & 1 \end{pmatrix}, \quad X_3 = \frac{1}{2}\begin{pmatrix} 0 & 1 \\ -1 & 0 \end{pmatrix}.$$

(i) Find the commutators $[X_1, X_2]$, $[X_2, X_3]$, $[X_3, X_1]$.
(ii) Let $z \in \mathbb{C}$. Find e^{zX_1}, e^{zX_2}, e^{zX_3}.
(iii) Let $u, v, r \in \mathbb{R}$. Show that

$$g(u, v, r) = e^{uX_3}e^{rX_2}e^{vX_3}$$

$$= e^{r/2}\begin{pmatrix} -\sin(u/2)\sin(v/2) & \cos(v/2)\sin(u/2) \\ -\cos(u/2)\sin(v/2) & \cos(u/2)\cos(v/2) \end{pmatrix}$$

$$+ e^{-r/2}\begin{pmatrix} \cos(u/2)\cos(v/2) & \cos(u/2)\sin(v/2) \\ -\cos(v/2)\sin(u/2) & -\sin(u/2)\sin(v/2) \end{pmatrix}.$$

(iv) Find $g(u, v, r)^{-1}$.

Solution 24. (i) We obtain $[X_1, X_2] = X_3$, $[X_2, X_3] = -X_1$, $[X_3, X_1] = -X_2$.
(ii) We find

$$e^{zX_1} = \begin{pmatrix} \cosh(z/2) & \sinh(z/2) \\ \sinh(z/2) & \cosh(z/2) \end{pmatrix}$$

$$e^{zX_2} = \begin{pmatrix} e^{-r/2} & 0 \\ 0 & e^{r/2} \end{pmatrix}$$

$$e^{zX_3} = \begin{pmatrix} \cos(z/2) & \sin(z/2) \\ -\sin(z/2) & \cos(z/2) \end{pmatrix}.$$

(iii) Using the result from (ii) we find the element $g(u, v, r)$ of the Lie group $SL(2, \mathbb{R})$.
(iv) We obtain

$$g(u, v, r)^{-1} = e^{-vX_3} e^{-rX_2} e^{-uX_3}.$$

Problem 25. Consider the semisimple Lie algebra $s\ell(n + 1, \mathbb{F})$. Let $E_{i,j}$ $(i, j \in \{1, 2, \ldots, n+1\})$ denote the standard basis, i.e. $(n+1) \times (n+1)$ matrices with all entries zero except for the entry in the i-th row and j-th column which is one. We can form a *Cartan-Weyl basis* with

$$H_j := E_{j,j} - E_{j+1,j+1}, \qquad j \in \{1, 2, \ldots, n\}.$$

Show that $E_{i,j}$ are *root vectors* for $i \neq j$, i.e. there exists $\lambda_{H,i,j} \in \mathbb{F}$ such that

$$[H, E_{i,j}] = \lambda_{H,i,j} E_{i,j}$$

for all $H \in \text{span}\{H_1, \ldots, H_n\}$.

Solution 25. Obviously $[H_j, H_k] = 0$ since H_j and H_k are diagonal. We have

$$[E_{i,j}, E_{k,l}] = E_{i,j}E_{k,l} - E_{k,l}E_{i,j} = \delta_{j,k}E_{i,l} - \delta_{l,i}E_{k,j}.$$

It follows that

$$[H_k, E_{i,j}] = [E_{k,k}, E_{i,j}] - [E_{k+1,k+1}, E_{i,j}]$$
$$= \delta_{i,k}E_{k,j} - \delta_{j,k}E_{i,k} - \delta_{i,k+1}E_{k+1,j} + \delta_{j,k+1}E_{i,k+1}.$$

Let

$$H = \sum_{j=1}^{n} \lambda_j H_j, \qquad \lambda_1, \ldots, \lambda_n \in \mathbb{F}.$$

Then, for $i \neq j$, we find

$$[H, E_{i,j}] = \sum_{k=1}^{n} \lambda_k (\delta_{i,k}E_{k,j} - \delta_{j,k}E_{i,k} - \delta_{i,k+1}E_{k+1,j} + \delta_{j,k+1}E_{i,k+1})$$

$$= \begin{cases} (\lambda_i - \lambda_j + \lambda_{j-1})E_{i,j} & i = 1 \\ (\lambda_i - \lambda_j - \lambda_{i-1})E_{i,j} & j = 1 \\ (\lambda_i - \lambda_j - \lambda_{i-1} + \lambda_{j-1})E_{i,j} & \text{otherwise} \end{cases}$$

so that $\lambda_{H,i,j}$ is given by

$$\lambda_{H,i,j} = \begin{cases} \lambda_i - \lambda_j + \lambda_{j-1} & i = 1 \\ \lambda_i - \lambda_j - \lambda_{i-1} & j = 1 \\ \lambda_i - \lambda_j - \lambda_{i-1} + \lambda_{j-1} & \text{otherwise} \end{cases}.$$

Problem 26. Consider the vector space $\mathbb{R}^3$ and the *vector product* $\times$. The vector product is not associative. The *associator* of three vectors $\mathbf{u}$, $\mathbf{v}$, $\mathbf{w}$ is defined by

$$\text{ass}(\mathbf{u} \times (\mathbf{v} \times \mathbf{w})) := (\mathbf{u} \times \mathbf{v}) \times \mathbf{w} - \mathbf{u} \times (\mathbf{v} \times \mathbf{w}).$$

The associator measures the failure of associativity.
(i) Consider the unit vectors

$$\mathbf{u} = \begin{pmatrix} 1 \\ 0 \\ 0 \end{pmatrix}, \quad \mathbf{v} = \begin{pmatrix} 0 \\ 1 \\ 0 \end{pmatrix}, \quad \mathbf{w} = \begin{pmatrix} 0 \\ 0 \\ 1 \end{pmatrix}.$$

Find the associator.
(ii) Consider the normalized vectors

$$\mathbf{u} \doteq \frac{1}{\sqrt{2}} \begin{pmatrix} 1 \\ 0 \\ 1 \end{pmatrix}, \quad \mathbf{v} = \begin{pmatrix} 0 \\ 1 \\ 0 \end{pmatrix}, \quad \mathbf{w} = \frac{1}{\sqrt{2}} \begin{pmatrix} 1 \\ 0 \\ -1 \end{pmatrix}.$$

Find the associator.

Solution 26. (i) Since $\mathbf{u} \times \mathbf{v} = \mathbf{w}$, $\mathbf{v} \times \mathbf{w} = \mathbf{u}$, $\mathbf{w} \times \mathbf{u} = \mathbf{v}$ we find that $\text{ass}(\mathbf{u} \times (\mathbf{v} \times \mathbf{w})) = \mathbf{0}$.
(ii) Since $\mathbf{u} \times \mathbf{v} = -\mathbf{w}$, $\mathbf{v} \times \mathbf{w} = -\mathbf{u}$, $\mathbf{w} \times \mathbf{u} = \mathbf{u}$ we find that the associator is the zero vector.

Problem 27. Consider vectors in the vector space $\mathbb{R}^3$ and the vector product $\times$. Consider the mapping of the vectors in $\mathbb{R}^3$ into 3×3 skew-symmetric matrices

$$\begin{pmatrix} a \\ b \\ c \end{pmatrix} \leftrightarrow \begin{pmatrix} 0 & c & -b \\ -c & 0 & a \\ b & -a & 0 \end{pmatrix}.$$

Calculate the vector product

$$\begin{pmatrix} a_2 \\ b_2 \\ c_2 \end{pmatrix} \times \begin{pmatrix} a_1 \\ b_1 \\ c_1 \end{pmatrix}$$

and the commutator $[M_1, M_2]$, where

$$M_1 = \begin{pmatrix} 0 & c_1 & -b_1 \\ -c_1 & 0 & a_1 \\ b_1 & -a_1 & 0 \end{pmatrix}, \quad M_2 = \begin{pmatrix} 0 & c_2 & -b_2 \\ -c_2 & 0 & a_2 \\ b_2 & -a_2 & 0 \end{pmatrix}.$$

Discuss.

Solution 27.   For the vector product we find

$$\begin{pmatrix} a_2 \\ b_2 \\ c_2 \end{pmatrix} \times \begin{pmatrix} a_1 \\ b_1 \\ c_1 \end{pmatrix} = \begin{pmatrix} b_2 c_1 - b_1 c_2 \\ a_1 c_2 - a_2 c_1 \\ a_2 b_1 - a_1 b_2 \end{pmatrix}.$$

For the commutator $[M_1, M_2]$ we obtain the skew-symmetric matrix

$$[M_1, M_2] = \begin{pmatrix} 0 & a_2 b_1 - a_1 b_2 & a_2 c_1 - a_1 c_2 \\ a_1 b_2 - a_2 b_1 & 0 & b_2 c_1 - b_1 c_2 \\ a_1 c_2 - a_2 c_1 & b_1 c_2 - b_2 c_1 & 0 \end{pmatrix}.$$

Thus we have isomorphic Lie algebras.

Problem 28.   A basis for the Lie algebra $su(3)$ is given by the eight 3×3 skew-hermitian matrices

$$X_1 = \begin{pmatrix} 0 & i & 0 \\ i & 0 & 0 \\ 0 & 0 & 0 \end{pmatrix}, \quad X_2 = \begin{pmatrix} 0 & 1 & 0 \\ -1 & 0 & 0 \\ 0 & 0 & 0 \end{pmatrix},$$

$$X_3 = \begin{pmatrix} i & 0 & 0 \\ 0 & -i & 0 \\ 0 & 0 & 0 \end{pmatrix}, \quad X_4 = \begin{pmatrix} 0 & 0 & i \\ 0 & 0 & 0 \\ i & 0 & 0 \end{pmatrix},$$

$$X_5 = \begin{pmatrix} 0 & 0 & 1 \\ 0 & 0 & 0 \\ -1 & 0 & 0 \end{pmatrix}, \quad X_6 = \begin{pmatrix} 0 & 0 & 0 \\ 0 & 0 & i \\ 0 & i & 0 \end{pmatrix},$$

$$X_7 = \begin{pmatrix} 0 & 0 & 0 \\ 0 & 0 & 1 \\ 0 & -1 & 0 \end{pmatrix}, \quad X_8 = \frac{1}{\sqrt{3}} \begin{pmatrix} i & 0 & 0 \\ 0 & i & 0 \\ 0 & 0 & -2i \end{pmatrix}.$$

Find

$$\sum_{j=1}^{8} (X_j \otimes X_j), \qquad \sum_{j=1}^{8} ((X_j X_k) \otimes X_j).$$

Express the results as Kronecker products of two 3×3 matrices.

Solution 28.   We obtain

$$\sum_{j=1}^{8} (X_j \otimes X_j) = \frac{1}{3}(I_3 \otimes I_3), \qquad \sum_{j=1}^{8} ((X_j X_k) \otimes X_j) = \frac{1}{3}(X_k \otimes I_3).$$

Problem 29.   Consider the non-commutative two-dimensional Lie algebra with $[A, B] = A$ and

$$A = \begin{pmatrix} 0 & 1 \\ 0 & 0 \end{pmatrix}, \qquad B = \begin{pmatrix} 0 & 0 \\ 0 & 1 \end{pmatrix}.$$

Show that the matrices

$$\{\, A \otimes I_2 + I_2 \otimes A, \ \ B \otimes I_2 + I_2 \otimes B \,\}$$

also form a non-commutative Lie algebra under the commutator. Discuss.

Solution 29.   We have

$$[A \otimes I_2 + I_2 \otimes A, B \otimes I_2 + I_2 \otimes B] = A \otimes I_2 + I_2 \otimes A.$$

Thus the two Lie algebras are isomorphic.

Problem 30.   Consider the three 3×3 matrices

$$A = \begin{pmatrix} 0 & 1 & 0 \\ -1 & 0 & 0 \\ 0 & 0 & 0 \end{pmatrix}, \quad B = \begin{pmatrix} 0 & 0 & -1 \\ 0 & 0 & 0 \\ 1 & 0 & 0 \end{pmatrix}, \quad C = \begin{pmatrix} 0 & 0 & 0 \\ 0 & 0 & 1 \\ 0 & -1 & 0 \end{pmatrix}$$

which satisfy the commutation relations $[A, B] = C$, $[B, C] = A$, $[C, A] = B$. Thus we have a basis of the simple Lie algebra $so(3)$. Calculate the commutators

$$[A \otimes I_3, B \otimes I_3], \quad [B \otimes I_3, C \otimes I_3], \quad [C \otimes I_3, A \otimes I_3].$$

Solution 30.   We have

$$[A \otimes I_3, B \otimes I_3] = [A, B] \otimes I_3 = C \otimes I_3.$$

Analogously we have $[B \otimes I_3, C \otimes I_3] = A \otimes I_3$, $[C \otimes I_3, A \otimes I_3] = B \otimes I_3$.

Problem 31.   (i) The *Heisenberg algebra* $h(3)$ is a three-dimensional non-commutative Lie algebra with basis E_1, E_2, E_3 and

$$[E_1, E_2] = E_3, \quad [E_1, E_3] = [E_2, E_3] = 0.$$

From the matrix representation

$$T = \begin{pmatrix} 0 & t_1 & t_3 \\ 0 & 0 & t_2 \\ 0 & 0 & 0 \end{pmatrix}, \quad t_1, t_2, t_3 \in \mathbb{R}$$

of an element $T = t_1 E_1 + t_2 E_2 + t_3 E_3$ in $h(3)$. Find $\exp(T)$.
(ii) The corresponding Lie group $H(3)$ is the group of 3×3 upper triangular matrices with 1's on the main diagonal

$$H(3) = \left\{\, g = \begin{pmatrix} 1 & \alpha_1 & \alpha_3 \\ 0 & 1 & \alpha_2 \\ 0 & 0 & 1 \end{pmatrix} \right\}.$$

Find g^{-1} and gTg^{-1}.

Solution 31.   (i) We have

$$\exp(T) = \begin{pmatrix} 1 & t_1 & t_1t_2 + t_3 \\ 0 & 1 & t_2 \\ 0 & 0 & 1 \end{pmatrix}.$$

(ii) We obtain

$$g^{-1} = \begin{pmatrix} 1 & -\alpha_1 & \alpha_1\alpha_2 - \alpha_3 \\ 0 & 1 & -\alpha_2 \\ 0 & 0 & 1 \end{pmatrix} \Rightarrow gTg^{-1} = \begin{pmatrix} 0 & t_1 & t_3 - \alpha_2 t_1 + \alpha_1 t_2 \\ 0 & 0 & t_2 \\ 0 & 0 & 0 \end{pmatrix}.$$

Problem 32.   Let $A_1, A_2, \ldots, A_r$ be $n \times n$ matrices which form a basis of a non-commutative Lie algebra. Let $B_1, B_2, \ldots, B_s$ be $n \times n$ matrices which form a basis of a commutative Lie algebra. Let $\otimes$ be the Kronecker product and form all elements

$$A_j \otimes B_k, \qquad j = 1, \ldots, r, \quad k = 1, \ldots, s.$$

Calculate the commutators

$$[A_j \otimes B_k, A_\ell \otimes B_m], \qquad j, \ell = 1, \ldots, r, \quad k, m = 1, \ldots, s.$$

Discuss.

Solution 32.   We have

$$[A_j \otimes B_k, A_\ell \otimes B_m] = (A_j A_\ell) \otimes (B_k B_m) - (A_\ell A_j) \otimes (B_m B_k).$$

Since $B_k B_m = B_m B_k$ we have

$$[A_j \otimes B_k, A_\ell \otimes B_m] = [A_j, A_\ell] \otimes (B_k B_m).$$

Thus all the elements $A_j \otimes B_k$ form a basis of a Lie algebra.

Problem 33.   Consider the Heisenberg algebra $h(1)$ with the generators given by $n \times n$ matrices H, X_1, X_2 and the commutation relations

$$[H, X_1] = 0_n, \quad [H, X_2] = 0_n, \quad [X_1, X_2] = H.$$

Find the commutators for the matrices $H \otimes I_n, X_1 \otimes I_n, X_2 \otimes I_n$.

Solution 33.   We find

$$[H \otimes I_n, X_1 \otimes I_n] = 0_{n^2}, \ [H \otimes I_n, X_2 \otimes I_n] = 0_{n^2}, \ [X_1 \otimes I_n, X_2 \otimes I_n] = H \otimes I_n.$$

Thus we find another representation of the Heisenberg algebra.

Problem 34.   (i) Let A, B be $n \times n$ matrices. Calculate the commutator

$$[A \otimes I_n + I_n \otimes A, B \otimes B]$$

and express the result using the commutator $[A, B]$.

(ii) Assume that $[A, B] = B$, i.e. A and B form a basis of a two-dimensional Lie algebra. Express the result from (i) using $[A, B] = B$. Discuss.

Solution 34.   (i) We find

$$[A \otimes I_n, I_n \otimes A, B \otimes B] = [A, B] \otimes B + B \otimes [A, B].$$

(ii) From (i) we find

$$[A \otimes I_n + I_n \otimes A, B \otimes B] = B \otimes B + B \otimes B = 2(B \otimes B).$$

Problem 35.   Let L be a finite dimensional Lie algebra. Let $C^\infty(S^1)$ be the set of all infinitely differentiable functions, where S^1 is the unit circle manifold. In the product space $L \otimes C^\infty(S^1)$ we define the Lie bracket ($g_1, g_2 \in L$ and $f_1, f_2 \in C^\infty(S^1)$)

$$[g_1 \otimes f_1, g_2 \otimes f_2] := [g_1, g_2] \otimes (f_1 f_2).$$

Calculate

$$[g_1 \otimes f_1, [g_2 \otimes f_2, g_3 \otimes f_3]] + [g_3 \otimes f_3, [g_1 \otimes f_1, g_2 \otimes f_2]] + [g_2 \otimes f_2, [g_3 \otimes f_3, g_1 \otimes f_1]].$$

Solution 35.   We have

$$\begin{aligned}
[g_1 \otimes f_1, [g_2 \otimes f_2, g_3 \otimes f_3]] &= [g_1, [g_2, g_3]] \otimes f_1 f_2 f_3 \\
[g_3 \otimes f_3, [g_1 \otimes f_1, g_2 \otimes f_2]] &= [g_3, [g_1, g_2]] \otimes f_3 f_2 f_1 \\
[g_2 \otimes f_2, [g_3 \otimes f_3, g_1 \otimes f_1]] &= [g_2, [g_3, g_1]] \otimes f_2 f_3 f_1.
\end{aligned}$$

Since $f_1 f_2 f_3 = f_3 f_1 f_2 = f_2 f_3 f_1$ we obtain

$$([g_1, [g_2, g_3]] + [g_3, [g_1, g_2]] + [g_2, [g_3, g_1]]) \otimes (f_1 f_2 f_3) = 0$$

where we applied the *Jacobi identity*.

Problem 36.   The elements (generators) $Z_1, Z_2, \ldots, Z_r$ of an r-dimensional Lie algebra satisfy the conditions

$$[Z_\mu, Z_\nu] = \sum_{\tau=1}^{r} c_{\mu\nu}^\tau Z_\tau$$

with $c_{\mu\nu}^\tau = -c_{\nu\mu}^\tau$, where the $c_{\mu\nu}^\tau$'s are called the *structure constants*. Let A be an arbitrary linear combination of the elements

$$A = \sum_{\mu=1}^{r} a^\mu Z_\mu.$$

Suppose that X is some other linear combination such that

$$X = \sum_{\nu=1}^{r} b^{\nu} Z_{\nu} \quad \text{and} \quad [A, X] = \rho X.$$

This equation has the form of an eigenvalue equation, where ρ is the corresponding eigenvalue and X the corresponding eigenvector. Assume that the Lie algebra is represented by matrices. Find the secular equation for the eigenvalues ρ.

Solution 36.   From $[A, X] = \rho X$ we have $AX - XA = \rho X$. Thus

$$\sum_{\mu=1}^{r} \sum_{\nu=1}^{r} (a^{\mu} b^{\nu} Z_{\mu} Z_{\nu} - a^{\mu} b^{\nu} Z_{\nu} Z_{\mu}) = \rho \sum_{\nu=1}^{r} b^{\nu} Z_{\nu}. \tag{1}$$

Inserting

$$Z_{\mu} Z_{\nu} = \sum_{\tau=1}^{r} c_{\mu\nu}^{\tau} Z_{\tau} + Z_{\nu} Z_{\mu}$$

into (1) yields

$$\sum_{\tau=1}^{r} \sum_{\nu=1}^{r} \sum_{\mu=1}^{r} a^{\mu} b^{\nu} c_{\mu\nu}^{\tau} Z_{\tau} = \rho \sum_{\tau=1}^{r} b^{\tau} Z_{\tau}.$$

It follows that

$$\sum_{\tau=1}^{r} \left(\sum_{\nu=1}^{r} \sum_{\mu=1}^{r} a^{\mu} b^{\nu} c_{\mu\nu}^{\tau} - \rho b^{\tau} \right) Z_{\tau} = 0.$$

Since Z_{τ}, with $\tau = 1, \ldots, r$ are linearly independent we have for a fixed $\tau = \tau^{*}$

$$\sum_{\nu=1}^{r} \left(\sum_{\mu=1}^{r} a^{\mu} b^{\nu} c_{\mu\nu}^{\tau^{*}} \right) - \rho b^{\tau^{*}} = 0.$$

Introducing the Kronecker delta $\delta_{\nu}^{\tau^{*}}$ with 1 if $\nu = \tau^{*}$ and 0 otherwise we can write

$$\sum_{\nu=1}^{r} \left(\sum_{\mu=1}^{r} a^{\mu} c_{\mu\nu}^{\tau^{*}} - \rho \delta_{\nu}^{\tau^{*}} \right) b^{\nu} = 0.$$

For $\nu, \tau^{*} = 1, \ldots, r$

$$\sum_{\mu=1}^{r} a^{\mu} c_{\mu\nu}^{\tau^{*}} - \rho \delta_{\nu}^{\tau^{*}}$$

defines an $r \times r$ matrix with the secular equation

$$\det \left(\sum_{\mu=1}^{r} a^{\mu} c_{\mu\nu}^{\tau^{*}} - \rho \delta_{\nu}^{\tau^{*}} \right) = 0$$

to find the eigenvalues ρ, where $\nu, \tau^{*} = 1, \ldots, r$.

Problem 37. Let $c_{\sigma\lambda}^{\tau}$ be the structure constants of a Lie algebra. We define

$$g_{\sigma\lambda} = g_{\lambda\sigma} = \sum_{\rho=1}^{r}\sum_{\tau=1}^{r} c_{\sigma\rho}^{\tau} c_{\lambda\tau}^{\rho} \qquad \text{and} \qquad g^{\sigma\lambda} g_{\sigma\lambda} = \delta_{\sigma}^{\lambda}.$$

A Lie algebra L is called *semisimple* if and only if $\det |g_{\sigma\lambda}| \neq 0$. We assume in the following that the Lie algebra is semisimple. We define

$$C := \sum_{\rho=1}^{r}\sum_{\sigma=1}^{r} g^{\rho\sigma} X_{\rho} X_{\sigma}.$$

The operator C is called *Casimir operator*. Let X_{τ} be an element of the Lie algebra L. Calculate the commutator $[C, X_{\tau}]$.

Solution 37. We have

$$[C, X_{\tau}] = \sum_{\rho=1}^{r}\sum_{\sigma=1}^{r} g^{\rho\sigma} [X_{\rho} X_{\sigma}, X_{\tau}]$$

$$= \sum_{\rho=1}^{r}\sum_{\sigma=1}^{r} g^{\rho\sigma} X_{\rho} [X_{\sigma}, X_{\tau}] + \sum_{\rho=1}^{r}\sum_{\sigma=1}^{r}\sum_{\lambda=1}^{r} g^{\rho\sigma} c_{\rho\tau}^{\lambda} X_{\lambda} X_{\sigma}$$

$$= \sum_{\rho=1}^{r}\sum_{\sigma=1}^{r}\sum_{\lambda=1}^{r} g^{\rho\sigma} c_{\sigma\tau}^{\lambda} X_{\rho} X_{\lambda} + \sum_{\rho=1}^{r}\sum_{\sigma=1}^{r}\sum_{\lambda=1}^{r} g^{\rho\sigma} c_{\rho\tau}^{\lambda} X_{\lambda} X_{\sigma}$$

$$= \sum_{\rho=1}^{r}\sum_{\sigma=1}^{r}\sum_{\lambda=1}^{r} g^{\rho\sigma} c_{\sigma\tau}^{\lambda} X_{\rho} X_{\lambda} + \sum_{\rho=1}^{r}\sum_{\sigma=1}^{r}\sum_{\lambda=1}^{r} g^{\sigma\rho} c_{\sigma\tau}^{\lambda} X_{\lambda} X_{\rho}$$

$$= \sum_{\rho=1}^{r}\sum_{\sigma=1}^{r}\sum_{\lambda=1}^{r} g^{\rho\sigma} c_{\sigma\tau}^{\lambda} (X_{\rho} X_{\lambda} + X_{\lambda} X_{\rho})$$

where we made a change in the variables σ and ρ. For semisimple Lie algebras we have

$$c_{\sigma\tau}^{\lambda} = \sum_{\nu=1}^{r} g^{\lambda\nu} c_{\nu\sigma\tau}$$

and therefore we obtain

$$[C, X_{\tau}] = \sum_{\rho=1}^{r}\sum_{\sigma=1}^{r}\sum_{\lambda=1}^{r}\sum_{\nu=1}^{r} g^{\rho\sigma} g^{\lambda\nu} c_{\nu\sigma\tau} (X_{\rho} X_{\lambda} + X_{\lambda} X_{\rho}).$$

Now $c_{\nu\sigma\tau}$ is antisymmetric, while the quantity in parentheses is symmetric in ρ and λ, and hence the right-hand side must vanish to give $[C, X_{\tau}] = 0$ for all $X_{\tau} \in L$.

Supplementary Problems

Problem 1.   Given the basis of the Lie algebra $s\ell(2, \mathbb{R})$

$$E = \begin{pmatrix} 0 & 1 \\ 0 & 0 \end{pmatrix}, \quad F = \begin{pmatrix} 0 & 0 \\ 1 & 0 \end{pmatrix}, \quad H = \begin{pmatrix} 1 & 0 \\ 0 & -1 \end{pmatrix}$$

with the commutation relations $[E, F] = H$, $[E, H] = -2E$, $[F, H] = -2F$.
(i) Let $z \in \mathbb{C}$. We define

$$\widetilde{E} = e^{zH/2} \otimes E + E \otimes e^{-zH/2}, \quad \widetilde{F} = e^{zH/2} \otimes F + F \otimes e^{-zH/2}, \quad \widetilde{H} = H \otimes I_2 + I_2 \otimes H.$$

Find the commutators $[\widetilde{E}, \widetilde{F}]$, $[\widetilde{E}, \widetilde{H}]$, $[\widetilde{F}, \widetilde{H}]$.
(ii) Let $U(s\ell(2, \mathbb{R}))$ be the universal enveloping algebra. Then any element of $U(s\ell(2, \mathbb{R}))$ can be expressed as a sum of product of the form $F^j H^k E^\ell$ where $j, k, \ell \geq 0$. Show that

$$EF^2 = -2F + 2FH + F^2E.$$

Hint: Utilize that $EF^2 \equiv [E, F^2] + F^2E$.

Problem 2.   Study the Lie algebra $s\ell(2, \mathbb{F})$, where $\text{char}(\mathbb{F}) = 2$.

Problem 3.   Do the matrices

$$E_+ = \begin{pmatrix} 0 & 1 & 0 \\ 0 & 0 & 1 \\ 0 & 0 & 0 \end{pmatrix}, \quad E_- = \begin{pmatrix} 0 & 0 & 0 \\ 1 & 0 & 0 \\ 0 & 1 & 0 \end{pmatrix}, \quad H = \begin{pmatrix} 1 & 0 & 0 \\ 0 & 0 & 0 \\ 0 & 0 & -1 \end{pmatrix}$$

form a basis for the Lie algebra $s\ell(2, \mathbb{R})$?

Problem 4.   Show that a basis of the Lie algebra $s\ell(2, \mathbb{C})$ is given by

$$\frac{1}{2} \begin{pmatrix} 0 & i \\ i & 0 \end{pmatrix}, \quad \frac{1}{2} \begin{pmatrix} 0 & 1 \\ -1 & 0 \end{pmatrix}, \quad \frac{1}{2} \begin{pmatrix} -i & 0 \\ 0 & i \end{pmatrix},$$

$$\frac{1}{2} \begin{pmatrix} 0 & 1 \\ 1 & 0 \end{pmatrix}, \quad \frac{1}{2} \begin{pmatrix} 0 & -i \\ i & 0 \end{pmatrix}, \quad \frac{1}{2} \begin{pmatrix} -1 & 0 \\ 0 & 1 \end{pmatrix}.$$

Problem 5.   Given the spin matrices

$$S_+ = \begin{pmatrix} 0 & 1 \\ 0 & 0 \end{pmatrix}, \quad S_- = \begin{pmatrix} 0 & 0 \\ 1 & 0 \end{pmatrix}.$$

Consider the 4×4 matrices

$$\begin{pmatrix} 0_2 & S_- \\ 0_2 & 0_2 \end{pmatrix}, \quad \begin{pmatrix} 0_2 & 0_2 \\ S_+ & 0_2 \end{pmatrix}, \quad \begin{pmatrix} S_+ & 0_2 \\ 0_2 & S_+ \end{pmatrix}, \quad \begin{pmatrix} S_- & 0_2 \\ 0_2 & S_- \end{pmatrix}.$$

Calculate the commutators of these matrices and extend the set so that one finds a basis of a Lie algebra.

Problem 6. Let L_1 and L_2 be two Lie algebras. Let $\varphi : L_1 \to L_2$ be a Lie algebra homomorphism. Show that im(φ) is a Lie subalgebra of L_2 and ker(φ) is an ideal in L_1.

Problem 7. Let A, B, C nonzero 3×3 matrices with the commutation relations

$$[A, B] = C, \quad [B, C] = A, \quad [C, A] = B$$

i.e. A, B, C form a basis of the Lie algebra $so(3)$. Find the Lie algebra generated by the 9×9 matrices

$$H = A \otimes I_3 + I_3 \otimes B + A \otimes B, \qquad K = A \otimes I_3 + I_3 \otimes B + B \otimes A.$$

Problem 8. Consider the Pauli spin matrices $\sigma_0 = I_2$, σ_1, σ_2, σ_3 and the sixteen 4×4 matrices ($j = 0, 1, 2, 3$)

$$\begin{pmatrix} \sigma_j & 0_2 \\ 0_2 & \sigma_j \end{pmatrix}, \quad \begin{pmatrix} \sigma_j & 0_2 \\ 0_2 & -\sigma_j \end{pmatrix}, \quad \begin{pmatrix} 0_2 & \sigma_j \\ \sigma_j & 0_2 \end{pmatrix}, \quad \begin{pmatrix} 0_2 & \sigma_j \\ -\sigma_j & 0_2 \end{pmatrix}.$$

Do the sixteen 4×4 matrices form a basis of a Lie algebra under the commutator?

Problem 9. Consider the Lie algebra of real-skew symmetric 3×3 matrices

$$A = \begin{pmatrix} 0 & a_3 & -a_2 \\ -a_3 & 0 & a_1 \\ a_2 & -a_1 & 0 \end{pmatrix}.$$

Let R be a real orthogonal 3×3 matrix, i.e. $RR^T = I_3$. Show that RAR^T is a real-skew symmetric matrix.

Problem 10. The Lie group $SU(2, 2)$ is defined as the group of transformation on the four dimensional complex space $\mathbb{C}^4$ leaving invariant the indefinite quadratic form

$$|z_1|^2 + |z_2|^2 - |z_3|^2 - |z_4|^2.$$

The Lie algebra $su(2, 2)$ is defined as the 4×4 matrices X with trace 0 and $X^*L + LX = 0_4$, where L is the 4×4 matrix

$$L = \begin{pmatrix} -I_2 & 0_2 \\ 0_2 & I_2 \end{pmatrix}.$$

Is

$$X = \begin{pmatrix} 0 & 0 & 0 & 1 \\ 0 & 0 & 1 & 0 \\ 0 & 1 & 0 & 0 \\ 1 & 0 & 0 & 0 \end{pmatrix}$$

an element of the Lie algebra $su(2, 2)$? Find $\exp(zX)$. Discuss.

Problem 11. The semisimple Lie algebra $s\ell(3, \mathbb{R})$ has dimension 8. The standard basis is given by

$$h_1 = \begin{pmatrix} 1 & 0 & 0 \\ 0 & -1 & 0 \\ 0 & 0 & 0 \end{pmatrix}, \quad h_2 = \begin{pmatrix} 0 & 0 & 0 \\ 0 & 1 & 0 \\ 0 & 0 & -1 \end{pmatrix},$$

$$e_1 = \begin{pmatrix} 0 & 1 & 0 \\ 0 & 0 & 0 \\ 0 & 0 & 0 \end{pmatrix}, \quad e_2 = \begin{pmatrix} 0 & 0 & 0 \\ 0 & 0 & 1 \\ 0 & 0 & 0 \end{pmatrix},$$

$$f_1 = \begin{pmatrix} 0 & 0 & 0 \\ 1 & 0 & 0 \\ 0 & 0 & 0 \end{pmatrix}, \quad f_2 = \begin{pmatrix} 0 & 0 & 0 \\ 0 & 0 & 0 \\ 0 & 1 & 0 \end{pmatrix},$$

$$e_{13} = \begin{pmatrix} 0 & 0 & 1 \\ 0 & 0 & 0 \\ 0 & 0 & 0 \end{pmatrix}, \quad f_{13} = \begin{pmatrix} 0 & 0 & 0 \\ 0 & 0 & 0 \\ 1 & 0 & 0 \end{pmatrix}.$$

Find the commutator table.

Problem 12. Let E_{jk} $(j, k = 1, 2, 3, 4)$ be the elementary matrices in the vector space of 4×4 matrices. Show that the 15 matrices

$$X_1 = E_{12}, \quad X_2 = E_{23}, \quad X_3 = E_{13}, \quad X_4 = E_{34}, \quad X_5 = E_{24}, \quad X_6 = E_{14}$$

$$Y_1 = E_{21}, \quad Y_2 = E_{32}, \quad Y_3 = E_{31}, \quad Y_4 = E_{43}, \quad Y_5 = E_{42}, \quad Y_6 = E_{41},$$

$$H_1 = \frac{1}{4}\mathrm{diag}(3, -1, -1, -1, -1), \quad H_2 = \frac{1}{2}\mathrm{diag}(1, 1, -1, -1),$$

$$H_3 = \frac{1}{4}\mathrm{diag}(1, 1, 1, -3)$$

form a basis (*Cartan-Weyl basis*) of the Lie algebra $s\ell(4, \mathbb{C})$.

Problem 13. Let A, B, C be 3×3 nonzero matrices such that (Lie algebra $so(3)$)

$$[A, B] = C, \quad [B, C] = A, \quad [C, A] = B.$$

Find the commutators of the 9×9 matrices

$$A \otimes I_3 + I_3 \otimes A, \quad B \otimes I_3 + I_3 \otimes B, \quad C \otimes I_3 + I_3 \otimes C.$$

Problem 14. The isomorphism of the Lie algebras $s\ell(2, \mathbb{C})$ and $so(3, \mathbb{C})$ has the form

$$\begin{pmatrix} a & b \\ c & -a \end{pmatrix} \leftrightarrow \begin{pmatrix} 0 & b-c & -i(b+c) \\ c-b & 0 & 2ia \\ i(b+c) & -2ia & 0 \end{pmatrix}.$$

Let $z \in \mathbb{C}$. Find

$$\exp\left(z\begin{pmatrix} a & b \\ c & -a \end{pmatrix}\right), \quad \exp\left(z\begin{pmatrix} 0 & b-c & -i(b+c) \\ c-b & 0 & 2ia \\ i(b+c) & -2ia & 0 \end{pmatrix}\right).$$

Problem 15. Consider a four-dimensional vector space with basis e_1, e_2, e_3, e_4. Assume that the non-zero commutators are given

$$[e_2, e_3] = e_1, \quad [e_1, e_4] = 2e_1, \quad [e_2, e_4] = e_2, \quad [e_3, e_4] = e_2 + e_3.$$

Do these relations define a Lie algebra? For example we have

$$[e_1, [e_2, e_3]] + [e_3, [e_1, e_2]] + [e_2, [e_3, e_1]] = [e_1, e_1] + [e_3, 0] + [e_2, 0] = 0.$$

If so find the adjoint representation.

Problem 16. Let A, B, C be nonzero $n \times n$ matrices. Assume that $[A, B] = 0_n$ and $[C, A] = 0_n$. Can we conclude that $[B, C] = 0_n$? Is the Jacobi identity

$$[A, [B, C]] + [C, [A, B]] + [B, [C, A]] = 0_n$$

of any use?

Problem 17. (i) Find the Lie algebra generated by the 2×2 matrices

$$A = \begin{pmatrix} 0 & 0 \\ 0 & 1 \end{pmatrix}, \quad B = \begin{pmatrix} 0 & 0 \\ -1 & 0 \end{pmatrix}.$$

Note that $[A, B] = A + B$.
(ii) Find the Lie algebra generated by the 3×3 matrices

$$A = \begin{pmatrix} 0 & 0 & 0 \\ 0 & 1 & 0 \\ 0 & 0 & -1 \end{pmatrix}, \quad B = \begin{pmatrix} 0 & 0 & 0 \\ -1 & 0 & 0 \\ 0 & 0 & 0 \end{pmatrix}, \quad C = \begin{pmatrix} 0 & 0 & 0 \\ 0 & 0 & 0 \\ 1 & 0 & 0 \end{pmatrix}.$$

Note that $[A, B] = B$, $[B, C] = 0_3$, $[C, A] = C$.

Problem 18. In the *decomposition* of the simple Lie algebra $s\ell(3, \mathbb{R})$ one finds the 3×3 matrices

$$A = \begin{pmatrix} a_{11} & a_{12} & 0 \\ a_{21} & a_{22} & 0 \\ 0 & 0 & -a_{11} - a_{22} \end{pmatrix}, \quad B = \begin{pmatrix} 0 & 0 & b_{13} \\ 0 & 0 & b_{23} \\ b_{31} & b_{32} & 0 \end{pmatrix}$$

where $a_{jk}, b_{jk} \in \mathbb{R}$. Find the commutators $[A, A']$, $[B, B']$ and $[A, B]$. Discuss. Note that

$$[A, B] = \begin{pmatrix} 0 & 0 & a_{12}b_{23} + a_{22}b_{13} \\ 0 & 0 & a_{21}b_{13} + a_{11}b_{23} \\ -a_{22}b_{31} - a_{21}b_{32} & -a_{11}b_{32} - a_{12}b_{31} & 0 \end{pmatrix}.$$

Thus the matrix of the commutator $[A, B]$ is of the form of matrix B.

Problem 19. Let σ_1, σ_2, σ_3 be the Pauli spin matrices with the commutation relations

$$[\sigma_1, \sigma_2] = 2i\sigma_3, \quad [\sigma_2, \sigma_3] = 2i\sigma_1, \quad [\sigma_3, \sigma_1] = 2i\sigma_2.$$

Thus we have $su(2) = \text{span}\{ i\sigma_1, i\sigma_2, i\sigma_3 \}$. Then

$$[\sigma_1 \otimes I_2 + I_2 \otimes \sigma_2, \sigma_2 \otimes I_2 + I_2 \otimes \sigma_3] = 2i(\sigma_3 \otimes I_2 + I_2 \otimes \sigma_1).$$

(i) Find the commutators

$$[\sigma_2 \otimes I_2 + I_2 \otimes \sigma_3, \sigma_3 \otimes I_2 + I_2 \otimes \sigma_1], \quad [\sigma_3 \otimes I_2 + I_2 \otimes \sigma_1, \sigma_2 \otimes I_2 + I_2 \otimes \sigma_3].$$

Discuss.

(ii) We know that the Pauli spin matrices σ_1, σ_2, σ_3 are elements of the Lie algebra $u(2)$, but not $su(2)$ since $\det(\sigma_j) = -1$ for $j = 1, 2, 3$. Are the nine 4×4 matrices

$$\sigma_j \otimes \sigma_k, \quad j, k = 1, 2, 3$$

elements of the Lie algebra $su(4)$?

(iii) Let $\star$ be the star operation. Find the commutators

$$[\sigma_1, \sigma_2], \quad [\sigma_2, \sigma_3], \quad [\sigma_3, \sigma_1], \quad [\sigma_1 \star \sigma_1, \sigma_2 \star \sigma_2], \quad [\sigma_2 \star \sigma_2, \sigma_3 \star \sigma_3], \quad [\sigma_3 \star \sigma_3, \sigma_1 \star \sigma_1].$$

Find the anti-commutators $[\sigma_1, \sigma_2]_+$, $[\sigma_2, \sigma_3]_+$, $[\sigma_3, \sigma_1]_+$,

$$[\sigma_1 \star \sigma_1, \sigma_2 \star \sigma_2]_+, \quad [\sigma_2 \star \sigma_2, \sigma_3 \star \sigma_3]_+, \quad [\sigma_3 \star \sigma_3, \sigma_1 \star \sigma_1]_+.$$

Problem 20. Let $(\alpha \in \mathbb{R})$. Show that the matrices

$$L_1 = \begin{pmatrix} \alpha + i & 0 \\ 0 & \alpha \end{pmatrix}, \quad L_2 = \begin{pmatrix} 0 & 1 \\ 0 & 0 \end{pmatrix}, \quad L_3 = \begin{pmatrix} 0 & i \\ 0 & 0 \end{pmatrix}$$

satisfy the commutation relations

$$[L_1, L_2] = L_3, \quad [L_2, L_3] = 0_2, \quad [L_3, L_1] = L_2.$$

Problem 21. Consider the four-dimensional real vector space with a basis e_0, e_1, e_2, e_3. The *Gödel quaternion algebra* $\mathbb{G}$ is defined by the non-commutative multiplication

$$e_0 e_k = e_k = e_k e_0, \quad (e_0)^2 = e_0$$
$$e_j e_k = (-1)^\ell \epsilon_{jk\ell} e_\ell - (-1)^k \delta_{jk} e_0, \quad j, k, \ell = 1, 2, 3.$$

Let q_0, q_1, q_2, q_3 be arbitrary real numbers. We call the vector

$$q = q_0 e_0 + q_1 e_1 + q_2 e_2 + q_3 e_3$$

a Gödel quaternion. We define the basis

$$e_{11} := \frac{1}{2}(e_0 + e_3), \quad e_{22} := \frac{1}{2}(e_0 - e_3),$$

$$e_{12} := \frac{1}{2}(e_1 + e_2), \quad e_{21} := \frac{1}{2}(e_1 - e_2).$$

Show that the e_{jk} satisfy the multiplication law

$$e_{jk}e_{rs} = \delta_{kr}e_{js}, \qquad j,k,r,s = 1,2.$$

Show that every Gödel quaternion q can be written as

$$q = \sum_{j,k=1}^{2} q_{jk}e_{jk}.$$

Show that

$$e_0 = \begin{pmatrix} 1 & 0 \\ 0 & 1 \end{pmatrix}, \quad e_1 = \begin{pmatrix} 0 & 1 \\ 1 & 0 \end{pmatrix}, \quad e_2 = \begin{pmatrix} 0 & 1 \\ -1 & 0 \end{pmatrix}, \quad e_3 = \begin{pmatrix} 1 & 0 \\ 0 & -1 \end{pmatrix}$$

is a matrix representation.

Chapter 20

Braid Group

Let $n \geq 3$. The *braid group* $\mathcal{B}_n$ of n strings (or group of n braids) has $n - 1$ generators $\{\sigma_1, \sigma_2, \ldots, \sigma_{n-1}\}$ (pairwise distinct) satisfying the relations

$$\sigma_j \sigma_{j+1} \sigma_j = \sigma_{j+1} \sigma_j \sigma_{j+1} \quad \text{for} \quad j = 1, 2, \ldots, n - 2 \quad (\textit{Yang-Baxter relation})$$
$$\sigma_j \sigma_k = \sigma_k \sigma_j \quad \text{for} \quad |j - k| \geq 2$$
$$\sigma_j \sigma_j^{-1} = \sigma_j^{-1} \sigma_j = e$$

where e is the identity element. Thus it is generated by elements σ_j (σ_j interchanges elements j and $j + 1$). Thus actually one should write σ_{12}, σ_{23}, $\ldots$, σ_{n-1n} instead of $\sigma_1, \sigma_2, \ldots, \sigma_{n-1}$. The braid group $\mathcal{B}_n$ is a generalization of the permutation group.

Let $n \geq 3$ and let σ_1, $\ldots$, σ_{n-1} be the generators. The braid group $\mathcal{B}_n$ on n-strings where $n \geq 3$ has a finite presentation of $\mathcal{B}_n$ given by

$$\langle \sigma_1, \ldots, \sigma_{n-1} : \sigma_i \sigma_j = \sigma_j \sigma_i, \quad \sigma_{i+1} \sigma_i \sigma_{i+1} = \sigma_i \sigma_{i+1} \sigma_i \rangle$$

where $1 \leq i, j < n - 1$, $|i - j| > 1$ or $j = n - 1$. Here $\sigma_i \sigma_j = \sigma_j \sigma_i$ and $\sigma_i \sigma_{i+1} \sigma_i = \sigma_{i+1} \sigma_i \sigma_{i+1}$ are called the braid relations.

The word written in terms of letters, generators from the set

$$\{\sigma_1, \ldots, \sigma_{n-1}, \sigma_1^{-1}, \ldots, \sigma_{n-1}^{-1}\}$$

gives a particular braid. The length of the braid is the total number of used letters, while the minimal irreducible length (referred sometimes as the primitive word) is the shortest non-contractible length of a particular braid which remains after applying all the group relations given above.

466

Problem 1. (i) Consider the braid group $\mathcal{B}_5$ with the generators σ_1, σ_2, σ_3, σ_4. Simplify

$$\sigma_1^{-1}\sigma_4^{-1}\sigma_3^{-1}\sigma_2\sigma_3\sigma_2\sigma_3^{-1}\sigma_4\sigma_1.$$

(ii) Consider the braid group $\mathcal{B}_3$ and with the generators σ_1 and σ_2. Let $a = \sigma_1\sigma_2\sigma_1$, $b = \sigma_1\sigma_2$. Show that $a^2 = b^3$.

Solution 1. (i) Utilizing $\sigma_2\sigma_3\sigma_2 = \sigma_3\sigma_2\sigma_3$ we find

$$\sigma_1^{-1}\sigma_4^{-1}\sigma_3^{-1}\sigma_2\sigma_3\sigma_2\sigma_3^{-1}\sigma_4\sigma_1 = \sigma_1^{-1}\sigma_4^{-1}\sigma_3^{-1}\sigma_3\sigma_2\sigma_3\sigma_3^{-1}\sigma_4\sigma_1 = \sigma_1^{-1}\sigma_4^{-1}\sigma_2\sigma_4\sigma_1$$
$$= \sigma_1^{-1}\sigma_2^{-1}\sigma_4^{-1}\sigma_4\sigma_1$$
$$= \sigma_1\sigma_2^{-1}\sigma_1.$$

(ii) Using the Yang-Baxter relation $\sigma_1\sigma_2\sigma_1 = \sigma_2\sigma_1\sigma_2$ we find

$$a^2 = \sigma_1\sigma_2\sigma_1\sigma_1\sigma_2\sigma_1 = \sigma_1\sigma_2\sigma_1\sigma_2\sigma_1\sigma_2 = b^3$$

Problem 2. Consider the braid group $\mathcal{B}_3$. A faithful representation for the *generators* σ_1 and σ_2 is

$$\sigma_1 = \begin{pmatrix} 1 & 0 \\ -1 & 1 \end{pmatrix}, \qquad \sigma_2 = \begin{pmatrix} 1 & 1 \\ 0 & 1 \end{pmatrix}.$$

Both are elements of the group $SL(2, \mathbb{Z})$ and the Lie group $SL(2, \mathbb{R})$.
(i) Find the inverse of σ_1 and σ_2. Find $\sigma_1\sigma_2$ and $\sigma_1^{-1}\sigma_2$.
(ii) Is $\sigma_1\sigma_2\sigma_1 = \sigma_2\sigma_1\sigma_2$?

Solution 2. (i) We find

$$\sigma_1^{-1} = \begin{pmatrix} 1 & 0 \\ 1 & 1 \end{pmatrix} = \sigma_2^T, \qquad \sigma_2^{-1} = \begin{pmatrix} 1 & -1 \\ 0 & 1 \end{pmatrix} = \sigma_1^T.$$

Obviously they are also elements of $SL(2, \mathbb{Z})$ and $SL(2, \mathbb{R})$. We find

$$\sigma_1\sigma_2 = \begin{pmatrix} 1 & 1 \\ -1 & 0 \end{pmatrix}, \qquad \sigma_1^{-1}\sigma_2 = \begin{pmatrix} 1 & 1 \\ 1 & 2 \end{pmatrix}.$$

(ii) Yes. We have

$$\sigma_1\sigma_2\sigma_1 = \begin{pmatrix} 0 & 1 \\ -1 & 0 \end{pmatrix} = \sigma_2\sigma_1\sigma_2.$$

Problem 3. Consider the unitary 2×2 matrices

$$A(\theta) = \begin{pmatrix} e^{-i\theta} & 0 \\ 0 & e^{i\theta} \end{pmatrix}, \qquad B(\theta) = \begin{pmatrix} \cos(\theta) & -i\sin(\theta) \\ -i\sin(\theta) & \cos(\theta) \end{pmatrix}$$

where $\theta \in \mathbb{R}$. Find the conditions on $\theta_1, \theta_2, \theta_3 \in \mathbb{R}$ such that (braid-like relation)

$$A(\theta_1)B(\theta_2)A(\theta_3) = B(\theta_3)A(\theta_2)B(\theta_1).$$

Solution 3. We obtain the condition

$$(e^{-2i\theta_2} + 1)(i - \sec(\theta_1 - \theta_3)\sin(\theta_1 + \theta_3)) = 2i$$

which can be written as

$$\theta_3 = \arctan\left(\frac{\sin(\theta_1 + \theta_3)}{\cos(\theta_1 - \theta_3)}\right).$$

Do the matrices $A(\theta)$ form a group under matrix multiplication?

Problem 4.   Can one find 2×2 matrices A and B with $[A, B] \neq 0_2$ and satisfying the braid-like relation $ABBA = BAAB$?

Solution 4. If $A = A^{-1}$ and $B = B^{-1}$ the relation holds. For example

$$A = \begin{pmatrix} 0 & 1 \\ 1 & 0 \end{pmatrix}, \qquad B = \frac{1}{\sqrt{2}}\begin{pmatrix} 1 & 1 \\ 1 & -1 \end{pmatrix}.$$

Problem 5.   Do the 2×2 unitary matrices

$$A = \begin{pmatrix} e^{-i\pi/4} & 0 \\ 0 & ie^{-i\pi/4} \end{pmatrix}, \qquad B = \frac{1}{\sqrt{2}}\begin{pmatrix} 1 & i \\ i & 1 \end{pmatrix}$$

satisfy the *braid-like relation* $ABA = BAB$?

Solution 5.   Yes the matrices A, B satisfy the braid-like relation.

Problem 6.   The 3×3 matrices g_1 and g_2 play a role for the matrix representation of the braid group $\mathcal{B}_4$

$$g_1 = \begin{pmatrix} -t & 1 & 0 \\ 0 & 1 & 0 \\ 0 & 1 & -t^{-1} \end{pmatrix}, \qquad g = \begin{pmatrix} 1-t & -t^{-1} & t^{-1} \\ 1-t^2 & -t^{-1} & 0 \\ 1 & -t^{-1} & 0 \end{pmatrix}$$

as generators. Let $g_2 = g_1 g^{-1}$. Find the eigenvalues and eigenvectors of g_1, g_2.

Solution 6.   We have

$$g_2 = g_1 g^{-1} = \begin{pmatrix} 0 & 0 & -t \\ 0 & -t^{-1} & t^{-1} - t \\ -1 & 0 & 1-t \end{pmatrix}.$$

The eigenvalues of g_1 and g_2 are $-t^{-1}, 1, -t$.

Problem 7.   Find the conditions on $a, b, c, d, e, f \in \mathbb{R}$ such that

$$\begin{pmatrix} a & b \\ 0 & c \end{pmatrix}\begin{pmatrix} d & e \\ 0 & f \end{pmatrix}\begin{pmatrix} a & b \\ 0 & c \end{pmatrix} = \begin{pmatrix} d & e \\ 0 & f \end{pmatrix}\begin{pmatrix} a & b \\ 0 & c \end{pmatrix}\begin{pmatrix} d & e \\ 0 & f \end{pmatrix}.$$

Find nontrivial solutions to these conditions.

Solution 7. We obtain the three conditions

$$ad(a - d) = 0, \qquad cf(c - f) = 0$$

$$ad(b - e) + cf(b - e) + ace - bdf = 0.$$

With the γ's arbitrary the eleven solutions are

$$a = 0, \ b = \gamma_1, \ c = \gamma_2, \ d = \gamma_3, \ e = \gamma_4, \ f = 0$$

$$a = \gamma_5, \ b = \gamma_6, \ c = 0, \ d = 0, \ e = \gamma_7, \ f = \gamma_8$$

$$a = \gamma_9, \ b = \gamma_{10}, \ c = \gamma_{11}, \ d = 0, \ e = 0, \ f = 0$$

$$a = \gamma_{13}, \ b = \gamma_{14}(\gamma_{13} - \gamma_{12}), \ c = \gamma_{12}, \ d = \gamma_{13}, \ e = \gamma_{14}, \ f = 0$$

$$a = 0 \ b = 0, \ c = 0, \ d = \gamma_{15}, \ e = \gamma_{16}, \ f = \gamma_{17}$$

$$a = \gamma_{18}, \ b = (\gamma_{18} - \gamma_{19})/(\gamma_{18} - \gamma_{20}), \ c = 0, \ d = \gamma_{18}, \ e = \gamma_{19}, \ f = \gamma_{20}$$

$$a = \gamma_{22}(\gamma_{23} - \gamma_{21})/\gamma_{23}, \ b = \gamma_{21}, \ c = \gamma_{22}, \ d = 0, \ e = \gamma_{23}, \ f = \gamma_{22}$$

$$a = 0, \ b = \gamma_{24}\gamma_{26}/(\gamma_{24} - \gamma_{25}), \ c = \gamma_{24}, \ d = \gamma_{25}, \ e = \gamma_{26}, \ f = \gamma_{24}$$

$$a = \gamma_{27}, b = \gamma_{28}, c = \frac{1}{2}\gamma_{27}(1 + i\sqrt{3}), d = \gamma_{27}, e = \gamma_{29}, f = \frac{1}{2}\gamma_{27}(1 + i\sqrt{3})$$

$$a = \gamma_{30}, b = \gamma_{31}, c = \frac{1}{2}\gamma_{30}(1 - i\sqrt{3}), d = \gamma_{30}, e = \gamma_{32}, f = \frac{1}{2}\gamma_{30}(1 - i\sqrt{3})$$

$$a = \gamma_{33}, \ b = \gamma_{34}, \ c = \gamma_{35}, \ d = \gamma_{33}, \ e = \gamma_{34}, \ f = \gamma_{35}.$$

Some of the solutions are trivial such as the first, third, fifth and last one.

Problem 8. Let A, B be $n \times n$ matrices with $ABA = BAB$.
(i) Show that $(A \otimes A)(B \otimes B)(A \otimes A) = (B \otimes B)(A \otimes A)(B \otimes B)$.
(ii) Show that $(A \oplus A)(B \oplus B)(A \oplus A) = (B \oplus B)(A \oplus A)(B \oplus B)$.

Solution 8. (i) We have

$$(A \otimes A)(B \otimes B)(A \otimes A) = (ABA) \otimes (ABA)$$

and

$$(B \otimes B)(A \otimes A)(B \otimes B) = ((BAB) \otimes (BAB)).$$

Thus since $ABA = BAB$ the statement is true.
(ii) We have

$$(A \oplus A)(B \oplus B)(A \oplus A) = (ABA) \oplus (ABA)$$

and

$$(B \oplus B)(A \oplus A)(B \oplus B) = ((BAB) \oplus (BAB)).$$

Thus since $ABA = BAB$ the statement is true.

Problem 9.   (i) Show that the matrices

$$S_1 = \begin{pmatrix} -t & 1 \\ 0 & 1 \end{pmatrix}, \qquad S_2 = \begin{pmatrix} 1 & 0 \\ t & -t \end{pmatrix}$$

satisfy the braid-like relation $S_1 S_2 S_1 = S_2 S_1 S_2$.
(ii) Show that the matrices $S_1 \otimes S_1$ and $S_2 \otimes S_2$ satisfy the braid-like relation

$$(S_1 \otimes S_1)(S_2 \otimes S_2)(S_1 \otimes S_1) = (S_2 \otimes S_2)(S_1 \otimes S_1)(S_2 \otimes S_2).$$

Solution 9.   The Maxima program will provide the proof

```
/* Note that . is matrix multiplication */
S1: matrix([-t,1],[0,1]);
S2: matrix([1,0],[t,-t]);
L1: S1 . S2 . S1;
R1: S2 . S1 . S2;
F: L1 - R1;
K1: kronecker_product(S1,S1);
K2: kronecker_product(S2,S2);
L2: K1 . K2 . K1;
R2: K2 . K1 . K2;
F: L2 - R2;
```

Problem 10.   Let T be an $n \times n$ matrix and R be an $n^2 \times n^2$ matrix. Consider the equation

$$R(T \otimes T) = (T \otimes T)R.$$

(i) Let $n = 2$ and

$$T = \begin{pmatrix} 0 & 1 \\ 1 & 0 \end{pmatrix}.$$

Find all 4×4 matrices R which satisfy the equation.
(ii) Let $n = 2$ and

$$R = \begin{pmatrix} 1 & 0 & 0 & 0 \\ 0 & 0 & 1 & 0 \\ 0 & 1 & 0 & 0 \\ 0 & 0 & 0 & 1 \end{pmatrix}.$$

Find all T which satisfy the equation.

Solution 10.   (i) We obtain the eight conditions

$$r_{11} = r_{44}, \quad r_{12} = r_{43}, \quad r_{13} = r_{42}, \quad r_{14} = r_{41},$$

$$r_{21} = r_{34}, \quad r_{22} = r_{33}, \quad r_{23} = r_{32}, \quad r_{24} = r_{31}$$

i.e.

$$R = \begin{pmatrix} r_{11} & r_{12} & r_{13} & r_{14} \\ r_{21} & r_{22} & r_{23} & r_{24} \\ r_{24} & r_{23} & r_{22} & r_{21} \\ r_{14} & r_{13} & r_{12} & r_{11} \end{pmatrix}.$$

(ii) All

$$T = \begin{pmatrix} t_{11} & t_{12} \\ t_{21} & t_{22} \end{pmatrix}$$

satisfy the condition $R(T \otimes T) = (T \otimes T)R$.

Problem 11. Find all 4×4 matrices A, B with $[A, B] \neq 0_4$ satisfying the conditions

$$ABA = BAB, \qquad ABBA = I_4.$$

The first condition is the braid relation and the second condition $ABBA = I_4$ runs under *Dirac game*.

Solution 11. We have $(ABA)^2 = (BAB)^2 = BABBAB = BB = B^2$ and

$$B^4 = (ABA)^4 = ABA(ABA)^2 ABA = ABAB^2 ABA = ABBA = I_4.$$

Problem 12. The *free group*, Γ_2, with two generators g_1 and g_2 admits the matrix representation

$$g_1 = \begin{pmatrix} 1 & 0 \\ 2 & 1 \end{pmatrix}, \qquad g_2 = \begin{pmatrix} 1 & 2 \\ 0 & 1 \end{pmatrix}.$$

(i) Find the inverse of the matrices.
(ii) Calculate $g_1 g_2 g_1$ and $g_2 g_1 g_2$. Discuss.

Solution 12. (i) We obtain

$$g_1^{-1} = \begin{pmatrix} 1 & 0 \\ -2 & 1 \end{pmatrix}, \qquad g_2^{-1} = \begin{pmatrix} 1 & -2 \\ 0 & 1 \end{pmatrix}.$$

(ii) We find

$$g_1 g_2 g_1 = \begin{pmatrix} 5 & 2 \\ 12 & 5 \end{pmatrix}, \qquad g_2 g_1 g_2 = \begin{pmatrix} 5 & 12 \\ 2 & 5 \end{pmatrix}.$$

Thus $g_1 g_2 g_1 = (g_2 g_1 g_2)^T$.

Problem 13. Let A, B be 2×2 matrices and

$$R_{12} = A \otimes B \otimes I_2, \quad R_{13} = A \otimes I_2 \otimes B, \quad R_{23} = I_2 \otimes A \otimes B.$$

Find the conditions on A and B such that $R_{12} R_{13} R_{23} = R_{23} R_{13} R_{12}$.

Solution 13. Since $R_{12} R_{13} R_{23} = A^2 \otimes BA \otimes B^2$ and

$$R_{23} R_{13} R_{12} = A^2 \otimes AB \otimes B^2$$

we have to solve $A^2 \otimes BA \otimes B^2 = A^2 \otimes AB \otimes B^2$.

Problem 14.   Let I_2 be the 2×2 unit matrix. Let R be a 4×4 matrix over $\mathbb{C}$. Then R satisfies the *Yang-Baxter equation* if

$$(R \otimes I_2)(I_2 \otimes R)(R \otimes I_2) = (I_2 \otimes R)(R \otimes I_2)(I_2 \otimes R).$$

Use computer algebra to show that

$$R = \begin{pmatrix} 1/\sqrt{2} & 0 & 0 & 1/\sqrt{2} \\ 0 & 1/\sqrt{2} & -1/\sqrt{2} & 0 \\ 0 & 1/\sqrt{2} & 1/\sqrt{2} & 0 \\ -1/\sqrt{2} & 0 & 0 & 1/\sqrt{2} \end{pmatrix}$$

satisfy the Yang-Baxter equation.

Solution 14.   A Maxima program which test the equation is

```
/* Note that . is matrix multiplication in Maxima */
I2: matrix([1,0],[0,1]);
R: matrix([1/sqrt(2),0,0,1/sqrt(2)],[0,1/sqrt(2),-1/sqrt(2),0],
          [0,1/sqrt(2),1/sqrt(2),0],[-1/sqrt(2),0,0,1/sqrt(2)]);
A1: kronecker_product(R,I2); A2: kronecker_product(I2,R);
A3: kronecker_product(R,I2);
B1: kronecker_product(I2,R); B2: kronecker_product(R,I2);
B3: kronecker_product(I2,R);
S1: A1 . A2 . A3; S2: B1 . B2 . B3;
F: S1 - S2;
```

Thus F is the 8×8 zero matrix.

Problem 15.   Show that the 2×2 matrix

$$A = \begin{pmatrix} 0 & 1 \\ 1 & 0 \end{pmatrix}$$

satisfies the braid-like relation

$$(A \otimes I_3)(I_3 \otimes A)(A \otimes I_3) = (I_3 \otimes A)(A \otimes I_3)(I_3 \otimes A).$$

Show that the 3×3 matrix

$$B = \begin{pmatrix} 0 & 0 & 1 \\ 0 & 1 & 0 \\ 1 & 0 & 0 \end{pmatrix}$$

satisfies the braid-like relation

$$(B \otimes I_2)(I_2 \otimes B)(B \otimes I_2) = (I_2 \otimes B)(B \otimes I_2)(I_2 \otimes B).$$

Solution 15.   The following Maxima program will test the equality

```
I2: matrix([1,0],[0,1]); I3: matrix([1,0,0],[0,1,0],[0,0,1]);
A: matrix([0,1],[1,0]);  B: matrix([0,0,1],[0,1,0],[1,0,0]);
T1: kronecker_product(A,I3); T2: kronecker_product(I3,A);
R1: T1 . T2 . T1; R2: T2 . T1 . T2;
R: R1 - R2;
U1: kronecker_product(B,I2); U2: kronecker_product(I2,B);
S1: U1 . U2 . U1; S2: U2 . U1 . U2;
S: S1-S2;
```

Problem 16. Consider the 2×2 matrices I_2 and σ_1 and the 4×4 matrices

$$R = \frac{1}{2}(I_2 \otimes I_2 + I_2 \otimes \sigma_1 + \sigma_1 \otimes I_2 - \sigma_1 \otimes \sigma_1), \quad F = \begin{pmatrix} 1 & 0 & 0 & 0 \\ 0 & 0 & 1 & 0 \\ 0 & 1 & 0 & 0 \\ 0 & 0 & 0 & 1 \end{pmatrix}.$$

Calculate the matrix $\tilde{R} = FR$. Find the determinant of $\tilde{R}$. Does the 4×4 matrix $\tilde{R}$ satisfy the braid-like relation

$$(I_2 \otimes \tilde{R})(\tilde{R} \otimes I_2)(I_2 \otimes \tilde{R}) = (\tilde{R} \otimes I_2)(I_2 \otimes \tilde{R})(\tilde{R} \otimes I_2)?$$

Consider the four *Bell states*

$$\frac{1}{\sqrt{2}}\begin{pmatrix} 1 \\ 0 \\ 0 \\ 1 \end{pmatrix}, \quad \frac{1}{\sqrt{2}}\begin{pmatrix} 0 \\ 1 \\ 1 \\ 0 \end{pmatrix}, \quad \frac{1}{\sqrt{2}}\begin{pmatrix} 0 \\ 1 \\ -1 \\ 0 \end{pmatrix}, \quad \frac{1}{\sqrt{2}}\begin{pmatrix} 1 \\ 0 \\ 0 \\ -1 \end{pmatrix}.$$

Apply the matrix $\tilde{R}$ to the Bell states. Discuss.

Solution 16. The following Maxima program will do the job. The determinant of $\tilde{R}$ is 1. The braid-like relation is satisfied and $\tilde{R}$ maps between the Bell states, i.e.

$$\tilde{R}\frac{1}{\sqrt{2}}\begin{pmatrix} 1 \\ 0 \\ 0 \\ 1 \end{pmatrix} = \frac{1}{\sqrt{2}}\begin{pmatrix} 0 \\ 1 \\ 1 \\ 0 \end{pmatrix}, \quad \tilde{R}\frac{1}{\sqrt{2}}\begin{pmatrix} 0 \\ 1 \\ 1 \\ 0 \end{pmatrix} = \frac{1}{\sqrt{2}}\begin{pmatrix} 1 \\ 0 \\ 0 \\ 1 \end{pmatrix}$$

$$\tilde{R}\frac{1}{\sqrt{2}}\begin{pmatrix} 1 \\ 0 \\ 0 \\ -1 \end{pmatrix} = \frac{1}{\sqrt{2}}\begin{pmatrix} 1 \\ 0 \\ 0 \\ -1 \end{pmatrix}, \quad \tilde{R}\frac{1}{\sqrt{2}}\begin{pmatrix} 0 \\ 1 \\ -1 \\ 0 \end{pmatrix} = \frac{1}{\sqrt{2}}\begin{pmatrix} 0 \\ -1 \\ 1 \\ 0 \end{pmatrix}.$$

```
/* braidrelation.mac */
I2: matrix([1,0],[0,1]);
s1: matrix([0,1],[1,0]);
T1: kronecker_product(I2,I2); T2: kronecker_product(s1,I2);
T3: kronecker_product(I2,s1); T4: kronecker_product(s1,s1);
R: (T1+T2+T3-T4)/2;
F: matrix([1,0,0,0],[0,0,1,0],[0,1,0,0],[0,0,0,1]);
RP: F . R;
```

```
detRP: determinant(RP);
V1: kronecker_product(I2,RP);  V2: kronecker_product(RP,I2);
/* braid relation */
Z: V1 . V2 . V1 - V2 . V1 . V2;
/* four Bell states */
B1: matrix([1],[0],[0],[1])/sqrt(2);
B2: matrix([0],[1],[1],[0])/sqrt(2);
B3: matrix([0],[1],[-1],[0])/sqrt(2);
B4: matrix([1],[0],[0],[-1])/sqrt(2);
/* applying BP to the Bell states */
RP . B1; RP . B2; RP . B3; RP . B4;
```

Problem 17.   Let

$$B(1) = \begin{pmatrix} 1 & 1 \\ 0 & 1 \end{pmatrix}$$

and $B(k)$ the $2^k \times 2^k$ matrix recursively defined by

$$B(k) := \begin{pmatrix} B(k-1) & B(k-1) \\ 0 & B(k-1) \end{pmatrix} = B(1) \otimes B(k-1), \qquad k = 2, 3, \ldots.$$

(i) Find $B(1)^{-1}$.
(ii) Find $B(k)^{-1}$ for $k = 2, 3, \ldots$.
(iii) Let $A(k)$ be the $2^k \times 2^k$ anti-diagonal matrix defined by $A_{ij}(k) = 1$ if $i = 2^k + 1 - j$ and 0 otherwise with $j = 1, 2, \ldots, 2^k$. Let

$$C(k) = A(k)B(k)^{-1}A(k).$$

Show that the matrices $B(k)$ and $C(k)$ have the braid-like relation

$$B(k)C(k)B(k) = C(k)B(k)C(k).$$

Solution 17.   (i) We have

$$B(1)^{-1} = \begin{pmatrix} 1 & -1 \\ 0 & 1 \end{pmatrix}.$$

(ii) We find $B(k)^{-1} = (B(1)^{-1})^{\otimes k}$.
(iii) We find

$$C(1) = \begin{pmatrix} 1 & 0 \\ -1 & 1 \end{pmatrix}.$$

The relation is true for $k = 1$ by a direct calculation. Since

$$B(k) = B(1)^{\otimes k}, \qquad C(k) = C(1)^{\otimes k}$$

and using the properties of the Kronecker product we find that the relation is true for every k.

Problem 18.   Let

$$B = \begin{pmatrix} 1 & 1 \\ 0 & 1 \end{pmatrix}, \qquad C = \begin{pmatrix} 1 & 0 \\ -1 & 1 \end{pmatrix}.$$

Then $BCB = CBC$. Is

$$(B \otimes B)(C \otimes C)(B \otimes B) = (C \otimes C)(B \otimes B)(C \otimes C)?$$

Solution 18. For the left-hand side we have

$$(B \otimes B)(C \otimes C)(B \otimes B) = (BC) \otimes (BC)(B \otimes B) = (BCB) \otimes (BCB).$$

For the right-hand side we have

$$(C \otimes C)(B \otimes B)(C \otimes C) = (CB) \otimes (CB)(C \otimes C) = (CBC) \otimes (CBC).$$

Since $BCB = CBC$ the identity holds.

Problem 19. Let $n \geq 2$. Let $u, v \in \mathbb{C}$ (u, v are called the *spectral parameters*). Consider the $n^2 \times n^2$ matrices $R(u)$ over $\mathbb{C}$ with the entries $(r_{k\ell}^{ij})(u)$ with $i, j, k, \ell = 1, \ldots, n$ and the entries are smooth functions of u. Let $\{\mathbf{e}_i : i = 1, \ldots, n\}$ be the standard basis in the vector space $\mathbb{C}^n$. Let $\otimes$ be the Kronecker product. Then

$$\{\mathbf{e}_i \otimes \mathbf{e}_j : i, j = 1, \ldots, n\}$$

is the standard basis in the vector space $\mathbb{C}^n \otimes \mathbb{C}^n$. One defines

$$R(\mathbf{e}_i \otimes \mathbf{e}_j)(u) := \sum_{\alpha=1}^{n} \sum_{\beta=1}^{n} r_{ij}^{\alpha\beta}(u) \mathbf{e}_\alpha \otimes \mathbf{e}_\beta, \quad i, j = 1, \ldots, n. \tag{1}$$

Thus

$$R(u) = \sum_{i,j=1}^{n} \sum_{\alpha,\beta=1}^{n} r_{ij}^{\alpha\beta}(u)(\mathbf{e}_\alpha \otimes \mathbf{e}_\beta)(\mathbf{e}_i \otimes \mathbf{e}_j)^T.$$

For $n = 2$ one finds that the matrix $R(u)$ is given by

$$R(u) = \begin{pmatrix} r_{11}^{11}(u) & r_{12}^{11}(u) & r_{21}^{11}(u) & r_{22}^{11}(u) \\ r_{11}^{12}(u) & r_{12}^{12}(u) & r_{21}^{12}(u) & r_{22}^{12}(u) \\ r_{11}^{21}(u) & r_{12}^{21}(u) & r_{21}^{21}(u) & r_{22}^{21}(u) \\ r_{11}^{22}(u) & r_{12}^{22}(u) & r_{21}^{22}(u) & r_{22}^{22}(u) \end{pmatrix}.$$

One defines the three $n^3 \times n^3$ matrices

$$R^{12}(u) := R(u) \otimes I_n, \qquad R^{23}(u) := I_n \otimes R(u)$$

and

$$R^{13}(\mathbf{e}_i \otimes \mathbf{e}_j \otimes \mathbf{e}_k)(u) := \sum_{\alpha=1}^{n} \sum_{\beta=1}^{n} r_{ik}^{\alpha\beta}(u) \mathbf{e}_\alpha \otimes \mathbf{e}_j \otimes \mathbf{e}_\beta.$$

(i) Write down the matrices $R^{12}(u)$, $R^{23}(u)$ and $R^{13}(u)$ for $n = 2$ and find the Yang-Baxter relation

$$R^{12}(u)R^{13}(u+v)R^{23}(v) = R^{23}(v)R^{13}(u+v)R^{12}(u).$$

Thus there are $2^6 = 64$ equations for $2^4 = 16$ unknowns.

(ii) Show that the matrix

$$R(u) = \begin{pmatrix} 1+u & 0 & 0 & 0 \\ 0 & u & 1 & 0 \\ 0 & 1 & u & 0 \\ 0 & 0 & 0 & 1+u \end{pmatrix}$$

satisfies the Yang-Baxter equation.

(iii) Let

$$P = \begin{pmatrix} 1 & 0 & 0 & 0 \\ 0 & 0 & 1 & 0 \\ 0 & 1 & 0 & 0 \\ 0 & 0 & 0 & 1 \end{pmatrix}.$$

This matrix is called the *swap matrix* with $P(\mathbf{x} \otimes \mathbf{y}) = \mathbf{y} \otimes \mathbf{x}$ where $\mathbf{x}, \mathbf{y} \in \mathbb{C}^2$. We define $\tilde{R}(u) := PR(u)$. Then the Yang-Baxter equation can be written as

$$(\tilde{R}(v) \otimes I_2)(I_2 \otimes \tilde{R}(u+v))(\tilde{R}(u) \otimes I_2) = (I_2 \otimes \tilde{R}(u))(\tilde{R}(u+v) \otimes I_2)(I_2 \otimes \tilde{R}(v)).$$

Show that $R(u)$ given in (ii) satisfies this equation. Note that without the spectral parameters u, v we obtain Artin's braid relation.

Solution 19.   (i) For $R^{12}(u) = R(u) \otimes I_2$ we have the 8×8 matrix (the dependence of the $r_{ij}^{\alpha\beta}$ on u is omitted)

$$R^{12}(u) = \begin{pmatrix} r_{11}^{11} & 0 & r_{12}^{11} & 0 & r_{21}^{11} & 0 & r_{22}^{11} & 0 \\ 0 & r_{11}^{11} & 0 & r_{12}^{11} & 0 & r_{21}^{11} & 0 & r_{22}^{11} \\ r_{11}^{12} & 0 & r_{12}^{12} & 0 & r_{21}^{12} & 0 & r_{22}^{12} & 0 \\ 0 & r_{11}^{12} & 0 & r_{12}^{12} & 0 & r_{21}^{12} & 0 & r_{22}^{12} \\ r_{11}^{21} & 0 & r_{12}^{21} & 0 & r_{21}^{21} & 0 & r_{22}^{21} & 0 \\ 0 & r_{11}^{21} & 0 & r_{12}^{21} & 0 & r_{21}^{21} & 0 & r_{22}^{21} \\ r_{11}^{22} & 0 & r_{12}^{22} & 0 & r_{21}^{22} & 0 & r_{22}^{22} & 0 \\ 0 & r_{11}^{22} & 0 & r_{12}^{22} & 0 & r_{21}^{22} & 0 & r_{22}^{22} \end{pmatrix}.$$

For $R^{23}(u) = I_2 \otimes R(u)$ we have

$$R^{23}(u) = \begin{pmatrix} r_{11}^{11} & r_{12}^{11} & r_{21}^{11} & r_{22}^{11} & 0 & 0 & 0 & 0 \\ r_{11}^{12} & r_{12}^{12} & r_{21}^{12} & r_{22}^{12} & 0 & 0 & 0 & 0 \\ r_{11}^{21} & r_{12}^{21} & r_{21}^{21} & r_{22}^{21} & 0 & 0 & 0 & 0 \\ r_{11}^{22} & r_{12}^{22} & r_{21}^{22} & r_{22}^{22} & 0 & 0 & 0 & 0 \\ 0 & 0 & 0 & 0 & r_{11}^{11} & r_{12}^{11} & r_{21}^{11} & r_{22}^{11} \\ 0 & 0 & 0 & 0 & r_{11}^{12} & r_{12}^{12} & r_{21}^{12} & r_{22}^{12} \\ 0 & 0 & 0 & 0 & r_{11}^{21} & r_{12}^{21} & r_{21}^{21} & r_{22}^{21} \\ 0 & 0 & 0 & 0 & r_{11}^{22} & r_{12}^{22} & r_{21}^{22} & r_{22}^{22} \end{pmatrix} = R(u) \oplus R(u)$$

where $\oplus$ denotes the direct sum. For $R^{13}(u)$ we obtain

$$R^{13}(u) = \begin{pmatrix} r^{11}_{11} & r^{11}_{12} & 0 & 0 & r^{11}_{21} & r^{11}_{22} & 0 & 0 \\ r^{12}_{11} & r^{12}_{12} & 0 & 0 & r^{12}_{21} & r^{12}_{22} & 0 & 0 \\ 0 & 0 & r^{11}_{11} & r^{11}_{12} & 0 & 0 & r^{11}_{21} & r^{11}_{22} \\ 0 & 0 & r^{12}_{11} & r^{12}_{12} & 0 & 0 & r^{12}_{21} & r^{12}_{22} \\ r^{21}_{11} & r^{21}_{12} & 0 & 0 & r^{21}_{21} & r^{21}_{22} & 0 & 0 \\ r^{22}_{11} & r^{22}_{12} & 0 & 0 & r^{22}_{21} & r^{22}_{22} & 0 & 0 \\ 0 & 0 & r^{21}_{11} & r^{21}_{12} & 0 & 0 & r^{21}_{21} & r^{21}_{22} \\ 0 & 0 & r^{22}_{11} & r^{22}_{12} & 0 & 0 & r^{22}_{21} & r^{22}_{22} \end{pmatrix}$$

where we utilized that

$$R^{13}(e_1 \otimes e_1 \otimes e_1)(u) = \sum_{\alpha=1}^{2} \sum_{\beta=1}^{2} r^{\alpha\beta}_{11}(u) e_\alpha \otimes e_1 \otimes e_\beta$$
$$= (r^{11}_{11} \ \ r^{12}_{11} \ \ 0 \ \ 0 \ \ r^{21}_{11} \ \ r^{22}_{11} \ \ 0 \ \ 0)^T$$

etc.

(ii) The Maxima program will test the Yang-Baxter relation for the given matrix $R(u)$.

```
I2: matrix([1,0],[0,1]);
Ru: matrix([1+u,0,0,0],[0,u,1,0],[0,1,u,0],[0,0,0,1+u]);
Rv: matrix([1+v,0,0,0],[0,v,1,0],[0,1,v,0],[0,0,0,1+v]);
Ruv: matrix([1+u+v,0,0,0],[0,u+v,1,0],[0,1,u+v,0],[0,0,0,1+u+v]);
R12u: kronecker_product(Ru,I2);
R23v: kronecker_product(I2,Rv);
R13uv: matrix([1+u+v,0,0,0,0,0,0,0],[0,u+v,0,0,1,0,0,0],
          [0,0,1+u+v,0,0,0,0,0],[0,0,0,u+v,0,0,1,0],
          [0,1,0,0,u+v,0,0,0],[0,0,0,0,0,1+u+v,0,0],
          [0,0,0,1,0,0,u+v,0],[0,0,0,0,0,0,0,1+u+v]);
LHS: R12u . R13uv . R23v;
RHS: R23v . R13uv . R12u;
Result: LHS - RHS;
Result: expand(Result);
```

(iii) The Maxima program will check the Yang-Baxter equation in braid form

```
I2: matrix([1,0],[0,1]);
P: matrix([1,0,0,0],[0,0,1,0],[0,1,0,0],[0,0,0,1]);
Ru: matrix([1+u,0,0,0],[0,u,1,0],[0,1,u,0],[0,0,0,1+u]);
Ru: P . Ru;
Rv: matrix([1+v,0,0,0],[0,v,1,0],[0,1,v,0],[0,0,0,1+v]);
Rv: P . Rv;
Ruv: matrix([1+u+v,0,0,0],[0,u+v,1,0],[0,1,u+v,0],[0,0,0,1+u+v]);
Ruv: P . Ruv;
T1u: kronecker_product(Ru,I2); T1v: kronecker_product(Rv,I2);
T2u: kronecker_product(I2,Ru); T2v: kronecker_product(I2,Rv);
T1uv: kronecker_product(Ruv,I2); T2uv: kronecker_product(I2,Ruv);
LHS: T1v . T2uv . T1u; RHS: T2u . T1uv . T2v;
D: LHS - RHS;
D: expand(D);
```

Problem 20.   Let $q = 4$ and $w := \exp(2\pi i/q)$, i.e. $w^4 = 1$ and $w = i$, $w^2 = -1$, $w^3 = -i$. Let D_1 and D_2 be the diagonal matrices

$$D_1 = \begin{pmatrix} 1 & 0 & 0 & 0 \\ 0 & w & 0 & 0 \\ 0 & 0 & w^2 & 0 \\ 0 & 0 & 0 & w^3 \end{pmatrix}, \quad D_2 = \begin{pmatrix} 1 & 0 & 0 & 0 \\ 0 & w^3 & 0 & 0 \\ 0 & 0 & w^2 & 0 \\ 0 & 0 & 0 & w \end{pmatrix}$$

and P the permutation matrix

$$P = \begin{pmatrix} 0 & 1 & 0 & 0 \\ 0 & 0 & 1 & 0 \\ 0 & 0 & 0 & 1 \\ 1 & 0 & 0 & 0 \end{pmatrix}.$$

Consider the 64×64 matrices

$$S_1 = D_1 \otimes D_2 \otimes I_4, \quad S_2 = I_4 \otimes P \otimes I_4, \quad S_3 = I_4 \otimes D_1 \otimes D_2, \quad S_4 = I_4 \otimes I_4 \otimes P.$$

Show that

$$S_j^q = I_{64} \qquad j = 1, 2, 3, 4$$
$$S_j S_{j+1} S_j^{-1} S_{j+1}^{-1} = w I_{64} \qquad j = 1, 2, 3$$
$$[S_1, S_3] = 0_{64}, \quad [S_1, S_4] = 0_{64}, \quad [S_2, S_4] = 0_{64}$$

where 0_{64} is the 64×64 zero matrix.

Solution 20.   This is a typical task for computer algebra to do. A Maxima program is

```
/* braid64.mac */
I4: matrix([1,0,0,0],[0,1,0,0],[0,0,1,0],[0,0,0,1]);
D1: matrix([1,0,0,0],[0,%i,0,0],[0,0,-1,0],[0,0,0,-%i]);
D2: matrix([1,0,0,0],[0,-%i,0,0],[0,0,-1,0],[0,0,0,%i]);
P: matrix([0,1,0,0],[0,0,1,0],[0,0,0,1],[1,0,0,0]);
S1: kronecker_product(D1,kronecker_product(D2,I4));
S2: kronecker_product(I4,kronecker_product(P,I4));
S3: kronecker_product(I4,kronecker_product(D1,D2));
S4: kronecker_product(I4,kronecker_product(I4,P));
R1: S1 . S2 . (S1^^-1) . (S2^^-1);
R2: S2 . S3 . (S2^^-1) . (S3^^-1);
R3: S3 . S4 . (S3^^-1) . (S4^^-1);
R4: S1 . S3 - S3 . S1;
R5: S1 . S4 - S4 . S1;
R6: S2 . S4 - S4 . S2;
```

Problem 21.   Consider the Pauli spin matrices $\sigma_0 = I_2$, σ_1, σ_2, σ_3.
(i) Find the 4×4 matrix

$$R(u) = \sum_{j=0}^{3} w_j(u)\sigma_j \otimes \sigma_j$$

with

$$w_0(u) = \sinh(u + \eta/2),$$
$$w_1(u) = w_2(u) = \sinh(\eta/2)\cosh(\eta/2),$$
$$w_3(u) = \sinh(\eta/2)\cosh(u + \eta/2)$$

where η is a fixed parameter.

Solution 21.   We have

$$
R(u) = \begin{pmatrix} w_0 + w_3 & 0 & 0 & w_1 - w_2 \\ 0 & w_0 - w_3 & w_1 + w_2 & 0 \\ 0 & w_1 + w_2 & w_0 - w_3 & 0 \\ w_1 - w_2 & 0 & 0 & w_0 + w_3 \end{pmatrix}
$$

$$
= \begin{pmatrix} \sinh(u + \eta) & 0 & 0 & 0 \\ 0 & \sinh(u) & \sinh(\eta) & 0 \\ 0 & \sinh(\eta) & \sinh(u) & 0 \\ 0 & 0 & 0 & \sinh(u + \eta) \end{pmatrix}.
$$

Problem 22.   Let $\mathcal{B}_n$ denote the braid group on n strands. $\mathcal{B}_n$ is generated by the elementary braids $\{\, b_1,\, b_2,\, \dots,\, b_{n-1}\,\}$ with the *braid relations*

$$b_j b_{j+1} b_j = b_{j+1} b_j b_{j+1}, \qquad 1 \le j < n - 1,$$

$$b_j b_k = b_k b_j, \qquad |j - k| \ge 2.$$

Let $\{\, e_1,\, e_2,\, \dots,\, e_n\,\}$ denote the standard basis in $\mathbb{R}^n$. Then $\mathbf{u} \in \mathbb{R}^n$ can be written as

$$\mathbf{u} = \sum_{k=1}^{n} c_k e_k, \qquad c_1, c_2, \dots, c_n \in \mathbb{R}.$$

Consider the linear operators B_j $(\alpha, \beta, \gamma, \delta \in \mathbb{R}$ and $\alpha, \gamma \ne 0)$ defined by

$$B_j \mathbf{u} := c_1 e_1 + \cdots + (\alpha c_{j+1} + \beta)e_j + (\gamma c_{j+1} + \delta)e_{j+1} + \cdots + c_n e_n$$

and the corresponding inverse operation

$$B_j^{-1} \mathbf{u} := c_1 e_1 + \cdots + \frac{1}{\gamma}(c_{j+1} - \delta)e_j + \frac{1}{\alpha}(\gamma c_j - \beta)e_{j+1} + \cdots + c_n e_n.$$

Use computer algebra to show that $B_1, B_2, \dots, B_{n-1}$ satisfy the braid condition

$$B_j B_{j+1} B_j \mathbf{u} = B_{j+1} B_j B_{j+1} \mathbf{u}$$

if $\gamma\beta + \delta = \alpha\delta + \beta$.

Solution 22.   We use SymbolicC++. The function B(k,u) implements B_k acting on $u \in \mathbb{R}^n$ expressed in terms of the standard basis e[1], ..., e[n].

```
// braid.cpp

#include <iostream>
#include "symbolicc++.h"
using namespace std;

Symbolic e = ~Symbolic("e");

Symbolic B(int j,const Symbolic &u)
{
UniqueSymbol x;
Symbolic r, alpha("alpha"), beta("beta"), gamma("gamma"), delta("delta");
r = u[e[j]==gamma*x];
r = r[e[j+1]==alpha*e[j]];
r = r[x==e[j+1]];
r += beta*e[j]+delta*e[j+1];
return r;
}


int main(void)
{
int n = 5, j, k, m;
Symbolic c("c"), u;
for(k=1;k<=n;k++) u += c[k]*e[k];
for(k=1;k<n-1;k++)
{
Symbolic r = B(k,B(k+1,B(k,u))) - B(k+1,B(k,B(k+1,u)));
cout << r << endl;
for(j=1;j<=n;j++) cout << e[j] << ": " << (r.coeff(e[j])==0) << endl;
}
return 0;
}
```

The program output is

```
beta*gamma*e[2]+delta*e[2]-delta*alpha*e[2]-beta*e[2]
e[1]: 0 == 0
e[2]: beta*gamma+delta-delta*alpha-beta == 0
e[3]: 0 == 0
e[4]: 0 == 0
e[5]: 0 == 0
beta*gamma*e[3]+delta*e[3]-delta*alpha*e[3]-beta*e[3]
e[1]: 0 == 0
e[2]: 0 == 0
e[3]: beta*gamma+delta-delta*alpha-beta == 0
e[4]: 0 == 0
e[5]: 0 == 0
beta*gamma*e[4]+delta*e[4]-delta*alpha*e[4]-beta*e[4]
e[1]: 0 == 0
e[2]: 0 == 0
e[3]: 0 == 0
e[4]: beta*gamma+delta-delta*alpha-beta == 0
e[5]: 0 == 0
```

Supplementary Problems

Problem 1. Let $n \geq 2$ and $j, k = 1, \ldots, 2n - 1$. Let Q be the 4×4 matrix

$$Q = \begin{pmatrix} 0 & 0 & 0 & 0 \\ 0 & s^4 & 1 & 0 \\ 0 & 1 & s^{-4} & 0 \\ 0 & 0 & 0 & 0 \end{pmatrix}.$$

Then $Q^2 = (s^4 + s^{-4})Q$. Consider the 2^{2n} matrices

$$Q_j = I_2 \otimes \cdots \otimes I_2 \otimes Q \otimes I_2 \otimes \cdots \otimes I_2$$

where the 4×4 matrix Q is at the j-th position. These matrices satisfy the relations (Temperley-Lieb algebra)

$$Q_j Q_j = (s^4 + s^{-4})Q_j$$
$$Q_j Q_{j\pm 1} Q_j = Q_j$$
$$Q_j Q_k = Q_k Q_j \quad \text{if} \quad |j - k| \geq 2.$$

These matrices play a role for the q-state Potts model. Consider the case $n = 2$ and the 16×16 matrices

$$Q_1 = Q \otimes I_2 \otimes I_2, \quad Q_2 = I_2 \otimes Q \otimes I_2, \quad Q_3 = I_2 \otimes I_2 \otimes Q.$$

Write computer algebra programs in SymbolicC++ and Maxima which check that these matrices satisfy

$$Q_1 Q_1 = (s^4 + s^{-4})Q_1, \quad Q_2 Q_2 = (s^4 + s^{-4})Q_2, \quad Q_3 Q_3 = (s^4 + s^{-4})Q_3$$

$$Q_1 Q_2 Q_1 = Q_1, \quad Q_2 Q_3 Q_2 = Q_2, \quad Q_3 Q_2 Q_3 = Q_3,$$
$$Q_2 Q_1 Q_2 = Q_2, \quad Q_1 Q_3 = Q_3 Q_1.$$

Problem 2. (i) Find the conditions on the 2×2 matrices over $\mathbb{C}$ such that $ABA = BAB$. Find solutions where $AB \neq BA$.
(ii) Find the conditions on the 2×2 matrices A and B such that

$$A \otimes B \otimes A = B \otimes A \otimes B.$$

Find solutions where $AB \neq BA$, i.e. $[A, B] \neq 0_2$.

Problem 3. Consider the 4×4 matrices

$$A = \begin{pmatrix} 1 & 0 & 0 & 0 \\ 1 & 1 & 0 & 0 \\ 1 & 2 & 1 & 0 \\ 1 & 3 & 3 & 1 \end{pmatrix}, \quad B = \begin{pmatrix} 1 & 3 & 3 & 1 \\ 0 & 1 & 2 & 1 \\ 0 & 0 & 1 & 1 \\ 0 & 0 & 0 & 1 \end{pmatrix}.$$

Is $AB^{-1}A = B^{-1}AB^{-1}$?

Problem 4.   (i) Consider the Pauli spin matrix $A = \sigma_1$. Does the matrix A satisfy the braid-like relation

$$(A \otimes I_2)(I_2 \otimes A)(A \otimes I_2) = (I_2 \otimes A)(A \otimes I_2)(I_2 \otimes A) ?$$

Does the matrix A satisfy the braid-like relation

$$(A \otimes I_3)(I_3 \otimes A)(A \otimes I_3) = (I_3 \otimes A)(A \otimes I_3)(I_3 \otimes A) ?$$

Ask the same question for the Pauli spin matrices σ_2 and σ_3.

(ii) Let $n \geq 2$ and $m \geq 2$. Let I_n be the $n \times n$ identity matrix. Let J_m be counter identity matrix, i.e. the $m \times m$ matrix whose entries are all equal to 0 except those on the counterdiagonal which are all equal to 1. Find the conditions on n and m such that

$$(J_n \otimes I_m)(I_m \otimes J_n)(J_n \otimes I_m) = (I_m \otimes J_n)(J_n \otimes I_m)(I_m \otimes J_n).$$

Problem 5.   Let

$$T = \begin{pmatrix} t_{11} & t_{12} \\ t_{21} & t_{22} \end{pmatrix}$$

be an element of the Lie group $SL(2, \mathbb{R})$, i.e. $\det(T) = 1$ and $t_{jk} \in \mathbb{R}$. Let

$$R = \begin{pmatrix} 1 & 0 & 0 & 0 \\ 0 & -1 & 0 & 0 \\ 0 & 4 & -1 & 0 \\ 0 & 0 & 0 & 1 \end{pmatrix}.$$

Find all $T \in SL(2, \mathbb{R})$ such that $[R, T \otimes T] = 0_4$.

Problem 6.   Let σ_1, σ_2, σ_3 be the Pauli spin matrices. Consider the 4×4 matrix

$$R = a(\lambda, \mu)\sigma_1 \otimes \sigma_1 + b(\lambda, \mu)(\sigma_2 \otimes \sigma_2 + \sigma_3 \otimes \sigma_3)$$

where

$$a(\lambda, \mu) = \frac{1}{4}\frac{\lambda^2 + \mu^2}{\lambda^2 - \mu^2}, \qquad b(\lambda, \mu) = \frac{1}{2}\frac{\lambda\mu}{\lambda^2 - \mu^2}.$$

Does R satisfy the braid-like relation

$$(R \otimes I_2)(I_2 \otimes R)(R \otimes I_2) = (I_2 \otimes R)(R \otimes I_2)(I_2 \otimes R) ?$$

Problem 7.   Consider the braid group $\mathcal{B}_3$ with the generators $\{\sigma_1, \sigma_2\}$ and the relation

$$\sigma_1\sigma_2\sigma_1 = \sigma_2\sigma_1\sigma_2.$$

Let $t \neq 0$. Show that a matrix representation is given by

$$\sigma_1 = \begin{pmatrix} -t & 1 \\ 0 & 1 \end{pmatrix}, \qquad \sigma_2 = \begin{pmatrix} 1 & 0 \\ t & -t \end{pmatrix}$$

with the inverse matrices

$$\sigma_1^{-1} = \begin{pmatrix} -1/t & 1/t \\ 0 & 1 \end{pmatrix}, \qquad \sigma_2^{-1} = \begin{pmatrix} 1 & 0 \\ 1 & -1/t \end{pmatrix}.$$

Let $\sigma = \sigma_1 \sigma_2 \sigma_1^{-1} \sigma_2^{-1} \sigma_1^{-1}$. Find $f(t) = \det(\sigma - I_2)$. Find minima and maxima of f.

Problem 8. (i) Let $a, b, c, d \in \mathbb{R}$ and $a \neq 0$. Does

$$R = \begin{pmatrix} a & 0 & 0 & 0 \\ 0 & 0 & c & 0 \\ 0 & d & a - cd/a & 0 \\ 0 & 0 & 0 & a \end{pmatrix}$$

satisfy $(R \otimes I_4)(I_4 \otimes R)(R \otimes I_4) = (I_4 \otimes R)(R \otimes I_4)(I_4 \otimes R)$? We have to check that $R^2 \otimes R = R \otimes R^2$.

(ii) Find the condition on the 4×4 matrix

$$S = \begin{pmatrix} s_{11} & 0 & 0 & s_{14} \\ 0 & s_{22} & s_{23} & 0 \\ 0 & s_{32} & s_{33} & 0 \\ s_{41} & 0 & 0 & s_{44} \end{pmatrix}$$

such that $(S \otimes I_4)(I_4 \otimes S)(S \otimes I_4) = (I_4 \otimes S)(S \otimes I_4)(I_4 \otimes S)$. We have to check that $S^2 \otimes S = S \otimes S^2$.

Problem 9. Let R be a 4×4 matrix written as

$$R = \begin{pmatrix} R_{11} & R_{12} \\ R_{21} & R_{22} \end{pmatrix}$$

where R_{12}, R_{21}, R_{22} are the 2×2 block matrices. Suppose that R satisfies the equation

$$(R \otimes I_2)(I_2 \otimes R)(R \otimes I_2) = (I_2 \otimes R)(R \otimes I_2)(I_2 \otimes R).$$

Consider the 8×8 matrix

$$\tilde{R} = R \star R = \begin{pmatrix} R_{11} & 0_2 & 0_2 & R_{12} \\ 0_2 & R_{11} & R_{12} & 0_2 \\ 0_2 & R_{21} & R_{22} & 0_2 \\ R_{12} & 0_2 & 0_2 & R_{22} \end{pmatrix}.$$

(i) Does $\tilde{R}$ satisfy $(\tilde{R} \otimes I_2)(I_2 \otimes \tilde{R})(\tilde{R} \otimes I_2) = (I_2 \otimes \tilde{R})(\tilde{R} \otimes I_2)(I_2 \otimes \tilde{R})$?
(ii) Does $\tilde{R}$ satisfy $(\tilde{R} \otimes I_4)(I_4 \otimes \tilde{R})(\tilde{R} \otimes I_4) = (I_4 \otimes \tilde{R})(\tilde{R} \otimes I_4)(I_4 \otimes \tilde{R})$?

Problem 10. Let R be a 4×4 matrix and

$$P = \begin{pmatrix} 1 & 0 & 0 & 0 \\ 0 & 0 & 1 & 0 \\ 0 & 1 & 0 & 0 \\ 0 & 0 & 0 & 1 \end{pmatrix}.$$

Let $\tilde{R} = PR$. Suppose that $\tilde{R}$ satisfies

$$(\tilde{R} \otimes I_2)(I_2 \otimes \tilde{R})(\tilde{R} \otimes I_2) = (I_2 \otimes \tilde{R})(\tilde{R} \otimes I_2)(I_2 \otimes \tilde{R}).$$

Show the following ten matrices satisfy this equation

$$R_0 = \begin{pmatrix} 1 & 0 & 0 & 0 \\ 0 & 0 & 1 & 0 \\ 0 & 1 & 0 & 0 \\ 0 & 0 & 0 & 1 \end{pmatrix}, \quad R_1 = \begin{pmatrix} 1 & 0 & 0 & i \\ 0 & 1 & 1 & 0 \\ 0 & 1 & -1 & 0 \\ i & 0 & 0 & 1 \end{pmatrix},$$

$$R_2 = \begin{pmatrix} 1+t & 0 & 0 & 1 \\ 0 & (1+t^2)^{1/2} & 1 & 0 \\ 0 & 1 & (1+t^2)^{1/2} & 0 \\ 1 & 0 & 0 & 1-t \end{pmatrix}, \quad R_3 = \begin{pmatrix} 1 & 0 & 0 & 0 \\ 0 & s & 0 & 0 \\ 0 & 1-t & st & 0 \\ 1 & 0 & 0 & -t \end{pmatrix},$$

$$R_4 = \begin{pmatrix} 1 & 0 & 0 & 0 \\ 0 & st & 1-t & 0 \\ 0 & 0 & s & 0 \\ 0 & 0 & 0 & -t \end{pmatrix}, \quad R_5 = \begin{pmatrix} 1 & 0 & 0 & 0 \\ 0 & r & 0 & 0 \\ 0 & 1-rt & t & 0 \\ 0 & 0 & 0 & 1 \end{pmatrix},$$

$$R_6 = \begin{pmatrix} 1 & 0 & 0 & 0 \\ 0 & r & 0 & 0 \\ 0 & 1-rt & t & 0 \\ 0 & 0 & 0 & -rt \end{pmatrix}, \quad R_7 = \begin{pmatrix} 1 & 0 & 0 & 0 \\ 0 & s & 0 & 0 \\ 0 & 0 & s & 0 \\ 1 & 0 & 0 & s' \end{pmatrix},$$

$$R_8 = \begin{pmatrix} 1 & 0 & 0 & 0 \\ 0 & r & 0 & 0 \\ 0 & 0 & p & 0 \\ 0 & 0 & 0 & t \end{pmatrix}, \quad R_9 = \begin{pmatrix} 0 & 0 & 0 & 1 \\ 0 & 0 & t & 0 \\ 0 & t & 0 & 0 \\ 1 & 0 & 0 & 0 \end{pmatrix}.$$

with $\tilde{R}_j = PR_j$ $(j = 0, 1, \dots, 9)$, $s^2 = 1$, $(s')^2 = 1$ and s, p, t, r are arbitrary.

Problem 11.   (i) Let A, B be invertible $n \times n$ matrices with $AB \neq BA$. Assume that

$$ABA = BAB \quad \text{and} \quad ABBA = I_n.$$

Show that $A^4 = B^4 = I_n$.
(ii) Find all 2×2 matrices A and B which satisfy the conditions given in (i).
(iii) Find all 3×3 matrices A and B which satisfy the conditions given in (i).
(iv) Find all 4×4 matrices A and B which satisfy the conditions given in (i).

Problem 12.   Let S be the unitary 4×4 matrix

$$S = \frac{1}{\sqrt{2}} \begin{pmatrix} 0 & 1 & i & 0 \\ 1 & 0 & 0 & 1 \\ -i & 0 & 0 & i \\ 0 & 1 & -i & 0 \end{pmatrix}.$$

Consider the 16×16 matrices $S \otimes I_2 \otimes I_2$, $I_2 \otimes S \otimes I_2$, $I_2 \otimes I_2 \otimes S$. Note that $I_4 = I_2 \otimes I_2$. Show that the braid-like relations

$$(S \otimes I_4)(I_2 \otimes S \otimes I_2)(S \otimes I_4) = (I_2 \otimes S \otimes I_2)(S \otimes I_4)(I_2 \otimes S \otimes I_2)$$
$$(I_2 \otimes S \otimes I_2)(I_4 \otimes S)(I_2 \otimes S \otimes I_2) = (I_4 \otimes S)(I_2 \otimes S \otimes I_2)(I_4 \otimes S)$$

are satisfied.

Problem 13. (i) Find all nonzero 4×4 matrices R such that

$$(R \otimes I_2)(I_2 \otimes R)(R \otimes I_2) = (I_2 \otimes R)(R \otimes I_2)(I_2 \otimes R).$$

We obtain $8 \times 8 = 64$ conditions.
(ii) Can one find R's with the additional condition $R^2 = I_4$?

Problem 14. (i) Let L_1, L_2 be 2×2 matrices and R a 4×4 matrix. Find the conditions on L_1, L_2, R such that

$$(L_1 \otimes L_2)R = R(L_2 \otimes L_1).$$

Simplify the system of equations using Gröbner bases.
(ii) Let

$$T = \sigma_2 = \begin{pmatrix} 0 & -i \\ i & 0 \end{pmatrix}$$

and $T_1 = T \otimes I_2$, $T_2 = I_2 \otimes T$. Find all invertible 4×4 matrices R over $\mathbb{C}$ such that

$$RT_1T_2 = T_2T_1R.$$

(iii) Let A, B be given nonzero 2×2 matrices. Find all 4×4 matrices R such that

$$R(A \otimes B) = (B \otimes A)R.$$

Find all 4×4 matrices R such that $R(A \otimes B) = (A \otimes B)R$.

Problem 15. Do the matrices

$$R_1 = \begin{pmatrix} a & 0 & 0 & 0 \\ 0 & 0 & b & 0 \\ 0 & c & 0 & 0 \\ 0 & 0 & 0 & d \end{pmatrix}, \quad R_2 = \begin{pmatrix} 0 & 0 & 0 & a \\ 0 & b & 0 & 0 \\ 0 & 0 & c & 0 \\ d & 0 & 0 & 0 \end{pmatrix}$$

satisfy the Yang-Baxter relation

$$(R \otimes I_2)(I_2 \otimes R)(R \otimes I_2) = (I_2 \otimes R)(R \otimes I_2)(I_2 \otimes R)?$$

Or what are the conditions on a, b, c, d so that the condition is satisfied?

Problem 16. The *braid linking matrix* B is a square symmetric $k \times k$ matrix defined by $B = (b_{ij})$ with b_{ii} the sum of half-twists in the i-th branch, b_{ij} the sum of the crossings between the i-th and the j-th branches of the ribbon graph with standard insertion. Thus the i-th diagonal element of B is the local torsion of the i-th branch. The off-diagonal elements of B are twice the linking numbers of the ribbon graph for the i-th and j-th branches. Consider the braid linking matrix

$$B = \begin{pmatrix} -1 & 0 & -1 \\ 0 & 2 & -1 \\ -1 & -1 & 0 \end{pmatrix}.$$

Discuss. Draw a graph.

Chapter 21

Graphs and Matrices

A *graph* consists of a non-empty set of points, called *vertices* (singular: vertex) or nodes, and a set of lines or curves, called *edges*, such that every edge is attached at each end to a vertex. We assume that $V = \{1, 2, \ldots, n\}$. A *digraph* (directed graph) is a diagram consisting of points, called vertices, joined by directed lines, called *arcs*. Thus each arc joins two vertices in a specified direction. To distinguish them from undirected graphs the edges of a digraph are called arcs.

The most frequently used representation schemes for graphs and digraphs are adjacency matrices. The *adjacency matrix* of an n-vertex graph $G = (V, E)$ is an $n \times n$ matrix A. The adjacency matrix $A(G)$ of a graph G with n vertices is an $n \times n$ matrix with the matrix element a_{ij} being the number of edges joining the vertices i and j.

The adjacency matrix $A(D)$ of a directed graph D with n vertices is an $n \times n$ matrix with the matrix element a_{ij} being the number of arcs from vertex i to vertex j.

A *walk* of length k in a graph is a succession of k edges joining two vertices. Edges can occur more than once in a walk. A *trail* is a walk in which all the edges (but not necessary all the vertices) are distinct. A *path* is a walk in which all the edges and all the vertices are distinct.

An *Eulerian graph* is a connected graph which contains a closed trail which includes every edge. The trail is called an Eulerian trail.

Problem 1. A walk of length k in a digraph is a succession of k arcs joining two vertices. A trail is a walk in which all the arcs (but not necessarily all the vertices) are distinct. A path is a walk in which all the arcs and all the vertices are distinct. Show that the number of walks of length k from vertex i to vertex j in a digraph D with n vertices is given by the ij-th element of the matrix A^k, where A is the adjacency matrix of the digraph.

Solution 1. We use *mathematical induction*. Assume that the result is true for $k \leq K - 1$. Consider any walk from vertex i to vertex j of length K. Such a walk consists of a walk of length $K - 1$ from vertex i to a vertex p which is adjacent to vertex j followed by a walk of length 1 from vertex p to vertex j. The number of such walks is $(A^{K-1})_{ip} \times A_{pj}$. The total number of walks of length k from vertex i to vertex j will then be the sum of the walks through any p, i.e.

$$\sum_{p=1}^{n}(A^{K-1})_{ip}A_{pj}.$$

This is just the expression for the ijth element of the matrix $A^{K-1}A = A^K$. Thus the result is true for $k = K$. The result is certainly true for the walks of length 1, i.e. $k = 1$ since this is the definition of the adjacency matrix A. Therefore it is true for all k.

Problem 2. Consider a digraph. The out-degree of a vertex v is the number of arcs incident from v and the in-degree of a vertex V is the number of arcs incident to v. Loops count as one of each.

Determine the in-degree and the out-degree of each vertex in the digraph given by the adjacency matrix

$$A = \begin{pmatrix} 0 & 1 & 0 & 0 & 0 \\ 0 & 0 & 1 & 0 & 0 \\ 1 & 0 & 0 & 0 & 1 \\ 0 & 0 & 1 & 0 & 0 \\ 0 & 0 & 0 & 1 & 0 \end{pmatrix}$$

and hence determine if it is an Eulerian graph. Display the digraph and determine an Eulerian trail.

Solution 2. The out-degree of each vertex is given by adding the entries in the corresponding row and the in-degree by adding the entries in the corresponding column.
Vertex 1 has out-degree 1 and in-degree 1
Vertex 2 has out-degree 1 and in-degree 1
Vertex 3 has out-degree 2 and in-degree 2
Vertex 4 has out-degree 1 and in-degree 1
Vertex 5 has out-degree 1 and in-degree 1
Hence each vertex has out-degree equal to in-degree and the digraph is Eulerian. An Eulerian trial is, for instance, 1235431.

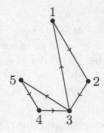

Problem 3. A digraph is strongly connected if there is a path between every pair of vertices. Show that if A is the adjacency matrix of a digraph D with n vertices and B is the matrix

$$B = A + A^2 + A^3 + \cdots + A^{n-1}$$

then D is strongly connected iff each non-diagonal element of B is greater than 0.

Solution 3. We must show both "if each non-diagonal element of B is greater than 0 then D is strongly connected" and "if D is strongly connected then each non-diagonal element of B is greater than 0".

Firstly, let D be digraph and suppose that each non-diagonal element of the matrix we have $b_{ij} > 0$, i.e. $b_{ij} > 0$ for $i \neq j$ and $i, j = 1, 2, \ldots, n$. Then $(A^k)_{ij} > 0$ for some $k \in [1, n-1]$, i.e. there is a walk of some length k between 1 and $n-1$ from every vertex i to every vertex j. Thus the digraph is strongly connected.

Secondly, suppose the digraph is strongly connected. Then, by definition, there is a path from every vertex i to every vertex j. Since the digraph has n vertices the path is of length no more than $n-1$. Hence, for all $i \neq j$, $(A^k)_{ij} > 0$ for some $k \leq n-1$. Hence, for all $i \neq j$, we have $b_{ij} > 0$.

Problem 4. Write down the adjacency matrix A for the digraph shown. Calculate the matrices A^2, A^3 and A^4. Consequently find the number of walks of length 1, 2, 3 and 4 from w to u. Is there a walk of length 1, 2, 3, or 4 from u to w? Find the matrix $B = A + A^2 + A^3 + A^4$ for the digraph and hence conclude whether it is strongly connected. This means finding out whether all off diagonal elements are nonzero.

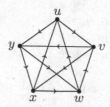

Solution 4.   We have

$$
A = \begin{matrix} & u & v & w & x & y \\ & \begin{pmatrix} 0 & 1 & 0 & 0 & 1 \\ 0 & 0 & 0 & 1 & 1 \\ 1 & 1 & 0 & 0 & 0 \\ 1 & 0 & 1 & 0 & 0 \\ 0 & 0 & 1 & 1 & 0 \end{pmatrix} \end{matrix}.
$$

Thus the powers of A are

$$
A^2 = \begin{pmatrix} 0 & 0 & 1 & 2 & 1 \\ 1 & 0 & 2 & 1 & 0 \\ 0 & 1 & 0 & 1 & 2 \\ 1 & 2 & 0 & 0 & 1 \\ 2 & 1 & 1 & 0 & 0 \end{pmatrix}, \quad A^3 = \begin{pmatrix} 3 & 1 & 3 & 1 & 0 \\ 3 & 3 & 1 & 0 & 1 \\ 1 & 0 & 3 & 3 & 1 \\ 0 & 1 & 1 & 3 & 3 \\ 1 & 3 & 0 & 1 & 3 \end{pmatrix}
$$

$$
A^4 = \begin{pmatrix} 4 & 6 & 1 & 1 & 4 \\ 1 & 4 & 1 & 4 & 6 \\ 6 & 4 & 4 & 1 & 1 \\ 4 & 1 & 6 & 4 & 1 \\ 1 & 1 & 4 & 6 & 4 \end{pmatrix}.
$$

The number of walks from w to u is given by the $(3, 1)$ element of each matrix so there is 1 walk of length 1, 0 walks of length 2, 1 walk of length 3 and 6 walks of length 4 from w to u. Walks from u to w are given by the $(1, 3)$ element of each matrix. Thus there are 0 walks of length 1, 1 walk of length 2, 3 walks of length 3 and 1 walk of length 4 from u to w. The matrix B is given by

$$
B = A + A^2 + A^3 + A^4 = \begin{pmatrix} 7 & 8 & 5 & 4 & 6 \\ 5 & 7 & 4 & 6 & 8 \\ 8 & 6 & 7 & 5 & 4 \\ 6 & 4 & 8 & 7 & 5 \\ 4 & 5 & 6 & 8 & 7 \end{pmatrix}.
$$

Therefore the digraph is strongly connected (all the off diagonal elements are nonzero).

Problem 5.   A graph $G(V, E)$ is a set of nodes V (points, vertices) connected by a set of links E (edges, lines). We assume that there are n nodes. The adjancy $(n \times n)$ matrix $A = A(G)$ takes the form with 1 in row i, column j if i is connected to j, and 0 otherwise. Thus A is a symmetric matrix. Associated with A is the degree distribution, a diagonal matrix with row-sums of A along the diagonal, and 0's elsewhere. We assume that $d_{ii} > 0$ for all $i = 1, 2, \ldots, n$. We define the Laplacian as $L := D - A$. Let

$$
A = \begin{pmatrix} 0 & 1 & 1 & 0 & 0 & 0 & 0 \\ 1 & 0 & 1 & 1 & 0 & 0 & 0 \\ 1 & 1 & 0 & 1 & 0 & 1 & 0 \\ 0 & 1 & 1 & 0 & 0 & 1 & 0 \\ 0 & 0 & 0 & 0 & 0 & 1 & 0 \\ 0 & 0 & 1 & 1 & 1 & 0 & 1 \\ 0 & 0 & 0 & 0 & 0 & 1 & 0 \end{pmatrix}.
$$

(i) Give an interpretation of A, A^2, A^3.

(ii) Find D and L.

(iii) Show that L admits the eigenvalue $\lambda_0 = 0$ (lowest eigenvalue) with eigenvector $\mathbf{x} = (1, 1, 1, 1, 1, 1, 1)^T$.

Solution 5.   (i) By definition A is the matrix of all pairs of nodes linked to each other. The matrix A^2 has a non-zero in row i column j if j is two steps away from i. Since i is 2 steps away from itself, the diagonal i, i entry counts the number of these 2-steps. The matrix A^3 has a non-zero entry in row i column j if j is 3 steps away from i.

(ii) We have

$$D = \begin{pmatrix} 2 & 0 & 0 & 0 & 0 & 0 & 0 \\ 0 & 3 & 0 & 0 & 0 & 0 & 0 \\ 0 & 0 & 4 & 0 & 0 & 0 & 0 \\ 0 & 0 & 0 & 3 & 0 & 0 & 0 \\ 0 & 0 & 0 & 0 & 1 & 0 & 0 \\ 0 & 0 & 0 & 0 & 0 & 4 & 0 \\ 0 & 0 & 0 & 0 & 0 & 0 & 1 \end{pmatrix}$$

and

$$L = D - A = \begin{pmatrix} 2 & -1 & -1 & 0 & 0 & 0 & 0 \\ -1 & 3 & -1 & -1 & 0 & 0 & 0 \\ -1 & -1 & 4 & -1 & 0 & -1 & 0 \\ 0 & -1 & -1 & 3 & 0 & -1 & 0 \\ 0 & 0 & 0 & 0 & 1 & -1 & 0 \\ 0 & 0 & -1 & -1 & -1 & 4 & -1 \\ 0 & 0 & 0 & 0 & 0 & -1 & 1 \end{pmatrix}.$$

(iii) With $L = D - A$ we have

$$L\mathbf{x} = \begin{pmatrix} 2 & -1 & -1 & 0 & 0 & 0 & 0 \\ -1 & 3 & -1 & -1 & 0 & 0 & 0 \\ -1 & -1 & 4 & -1 & 0 & -1 & 0 \\ 0 & -1 & -1 & 3 & 0 & -1 & 0 \\ 0 & 0 & 0 & 0 & 1 & -1 & 0 \\ 0 & 0 & -1 & -1 & -1 & 4 & -1 \\ 0 & 0 & 0 & 0 & 0 & -1 & 1 \end{pmatrix} \begin{pmatrix} 1 \\ 1 \\ 1 \\ 1 \\ 1 \\ 1 \\ 1 \end{pmatrix} = \begin{pmatrix} 0 \\ 0 \\ 0 \\ 0 \\ 0 \\ 0 \\ 0 \end{pmatrix}.$$

Problem 6.   A graph $G(V, E)$ is a set of nodes V (points, vertices) connected by a set of links E (edges, lines). We assume that there are n nodes. The adjacency $(n \times n)$ matrix $A = A(G)$ takes the form with 1 in row i, column j if i is connected to j, and 0 otherwise. Thus A is a symmetric matrix. Associated with A is the degree distribution D, a diagonal matrix with row-sums of A along the diagonal, and 0's elsewhere. D describes how many connections each node has. We define the *Laplacian* as $L := D - A$. Let $A = (a_{ij})$, i.e. a_{ij} are the entries of adjacency matrix. Find the minimum of the weighted sum

$$S = \frac{1}{2} \sum_{i,j=1}^{n} (x_i - x_j)^2 a_{ij}$$

with the constraint $\mathbf{x}^T\mathbf{x} = 1$, where $\mathbf{x}^T = (x_1, x_2, \ldots, x_n)$. Use the Lagrange multiplier method. The sum is over all pairs of squared distances between nodes which are connected, and so the solution should result in nodes with large numbers of inter-connections being clustered together.

Solution 6. Taking into account that A is a symmetric matrix we obtain

$$S = \frac{1}{2}\sum_{i,j=1}^{n}(x_i - x_j)^2 a_{ij} = \frac{1}{2}\sum_{i,j=1}^{n}(x_i^2 - 2x_i x_j + x_j^2)a_{ij}$$
$$= \mathbf{x}^T D\mathbf{x} - \mathbf{x}^T A\mathbf{x} = \mathbf{x}^T(D - A)\mathbf{x}$$
$$= \mathbf{x}^T L\mathbf{x}$$

where $L = D - A$. Let λ be the Lagrange multiplier. Then we have to minimize the expression

$$Z = \mathbf{x}^T L\mathbf{x} - \lambda \mathbf{x}^T \mathbf{x}.$$

This leads to the eigenvalue equation $L\mathbf{x} = \lambda\mathbf{x}$. The lowest eigenvalue is $\lambda_0 = 0$ with the (trivial) eigenvector

$$\mathbf{x}_0 = (1 \quad 1 \quad 1 \quad 1 \quad 1 \quad 1 \quad 1)^T.$$

Thus we can order $0 = \lambda_0 \leq \lambda_1 \leq \cdots \leq \lambda_{n-1}$. The eigenvector $\mathbf{x}_1$ corresponding to the second smallest eigenvalue λ_1 can be used to solve the *min-cut* problem: separate the network into two approximately equal sets of nodes with the fewest number of connections between them, based on the signs of the entries of $\mathbf{x}_1$. The eigenvector $\mathbf{x}_1$ is called the *Fiedler vector*.

Problem 7. Find the eigenvalues of the three adjacent matrices

$$A = \begin{pmatrix} 1 & 0 & 1 \\ 1 & 1 & 1 \\ 1 & 0 & 1 \end{pmatrix}, \quad B = \begin{pmatrix} 1 & 0 & 1 \\ 0 & 1 & 0 \\ 1 & 0 & 1 \end{pmatrix}, \quad C = \begin{pmatrix} 1 & 1 & 0 \\ 0 & 1 & 0 \\ 0 & 1 & 1 \end{pmatrix}$$

provided by three simple graphs. Find the "energy" $E(G)$ of each graph defined by

$$E(G) = \sum_{j=1}^{3}|\lambda_j|.$$

Solution 7. The eigenvalues of A are 1, 2, 0 with the energy $E = 3$. The eigenvalues of B are 1, 2, 0 with the energy $E = 3$. The eigenvalues of C are 1 (three-times) with the energy $E = 3$.

Problem 8. Consider the two directed graphs with the adjacency matrices

$$A_1 = \begin{pmatrix} 0 & 0 & 1 & 0 & 0 \\ 1 & 0 & 0 & 1 & 0 \\ 0 & 1 & 0 & 0 & 0 \\ 0 & 0 & 0 & 0 & 1 \\ 1 & 0 & 0 & 1 & 0 \end{pmatrix}, \quad A_2 = \begin{pmatrix} 0 & 0 & 0 & 0 & 1 \\ 1 & 0 & 0 & 1 & 1 \\ 0 & 1 & 0 & 0 & 0 \\ 0 & 1 & 1 & 0 & 0 \\ 1 & 0 & 0 & 1 & 0 \end{pmatrix},$$

(i) Find the characteristic polynomials and the eigenvalues of A_1 and A_2.
(ii) Do the directed graphs admit a Hamilton cycle?
(iii) Can one remove some of the 1's in A_1 so that one has a permutation matrix?

Solution 8.   (i) One finds the same characteristic polynomial for both matrices, namely $\lambda^5 - \lambda^3 - \lambda = 0$ with the eigenvalues $\lambda_1 = 0$, $\lambda_2 = 0$, ...
(ii) For A_1 we find the Hamilton cycle $1 - 3 - 2 - 4 - 5 - 1$. The adjacency matrix A_2 admits no Hamilton cycle.
(iii) If one sets $a_{21} = 0$ we obtain a permutation matrix.

Problem 9.   Consider the directed graph G

(i) Find the adjacency matrix A.
(ii) Calculate $\frac{1}{2}\text{tr}(A^2)$. How many 2-vertex cycles (loops) does G have?
(iii)] Calculate $\frac{1}{3}\text{tr}(A^3)$. How many 3-vertex cycles (triangles) does G have?

Solution 9.   The order of the vertices determine the structure of the adjacency matrix. Consider the ordering

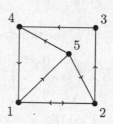

For any other ordering of the vertices, the adjacency matrix B satisfies $B = PAP^T$ for some permutation matrix P. Consequently $\text{tr}(A^2) = \text{tr}(B^2)$ and $\text{tr}(A^3) = \text{tr}(B^3)$.
(i) We have

$$A = \begin{pmatrix} 0 & 1 & 0 & 0 & 1 \\ 1 & 0 & 1 & 0 & 0 \\ 0 & 0 & 0 & 1 & 0 \\ 1 & 0 & 0 & 0 & 0 \\ 0 & 1 & 0 & 1 & 0 \end{pmatrix}.$$

(ii) We have

$$A^2 = \begin{pmatrix} 1 & 1 & 1 & 1 & 0 \\ 0 & 1 & 0 & 1 & 1 \\ 1 & 0 & 0 & 0 & 0 \\ 0 & 1 & 0 & 0 & 1 \\ 2 & 0 & 1 & 0 & 0 \end{pmatrix}$$

and $\frac{1}{2}\mathrm{tr}(A^2) = 1$. There is one loop: $(1,2)$.

(iii) We have

$$A^3 = \begin{pmatrix} 2 & 1 & 1 & 1 & 1 \\ 2 & 1 & 1 & 1 & 0 \\ 0 & 1 & 0 & 0 & 1 \\ 1 & 1 & 1 & 1 & 0 \\ 0 & 2 & 0 & 1 & 2 \end{pmatrix}$$

and $\frac{1}{3}\mathrm{tr}(A^3) = 2$. There are 2 triangles: $(1,5,4)$ and $(1,5,2)$.

Problem 10. Let G be a graph with vertices V and edges E. The graph G is *transitive* if $(v_1, v_2), (v_2, v_3) \in E$ implies that $(v_1, v_3) \in E$. The *transitive closure* $\widetilde{G}$ of G, is the smallest transitive graph $\widetilde{G}$ which has G as a subgraph. Let H be the adjacency matrix for G and $\widetilde{H}$ be the adjacency matrix for $\widetilde{G}$. If $(H)_{ij} = 1$ and $(H)_{jk} = 1$ then $(\widetilde{H})_{ij} = (\widetilde{H})_{jk} = (\widetilde{H})_{ik} = 1$. Write a C++ program which finds $\widetilde{H}$ for any given adjacency matrix H. Find the adjacency matrix for the transitive closure of the graph given by the adjacency matrix

$$H = \begin{pmatrix} 0 & 1 & 0 & 0 & 0 \\ 1 & 0 & 0 & 0 & 1 \\ 0 & 0 & 0 & 0 & 0 \\ 0 & 0 & 0 & 1 & 1 \\ 0 & 1 & 0 & 1 & 0 \end{pmatrix}.$$

Solution 10. We search for pairs $(H)_{ij} = 1$ and $(H)_{jk} = 1$ over all values of i, j and k and set $(H)_{ik}$ to 1 if such a pair is found. We have to repeat this procedure again for each of the nodes because new pairs may have been introduced with $(H)_{ij} = 1$ and $(H)_{jk} = 1$. Here $||$ denotes the logical OR operation and && denotes the logical and operation. The program outputs $\widetilde{H}$

$$\widetilde{H} = \begin{pmatrix} 1 & 1 & 0 & 1 & 1 \\ 1 & 1 & 0 & 1 & 1 \\ 0 & 0 & 0 & 0 & 0 \\ 1 & 1 & 0 & 1 & 1 \\ 1 & 1 & 0 & 1 & 1 \end{pmatrix}.$$

```cpp
// closure.cpp

#include <iostream>
using namespace std;

int main(void)
{
  const int n = 5;
  bool H[n][n] = { { 0, 1, 0, 0, 0 },{ 1, 0, 0, 0, 1 },
                   { 0, 0, 0, 0, 0 },{ 0, 0, 0, 1, 1 },
                   { 0, 1, 0, 1, 0 } };
  int i, j, k, l;
  cout << "Adjacency matrix" << endl;
```

```
for(i=0;i<n;++i)
{
cout << "[ ";
for(j=0;j<n;++j) cout << H[i][j] << " ";
cout << "]" << endl;
}
cout << endl;
for(l=0;l<n;++l)
 for(k=0;k<n;++k)
  for(j=0;j<n;++j)
   for(i=0;i<n;++i) H[i][j] = (H[i][j] || (H[i][k] && H[k][j]));
cout << "Transitive closure" << endl;
for(i=0;i<n;++i)
{
cout << "[ ";
for(j=0; j<n; ++j) cout << H[i][j] << " ";
cout << "]" << endl;
}
return 0;
}
```

Supplementary Problems

Problem 1.   Compare the two adjacency matrices

$$A = \begin{pmatrix} 0 & 1 & 1 & 0 \\ 1 & 0 & 0 & 1 \\ 1 & 0 & 0 & 1 \\ 0 & 1 & 1 & 0 \end{pmatrix}, \quad B = \begin{pmatrix} 1 & 0 & 0 & 1 \\ 0 & 1 & 1 & 0 \\ 0 & 1 & 1 & 0 \\ 1 & 0 & 0 & 1 \end{pmatrix}.$$

Do they admit an Euler path?

Problem 2.   Consider the adjacency matrix

$$A = \begin{pmatrix} 0 & 1 & 1 & 1 & 0 \\ 1 & 0 & 0 & 0 & 1 \\ 1 & 0 & 0 & 0 & 1 \\ 1 & 0 & 0 & 0 & 1 \\ 0 & 1 & 1 & 1 & 0 \end{pmatrix}.$$

Is there a Hamilton cycle?

Problem 3.   Let G be a graph with vertices $V_G = \{1, \ldots, n_G\}$ and edges $E_G \subseteq V_G \times V_G$. The $n_G \times n_G$ adjacency matrix A_G is given by $(A_G)_{ij} = \chi_{E_G}(i, j) \in \{0, 1\}$ where $i, j \in V$. In the following products of graphs G_1 and G_2 with $n_{G_1} n_{G_2} \times n_{G_2} n_{G_2}$ adjacency matrices. The rows and columns are indexed by $V_{G_1} \times V_{G_2}$, where we use the ordering $(i, j) \leq (k, l)$ if $i < k$ or, $i = k$ and $j \leq l$.

The cartesian product $G_1 \times G_2$ of two graphs G_1 and G_2 is the graph with vertices

$$V_{G_1 \times G_2} = V_{G_1} \times V_{G_2}$$

and edges

$$E_{G_1 \times G_2} = \{\, ((a,b),(a,b')) \,:\, a \in V_{G_1} \text{ and } (b,b') \in E_{G_2} \,\}$$
$$\cup \{\, ((a,b),(a',b)) \,:\, (a,a') \in E_{G_1} \text{ and } b \in V_{G_2} \,\}.$$

It follows that

$$(A_{G_1 \times G_2})_{(i,j),(k,l)} = \delta_{ij}(A_{G_2})_{(k,l)} \boxplus \delta_{kl}(A_{G_1})_{(i,j)}$$

where $\boxplus$ is the usual addition with the convention that $1 \boxplus 1 = 1$. Thus

$$A_{G_1 \times G_2} = (A_{G_1} \otimes I_{n_{G_2}}) \boxplus (I_{n_{G_2}} \otimes A_{G_2}).$$

The lexicographic product $G_1 \bullet G_2$ of two graphs G_1 and G_2 is the graph with vertices

$$V_{G_1 \bullet G_2} = V_{G_1} \times V_{G_2}$$

and edges

$$E_{G_1 \bullet G_2} = \{\, ((a,b),(a,b')) \,:\, a \in V_{G_1} \text{ and } (b,b') \in E_{G_2} \,\}$$
$$\cup \{\, ((a,b),(a',b')) \,:\, (a,a') \in E_{G_1} \text{ and } b, b' \in V_{G_2} \,\}.$$

Thus $E_{G_1 \times G_2} \subseteq E_{G_1 \bullet G_2}$. It follows that

$$(A_{G_1 \times G_2})_{(i,j),(k,l)} = \delta_{ij}(A_{G_2})_{(k,l)} \boxplus (A_{G_1})_{(i,j)}.$$

Hence

$$A_{G_1 \times G_2} = (A_{G_1} \otimes 1_{n_{G_2}}) \boxplus (I_{n_{G_2}} \otimes A_{G_2}).$$

Here $1_{n_{G_1}}$ is the $n_{G_2} \times n_{G_2}$ with every entry equal to 1.
The tensor product $G_1 \otimes G_2$ of two graphs G_1 and G_2 is the graph with vertices

$$V_{G_1 \otimes G_2} = V_{G_1} \times V_{G_2}$$

and edges

$$E_{G_1 \otimes G_2} = \{\, ((a,b),(a',b')) \,:\, (a,a') \in E_{G_1} \text{ and } (b,b') \in E_{G_2} \,\}.$$

It follows that

$$(A_{G_1 \otimes G_2})_{(i,j),(k,l)} = (A_{G_1})_{(i,j)}(A_{G_2})_{(k,l)} = (A_{G_1} \otimes A_{G_2})_{(i-1)n_{G_2}+k,(j-1)n_{G_2}+l}.$$

Thus $A_{G_1 \otimes G_2} = A_{G_1} \otimes A_{G_2}$ where $\otimes$ is the Kronecker product of matrices.
The normal product $G_1 \star G_2$ of two graphs G_1 and G_2 is the graph with vertices

$$V_{G_1 \star G_2} = V_{G_1} \times V_{G_2}$$

and edges

$$E_{G_1 \star G_2} = E_{G_1 \times G_2} \cup E_{G_1 \otimes G_2}.$$

Thus

$$A_{G_1 \star G_2} = A_{G_1 \times G_2} \boxplus A_{G_1 \otimes G_2} = (A_{G_1} \otimes I_{n_{G_2}}) \boxplus (I_{n_{G_2}} \otimes A_{G_2}) \boxplus A_{G_1} \otimes A_{G_2}.$$

(i) Show that the Euler path is not preserved (in general) under these operations.
(ii) Show that the Hamilton path is not preserved (in general) under these operations.

Chapter 22

Hilbert Spaces and Mutually Unbiased Bases

A complex inner product vector space (also called pre-Hilbert space) is a complex vector space together with an inner product. An inner product vector space which is complete with respect to the norm induced by the inner product is called a Hilbert space. Two finite dimensional Hilbert spaces are considered. The Hilbert space $\mathbb{C}^d$ with the scalar product

$$\langle \mathbf{v}_1, \mathbf{v}_2 \rangle = \mathbf{v}_1^* \mathbf{v}_2$$

and the Hilbert space of $m \times m$ matrices X, Y over $\mathbb{C}$ with the scalar product

$$\langle X, Y \rangle := \operatorname{tr}(XY^*).$$

Let $\mathcal{H}_1$ and $\mathcal{H}_2$ be two finite dimensional Hilbert spaces with

$$\dim(\mathcal{H}_1) = \dim(\mathcal{H}_2) = d.$$

Let $\mathcal{B}_{1,1} = \{|j_1\rangle\}$ and $\mathcal{B}_{1,2} = \{|j_2\rangle\}$ $(j_1, j_2 = 1, \ldots, d$) be two orthonormal bases in the Hilbert space $\mathcal{H}_1$. They are called *mutually unbiased* iff

$$|\langle j_1 | j_2 \rangle|^2 = \frac{1}{d} \quad \text{for all} \quad j_1, j_2 = 1, \ldots, d.$$

Let $\mathcal{B}_{2,1} = \{|k_1\rangle\}$ and $\mathcal{B}_{2,2} = \{|k_2\rangle\}$ $(k_1, k_2 = 1, \ldots, d_2)$ be orthonormal bases in the Hilbert space $\mathcal{H}_2$. These means that two orthonormal basis in the Hilbert space $\mathbb{C}^d$

$$\mathcal{A} = \{\mathbf{e}_1, \ldots, \mathbf{e}_d\}, \quad \mathcal{B} = \{\mathbf{f}_1, \ldots, \mathbf{f}_d\}$$

are called mutually unbiased if for every $1 \leq j, k \leq d$

$$|\langle \mathbf{e}_j, \mathbf{f}_k \rangle| = \frac{1}{\sqrt{d}}.$$

Problem 1. The four *Dirac matrices* are given by

$$\gamma_1 = \begin{pmatrix} 0 & 0 & 0 & -i \\ 0 & 0 & -i & 0 \\ 0 & i & 0 & 0 \\ i & 0 & 0 & 0 \end{pmatrix}, \quad \gamma_2 = \begin{pmatrix} 0 & 0 & 0 & -1 \\ 0 & 0 & 1 & 0 \\ 0 & 1 & 0 & 0 \\ -1 & 0 & 0 & 0 \end{pmatrix},$$

$$\gamma_3 = \begin{pmatrix} 0 & 0 & -i & 0 \\ 0 & 0 & 0 & i \\ i & 0 & 0 & 0 \\ 0 & -i & 0 & 0 \end{pmatrix}, \quad \gamma_4 = \begin{pmatrix} 1 & 0 & 0 & 0 \\ 0 & 1 & 0 & 0 \\ 0 & 0 & -1 & 0 \\ 0 & 0 & 0 & -1 \end{pmatrix}$$

and

$$\gamma_5 = \gamma_1\gamma_2\gamma_3\gamma_4 = \begin{pmatrix} 0 & 0 & -1 & 0 \\ 0 & 0 & 0 & -1 \\ -1 & 0 & 0 & 0 \\ 0 & -1 & 0 & 0 \end{pmatrix}.$$

We define the 4×4 matrices

$$\sigma_{jk} := \frac{i}{2}[\gamma_j, \gamma_k], \quad j < k$$

where $j = 1, 2, 3$, $k = 2, 3, 4$.
(i) Calculate $\sigma_{12}, \sigma_{13}, \sigma_{14}, \sigma_{23}, \sigma_{24}, \sigma_{34}$.
(ii) Do the 16 matrices

$$I_4, \gamma_1, \gamma_2, \gamma_3, \gamma_4, \gamma_5, \gamma_5\gamma_1, \gamma_5\gamma_2, \gamma_5\gamma_3, \gamma_5\gamma_4, \sigma_{12}, \sigma_{13}, \sigma_{14}, \sigma_{23}, \sigma_{24}, \sigma_{34}$$

form a basis in the Hilbert space $M_4(\mathbb{C})$? If so is the basis orthogonal?

Solution 1. (i) We have

$$\sigma_{12} = \frac{i}{2}[\gamma_1, \gamma_2] = \mathrm{diag}(-1, +1, -1, +1)$$

$$\sigma_{13} = \frac{i}{2}[\gamma_1, \gamma_3] = \begin{pmatrix} 0 & -i & 0 & 0 \\ i & 0 & 0 & 0 \\ 0 & 0 & 0 & -i \\ 0 & 0 & i & 0 \end{pmatrix}$$

$$\sigma_{14} = \frac{i}{2}[\gamma_1, \gamma_4] = \begin{pmatrix} 0 & 0 & 0 & -1 \\ 0 & 0 & -1 & 0 \\ 0 & -1 & 0 & 0 \\ -1 & 0 & 0 & 0 \end{pmatrix}$$

$$\sigma_{23} = \frac{i}{2}[\gamma_2, \gamma_3] = \begin{pmatrix} 0 & -1 & 0 & 0 \\ -1 & 0 & 0 & 0 \\ 0 & 0 & 0 & -1 \\ 0 & 0 & -1 & 0 \end{pmatrix}$$

$$\sigma_{24} = \frac{i}{2}[\gamma_2, \gamma_4] = \begin{pmatrix} 0 & 0 & 0 & i \\ 0 & 0 & -i & 0 \\ 0 & i & 0 & 0 \\ -i & 0 & 0 & 0 \end{pmatrix}$$

$$\sigma_{34} = \frac{i}{2}[\gamma_3, \gamma_4] = \begin{pmatrix} 0 & 0 & -1 & 0 \\ 0 & 0 & 0 & 1 \\ -1 & 0 & 0 & 0 \\ 0 & 1 & 0 & 0 \end{pmatrix}.$$

(ii) The 16 matrices are linearly independent since from the equation

$$c_0 I_4 + \sum_{j=1}^{5} c_j \gamma_j + \sum_{j=1}^{4} d_j \gamma_5 \gamma_j + \sum_{j<k}^{4} e_{jk} \sigma_{jk} = 0_4$$

it follows that all the coefficients c_j, d_j, e_{jk} are equal to 0. Calculating the scalar product of all possible pairs of matrices we find 0. For example

$$\langle \gamma_1, \gamma_2 \rangle = \text{tr}(\gamma_1 \gamma_2^*) = \text{tr} \begin{pmatrix} i & 0 & 0 & 0 \\ 0 & -i & 0 & 0 \\ 0 & 0 & i & 0 \\ 0 & 0 & 0 & -i \end{pmatrix} = 0.$$

Problem 2.   Let P be the $n \times n$ primary permutation matrix

$$P := \begin{pmatrix} 0 & 1 & 0 & \cdots & 0 \\ 0 & 0 & 1 & \cdots & 0 \\ \vdots & \vdots & \vdots & \ddots & \vdots \\ 0 & 0 & 0 & \cdots & 1 \\ 1 & 0 & 0 & \cdots & 0 \end{pmatrix}$$

and V be the $n \times n$ unitary diagonal matrix ($\zeta \in \mathbb{C}$)

$$V := \begin{pmatrix} 1 & 0 & 0 & \cdots & 0 \\ 0 & \zeta & 0 & \cdots & 0 \\ 0 & 0 & \zeta^2 & \cdots & 0 \\ \vdots & \vdots & \vdots & \ddots & \vdots \\ 0 & 0 & 0 & \cdots & \zeta^{n-1} \end{pmatrix}$$

where $\zeta^n = 1$. Then the set of matrices

$$\{ P^j V^k \ : \ j, k = 0, 1, \ldots, n-1 \}$$

provide a basis in the Hilbert space for all $n \times n$ matrices with the *scalar product*

$$\langle A, B \rangle := \frac{1}{n} \text{tr}(AB^*)$$

for $n \times n$ matrices A and B. Write down the basis for $n = 2$.

Solution 2.   For $n = 2$ we have the combinations

$$(jk) \in \{ (00), (01), (10), (11) \}.$$

This yields the orthonormal basis in the vector space of 2×2 matrices

$$I_2 = \begin{pmatrix} 1 & 0 \\ 0 & 1 \end{pmatrix}, \quad \sigma_3 = \begin{pmatrix} 1 & 0 \\ 0 & -1 \end{pmatrix}, \quad \sigma_1 = \begin{pmatrix} 0 & 1 \\ 1 & 0 \end{pmatrix}, \quad -i\sigma_2 = \begin{pmatrix} 0 & -1 \\ 1 & 0 \end{pmatrix}.$$

Problem 3. Consider the two normalized vectors

$$\mathbf{v}(\alpha) = \begin{pmatrix} \cos(\alpha) \\ \sin(\alpha) \end{pmatrix}, \quad \mathbf{u}(\beta) = \begin{pmatrix} \cos(\beta) \\ \sin(\beta) \end{pmatrix}$$

in $\mathbb{R}^2$. Find the condition on $\alpha, \beta \in \mathbb{R}$ such that

$$|\mathbf{v}^*(\alpha)\mathbf{u}(\beta)|^2 = \frac{1}{2}.$$

Solution 3. Owing to the identity

$$\cos(\alpha)\cos(\beta) + \sin(\alpha)\sin(\beta) \equiv \cos(\alpha - \beta)$$

we obtain $\alpha - \beta \in \{ \pi/4, \ 3\pi/4, \ 5\pi/4, \ 7\pi/4 \} \bmod 2\pi$.

Problem 4. Consider the four nonnormal 2×2 matrices

$$Q = \frac{1}{\sqrt{2}} \begin{pmatrix} 1 & 1 \\ 0 & 0 \end{pmatrix}, \quad R = \frac{1}{\sqrt{2}} \begin{pmatrix} 0 & 0 \\ 1 & -1 \end{pmatrix},$$

$$S = \frac{1}{\sqrt{2}} \begin{pmatrix} 1 & -1 \\ 0 & 0 \end{pmatrix}, \quad T = \frac{1}{\sqrt{2}} \begin{pmatrix} 0 & 0 \\ 1 & 1 \end{pmatrix}.$$

Show that they form an orthonormal basis in the Hilbert space of the 2×2 matrices with the scalar product $\langle X, Y \rangle = \mathrm{tr}(XY^*)$.

Solution 4. The matrices are nonzero

$$\langle Q, R \rangle = 0, \ \langle Q, S \rangle = 0, \ \langle Q, T \rangle = 0, \ \langle R, S \rangle = 0, \ \langle R, T \rangle = 0, \ \langle S, T \rangle = 0.$$

Furthermore $\langle Q, Q \rangle = \langle R, R \rangle = \langle S, S \rangle = \langle T, T \rangle = 1$.

Problem 5. Consider the Hilbert space $M_d(\mathbb{C})$ of $d \times d$ matrices with scalar product $\langle A, B \rangle := \mathrm{tr}(AB^*)$, $A, B \in M_d(\mathbb{C})$. Consider an orthogonal basis of d^2 $d \times d$ hermitian matrices $B_1, B_2, \ldots, B_{d^2}$, i.e.

$$\langle B_j, B_k \rangle = \mathrm{tr}(B_j B_k) = d\delta_{jk}$$

since $B_k^* = B_k$ for a hermitian matrix. Let M be a $d \times d$ hermitian matrix. Let

$$m_j = \mathrm{tr}(B_j M) \qquad j = 1, \ldots, d^2.$$

Given m_j and B_j $(j = 1, \ldots, d^2)$. Find M.

Solution 5. We find

$$M = \frac{1}{d} \sum_{j=1}^{d^2} m_j B_j$$

since

$$MB_k = \frac{1}{d} \sum_{j=1}^{d^2} m_j B_j B_k.$$

Thus

$$\mathrm{tr}\left(\frac{1}{d} \sum_{j=1}^{d^2} m_j B_j B_k\right) = \frac{1}{d} \sum_{j=1}^{d^2} m_j \mathrm{tr}(B_j B_k) = \sum_{j=1}^{d^2} m_j \delta_{jk} = m_k.$$

Problem 6.   (i) Consider the Hilbert space $\mathcal{H}$ of the 2×2 matrices over the complex numbers. Show that the rescaled Pauli spin matrices $\mu_j = \frac{1}{\sqrt{2}} \sigma_j$, $j = 1, 2, 3$

$$\mu_1 = \frac{1}{\sqrt{2}} \begin{pmatrix} 0 & 1 \\ 1 & 0 \end{pmatrix}, \quad \mu_2 = \frac{1}{\sqrt{2}} \begin{pmatrix} 0 & -i \\ i & 0 \end{pmatrix}, \quad \mu_3 = \frac{1}{\sqrt{2}} \begin{pmatrix} 1 & 0 \\ 0 & -1 \end{pmatrix}$$

plus the rescaled 2×2 identity matrix

$$\mu_0 = \frac{1}{\sqrt{2}} \begin{pmatrix} 1 & 0 \\ 0 & 1 \end{pmatrix}$$

form an orthonormal basis in the Hilbert space $\mathcal{H}$.
(ii) Find an orthonormal basis given by hermitian matrices in the Hilbert space $\mathcal{H}$ of 4×4 matrices over $\mathbb{C}$. Hint. Start with hermitian 2×2 matrices and then use the Kronecker product.

Solution 6.   (i) The Hilbert space $\mathcal{H}$ is four dimensional. Since

$$\langle \mu_j, \mu_k \rangle = \delta_{jk}$$

and $\mu_0, \mu_1, \mu_2, \mu_3$ are nonzero matrices we have an orthonormal basis.
(ii) Consider the rescaled Pauli spin matrices μ_1, μ_2, μ_3 and the rescaled 2×2 identity matrix μ_0. Now the Kronecker product of two hermitian matrices is again hermitian. Thus forming all possible 16 Kronecker products

$$\mu_j \otimes \mu_k, \qquad j, k = 0, 1, 2, 3$$

we find an orthonormal basis of the Hilbert space of the 4×4 matrices, where we used

$$\mathrm{tr}((\mu_j \otimes \mu_k)(\mu_m \otimes \mu_n)) = \mathrm{tr}((\mu_j \mu_m) \otimes (\mu_k \mu_n)) = (\mathrm{tr}(\mu_j \mu_m))(\mathrm{tr}(\mu_k \mu_n))$$

with $j, k, m, n = 0, 1, 2, 3$. We applied that $\mathrm{tr}(A \otimes B) = \mathrm{tr}(A)\mathrm{tr}(B)$, where A and B are $n \times n$ matrices.

Problem 7.   Let A, B be hermitian matrices. Find the scalar product

$$\langle A \otimes B, B \otimes A \rangle.$$

Is $\langle A \otimes B, B \otimes A \rangle \geq 0$. Prove or disprove.

Solution 7. Since $B \otimes A$ is hermitian we find

$$\langle A \otimes B, B \otimes A \rangle = \text{tr}((AB) \otimes (BA)) = \text{tr}(AB)\text{tr}(BA) = (\text{tr}(AB))^2.$$

Problem 8. Consider the Hilbert space of the $n \times n$ matrices over $\mathbb{C}$. Let A, B be two $n \times n$ matrices over $\mathbb{C}$. Assume that $\langle A, B \rangle = 0$. Calculate

$$\langle A \otimes A, B \otimes B \rangle, \qquad \langle A \otimes B, A \otimes B \rangle, \qquad \langle A \otimes B, B \otimes A \rangle.$$

Solution 8. We have

$$\langle A \otimes A, B \otimes B \rangle = \text{tr}((AB^*) \otimes (AB^*)) = \text{tr}(AB^*)\text{tr}(AB^*) = 0$$
$$\langle A \otimes B, A \otimes B \rangle = \text{tr}((AA^*) \otimes (BB^*)) = \text{tr}(AA^*)\text{tr}(BB^*)$$
$$\langle A \otimes B, B \otimes A \rangle = \text{tr}((AB^*) \otimes (BA^*)) = \text{tr}(AB^*)\text{tr}(BA^*) = 0.$$

Problem 9. Consider the Hilbert space $\mathcal{H} = \mathbb{C}^4$ and the hermitian matrices

$$H = \begin{pmatrix} 2 & -1 & 0 & -1 \\ -1 & 2 & -1 & 0 \\ 0 & -1 & 2 & -1 \\ -1 & 0 & -1 & 2 \end{pmatrix}$$

$$A = \begin{pmatrix} 1 & 0 & 0 & 0 \\ 0 & 0 & 0 & 0 \\ 0 & 0 & -1 & 0 \\ 0 & 0 & 0 & 0 \end{pmatrix}, \qquad B = \begin{pmatrix} 0 & 0 & 0 & 0 \\ 0 & 1 & 0 & 0 \\ 0 & 0 & 0 & 0 \\ 0 & 0 & 0 & -1 \end{pmatrix}.$$

We consider the 16×16 matrix

$$K = \frac{1}{2}(H \otimes I_4 + I_4 \otimes H) - \lambda A \otimes A - \mu B \otimes B$$

acting in the Hilbert space $\mathbb{C}^{16}$, where $\lambda, \mu \geq 0$.
(i) Consider the permutation matrix

$$R = \begin{pmatrix} 0 & 1 & 0 & 0 \\ 0 & 0 & 1 & 0 \\ 0 & 0 & 0 & 1 \\ 1 & 0 & 0 & 0 \end{pmatrix}.$$

Thus the inverse of R exists. Calculate $(R \otimes R)^{-1} K (R \otimes R)$ and the commutator $[R \otimes R, K]$. Discuss.
(ii) Consider the permutation matrices

$$S = \begin{pmatrix} 0 & 0 & 1 & 0 \\ 0 & 1 & 0 & 0 \\ 1 & 0 & 0 & 0 \\ 0 & 0 & 0 & 1 \end{pmatrix}, \qquad T = \begin{pmatrix} 1 & 0 & 0 & 0 \\ 0 & 0 & 0 & 1 \\ 0 & 0 & 1 & 0 \\ 0 & 1 & 0 & 0 \end{pmatrix}.$$

Find the commutators $[S \otimes S, K]$ and $[T \otimes T, K]$. Discuss.

Solution 9.   (i) We find $(R \otimes R)^{-1} K (R \otimes R) = K$ and $[R \otimes R, K] = 0_4$ if $\lambda = \mu$.
(ii) We find $[S \otimes S, K] = 0_4$ and $[T \otimes T, K] = 0_4$ for any $\mu, \lambda \geq 0$.

Problem 10.   Consider the Hilbert space of 2×2 matrices. Let $\alpha, \beta \in \mathbb{R}$ and

$$
A(\alpha) = \frac{1}{\sqrt{2}} \begin{pmatrix} \cos(\alpha) & \sin(\alpha) \\ \sin(\alpha) & -\cos(\alpha) \end{pmatrix}, \qquad B(\beta) = \frac{1}{\sqrt{2}} \begin{pmatrix} \cos(\beta) & \sin(\beta) \\ \sin(\beta) & -\cos(\beta) \end{pmatrix}.
$$

Find the conditions on α and β such that $\langle A(\alpha), B(\beta) \rangle = \frac{1}{2}$. The matrices $A(\alpha)$, $B(\beta)$ contain $\frac{1}{\sqrt{2}} \sigma_3$ with $\alpha = 0$ and $\frac{1}{\sqrt{2}} \sigma_1$ with $\alpha = \pi/2$.

Solution 10.   Note that $B^*(\beta) = B(\beta)$. We find

$$
\langle A(\alpha), B(\beta) \rangle = \text{tr}(A(\alpha) B^*(\beta)) = \cos(\alpha - \beta) = \frac{1}{2}
$$

with the solution $\alpha - \beta = \pi/3, 5\pi/3 \pmod{2\pi}$. Study the case

$$
\langle A(\alpha) \otimes A(\alpha) \rangle = B(\beta) \otimes B(\beta) \rangle = \frac{1}{4}.
$$

Extend the matrix $A(\alpha)$ to

$$
A(\alpha, \phi) = \frac{1}{\sqrt{2}} \begin{pmatrix} \cos(\alpha) & e^{i\phi} \sin(\alpha) \\ e^{-i\phi} \sin(\alpha) & -\cos(\alpha) \end{pmatrix}.
$$

Problem 11.   Consider the Pauli spin matrices σ_3, σ_1, σ_2. Show that the normalized eigenvectors of σ_3, σ_1, σ_2 provide a set of mutually unbiased bases.

Solution 11.   The eigenvalues of the three matrices are $+1$ and -1. The normalized eigenvectors of σ_3 are given by

$$
\mathcal{B}_3 = \left\{ \begin{pmatrix} 1 \\ 0 \end{pmatrix}, \begin{pmatrix} 0 \\ 1 \end{pmatrix} \right\}.
$$

The normalized eigenvectors of σ_1 are

$$
\mathcal{B}_1 = \left\{ \frac{1}{\sqrt{2}} \begin{pmatrix} 1 \\ 1 \end{pmatrix}, \frac{1}{\sqrt{2}} \begin{pmatrix} 1 \\ -1 \end{pmatrix} \right\}.
$$

The normalized eigenvectors of σ_2 are

$$
\mathcal{B}_3 = \left\{ \frac{1}{\sqrt{2}} \begin{pmatrix} 1 \\ i \end{pmatrix}, \frac{1}{\sqrt{2}} \begin{pmatrix} 1 \\ -i \end{pmatrix} \right\}.
$$

This is a set of three mutually unbiased basis.

Problem 12. Consider the three 4×4 matrices $\sigma_3 \otimes \sigma_3$, $\sigma_1 \otimes \sigma_1$, $\sigma_2 \otimes \sigma_2$. Find mutually unbiased bases for these matrices applying the result from the previous problem.

Solution 12. Utilizing the Kronecker product and the basis given above the sets

$$\mathcal{B}_1 = \left\{ \begin{pmatrix} 1 \\ 0 \end{pmatrix} \otimes \begin{pmatrix} 1 \\ 0 \end{pmatrix}, \ \begin{pmatrix} 1 \\ 0 \end{pmatrix} \otimes \begin{pmatrix} 0 \\ 1 \end{pmatrix}, \ \begin{pmatrix} 0 \\ 1 \end{pmatrix} \otimes \begin{pmatrix} 1 \\ 0 \end{pmatrix}, \ \begin{pmatrix} 0 \\ 1 \end{pmatrix} \otimes \begin{pmatrix} 0 \\ 1 \end{pmatrix} \right\}$$

$$\mathcal{B}_2 = \left\{ \frac{1}{2} \begin{pmatrix} 1 \\ 1 \end{pmatrix} \otimes \begin{pmatrix} 1 \\ 1 \end{pmatrix}, \ \frac{1}{2} \begin{pmatrix} 1 \\ 1 \end{pmatrix} \otimes \begin{pmatrix} 1 \\ -1 \end{pmatrix}, \right.$$
$$\left. \frac{1}{2} \begin{pmatrix} 1 \\ -1 \end{pmatrix} \otimes \begin{pmatrix} 1 \\ 1 \end{pmatrix}, \ \frac{1}{2} \begin{pmatrix} 1 \\ -1 \end{pmatrix} \otimes \begin{pmatrix} 1 \\ -1 \end{pmatrix} \right\}$$

$$\mathcal{B}_3 = \left\{ \frac{1}{2} \begin{pmatrix} 1 \\ i \end{pmatrix} \otimes \begin{pmatrix} 1 \\ i \end{pmatrix}, \ \frac{1}{2} \begin{pmatrix} 1 \\ i \end{pmatrix} \otimes \begin{pmatrix} 1 \\ -i \end{pmatrix}, \right.$$
$$\left. \frac{1}{2} \begin{pmatrix} 1 \\ -i \end{pmatrix} \otimes \begin{pmatrix} 1 \\ i \end{pmatrix}, \ \frac{1}{2} \begin{pmatrix} 1 \\ -i \end{pmatrix} \otimes \begin{pmatrix} 1 \\ -i \end{pmatrix} \right\}$$

provide mutually unbiased bases in the Hilbert space $\mathbb{C}^4$.

Extend the results from problem 11 and problem 12 the case in the Hilbert space $\mathbb{C}^8$ with $\sigma_3 \otimes \sigma_3 \otimes \sigma_3$, $\sigma_1 \otimes \sigma_1 \otimes \sigma_1$, $\sigma_2 \otimes \sigma_2 \otimes \sigma_2$. Extend to n Kronecker products of the Pauli spin matrices.

Problem 13. Let $d \geq 2$. Consider the Hilbert space $\mathbb{C}^d$ and $\mathbf{e}_0$, $\mathbf{e}_1$, ..., $\mathbf{e}_{d-1}$ be the standard basis

$$\mathbf{e}_0 = \begin{pmatrix} 1 \\ 0 \\ 0 \\ \vdots \\ 0 \end{pmatrix}, \quad \mathbf{e}_1 = \begin{pmatrix} 0 \\ 1 \\ 0 \\ \vdots \\ 0 \end{pmatrix}, \quad \ldots, \quad \mathbf{e}_{d-1} = \begin{pmatrix} 0 \\ 0 \\ \vdots \\ 0 \\ 1 \end{pmatrix}.$$

Let $\omega_d := e^{i2\pi/d}$. Then we can form new orthonormal bases via

$$\mathbf{v}_{m;b} \equiv |m; b\rangle = \frac{1}{\sqrt{d}} \sum_{n=0}^{d-1} \omega^{bn(n-1)/2 - nm} \mathbf{e}_n, \quad b, m = 0, 1, \ldots, d-1$$

where the d labelled by the b are the bases and m labels the state within a basis. Thus we find mutually unbiased bases.
(i) Consider $d = 2$ and $b = 0$. Find the basis $|0; 0\rangle$, $|1; 0\rangle$.
(ii) Consider $d = 3$ and $b = 0$. Find the basis $|1; 0\rangle$, $|2; 0\rangle$, $|3; 0\rangle$.
(iii) Consider $d = 4$ and $b = 0$. Find the basis $|0; 0\rangle$, $|1; 0\rangle$, $|2; 0\rangle$, $|3; 0\rangle$. Find out whether the states can be written as Kronecker product of two vectors in $\mathbb{C}^2$, i.e. whether the states are entangled or not?

Solution 13.   (i) We have $\omega_2 = e^{i\pi} = -1$ and $e^{-i\pi} = -1 = \omega_2^{-1}$. Thus

$$|0;0\rangle = \frac{1}{\sqrt{2}} \sum_{n=0}^{1} \mathbf{e}_n = \frac{1}{\sqrt{2}} \begin{pmatrix} 1 \\ 1 \end{pmatrix}$$

and

$$|1;0\rangle = \frac{1}{\sqrt{2}} \sum_{n=0}^{1} \omega^{-n} \mathbf{e}_n = \frac{1}{\sqrt{2}}(\mathbf{e}_0 + \omega^{-1}\mathbf{e}_1) = \frac{1}{\sqrt{2}} \begin{pmatrix} 1 \\ -1 \end{pmatrix}.$$

These are the normalized eigenvectors of σ_1.

(ii) We have $\omega_3 = e^{i2\pi/3} = (-1 + i\sqrt{3})/2$ and $\omega_3^{-1} = e^{-i2\pi/3} = (-1 - i\sqrt{3})/2$ and

$$|0;0\rangle = \frac{1}{\sqrt{3}} \begin{pmatrix} 1 \\ 1 \\ 1 \end{pmatrix}, \qquad |1;0\rangle = \frac{1}{\sqrt{3}} \begin{pmatrix} 1 \\ e^{-i2\pi/3} \\ e^{-i4\pi/3} \end{pmatrix}$$

$$|2;0\rangle = \frac{1}{\sqrt{3}} \begin{pmatrix} 1 \\ e^{-i4\pi/3} \\ e^{-i2\pi/3} \end{pmatrix}.$$

(iii) We have $\omega_4 = e^{i\pi/2} = i$ and $\omega_4^{-1} = e^{-i\pi/2} = -i$. Thus

$$|0;0\rangle = \frac{1}{2} \begin{pmatrix} 1 \\ 1 \\ 1 \\ 1 \end{pmatrix} = \frac{1}{\sqrt{2}} \begin{pmatrix} 1 \\ 1 \end{pmatrix} \otimes \frac{1}{\sqrt{2}} \begin{pmatrix} 1 \\ 1 \end{pmatrix}$$

$$|1;0\rangle = \frac{1}{2} \begin{pmatrix} 1 \\ -i \\ -1 \\ i \end{pmatrix} = \frac{1}{\sqrt{2}} \begin{pmatrix} 1 \\ -1 \end{pmatrix} \otimes \frac{1}{\sqrt{2}} \begin{pmatrix} 1 \\ -i \end{pmatrix}$$

$$|2;0\rangle = \frac{1}{2} \begin{pmatrix} 1 \\ -1 \\ 1 \\ -1 \end{pmatrix} = \frac{1}{\sqrt{2}} \begin{pmatrix} 1 \\ 1 \end{pmatrix} \otimes \frac{1}{\sqrt{2}} \begin{pmatrix} 1 \\ -1 \end{pmatrix}$$

$$|3;0\rangle = \frac{1}{2} \begin{pmatrix} 1 \\ i \\ -1 \\ -i \end{pmatrix} = \frac{1}{\sqrt{2}} \begin{pmatrix} 1 \\ -1 \end{pmatrix} \otimes \frac{1}{\sqrt{2}} \begin{pmatrix} 1 \\ i \end{pmatrix}.$$

Thus none of the states are entangled.

Supplementary Problems

Problem 1.   Assume in the following that $\mathcal{B}_{1,1}$ and $\mathcal{B}_{1,2}$ are mutually unbiased bases in the Hilbert space $\mathcal{H}_1$ and $\mathcal{B}_{2,1}$ and $\mathcal{B}_{2,2}$ are mutually unbiased bases in the Hilbert space $\mathcal{H}_2$, respectively. Show that

$$\{\, |j_1\rangle \otimes |k_1\rangle \,\}, \quad \{\, |j_2\rangle \otimes |k_2\rangle \,\}, \quad j_1, j_2 = 1, \ldots, d_1, \ \ k_1, k_2 = 1, \ldots, d_2$$

are mutually unbiased bases in the finite dimensional product Hilbert space $\mathcal{H}_1 \otimes \mathcal{H}_2$ with $\dim(\mathcal{H}_1 \otimes \mathcal{H}_2) = d_1 \cdot d_2$ and the scalar product in the product Hilbert space

$$((\langle j_1| \otimes \langle k_1|)(|j_2\rangle \otimes |k_2\rangle) = \langle j_1|j_2\rangle\langle k_1|k_2\rangle.$$

Apply this result to $\mathcal{H}_2 = \mathbb{C}^3$ with

$$\mathcal{B}_{2,1} = \{ |1\rangle_{2,1} = \begin{pmatrix} 1 \\ 0 \\ 0 \end{pmatrix}, \ |2\rangle_{2,1} = \begin{pmatrix} 0 \\ 1 \\ 0 \end{pmatrix}, \ |3\rangle_{2,1} = \begin{pmatrix} 0 \\ 0 \\ 1 \end{pmatrix} \}$$

$$\mathcal{B}_{2,2} = \{ |1\rangle_{2,2} = \frac{1}{\sqrt{3}} \begin{pmatrix} 1 \\ 1 \\ 1 \end{pmatrix}, \ |2\rangle_{2,2} = \begin{pmatrix} 1/\sqrt{3} \\ -i/2 - 1/(2\sqrt{3}) \\ i/2 - 1/(2\sqrt{3}) \end{pmatrix},$$

$$|3\rangle_{2,2} = \begin{pmatrix} 1/\sqrt{3} \\ i/2 - 1/(2\sqrt{3}) \\ i/2 - 1/(2\sqrt{3}) \end{pmatrix} \}.$$

Problem 2. The vector space of all $n \times n$ matrices over $\mathbb{C}$ form a Hilbert space with the scalar product defined by $\langle A, B \rangle := \mathrm{tr}(AB^*)$. This implies a norm $\|A\|^2 = \mathrm{tr}(AA^*)$.
(i) Consider the Lie group $U(n)$. Find two unitary 2×2 matrices U_1, U_2 such that $\|U_1 - U_2\|$ takes a maximum.
(ii) Are the matrices

$$U_1 = \begin{pmatrix} 0 & 1 \\ 1 & 0 \end{pmatrix}, \qquad U_2 = \frac{1}{\sqrt{2}} \begin{pmatrix} 1 & 1 \\ 1 & -1 \end{pmatrix}$$

such a pair?

Problem 3. Consider the two normalized vectors

$$\mathbf{v}(\alpha, \phi) = \begin{pmatrix} e^{i\phi}\cos(\alpha) \\ \sin(\alpha) \end{pmatrix}, \qquad \mathbf{u}(\beta, \psi) = \begin{pmatrix} e^{i\psi}\cos(\beta) \\ \sin(\beta) \end{pmatrix}$$

in $\mathbb{C}^2$. Find the condition on $\alpha, \beta, \phi, \psi \in \mathbb{R}$ such that

$$|\mathbf{v}^*(\alpha, \phi)\mathbf{u}(\beta, \psi)|^2 = \frac{1}{2}.$$

Note that

$$\mathbf{v}^*(\alpha, \phi)\mathbf{u}(\beta, \psi) = e^{i(\psi - \phi)}\cos(\alpha)\cos(\beta) + \sin(\alpha)\sin(\beta)$$

and

$$|\mathbf{v}^*(\alpha, \phi)\mathbf{u}(\beta, \psi)|^2 = \cos^2(\alpha) + 2\cos(\psi - \phi)\cos(\alpha)\cos(\beta) + \sin^2(\beta).$$

Problem 4. Consider the two normalized vectors

$$\mathbf{v}_1(\phi_1, \theta_1) = \begin{pmatrix} \cos(\phi_1)\sin(\theta_1) \\ \sin(\phi_1)\sin(\theta_1) \\ \cos(\theta_1) \end{pmatrix}, \qquad \mathbf{v}_2(\phi_2, \theta_2) = \begin{pmatrix} \cos(\phi_2)\sin(\theta_2) \\ \sin(\phi_2)\sin(\theta_2) \\ \cos(\theta_2) \end{pmatrix},$$

Find the conditions on ϕ_1, θ_1, ϕ_2, θ_2 such that $|\mathbf{v}_1^*(\phi_1, \theta_1)\mathbf{v}_2(\phi_2, \theta_2)| = \frac{1}{\sqrt{3}}$.

Problem 5. Consider the Hilbert space $M(2, \mathbb{C})$ of the 2×2 matrices over the complex numbers with the scalar product $\langle A, B \rangle = \text{tr}(AB^*)$ and $A, B \in M(2, \mathbb{C})$. Two mutually unbiased bases are given by

$$\begin{pmatrix} 1 & 0 \\ 0 & 0 \end{pmatrix}, \quad \begin{pmatrix} 0 & 1 \\ 0 & 0 \end{pmatrix}, \quad \begin{pmatrix} 0 & 0 \\ 1 & 0 \end{pmatrix}, \quad \begin{pmatrix} 0 & 0 \\ 0 & 1 \end{pmatrix}$$

and

$$\frac{1}{2}\begin{pmatrix} 1 & 1 \\ 1 & 1 \end{pmatrix}, \quad \frac{1}{2}\begin{pmatrix} 1 & -1 \\ 1 & -1 \end{pmatrix}, \quad \frac{1}{2}\begin{pmatrix} 1 & 1 \\ -1 & -1 \end{pmatrix}, \quad \frac{1}{2}\begin{pmatrix} 1 & -1 \\ -1 & 1 \end{pmatrix}.$$

Construct mutually unbiased bases applying the Kronecker product for the Hilbert space $M(4, \mathbb{C})$.

Chapter 23

Linear Differential Equations

Let A be an $n \times n$ matrix over $\mathbb{R}$. Homogeneous systems of linear differential equations with constant coefficients are given by

$$\frac{d\mathbf{x}}{dt} = A\mathbf{x}, \qquad \mathbf{x} = (x_1, x_2, \dots, x_n)^T \tag{1}$$

with the initial condition $\mathbf{x}(t = 0) = \mathbf{x}_0$. The solution is given by

$$\mathbf{x}(t) = \exp(tA)\mathbf{x}_0. \tag{2}$$

The nonhomogeneous system of linear differential equations with constant coefficients are given by

$$\frac{d\mathbf{x}}{dt} = A\mathbf{x} + \mathbf{y}(t) \tag{3}$$

where $y_1(t), y_2(t), \dots, y_n(t)$ are analytic functions. The solution can be found with the method called *variation of constants*. The solution which vanishes at $t = 0$ is given by

$$\mathbf{x}(t) = \int_0^t \exp((t - \tau)A)\mathbf{y}(\tau)d\tau = \exp(tA) \int_0^t \exp(-\tau A)\mathbf{y}(\tau)d\tau. \tag{4}$$

Thus the solution of a system of differential equations (3) is the sum of the general solution of (1) given by (2) and the particular solution given by (4).

Problem 1.   Solve the initial value problem of $dx/dt = Ax$, where

$$A = \begin{pmatrix} 0 & 1 \\ 1 & 0 \end{pmatrix}.$$

Solution 1.   Since

$$\exp(tA) = \begin{pmatrix} \cosh(t) & \sinh(t) \\ \sinh(t) & \cosh(t) \end{pmatrix}$$

we obtain

$$\begin{pmatrix} x_1(t) \\ x_2(t) \end{pmatrix} = \exp(tA) \begin{pmatrix} x_1(0) \\ x_2(0) \end{pmatrix} = \begin{pmatrix} x_1(0)\cosh(t) + x_2(0)\sinh(t) \\ x_1(0)\sinh(t) + x_2(0)\cosh(t) \end{pmatrix}.$$

Problem 2.   (i) Solve the initial value problem of $dx/dt = Ax$, where

$$A = \begin{pmatrix} a & c \\ 0 & b \end{pmatrix}, \quad a, b, c \in \mathbb{R}.$$

(ii) Let

$$A = \begin{pmatrix} 1 & 1 \\ 0 & 1 \end{pmatrix}.$$

Calculate $\exp(tA)$, where $t \in \mathbb{R}$. Find the solution of the initial value problem of the differential equation

$$\begin{pmatrix} du_1/dt \\ du_2/dt \end{pmatrix} = A \begin{pmatrix} u_1 \\ u_2 \end{pmatrix}$$

with the initial conditions $u_1(t = 0) = u_{10}$, $u_2(t = 0) = u_{20}$. Use the result from (i).

Solution 2.   (i) We have

$$\exp(tA) = \begin{pmatrix} e^{at} & d(t)t \\ 0 & e^{bt} \end{pmatrix}$$

where

$$d(t) = \frac{c(e^{bt} - e^{at})}{b - a} \quad \text{for} \quad b \neq a$$

and $d(t) = ce^{at}$ for $b = a$. Thus

$$\begin{pmatrix} x_1(t) \\ x_2(t) \end{pmatrix} = \begin{pmatrix} e^{at} & d(t)t \\ 0 & e^{bt} \end{pmatrix} \begin{pmatrix} x_1(0) \\ x_2(0) \end{pmatrix}.$$

(ii) We have

$$\exp(tA) = \begin{pmatrix} e^t & te^t \\ 0 & e^t \end{pmatrix}.$$

(ii) Using (i) we obtain

$$\begin{pmatrix} u_1(t) \\ u_2(t) \end{pmatrix} = \exp(tA) \begin{pmatrix} u_{10} \\ u_{20} \end{pmatrix}.$$

Thus the solution of the initial value problem is

$$u_1(t) = e^t u_{10} + t e^t u_{20}, \qquad u_2(t) = e^t u_{20}.$$

Problem 3. Show that the n-th order differential equation

$$\frac{d^n x}{dt^n} = c_0 x + c_1 \frac{dx}{dt} + \cdots + c_{n-1} \frac{d^{n-1} x}{dt^{n-1}}, \qquad c_j \in \mathbb{R}$$

can be written as a system of first-order differential equation.

Solution 3. We set $x_1 = x$ and

$$\frac{dx_1}{dt} := x_2, \qquad \frac{dx_2}{dt} := x_3, \qquad \cdots \qquad , \frac{dx_n}{dt} := c_0 x_1 + c_1 x_2 + \cdots + c_{n-1} x_n.$$

Thus we have the system $dx/dt = Ax$ with the matrix

$$A = \begin{pmatrix} 0 & 1 & 0 & \cdots & 0 \\ 0 & 0 & 1 & \cdots & 0 \\ \cdot & \cdot & \cdot & \cdots & \cdot \\ 0 & 0 & 0 & \cdots & 1 \\ c_0 & c_1 & c_2 & \cdots & c_{n-1} \end{pmatrix}.$$

Problem 4. Let A, X, F be $n \times n$ matrices. Assume that the matrix elements of X and F are differentiable functions of t. Consider the initial-value linear matrix differential equation with an inhomogeneous part

$$\frac{dX(t)}{dt} = AX(t) + F(t), \qquad X(t_0) = C.$$

Find the solution of this matrix differential equation.

Solution 4. The solution is given by

$$X(t) = e^{(t-t_0)A} C + e^{tA} \int_{t_0}^t e^{-sA} F(s) ds \Leftrightarrow X(t) = e^{(t-t_0)A} C + \int_{t_0}^t e^{(t-s)A} F(s) ds.$$

Problem 5. Let A, B, C, Y be $n \times n$ matrices. We know that $AY + YB = C$ can be written as

$$((I_n \otimes A) + (B^T \otimes I_n)) \mathrm{vec}(Y) = \mathrm{vec}(C)$$

where $\otimes$ denotes the Kronecker product. The *vec operation* is defined as

$$\text{vec}(Y) := (y_{11}, \ldots, y_{n1}, y_{12}, \ldots, y_{n2}, \ldots, y_{1n}, \ldots, y_{nn})^T.$$

Apply the vec operation to the matrix differential equation

$$\frac{d}{dt}X(t) = AX(t) + X(t)B$$

where A, B are $n \times n$ matrices and the initial matrix $X(t=0) \equiv X(0)$ is given. Find the solution of this differential equation.

Solution 5. The vec operation is linear and using the result from above we obtain

$$\frac{d}{dt}\text{vec}(X(t)) = (I_n \otimes A + B^T \otimes I_n)\text{vec}(X(t)).$$

The solution of the linear differential equation is given by

$$\text{vec}(X(t)) = e^{t(I_n \otimes A + B^T \otimes I_n)}\text{vec}(X(0)).$$

Since $[I_n \otimes A, B^T \otimes I_n] = 0_{n^2}$ this can be written as

$$\text{vec}(X(t)) = (e^{tB^T} \otimes e^{tA})\text{vec}(X(0)).$$

Problem 6. The *motion of a charge q* in an electromagnetic field is given by

$$m\frac{d\mathbf{v}}{dt} = q(\mathbf{E} + \mathbf{v} \times \mathbf{B}) \tag{1}$$

where m denotes the mass and $\mathbf{v}$ the velocity. Assume that

$$\mathbf{E} = \begin{pmatrix} E_1 \\ E_2 \\ E_3 \end{pmatrix}, \qquad \mathbf{B} = \begin{pmatrix} B_1 \\ B_2 \\ B_3 \end{pmatrix} \tag{2}$$

are constant fields. Find the solution of the initial value problem.

Solution 6. Equation (1) can be written in the form

$$\begin{pmatrix} dv_1/dt \\ dv_2/dt \\ dv_3/dt \end{pmatrix} = \frac{q}{m}\begin{pmatrix} 0 & B_3 & -B_2 \\ -B_3 & 0 & B_1 \\ B_2 & -B_1 & 0 \end{pmatrix}\begin{pmatrix} v_1 \\ v_2 \\ v_3 \end{pmatrix} + \frac{q}{m}\begin{pmatrix} E_1 \\ E_2 \\ E_3 \end{pmatrix}. \tag{3}$$

We set

$$B_j \to \frac{q}{m}B_j, \qquad E_j \to \frac{q}{m}E_j. \tag{4}$$

Thus

$$\begin{pmatrix} dv_1/dt \\ dv_2/dt \\ dv_3/dt \end{pmatrix} = \begin{pmatrix} 0 & B_3 & -B_2 \\ -B_3 & 0 & B_1 \\ B_2 & -B_1 & 0 \end{pmatrix}\begin{pmatrix} v_1 \\ v_2 \\ v_3 \end{pmatrix} + \begin{pmatrix} E_1 \\ E_2 \\ E_3 \end{pmatrix}. \tag{5}$$

Equation (5) is a system of nonhomogeneous linear differential equations with constant coefficients. The solution to the homogeneous equation

$$\begin{pmatrix} dv_1/dt \\ dv_2/dt \\ dv_3/dt \end{pmatrix} = \begin{pmatrix} 0 & B_3 & -B_2 \\ -B_3 & 0 & B_1 \\ B_2 & -B_1 & 0 \end{pmatrix} \begin{pmatrix} v_1 \\ v_2 \\ v_3 \end{pmatrix} \tag{6}$$

is given by

$$\begin{pmatrix} v_1(t) \\ v_2(t) \\ v_3(t) \end{pmatrix} = e^{tM} \begin{pmatrix} v_1(0) \\ v_2(0) \\ v_3(0) \end{pmatrix} \tag{7}$$

where M is the matrix of the right-hand side of (6) and $v_j(0) = v_j(t=0)$. The solution of the system of nonhomogeneous linear differential equations (5) can be found with the help of the method called *variation of constants*. One sets

$$\mathbf{v}(t) = e^{tM} \mathbf{f}(t) \tag{8}$$

where $\mathbf{f} : \mathbb{R} \to \mathbb{R}^3$ is some differentiable curve. Then

$$\frac{d\mathbf{v}}{dt} = M e^{tM} \mathbf{f}(t) + e^{tM} \frac{d\mathbf{f}}{dt}. \tag{9}$$

Inserting (9) into (5) yields

$$M\mathbf{v}(t) + \mathbf{E} = M e^{tM} \mathbf{f}(t) + e^{tM} \frac{d\mathbf{f}}{dt} = M\mathbf{v}(t) + e^{tM} \frac{d\mathbf{f}}{dt}. \tag{10}$$

Consequently

$$\frac{d\mathbf{f}}{dt} = e^{-tM} \mathbf{E}. \tag{11}$$

By integration we obtain

$$\mathbf{f}(t) = \int_0^t e^{-sM} \mathbf{E}\, ds + \mathbf{K} \tag{12}$$

where

$$\mathbf{K} = \begin{pmatrix} K_1 & K_2 & K_3 \end{pmatrix}^T.$$

Therefore we obtain the general solution of the initial value problem of the nonhomogeneous system (5), namely

$$\mathbf{v}(t) = e^{tM} \left(\int_0^t e^{-sM} \mathbf{E}\, ds + \mathbf{K} \right)$$

where

$$\mathbf{K} = \begin{pmatrix} v_1(0) \\ v_2(0) \\ v_3(0) \end{pmatrix}.$$

We now have to calculate e^{tM} and e^{-sM}. We find that

$$M^2 = \begin{pmatrix} -B_2^2 - B_3^2 & B_1 B_2 & B_1 B_3 \\ B_1 B_2 & -B_1^2 - B_3^2 & B_2 B_3 \\ B_1 B_3 & B_2 B_3 & -B_1^2 - B_2^2 \end{pmatrix}$$

and $M^3 = M^2 M = -B^2 M$, where $B^2 := B_1^2 + B_2^2 + B_3^2$. Hence $M^4 = -B^2 M^2$. Since

$$e^{tM} := \sum_{k=0}^{\infty} \frac{(tM)^k}{k!} = I + \frac{tM}{1!} + \frac{t^2 M^2}{2!} + \frac{t^3 M^3}{3!} + \frac{t^4 M^4}{4!} + \cdots$$

where I denotes the 3×3 unit matrix, we obtain

$$e^{tM} = I + M \left(t - \frac{t^3}{3!} B^2 + \frac{t^5}{5!} B^4 - \cdots \right) + M^2 \left(\frac{t^2}{2!} - \frac{t^4}{4!} B^2 + \frac{t^6}{6!} B^4 - \cdots \right).$$

Thus

$$e^{tM} = I + \frac{M}{B} \left(tB - \frac{t^3}{3!} BB^2 + \cdots \right) - \frac{M^2}{B^2} \left(1 - 1 - \frac{t^2 B^2}{2!} + \frac{t^4 B^4}{4!} - \cdots \right)$$

$$= I + \frac{M}{B} \sin(Bt) + \frac{M^2}{B^2} (1 - \cos(Bt)).$$

Therefore

$$e^{tM} = I + \frac{M}{B} \sin(Bt) + \frac{M^2}{B^2} (1 - \cos(Bt))$$

and

$$e^{-sM} = I - \frac{M}{B} \sin(Bs) + \frac{M^2}{B^2} (1 - \cos(Bs))$$

using the substitution $t \to -s$. Since

$$\int_0^t e^{-sM} \mathbf{E} ds = \int_0^t \left(I - \frac{M}{B} \sin(Bs) + \frac{M^2}{B^2} (1 - \cos(Bs)) \right) \mathbf{E} ds$$

$$= \mathbf{E} t + \frac{M\mathbf{E}}{B^2} \cos(Bt) - \frac{M\mathbf{E}}{B^2} + \frac{M^2 \mathbf{E} t}{B^2} - \frac{M^2 \mathbf{E}}{B^2 B} \sin(Bt)$$

we find as the solution of the initial value problem of system (5)

$$\mathbf{v}(t) = \mathbf{E} t \left(1 + \frac{M^2}{B^2} \right) - \frac{M^2 \mathbf{E}}{B^2 B} \sin(Bt) + \frac{M\mathbf{v}(0)}{B} \sin(Bt)$$

$$+ \frac{M\mathbf{E}}{B^2} (1 - \cos(Bt)) + \frac{M^2 \mathbf{v}(0)}{B^2} (1 - \cos(Bt)) + \mathbf{v}(0)$$

where $\mathbf{v}(t = 0) \equiv \mathbf{v}(0)$.

Problem 7. Consider a system of linear ordinary differential equations with periodic coefficients

$$\frac{d\mathbf{u}}{dt} = A(t)\mathbf{u} \tag{1}$$

where $A(t)$ is an $n \times n$ matrix of periodic functions with a period T. From *Floquet theory* we know that any fundamental $n \times n$ matrix $\Phi(t)$, which is defined as a nonsingular matrix satisfying the matrix differential equation

$$\frac{d\Phi(t)}{dt} = A(t)\Phi(t)$$

can be expressed as

$$\Phi(t) = P(t)\exp(tR).\tag{2}$$

Here $P(t)$ is a nonsingular $n \times n$ matrix of periodic functions with the same period T, and R, a constant matrix, whose eigenvalues are called the *characteristic exponents* of the periodic system (1). Let

$$\mathbf{v}(t) = P^{-1}(t)\mathbf{u}(t).$$

Show that $\mathbf{y}$ satisfies the system of linear differential equations with constant coefficients

$$\frac{d\mathbf{v}(t)}{dt} = R\mathbf{v}(t).$$

Solution 7. From $P(t)P^{-1}(t) = I_n$ we have

$$\frac{dP^{-1}(t)}{dt} = -P^{-1}(t)\frac{dP(t)}{dt}P^{-1}(t)$$

and from (2) it follows that

$$\frac{d\Phi(t)}{dt} = \frac{dP(t)}{dt}e^{tR} + P(t)Re^{tR}$$

or

$$\frac{dP(t)}{dt} = \frac{d\Phi(t)}{dt}e^{-tR} - P(t)R.$$

Using these two results we obtain from $\mathbf{v}(t) = P^{-1}(t)\mathbf{u}(t)$

$$\frac{d\mathbf{v}}{dt} = \frac{dP^{-1}(t)}{dt}\mathbf{u} + P^{-1}(t)\frac{d\mathbf{u}}{dt} = -P^{-1}(t)\frac{dP(t)}{dt}P^{-1}(t)\mathbf{u} + P^{-1}(t)A(t)\mathbf{u}$$

$$= -P^{-1}(t)\left(\frac{d\Phi(t)}{dt}e^{-tR} - P(t)R\right)\mathbf{v} + P^{-1}(t)A(t)\mathbf{u}$$

$$= -P^{-1}(t)\frac{d\Phi(t)}{dt}e^{-tR}\mathbf{v} + R\mathbf{v} + P^{-1}(t)A(t)P(t)\mathbf{v}$$

$$= -P^{-1}(t)A(t)\Phi(t)e^{-tR}\mathbf{v} + P^{-1}(t)A(t)P(t) + R\mathbf{v}$$

$$= P^{-1}(t)A(t)(-\Phi(t)e^{-tR} + P(t))\mathbf{v} + R\mathbf{v}$$

$$= R\mathbf{v}.$$

Problem 8. Consider a system of linear ordinary differential equations with periodic coefficients

$$\frac{d\mathbf{u}}{dt} = A(t)\mathbf{u}, \qquad \mathbf{u} = \begin{pmatrix} u_2 \\ u_2 \end{pmatrix}$$

where $A(t)$ is a 2×2 matrix of periodic functions with period T. By the classical *Floquet theory*, any fundamental matrix $\Phi(t)$, which is defined as a nonsingular matrix satisfying the matrix differential equation

$$\frac{d\Phi}{dt} = A(t)\Phi(t)$$

can be expressed as

$$\Phi(t) = P(t)\exp(TR).$$

Here $P(t)$ is nonsingular matrix of periodic functions with the same period T, and R, a constant matrix, whose eigenvalues λ_1 and λ_2 are called the characteristic exponents of the periodic system (1). For a choice of fundamental matrix $\Phi(t)$, we have

$$\exp(TR) = \Phi(t_0)\Phi(t_0 + T)$$

which does not depend on the initial time t_0. The matrix $\exp(TR)$ is called the *monodromy matrix* of the periodic system (1). Calculate

$$\mathrm{tr}(\exp(TR)).$$

Solution 8.   Using $P(t + T) = P(t)$, $\Phi^{-1}(t) = \exp(-tR)P^{-1}(t)$ we have

$$\mathrm{tr}(\Phi^{-1}(t_0)\Phi(t_0 + T) = \mathrm{tr}(e^{-t_0 R}P^{-1}(t_0)P(t_0 + T)e^{(t_0+T)R})$$
$$= \mathrm{tr}(e^{TR}P^{-1}(t_0)P(t_0 + T)) = \mathrm{tr}(e^{TR}P^{-1}(t_0)P(t_0))$$
$$= \mathrm{tr}(e^{TR})$$
$$= e^{\lambda_1 T} + e^{\lambda_2 T}.$$

Problem 9.   Consider the autonomous system of nonlinear first-order ordinary differential equations

$$\frac{dx_1}{dt} = a(x_2 - x_1) = f_1(x_1, x_2, x_3)$$

$$\frac{dx_2}{dt} = (c - a)x_1 + cx_2 - x_1 x_3 = f_2(x_1, x_2, x_3)$$

$$\frac{dx_3}{dt} = -bx_3 + x_1 x_2 = f_3(x_1, x_2, x_3)$$

where $a > 0$, $b > 0$ and c are real constants with $2c > a$.
(i) The *fixed points* are defined as the solutions of the system of equations

$$f_1(x_1^*, x_2^*, x_3^*) = a(x_2^* - x_1^*) = 0$$
$$f_2(x_1^*, x_2^*, x_3^*) = (c - a)x_1^* + cx_2^* - x_1^* x_3^* = 0$$
$$f_3(x_1^*, x_2^*, x_3^*) = -bx_3^* + x_1^* x_2^* = 0.$$

Find the fixed points. Obviously $(0, 0, 0)$ is a fixed point.
(ii) The *linearized equation* (or variational equation) is given by

$$\begin{pmatrix} dy_1/dt \\ dy_2/dt \\ dy_3/dt \end{pmatrix} = A \begin{pmatrix} y_1 \\ y_2 \\ y_3 \end{pmatrix}$$

where the 3×3 matrix A is given by

$$A_{\mathbf{x}=\mathbf{x}^*} = \begin{pmatrix} \partial f_1/\partial x_1 & \partial f_1/\partial x_2 & \partial f_1/\partial x_3 \\ \partial f_2/\partial x_1 & \partial f_2/\partial x_2 & \partial f_2/\partial x_3 \\ \partial f_3/\partial x_1 & \partial f_3/\partial x_2 & \partial f_3/\partial x_3 \end{pmatrix}_{\mathbf{x}=\mathbf{x}^*}$$

where $\mathbf{x} = \mathbf{x}^*$ indicates to insert one of the fixed points into A. Calculate A and insert the first fixed point $(0,0,0)$. Calculate the eigenvalues of A. If all eigenvalues have negative real part then the fixed point is stable. Thus study the stability of the fixed point.

Solution 9. (i) Obviously $x_1^* = x_2^* = x_3^* = 0$ is a solution. The other two solutions we find by eliminating x_3^* from the third equation and x_2^* from the first equation. We obtain

$$x_1^* = \sqrt{b(2c - a)}, \quad x_2^* = \sqrt{b(2c - a)}, \quad x_3^* = 2c - a$$

and

$$x_1^* = -\sqrt{b(2c - a)}, \quad x_2^* = -\sqrt{b(2c - a)}, \quad x_3^* = 2c - a.$$

(ii) We obtain

$$\begin{pmatrix} \partial f_1/\partial x_1 & \partial f_1/\partial x_2 & \partial f_1/\partial x_3 \\ \partial f_2/\partial x_1 & \partial f_2/\partial x_2 & \partial f_2/\partial x_3 \\ \partial f_3/\partial x_1 & \partial f_3/\partial x_2 & \partial f_3/\partial x_3 \end{pmatrix} = \begin{pmatrix} -a & a & 0 \\ c - a - x_3 & c & -x_1 \\ x_2 & x_1 & -b \end{pmatrix}.$$

Inserting the fixed point $(0,0,0)$ yields the matrix

$$A_{\mathbf{x}^*=(0,0,0)} = \begin{pmatrix} -a & a & 0 \\ c - a & c & 0 \\ 0 & 0 & -b \end{pmatrix}.$$

The eigenvalues are

$$\lambda_{1,2} = -\frac{a - c}{2} \pm \sqrt{a(2c - a)}, \qquad \lambda_3 = -b.$$

Depending on c the fixed point $(0,0,0)$ is unstable.

Problem 10. Let A be an $n \times n$ matrix over $\mathbb{R}$. Consider the initial value problem of the system of linear differential equations

$$\frac{d\mathbf{u}(t)}{dt} + A\mathbf{u}(t) = \mathbf{g}(t), \qquad \mathbf{u}(0) = \mathbf{u}_0 \tag{1}$$

where $\mathbf{g}(t) = (g_1(t), g_2(t), \ldots, g_n(t))^T$. The solution of the initial value problem is

$$\mathbf{u}(t) = e^{-tA}\mathbf{u}_0 + \int_0^t e^{-(t-\tau)A}\mathbf{g}(\tau)d\tau. \tag{2}$$

Apply it to the 3×3 matrices

$$A = \begin{pmatrix} 0 & 0 & 1 \\ 0 & 1 & 0 \\ 1 & 0 & 0 \end{pmatrix}$$

with initial values $\mathbf{u}_0 = (1 \quad 1 \quad 1)^T$ and $\mathbf{g}(t) = (1 \quad 0 \quad 1)^T$.

Solution 10.   Since $e^{cA} = I_3 \cosh(c) + A \sinh(c)$ $(c \in \mathbb{R})$ we find for the solution

$$\begin{pmatrix} u_1(t) \\ u_2(t) \\ u_3(t) \end{pmatrix} = \begin{pmatrix} \cosh(t) & 0 & -\sinh(t) \\ 0 & \cosh(t) - \sinh(t) & 0 \\ -\sinh(t) & 0 & \cosh(t) \end{pmatrix} \begin{pmatrix} 1 \\ 1 \\ 1 \end{pmatrix} + \begin{pmatrix} e^{t-\tau} \\ 0 \\ e^{t-\tau} \end{pmatrix}.$$

Discretize the system with the implicit Euler method with step size h and $h = 0.1$. One has

$$(I_n + hA)\mathbf{u}_j = \mathbf{u}_{j-1} + h\mathbf{g}(t_j), \qquad j = 1, \ldots, M$$

with

$$(A + hI_3)^{-1} = \frac{1}{h^2(h+1) - h - 1} \begin{pmatrix} h(h+1) & 0 & -h-1 \\ 0 & h^2 - 1 & 0 \\ -h-1 & 0 & h(h^2+1) \end{pmatrix}.$$

Compare the two solutions for the matrix A.

Problem 11.   Let L and K be two $n \times n$ matrices. Assume that the entries depend on a parameter t and are differentiable with respect to t. Assume that $K^{-1}(t)$ exists for all t. Assume that the time-evolution of L is given by

$$L(t) = K(t)L(0)K^{-1}(t).$$

(i) Show that $L(t)$ satisfies the matrix differential equation

$$\frac{dL}{dt} = [L, B](t)$$

where $[\,,\,]$ denotes the commutator and

$$B = -\frac{dK}{dt}K^{-1}(t).$$

(ii) Show that if $L(t)$ is hermitian and $K(t)$ is unitary, then the matrix $B(t)$ is skew-hermitian.

Solution 11.   (i) Differentiation of $K(t)K^{-1}(t) = I_n$ with respect to t yields

$$\frac{dK^{-1}}{dt} = -K^{-1}(t)\frac{dK}{dt}K^{-1}(t).$$

Differentiation of $L(t)$ with respect to t provides

$$\frac{dL}{dt} = \frac{dK}{dt}L(0)K^{-1}(t) + K(t)L(0)\frac{dK^{-1}}{dt}.$$

Inserting dK^{-1}/dt gives

$$\begin{aligned}
\frac{dL}{dt} &= \frac{dK}{dt}L(0)K^{-1}(t) - K(t)L(0)K^{-1}(t)\frac{dK}{dt}K^{-1}(t) \\
&= \frac{dK}{dt}K^{-1}(t)K(t)L(0)K^{-1}(t) - L(t)\frac{dK}{dt}K^{-1}(t) \\
&= \frac{dK}{dt}K^{-1}(t)L(t) - L(t)\frac{dK}{dt}K^{-1}(t) \\
&= -B(t)L(t) + L(t)B(t) \\
&= [L, B](t).
\end{aligned}$$

(ii) We have

$$(B(t))^* = -\left(\frac{dK}{dt}K^{-1}(t)\right)^* = -K(t)\left(\frac{dK}{dt}\right)^* = \frac{dK}{dt}K(t)^* = -B.$$

Problem 12. Solve the initial value problem for the matrix differential equation

$$[B, A(\epsilon)] = \frac{dA(\epsilon)}{d\epsilon}$$

where $A(\epsilon)$ and $B = \sigma_1$ are 2×2 matrices.

Solution 12. Inserting $B = \sigma_1$ into the matrix differential equation yields

$$\frac{da_{11}}{d\epsilon} = a_{21} - a_{12}, \quad \frac{da_{12}}{d\epsilon} = a_{22} - a_{11}, \quad \frac{da_{21}}{d\epsilon} = a_{11} - a_{22}, \quad \frac{da_{22}}{d\epsilon} = a_{12} - a_{21}.$$

Thus

$$\frac{d}{d\epsilon}(a_{11} + a_{22}) = 0, \quad \frac{d}{d\epsilon}(a_{12} + a_{21}) = 0.$$

In matrix form we can write

$$\begin{pmatrix} da_{11}/d\epsilon \\ da_{12}/d\epsilon \\ da_{21}/d\epsilon \\ da_{22}/d\epsilon \end{pmatrix} = \begin{pmatrix} 0 & -1 & 1 & 0 \\ -1 & 0 & 0 & 1 \\ 1 & 0 & 0 & -1 \\ 0 & 1 & -1 & 0 \end{pmatrix} \begin{pmatrix} a_{11} \\ a_{12} \\ a_{21} \\ a_{22} \end{pmatrix}.$$

The eigenvalues of the 4×4 matrix are $0, 0, 2, -2$.

Problem 13. Let $a, b \in \mathbb{R}$. Consider the linear matrix differential equation

$$\frac{d^2X}{dt^2} + a\frac{dX}{dt} + bX = 0.$$

Find the solution of the initial value problem.

Solution 13. The general solution of the initial-value problem is given by

$$X(t) = \exp(\alpha t)\left(X(0)\cos(\beta t) + \frac{1}{\beta}\left(\frac{dX(0)}{dt} - \alpha X(0)\right)\sin(\beta t)\right)$$

where $\alpha = -\frac{a}{2}$, $\beta = \sqrt{b - a^2/4}$.

Problem 14. Let A be an $n \times n$ matrix over $\mathbb{R}$. The autonomous system of first-order differential equations $du/dt = Au$ admits the solution of the initial value problem $u(t) = \exp(A)u(0)$. Differentiation of the differential equations yields the second-order system

$$\frac{d^2u}{dt^2} = A\frac{du}{dt} = A^2u.$$

Thus we can write

$$\frac{d\mathbf{u}}{dt} = \mathbf{v} = A\mathbf{u}, \quad \frac{d\mathbf{v}}{dt} = A^2\mathbf{u} = A\mathbf{v}$$

or in matrix form

$$\begin{pmatrix} d\mathbf{u}/dt \\ d\mathbf{v}/dt \end{pmatrix} = \begin{pmatrix} 0_n & I_n \\ A^2 & 0_n \end{pmatrix} \begin{pmatrix} \mathbf{u}(0) \\ \mathbf{v}(0) \end{pmatrix}.$$

Find the solution of the initial value problem. Assume that A is invertible.

Solution 14.   We have

$$\exp\begin{pmatrix} 0_n & I_n \\ A^2 & 0_n \end{pmatrix} = \begin{pmatrix} \cosh(A) & A^{-1}\sinh(A) \\ A\sinh(A) & \cosh(A) \end{pmatrix}$$

and therefore the solution of the initial value problem is

$$\begin{pmatrix} \mathbf{u}(t) \\ \mathbf{v}(t) \end{pmatrix} = \begin{pmatrix} \cosh(A) & A^{-1}\sinh(A) \\ A\sinh(A) & \cosh(A) \end{pmatrix} \begin{pmatrix} \mathbf{u}(0) \\ \mathbf{v}(0) \end{pmatrix}.$$

Problem 15.   (i) Let $\mathbf{u} = (u_1, \ldots, u_n)^T$ and A be an $n \times n$ matrix over $\mathbb{R}$. Solve the initial value problem of the system of linear differential equations

$$\frac{d\mathbf{u}}{dt} = A\mathbf{u}$$

using the *Laplace transform*, where $\mathbf{u}(t = 0) = \mathbf{u}(0)$. Note that (applying integration by parts)

$$\mathcal{L}(d\mathbf{u}/dt)(s) = \int_0^\infty e^{-st}\frac{d\mathbf{u}}{dt}dt = e^{-st}\mathbf{u}(t)\big|_{t=0}^{t=\infty} + s\int_0^\infty e^{-st}\mathbf{u}(t)dt = s\mathbf{U}(s) - \mathbf{u}(0).$$

(ii) Apply it to the 2×2 Hadamard matrix

$$H = \frac{1}{\sqrt{2}}\begin{pmatrix} 1 & 1 \\ 1 & -1 \end{pmatrix}.$$

Solution 15.   (i) Taking the Laplace transform of $d\mathbf{u}/dt = A\mathbf{u}$ yields $s\mathbf{U}(s) - \mathbf{u}(0) = A\mathbf{U}(s)$. Thus

$$\mathbf{U}(s) = (sI_n - A)^{-1}\mathbf{U}(0).$$

One calls $(sI_n - A)^{-1}$ the resolvent of the matrix A. The resolvent is defined for $s \in \mathbb{C}$ except the eigenvalues of A. Taking the inverse Laplace transform we find the solution

$$\mathbf{u}(t) = \mathcal{L}^{-1}((sI_n - A)^{-1}\mathbf{u}(0)).$$

(ii) For the Hadamard matrix we have

$$sI_2 - H = \begin{pmatrix} s - 1/\sqrt{2} & -1/\sqrt{2} \\ -1/\sqrt{2} & s + 1/\sqrt{2} \end{pmatrix}.$$

Thus

$$(sI_2 - H)^{-1} = \frac{1}{s^2 + 1} \begin{pmatrix} s + 1/\sqrt{2} & 1/\sqrt{2} \\ 1/\sqrt{2} & s - 1/\sqrt{2} \end{pmatrix}.$$

Since

$$\mathcal{L}^{-1}\left(\frac{s}{s^2 - 1}\right) = \cosh(t), \qquad \mathcal{L}^{-1}\left(\frac{1}{s^2 - 1}\right) = \sinh(t)$$

we obtain

$$\begin{pmatrix} u_1(t) \\ u_2(t) \end{pmatrix} = \begin{pmatrix} \cosh(t) + \frac{1}{\sqrt{2}}\sinh(t) & \frac{1}{\sqrt{2}}\sinh(t) \\ \sinh(t)/\sqrt{2} & \cosh(t) - \frac{1}{\sqrt{2}}\sinh(t) \end{pmatrix} \begin{pmatrix} u_1(0) \\ u_2(0) \end{pmatrix}.$$

Supplementary Problems

Problem 1. Consider the *Lotka-Volterra model*

$$\frac{du_1}{dt} = f_1(u_1, u_2) = u_1 - u_1 u_2, \qquad \frac{du_2}{dt} = f_2(u_1, u_2) = -u_2 + u_1 u_2$$

with the fixed point $(u_1^* = 1, u_2^* = 1)$. The linearized equation (variational equation) with

$$\frac{\partial f_1}{\partial x_1} = 1 - u_2, \quad \frac{\partial f_1}{\partial x_2} = -u_1, \quad \frac{\partial f_2}{\partial x_1} = u_2, \quad \frac{\partial f_2}{\partial x_2} = -1 + u_1$$

is given by

$$\begin{pmatrix} dy_1/dt \\ dy_2/dt \end{pmatrix} = \begin{pmatrix} 0 & -1 \\ 1 & 0 \end{pmatrix} \begin{pmatrix} y_1 \\ y_2 \end{pmatrix}.$$

Solve this system of linear differential equations. First show that the eigenvalues of the 2×2 matrix are given by $\pm i$.

Problem 2. Consider the initial problem of the matrix differential equation

$$\frac{dX}{dt} = A(t)X, \qquad X(0) = I_n$$

where $A(t)$ is an $n \times n$ matrix which depends smoothly on t. It is known that the solution of this matrix differential equation can locally be written as

$$X(t) = \exp(\Omega(t))$$

where $\Omega(t)$ is obtained as an infinite series

$$\Omega(t) = \sum_{k=1}^{\infty} \Omega_k(t).$$

This is the so-called *Magnus expansion*. Implement this recursion in Symbolic++ and apply it to

$$A(t) = \begin{pmatrix} \cos(\omega t) & -\sin(\omega t) \\ \sin(\omega t) & \cos(\omega t) \end{pmatrix}$$

where ω is a constant frequency.

Chapter 24

Differentiation and Matrices

Consider the $m \times n$ matrix

$$F(x) = \begin{pmatrix} f_{11}(x) & f_{12}(x) & f_{13}(x) & \cdots & f_{1n}(x) \\ f_{21}(x) & f_{22}(x) & f_{23}(x) & \cdots & f_{2n}(x) \\ \vdots & \vdots & \vdots & \ddots & \vdots \\ f_{m1}(x) & f_{m2}(x) & f_{m3}(x) & \cdots & f_{mn}(x) \end{pmatrix}$$

with f_{jk} are differentiable function of x. Then $dF(x)/dx$ is defined entrywise. For example

$$\frac{d}{dx} \begin{pmatrix} \cos(x) & -\sin(x) \\ \sin(x) & \cos(x) \end{pmatrix} = \begin{pmatrix} -\sin(x) & -\cos(x) \\ \cos(x) & -\sin(x) \end{pmatrix}.$$

Let A, B be $n \times n$ matrices. Assume that the entries of A and B are analytic functions of x. Then

$$\frac{d}{dx}(AB) = \left(\frac{d}{dx}A \right) B + A \left(\frac{d}{dx}B \right)$$

$$\frac{d}{dx}(A \otimes B) = \left(\frac{d}{dx}A \right) \otimes B + A \otimes \left(\frac{d}{dx}B \right)$$

where $\otimes$ denotes the Kronecker product and

$$\frac{d}{dx}(A \bullet B) = \left(\frac{d}{dx}A \right) \bullet B + A \bullet \left(\frac{d}{dx}B \right)$$

where $\bullet$ denotes the Hadamard product (Schur product, entrywise product).

Problem 1. (i) Consider the 2×2 rotation matrix

$$R(\alpha) = \begin{pmatrix} \cos(\alpha) & -\sin(\alpha) \\ \sin(\alpha) & \cos(\alpha) \end{pmatrix}$$

with $\det(R(\alpha)) = 1$. Is the determinant preserved under repeated differentiation of $R(\alpha)$ with respect to α?
(ii) Consider the 2×2 matrix

$$S(\alpha) = \begin{pmatrix} \cos(\alpha) & \sin(\alpha) \\ \sin(\alpha) & -\cos(\alpha) \end{pmatrix}$$

with $\det(S(\alpha)) = -1$. Is the determinant preserved under repeated differentiation of $S(\alpha)$ with respect to α?

Solution 1. (i) Note that $d\sin(\alpha)/d\alpha = \cos(\alpha)$, $d\cos(\alpha)/d\alpha = -\sin(\alpha)$. We find

$$\frac{dR(\alpha)}{d\alpha} = \begin{pmatrix} -\sin(\alpha) & -\cos(\alpha) \\ \cos(\alpha) & -\sin(\alpha) \end{pmatrix}.$$

We have $\det(dR/d\alpha) = 1$. Thus the determinant is preserved.
(ii) We obtain

$$\frac{dS(\alpha)}{d\alpha} = \begin{pmatrix} -\sin(\alpha) & \cos(\alpha) \\ \cos(\alpha) & \sin(\alpha) \end{pmatrix}$$

with $\det(dS/d\alpha) = -1$. Thus the determinant is preserved.

Problem 2. Let $f_{jk} : \mathbb{R} \to \mathbb{R}$ be analytic functions, where $j, k = 1, 2$ and $F(\epsilon) = (f_{jk}(\epsilon))$. Find the differential equations for $F(\epsilon)$ such that

$$F(\epsilon)\frac{dF(\epsilon)}{d\epsilon} = \frac{dF(\epsilon)}{d\epsilon}F(\epsilon).$$

Solution 2. We obtain the three conditions $f_{12}f'_{21} = f_{21}f'_{12}$,

$$f_{11}f'_{12} + f_{12}f'_{22} = f_{12}f'_{11} + f_{22}f'_{12}, \quad f_{11}f'_{21} + f_{21}f'_{22} = f_{21}f'_{11} + f_{22}f'_{21}.$$

Note that

$$f_{11}(\epsilon) = \cos(\epsilon), \quad f_{12}(\epsilon) = -\sin(\epsilon), \quad f_{21}(\epsilon) = \sin(\epsilon), \quad f_{22}(\epsilon) = \cos(\epsilon)$$

satisfy the three conditions.

Problem 3. Consider the invertible 2×2 matrix

$$A(x) = \begin{pmatrix} \cos(x) & \sin(x) \\ -\sin(x) & \cos(x) \end{pmatrix}.$$

Show that $d(\ln(\det(A(x)))) = \text{tr}(A^{-1}(x)dA(x))$ where d denotes the exterior derivative.

Solution 3.   From $A(x)$ we obtain

$$A^{-1}(x) = \begin{pmatrix} \cos(x) & -\sin(x) \\ \sin(x) & \cos(x) \end{pmatrix}, \quad dA(x) = \begin{pmatrix} -\sin(x)dx & \cos(x)dx \\ -\cos(x)dx & -\sin(x)dx \end{pmatrix}.$$

Thus

$$A^{-1}(x)dA(x) = \begin{pmatrix} 0 & * \\ * & 0 \end{pmatrix}$$

and therefore for the right-hand side we have $\operatorname{tr}(A^{-1}dA) = 0$. For the left-hand side we have $\det(A) = 1$ and $\ln(1) = 0$. Thus $d(\ln(\det(A))) = 0$.

Problem 4.   (i) Consider the analytic function $\mathbf{f} : \mathbb{R}^2 \to \mathbb{R}^2$

$$f_1(x_1, x_2) = \sinh(x_2), \quad f_2(x_1, x_2) = \sinh(x_1).$$

Show that this function admits the (only) fixed point $(0,0)$. Find the functional matrix at the fixed point

$$\begin{pmatrix} \partial f_1/\partial x_1 & \partial f_1/\partial x_2 \\ \partial f_2/\partial x_1 & \partial f_2/\partial x_2 \end{pmatrix}\Bigg|_{(0,0)}.$$

(ii) Consider the analytic function $\mathbf{g} : \mathbb{R}^2 \to \mathbb{R}^2$

$$g_1(x_1, x_2) = \sinh(x_1), \quad g_2(x_1, x_2) = -\sinh(x_2).$$

Show that this function admits the (only) fixed point $(0,0)$. Find the functional matrix at the fixed point

$$\begin{pmatrix} \partial g_1/\partial x_1 & \partial g_1/\partial x_2 \\ \partial g_2/\partial x_1 & \partial g_2/\partial x_2 \end{pmatrix}\Bigg|_{(0,0)}.$$

(iii) Multiply the two matrices found in (i) and (ii).
(iv) Find the composite function $\mathbf{h} : \mathbb{R}^2 \to \mathbb{R}^2$

$$\mathbf{h}(\mathbf{x}) = (\mathbf{f} \circ \mathbf{g})(\mathbf{x}) = \mathbf{f}(\mathbf{g}(\mathbf{x})).$$

Show that this function also admits the fixed point $(0,0)$. Find the functional matrix at this fixed point

$$\begin{pmatrix} \partial h_1/\partial x_1 & \partial h_1/\partial x_2 \\ \partial h_2/\partial x_1 & \partial h_2/\partial x_2 \end{pmatrix}\Bigg|_{(0,0)}.$$

Compare this matrix with the matrix found in (iii).

Solution 4.   (i) Since $\sinh(0) = 0$ we find from $\sinh(x_2) = x_1$, $\sinh(x_1) = x_2$ that $x_1 = 0$, $x_2 = 0$. Since $d\sinh(x)/dx = \cosh(x)$ the functional matrix at $(0,0)$ is

$$\begin{pmatrix} 0 & 1 \\ 1 & 0 \end{pmatrix}.$$

(ii) Since $\sinh(0) = 0$ we find from $\sinh(x_1) = x_1$, $-\sinh(x_2) = x_2$ that $x_1 = 0$, $x_2 = 0$. Since $d\sinh(x)/dx = \cosh(x)$ the functional matrix at $(0,0)$ is

$$\begin{pmatrix} 1 & 0 \\ 0 & -1 \end{pmatrix}.$$

(iii) Multiplication of the two matrices yields

$$\begin{pmatrix} 0 & 1 \\ 1 & 0 \end{pmatrix} \begin{pmatrix} 1 & 0 \\ 0 & -1 \end{pmatrix} = \begin{pmatrix} 0 & -1 \\ 1 & 0 \end{pmatrix}.$$

(iv) We have $\mathbf{h}(\mathbf{x}) = (f_1(g_1, g_2), f_2(g_1, g_2))$. Thus

$$h_1(x_1, x_2) = -\sinh(\sinh(x_2)), \quad h_2(x_1, x_2) = \sinh(\sinh(x_1))$$

which admits the fixed point $(0,0)$. For the functional matrix of $\mathbf{h}$ at $(0,0)$ we find

$$\begin{pmatrix} 0 & -1 \\ 1 & 0 \end{pmatrix}.$$

This is the matrix from (iii).

Problem 5. Consider the 2×2 matrix

$$V(t) = \begin{pmatrix} \cos(\omega t) & \sin(\omega t) \\ -\sin(\omega t) & \cos(\omega t) \end{pmatrix}.$$

Calculate $dV(t)/dt$ and then find the commutator $[dV(t)/dt, V(t)]$. Discuss.

Solution 5. We have

$$\frac{dV(t)}{dt} = \begin{pmatrix} -\omega\sin(\omega t) & \omega\cos(\omega t) \\ -\omega\cos(\omega t) & -\omega\sin(\omega t) \end{pmatrix};$$

For the commutator we find $[dV(t)/dt, V(t)] = 0_2$.

Problem 6. Let A be an $n \times n$ matrix. Assume that the inverse of A exists, i.e. $\det(A) \neq 0$. Then the inverse $B = A^{-1}$ can be calculated as

$$\frac{\partial}{\partial a_{jk}} \ln(\det(A)) = b_{kj}.$$

Apply this formula to the 2×2 matrix A with $\det(A) = a_{11}a_{22} - a_{12}a_{21} \neq 0$.

Solution 6. We have

$$b_{11} = \frac{\partial}{\partial a_{11}} \ln(a_{11}a_{22} - a_{12}a_{21}) = \frac{a_{22}}{\det(A)}$$

$$b_{12} = \frac{\partial}{\partial a_{21}} \ln(a_{11}a_{22} - a_{12}a_{21}) = \frac{-a_{12}}{\det(A)}$$

$$b_{21} = \frac{\partial}{\partial a_{12}} \ln(a_{11}a_{22} - a_{12}a_{21}) = \frac{-a_{21}}{\det(A)}$$

$$b_{22} = \frac{\partial}{\partial a_{22}} \ln(a_{11}a_{22} - a_{12}a_{21}) = \frac{a_{11}}{\det(A)}.$$

Thus the inverse is given by

$$A^{-1} = \frac{1}{\det(A)} \begin{pmatrix} a_{22} & -a_{12} \\ -a_{21} & a_{11} \end{pmatrix}.$$

Problem 7. Let Q and P be $n \times n$ symmetric matrices over $\mathbb{R}$, i.e. $Q = Q^T$ and $P = P^T$. Assume that P^{-1} exists. Find the maximum of the function $f : \mathbb{R}^n \to \mathbb{R}$

$$f(\mathbf{x}) = \mathbf{x}^T Q \mathbf{x}$$

subject to $\mathbf{x}^T P \mathbf{x} = 1$. Use the *Lagrange multiplier method*.

Solution 7. The Lagrange function is

$$L(\mathbf{x}, \lambda) = \mathbf{x}^T Q \mathbf{x} + \lambda(1 - \mathbf{x}^T P \mathbf{x})$$

where λ is the Lagrange multiplier. The partial derivatives lead to

$$\frac{\partial L}{\partial \mathbf{x}}(\widetilde{\mathbf{x}}, \widetilde{\lambda}) = 2\widetilde{\mathbf{x}}^T Q - 2\widetilde{\lambda}\widetilde{\mathbf{x}}^T P$$

$$\frac{\partial L}{\partial \lambda}(\widetilde{\mathbf{x}}, \widetilde{\lambda}) = 1 - \widetilde{\mathbf{x}}^T P \widetilde{\mathbf{x}}.$$

Thus the necessary conditions are

$$Q\widetilde{\mathbf{x}} = \widetilde{\lambda} P \widetilde{\mathbf{x}}, \qquad \widetilde{\mathbf{x}}^T P \widetilde{\mathbf{x}} = 1.$$

Since P is nonsingular we obtain

$$P^{-1} Q \widetilde{\mathbf{x}} = \widetilde{\lambda} \widetilde{\mathbf{x}}.$$

Since $\mathbf{x}^T P \mathbf{x} = 1$ the (column) vector $\mathbf{x}$ cannot be the zero vector. Thus the above equation is an eigenvalue equation, i.e. $\widetilde{\lambda}$ is an eigenvalue of the $n \times n$ matrix $P^{-1} Q$. Multiplying this equation with the row vector $\widetilde{\mathbf{x}}^T P$ we obtain the scalar equation

$$\widetilde{\mathbf{x}}^T Q \widetilde{\mathbf{x}} = \widetilde{\lambda} \widetilde{\mathbf{x}}^T P \widetilde{\mathbf{x}} = \widetilde{\lambda}.$$

Thus we conclude that the maximizer of $\mathbf{x}^T Q \mathbf{x}$ subject to the constraint $\mathbf{x}^T P \mathbf{x} = 1$ is the eigenvector of the matrix $P^{-1} Q$ corresponding to the largest eigenvalue.

Problem 8. Let $\Phi : \mathbb{R}^n \times \mathbb{R}^n \to \mathbb{R}$ be an analytic function, where $(\mathbf{x}, \mathbf{y}) = (x_1, \ldots, x_n, y_1, \ldots, y_n) \in \mathbb{R}^n \times \mathbb{R}^n$. The *Monge-Ampere determinant* $M(\Phi)$ is defined by

$$M(\Phi) := \det \begin{pmatrix} \Phi & \partial\Phi/\partial x_1 & \cdots & \partial\Phi/\partial x_n \\ \partial\Phi/\partial y_1 & \partial^2\Phi/\partial x_1 \partial y_1 & \cdots & \partial^2\Phi/\partial x_n \partial y_1 \\ \vdots & \vdots & \ddots & \vdots \\ \partial\Phi/\partial y_n & \partial^2\Phi/\partial x_1 \partial y_n & \cdots & \partial^2\Phi/\partial x_n \partial y_n \end{pmatrix}.$$

Let $n = 2$ and

$$\Phi(x_1, x_2, y_1, y_2) = x_1^2 + x_2^2 + (x_1 y_1)^2 + (x_2 y_2)^2 + y_1^2 + y_2^2.$$

Find the Monge-Ampere determinant and the conditions on x_1, x_2, y_1, y_2 such that $M(\Phi) = 0$.

Solution 8. We find $M(\Phi) = -32x_1x_2y_1y_2$. For $M(\Phi) = 0$ we need one of the variables to be equal to 0.

Problem 9. Consider the $m \times m$ matrix $F(\mathbf{x}) = (f_{jk}(\mathbf{x}))$ $(j, k = 1, \ldots, m)$, where $f_{jk} : \mathbb{R}^n \to \mathbb{R}$ are analytic functions. Assume that $F(\mathbf{x})$ is invertible for all $\mathbf{x} \in \mathbb{R}^n$. Then we have the identities $(j = 1, \ldots, m)$

$$\frac{\partial(\det(F(\mathbf{x})))}{\partial x_j} \equiv \det(F(\mathbf{x}))\mathrm{tr}\left(F^{-1}(\mathbf{x})\frac{\partial F(\mathbf{x})}{\partial x_j} \right)$$

and

$$\frac{\partial F^{-1}(\mathbf{x})}{\partial x_j} \equiv -F^{-1}(\mathbf{x})\frac{\partial F(\mathbf{x})}{\partial x_j}F^{-1}(\mathbf{x}).$$

The differentiation is understood entrywise. Apply the identities to the matrix $(m = 2, n = 1)$

$$F(x) = \begin{pmatrix} \cos(x) & \sin(x) \\ -\sin(x) & \cos(x) \end{pmatrix}.$$

Solution 9. For the left-hand side of the first identity we have $\det(F(x)) = 1$ and thus $d(\det(F(x)))/dx = 0$. Since

$$F^{-1}(x) = \begin{pmatrix} \cos(x) & -\sin(x) \\ \sin(x) & \cos(x) \end{pmatrix}$$

and therefore

$$F^{-1}(x)\frac{dF(x)}{dx} = \begin{pmatrix} \cos(x) & -\sin(x) \\ \sin(x) & \cos(x) \end{pmatrix}\begin{pmatrix} -\sin(x) & \cos(x) \\ -\cos(x) & -\sin(x) \end{pmatrix} = \begin{pmatrix} 0 & * \\ * & 0 \end{pmatrix}$$

we obviously also find 0.
The left-hand side for the second identity is

$$\frac{dF^{-1}(x)}{dx} = \begin{pmatrix} -\sin(x) & -\cos(x) \\ \cos(x) & -\sin(x) \end{pmatrix}$$

and for the right-hand side we have

$$F^{-1}(x)\frac{dF(x)}{dx}F^{-1}(x) = \begin{pmatrix} \sin(x) & \cos(x) \\ -\cos(x) & \sin(x) \end{pmatrix}.$$

Problem 10. Let A be an invertible $n \times n$ matrix over $\mathbb{R}$. Consider the functions

$$E_j = \frac{1}{2}(A\mathbf{c}_j - \mathbf{e}_j)^T(A\mathbf{c}_j - \mathbf{e}_j)$$

where $j = 1, \ldots, n$, $\mathbf{c}_j$ is the j-th column of the inverse matrix of A, $\mathbf{e}_j$ is the j-th column of the $n \times n$ identity matrix. This means $\mathbf{e}_1, \ldots, \mathbf{e}_n$ is the standard basis (as column vectors) in $\mathbb{R}^n$. The c_j are determined by minimizing the E_j with respect to the c_j. Apply this method to find the inverse of the 3×3 matrix

$$A = \begin{pmatrix} 1 & 0 & 1 \\ 0 & 1 & 0 \\ 1 & 0 & -1 \end{pmatrix}.$$

Solution 10.   For the matrix A we have

$$E_1 = c_{1,1}^2 + c_{1,3}^2 - c_{1,1} - c_{1,3} + \frac{1}{2}c_{1,2}^2 + \frac{1}{2}$$

$$E_2 = c_{2,1}^2 + c_{2,3}^2 + \frac{1}{2}c_{2,2}^2 - c_{2,2} + \frac{1}{2}$$

$$E_3 = c_{3,1}^2 + c_{3,3}^2 - c_{3,1} + c_{3,3} + \frac{1}{2}c_{3,2}^2 + \frac{1}{2}.$$

From $\partial E_1/\partial c_{1,1} = 0$, $\partial E_1/\partial c_{1,2} = 0$, $\partial E_1/\partial c_{1,3} = 0$ we obtain

$$c_{1,1} = \frac{1}{2}, \quad c_{1,2} = 0, \quad c_{1,3} = 1/2.$$

From $\partial E_2/\partial c_{2,1} = 0$, $\partial E_2/\partial c_{2,2} = 0$, $\partial E_1/\partial c_{2,3} = 0$ we obtain

$$c_{2,1} = 0, \quad c_{2,2} = 1, \quad c_{2,3} = 0.$$

From $\partial E_3/\partial c_{3,1} = 0$, $\partial E_3/\partial c_{3,2} = 0$, $\partial E_3/\partial c_{3,3} =$ we obtain

$$c_{3,1} = \frac{1}{2}, \quad c_{3,2} = 0, \quad c_{3,3} = -1/2.$$

Thus the inverse matrix is

$$A^{-1} = \begin{pmatrix} 1/2 & 0 & 1/2 \\ 0 & 1 & 0 \\ 1/2 & 0 & -1/2 \end{pmatrix}.$$

Problem 11.   Let $\mathbf{f}$ be a function from U, an open subset of $\mathbb{R}^m$, to $\mathbb{R}^n$. Assume that the component function f_j $(j = 1, \ldots, n)$ possess first order partial derivatives. Then we can associate the $n \times m$ matrix

$$\left(\left. \frac{\partial f_j}{\partial x_k} \right|_{\mathbf{p}} \right), \qquad j = 1, \ldots, n \quad k = 1, \ldots, m$$

where $\mathbf{p} \in U$. The matrix is called the *Jacobian matrix* of $\mathbf{f}$ at the point $\mathbf{p}$. When $m = n$ the determinant of the square matrix $\mathbf{f}$ is called the *Jacobian* of $\mathbf{f}$. Let

$$A = \{ r \in \mathbb{R} : r > 0 \}, \qquad B = \{ \theta \in \mathbb{R} : 0 \le \theta < 2\pi \}$$

and $\mathbf{f} : A \times B \to \mathbb{R}^2$ with $f_1(r, \theta) = r\cos(\theta)$, $f_2(r, \theta) = r\sin(\theta)$. Find the Jacobian matrix and the Jacobian.

Solution 11. For the Jacobian matrix we find

$$\begin{pmatrix} \partial f_1/\partial r & \partial f_1/\partial \theta \\ \partial f_2/\partial r & \partial f_2/\partial \theta \end{pmatrix} = \begin{pmatrix} \cos(\theta) & -r\sin(\theta) \\ \sin(\theta) & r\cos(\theta) \end{pmatrix}.$$

Thus the Jacobian is r.

Problem 12. Consider the rotation matrix

$$R(t) = \begin{pmatrix} \cos(\omega t) & \sin(\omega t) \\ -\sin(\omega t) & \cos(\omega t) \end{pmatrix}$$

where ω is a fixed frequency. Find the 2×2 matrix

$$H(t) = i\hbar \frac{dR(t)}{dt} R^T(t)$$

and show it is hermitian.

Solution 12. We have

$$\frac{dR(t)}{dt} = \begin{pmatrix} -\omega\sin(\omega t) & \omega\cos(\omega t) \\ -\omega\cos(\omega t) & -\omega\sin(\omega t) \end{pmatrix}, \quad R(t)^T = \begin{pmatrix} \cos(\omega t) & -\sin(\omega t) \\ \sin(\omega t) & \cos(\omega t) \end{pmatrix}.$$

Thus

$$\frac{dR(t)}{dt} R(t)^T = \begin{pmatrix} 0 & \omega \\ -\omega & 0 \end{pmatrix}$$

and therefore

$$H(t) = \hbar\omega \begin{pmatrix} 0 & i \\ -i & 0 \end{pmatrix} \equiv -\hbar\omega\sigma_2.$$

Problem 13. Let $GL(m, \mathbb{C})$ be the general linear group over $\mathbb{C}$. This Lie group consists of all nonsingular $m \times m$ matrices. Let G be a Lie subgroup of $GL(m, \mathbb{C})$. Suppose $u_1, u_2, \ldots, u_n$ is a coordinate system on G in some neighborhood of I_m, the $m \times m$ identity matrix, and that $X(u_1, u_2, \ldots, u_m)$ is a point in this neighborhood. The matrix dX of differential one-forms contains n linearly independent differential one-forms since the n-dimensional Lie group G is smoothly embedded in $GL(m, \mathbb{C})$. Consider the matrix of differential one forms

$$\Omega := X^{-1}dX, \qquad X \in G.$$

The matrix Ω of differential one forms contains n-linearly independent ones.
(i) Let A be any fixed element of G. The *left-translation* by A is given by

$$X \to AX.$$

Show that $\Omega = X^{-1}dX$ is left-invariant.
(ii) Show that

$$d\Omega + \Omega \wedge \Omega = 0$$

where $\wedge$ denotes the *exterior product* for matrices, i.e. we have matrix multiplication together with the exterior product. The exterior product is linear and satisfies

$$du_j \wedge du_k = -du_k \wedge du_j.$$

Therefore $du_j \wedge du_j = 0$ for $j = 1, \ldots, n$. The exterior product is also associative.
(iii) Find dX^{-1} using $XX^{-1} = I_m$.

Solution 13.   (i) We have

$$(AX)^{-1}d(AX) = X^{-1}A^{-1}(AdX) = X^{-1}(A^{-1}A)dX = X^{-1}dX$$

since $d(AX) = AdX$.
(ii) From $\Omega = X^{-1}dX$ we obtain

$$X\Omega = dX$$
$$d(X\Omega) = ddX = 0$$
$$dX \wedge \Omega + Xd\Omega = 0$$
$$(X\Omega) \wedge \Omega + Xd\Omega = 0$$
$$\Omega \wedge \Omega + d\Omega = 0.$$

(iii) From $XX^{-1} = I_m$ we obtain

$$d(XX^{-1}) = 0$$
$$(dX)X^{-1} + XdX^{-1} = 0$$
$$X^{-1}(dX)X^{-1} + dX^{-1} = 0.$$

Thus $dX^{-1} = -X^{-1}(dX)X^{-1}$.

Problem 14.   Consider $GL(m, \mathbb{R})$ and a Lie subgroup of it. We interpret each element X of G as a linear transformation on the vector space $\mathbb{R}^m$ of row vectors $\mathbf{v} = (v_1, v_2, \ldots, v_n)$. Thus $\mathbf{v} \to \mathbf{w} = \mathbf{v}X$. Show that $d\mathbf{w} = \mathbf{w}\Omega$.

Solution 14.   We have

$$\mathbf{w} = \mathbf{v}X$$
$$d\mathbf{w} = d(\mathbf{v}X) = \mathbf{v}dX$$
$$d\mathbf{w} = (\mathbf{w}X^{-1})dX$$
$$d\mathbf{w} = \mathbf{w}(X^{-1}dX)$$
$$d\mathbf{w} = \mathbf{w}\Omega.$$

This means Ω can be interpreted as an "infinitesimal groups element".

Problem 15.   Consider the compact Lie group $SO(2)$ consisting of the matrices

$$X(u) = \begin{pmatrix} \cos(u) & -\sin(u) \\ \sin(u) & \cos(u) \end{pmatrix}.$$

Calculate dX and $X^{-1}dX$.

Solution 15. We have

$$dX = \begin{pmatrix} -\sin(u)du & -\cos(u)du \\ \cos(u)du & -\sin(u)du \end{pmatrix} = \begin{pmatrix} -\sin(u) & -\cos(u) \\ \cos(u) & -\sin(u) \end{pmatrix} du.$$

Since

$$X^{-1} = \begin{pmatrix} \cos(u) & \sin(u) \\ -\sin(u) & \cos(u) \end{pmatrix}$$

and using $\cos^2(u) + \sin^2(u) = 1$ we obtain

$$X^{-1}dX = \begin{pmatrix} 0 & -1 \\ 1 & 0 \end{pmatrix} du.$$

Problem 16. Let n be the dimension of the Lie group G. Since the vector space of differential one-forms at the identity element is an n-dimensional vector space, there are exactly n linearly independent left invariant differential one-forms in G. Let $\sigma_1, \sigma_2, \ldots, \sigma_n$ be such a system. Consider the Lie group

$$G := \left\{ \begin{pmatrix} u_1 & u_2 \\ 0 & 1 \end{pmatrix} : u_1, u_2 \in \mathbb{R}, \ u_1 > 0 \right\}.$$

Let

$$X = \begin{pmatrix} u_1 & u_2 \\ 0 & 1 \end{pmatrix}.$$

(i) Find X^{-1} and $X^{-1}dX$. Calculate the left-invariant differential one-forms. Calculate the left-invariant volume element.
(ii) Find the right-invariant forms.

Solution 16. (i) Since $\det(X) = u_1$ we obtain

$$X^{-1} = \frac{1}{u_1} \begin{pmatrix} 1 & -u_2 \\ 0 & u_1 \end{pmatrix}, \quad dX = \begin{pmatrix} du_1 & du_2 \\ 0 & 0 \end{pmatrix}.$$

It follows that

$$X^{-1}dX = \frac{1}{u_1} \begin{pmatrix} du_1 & du_2 \\ 0 & 0 \end{pmatrix}.$$

Hence $\sigma_1 = du_1/u_1$, $\sigma_2 = du_2/u_1$ are left-invariant. Then the left-invariant volume element follows as

$$\sigma_1 \wedge \sigma_2 = \frac{du_1 \wedge du_2}{u_1^2}.$$

(ii) We have

$$(dX)X^{-1} = \frac{1}{u_1} \begin{pmatrix} du_1 & du_2 \\ 0 & 0 \end{pmatrix} \begin{pmatrix} 1 & -u_2 \\ 0 & u_1 \end{pmatrix} = \frac{1}{u_1} \begin{pmatrix} du_1 & -u_2 du_1 + u_1 du_2 \\ 0 & 0 \end{pmatrix}.$$

Thus

$$\sigma_1 = \frac{du_1}{u_1}, \qquad \sigma_2 = \frac{1}{u_1}(-u_2 du_1 + u_1 du_2).$$

It follows that

$$\sigma_1 \wedge \sigma_2 = \frac{du_1 \wedge du_2}{u_1}$$

since $du_1 \wedge du_1 = 0$. The left- and right-invariant volume forms are different.

Problem 17. Consider the Lie group consisting of the matrices

$$X = \begin{pmatrix} u_1 & u_2 \\ 0 & u_1 \end{pmatrix}, \quad u_1, u_2 \in \mathbb{R}, \quad u_1 > 0.$$

Calculate X^{-1} and $X^{-1}dX$. Find the left-invariant differential one-forms and the left-invariant volume element.

Solution 17. Since $\det(X) = u_1^2$ we obtain

$$X^{-1} = \frac{1}{u_1^2}\begin{pmatrix} u_1 & -u_2 \\ 0 & u_1 \end{pmatrix}.$$

Moreover

$$dX = \begin{pmatrix} du_1 & du_2 \\ 0 & du_1 \end{pmatrix}.$$

From $\Omega = X^{-1}dX$ we have

$$\Omega = \frac{1}{u_1^2}\begin{pmatrix} u_1 & -u_2 \\ 0 & u_1 \end{pmatrix}\begin{pmatrix} du_1 & du_2 \\ 0 & du_1 \end{pmatrix} = \frac{1}{u_1^2}\begin{pmatrix} u_1 du_1 & u_1 du_2 - u_2 du_1 \\ 0 & u_1 du_1 \end{pmatrix}.$$

Thus we may take

$$\sigma_1 = \frac{du_1}{u_1}, \qquad \sigma_2 = \frac{u_1 du_2 - u_2 du_1}{u_1^2}.$$

Therefore the volume form is

$$\sigma_1 \wedge \sigma = \frac{du_1 \wedge du_2}{u_1^2}$$

where we used $du_1 \wedge du_1 = 0$.

Problem 18. Let A be a 2×2 symmetric matrix

$$A = \begin{pmatrix} a_{11} & a_{12} \\ a_{12} & a_{22} \end{pmatrix}$$

over $\mathbb{R}$. We define

$$\frac{\partial}{\partial a_{12}}A := \begin{pmatrix} 0 & 1 \\ 1 & 0 \end{pmatrix}.$$

Show that

$$\frac{\partial}{\partial a_{12}}\text{tr}(A^2) = \text{tr}\left(\frac{\partial}{\partial a_{12}}A^2\right) = \text{tr}\left(2A\frac{\partial A}{\partial a_{12}}\right).$$

Solution 18. We have

$$\frac{\partial}{\partial a_{12}}\text{tr}(A^2) = \frac{\partial}{\partial a_{12}}(a_{11}^2 + a_{12}^2 + a_{12}^2 + a_{22}^2) = \frac{\partial}{\partial a_{12}}2a_{12}^2 = 4a_{12}.$$

On the other hand we have

$$\text{tr}\left(\frac{\partial}{\partial a_{12}}\begin{pmatrix} a_{11}^2 + a_{12}^2 & a_{11}a_{12} + a_{12}a_{22} \\ a_{11}a_{12} + a_{12}a_{22} & a_{12}^2 + a_{22}^2 \end{pmatrix}\right) = \text{tr}\begin{pmatrix} 2a_{12} & a_{11} + a_{22} \\ a_{11} + a_{22} & 2a_{12} \end{pmatrix}$$
$$= 4a_{12}$$

and

$$\text{tr}\left(2A\frac{\partial A}{\partial a_{12}}\right) = 2\text{tr}\left(A\begin{pmatrix} 0 & 1 \\ 1 & 0 \end{pmatrix}\right) = 2\text{tr}\begin{pmatrix} a_{12} & a_{11} \\ a_{22} & a_{12} \end{pmatrix} = 4a_{12}.$$

Problem 19. Consider the matrix

$$A = \begin{pmatrix} 0 & -i \\ i & 0 \end{pmatrix}.$$

Find the function (characteristic polynomial) $p(\lambda) = \det(A - \lambda I_2)$. Find the eigenvalues of A by solving $p(\lambda) = 0$. Find the minima of the function $f(\lambda) = |p(\lambda)|$. Discuss.

Solution 19. We find $p(\lambda) = \lambda^2 - 1$ with the eigenvalues $+1$ and -1. Now

$$f(\lambda) = |p(\lambda)| = |\lambda^2 - 1|$$

which has minima at $+1$ and -1 and $f(+1) = 0$, $f(-1) = 0$.

Problem 20. Let $\alpha, \beta \in \mathbb{R}$. Consider the 2×2 matrix

$$B = \begin{pmatrix} -i\alpha & -\beta \\ -\beta & i\alpha \end{pmatrix}.$$

Find $\exp(tB)$, where $t \in \mathbb{R}$ and thus solve the initial value problem of the matrix differential equation

$$\frac{dA}{dt} = BA(t).$$

Solution 20. We find

$$\exp(tB) = \begin{pmatrix} \cos(\kappa t) - i\frac{\alpha}{\kappa}\sin(\kappa t) & -\frac{\beta}{\kappa}\sin(\kappa t) \\ -\frac{\beta}{\kappa}\sin(\kappa t) & \cos(\kappa t) + i\frac{\alpha}{\kappa}\sin(\kappa t) \end{pmatrix}$$

where $\kappa := \sqrt{\alpha^2 - \beta^2}$. Then the solution of the initial value problem is $A(t) = \exp(tB)A(0)$.

Problem 21. The *Fréchet derivative* of a matrix function $f : \mathbb{C}^{n \times n}$ at a point $X \in \mathbb{C}^{n \times n}$ is a linear mapping $L_X : \mathbb{C}^{n \times n} \to \mathbb{C}^{n \times n}$ such that for all $Y \in \mathbb{C}^{n \times n}$

$$f(X + Y) - f(X) - L_X(Y) = o(\|Y\|).$$

Calculate the Fréchet derivative of $f(X) = X^2$.

Solution 21.   We have $f(X+Y) - f(X) = XY + YX + Y^2$. Thus

$$L_X(Y) = XY + YX.$$

The right-hand side is the anti-commutator of X and Y.

Supplementary Problems

Problem 1.   Let A, B be $n \times n$ matrices. Assume that the entries of A and B are analytic functions of x.
(i) Show that

$$\frac{d}{dx}(AB) = \left(\frac{d}{dx}A\right)B + A\left(\frac{d}{dx}B\right).$$

(ii) Show that

$$\frac{d}{dx}(A \otimes B) = \left(\frac{d}{dx}A\right) \otimes B + A \otimes \left(\frac{d}{dx}B\right).$$

(iii) Show that

$$\frac{d}{dx}(A \bullet B) = \left(\frac{d}{dx}A\right) \bullet B + A \bullet \left(\frac{d}{dx}B\right)$$

where $\bullet$ denotes the Hadamard product.

Problem 2.   Let A be an $m \times n$ matrix over $\mathbb{R}$ and

$$\mathbf{x} := (1, x, x^2, \dots, x^{m-1})^T, \qquad \mathbf{y} := (1, y, y^2, \dots, y^{n-1})^T.$$

Find the extrema of the function $p(\mathbf{x}, \mathbf{y}) = \mathbf{x}^T A \mathbf{y}$.

Problem 3.   (i) Let $\epsilon \in \mathbb{R}$. Let $A(\epsilon)$ be an invertible $n \times n$ matrix. Assume that the entries a_{jk} are analytic functions of ϵ. Show that

$$\operatorname{tr}\left(A^{-1}(\epsilon)\frac{d}{d\epsilon}A(\epsilon)\right) = \frac{1}{\det(A(\epsilon))}\frac{d}{d\epsilon}\det(A(\epsilon)).$$

(ii) Consider the 2×2 matrix

$$M(\epsilon) = \begin{pmatrix} f_1(\epsilon) & f_2(\epsilon) \\ f_3(\epsilon) & f_4(\epsilon) \end{pmatrix}$$

where f_j $(j = 1, 2, 3, 4)$ are smooth functions and $\det(M(\epsilon)) > 0$ for all ϵ. Show that

$$\operatorname{tr}((dM(\epsilon))M(\epsilon)^{-1}) = d(\ln(\det(M(\epsilon))))$$

where d is the exterior derivative.

Problem 4. Let $f_j : \mathbb{R} \to \mathbb{R}$ $(j = 1, \ldots, n)$ be analytic functions. Consider the determinant (*Wronskian*)

$$W(x) = \det \begin{pmatrix} f_1(x) & f_2(x) & \cdots & f_n(x) \\ f_1'(x) & f_2'(x) & \cdots & f_n'(x) \\ \vdots & \vdots & \ddots & \\ f_1^{(n-1)}(x) & f_2^{(n-1)}(x) & \cdots & f_n^{(n-1)}(x) \end{pmatrix}$$

where $f_j(n-1)$ denotes the $(n-1)$ derivative of f_j. If the Wronskian of these functions is not identically 0 on $\mathbb{R}$, then these functions form a linearly independent set. Let $n = 3$ and $f_1(x) = x$, $f_2(x) = e^x$, $f_3(x) = e^{2x}$. Find the Wronskian. Discuss.

Problem 5. Let $f_1, f_2, f_3 : \mathbb{R}^3 \to \mathbb{R}$ be continuously differentiable functions. Find the determinant of the 3×3 matrix $A = (a_{jk})$

$$a_{jk} := \frac{\partial f_j}{\partial x_k} - \frac{\partial f_k}{\partial x_j}.$$

Problem 6. Let A, B be $n \times n$ matrices over $\mathbb{C}$. Let $\mathbf{u} \in \mathbb{C}^n$ be considered as column vector and

$$\partial := (\partial/\partial x_1 \quad \cdots \quad \partial/\partial x_n)^T.$$

Is

$$[\mathbf{u}^* A \partial, \mathbf{u}^* B \partial] = \mathbf{u}^* [A, B] \partial?$$

Problem 7. Let $V(\alpha)$ be a 2×2 matrix where all the entries are smooth functions of α. Calculate $dV(\alpha)/d\alpha$ and then find the conditions on the entries such that $[dV(\alpha)/d\alpha, V(\alpha)] = 0_2$.

Problem 8. Let j be a positive integer. Let A, B be $n \times n$ matrices over $\mathbb{R}$. Calculate

$$\lim_{\epsilon \to 0} \frac{1}{\epsilon} ((A + \epsilon B)^j - A^j), \qquad \frac{d}{d\epsilon} \operatorname{tr}(A + \epsilon B)^j \big|_{\epsilon=0}.$$

Problem 9. Find the partial differential equation given by the condition

$$\det \begin{pmatrix} 0 & \partial u/\partial x_1 & \partial u/\partial x_2 \\ \partial u/\partial x_1 & \partial^2 u/\partial x_1^2 & \partial^2 u/\partial x_1 \partial x_2 \\ \partial u/\partial x_2 & \partial^2 u/\partial x_2 \partial x_1 & \partial^2 u/\partial x_2^2 \end{pmatrix} = 0.$$

Find a nontrivial solution of the partial differential equation.

Problem 10. Let $\phi : \mathbb{R} \to \mathbb{R}$ be an analytic functions. Consider the matrices

$$A(t) = \begin{pmatrix} e^{i\phi(t)} & 1 \\ 1 & e^{id\phi(t)/dt} \end{pmatrix}, \qquad B(t) = \begin{pmatrix} 1 & e^{i\phi(t)} \\ e^{id\phi(t)/dt} & 1 \end{pmatrix}.$$

(i) Find the differential equation for ϕ from the condition $\mathrm{tr}(AB) = 0$.

(ii) Find the differential equation for ϕ from the condition $\det(AB) = 0$.

Problem 11.   Consider the four 6×6 matrices

$$
\beta_0 = \begin{pmatrix} -1 & 0 & 0 & 0 & 0 & 0 \\ 0 & -1 & 0 & 0 & 0 & 0 \\ 0 & 0 & -1 & 0 & 0 & 0 \\ 0 & 0 & 0 & -1 & 0 & 0 \\ 0 & 0 & 0 & 0 & -1 & 0 \\ 0 & 0 & 0 & 0 & 0 & -1 \end{pmatrix}, \quad
\beta_1 = \begin{pmatrix} 0 & 0 & 0 & 0 & 0 & 0 \\ 0 & 0 & 0 & 0 & 0 & -1 \\ 0 & 0 & 0 & 0 & 1 & 0 \\ 0 & 0 & 0 & 0 & 0 & 0 \\ 0 & 0 & 1 & 0 & 0 & 0 \\ 0 & -1 & 0 & 0 & 0 & 0 \end{pmatrix},
$$

$$
\beta_2 = \begin{pmatrix} 0 & 0 & 0 & 0 & 0 & 1 \\ 0 & 0 & 0 & 0 & 0 & 0 \\ 0 & 0 & 0 & -1 & 0 & 0 \\ 0 & 0 & -1 & 0 & 0 & 0 \\ 0 & 0 & 0 & 0 & 0 & 0 \\ 1 & 0 & 0 & 0 & 0 & 0 \end{pmatrix}, \quad
\beta_3 = \begin{pmatrix} 0 & 0 & 0 & 0 & -1 & 0 \\ 0 & 0 & 0 & 1 & 0 & 0 \\ 0 & 0 & 0 & 0 & 0 & 0 \\ 0 & 1 & 0 & 0 & 0 & 0 \\ -1 & 0 & 0 & 0 & 0 & 0 \\ 0 & 0 & 0 & 0 & 0 & 0 \end{pmatrix}
$$

and $\psi = (\, E_1 \quad E_2 \quad E_3 \quad cB_1 \quad cB_2 \quad cB_3 \,)^T$. Show that

$$
\beta_0 \frac{\partial}{c \partial t}\psi + \beta_1 \frac{\partial}{\partial x_1}\psi + \beta_2 \frac{\partial}{\partial x_2}\psi + \beta_3 \frac{\partial}{\partial x_3}\psi = (0 \ \ 0 \ \ 0 \ \ 0 \ \ 0 \ \ 0)^T
$$

are Maxwell's equations

$$
\frac{1}{c^2}\frac{\partial \mathbf{E}}{\partial t} = \nabla \times \mathbf{B}, \qquad -\frac{\partial \mathbf{B}}{\partial t} = \nabla \times \mathbf{E}.
$$

Problem 12.   Let σ_1, σ_2, σ_3 be the Pauli spin matrices and $f_j(x_1, x_2)$ ($j = 1, 2, 3$) be real-valued smooth functions $f_j : \mathbb{R}^2 \to \mathbb{R}$. Consider the 2×2 matrix

$$
N(x_1, x_2) = f_1\sigma_1 + f_2\sigma_2 + \sigma_3 f_3 \equiv \begin{pmatrix} f_3 & -if_2 + f_1 \\ if_2 + f_1 & -f_3 \end{pmatrix}.
$$

Find dN, N^*. Then calculate $d(N^* dN)$. Find the conditions of f_1, f_2, f_3 such that $d(N^* dN) = 0_2$.

Problem 13.   Let f_{11}, f_{22}, f_{33} be analytic functions $f_{jj} : \mathbb{R} \to \mathbb{R}$. Consider the 4×4 matrix

$$
M(x) = \begin{pmatrix} f_{11}(x) & 0 & 0 & 0 \\ 1 & f_{22}(x) & 0 & 1 \\ 0 & 0 & f_{33}(x) & 1 \\ f'_{11}(x) & f'_{22}(x) & f'_{33}(x) & 0 \end{pmatrix}
$$

where $'$ denotes differentiation with respect to x. Find the determinant of the matrix and write down the ordinary differential equation which follows from $\det(M(x)) = 0$. Find solutions of the differential equation.

Chapter 25

Integration and Matrices

Let $A(t)$ be an $n \times n$ matrix

$$A(t) = \begin{pmatrix} a_{11}(t) & a_{12}(t) & \cdots & a_{1n}(t) \\ a_{21}(t) & a_{22}(t) & \cdots & a_{2n}(t) \\ \vdots & \vdots & \ddots & \vdots \\ a_{n1}(t) & a_{n2}(t) & \cdots & a_{nn}(t) \end{pmatrix}$$

with $a_{jk}(t)$ are integrable functions. The integration is defined entrywise. For example

$$\int_0^t \begin{pmatrix} \cos(s) & -\sin(s) \\ \sin(s) & \cos(s) \end{pmatrix} ds = \begin{pmatrix} \sin(t) & \cos(t) - 1 \\ -\cos(t) + 1 & \sin(t) \end{pmatrix}.$$

Let A be an $n \times n$ matrix over $\mathbb{C}$. Suppose f is an analytic function inside on a closed contour Γ which encircles $\lambda(A)$, where $\lambda(A)$ denotes the eigenvalues of A. We define $f(A)$ to be the $n \times n$ matrix

$$f(A) = \frac{1}{2\pi i} \oint_\Gamma f(z)(zI_n - A)^{-1} dz.$$

This is a matrix version of the *Cauchy integral theorem*. The integral is defined on an element-by-element basis $f(A) = (f_{jk})$, where

$$f_{jk} = \frac{1}{2\pi i} \oint_\Gamma f(z) e_j^T (zI_n - A)^{-1} e_k dz$$

where e_j $(j = 1, \ldots, n)$ is the standard basis in $\mathbb{C}^n$.

Problem 1.   Consider the 2×2 matrix over $\mathbb{R}$

$$F(s) = \begin{pmatrix} \cosh(s) & \sinh(s) \\ \sinh(s) & \cosh(s) \end{pmatrix}.$$

Find

$$G(t) = \int_0^t F(s)ds.$$

Then calculate the commutator $[F(s), G(t)]$.

Solution 1.   We do the task with the following Maxima program

```
/* integration2.mac */
F: matrix([cosh(s),sinh(s)],[sinh(s),cosh(s)]);
assume(t > 0);
g11(t) := ''(integrate(F[1,1],s,0,t));
g12(t) := ''(integrate(F[1,2],s,0,t));
g21(t) := ''(integrate(F[2,1],s,0,t));
g22(t) := ''(integrate(F[2,2],s,0,t));
G: matrix([g11(t),g12(t)],[g21(t),g22(t)]);
C: F . G - G . F;
C: expand(C);
```

The output is

$$\begin{pmatrix} \sinh(t) & \cosh(t) - 1 \\ \cosh(t) - 1 & \sinh(t) \end{pmatrix}$$

and the commutator $[F(s), G(t)]$ vanishes.

Problem 2.   Consider the analytic function $f : \mathbb{C} \to \mathbb{C}$, $f(z) = z^2$ and the 2×2 matrix

$$A = \begin{pmatrix} 0 & 1 \\ 1 & 0 \end{pmatrix}.$$

Calculate $f(A)$ applying the Cauchy integral theorem.

Solution 2.   The eigenvalues of A are $+1$ and -1. We consider a circle as the contour with radius $r = 2$. Now we have

$$(zI_2 - A)^{-1} = \begin{pmatrix} z & -1 \\ -1 & z \end{pmatrix}^{-1} = \frac{1}{z^2 - 1} \begin{pmatrix} z & 1 \\ 1 & z \end{pmatrix}.$$

Thus

$$f_{11} = \frac{1}{2\pi i} \oint_\Gamma \frac{z^3}{z^2 - 1}dz, \qquad f_{12} = \frac{1}{2\pi i} \oint_\Gamma \frac{z^2}{z^2 - 1}dz,$$

$$f_{21} = \frac{1}{2\pi i} \oint_\Gamma \frac{z^2}{z^2 - 1}dz, \qquad f_{22} = \frac{1}{2\pi i} \oint_\Gamma \frac{z^3}{z^2 - 1}dz.$$

Using the *residue theorem* these integrals with the singularities at $+1$ and -1 can be easily calculated. One finds $f_{11} = f_{22} = 1$, $f_{12} = f_{21} = 0$. Obviously, $f(A)$ is the identity matrix.

Problem 3. Let A be an $n \times n$ matrix over $\mathbb{C}$. If λ is not an eigenvalue of A, then the matrix $(A - \lambda I_n)$ has an inverse, namely the *resolvent*

$$R_\lambda = (A - \lambda I_n)^{-1}.$$

Let λ_j be the eigenvalues of A. For $|\lambda| \geq a$, where a is any positive constant greater than all the $|\lambda_j|$ the resolvent can be expanded as

$$R_\lambda = -\frac{1}{\lambda}\left(I_n + \frac{1}{\lambda}A + \frac{1}{\lambda^2}A^2 + \cdots\right).$$

Calculate

$$-\frac{1}{2\pi i}\oint_{|\lambda|=a} \lambda^m R_\lambda d\lambda, \qquad m = 0, 1, 2, \ldots.$$

Solution 3. We obtain

$$-\frac{1}{2\pi i}\oint_{|\lambda|=a} \lambda^m R_\lambda d\lambda = A^m, \qquad m = 0, 1, 2, \ldots.$$

Problem 4. Let $\mathbb{C}^{n \times N}$ be the vector space of all $n \times N$ complex matrices. Let $Z \in \mathbb{C}^{n \times N}$. Then $Z^* \equiv \bar{Z}^T$, where T denotes transpose. One defines a *Gaussian measure* μ on $\mathbb{C}^{n \times N}$ by

$$d\mu(Z) := \frac{1}{\pi^{nN}}\exp(-\text{tr}(ZZ^*))dZ$$

where dZ denotes the Lebesgue measure on $\mathbb{C}^{n \times N}$. The *Fock space* $\mathcal{F}(\mathbb{C}^{n \times N})$ consists of all entire functions on $\mathbb{C}^{n \times N}$ which are square integrable with respect to the Gaussian measure $d\mu(Z)$. With the scalar product

$$\langle f|g \rangle := \int_{\mathbb{C}^{n \times N}} f(Z)\overline{g(Z)}d\mu(Z), \qquad f, g \in \mathcal{F}(\mathbb{C}^{n \times N})$$

one has a Hilbert space. Show that this Hilbert space has a *reproducing kernel* K. This means a continuous function $K(Z, Z') : \mathbb{C}^{n \times N} \times \mathbb{C}^{n \times N} \to \mathbb{C}$ such that

$$f(Z) = \int_{\mathbb{C}^{n \times N}} K(Z, Z')f(Z')d\mu(Z')$$

for all $Z \in \mathbb{C}^{n \times N}$ and $f \in \mathcal{F}(\mathbb{C}^{n \times N})$.

Solution 4. We find

$$K(Z, Z') = \exp(\text{tr}(Z(Z')^*)).$$

Problem 5. Let A be an $n \times n$ positive definite matrix over $\mathbb{R}$, i.e. $\mathbf{x}^T A\mathbf{x} > 0$ for all $\mathbf{x} \in \mathbb{R}^n$. Calculate

$$\int_{\mathbb{R}^n} \exp(-\mathbf{x}^T A\mathbf{x})d\mathbf{x}.$$

Solution 5.   Since A is positive definite it can be written in the form $A = S^T S$, where ^{T} denotes transpose and $S \in GL(n, \mathbb{R})$. Thus

$$|\det(S)| = \sqrt{\det(A)}.$$

Let $\mathbf{y} = S\mathbf{x}$. Now

$$\int_{\mathbb{R}^n} \exp(-\mathbf{x}^T A \mathbf{x}) d\mathbf{x} = \int_{\mathbb{R}^n} \exp(-(S\mathbf{x})^T (S\mathbf{x})) d\mathbf{x} = \int_{\mathbb{R}^n} \exp(-\mathbf{y}^T \mathbf{y}) d(S^{-1}\mathbf{y})$$

$$= |\det(S^{-1})| \int_{\mathbb{R}^n} \exp(-\mathbf{y}^T \mathbf{y}) d\mathbf{y}$$

$$= \frac{\pi^{n/2}}{\sqrt{\det(A)}}.$$

where we used that

$$\int_{\mathbb{R}^n} e^{-\mathbf{y}^T \mathbf{y}} d\mathbf{y} = \pi^{n/2}.$$

Problem 6.   Let V be an $N \times N$ unitary matrix, i.e. $VV^* = I_N$. The eigenvalues of V lie on the unit circle; that is, they may be expressed in the form $\exp(i\theta_n)$, $\theta_n \in \mathbb{R}$. A function $f(V) = f(\theta_1, \ldots, \theta_N)$ is called a *class function* if f is symmetric in all its variables. Weyl gave an explicit formula for averaging class functions over the circular unitary ensemble

$$\int_{U(N)} f(V) dV$$

$$= \frac{1}{(2\pi)^N N!} \int_0^{2\pi} \cdots \int_0^{2\pi} f(\theta_1, \ldots, \theta_N) \prod_{1 \le j < k \le N} |e^{i\theta_j} - e^{i\theta_k}|^2 d\theta_1 \cdots d\theta_N.$$

Thus we integrate the function $f(V)$ over $U(N)$ by parametrizing the group by the θ_i and using *Weyl's formula* to convert the integral into an N-fold integral over the θ_i. By definition the Haar measure dV is invariant under $V \to \tilde{U} V \tilde{U}^*$, where $\tilde{U}$ is any $N \times N$ unitary matrix. The matrix V can always be diagonalized by a unitary matrix, i.e.

$$V = W \begin{pmatrix} e^{i\theta_1} & \cdots & 0 \\ \vdots & \ddots & \vdots \\ 0 & \cdots & e^{i\theta_N} \end{pmatrix} W^*$$

where W is an $N \times N$ unitary matrix. Thus the integral over V can be written as an integral over the matrix elements of W and the eigenphases θ_n. Since the measure is invariant under unitary transformations, the integral over the matrix elements of U can be evaluated straightforwardly, leaving the integral over the eigenphases. Show that for f a class function we have

$$\int_{U(N)} f(V) dV = \frac{1}{(2\pi)^N} \int_0^{2\pi} \cdots \int_0^{2\pi} f(\theta_1, \ldots, \theta_N) \det(e^{i\theta_n (n-m)}) d\theta_1 \cdots d\theta_N.$$

Solution 6. From Weyl's formula we know that

$$\int_{U(N)} f(V)dV$$

can be written as

$$\frac{1}{(2\pi)^N N!} \int_0^{2\pi} \cdots \int_0^{2\pi} f(\theta_1,\ldots,\theta_N) \prod_{1\leq j<k\leq N} |e^{i\theta_j} - e^{i\theta_k}|^2 d\theta_1 \cdots d\theta_N.$$

We first note that

$$\prod_{1\leq j<k\leq N} |e^{i\theta_j} - e^{i\theta_k}|^2 = \det\left(\begin{pmatrix} 1 & 1 & \cdots \\ e^{i\theta_1} & e^{i\theta_2} & \cdots \\ \vdots & & \ddots & \cdots \\ e^{i(N-1)\theta_1} & e^{i(N-1)\theta_2} & \cdots \end{pmatrix}\right.$$

$$\left.\times \begin{pmatrix} 1 & e^{-i\theta_1} & e^{-2i\theta_1} & \cdots \\ 1 & e^{-i\theta_2} & e^{-2i\theta_2} & \cdots \\ \vdots & \vdots & & \ddots & \cdots \\ 1 & e^{-i\theta_N} & e^{-2i\theta_N} & \cdots \end{pmatrix}\right)$$

$$= \det\left(\sum_{\ell=1}^{N} e^{i\theta_\ell(n-m)}\right).$$

Therefore

$$\int_{U(N)} f(\theta_1,\ldots,\theta_N)dV = \frac{1}{(2\pi)^N N!} \int_0^{2\pi} \cdots \int_0^{2\pi} f(\theta_1,\ldots,\theta_N)$$

$$\times \det \begin{pmatrix} \sum_{\ell=1}^N 1 & \sum_{\ell=1}^N e^{-i\theta_\ell} & \cdots & \sum_{\ell=1}^N e^{-(N-1)i\theta_\ell} \\ \sum_{\ell=1}^N e^{i\theta_\ell} & \sum_{\ell=1}^N 1 & \cdots & \sum_{\ell=1}^N e^{-i(N-2)\theta_\ell} \\ \vdots & \vdots & \ddots & \vdots \\ \sum_{\ell=1}^N e^{i(N-1)\theta_\ell} & \sum_{\ell=1}^N e^{i(N-2)\theta_\ell} & \cdots & \sum_{\ell=1}^N 1 \end{pmatrix} d\theta_1 \cdots d\theta_N.$$

Using that f is symmetric in its arguments provides

$$\int_{U(N)} f(\theta_1,\ldots,\theta_N)dV = \frac{1}{(2\pi)^N N!} \int_0^{2\pi} \cdots \int_0^{2\pi} f(\theta_1,\ldots,\theta_N)$$

$$\times \det \begin{pmatrix} 1 & e^{-i\theta_1} & \cdots & e^{-(N-1)i\theta_1} \\ \sum_{\ell=1}^N e^{i\theta_\ell} & \sum_{\ell=1}^N 1 & \cdots & \sum_{\ell=1}^N e^{-i(N-2)\theta_\ell} \\ \vdots & \vdots & \ddots & \vdots \\ \sum_{\ell=1}^N e^{i(N-1)\theta_\ell} & \sum_{\ell=1}^N e^{i(N-2)\theta_\ell} & \cdots & \sum_{\ell=1}^N 1 \end{pmatrix} d\theta_1 \cdots d\theta_N.$$

Subtracting $e^{i\theta_1}$ times the first row from the second row then gives

$$\int_{U(N)} f(\theta_1,\ldots,\theta_N)dV = \frac{1}{(2\pi)^N N!} \int_0^{2\pi} \cdots \int_0^{2\pi} f(\theta_1,\ldots,\theta_N)\times$$

$$\det \begin{pmatrix} \sum_{\ell=2}^{N} e^{i\theta_\ell} & \dfrac{e^{-i\theta_1}}{\sum_{\ell=2}^{N} 1} & \cdots & \dfrac{e^{-(N-1)i\theta_1}}{\sum_{\ell=2}^{N} e^{-i(N-2)\theta_\ell}} \\ \vdots & \vdots & \ddots & \vdots \\ \sum_{\ell=2}^{N} e^{i(N-1)\theta_\ell} & \sum_{\ell=2}^{N} e^{i(N-2)\theta_\ell} & \cdots & \sum_{\ell=2}^{N} 1 \end{pmatrix} d\theta_1 \cdots d\theta_N.$$

This process is continued reducing the second row to $e^{i\theta_2}, 1, e^{-i\theta_2}, \ldots, e^{-(N-2)i\theta_2}$ and thus pulling out a factor of $N-1$. Then doing the same to the third row and so on. The factor of $N!$ resulting from these row manipulations cancels the $N!$ in the normalization constant of Weyl's formula.

Problem 7. Let A be an $n \times n$ positive definite matrix over $\mathbb{R}$. Let $\mathbf{q}$ and $\mathbf{J}$ be column vectors in $\mathbb{R}^n$. Calculate

$$Z(\mathbf{J}) = \int_{-\infty}^{\infty} \cdots \int_{-\infty}^{\infty} dq_1 \cdots dq_n \exp\left(-\frac{1}{2}\mathbf{q}^T A \mathbf{q} + \mathbf{J}^T \mathbf{q}\right).$$

Note that

$$\int_{-\infty}^{\infty} dq e^{-(aq^2 + bq + c)} = \sqrt{\frac{\pi}{a}} e^{(b^2 - 4ac)/(4a)}. \tag{1}$$

Solution 7. Since A is positive definite the inverse A^{-1} of A exists and we have $\det A \neq 0$. Using (1) we find

$$Z(\mathbf{J}) = \frac{(2\pi)^{n/2}}{\sqrt{\det A}} \exp\left(\frac{1}{2}\mathbf{J}^T A^{-1} \mathbf{J}\right).$$

Problem 8. Let A be an $n \times n$ positive definite matrix over $\mathbb{R}$, i.e. all the eigenvalues, which are real, are positive. We also have $A^T = A$. Consider the analytic function $f : \mathbb{R}^n \to \mathbb{R}$

$$f(\mathbf{x}) = \exp\left(-\frac{1}{2}\mathbf{x}^T A^{-1} \mathbf{x}\right).$$

Calculate the *Fourier transform* of f. The Fourier transform is defined by

$$\hat{f}(\mathbf{k}) := \int_{\mathbb{R}^n} f(\mathbf{x}) e^{i\mathbf{k}\cdot\mathbf{x}} d\mathbf{x}$$

where $\mathbf{k}\cdot\mathbf{x} \equiv \mathbf{k}^T\mathbf{x} \equiv k_1 x_1 + \cdots + k_n x_n$ and $d\mathbf{x} = dx_1 \ldots dx_n$. The inverse Fourier transform is given by

$$f(\mathbf{x}) = \frac{1}{(2\pi)^n} \int_{\mathbb{R}^n} \hat{f}(\mathbf{k}) e^{-i\mathbf{k}\cdot\mathbf{x}} d\mathbf{k}$$

where $d\mathbf{k} = dk_1 \ldots dk_n$. With $a > 0$ we have

$$\int_{\mathbb{R}} e^{-(ax^2 + bx + c)} dx = \sqrt{\frac{\pi}{a}} e^{(b^2 - 4ac)/(4a)}.$$

Solution 8. Consider first the case $n = 1$. Thus we have

$$\int_{\mathbb{R}} e^{-x^2/(2\lambda)} e^{ikx} dx = \int_{\mathbb{R}} e^{-(x^2/(2\lambda)-ikx)} dx = \sqrt{2\lambda\pi} e^{-\lambda k^2/2}$$

where λ is the (positive) eigenvalue of the 1×1 matrix A. Consider now the general case. Since A is positive definite we can find an orthogonal matrix O such that OAO^{-1} is diagonal, where $O^{-1} = O^T$. Also $OA^{-1}O^{-1}$ is diagonal. Thus we introduce $\mathbf{y} = O\mathbf{x}$. It follows that

$$\mathbf{x}^T A^{-1} \mathbf{x} = \mathbf{y}^T OA^{-1}O^{-1}\mathbf{y} = \mathbf{y}^T D\mathbf{y}$$

where D is the diagonal matrix $D = \text{diag}(1/\lambda_1, \ldots, 1/\lambda_n)$. Thus the multidimensional integral reduces to a product. Using the result for the case $n = 1$ we find

$$\hat{f}(\mathbf{k}) = \sqrt{(2\pi)^n \det(A)} \exp\left(-\frac{1}{2}\mathbf{k}^T A\mathbf{k}\right).$$

Problem 9. Let A be an $n \times n$ matrix. Let $\omega, \mu \in \mathbb{R}$. Assume that

$$\|e^{tA}\| \leq Me^{\omega t}, \qquad t \geq 0$$

and $\mu > \omega$. Then we have

$$(\mu I_n - A)^{-1} \equiv \int_0^\infty e^{-\mu t} e^{tA} dt. \tag{1}$$

Calculate the left- and right-hand side of (1) for the matrix $A = \sigma_1$.

Solution 9. For the left-hand side we have

$$(\mu I_2 - A)^{-1} = \begin{pmatrix} \mu & -1 \\ -1 & \mu \end{pmatrix}^{-1} = \frac{1}{1-\mu^2}\begin{pmatrix} -\mu & -1 \\ -1 & -\mu \end{pmatrix}.$$

Since

$$\exp(tA) = I_2 \cosh(t) + \begin{pmatrix} 0 & 1 \\ 1 & 0 \end{pmatrix}\sinh(t) = \begin{pmatrix} \cosh(t) & \sinh(t) \\ \sinh(t) & \cosh(t) \end{pmatrix}$$

and

$$\exp(-\mu t I_2) = \begin{pmatrix} e^{-\mu t} & 0 \\ 0 & e^{-\mu t} \end{pmatrix}$$

we find for the right-hand side

$$\int_0^\infty e^{-\mu t} e^{tA} dt = \int_0^\infty e^{-\mu t I_2} e^{tA} dt$$

$$= \int_0^\infty \begin{pmatrix} e^{-\mu t} & 0 \\ 0 & e^{-\mu t} \end{pmatrix}\begin{pmatrix} \cosh(t) & \sinh(t) \\ \sinh(t) & \cosh(t) \end{pmatrix} dt$$

$$= \int_0^\infty \begin{pmatrix} e^{-\mu t}\cosh(t) & e^{-\mu t}\sinh(t) \\ e^{-\mu t}\sinh(t) & e^{-\mu t}\cosh(t) \end{pmatrix} dt.$$

Since

$$\int_0^\infty \cosh(t)e^{-\mu t}\,dt = -\frac{\mu}{1-\mu^2}, \qquad \int_0^\infty \sinh(t)e^{-\mu t}\,dt = -\frac{1}{1-\mu^2}$$

we obtain

$$\int_0^\infty e^{-\mu t}e^{tA}\,dt = \frac{1}{1-\mu^2}\begin{pmatrix} -\mu & -1 \\ -1 & -\mu \end{pmatrix}.$$

Problem 10. Consider the differentiable manifold

$$S^3 = \{\,(x_1, x_2, x_3, x_4) : x_1^2 + x_2^2 + x_3^2 + x_4^2 = 1\,\}.$$

(i) Show that the matrix

$$U(x_1, x_2, x_3, x_4) = -i\begin{pmatrix} x_3 + ix_4 & x_1 - ix_2 \\ x_1 + ix_2 & -x_3 + ix_3 \end{pmatrix}$$

is unitary. Show that the matrix is an element of $SU(2)$.
(ii) Consider the parameters (θ, ψ, ϕ) with $0 \le \theta < \pi$, $0 \le \psi < 4\pi$, $0 \le \phi < 2\pi$. Show that

$$x_1(\theta, \psi, \phi) + ix_2(\theta, \psi, \phi) = \cos(\theta/2)e^{i(\psi+\phi)/2}$$
$$x_3(\theta, \psi, \phi) + ix_4(\theta, \psi, \phi) = \sin(\theta/2)e^{i(\psi-\phi)/2}$$

is a parametrization. Thus the matrix given in (i) takes the form

$$-i\begin{pmatrix} \sin(\theta/2)e^{i(\psi-\phi)/2} & \cos(\theta/2)e^{-i(\psi+\phi)/2} \\ \cos(\theta/2)e^{i(\psi+\phi)/2} & -\sin(\theta/2)e^{-i(\psi-\phi)/2} \end{pmatrix}$$

and

$$U^* = U^{-1} = \begin{pmatrix} ie^{-i(\psi-\phi)/2}\sin(\theta/2) & e^{-i(\psi+\phi)/2}\cos(\theta/2) \\ e^{i(\psi+\phi)/2}\cos(\theta/2) & -ie^{i(\psi-\phi)/2}\sin(\theta/2) \end{pmatrix}.$$

(iii) Let $(\xi_1, \xi_2, \xi_3) = (\theta, \psi, \phi)$ with $0 \le \theta < \pi$, $0 \le \psi < 4\pi$, $0 \le \phi < 2\pi$. Show that

$$\frac{1}{24\pi^2}\int_0^\pi d\theta \int_0^{4\pi} d\psi \int_0^{2\pi} d\phi \sum_{j,k,\ell=1}^3 \epsilon_{jk\ell}\mathrm{tr}\left(U^{-1}\frac{\partial U}{\partial \xi_j}U^{-1}\frac{\partial U}{\partial \xi_k}U^{-1}\frac{\partial U}{\partial \xi_\ell}\right) = 1$$

where $\epsilon_{123} = \epsilon_{321} = \epsilon_{132} = +1$, $\epsilon_{213} = \epsilon_{321} = \epsilon_{132} = -1$ and 0 otherwise.
(iv) Consider the metric tensor field

$$g = dx_1 \otimes dx_1 + dx_2 \otimes dx_2 + dx_3 \otimes dx_3 + dx_4 \otimes dx_4.$$

Using the parametrization show that

$$g_{S^3} = \frac{1}{4}(d\theta \otimes d\theta + d\psi \otimes d\psi + d\phi \otimes d\phi + \cos(\theta)d\psi \otimes d\phi + \cos(\theta)d\phi \otimes d\psi).$$

(v) Consider the three differential one forms e_1, e_2, e_3 defined by

$$\begin{pmatrix} e_1 \\ e_2 \\ e_3 \end{pmatrix} = \begin{pmatrix} -x_4 & -x_3 & x_2 & x_1 \\ x_3 & -x_4 & -x_1 & x_2 \\ -x_2 & x_1 & -x_4 & x_3 \end{pmatrix}\begin{pmatrix} dx_1 \\ dx_2 \\ dx_3 \\ dx_4 \end{pmatrix}.$$

Show that $g_{S^3} = de_1 \otimes de_1 + de_2 \otimes de_2 + de_3 \otimes de_3$.

(vi) Show that

$$de_j = \sum_{k,\ell=1}^{3} \epsilon_{jk\ell} e_k \wedge e_\ell$$

i.e. $de_1 = 2e_2 \wedge e_3$, $de_2 = 2e_3 \wedge e_1$, $de_3 = 2e_1 \wedge e_2$.

Solution 10. (i) We have $UU^* = I_4$. Thus the matrix is unitary. We have $\det(U) = 1$. Thus U is an element of $SU(2)$.

(ii) Since

$$x_1 = \cos(\theta/2)\cos((\psi + \phi)/2), \quad x_2 = \cos(\theta/2)\sin((\psi + \phi)/2),$$

$$x_3 = \sin(\theta/2)\cos((\psi - \phi)/2), \quad x_4 = \sin(\theta/2)\sin((\psi - \phi)/2)$$

we obtain

$$x_1^2 + x_2^2 = \cos^2(\theta/2), \qquad x_3^2 + x_4^2 = \sin^2(\theta/2).$$

Thus $x_1^2 + x_2^2 + x_3^2 + x_4^2 = 1$.

(iii) We have

$$\frac{\partial U}{\partial \xi_1} = -\frac{i}{2}\begin{pmatrix} e^{i(\psi+\phi)/2}\cos(\theta/2) & -e^{-i(\psi+\phi)/2}\sin(\theta/2) \\ -e^{i(\psi+\phi)/2}\sin(\theta/2) & -e^{-i(\psi-\phi)/2}\cos(\theta/2) \end{pmatrix}$$

$$\frac{\partial U}{\partial \xi_2} = -\frac{1}{2}\begin{pmatrix} \sin(\theta/2)e^{i(\psi-\phi)/2} & -\cos(\theta/2)e^{-i(\psi+\phi)/2} \\ \cos(\theta/2)e^{i(\psi+\phi)/2} & \sin(\theta/2)e^{-i(\psi-\phi)/2} \end{pmatrix}$$

$$\frac{\partial U}{\partial \xi_3} = \frac{1}{2}\begin{pmatrix} \sin(\theta/2)e^{i(\psi-\phi)/2} & -\cos(\theta/2)e^{-i(\psi+\phi)/2} \\ \cos(\theta/2)e^{i(\psi+\phi)/2} & -\sin(\theta/2)e^{-i(\psi-\phi)/2} \end{pmatrix}.$$

(vi) We show that $de_3 = 2e_1 \wedge e_2$. We have

$$de_3 = 2(dx_1 \wedge dx_2 + dx_3 \wedge dx_4).$$

Now

$$\begin{aligned}
e_1 \wedge e_2 &= (x_3^2 + x_4^2)dx_1 \wedge dx_2 + (x_1x_4 - x_2x_3)dx_1 \wedge dx_3 \\
&\quad + (-x_1x_3 - x_2x_4)dx_1 \wedge dx_4 + (x_1x_3 + x_2x_4)dx_2 \wedge dx_3 \\
&\quad + (-x_2x_3 + x_1x_4)dx_2 \wedge dx_4 + (x_1^2 + x_2^2)dx_3 \wedge dx_4.
\end{aligned}$$

Using

$$x_1dx_1 = -x_2dx_2 - x_3dx_3 - x_4dx_4, \quad x_3dx_3 = -x_1dx_1 - x_2dx_2 - x_4dx_4$$

we obtain

$$(x_1x_4 - x_2x_3)dx_1 \wedge dx_3 = x_2^2dx_1 \wedge dx_2 + x_4^2dx_3 \wedge dx_4 - x_2x_4dx_2 \wedge dx_3 + x_2x_4dx_1 \wedge dx_4.$$

Using

$$x_2dx_2 = -x_1dx_1 - x_3dx_3 - x_4dx_4, \quad x_4dx_4 = -x_1dx_1 - x_2dx_2 - x_3dx_3$$

we obtain

$$(x_1x_4 - x_2x_3)dx_2 \wedge dx_4 = x_1^2 dx_1 \wedge dx_2 + x_3^2 dx_3 \wedge dx_4 + x_1 x_3 dx_1 \wedge dx_4 - x_1 x_3 dx_2 \wedge dx_3.$$

Inserting these terms into $e_1 \wedge e_2$ we arrive at

$$e_1 \wedge e_2 = (x_1^2 + x_2^2 + x_3^2 + x_4^2)dx_1 \wedge dx_2 + (x_1^2 + x_2^2 + x_3^2 + x_4^2)dx_3 \wedge dx_4$$
$$= dx_1 \wedge dx_2 + dx_3 \wedge dx_4.$$

Analogously we show that $de_1 = e_2 \wedge e_3$ and $de_3 = e_2 \wedge e_1$.

Supplementary Problems

Problem 1. Let A, B be $n \times n$ matrices over $\mathbb{C}$. Let $\beta, \epsilon \in \mathbb{R}$. Show that

$$\exp(\beta(A + B)) \equiv \exp(\beta A)\left(I_n + \int_0^\beta d\epsilon\, e^{-\epsilon A} B e^{\epsilon(A+B)}\right).$$

Multiply both sides from the left with $\exp(-\beta A)$ and differentiate with respect to β.

Problem 2. Let $\alpha \in \mathbb{R}$. Consider an $n \times n$ matrix $A(\alpha)$, where the entries of A depends smoothly on α. Then one has the identity

$$\frac{d}{d\alpha} e^{A(\alpha)} \equiv \int_0^1 e^{(1-s)A(\alpha)} \frac{dA(\alpha)}{d\alpha} e^{sA(\alpha)} ds \equiv \int_0^1 e^{sA(\alpha)} \frac{dA(\alpha)}{d\alpha} e^{(1-s)A(\alpha)} ds.$$

Let $n = 2$ and

$$A(\alpha) = \begin{pmatrix} \cos(\alpha) & -\sin(\alpha) \\ \sin(\alpha) & \cos(\alpha) \end{pmatrix} \Rightarrow \frac{dA(\alpha)}{d\alpha} = \begin{pmatrix} -\sin(\alpha) & -\cos(\alpha) \\ \cos(\alpha) & -\sin(\alpha) \end{pmatrix}.$$

Calculate the right-hand sides of the identities.

Problem 3. Let A, B be positive definite matrices. Then we have the integral representation ($x \geq 0$)

$$\ln(A + xB) - \ln(A) \equiv \int_0^\infty (A + uI_n)^{-1} xB(A + xB + uI_n)^{-1} du.$$

Let

$$A = \begin{pmatrix} 2 & -1 \\ -1 & 1 \end{pmatrix}, \qquad B = \begin{pmatrix} 1 & 1 \\ 1 & 2 \end{pmatrix}.$$

Calculate the left- and right-hand side of the integral representation.

Problem 4. Consider the 2×2 matrices

$$A(t) = \begin{pmatrix} \cos(t) & -\sin(t) \\ \sin(t) & \cos(t) \end{pmatrix}, \qquad B(t) = \int_0^t A(s)ds.$$

Find the commutator $[A(t), B(t)]$. Discuss. What is the condition such that $[A(t), B(t)] = 0_2$?

Problem 5. Let P_j $(j = 0, 1, 2, \ldots)$ be the *Legendre polynomials*

$$P_0(x) = 1, \quad P_1(x) = x, \quad P_2(x) = \frac{1}{2}(3x^2 - 1), \quad \ldots .$$

Show that the infinite dimensional matrix $A = (a_{jk})$

$$a_{jk} := \int_{-1}^{+1} P_j(x) \frac{dP_k(x)}{dx} dx$$

where $j, k = 0, 1, \ldots$ is given by

$$A = \begin{pmatrix} 0 & 0 & 0 & 0 & 0 & 0 & 0 & \cdots \\ 2 & 0 & 0 & 0 & 0 & 0 & 0 & \cdots \\ 0 & 2 & 0 & 0 & 0 & 0 & 0 & \cdots \\ 2 & 0 & 2 & 0 & 0 & 0 & 0 & \cdots \\ 0 & 2 & 0 & 2 & 0 & 0 & 0 & \cdots \\ 2 & 0 & 2 & 0 & 2 & 0 & 0 & \cdots \\ 0 & 2 & 0 & 2 & 0 & 2 & 0 & \cdots \\ \vdots & \vdots & \vdots & \vdots & \vdots & \vdots & \vdots & \ddots \end{pmatrix}.$$

Note that the matrix is not symmetric. The matrix A can be considered as a linear operator in the Hilbert space $\ell_2(\mathbb{N}_0)$. Is $\|A\| < \infty$?

Problem 6. Let A, B, C be positive definite $n \times n$ matrices. Then (*Lieb inequality*)

$$\operatorname{tr}(e^{\ln(A) - \ln(B) + \ln(C)}) \leq \operatorname{tr} \int_0^\infty A(B + uI_n)^{-1} C(B + uI_n)^{-1} du.$$

(i) Let

$$A = \begin{pmatrix} 2 & -1 \\ -1 & 1 \end{pmatrix}, \quad B = \begin{pmatrix} 1 & 1 \\ 1 & 2 \end{pmatrix}, \quad C = \begin{pmatrix} 4 & 2 \\ 2 & 5 \end{pmatrix}.$$

Calculate the left-hand side and right-hand side of the inequality.
(ii) Show that a sufficient condition such that one has an equality is that A, B, C commute. However this condition is not necessary.

Bibliography

Aldous J. M. and Wilson R. J.
Graphs and Applications: An Introductory Approach, Springer (2000)

Armstrong M. A.
Groups and Symmetry, Springer (1988)

Beezer R. A.
A First Course in Linear Algebra, Congruent Press, Washington (2013)

Bredon G. E.
Introduction to Compact Transformation Groups, Academic Press (1972)

Bronson R.
Matrix Operations, Schaum's Outlines, McGraw-Hill (1989)

Bump D.
Lie Groups, Springer (2000)

Burnside W.
Theory of Groups of Finite Order, Dover, New York (1955)

Carter R. W.
Simple Groups of Lie Type, John Wiley (1972)

Cullen C. G.
Matrices and Linear Transformations, Second edition, Dover, New York (1990)

Davis P. J.
Circulant Matrices, John Wiley (1997)

de Souza P. N. and Silva J.-N.
Berkeley Problems in Mathematics, Springer (1998)

Dixmier J.
Enveloping Algebras, North-Holland (1974)

Erdmann K. and Wildon M.
Introduction to Lie Algebras, Springer (2006)

Flanders H.
Differential Forms with Applications to the Physical Sciences, Academic Press (1963)

Fuhrmann, P. A.
A Polynomial Approach to Linear Algebra, Springer (1996)

Fulton W. and Harris J.
Representation Theory, Springer (1991)

Gallian J. A.
Contemporary Abstract Algebra, Sixth edition, Houghton Mifflin (2006)

Gantmacher F.
Theory of Matrices, AMS Chelsea Publishing (1953)

Golan J. S.
The Linear Algebra a Beginning Graduate Student Ought to Know, Springer (2012)

Harville D. A.
Matrix Algebra from a Statistician's Perspective, Springer (1997)

Golub G. H. and Van Loan C. F.
Matrix Computations, Third edition, Johns Hopkins University Press (1996)

Grossman S. I.
Elementary Linear Algebra, Third edition, Wadsworth Publishing, Belmont (1987)

Hall B. C.
Lie Groups, Lie Algebras, and Representations: An elementary introduction, Springer (2003)

Handbook of Linear Algebra, Second edition, edited by L. Hogben, CRC Press, Boca Raton (2014)

Horn R. A. and Johnson C. R.
Topics in Matrix Analysis, Cambridge University Press (1999)

Humphreys J. E.
Introduction to Lie Algebras and Representation Theory, Springer (1972)

Inui T., Tanabe Y. and Onodera Y.
Group Theory and its Applications in Physics, Springer (1990)

James G. and Liebeck M.
Representations and Characters of Groups, Second edition, Cambridge University Press (2001)

Johnson D. L.
Presentation of Groups, Cambridge University Press (1976)

Jones H. F.
Groups, Representations and Physics, Adam Hilger, Bristol (1990)

Kedlaya K. S., Poonen B. and Vakil R.
The William Lowell Putnam Mathematical Competition 1985–2000, The Mathematical Association of America (2002)

Lang S.
Linear Algebra, Addison-Wesley, Reading (1968)

Lee Dong Hoon
The Structure of Complex Lie Groups, Chapman and Hall/CRC (2002)

Marcus M. and Minc H.
A Survey of Matrix Theory and Matrix Inequalities, Dover Publications, New York (1992)

Miller W.
Symmetry Groups and Their Applications, Academic Press, New York (1972)

Schneider H. and Barker G. P.
Matrices and Linear Algebra, Dover Publications, New York (1989)

Searle S. R.
Matrix Algebra Useful in Statistics, John Wiley (2006)

Seber G. A. F.
A Matrix Handbook for Statisticians, Wiley Series in Probability and Statistics (2008)

Steeb W.-H.
Matrix Calculus and Kronecker Product with Applications and C++ Programs, World Scientific Publishing, Singapore (1997)

Steeb W.-H.
Continuous Symmetries, Lie Algebras, Differential Equations and Computer Algebra, World Scientific Publishing, Singapore (1996)

Steeb W.-H.
Hilbert Spaces, Wavelets, Generalized Functions and Quantum Mechanics, Kluwer Academic Publishers, Dordrecht (1998)

Steeb W.-H.
Problems and Solutions in Theoretical and Mathematical Physics, Second edition, Volume I: Introductory Level, World Scientific Publishing, Singapore (2003)

Steeb W.-H.
Problems and Solutions in Theoretical and Mathematical Physics, Second edition, Volume II: Advanced Level, World Scientific Publishing, Singapore (2003)

Steeb W.-H., Hardy Y., Hardy A. and Stoop R.
Problems and Solutions in Scientific Computing with C++ and Java Simulations, World Scientific Publishing, Singapore (2004)

Sternberg S.
Lie Algebras www.math.harvard.edu/~shlomo/docs/lie_algebras.pdf

Varadarajan V. S.
Lie Groups, Lie Algebras and Their Representations, Springer (2004)

Weyl H.
Classical Groups, Princeton, UP, 1946

Watkins D. S.
Fundamentals of Matrix Computations, John Wiley (1991)

Wawrzynczyk A.
Group Representations and Special Functions, D. Reidel (1984)

Wybourne B. G.
Classical Groups for Physicists, John Wiley (1974)

Zhang F.
Matrix Theory, Basic Results and Techniques, Second edition, Springer (2011)

Index

Printed in the United States
By Bookmasters